Springer Proceedings in Physics

Volume 143

For further volumes:
http://www.springer.com/series/361

P. K. Giri · D. K. Goswami
A. Perumal
Editors

Advanced Nanomaterials and Nanotechnology

Proceedings of the 2nd International Conference on Advanced Nanomaterials and Nanotechnology, Dec 8–10, 2011, Guwahati, India

With 332 Figures

Springer

Editors
P. K. Giri
D. K. Goswami
A. Perumal
Department of Physics
Indian Institute of Technology Guwahati
Guwahati
India

ISSN 0930-8989 ISSN 1867-4941 (electronic)
ISBN 978-3-642-34215-8 ISBN 978-3-642-34216-5 (eBook)
DOI 10.1007/978-3-642-34216-5
Springer Heidelberg New York Dordrecht London

Library of Congress Control Number: 2012954626

Printed on acid-free paper

Springer is part of Springer Science+Business Media (www.springer.com)

Contents

Nanoemulsion as a Carrier for Topical Delivery of Aceclofenac

S. Choudhury, S. Dasgupta, D. K. Patel, Y. R. Ramani, S. K. Ghosh and B. Mazumder

Abstract The aim of the present study was to investigate the potential of a nanoemulsion formulated by excipients, selected for their emulsification ability rather than their solubilising ability of drug, for transdermal delivery of aceclofenac. The oil in water nanoemulsions was prepared by screening the excipients and by identifying nanoemulsion area after constructing pseudoternary phase diagram with the help of the selected excipients. The prepared nanoemulsions were subjected to different thermodynamic stability tests. The nanoemulsion formulations that passed thermodynamic stability tests were characterized for viscosity, droplet size, transmission electron microscopy and refractive index. *In vitro* release rate and permeation rate of aceclofenac nanoemulsion were determined. The *in vitro* release profile of optimized formulations (NE31) was compared with that of nanoemulsion gel (NG31) and marketed aceclofenac gel (HIG). *In vivo* anti-inflammatory efficacy study was also carried out for NE31, NG31 and HIG. Transmission Electron Microscopy (TEM) of the passed samples was carried out, which correlated with the particle sizes (390.48 nm) observed. A significant increase in release rate was observed in optimized nanoemulsion formulation NE31 (23.85 % Cremophor EL, 7.95 % PEG400, 13.6 % Triacetin). An increase in permeability parameters such as steady state flux (J_{ss}), permeability coefficient (K_p) and enhancement ratio (E_r) was observed in optimized nanoemulsion formulation NE31. On the basis of these parameters NE31was selected for *in vivo* studies, which in turn revealed a significant increase in anti-inflammatory effects as compared with marketed aceclofenac gel. The anti-inflammatory effects of

S. Choudhury · S. Dasgupta · D. K. Patel · S. K. Ghosh · B. Mazumder (✉)
Department of Pharmaceutical Sciences, Dibrugarh University, Dibrugarh, Assam 786004, India
e-mail: pharmadu10@gmail.com

Y. R. Ramani
Department of Pharmacology, M.K.C.G Medical College and Hospital, Brahmapur, Orissa, India

P. K. Giri et al. (eds.), *Advanced Nanomaterials and Nanotechnology*,
Springer Proceedings in Physics 143, DOI: 10.1007/978-3-642-34216-5_1,

formulation NE31 showed a significant increase in percent inhibition value when compared with control, this difference was found to be highly significant ($p < 0.001$). These results suggested that the aim to formulate a nanoemulsion by choosing the excipients on the basis of their emulsification capabilities was proved to have been accomplished.

Keywords Nanoemulsion · Emulsification capability · Aceclofenac · Topical delivery

1 Introduction

In the last few years, nanoemulsions have become one of the most researched subjects of formulation development. It has attained such popularity due to advantages like ease of formulation, non-toxicity and acting as an effective transport system topically.

Aceclofenac is a Non-steroidal anti-inflammatory drugs (*NSAID*) which is used for the treatment of osteoarthritis and rheumatoid arthritis [1–3]. It is a *Biopharmaceutics classification system class II* (*BCS class II*) drug which has an intermediate half-life of 3–4 h and undergoes substantial first pass metabolism. In the present day pharmaceutical market, aceclofenac is available either in oral form (tablet) or in topical form (gel). Oral administration of aceclofenac causes problems like gastrointestinal ulcer and bleeding upon chronic use which ultimately leads to anaemia. Thus topical formulations like gels were developed which not only negated these problems but also avoided first pass metabolism. But the drug release from these formulations was very low compared to those administered orally. So there was a need to develop a topical formulation with a better release profile. Nanoemulsions, owing to their small size, provide a large interfacial area for rapid drug release, and so the drug should exhibit an enhanced bioavailability, enabling a reduction in dose, more consistent temporal profiles of drug absorption, and the protection of drugs from the hostile environment of the body. *In addition, increasing* the rate of drug release, nanoemulsion can also be used as a reservoir and actually slow the release of drug and prolong its effect, thereby avoiding high concentrations in the blood. This improvement in transdermal delivery properties have been proved in many studies [4–19]. Though nanoemulsion formulations of aceclofenac have been already established but the process of selection of excipients is based on multiple parameters. This article describes the potential of nanoemulsion systems in transdermal delivery of aceclofenac using nonirritating, pharmaceutically acceptable ingredients selected for their emulsification ability rather than their solubilising ability of drug, for transdermal delivery of aceclofenac.

2 Experimental

2.1 Materials

Aceclofenac was a gift sample from Cipla ltd (Sikkim, India). Isopropyl myristate (IPM), oleic acid, castor oil, ethyl oleate, Tween 80 and polyoxy-35-castor oil (Cremophor EL) were purchased from HI MEDIA (Mumbai, India). Span 80 was purchased from Loba Chemicals. Glycerol triacetate (Triacetin) was purchased from E-Merck (Mumbai, India). All other chemicals used in the study were of analytical grade.

2.2 Screening of Oil

The drug chosen for the experiment was aceclofenac which is a BCS class II type of drug, which means it is a highly permeable but less soluble in water. *The molecular formula of aceclofenac is* $C_{16}H_{13}C_{12}NO_4$ *and structure is shown in* Fig. 1. The only way to deliver such a drug would be to encapsulate the drug in oil, which in turn should be so minute in size that passive diffusion of the droplets through skin barriers becomes a possibility. Since the aim of this study is to develop a transdermal formulation, therefore, solubility of drug in oils is more important as the ability of nanoemulsion to maintain the drug in solubilised form is greatly influenced by the solubility of the drug in oil phase. The solubility of aceclofenac in various oils was determined by adding an excess amount of drug in 5 ml of selected oils (Castor oil, Triacetin, IPM, Ethyl oleate, Oleic acid), in 15 ml capacity stoppered vials, and mixed using a magnetic stirrer for 24 h. The mixture vials were then kept at 25 ± 1.0 °C in an isothermal shaker for 72 h to reach equilibrium. The equilibrated samples were removed from shaker and centrifuged at 3,000 rpm for 15 min. The supernatant was taken and filtered. The concentration of aceclofenac was determined in the oils using UV spectrophotometer at 273 nm.

Fig. 1 Structure of aceclofenac

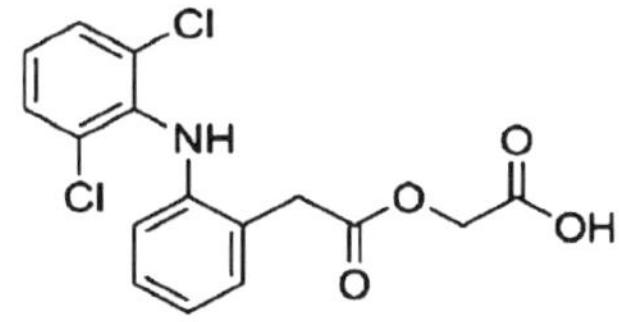

2.3 Screening of Surfactants

Emulsions are dispersions of one liquid phase in another immiscible liquid phase that are made using mechanical shear [20]. Due to differences in attractive interactions between the molecules of the two liquid phases, an interfacial tension, σ, exists between the two liquids which can be reduced by incorporating surfactants into the system. So to select the best surfactant from a pool, the emulsification ability of the surfactants was screened. 300 mg of surfactant was added to 300 mg of the selected oily phase. The mixture was gently heated at 45–60 °C for homogenizing the components. The isotropic mixture, 50 mg, was accurately weighed and diluted with distilled water to 50 ml to yield fine emulsion. The ease of formation of emulsions was monitored by noting the number of volumetric flask inversions required to give uniform emulsion. The resulting emulsions were observed visually for the relative turbidity. The emulsions were allowed to stand for 2 h to note for any change in turbidity and their transmittance was assessed at 273 nm by colorimeter using distilled water as blank.

2.4 Screening of Cosurfactants

Cosurfactants are the excipients added to the emulsion to further stabilise the interfacial film and prevent coalescence of the droplets. 100 mg of cosurfactant was added to 200 mg of the surfactant selected previously and the S_{mix} was added to the selected oily phase. The mixture was gently heated at 45–60 °C for homogenizing the components. The isotropic mixture, 50 mg, was accurately weighed and diluted with distilled water to 50 ml to yield fine emulsion. The ease of formation of emulsions was monitored by noting the number of volumetric flask inversions required to give uniform emulsion. The resulting emulsions were observed visually for the relative turbidity. The emulsions were allowed to stand for 2 h to note for any change in turbidity and their transmittance was assessed at 273 nm by colorimeter using distilled water as blank. As the ratio of co-surfactants to surfactant/s is the same, the turbidity of resulting nanoemulsions will help in assessing the relative efficacy of the co-surfactants to improve the nanoemulsification ability of surfactant/s.

2.5 Pseudoternary Phase Diagram

On the basis of the solubility studies of drug, Triacetin was selected as the oil phase. Cremophor EL was used as surfactant and PEG 400 used as cosurfactant. Distilled water was used as an aqueous phase for the construction of phase diagrams. Surfactant and cosurfactant (S_{mix}) were mixed in different weight ratios

[1:3, 0.5:1 (1:2), 1:1, 2:1 (1:0.5), 3:1]. These S_{mix} ratios were chosen in increasing concentration of surfactant with respect to cosurfactant and increasing concentration of cosurfactant with respect to surfactant for detailed study of the phase diagrams for formulation of nanoemulsion. For each phase diagram, oil and specific S_{mix} ratio was mixed thoroughly in different weight ratios from 1:9 to 9:1 in different glass vials. Sixteen different combinations of oil and S_{mix}, 1:9, 1:8, 1:7, 1:6, 1:5, 2:8 (1:4), 1:3.5, 1:3, 3:7 (1:2.3), 1:2, 4:6 (1:1.5), 5:5 (1:1), 6:4 (1:0.7), 7:3 (1:0.43), 8:2(1:0.25), 9:1 (1:0.1), were made so that maximum ratios were covered for the study to delineate the boundaries of phases precisely formed in the phase diagrams. Pseudo-ternary phase diagrams were developed using aqueous titration method. Slow titration with aqueous phase was done to each weight ratio of oil and S_{mix} and visual observation was carried out for transparent and easily flowable o/w nanoemulsions. The physical state of the nanoemulsion was marked on a pseudo-three-component phase diagram with one axis representing aqueous phase, the other representing oil and the third representing a mixture of surfactant and cosurfactant at fixed weight ratios (S_{mix} ratio). Care must be taken to ensure that observations are not made on metastable systems.

2.6 *Selection of Formulations from Phase Diagrams*

From each phase diagram constructed, different formulations were selected from the nanoemulsion region so that the drug could be incorporated into the oil phase. The formulation was chosen with the criteria of; maximum oil being emulsified with minimum amount of S_{mix}. Exactly 1.5 % wt/wt of aceclofenac, which was kept constant in all the selected formulations, was dissolved in the oil phase of the emulsion formulation. Selected formulations were subjected to different thermodynamic stability tests.

2.7 *Formulation of Nanoemulsions*

1.5 % wt/wt of aceclofenac was dissolved in the oil phase of the emulsion formulation. Distilled water was then added to the formulation drop by drop and the whole mixture was stirred at 200 rpm for 5 min with the help of a magnetic stirrer to formulate the nanoemulsion.

2.8 *Thermodynamic Stability Studies*

To overcome the problem of metastable formation, due to time constraints thermodynamic stability tests were performed [21], which were as follows:

1. Heating cooling cycle: Six cycles between refrigerator temperature (4 °C) and 45 °C with storage at each temperature of not less than 48 h was studied. Those formulations, which were stable at these temperatures, were subjected to centrifugation test.
2. Centrifugation: Passed formulations were centrifuged at 3,500 rpm for 30 min. Those formulations that did not show any phase separation were taken for the freeze thaw stress test.
3. Freeze thaw cycle: Three freeze thaw cycles between −21 and +25 °C with storage at each temperature for not less than 48 h was done for the formulations.

Those formulations, which passed these thermodynamic stress tests, were selected for further study.

3 Characterization of Nanoemulsions

3.1 Globule Size Analysis

One of the most important criterion of a nanoemulsion is the globule size of the emulsion being formed. Globule size of the nanoemulsion was determined by photon correlation spectroscopy that analyzes the fluctuations in light scattering due to Brownian motion of the particles [22], using a Zetasizer 1000 HS (Malvern Instruments, UK). Light scattering was monitored at 25 °C at a 90° angle.

3.2 Viscosity

The viscosity of the formulations (0.5 g) was determined as such without dilution using Brookfield DVE viscometer (Brookfield Engineering Laboratories, Inc., Middleboro, MA) using spindle no. 63 at different rpm at 25 ± 0.5 °C.

3.3 Transmission Electron Microscopy

Morphology and structure of the nanoemulsion were studied using transmission electron microscopy (TEM) JEM CX 100 II operating at 200 kV capable of point to point resolution. Combination of bright field imaging at increasing magnification and of diffraction modes was used to reveal the form and size of the nanoemulsion.

3.4 Refractive Index

The refractive index of placebo formulations and drug loaded formulations was determined using an Abbe-type refractometer (Macro Scientific Works, Delhi).

3.5 In Vitro Study

In vitro skin permeation studies were carried out on a Franz diffusion cell with an effective diffusional area of 3.14 cm^2 and 100 ml receiving chamber capacity using dialysis membrane 70. Formulations containing 1.5 % wt/wt drug were placed in the donor compartment and diffusion of the drug to the receiver compartment containing phosphate buffer (pH 7.4) was observed for 24 h at 37.5 ± 0.5 °C. Samples were withdrawn at regular intervals (0.5, 1, 2, 4, 8, 12, 18, 24 h), filtered and analyzed for drug content by UV spectrophotometer at λmax of 273 nm. The formulation NE31 provided the highest release as compared with other nanoemulsion formulation.

Thus the formulation NE31 was also converted into nanoemulsion gel formulation. 1 % wt/wt Carbopol 934 was dissolved in sufficient quantity of distilled water and to it Triacetin was added. The gel was neutralised using 0.5 % wt/wt Triethanolamine and to the subsequent gel Cremophor EL and PEG 400 was incorporated slowly in the same amount as present in NE31. After the addition of the S_{mix} rest of the water was added and the gel so formed was coded as NG31. It's *in vitro* release profile was compared with NE31 and with marketed formulation Hifenac gel.

3.6 Permeation Data Analysis

The permeation profiles were constructed by plotting the cumulative amount of aceclofenac permeated per unit dialysis membrane area (μg/cm^2) versus time. The steady state flux (J_{ss}, μg/cm^2/h) [23] of aceclofenac was calculated from the slope of the plot using linear regression analysis. The permeability co-efficient (K_p) of the drug through the membrane was calculated using Eq. (1) [24].

$$K_p = \frac{J_{ss}}{C} \tag{1}$$

where, C is the initial concentration of the drug in the donor compartment.

The penetration enhancing effect was calculated in terms of enhancement ratio (ER) by using Eq. (2)

$$E_r = \frac{J_{ss}\text{ of formulation}}{J_{ss}\text{ of control}}. \tag{2}$$

3.7 Skin Irritation Study

In skin irritation study 4 rats were used. Hairs were depleted from the back of rats with the help of depilatories and area was marked on both the side. One side served as control while other as test and animal were used after 24 h. After hair depletion gel was applied (500 mg/rats) once a day for 7 days, observations were made for any sensitivity and the reaction if any was graded as [25, 26]. (a) No reaction, (b) Slight, patchy erythema, (c) Moderate but patchy erythema, (d) Moderate erythema, (e) Severe erythema with or without edema.

3.8 In Vivo Efficacy Study

Approval to carry out *in vivo* studies was obtained from Institutional Animal Ethical Committee, M.K.C.G. Medical College and Hospital, Orissa, India. The anti-inflammatory and the sustaining action of the optimized formulation NE31 was evaluated using the carrageenan-induced hind paw edema method developed by Winter et al. in Wistar rats [27]. Young Wistar rats weighing 120–150 g were randomly divided into 4 groups; control, nanoemulsion (NE31), nanoemulsion gel (NG31) and marketed formulation (HIG), each containing 6 rats. The animals were kept in standard laboratory conditions and were fed a standard laboratory diet. NE31, NG31, HIG were applied to the hind left paws of the rats half an hour before the subplanter injection (1 % wt/wt carrageenan solution) was administered to those paws. The volume of the paw of the rats was measured using a digital plethysmometer (Orchid Scientifics, Nasik) before injection and at regular intervals (1, 2, 4, 8, 12, 24 h) after injection. The amount of paw swelling was determined for 24 h and expressed as percentage edema relative to initial hind paw volume. Percentage inhibition of edema produced by each formulation treated group was calculated against the respective control group.

$$\text{Edema rate percentage} = (V_t - V_0)/100 \times 1/100$$

where, $V_t =$ volume of paw at time t

$V_0 =$ volume of paw at start ($t = 0$)

$$\text{And, Inhibition rate percentage} = (E_c - E_t)/E_t \times 100 \times 1/100$$

where, $E_c =$ edema rate of the control group at time t

$E_t =$ edema rate of the treated group at time t

Results of the anti-inflammatory activity were compared using the Dunnett's t test of 1 way ANOVA.

4 Result and Discussion

4.1 Selection of Excipients

Solubility studies were aimed at identifying suitable oily phase and surfactant/s for the development of nanoemulsions. Identifying the suitable oil having the maximal solubilizing potential for drug under investigation is very important to achieve optimum drug loading [28, 29]. It is even more important for aceclofenac, as the target dose is substantially high.

4.2 Selection of Oil

Solubility of aceclofenac in oils is presented in Table 1. Amongst the various oils that were screened, triacetin could solubilise the maximum amount of drug. The selection of surfactant or co-surfactant in the further study was governed by their emulsification efficiency rather than their ability to solubilize aceclofenac.

4.3 Screening of Surfactants for Emulsifying Ability

The %transmittance values of various dispersions are given in Table 2. Emulsification studies clearly distinguished the ability of various surfactants to emulsify aceclofenac. These studies indicated that Cremophor EL had very good ability to emulsify aceclofenac followed by PEG 400, Tween 80 and Span 80. Although, the HLB values of the surfactants used in the investigation were above 10 except for span 80, there was a great difference in their emulsification ability. This observation is in line with the investigations reported by Malcolmson et al. [30] and Warisnoicharoen et al. [31] who concluded that microemulsification is also influenced by the structure and chain length of the surfactant. Cremophor-EL rendered very good nanoemulsion requiring short time for nanoemulsification and was selected for further investigation.

Table 1 Solubility of aceclofenac in oils

Oils	Mean concentration of drug (mg/ml) ± standard deviation (n = 3)
Castor oil	45.117 ± 0.973
Ethyl oleate	22.187 ± 0.160
IPM	3.418 ± 0.097
Oleic acid	4.703 ± 0.049
Triacetin	49.894 ± 0.184

Table 2 Emulsification capability of surfactants and cosurfactants

Surfactants	Maximum number of inversions	%transmittance	Cosurfactants + cremophor EL	Maximum number of inversions	%transmittance
Cremophor EL	21	90	PEG 400	11	98
Span 80	42	15	Span 80	36	42
Tween 80	37	47	Tween 80	29	64
PEG 400	30	62			

4.4 Screening of Co-surfactants

The investigations clearly distinguished the ability of various co-surfactants, both hydrophilic and lipophilic, to improve the nanoemulsification of selected surfactant/s. All the cosurfactants increased the spontaneity of the nanoemulsion formation. It was however interesting to note that addition of a polymeric cosurfactant improved the nanoemulsification ability of Cremophor EL (Table 2). In case of lipophilic surfactants the transmittance value was seen to decrease leading to the hypothesis of an inferior quality of emulsion being formed. Thus PEG 400 was selected as the cosurfactant for further investigation.

4.5 Construction of Pseudoternary Phase Diagram

Constructing phase diagrams is time-consuming, particularly when the aim is to accurately delineate a phase boundary [32]. Care was taken to ensure that observations were not made on metastable systems—although the free energy required to form an emulsion is very low, the formation is thermodynamically spontaneous [33]. The relationship between the phase behavior of a mixture and its composition can be captured with the aid of a phase diagram [34].

As per the earlier results

- Triacetin was taken as oil
- Cremophor EL was taken as surfactant
- PEG 400 was taken as cosurfactant.

Cremophor EL and PEG 400 were mixed in different volume ratios (3:1, 2:1, 1:1, 1:2, 1:3) to form the S_{mix} [21]. These S_{mix} ratios were chosen to reflect increasing concentrations of cosurfactant with respect to surfactant and increasing concentrations of surfactant with respect to cosurfactant for detailed study of the phase diagrams in the nanoemulsion formation.

For each phase diagram, oil (Triacetin) and specific S_{mix} ratio were mixed thoroughly in different volume ratios from 1:9 to 9:1 in different glass vials. Sixteen different combinations of oil and S_{mix} [1:9, 1:8, 1:7, 1:6, 1:5, 1:4, 1:3.5, 1:3, 1:2.33 (3:7), 1:2, 1:1.5 (4:6), 1:1, 1:0.66 (6:4), 1:0.43 (7:3), 1:0.25 (8:2),

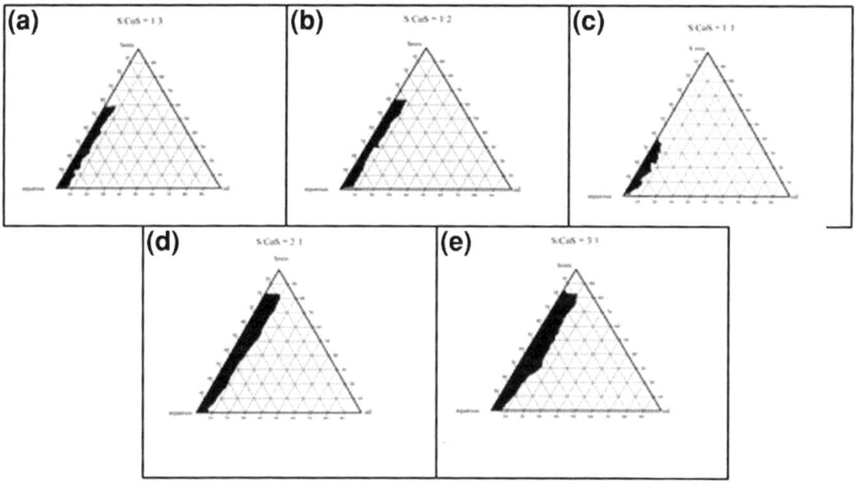

Fig. 2 a–e Pseudoternary phase diagrams of different S_{mix}

and 1:0.11 (9:1)] were made for the study to delineate the boundaries of phases precisely formed in the phase diagrams [21].

Only those mixtures which were found to be transparent even after being left standing for 24 h were designated to have formed the requisite nanoemulsion so that the system was given time to equilibrate.

From the pseudoternary Fig. 2a–e it was deciphered that the nanoemulsion region represented by the shaded region increased with an increase in the concentration of the surfactant. The maximum nanoemulsion region was obtained with the S_{mix} 3:1 and the minimum was obtained with the S_{mix} 1:1. As the surfactant concentration was increased in the S_{mix} ratio 2:1 and 3:1, a higher nanoemulsion region was observed, perhaps because of further reduction of the interfacial tension, increasing the fluidity of the interface, thereby increasing the entropy of the system. There may be greater penetration of the oil phase in the hydrophobic region of the surfactant monomers [32, 35]. The nanoemulsion was mostly formed at the water rich apex of the diagram further strengthening the fact that an o/w emulsion was being formed. The nanoemulsion region formed by S_{mix} ratio 4:1 was very small which may be due to absence of sufficient quantity of cosurfactant. Hence the S_{mix} ratio of 4:1 was avoided.

4.6 Selection of Nanoemulsion Formulations

It is well known that large amounts of surfactants cause skin irritation [34, 36]; therefore, it is important to determine the surfactant concentration properly and use the optimum concentration of surfactant in the formulation. The pseudoternary

Table 3 Composition of selected nanoemulsion formulations

% Wt/Wt of components in nanoemulsion formulation						
S_{mix}	Oil (%)	Surfactant (%)	Cosurfactant (%)	Water (%)	Oil:S_{mix} ratio	Code
1:3	6.6	3.9	11.6	77.9	1:2.34	NE13
1:2	7.4	8.6	17.3	66.7	1:3.5	NE12
1:1	7.7	11.7	11.7	68.9	1:3.03	NE11
2:1	11.11	36.7	18.3	33.89	1:4.95	NE21
3:1	13.6	23.9	7.9	54.6	1:2.33	NE31

phase diagrams were carefully observed and from them the formulations in which the amount of oil phase completely solubilized the drug and which could accommodate the optimum quantity of S_{mix} and distilled water were selected for the study.

From each phase diagram constructed, different formulations were selected from the nanoemulsion (Table 3) region so that the drug could be incorporated into the oil phase. Exactly 1.5 % wt/wt of aceclofenac, which was kept constant in all the selected formulations, was dissolved in the oil phase of the nanoemulsion formulation. Selected formulations were subjected to different thermodynamic stability tests.

4.7 Thermodynamic Stability Tests

Care must be taken to ensure that observations are not made on metastable systems. Clearly, however, time constraints impose a physical limit on the length of time systems can be left to equilibrate, and consequently the elimination of metastable states can be difficult to ensure in practice [29, 32, 34]. To overcome this problem of metastable formation, thermodynamic stability tests were performed. The formulations selected were subjected to different stress tests. If the nanoemulsions are stable over these conditions, metastable formulations are thus avoided.

The formulations that survived thermodynamic stability tests were selected for further study.

5 Characterization of Nanoemulsions

5.1 TEM

Morphology and structure of the nanoemulsion were studied using transmission electron microscopy (TEM), with JEM 100 CX II and capable of point-to-point resolution. To perform the TEM observations, 1 ml of the nanoemulsion was

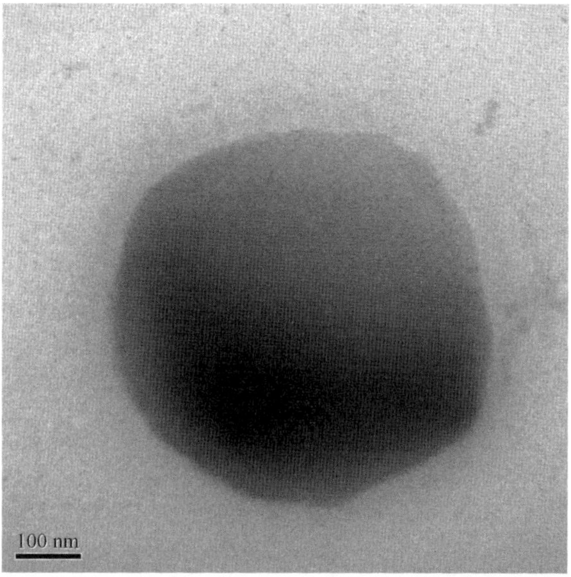

Fig. 3 TEM images of nanoemulsion formulation

diluted to 100 ml using distilled water and a drop of the resultant system was deposited on the holey film grid and observed after drying.

In the TEM positive image, the nanoemulsion appeared dark and the surroundings were bright (Fig. 3). Some droplet sizes were measured, as TEM is capable of point-to-point resolution. These sizes were in agreement with the droplet size distribution measured using photon correlation spectroscopy.

5.2 *Globule Size Analysis*

The droplet/globule size was found to increase with the increase in the concentration of the surfactants. The sample NE31 was found to have the droplets of the lowest size, followed by NE21 and NE11 (Table 4).

5.3 *Viscosity Determination*

The viscosity of the selected formulations was determined using spindle no. 63 at 50 rpm. The viscosity of NE31 was found to be the highest and with decreasing concentration of cosurfactant the viscosity was found to decrease (Table 4).

This deferred from one characteristic of nanoemulsion which is that they are usually of low viscosity [34]. Instead a hypothesis can be put forward that due to the higher viscosity of the nanoemulsions the droplets were not able to coalesce and thus retain their nanosize.

Table 4 Test result for nanoemulsion formulation

Code	Centrifugation	Heating cooling cycle	Freeze thaw cycle	Average droplet size (nm)	Viscosity at 50 RPM (cps)	Refractive index of formulations	
						Fresh	Placebo
NE13	✔	✔	–				
NE12	✔	✔	✔	649.75	98.8	1.372 ± 0.009	1.374 ± 0.004
NE11	✔	✔	✔	585.17	130.6	1.359 ± 0.002	1.353 ± 0.002
NE21	✔	✔	✔	412.45	221.4	1.389 ± 0.010	1.384 ± 0.001
NE31	✔	✔	✔	390.48	339.5	1.375 ± 0.003	1.372 ± 0.002

✔ Test passed, – Test failed

5.4 Refractive Index

The refractive index of placebo formulations and drug loaded formulations was determined using an Abbe-type refractometer (Macro Scientific Works). The values of the refractive index of drug-loaded formulations and placebo formulations are given in Table 4. When the refractive index values for formulations were compared with those of the placebo, it was found that there were no significant differences between the values. Therefore, it can be concluded that the nanoemulsion formulations were not only thermodynamically stable but also chemically stable and remained isotropic; thus, there were no interactions between nanoemulsion excipients and drug.

5.5 In Vitro Studies

In vitro studies were performed to compare the release rate of the drug from the various nanoemulsion formulations (NE31, NE21, NE11, NE12) and Hifenac gel manufactured by Intas pharmaceuticals limited, which was code named as HIG, all having the same quantity of aceclofenac (1.5 % wt/wt). *The release rate of NE*31 *was found to be the best* (Fig. 4) *and thus nanoemulsion gel* (*NG*31) was prepared and studied *in vitro*. The comparison between NE31, NG31 and HIG showed that even though the release rate of NG31 was less than NE31 but it was significantly more than that of HIG. Thus NE31, NG31 and HIG were used for *in vivo* efficacy study.

5.6 Permeation Data Analysis

Permeability parameters like steady-state flux (J_{ss}), permeability coefficient (K_p), and enhancement ratio (E_r) were significantly increased in nanoemulsions and the NG31 formulation as compared with marketed formulation HIG. This is because nanoemulsions and NG31 excipients contain permeation enhancers like Triacetin,

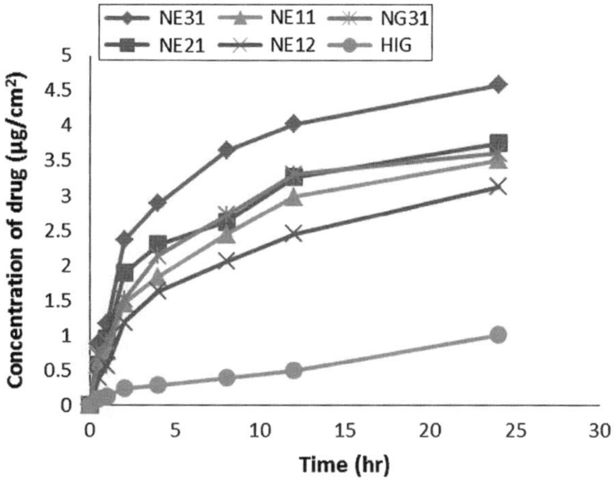

Fig. 4 Comparative drug release profile of NE31, NE21, NE11, NE12, NG31 and HIG

Table 5 Permeability parameters of nanoemulsion formulations

Sample code	J_{ss} ± SD (μg/cm^2/hr)	K_p ± SD (cm/hr) × 10^{-2}	Enhancement ratio (E_r)
HIG	43.667 ± 2.108	0.291 ± 0.014	–
NE31	254.900 ± 1.253	1.699 ± 0.008	5.837
NE21	204.033 ± 4.650	1.360 ± 0.031	4.672
NE11	187.267 ± 3.262	1.248 ± 0.022	4.289
NE12	162.033 ± 3.667	1.080 ± 0.024	3.711
NG31	199.600 ± 6.928	1.331 ± 0.046	4.571

Cremophor EL, and PEG 400. The permeability parameters of different formulations are given in Table 5.

5.7 Skin Irritation Study

Following seven days application of the gel, the results of skin irritation test were observed. The results indicated that NE31, NG31, Hifenac gel (HIG) (marketed product) and Blank NE31 gel did not cause any skin allergy or any other *reactions* (*grade A-no reaction*). It was assured that prepared gel did not cause any skin irritation and could be used in the gel formulation for topical application.

5.8 In Vivo Efficacy Study

Based on higher drug release and lowest droplet size formulation NE31 was chosen for the study of *in vivo* anti-inflammatory effects. The anti-inflammatory and sustaining action of the optimized formulation was evaluated by carrageenan

Table 6 Stastistical analysis of *in vivo* results

Groups	Edema rate (%) ± SD [Inhibition rate (%)]						
	0 h	1 h	2 h	4 h	8 h	12 h	24 h
Control	0 ± 0.01	36.50 ± 0.02	44.47 ± 0.02	46.79 ± 0.02	47.81 ± 0.01	43.70 ± 0.02	37.28 ± 0.02
HIG	0 ± 0.01 [0]	34.09 ± 0.02* [6.61]	38.89 ± 0.02* [12.56]	36.62 ± 0.03* [21.74]	36.87 ± 0.02* [22.89]	28.28 ± 0.02* [35.28]	21.72 ± 0.02* [41.74]
NE31	0 ± 0.03 [0]	29.90 ± 0.01* [18.09]	28.39 ± 0.02* [36.16]	23.12 ± 0.02* [50.59]	18.09 ± 0.02* [62.17]	14.82 ± 0.02* [66.08]	10.05 ± 0.02* [73.04]
NG31	0 ± 0.01 [0]	33.16 ± 0.02* [9.15]	30.89 ± 0.01* [30.55]	27.59 ± 0.02* [41.02]	22.28 ± 0.02* [53.41]	18.73 ± 0.02* [57.13]	12.91 ± 0.03* [65.36]

*Edema rate (%) values were statistically significant from the saline control using one way ANOVA followed by Dunnett's t- test at $p < 0.001$ (*)

induced hind paw inflammation method developed by Winter et al. in Wistar rats. The rate percentage inhibition of inflammation of rat paws by NE31, NG31 and HIG was evaluated. 0.1 ml of carrageenan solution (1 % wt/wt) was administered to the hind paws of the rats 2 h after the formulations were applied. Both the formulation NE31 and NG31 under study not only decreased the inflammation to the larger magnitude, but also sustaining the magnitude. *In the first hour, the percent edema for NE*31 *and NG*31 *were* 29.90 *and* 33.16 % *respectively* (shown in Table 6). The percentage edema inhibition for NE31 and NG31 was 18.09 and 09.15 % respectively while in case of Hifenac gel percentage edema was 34.09 % and percentage inhibition was 6.61 %. Hence both NE31 and NG31 formulation of aceclofenac remained superior to the marketed product in its ability to suppress edema starting from third hour and throughout the whole study time. The enhanced anti-inflammatory effect of formulation NE31 could be due to the enhanced permeation of aceclofenac through the skin.

6 Conclusion

From the results as obtained and discussed earlier, we can project that the formulation of nanoemulsion of aceclofenac using the excipients, that is, the surfactant and cosurfactant chosen on the basis of their emulsifying capability, is a possibility. The permeation of the drug from the formulation was also found to be enhanced in comparison to the marketed formulation, making the formulation suitable for transdermal application.

On the basis of highest drug release, lowest droplet size and optimum surfactant and cosurfactant concentration NE31(23.85 % Cremophor EL, 7.95 % PEG400, 13.6 % Triacetin) was selected for *in vivo* studies, which in turn revealed a significant increase in anti- inflammatory effects as compared with marketed aceclofenac gel. Thus the aim to formulate a nanoemulsion, by choosing the excipients on the basis of their emulsification capabilities was proved to have been established. Scaling up of such a protocol is also a possibility since it can be easily mechanised and the formation of such a nanoemulsion would only require a brisk mixing of oil and water phases in presence of excipients.

Thus, such a protocol can be followed for the development of other BCS class II drugs in future.

References

1. E. Escribano, A.C. Calpena, J. Queralt, R. Obach, J. Domenech, Assessment of diclofenac permeation with different formulations: anti-inflammatory study of a selected formula. Eur. J. Pharm. Sci. **19**, 203–210 (2003)
2. R. Yamazaki, S. Kawai, T. Mastsuzaki, Aceclofenac blocks prostaglandin E2 production following its intracellular conversion into cyclooxygenase inhibitors. Eur. J. Pharmacol. **329**, 181–187 (1997)

3. E. Gonzalez, C. Cruz, R. Nicolas, J. Egido, G. Herrero-Beaumont, Long-term effects of nonsteroidal anti-inflammatory drugs on the production of cytokines and other inflammatory mediators by blood cells of patients with osteoarthritis. Agents Actions **41**, 171–178 (1994)
4. D.W. Osborne, A.J. Ward, K.J. Neil, Microemulsions as topical delivery vehicles: in vitro transdermal studies of a model hydrophilic drug. J. Pharm. Pharmacol. **43**, 450–454 (1991)
5. M. Trotta, F. Pattarino, M.R. Gasco, Influence of counter ions on the skin permeation of methotrexate from water-oil microemulsions. Pharm. Acta Helv. **71**, 135–140 (1996)
6. M.B. Delgado-Charro, G. Iglesias-Vilas, J. Blanco-Mendez, M.J. Lopez-Quintela, M.A. Marty, J.P. Guy, Delivery of a hydrophilic solute through the skin from novel microemulsion systems. Eur. J. Pharm. Biopharm. **43**, 37–42 (1997)
7. F. Dreher, P. Walde, P. Walter, E. Wehrli, Interaction of a lecithin microemulsion gel with human stratum corneum and its effect on transdermal transport. J. Control Rel. **45**, 131–140 (1997)
8. U. Schmalfus, R. Neubart, W. Wohlrab, Modification of drug penetration into human skin using microemulsions. J. Control Rel. **46**, 279–285 (1997)
9. M. Kreilgaard, E.J. Pedersen, J.W. Jaroszewski, NMR characterization and transdermal drug delivery potentials of microemulsion systems. J Control Rel. **69**, 421–433 (2000)
10. M.J. Alvarez-Figueroa, J. Blanco-Mendez, Transdermal delivery of methotrexate: iontophoretic delivery from hydrogels and passive delivery from microemulsions. Int. J. Pharm. **215**, 57–65 (2001)
11. Y.S. Rhee, J.G. Choi, E.S. Park, S.C. Chi, Transdermal delivery of ketoprofen using microemulsions. Int. J. Pharm. **228**, 161–170 (2001)
12. P.J. Lee, R. Langer, V.P. Shastri, Novel microemulsion enhancer formulation for simultaneous transdermal delivery of hydrophilic and hydrophobic drugs. Pharm. Res. **20**, 264–269 (2003)
13. J. Kemken, A. Ziegler, B.W. Muller, Influence of supersaturation on the pharmacodynamic effect of bupranolol after dermal administration using microemulsions as vehicle. Pharm. Res. **9**, 554–558 (1992)
14. M. Kreilgaard, Dermal pharmacokinetics of microemulsion formulations determined by in vitro microdialysis. Pharm. Res. **18**, 367–373 (2001)
15. M. Kreilgaard, M.J.B. Kemme, J. Burggraaf, R.C. Schoemaker, A.F. Cohen, Influence of a microemulsion vehicle on cutaneous bioequivalence of a lipophilic model drug assessed by microdialysis and pharmacodynamics. Pharm. Res. **18**(593–599), 11 (2001)
16. G. Ktistis, I. Niopas, A study on the in vitro percutaneous absorption of propranolol from disperse systems. J. Pharm. Pharmacol. **50**, 413–419 (1998)
17. M.R. Gasco, M. Gallarate, F. Pattarino, In vitro permeation of azelaic acid from viscosized microemulsions. Int. J. Pharm. **69**, 193–196 (1991)
18. K. Kriwet, C.C. Muller-Goymann, Diclofenac release from phospholipid drug systems and permeation through excised human stratum corneum. Int. J. Pharm. **125**, 231–242 (1995)
19. M. Trotta, Influence of phase transformation on indomethacin release from microemulsions. J. Control Rel. **60**, 399–405 (1999)
20. S.J. Carter, in *Dispensing for pharmaceutical students*, 12 edn. (2000), p. 120
21. S. Shafiq-un-Nabi, F. Shakeel, S. Talegaonkar, Development and bioavailability assessment of ramipril nanoemulsion formulation. Eur. J. Pharm. Biopharm. **66**, 227–243 (2007)
22. D. Attwood, L.R.J. Currie, P.H. Elworthy, Studies of Solubilized miceller solutions: phase studies and particle-size analysis of solution formed with non-ionic surfactants. J. Coll. Interf. Sci. **46**, 249–254 (1974)
23. D.A. Edwards, R. Langer, A linear theory of transdermal transport phenomena. J. Pharm. Sci. **83**, 1315–1334 (1994)
24. M.A. Yumune, A.C. Williams, B.W. Barry, Terpenes penetration enhancers in propylene glycol/water co-solvent system effectiveness and mechanism of action. J. Pharm. Pharmacol. **47**, 978–989 (1995)
25. P.R. Prakash, N.R. Rao, S. Chowdary, Formulation, evaluation and anti-inflammatory activity of topical etoricoxib gel. Asian J. Pharm. Clin. Res. **3**(2), 126–129 (2010)

26. S.K. Kulkarni, N.K. Jain, Pharmacological and Pharmacokinetic studies on marketed gel formulations of nimesulide. Indian Drugs **38**(2), 63–66 (2001)
27. C.A. Winter, Anti-inflammatory testing methods: comparative evaluation of indomethacin and other agents. Nonsteroidal Anti-inflamm. Drugs **52**, 190–202 (1965)
28. C.W. Pouton, Formulation of self-emulsifying drug delivery systems. Adv. Drug Del. Rev. **25**, 47–58 (1997)
29. C.W. Pouton, Lipid formulations for oral administration of drugs: nonemulsifying, self-emulsifying and 'self-microemulsifying' drug delivery systems. Eur. J. Pharm. Sci. **11**, S93–S98 (2000)
30. C. Malcolmson, A. Sidhu, C. Satra, S. Kantaria, M.J. Lawrence, Effect of the nature of oil on the incorporation of testosterone propionate into non-ionic oil-in-water microemulsions. J. Pharm. Sci. **87**, 109–116 (1998)
31. W. Warisnoicharoen, A.B. Lansley, M.J. Lawrence, Nonionic oil-in water microemulsions: the effect of oil type on phase behavior. Int. J. Pharm. **198**, 7–27 (2000)
32. J. Eccleston, Microemulsions, in *Encyclopedia of Pharmaceutical Technology*, vol. 9, ed. by J. Swarbrick, J.C. Boylan (Marcel Dekker, New York, 1994), pp. 375–421
33. D.Q.M. Craig, S.A. Barker, D. Banning, S.W. Booth, An investigation into the mechanisms of self-emulsification using particle size analysis and low frequency dielectric spectroscopy. Int. J. Pharm. **114**, 103–110 (1995)
34. M.J. Lawrence, G.D. Rees, Microemulsion-based media as novel drug delivery systems. Adv. Drug Deliv. Rev. **45**, 89–121 (2000)
35. K. Kawakami, T. Yoshikawa, Y. Moroto et al., Microemulsion formulation for enhanced absorption of poorly soluble drugs, I: prescription design. J. Control Rel. **81**, 65–74 (2002)
36. P. Li, A. Ghosh, R.F. Wagner, S. Krill, Y.M. Joshi, A.T.M. Serajuddin, Effect of combined use of non-ionic surfactant on formation of oil-in water microemulsions. Int. J. Pharm. **288**, 27–34 (2005)

Silver Tip Formation on Colloidal CdSe Nanorods by a Facile Phase Transfer Protocol

Tanushree Bala and Kevin M. Ryan

Abstract A facile phase transfer procedure is described for the formation of uniform silver metal tips on II–VI semiconductor nanorods. Judicious choice of a functional ligand dimethyl phenol (DMP) which binds to the semiconductor rod in the organic phase enables the transfer of metal ions from the aqueous phase and their reduction onto the nanorod. The nanorod hybrids can be assembled into perpendicularly aligned arrays by simple solvent evaporation.

Keywords CdSe nanorods · Ag tipping · Phase transfer

1 Introduction

Colloidal semiconductor nanorods synthesised by precise shape control of nanocrystals have found significant application in photovoltaics, photocatalysis and field emission devices [1–5]. These bulk applications are possible as the nanorods can be readily assembled into scalable perpendicular architectures where length dependent properties such as total absorption or emission can be independently tuned from diameter dependent properties such as band gap [2, 6–8]. Recent advances allow the formation of gold tips on the ends of the semiconductor nanorods as a sequential synthetic step in solution or by post-synthetic spin casting

T. Bala (✉)
Department of Chemistry, University of Calcutta, Kolkata, India
e-mail: tanushreebala@gmail.com

T. Bala · K. M. Ryan
Materials and Surface Science Institute, University of Limerick, Limerick, Ireland
e-mail: kevin.m.ryan@ul.ie

P. K. Giri et al. (eds.), *Advanced Nanomaterials and Nanotechnology*,
Springer Proceedings in Physics 143, DOI: 10.1007/978-3-642-34216-5_2,

on a preformed assembly [9–21]. In Au–CdSe, enhanced photocatalytic reactions are possible in solution through charge separation at the M–SC interface [6, 20, 22]. More recently, electron transfer via redox processes in a protein immobilised on metal tipped nanorod arrays was detected at a substrate demonstrating potential for sensor applications [23]. A facile route to Au tip formation has been achieved where the metal precursor either ($AuCl_3$) or more recently (AuCl) and a reducing agent are solubilized in the organic phase using surfactants. Control of the reaction time, temperature and light activation can tune the structure from double tipped (both ends) to single tip to enlarged single tip [8–14]. The organic phase reduction methods while extremely successful with Au-tipping limit the extension of the tipping protocol to other metals due to the insolubility of their metal salts in non-polar solvents. Tipping using organometallic precursors has been successful with Pt and Co although complex reaction schemes are necessary for these non-commercially available precursors [24–27]. A successful approach allowed Au and Pt tipping on rods from their water soluble salts whereby the nanorod was transferred to the aqueous phase using ligand exchange followed by reduction of the water soluble metal precursor. The tip growth is pH dependent and reaction times of up to 2 days are required [24, 28].

Here we describe a simplified protocol to yield instantaneous Ag–semiconductor nanorod hybrid structures where instead of transferring the nanorod to the aqueous phase, the metal ions are phase transferred from the aqueous to the organic phase. This we achieve by the introduction of suitable organic ligand molecules onto the surface of the nanorods possessing the bi-functionality of phase transfer and reducing capability. Tipping by this phase transfer protocol occurs at room temperature on CdSe nanorods. This method avoids the preference for the Ag ion to cation exchange with Cd forming Ag_2Se. The tipped nanorods also readily assemble into close packed super lattices from solution with the c-axis of the rod perpendicular to the substrate. The density of packing and proper assembly [29] will likely find significant application where silver metal–semiconductor hybrids are needed in high density arrays such as photovoltaics or photocatalysis [8–10].

2 Experimental Section

2.1 Synthesis of CdSe Nanorods

Materials: Cadmium oxide (CdO) was purchased from Fluka, trioctylphosphine (TOP, 90 %), trioctylphosphine oxide (TOPO, 99 %), sulfur (S), silver nitrate ($AgNO_3$), dimethyl phenol ($C_8H_{16}O$, mentioned as DMP in this paper), and toluene were purchased from Aldrich, and n-octadecylphosphonic acid (ODPA) was obtained from Poly Carbon Industries Inc. (PCI).

CdSe nanorods were synthesised with slight modifications to the published procedure [30].

2.2 Synthesis of Ag–Semiconductor Hybrids

In a typical reaction, 10 ml of 10^{-1} M DMP in toluene was injected into 15 ml of nanorod solution and the resulting mixture was sonicated before precipitating with acetone/toluene mixture. The solution was then centrifuged at 5,000 rpm for 10 min. The colorless supernatant was decanted off and the above process was repeated for 3 times. The precipitate was then re-dispersed in toluene maintaining a concentration of 0.06 mg ml^{-1}, a part of which ($\sim$5 ml) was then stirred with 5×10^{-3} M aqueous $AgNO_3$ in the presence of 5×10^{-5} M KOH solution. Aliquots of the top toluene layer were separated from the biphasic mixture at different time intervals for characterization.

2.3 Characterization Techniques

Samples were characterized using Transmission electron microscopy (JEOL 2011) at an accelerating voltage of 200 kV, UV–Vis spectroscopy (Cary-300 Bio UV–Vis spectrophotometer) operated at a resolution of 1 nm, fluorescence spectroscopy (Varian Cary Eclipse Fluorescence Spectrophotometer), X-Ray diffraction (PANalyticalX'Pert MPD Pro) using Cu Ka radiation with a 1-DX' Celerator strip detector, Fourier transform infrared spectroscopy (Perkin-Elmer-Spectrum One FTIR), X-Ray photoelectron spectroscopy (Kratos Axis 165 spectrometer). XPS high resolution spectra were taken using monochromated Al Kα radiation of an energy of 1,486.6 eV at fixed pass energy of 20 eV. For peak synthesis, a mixed Gaussian–Lorenzian function with a Shirley type background subtraction was used. Samples were flooded with low energy electrons for efficient charge neutralisation. Binding energies (BE) were determined using C 1 s at 284.8 eV as charge reference.

3 Results and Discussion

Stirring a biphasic mixture of organically soluble di-methyl phenol (DMP) with aqueous $AgNO_3$ in the presence of 5×10^{-5} M KOH resulted in the formation of Ag nanocrystals (Fig. 1a) in the organic medium (yellow colour). Nanocrystals were FCC Ag of the order of 5–10 nm in size with the density dependent on the concentration of the DMP (Fig. 1b). The introduction of CdSe nanorods that were pre-treated with 10^{-1} M DMP into the organic phase, with similar concentrations of $AgNO_3$ and KOH in the aqueous phase resulted in the nucleation of metal dots (<1 nm) onto the rod surface without additional free metal nanocrystal formation. The formation of the hybrid was schematically represented in Fig. 1c. The growth of these Ag dots was reaction time dependent with multiple grains occurring after

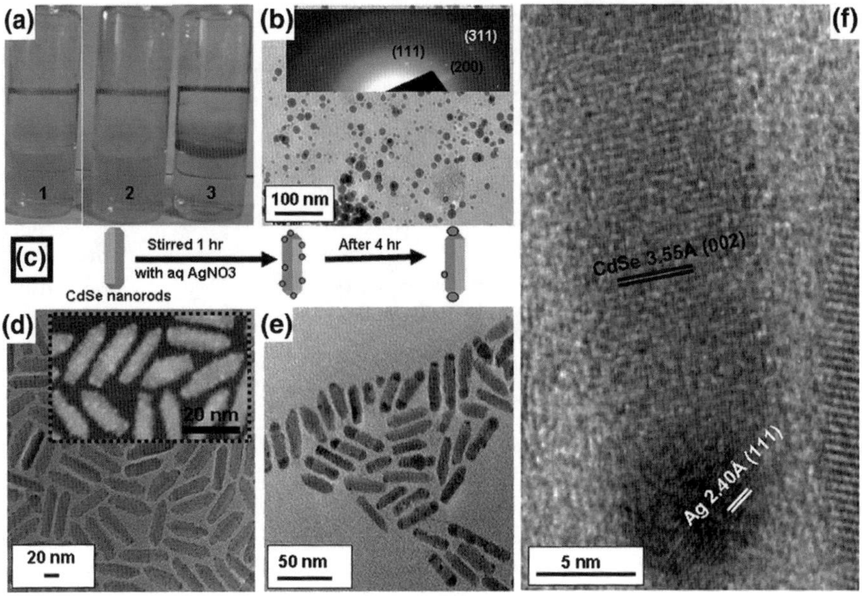

Fig. 1 **a** Photographs of the Ag nanoparticles formed using 1×10^{-4} (*photo 1*), 1×10^{-3} (*photo 2*) and 1×10^{-2} M (*photo 3*) aqueous Ag + and 10^{-2} M DMP in toluene. **b** TEM and SAED of Ag nanoparticles. **c** Schematic representation of the formation of Ag tips on bare CdSe nanorods. **d** TEM of the anisotropic growth of Ag nanoparticles all over the surface of the nanorods at t = 1 h. *The inset* shows STEM of Ag tipped CdSe nanorods. **e** Ag tips only at the ends of CdSe nanorods at t = 4 h. **f** HRTEM of Ag tipped CdSe nanorods aligned horizontally revealing the planes corresponding both to CdSe and Ag

1 h stirring of the biphasic mixture which was clearly evidenced in the TEM image, Fig. 1d, and the corresponding STEM image in the inset of the same.

The Ag dots showed a strong preference for migration towards the ends of the CdSe rods on prolonged reaction (~4 h), ripening to a larger metal tip (4–7 nm) on one or both ends (TEM image, Fig. 1e). The persistence of some unripened dots on the lateral sides of the CdSe nanorods was also evident. HRTEM of a single tipped rod allowed characterisation of the CdSe (002) and Ag (111) lattice spacings at 3.55 and 2.40 A respectively conclusively proving the growth of Ag metal on CdSe nanorods, Fig. 1f. The presence of the semiconductor nanorods in the organic phase (toluene) and the surface attachment of the DMP molecules clearly enabled the selective transfer of Ag^+ from the aqueous phase and its reduction onto the nanorod. Though DMP molecules were not common for reduction of Ag^+ to form Ag nanoparticles, phenolic OH groups had previously been shown to lead to the reduction of Ag^+ in an alkaline medium [31]. Crucially, the attachment of the DMP to nanorods resulted in metal formation on the nanorod structure whereas in solution it capped the Ag nanoparticles forming a stable dispersion. The phenomenon of multiple Ag dots formation and their Oswald ripening to form one larger single tip could be explained in terms of significant lattice mismatch for Ag and the semiconductor materials used in this study due to a strain relief mechanism [9, 32–34]. The lattice mismatch and associated strain relief had already been reported to grow small islands of noble metals of a critical atom density which ripen into larger islands as the reaction continued. Robinson and co-workers observed similar strain-dependent island formation on CdS nanorods during the partial cation exchange of Cd^{2+} with Ag^{+} [33]. However, as our protocol favoured reduction of the metal under conditions not conducive to Cd ion removal, cation exchange was not expected as confirmed by detailed characterisation.

The UV–Vis spectrum of as synthesized CdSe NRs showed two peaks at 552 and 674 nm (Fig. 2a, curve 1), which were characteristic of the excitonic band transitions [35]. The absorption spectrum of the Ag–CdSe nanorod hybrid retained the peak at 674 nm and showed a peak at 421 nm. The latter corresponded to the surface Plasmon resonance peak of Ag nanoparticles (Fig. 2a, curve 2). The appearance of a surface plasmon peak characteristic of Ag nanoparticles clearly indicated the reduction of Ag^{+} to form Ag nano-dots onto the semiconductor nanoparticles [36, 37]. The PL peak for CdSe nanorods was found to be centred at ~684 nm, after exciting the sample at 400 nm (Fig. 2b, curve 1). The emission was found to be quenched heavily after the formation of Ag–CdSe nanorod hybrid structures (Fig. 2b, curve 2). This kind of quenching of emission peak was probably due to hybrid formation and had previously been reported. It could be explained due to efficient electron transfer from semiconductors to metal nanoparticles [9–11, 38]. The dramatic quenching of emission was suggestive of charge separation at a silver metal–CdSe semiconductor interface. The CdSe–Ag_2Se interface in contrast is an abrupt n–n heterojunction which had a limited effect on the PL intensity as observed in previous studies [39].

The X-ray diffractograms of CdSe nanorods after reaction with $AgNO_3$ revealed the presence of Ag (111), (200) and (220) peaks (Fig. 3a), confirming the presence of fcc Ag with a lattice parameter estimated to be 4.08 A (PCPDF no. 04-0783). It also showed the basic pattern of wurtzite CdSe remained unchanged indicating equally good crystallinity even after the tip formation. The comparatively sharp (002) peak of CdSe indicated the growth of the nanorods

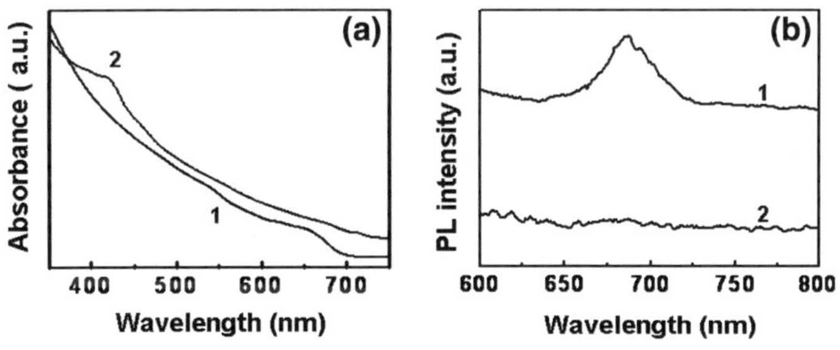

Fig. 2 **a** UV–Vis spectra and **b** photoluminescence spectra of CdSe nanorods, before and after the hybrid formation. In both cases, *curve 1* indicates the as synthesized nanorods whereas *curve 2* corresponds to semiconductor–Ag hybrids

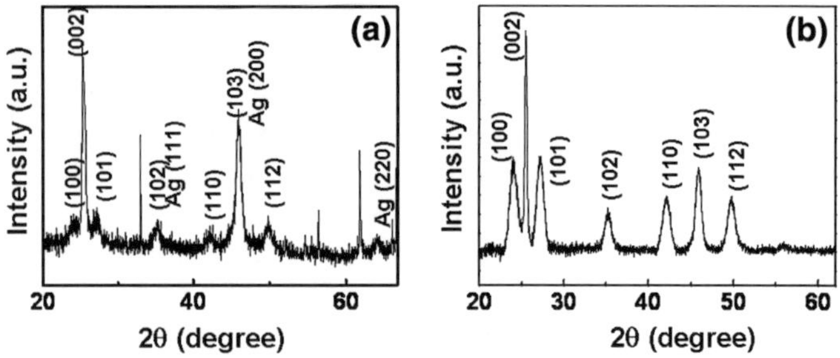

Fig. 3 **a** X-Ray diffractograms of CdSe–Ag hybrids. The peaks for Ag were mentioned in the curve and the rest were matched with wurtzite CdSe. **b** XRD of pristine CdSe nanorods

along this direction which was in exact agreement with the HRTEM in Fig. 1. XRD analysis of pristine CdSe was also performed for comparison which matched well with the wurtzite CdSe (Fig. 3b).

A systematic XPS analysis was carried out on the pristine CdSe nanorods as well as on the hybrid (Fig. 4a–e). In both the cases Cd 3d peak was fitted with two components corresponding to $3d_{5/2}$ and $3d_{3/2}$ at 405.0 and 411.7 eV respectively, Fig. 4a, c. The fitting clearly showed the presence of single species for Cd coming from CdSe. The deconvolution of Se 3d peak showed the presence of two decoupled components at 53.5 and 54.6 eV. (Fig. 4b, d) due to $3d_{5/2}$ and $3d_{3/2}$ which appeared from CdSe. Absence of any extra components in the fitted curve ruled out the probability of the presence of un-reacted Se. The Ag 3d doublet obtained from the hybrids appears at 367.9 and 373.9 eV, with a separation of 6.0 eV and full width at half maximum of 0.9 eV, characteristic of metallic silver [40] (Fig. 4e). However, since the energy differences between the different chemical states of Ag were <0.5 eV, formation of Ag compounds could not be ruled out entirely solely depending on XPS [39]. However, in combination with HRTEM, UV–Vis, and XRD, the characterizations were indicative of metallic Ag tips on rods instead of Ag_2X (X = S, Se) [41] suggesting the phase transfer of Ag^+ ions and their reduction by DMP occurred faster than cation exchange processes [42].

FTIR analyses were also carried out on CdSe rods capped with ligands (Fig. 5) showing the characteristic peaks at 1,463, 1,376, 1,151 and 1,087 cm^{-1} which were related to $-CH_3$ groups of the capping ligand, P = O of alkyl phosphonic acids [43], P = O of TOPO [44] and P–O–H of alkyl phosphonic acids [44] respectively, curve 1, Fig. 5. After modification of the surface with DMP, curve 2, Fig. 5, a new peak at 1,409 cm^{-1} and a broad band at 3,250 cm^{-1} emerged corresponding to –OH in plane bending vibration and stretching vibration respectively of the molecule. Though it was difficult to conclude on the surface ligand composition solely based on these FTIR analyses [45], it was observed that the –OH stretching frequency was damped out after the formation of Ag tips on the particles (Fig. 5, curve 3). This further supported the proposed role of phenolic –OH of DMP in the formation

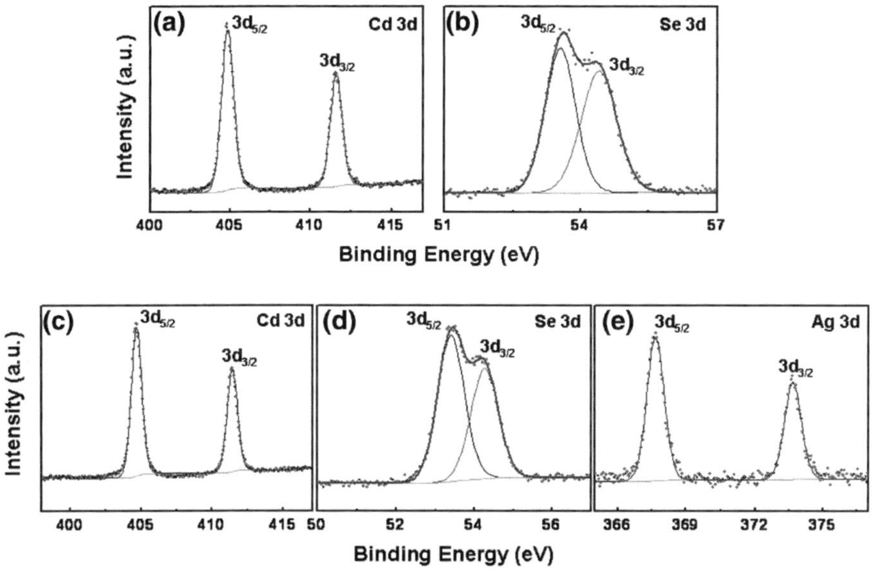

Fig. 4 **a, b** X-Ray photoelectron spectroscopic analysis of Cd 3d, Se 3d from neat CdSe nanorods. **c–e** Same for Cd 3d, Se 3d and Ag 3d from Ag tipped CdSe nanorods

Fig. 5 FTIR analysis of the pristine CdSe rods (*curve 1*), CdSe nanorods after modifying the surface with DMP (*curve 2*) and after the formation of CdSe rod–Ag hybrid (*curve 3*)

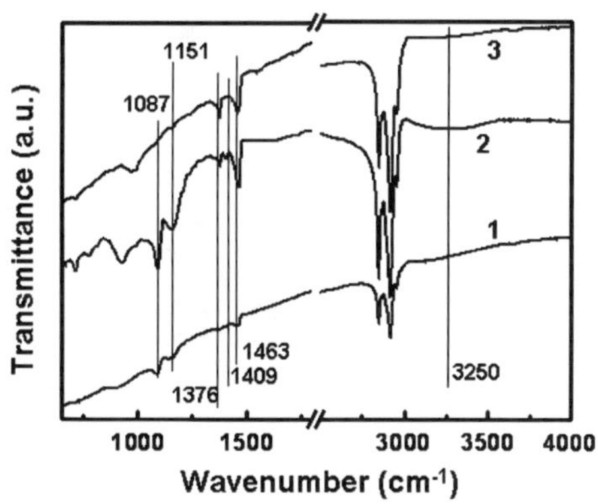

of Ag tips on the CdSe rods. It was interesting to note that in the absence of DMP molecules none of the nanorods allowed the growth of any Ag dots which again validated our proposition that DMP played the important roles of phase transfer and reduction of the metal ions site selectively on the surface of the semiconductor materials to synthesize these hybrids.

Through careful control of concentration, Ag-tipped CdSe rods readily organized into close packed vertical assemblies from solution, as observed from TEM and STEM analyses (Fig. 6a and inset). Arrows highlighted the non-assembled

Ag–CdSe rods at the edge of the super lattice. A further magnified TEM image clearly showed the vertically arranged Ag tipped wurtzite rods adopted a hexagonal close packed super-structure with individual rods separated at 1–2 nm by interdigitation of the surfactant ligands, Fig. 6b. The planes corresponding to both CdSe and Ag could be distinguished on the top end of a vertical rod through HRTEM imaging (Fig. 6c). The tipped rods adopted an orthogonal orientation with the CdSe component in contact with the substrate similar to Au-tipped CdSe structures reported by Zhao et al. [17] and O'Sullivan et al. [9, 23]. Brighter contrast due to Ag tips was also clearly visible on vertically aligned CdSe rods under a scanning electron microscope, Fig. 6d. The SEM image of some of the laterally aligned rods is shown in the inset, Fig. 6d.

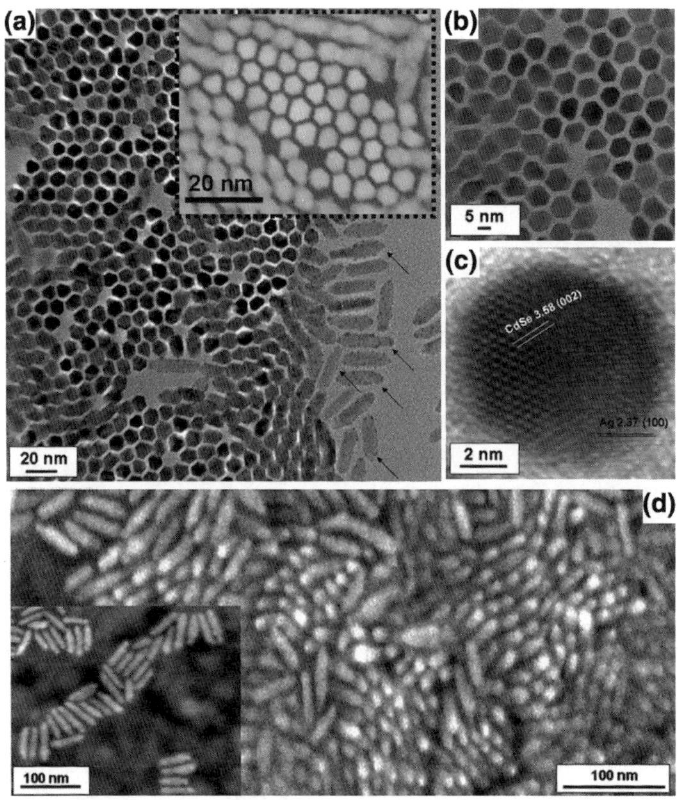

Fig. 6 Vertically assembled Ag-tipped CdSe nanorods at lower **a** and at higher magnification **b**. The STEM image of these assembled hybrids are shown in the inset of (**a**). **c** HRTEM of the vertically aligned hybrid nanorod showing the planes corresponding to both CdSe and Ag. **d** The SEM image of the Ag-tipped CdSe nanorods mainly aligned vertically on the substrate along with few horizontally aligned rods in the inset

4 Conclusion

In summary, we demonstrated a novel synthetic protocol for the formation of Ag noble metal tips on semiconductor nanorods where water soluble metal precursors were transferred to organic media and selectively reduced on nanorods in a single step. This facile route was mediated by an organic ligand (phenolic-OH) which could mediate phase transfer in addition to metal ion reduction at room temperature, thus avoiding the use of costly organometallic precursors and/or toxic organic solvents. The procedure can be extended to a generalized method for different noble metal tipping on semiconductor nanorods through judicious selection of suitable precursors and ligands. The facile hierarchical assembly of the metal tipped nanorods may allow scalable application for bulk devices where nanoscale metal–semiconductor interfaces are required in high density.

Acknowledgments TB acknowledges University of Calcutta for financial support (UGC/1039/Nano Travel Grant) to participate in the conference. The financial supports by Science Foundation Ireland (SFI, Contract No. 06/IN.1/I85 at UL), Institutional funding from INSPIRE are gratefully acknowledged.

Notes and References

1. A.I. Hochbaum, P. Yang, Semiconductor nanowires for energy conversion. Chem. Rev. **110**, 527 (2010)
2. D.V. Talapin, J.-S. Lee, M.V. Kovalenko, E.V. Shevchenko, Prospects of colloidal nanocrystals for electronic and optoelectronic applications. Chem. Rev. **110**, 389 (2010)
3. J. Liqiang, Q. Yichun, W. Baiqi, L. Shudan, J. Baojiang, Y. Libin, F. Wei, F. Honggang, S. Jiazhong, Review of photoluminescence performance of nano-sized semiconductor materials and its relationships with photocatalytic activity. Sol. Energy. Mater. Sol. Cells. **90**, 1773 (2006)
4. R. Costi, A.E. Saunders, E. Elmalem, A. Salant, U. Banin, Visible Light-Induced Charge Retention and Photocatalysis with Hybrid CdS–Au Nano dumbbells. Nano Lett. **8**, 637 (2008)
5. X. Chen, S.S. Mao, Titanium Dioxide nanomaterials: synthesis, properties, modifications and applications. Chem. Rev. **107**, 2891 (2007)
6. D.V. Talapin, I. Mekis, S. Gotzinger, A. Kornowski, O. Benson, H. Weller, CdSe/CdS/ZnS and CdSe/ZnSe/ZnS core−shell − shell nanocrystals. J. Phys. Chem. B **108**, 18826 (2004)
7. C. O'Sullivan, R.D. Gunning, A. Sanyal, C.A. Barrett, H. Geaney, F.R. Laffir, S. Ahmed, K.M. Ryan, Spontaneous room temperature elongation of CdS and Ag2S nanorods via oriented attachment. J. Am. Chem. Soc. **131**, 12250 (2009)
8. S. Ahmed, K.M. Ryan, Centimeter scale assembly of vertically aligned and close packed semiconductor nanorods from solution. Chem. Commun. **42**, 6421 (2009)
9. C. O'Sullivan, S. Ahmed, K.M. Ryan, Gold tip formation on perpendicularly aligned semiconductor nanorods assemblies. J. Mater. Chem. **18**, 5218 (2008)
10. C. O'Sullivan, R.D. Gunning, C.A. Barrett, A. Singh, K.M. Ryan, Size controlled gold tip growth onto II-VI nanorods. J. Mater. Chem. **20**, 7875 (2010)
11. T. Mokari, E. Rothenberg, I. Popov, R. Costi, U. Banin, Selective growth of metal tips onto semiconductor quantum rods and tetrapods. Science **304**, 1787 (2004)

12. T. Mokari, C.G. Sztrum, A. Salant, E. Rabani, U. Banin, Formation of asymmetric one sided metal tipped semiconductor naocrystals dots and rods. Nat. Mater. **4**, 855 (2005)
13. A.E. Saunders, I. Popov, U. Banin, Synthesis of hybrid CdS–Au colloidal nanostructures. J. Phys. Chem. B **110**, 25421 (2006)
14. L. Carbone, S. Kudera, C. Giannini, G. Ciccarella, R. Cingolani, P.D. Cozzoli, L. Manna, Selective reactions on tips of semiconductor nanorods. J. Mater. Chem. **16**, 3952 (2006)
15. W.L. Shi, H. Zeng, Y. Sahoo, T.Y. Ohulchansky, Y. Ding, Z.L. Wang, M. Swihart, P.N. Prasad, A general approach to binary and ternary hybrid nanocrystals. Nano Lett. **6**, 875 (2006)
16. D.V. Talapin, E.V. Shevchenko, C.B. Murray, A. Kornowski, S. Forster, H. Weller, CdSe and CdSe/CdS nanorod solids. J. Am. Chem. Soc. **126**, 12984 (2004)
17. N. Zhao, K. Liu, J. Greener, Z. Nie, E. Kumacheva, Close-packed super lattices of side-by-side assembled Au-CdSe nanorods. Nano Lett. **9**, 3077 (2009)
18. A. Salant, E.A. Sadovsky, U. Banin, Directed self-assembly of gold-tipped CdSe nanorods. J. Am. Chem. Soc. **128**, 10006 (2006)
19. C.A.J. Lin, T.Y. Yang, C.H. Lee, S.H. Huang, R.A. Sperling, M. Zanella, J.K. Li, J.L. Shen, H.H. Wang, H.I. Yeh, W.J. Parak, W.H. Chang, Synthesis, characterization, and bioconjugation of fluorescent gold nanoclusters toward biological labeling applications. ACS Nano **3**, 395 (2009)
20. V.R. Reddy, N.R. Reddy, C.-J. Choi, Electrical and structural properties of low-resistance Pt/Ag/Au ohmic contacts to p-type GaN. Solid-State Electron. **49**, 1213 (2005)
21. M.T. Sheldon, P.-E. Trudeau, T. Mokari, L.-W. Wang, A.P. Alivisatos, Enhanced semiconductor nanocrystal conductance via solution grown contacts. Nano Lett. **9**, 3676 (2009)
22. K.T. Yong, Y. Sahoo, K.R. Choudhury, M.T. Swihart, J.R. Minter, P.N. Prasad, Shape control of PbSe nanocrystals using noble metal seed particles. Nano Lett. **6**, 709 (2006)
23. C. O'Sullivan, S. Crilly, F.R. Laffir, A. Singh, E. Magner, K.M. Ryan, Protein immobilization on perpendicularly aligned gold tipped nanorods assemblies. Chem. Commun. **47**, 2655 (2011)
24. S.E. Habas, P. Yang, T. Mokari, Selective growth of metal and binary metal tips on CdS nanorods. J. Am. Chem. Soc. **130**, 3294 (2008)
25. B.D. Yuhas, S.E. Habas, S.C. Fakra, T. Mokari, Probing compositional variation within hybrid nanostructures. ACS Nano **3**, 3369 (2009)
26. L.-Y. Fan, S.-H. Yu, ZnO@Co hybrid nanotube arrays growth from electrochemical deposition: structural, optical, photocatalytic and magnetic properties. Phys. Chem. Chem. Phys. **11**, 3710 (2009)
27. J. Maynadie, A. Salant, A. Falqui, M. Respaud, E. Shaviv, U. Banin, K. Soulantica, B. Chaudret, Cobalt growth on tips of CdSe nanorods. Angew. Chem. Int. Ed. **48**, 1814 (2009)
28. E. Elmalem, A.E. Saunders, R. Costi, A. Salant, U. Banin, Growth of photocatalytic CdSe-Pt nanorods and nanonets. Adv. Mater. **20**, 4312 (2008)
29. A. Singh, R.D. Gunning, A. Sanyal, K.M. Ryan, Directing semiconductor nanorods assembly into 1D or 2D super crystals by altering the surface charge. Chem. Commun. **46**, 7193 (2010)
30. I. Gur, N.A. Fromer, M.L. Geier, A.P. Alivisatos, Air-stable all-inorganic nanocrystal solar cells processed from solution. Science **310**, 462 (2005)
31. P.R. Selvakannan, A. Swami, D. Srisathiyanarayanan, P.S. Shirude, R. Pasricha, A.B. Mandale, M. Sastry, Synthesis of aqueous Au Core–Ag shell nanoparticles using tyrosine as a pH-dependent reducing agent and assembling phase-transferred silver nanoparticles at the air–water interface. Langmuir **20**, 7825 (2004)
32. B. Muller, L.P. Nedelmann, B. Fischer, H. Brune, J.V. Barth, K. Kern, D. Erdos, J. Wollschlager, Strain relief via island ramification in sub monolayer herero epitaxy. Surf. Rev. Lett. **5**, 769 (1998)
33. R.D. Robinson, B. Sadtler, D.O. Demchenko, C.K. Erdonmez, L.W. Wang, A.P. Alivisatos, Spontaneous superlattice formation in nanorods through partial cation exchange. Science **317**, 355 (2007)

34. D.H. Son, S.M. Hughes, Y. Yin, A.P. Alivisatos, Cation Exchange Reactions in Ionic Nanocrystals. Science **306**, 1009 (2004)
35. G. Ramalingam, N. Melikechi, P. DennisChristy, S. Selvakumar, P. Sagayaraj, Structural and optical property studies of CdSe crystalline nanorods synthesized by a solvothermal method. J. Cryst. Growth **311**, 3138 (2009)
36. J.A. Creighton, D.G. Eadon, Ultraviolet–visible absorption spectra of the colloidal metallic elements. J. Chem. Soc., Faraday Trans. **87**, 3881 (1991)
37. A. Kumar, V. Chaudhary, Optical and photophysical properties of Ag/CdS nanocomposites—An analysis of relaxation kinetics of the charge carriers. J. Photochem. Photobiol. A **189**, 272 (2007)
38. Y.P. Hsieh, C.T. Liang, Y.F. Chen, C.W. Lai, P.T. Chou, Mechanism of giant enhancement of light emission from Au/CdSe nanocomposites. Nanotechnology **18**, 415707 (2007)
39. L.K. Leung, N.J. Komplin, A.B. Ellis, N. Tabatabaie, Photoluminescence studies of silver-exchanged cadmium selenide crystals: Modification of a chemical sensor for aniline derivatives by heterojunction formation. J. Phys. Chem. **95**, 5918 (1991)
40. NIST X-ray Photoelectron Spectroscopy Database, Version 3.5, http://srdata.nist.gov/xps/
41. J. Yang, E. Sargent, S. Kelley, J. Y. Ying, A general phase-transfer protocol for metal ions and its application in nanocrystal synthesis. Nat. Mater. **8**, 683 (2009)
42. S. Chakrabortty, J.A. Yang, Y.M. Tan, N. Mishra, Y. Chan, Asymmetric Dumbbells from Selective Deposition of Metals on Seeded Semiconductor Nanorods. Angew. Chem. **122**, 2950 (2010)
43. R. Luschtinetz, G. Seifert, E. Jaehne, H. J. P. Adler, Infrared Spectra of Alkylphosphonic Acid Bound to Aluminium Surfaces. Macromol. Symp. **254**, 248 (2007)
44. P.K. Sahoo, S.S.K. Kamal, M. Premkumar, T.J. Kumar, B. Sreedhar, A.K. Singh, S.K. Srivastava, K.C. Sekhar, Synthesis of tungsten nanoparticles by solvothermal decomposition of tungsten hexacarbonyl. Int. J. Refract. Met. Hard Mater. **27**, 784 (2009)
45. A.J. Morris-Cohen, M.D. Donakowski, K.E. Knowles, E.A. Weiss, The Effect of a Common Purification Procedure on the Chemical Composition of the Surfaces of CdSe Quantum Dots Synthesized with Trioctylphosphine Oxide. J. Phys. Chem. C **114**, 897 (2010)

Comparative Study of Lanthanum Based Perovskites Synthesized by Different Methods

Shrikant K. Nikam, Dipti V. Dharmadhikari and Anjali A. Athawale

Abstract A series of lanthanum perovskites $LaMO_3$ (M = Cr, Mn, Fe, Co, Ni and Cu) have been synthesized by different methods: heating a mixture of precursor salts at 800 °C, hydrothermal activation and microwave combustion methods. The hydrothermal synthesis was carried out at 160 °C under autogenous pressure while microwave combustion was done at 80 % microwave power. However, pure phase of perovskites were formed only after heating the products of the above synthesis. The precursor salt mixtures and products of hydrothermal activation were heated at 800 °C for 7 h. The microwave combustion products required sintering at 200 °C for minimum of 6 h to yield pure phase products. The formation of products were assured from the Fourier transform infrared (FTIR) analysis of the samples. The band centered at $550 \pm 50\ cm^{-1}$ in the FT-IR spectra corresponds to the metal oxygen bond indicating formation of products. Phase purity of the samples were confirmed from the X-ray diffraction (XRD) analysis. The phases were identified by comparing the results with the standard Joint Committee on Powder Diffraction Standards (JCPDS) data. The products of simple heating and hydrothermal activation are of $LaMO_3$ type except lanthanum copper oxide (La_2CuO_4). In case of microwave combustion, copper and nickel exhibits La_2MO_4 type phase while cobalt shows mixed phase. The micrographs reveal the presence of rod like morphology of the products obtained by hydrothermal synthesis, while, samples obtained after simple heating and microwave combustion synthesis yielded nanoparticles with irregular shape and a size distribution of 16–50 nm.

S. K. Nikam · D. V. Dharmadhikari · A. A. Athawale (✉)
Department of Chemistry, University of Pune, Pune 411007, India
e-mail: agbed@chem.unipune.ac.in

A. A. Athawale
Centre for Nanoscience and Quantum Systems, Pune 411007, India

S. K. Nikam
Anantrao Thopte College, Bhor, Pune 412206, India

P. K. Giri et al. (eds.), *Advanced Nanomaterials and Nanotechnology*,
Springer Proceedings in Physics 143, DOI: 10.1007/978-3-642-34216-5_3,

Keywords Perovskites · FTIR analysis · Transmission electron microscopy · Nanoparticles

1 Introduction

Ceramic oxides based on the perovskite structure ABO_3 have proven to be a valuable source of strategic materials for manifold applications [1]. The vast majority of catalysts used in modern chemical industry are based on mixed metal oxides. $LaMO_3$ type perovskites exhibit good electronic, magnetic, and optical properties together with catalytic activity [2, 3]. An additional advantage of perovskite materials is their high thermal stability in oxidizing environment, which is a requirement for many catalytic applications. The thermal and thermodynamic stability of the lanthanum based transition metal perovskites are of importance as these materials are used as electrodes in magneto hydrodynamic generators and as catalysts in pollution control [4, 5].

These perovskites are commonly prepared by solid-state method which requires high reaction temperature, intermediate grinding, long preparative time and the products thus obtained have limited degree of chemical homogeneity with large particle size. Mixed metal oxides are stable in the perovskite structure provided that their ionic radii fit well in the 12—coordinated A and 6—coordinated B sites e.g. $r_A > 0.90$ Å and $r_B > 0.51$ Å. Much attention is given to the preparation of lanthanum perovskites since their performances are strongly dependent on the synthesis procedures [2]. Now a days, numerous wet-chemical methods are being adopted to prepare fine sized, morphology oriented homogeneous perovskites such as sol–gel, citrate complexation, microemulsion, sonochemical, combustion, hydrothermal, co-precipitation, microwave combustion and physical mixing of metal oxides followed by calcination at high temperatures [3, 6]. The magnetic study of lanthanum based perovskites shows that chromite, manganite, ferrite and cuprate are antiferromagnetic while cobaltite and nickelate are nonmagnetic in nature [7, 8].

The present work is directed towards the synthesis of lanthanum perovskites of transition metals (Cr, Mn, Fe, Co, Ni and Cu). $LaMO_3$ or La_2MO_4 phases were obtained by heating homogeneous mixtures of lanthanum nitrate and transition metal nitrates with and without KOH, hydrothermal assisted synthesis and microwave combustion synthesis followed by sintering. The minimum phase formation temperature for the lanthanum manganite is higher than other perovskites [1, 4]. To obtain pure phase, sintering temperature was confirmed from thermal analysis of the samples and microstructure was analyzed from transmission electron microscopy (TEM). Phase formation temperature of LM-1, LM-2 and LM-4 were ≥760 °C and that of microwave treated LM-6 was ≥200 °C, hence all the samples were sintered at respective temperatures.

2 Experimental

All the chemicals used were of A.R. grade. La $(NO_3)_3$, $Cr(NO_3)_3$, $Co(NO_3)_2$, $Ni(NO_3)_2$, $Cu(NO_3)_2$ and Urea were from Loba Chemie Pvt. Ltd., $Fe(NO_3)_2$ from Qualigens Fine Chemicals, Mn $(NO_3)_2$ from Otto Kemi and KOH from SRL. Double distilled water was used wherever necessary.

In the first set of samples, the precursors (metal nitrates, 1:1 proportion) without/with KOH (5 times) (LM-1 and 2) were heated at 800 °C for 7 h. In the second set, the precursors were taken in 1:1 proportion with KOH (5 times) and subjected to hydrothermal activation at 160 °C under autogenous pressure for 6 h with appropriate amount of water in a Teflon lined stainless steel autoclave (LM-3) which was further heated as above to obtain pure phase perovskite (LM-4). The third set involved microwave combustion synthesis, the precursor salts and fuel (urea) were taken in 1:1:1 proportion and subjected to 80 % microwave power for 10 min (LM-5) and the sample LM-6 was obtained by heating LM-5 at 200 °C for 6 h. Instant combustion reaction takes place in microwave oven within 2 min to form the product.

3 Characterization

The products were characterized using Fourier transmission infrared (FTIR) analysis, X-ray diffraction (XRD) analysis, and Transmission electron microscopy (TEM). FTIR spectra of the dry powders were recorded on Thermo Scientific Spectrophotometer Model—NICOLET—6700 over a frequency range of 400–4,000 cm^{-1}. XRD analysis of the powders were done on a Bruker AXS D-8 Advanced X-ray Diffractometer with monochromatic CuKα radiation ($\lambda = 1.54$ Å), silicon was used as an external standard for correction due to instrumental broadening. Diffraction data were collected from 20 to 80° 2θ angle. TEM of the samples were observed under the Philips CM-200 instrument at an accelerating voltage of 200 kV. Suspension of the samples in ethanol were well dispersed and loaded on carbon coated copper grids of 200 mesh size. The grids were then dried under IR lamp and viewed under the microscope.

4 Results and Discussion

4.1 FTIR Analysis

The FTIR spectra of the samples synthesized by different methods (A-F) are shown in Fig. 1. The curves a, b, c, d, e and f represent the products of chromite, manganite, ferrite, cobaltite, nickelate and cuprate respectively.

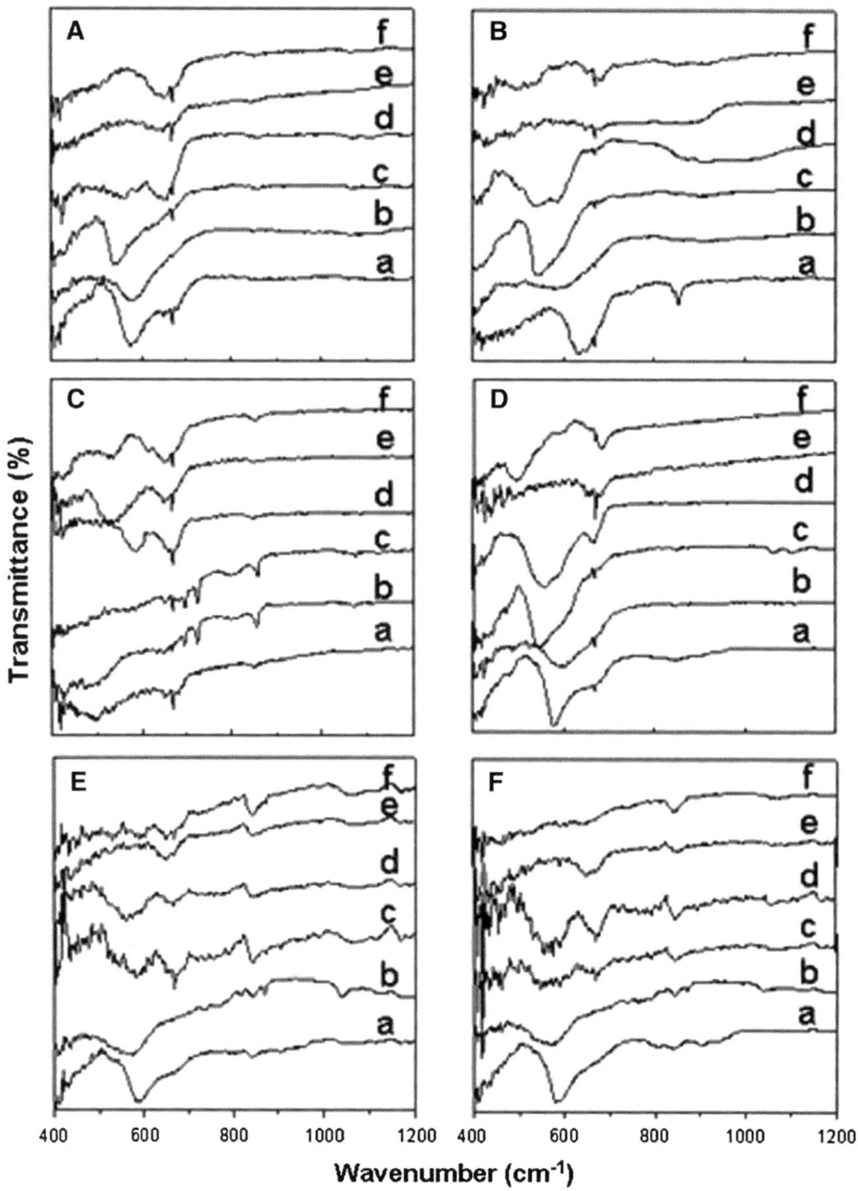

Fig. 1 FTIR spectra of samples prepared by different methods **A** LM-1, **B** LM-2, **C** LM-3, **D** LM-4, **E** LM-5, **F** LM-6 *a* chromite, *b* manganite, *c* ferrite, *d* cobaltite, *e* nickelate *f* cuprate

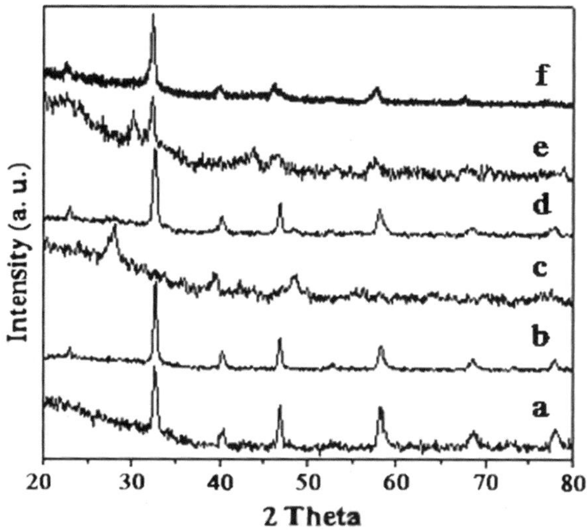

Fig. 2 XRD's of lanthanum manganites prepared by different methods *a* LM-1, *b* LM-2, *c* LM-3, *d* LM-4, *e* LM-5, *f* LM-6

The stretching bands observed in the range of 400–850 cm^{-1} are mainly due to the symmetric and antisymmetric stretching modes of the metal oxygen bonds [9]. A comparison of the results with respect to the methods of preparation reveals that oxide phase of the products are obtained when the salt precursors are heated in absence of KOH (Fig. 1A, LM-1) with the stretching frequency appearing at 550 ± 50 cm^{-1}. Similar observations are noted for hydrothermally activated and microwave combustion products after heating at 800 °C for 7 h and 200 °C for 6 h respectively (Fig. 1D and F, LM-4 and LM-6). On the contrary, the precursor salts heated in presence of KOH (LM-2) and as synthesized products of hydrothermal activation (LM-3) and microwave combustion (LM-5) exhibit formation of hydroxides ($\nu_s = 650\,cm^{-1}$) as the predominant phase. The sharpness of the ν_s band could be attributed to the narrow size distribution of the particles. The shift in ν_s band from lower to higher frequency can be correlated to the change in the morphology of the sample (particles—rod as observed in Fig. 3). Further observations drawn from the FTIR spectra are that the hydroxide ν_s band does not show shift with change in metal ions while the oxide ν_s band shows a shift from 500 to 600 cm^{-1}, O–M–O deformation vibrations (Cu-510 and 680 cm^{-1}, Ni-530 cm^{-1}, Fe-545 cm^{-1}, Co-563 cm^{-1}, Cr-580 cm^{-1}, Mn-590 cm^{-1}) indicating that the hydroxide phase is governed by the A site i.e. lanthanum ion, while the oxide phase is governed by the B site i.e. transition metal ion of the perovskite.

Thus, from the overall results it appears that sufficient energy to form perovskite phase is obtained from thermal heating. The base KOH and water serve as mineralizer and hydrolyzing agents respectively, affecting the morphology of the samples.

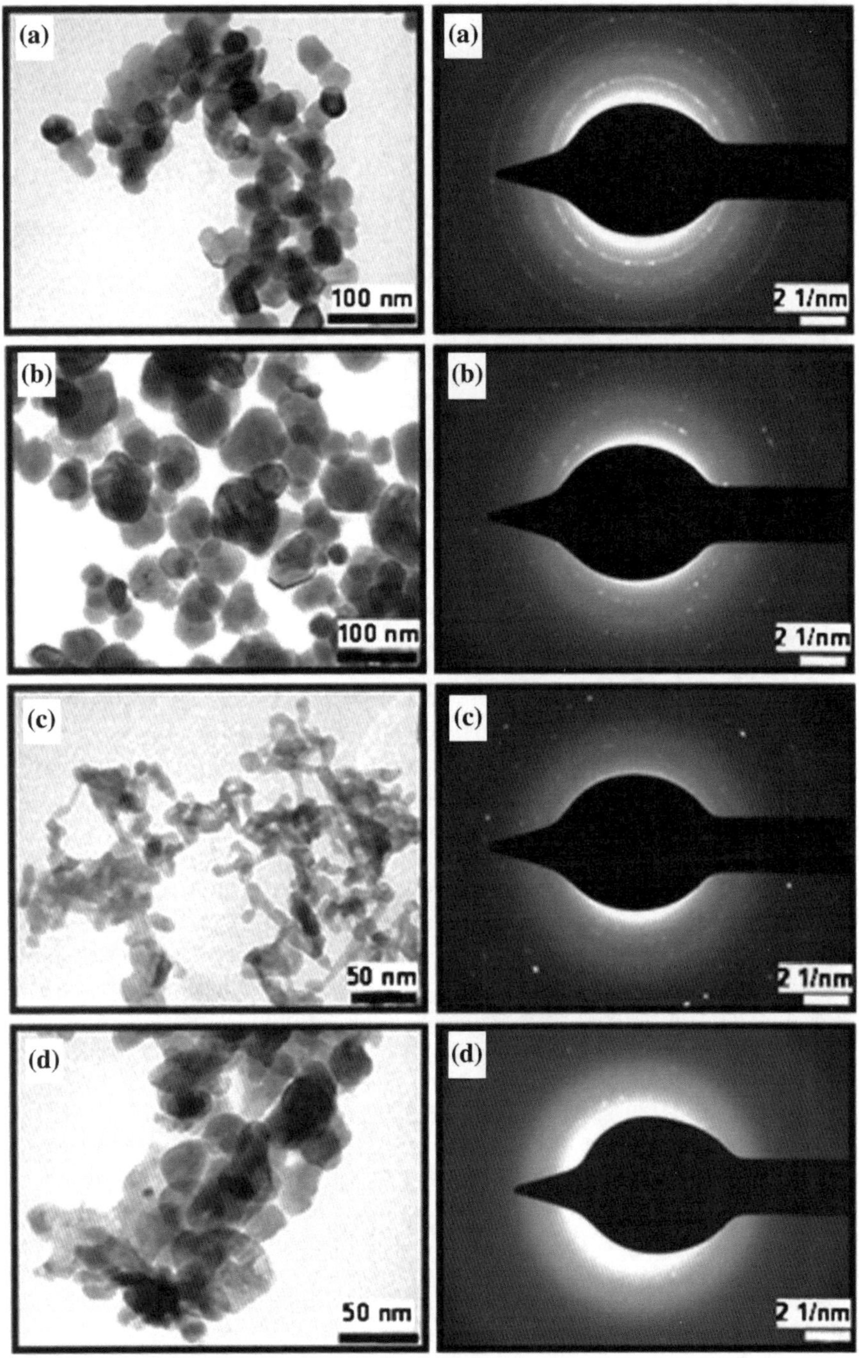

Fig. 3 TEM and SAED of lanthanum manganites prepared by different methods **a** LM-1, **b** LM-2, **c** LM-4, **d** LM-6

4.2 XRD

Figure 2 depicts the XRD of lanthanum manganite samples as a representative system, a–f correspond to LM-1, LM-2, LM-3, LM-4, LM-5 and LM-6 respectively. The peaks observed at 2θ values 22.9, 32.6, 40.2, 46.9, 52.8, 58.4, 68.4, 78.1 represent the perovskite phase, a peak at 2θ value 28.0 correspond to the hydroxide phase while, the oxide phase shows a peak at 2θ value $\sim$30.2. Pure phase products are obtained for samples LM-1, LM-2, LM-4 and LM-6, hydroxide phase for LM-3 and oxide phase for LM-5.

The XRD peak values conform to those reported in the JCPDS data (Card number PCPDF-75-0440). The three most intense reflections in the XRD were used to calculate the particle size with the help of Scherrer formula and the average value were found to be <50 nm. Similar observations were noted for all the systems. The lanthanum copper oxide shows Ruddlesden-Popper (A_2BO_4) type phase, nickelate yields this type of phase after microwave combustion synthesis while lanthanum cobalt oxide shows the formation of both ABO_3 along with A_2BO_4 type phase.

4.3 Morphological Investigations

The microstructure and particle size distribution of the samples were studied by TEM. Figure 3a, b, c and d are the TEM images and Selected Area Electron Diffraction (SAED) patterns of the samples LM-1, LM-2, LM-4 and LM-6 respectively for lanthanum manganite.

As observed from the TEM images, the morphology of the samples seem to differ with the method of synthesis. Octagonal shaped particles with well defined boundaries and a narrow particle size distribution (24–35 nm) (Fig. 3a) are seen for LM-1 sample. Figure 3b shows the TEM image of LM-2 sample with spherical particles having a broad size distribution (16–70 nm). Whereas, the sample LM-4 reveals mixed morphology (Fig. 3c) including particles and rod like structure (unidirectional particle stacking). Figure 3d shows TEM image of LM-6, with oval particles and diffused boundaries. Average particle size distribution is about 17–42 nm. The SAED patterns of each sample exhibit polycrystalline nature.

5 Conclusions

$LaMO_3$ (M = Cr, Mn, Fe, Co, Ni and Cu) were synthesized by different methods. The products obtained by heating mixture of precursors are nanosized and homogenous with well defined boundaries while those obtained by microwave combustion synthesis are porous in nature with diffused boundaries. Hydrothermal products show tendency to form rod like structures. The phase formation of the cobaltite, nickelate and cuprate are a function of hydrothermal activation time and sintering temperature. The shift and sharpening of the metal oxygen stretching band observed in FTIR correspond to phase formation, purity, microstructure and the type of B site cation.

Acknowledgments Shrikant K. Nikam gratefully acknowledges UGC for providing the teacher's fellowship and financial support, Sangram Thopte, President, Rajgad Dnyanpeeth—Bhor. Dipti V. Dharmadhikari gratefully acknowledges CNQS for providing the financial support. Thanks to DST unit on Nanoscience and CNQS, Department of Physics, University of Pune for providing XRD facility, SAIF, IIT Powai, Mumbai is acknowledged for TEM analysis.

References

1. E. Krupicka, A. Reller, A. Weidenkaff, Morphology of nanoscaled $LaMO_3$- particles (M = Mn, Fe, Co, Ni) derived by citrate precursors in aqueous and alcoholic solvents. Cryst. Eng. **5**, 195 (2002)
2. P. Herve, T. Ngamou, N. Bahlawane, Chemical vapor deposition and electric characterization of perovskite oxides $LaMO_3$ (M = Co, Fe, Cr and Mn) thin films. J. Solid State Chem. **182**, 849 (2009)
3. Y. Wang, X. Yang, L. Lu, X. Wang, Experimental study on preparation of $LaMO_3$ (M = Fe, Co, Ni) nanocrystals and their catalytic activity. Thermochemica Acta **443**, 225 (2006)
4. O.M. Sheedharan, M.S. Chandrashekharaiah, Standard Gibbs' energy of formation of $LaFeO_3$ and comparison of stability of $LaMO_3$ (M = Mn, Fe, Co or Ni) compounds. J. Mater. Sci. **21**, 2581 (1986)
5. M. Panneerselvam, K.J. Rao, Microwave preparation and sintering of industrially important perovskite oxides: $LaMO_3$ (M = Cr, Co, Ni). J. Mater. Chem. **13**, 596 (2003)
6. M. Wallin, N. Cruise, U. Klement, A. Palmqvist, M. Skoglundh, Preparation of Mn, Fe and Co based perovskite catalysts using microemulsions. Colloids Surf. A: Physicochem. Eng. Aspects **238**, 27 (2004)
7. D.D. Sarma, N. Shanthi, S.R. Barman, N. Hemade, H. Sawada, K. Terakura, Band theory for ground-state properties and excitation spectra of Perovskite $LaMO_3$ (M = Mn, Fe, Co, Ni). Phys. Rev. Lett. **75**, 1126 (1995)
8. G. Pari, S. Mathi Jaya, G. Subramoniam, R. Asokamani, Density-functional description of the electronic structure of $LaMO_3$ (M = Sc, Ti, V, Cr, Mn, Fe, Co, Ni). Phys. Rev. B **51**, 16575 (1995)
9. Y. Gong, C. Ding, M. Zhou, Infrared spectra of oxygen-rich Yttrium and Lanthanum Dioxygen/Ozonide complexes in solid Argon. J. Phys. Chem. A **113**, 8569 (2009)

Novel, Fast, Bio-Derivatized Sonochemical Synthesis of Gold Nanoparticles by Using *Piper betle* Leaf Broth as a Reducing and Capping Agent

K. Mallikarjuna, N. John Sushma and B. Deva Prasad Raju

Abstract The use of biomolecules hosted products for synthesis of nano sized materials of increasing interest to benign medicinal field. We report a biomolecules hosting, rapid, environmentally benign, bio-degradable, non-toxic, and sonochemical synthesis of gold nanoparticles by using *piper betle* leaf broth as a reducing and capping agent. The aqueous reaction of medium containing gold nanoparticles showed a peak at 540 nm was studied by UV-Vis spectra. In addition, we determined the size of the nanoparticles and optical energy gap from the surface plasmon absorption peak. The crystalline structural characteristics of a biomolecules hosted gold nanoparticles were studied by X-ray diffraction. The morphology of nanoparticles was analyzed by scanning electron microscopy. The stoichiometric chemical composition of elemental presence in the medium was determined by energy dispersive spectrum. The presences of biomolecules which are act as capping agents around the nanoparticles were studied by fourier transform infrared spectroscopy.

Keywords Sonochemical synthesis · UV-Vis spectra · XRD · FTIR

K. Mallikarjuna
Department of Physics, Sri Venkateswara University, Tirupati 517502, India

B. Deva Prasad Raju (✉)
Department of Future Studies, Sri Venkateswara University, Tirupati 517502, India
e-mail: drdevaprasadraju@gmail.com

N. J. Sushma
Department of Biotechnology, Sri Padmavati Women's University, Tirupati 517502, India

P. K. Giri et al. (eds.), *Advanced Nanomaterials and Nanotechnology*,
Springer Proceedings in Physics 143, DOI: 10.1007/978-3-642-34216-5_4,

1 Introduction

Proliferation of green chemical nanoscience research has opened up novel fundamental and frontiers in material science medicinal research. Integration of green chemistry principles to nano is one of the key issues of nanoscience and technology research [1, 2]. Interdisciplinary nanoscience research involving physicists, chemists, biologists and material engineers are concerned about the need for developing eco-friendly methods to synthesize of nano materials [1, 2]. Among the noble metal nanoparticles, gold nanoparticles has fascinating chemical and physical properties and has important value in the fundamental and application research fields [3, 4]. Traditionally nanoparticles have been prepared by a number of physical and chemical methods like laser ablation, pyrolysis, lithography; chemical vapor deposition, sol–gel technique and electro deposition techniques are slow, expensive and often involve the use of toxic, hazardous chemicals which may pose environmental risks and human health [5]. Among these synthesis methods, the use of ultrasound offers the sonochemical reduction drew particular attention due to the generation of nanoparticles with smaller size [6]. Nature provides a clue in the form of biomolecules that can be effectively utilized in capping nanoparticles. For instance, a great deal of effort has put into the green synthesis of inorganic materials, especially metal nanoparticles [7]. Consequently there have been interesting reports on the preparation of metal nanoparticles employing biomimetic entities such as enzymes, bacteria, viruses, fungi, yeast and plant or plant leaf extracts [8–13].

Green chemistry is reducing or eliminating the generation of substance hazardous to human health and the environment strategies of addressing the mounting environmental concerns with current approaches include the use of environmentally benign solvents, biodegradable polymers and non-toxic chemical [14]. The *Piper betle* is a traditional medicinal plant of India, which is a source of diosgenin, triterpenes, alkaloids, nitrosoguvacoline, propenylphenols and piperbetol [15]. *Piper betle* leaves have long been studied for their diverse pharmalogical actions such as anti-inflammatory, anti-oxidant, radio protective, anti-fungal, anti-bacterial and anti-allergic activities [16, 17]. So far, there have been no reports on the synthesis of nanoparticles from *piper betle* leaf extract by using sonochemical synthesis. In this paper we report on the sonochemical synthesis of gold nanoparticles using *piper betle* leaf extract as a reducing agent and stabilizer.

2 Experimental

2.1 Materials

Hydrogen tetrachloroaurate (III) hydrate ($HAuCl_4$. $3H_2O$) was obtained from Sigma Aldrich (USA). The fresh *Piper betle* leaves were collected from the retail shop near by the Sri Venkateswara University campus, Tirupati, Andhra Pradesh, India. The solutions were prepared with nanopure water which was obtained by passing twice distilled water through a milli-Q system.

2.2 Preparation of Piper betle Leaves Extract

The fresh *piper betle* leaf extract used for the reduction of gold ions was prepared by taking 20 gm of thoroughly washed, finely cut leaves in 500 ml Erlenmeyer flask along with 100 ml of deionized water and then boiling the mixture for 2 min before decanting it. Further, the extract was filtered with Whatman No. 1 filter paper and stored at 4 °C and used for further experiments.

2.3 Sonochemical Synthesis of Gold Nanoparticles

In a typical experiment, 0.5 ml of the leaf extract was added to 10 ml of 1 mM $HAuCl_4$ aqueous solution. Synthesis of gold nanoparticles was prepared by sonication of the solution was a high power ultrasonicator (POWER SONIC 300). The bioreduced aqueous component (0.5 ml) was used to measure UV-Vis spectra of the solution. The particle suspension was diluted in the order 1:10 ratio with nanopure water, to avoid the errors that could occur due to the high optical density of the solution.

3 Results and Discussion

3.1 UV-Visible Absorption Spectroscopy

The UV-Visible absorption spectroscopy is a valuable tool for understanding the kinetics coagulation of aqueous chloroaurate ions during its reaction with leaf extract under ultrasound was examine the bioreduced aqueous solution. The visual inspection of gold reduced solution gradually turned from yellow to pink. Subsequently, we observed that the reaction time is increases continuously the colour changes from pink to deep red. Under ultrasound, the progress of reaction between the *piper betle* leaf broth and the aqueous chloroauric acid as a function of time at room temperature was recorded on spectrophotometer (PERKIN-ELMER LAMBDA 25) as shown in Fig. 1. With increasing time, the full width half maximum decreases supporting the reduction of size of the particle. Strong absorption band is observed at 540 nm supporting the formation of spherical nanoparticles in aqueous solution. In recent years, the gold nanoparticles prepared by eco-friendly

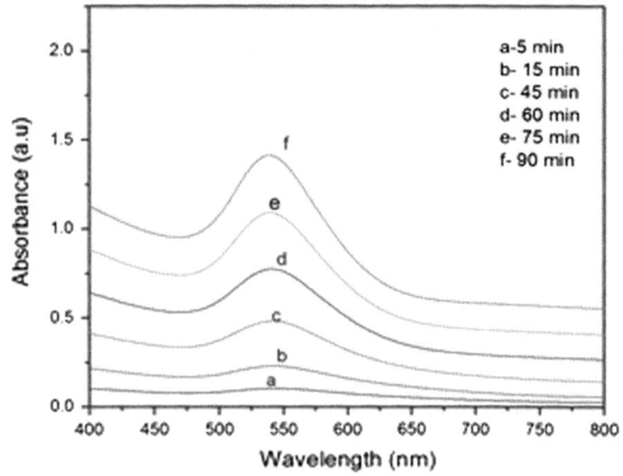

Fig. 1 UV-Vis spectra of gold nanoparticles

procedure have widely applied. The calculation of size of the particles by electron microscoptic techniques is good, but it is not simple and fast. It is not reliable calculation of size of the nanoparticles at kinetics of reaction. A simple, fast spectroscopic method used to calculate size of the nanoparticles. The average radii of the gold nanoparticles can be estimate by the equation $R = V_f/\Delta\varpi_{1/2}$ [18]. Where V_f is the Fermi velocity of the electrons in the metal, $\Delta\varpi_{1/2} = 2\Pi C\ (\Delta\lambda/\lambda_p^2)$. Where $\Delta\lambda$ is full width half maximum wavelength and λ_p is plasmon peak wavelength.

3.2 Calculation of Optical Band Gap

The solid state materials are illustrated with optical band gap as an important parameter. The band gap of material is defined as the gap between the valance and conduction bands. There are two kinds of optical transitions at the fundamental absorption edge, direct and indirect transitions both of which involve the interaction of an electromagnetic wave with an electron in the valence band. The group energy levels are quite different between nanomaterials and bulk materials in that the nanomaterials have larger band gaps and consist of more discrete energy levels. This observation may be attributed to the various quantum confinement effects within nanoparticles. We calculated the optical band gap energy for gold nanostructures from the UV-Vis absorption spectrum. The relation between optical absorption coefficient α and $h\nu$ is expressed as $(\alpha h\nu)^2 = C\ (h\nu - E_g)$ where C is a constant. Figure 2 shows the plot of $(\alpha h\nu)^2$ versus $h\nu$ for the gold nanostructures synthesized from the *piper betle* leaf broth. The energy gap is estimated from the extrapolated linear portion of the plot. In the present case, the energy gap is found to be 2.0 eV and it is in good agreement with the reported values in the literature [19].

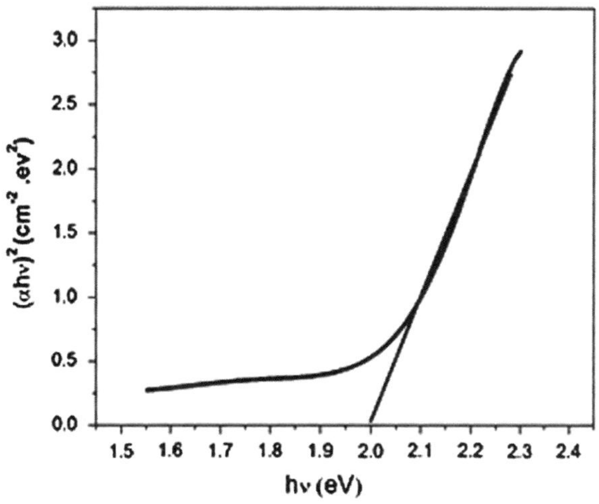

Fig. 2 Plot of $(\alpha h\upsilon)^2$ versus photon energy ($h\upsilon$) of gold nanoparticles

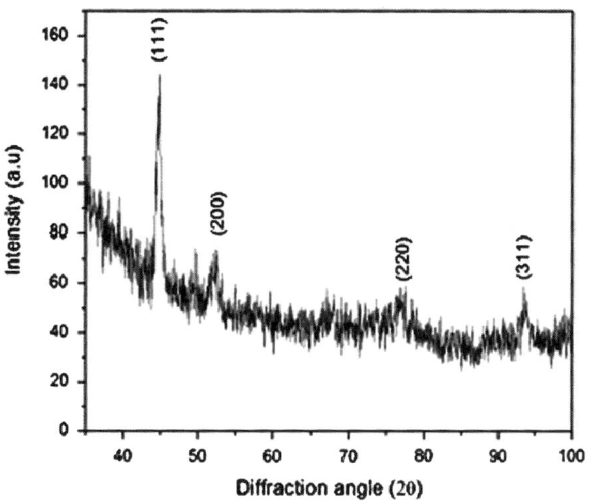

Fig. 3 X-ray diffraction spectrum of synthesized gold nanoparticles

3.3 Crystalline Structural Analysis of Gold Nanoparticles

An X-ray diffraction (XRD) measurement of a thin film of the bioreduced gold ions aqueous solution was drop coated onto a glass slide and an INEL X-ray diffractometer. The diffraction pattern was recorded by Co-kα_1 radiation with λ of 1.78A° in the region of 2θ from 20 to 90° at 0.02°/min and the time constant was 2 s. The size of the nanoparticles was calculated through the Scherer's equation. The crystalline nature of gold nanoparticles was studied with the aid of an X-ray diffraction as shown in Fig. 3. The number of strong Bragg's diffracted peaks was observed at 44.6, 52.1, 76.8 and 93.4° corresponding to the (111), (200), (220) and

Fig. 4 Scanning electron microscopy image of gold nanoparticles

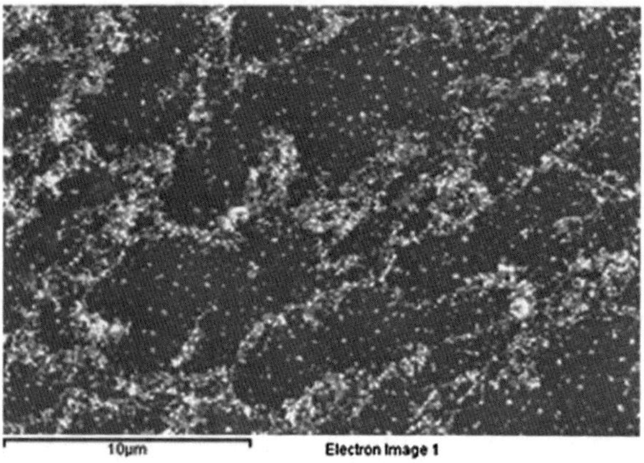

(311) facets of the face centered cubic lattice of gold were obtained. The average domain size of the biomolecules hosted gold nanoparticles was found to be 3.2 nm; and it was done by using the width of the (111) Bragg's reflection which was in consonance with the size of the particle calculated from UV-Vis analysis.

3.4 Scanning Electron Microscopy Studies

The morphology of the nanoparticles was studied with scanning electron microscopy. The SEM samples of the aqueous suspension of gold nanoparticles were prepared by dropping the suspension onto clean glass plate and allowing water to completely evaporate. SEM observations were carried out on a PHILLIPS FEI TECNAI G2S TWIN instrument. The SEM image shows that the gold nanoparticles are predominantly spherical in shape as shown in Fig. 4.

3.5 Energy Dispersive X-Ray Measurements

The energy dispersive X-ray analysis (EDS) reveals the stoichiometric elemental chemical composition in fabricated nanoparticles. The EDS revealed the formation of a strong signal in the gold region and confirmed the formation of gold nanoparticles as shown in Fig. 5.

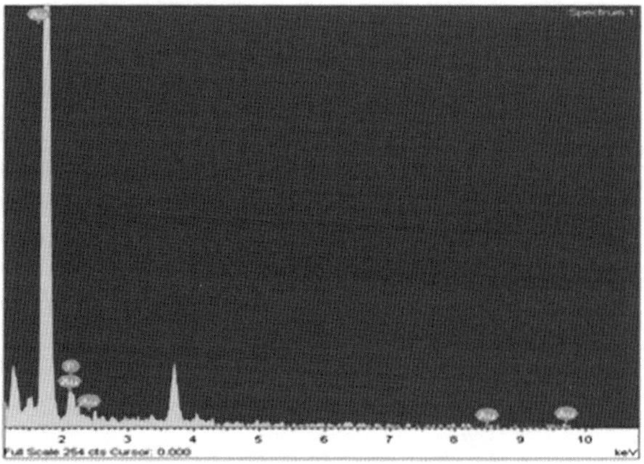

Fig. 5 Energy dispersive X-ray spectrum of gold nanoparticles

3.6 Fourier Transform Infra-Red Spectroscopy

FTIR has emerged as a valuable tool for understanding the involvement of surface functional biological groups in metal interactions. For fourier transformed infrared (FTIR) measurements, the bioreduced aurate ions aqueous component was centrifuged at 15,000 rpm for 20 min. The dried sample was grinded with KBr pellets and analyzed on THERMO-NICOLET IR200 FT-IR spectrometer which was operated at a resolution 4 cm^{-1} in the region of 4,000–400 cm^{-1}. Figure 6 shows the FTIR spectrum of synthesized gold nanoparticles reduced by *Piper betle* leaf broth. It confirmed the fact that, the identified biomolecules act as the reducing and efficient stabilizing agents of the gold nanoparticles. The band 3,427 cm^{-1} due to O–H stretching as also the H-boned alcohols and phenols. The peak at 2,920 cm^{-1} corresponds to aldehydic C–H stretching. The band at 2,359 cm^{-1} indicates the C$\equiv$N stretch nitriles. The peak at 1,620 cm^{-1} can be represents to the vibrational modes of C=C double bonds of these molecules. The 1,549 cm^{-1} can be assigned to the N–H bend secondary amines. The band at 1,365 cm^{-1} indicates the N=O bend nitro groups and 1,060 cm^{-1} corresponds the vibrational modes of esters and ethers. From the FTIR spectrum it may be confirm that phenolic compounds was probable capping agent on the nanoparticles [20]. When the metal nanoparticles in the form of solution, they must be stabilized against the Vander Waals forces of attraction which may otherwise cause coagulation. It is known that *Piper betle* leaf broth is source of tannins, alkaloids, triperpens and phenolic compounds. To understand the trends of phenolic compounds in enabling nanoparticles nucleation, phenolic compounds have –OH groups in its structures and it has O-dihydroxyl-phenyl groups are capable of taking part in redox reactions to form quinones. Phenolic compounds can be used as both reducing and stabilizing agents and it is difficult to distinguish between the reducing and stabilizing roles of the phenolic compounds. In this procedure for the production of gold nanoparticles using leaf extract and ultrasound is novel because the formation of nanoparticles can be achieved within few minutes.

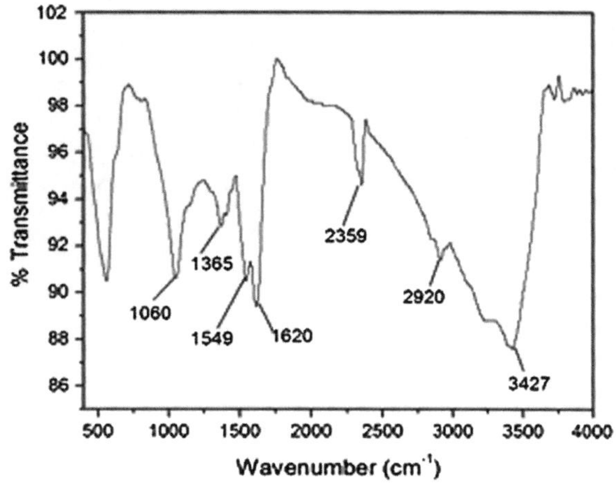

Fig. 6 FTIR spectrum of phytofabricated gold nanoparticles

4 Conclusion

The synthesis of spherical gold nanoparticles using leaf broth of *piper betle* provides an environment friendly, simple, cost effective and efficient route for the benign nanoparticles. In this synthesis protocol ultrasound used for reduced the gold nanoparticles in few minutes. From the absorption maxima obtained by UV-visible spectrum, the theoretical particle size was calculated and it is compared with calculated particle size from the XRD studies. From a technological point of view, these obtained gold nanoparticles have many potential applications in the medicinal fields, catalysis, optoelectronics, modern nanobiotechnology, materials science and this novel procedure has several advantages such as the compatibility for medical and pharmaceutical applications, and large scale commercial productions as well.

Acknowledgments The authors are thankful to Dr. B.V. Subba Reddy, IICT, Hyderabad for their permission to use the facilities therein for characterizing the sample. They would similarly like to express their gratitude to Mr. Ravi Shankar, for the XRD characterization at the School of Physics, University of Hyderabad, Hyderabad.

References

1. W.C.W. Chan, S. Nie, Quantum dot bioconjugates for ultrasensitive non isotropic detection. Science **281**, 2016 (1998)
2. S.J. Park, T.A. Taton, C.A. Mirkin, Array based electrical detection of DNA with nanoparticles probe. Science **295**, 1503 (2002)
3. H. Klefenz, Nanotechnology: From molecules to systems. Eng. Life Sci. **4**, 211 (2004)
4. J. Zhiliang, F. Zhongwei, L. Tingsheng, L. Fang, Z. Fuxin, X. Jiyun, Y. Xianghui, Resonance scattering spectroscopy of gold nanoparticles. Sci. China **44**, 175 (2001)

5. A. Tripathy, A.M. Raichur, N. Chanrasekaran, T.C. Prathna, A. Mukherjee, Process variables in biomimetic synthesis of silver nanoparticles by aqueous extract of Azadirechta indica (neem) leaves. J. Nanopart. Res. **12**, 237 (2010)
6. L.M.C. Aguilera, M.F. Romano, M.L.A. Gil, I.N. Rodriguez, J.L.H.H. Cisneros, J.M.P. Santander, New, fast and green procedure of gold nanoparticles based on sonocatalysis. Ultrasonic Sonochem. **18**, 789 (2011)
7. R. Satyavathi, M.B. Krishna, S.V. Rao, R. Saritha, D.N. Rao, Biosynthesis of silver nanoparticles using coriandrum sativum leaf extract and their application in nonlinear optics. Adv. Sci. Lett. **3**, 1 (2010)
8. I. Willner, R. Baron, B. Willner, Growing metals nanoparticles by enzymes. Adv. Mater. **18**, 1109 (2006)
9. K. Sen, P. Sinha, S. Lahari, Time dependent formation of gold nanoparticles in yeast cells: A comparative study. Biochem. Eng. J. **55**, 1 (2011)
10. J.M. Slocik, R.R. Naik, M.O. Stone, D.W. Wright, Viral templates for gold nanoparticles synthesis. J. Mater. Chem. **15**, 749 (2005)
11. R. Venu, T.S. Ramulu, S. Ananakumar, V.S. Rani, C.G. Kim, Bio-directed synthesis of platinum nanoparticles using aqueous honey solutions and their catalytic applications. Colloid. Surf A. **384**, 733 (2011)
12. D.S. Sheny, J. Mathew, D. Philip, Phytosynthesis of Au, Ag and Au-Ag bimetallic nanoparticles using aqueous extract and dried leaf of Anacardium occientale. Spectrochim. Acta A. **79**, 254 (2011)
13. D. Philip, Mangifera indica leaf assisted biosynthesis of well dispersed silver nanoparticles. Spectrochim. Acta A. **78**, 327 (2011)
14. P. Raveendran, J. Fu, S.L. Wallen, Completely green synthesis and stabilization of metal nanoparticles. J. Am. Chem. Soc. **125**, 13940 (2003)
15. R. Shukla, V. Satish, V.K. Singh, S. Kumar, S. Gupta, U. Gavani, Antibacterial activity of fresh leaves of Piper betle linn. Pharma Res. **1**, 110 (2009)
16. J.S. Rathee, B.S. Patro, M. Soumyaditya, G. Sunita, C. Subratha, Antioxident activity of Piper betle leaf extract and its constituents. J. Agric. Foochem. **54**, 9046 (2006)
17. L.R. Jaidev, G. Narasimha, Fungus mediated biosynthesis of silver nanoparticles, characterization and antimicrobial activity. Colloid. surf B. **81**, 430 (2010)
18. L.H. Zhou, H.H. Zhang, Y.T. Yang, B.S. Li, L.Q. Zhang, Y.C. Fu, H.H. Zhang, Formation of Au nanoparticles in sapphire y using Ar ion implantation and thermal annealing. Nuclear. Instru. meth B. **267**, 58 (2009)
19. A.S. Zadeh, M. Ghasemifard, G.A. Mansoori, Structural and optical characterization of folate conjugated gold nanoparticles. Physica E **42**, 1272 (2010)
20. M.G. Sayed, B. Mohsen, K. Maryan, Green synthesis of silver and gold nanoparticles using rosa damascene and its primary application in electro chemistry. Physica E **44**, 97 (2011)

Understanding the Role of Mode of Heating on Phase Formation of Fe–Pt Nanoparticles

S. A. Acharya, A. Dani, S. K. Sayyed and V. M. Gaikawad

Abstract In the present work, microwave heating route (MH) is used to prepare Fe–Pt nanoparticles and the results are compared with the materials obtained by conventional heating routes (CH). The effects of conditions of heating (microwave power, irradiation time, temperature) on the growth process of Fe–Pt nanoparticles are systematically investigated. The as-prepared Fe–Pt NPs by microwave heating route are observed in ordered face centered tetragonal (fct) $L1_0$ phase without any post-synthesis treatment, while conventional heating route gives rise to disordered face centered cubic (fcc) phase. The hysteresis measurements are performed at 300 K to study magnetic properties of as-synthesized Fe–Pt as a function of mode of heating and crystallite size. Conventionally, synthesized Fe–Pt (NPs) shows super paramagnetic behavior, while microwave-assisted sample exhibits ferromagnetic behavior. The particle size and magnetic properties of the as-prepared Fe–Pt are observed to be very sensitive to the preparative parameters such as microwave irradiation power and temperature, while influence of reaction time is insignificant. Size dependent magnetic behavior of microwave-assisted synthesized samples shows that coercivity and saturation magnetizations are observed to be decreasing with particle size. The microwave-assisted route is found to be a simple technique for direct synthesis of metal alloys and tuning particles size at nanoscale may prove to be a potential tool of high density data storage materials such as Fe–Pt.

S. A. Acharya (✉) · S. K. Sayyed · V. M. Gaikawad
Department of Physics, RTM Nagpur University, Amaravati Road,
Nagpur 440033, Maharashtra, India
e-mail: saha275@yahoo.com

A. Dani
Department of Applied Physics, Priyadarshini College of Engineering,
Nagpur, India

P. K. Giri et al. (eds.), *Advanced Nanomaterials and Nanotechnology*,
Springer Proceedings in Physics 143, DOI: 10.1007/978-3-642-34216-5_5,

Keywords Fe–Pt nanoparticles · Microwave-heating · *L*1*o*-phase · Chemical reduction and tuning of particles size

1 Introduction

The chemically ordered nanoparticles (NPs) of FePt alloy in $L1_0$-phase have been identified as promising materials for a new generation of ultrahigh density magnetic recording media. Large magnetocrystalline anisotropy (Ku $= 7 \times 10^6$ J/cm^3) and excellent hard magnetic properties of Fe–Pt NPs in $L1_0$-phase have been demonstrated [1–6]. Unfortunately, as-synthesized Fe–Pt NPs assume a disordered face-centered cubic (fcc) structure that has low magnetocrystalline anisotropy and reveal superparamagnetic behavior. Heat treatment is necessary to convert the fcc structure to the ordered face-centered tetragonal (fct) structure ($L1_0$-phase). The high temperature treatments have been observed to lead to the coalescence and agglomeration of NPs resulting in suppressing the recording density evolution [7, 8]. The direct synthesis of Fe–Pt NPs in ordered fct structure ($L1_0$-phase) is highly desirable to avoid high temperature treatment. Recently, different methods have been used to attempt to direct synthesis of chemically ordered fct-Fe–Pt NPs. The direct synthesis route for chemically ordered fct-Fe–Pt NPs by co-reduction of iron and platinum ions in polyol has been attempted [9–12]. The controlled synthesis conditions such as reaction temperature, time and type of precursor are found to be the effective parameters in predicting structure of as-synthesized Fe–Pt NPs. Howard et al. [13] reports the direct synthesis of fact Fe–Pt NPs using collman's reagent, $Na_2Fe(CO_4)$, as reducing agent for Pt^{2+}. The as-made Fe–Pt NPs are partially ordered. Others approaches for obtaining monodisperse fct NPs include PVP as protective reagent by Iwamoto et al. [14], annealing with salt matrix [15] to avoid particle agglomeration during annealing. Zafiropoulon et al. [16] reports the direct synthesis of fct phase without annealing by thermal decomposition of $Fe(acac)_3$ and $Pt(acac)_2$ in presence of oleic acid as surfactant but the synthesis temperature is very high. Microwave-assisted polyol method to directly synthesis of Fe–Pt NPs has been reported [17]. However, the Fe–Pt NPs that has been synthesized by this method was observed to be highly coalescence and agglomerated. Most of the efforts on direct synthesis of fct-Fe–Pt result out the partially ordering and there are required high temperature annealing for complete ordering. Very few references are found on the direct synthesis of fct-Fe–Pt phase without annealing.

The synthesis methodology and kinetics of reaction play very important role in the development of structure of Fe–Pt. Mostly atomic disorder in the crystal structure are the consequences of non-uniformity in the processing conditions during synthesis. From the literature [18–20] it can be revealed that the disorder structure in fcc-Fe–Pt is a result of fast kinetics of the reaction during processing. By controlling the rate of co-reduction of Fe and Pt ions during chemical synthesis, gives rise to atomic ordering. In the chemical synthesis, process parameters such as the injection temperature of the precursors, the injection time of the surfactants, the reaction solvent, the refluxing temperature and heating rate to the refluxing temperature etc. have strongly influenced in deciding kinetics of reaction and characteristics of final product. With this basic perceptive, in the present work we have used two mode of heating conventional and microwave to process the synthesis. So that, kinetics of reaction can be easily manifested. The roles of different conditions of synthesis such as microwave power, irradiation time, temperature etc. on the structure and magnetic behavior of final product are systematically investigated. The results are compared taking into account the characteristics of the product obtained.

2 Experimental Details

2.1 Experimental Procedure

In the present work, we have used two mode of heating conventional and microwave to process the chemical reaction. The nomenclatures CH-Fe–Pt and MH-Fe–Pt are used for the samples proceed by conventional and microwave mode of heating, respectively. The syntheses are carried out using standard airless techniques in argon atmosphere. In a typical procedure, 1 mmol of $Pt(acac)_2$ was added to 25 ml benzyl ether. After purging with argon for 30 min at room temperature the flask was heated up to 100 °C for 15 min and a designated amount of oleic acid and oleyl amine was added. $Fe(CO)_5$ (2 m mol) was added when the platinum precursor dissolved completely, the mixture was made to react in a 250 cm^3 round bottom flask with condenser attachment kept in a microwave refluxing system. This was a microwave chamber of 360 × 210 × 430 mm dimensions with a 2.45 GHz frequency multimode source. Maximum deliverable power output was 700 W. We prepared the samples for different microwave synthesis conditions by varying the scale of microwave power level from 120 to 420 W and irradiation time 15–65 min under the argon atmosphere, respective temperature was observed to be varied from 150 to 300 °C. The temperatures were monitored during the processing in order to know the actual energy input. Temperature was recorded using optical fibre thermocouple inserting into the solution. Time versus temperature graph for different microwave power is shown in Fig. 1. Same procedure was repeated for conventional processing simply placing the flask with condenser attachment on the heating plate with magnetic stirrer of temperature scale from 200 to 450 °C and time 30–180 min under argon of temperature. The reaction products were cooled down at room temperature and 80 ml of anhydrous ethanol was then added to precipitate the particles. The precipitates were centrifuged four times, and then dispersed in toluene.

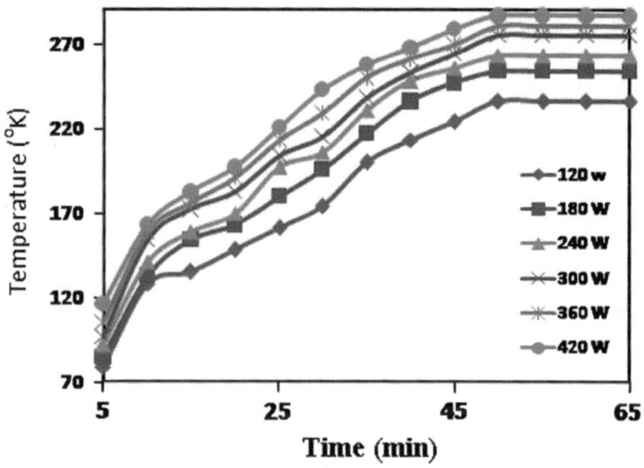

Fig. 1 Rate of heating microwave at different power level during synthesis of Fe–Pt

2.2 Structural Characterization

The measurements of X-ray diffraction (XRD) of as-synthesized Fe–Pt NPs were carried out in the reflection geometry using an X-ray diffractometer D8 Advanced Bruker instrument with Cu Kα radiation (1.54 A°) at a scanning rate of 10 min^{-1}. The transmission electron microscopy (TEM) was studied using a Technai 20 G^2. Energy-dispersive X-ray spectroscopy measurements were obtained using a scanning electron microscopy JEOL JSM-6360A. The magnetic properties of the Fe–Pt NPs sample were determined using a quantum design DC superconducting Quantum Interference Device magnetometer (SQUID).

For the X-ray diffraction and magnetic characterization, the NPs were drop-casted from the original solution (without dilution) on the Si-substrate. For the transmission electron microscopy (TEM), the solutions of NPs were diluted and used to deposit the particles on carbon coated TEM grids for subsequent imaging and analysis.

3 Results and Discussion

XRD patterns of as-synthesized Fe–Pt NPs obtained by conventional and microwave mode of heating is shown in Fig. 2a, b, respectively. XRD peaks of Fig. 2a match with fcc-structure of Fe–Pt indicate as-synthesized CH-Fe–Pt has fcc-phase. However, sharp diffraction at (001) and (110) (Fig. 2b) shows the characteristics super-lattice peaks of the chemically ordered fct phase of Fe–Pt. It indicates that MH-Fe–Pt has directly developed into chemically ordered fct-phase. The XRD patterns of as-synthesized MH and CH Fe–Pt precede at different microwave power and temperature, respectively are shown in Fig. 3a, b. The direct formation of $L1_0$ phase in as-synthesized MH-Fe–Pt can be confirmed from the XRD spectra at different microwave power (Fig. 3a). This is clear evidence for the direct

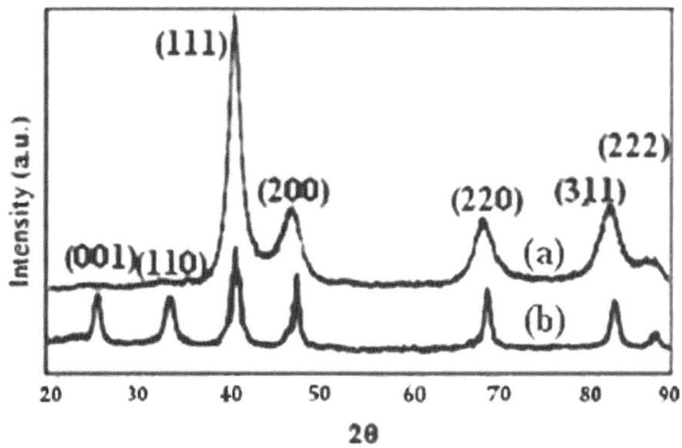

Fig. 2 XRD of Fe–Pt NPs **a** conventional and **b** microwave

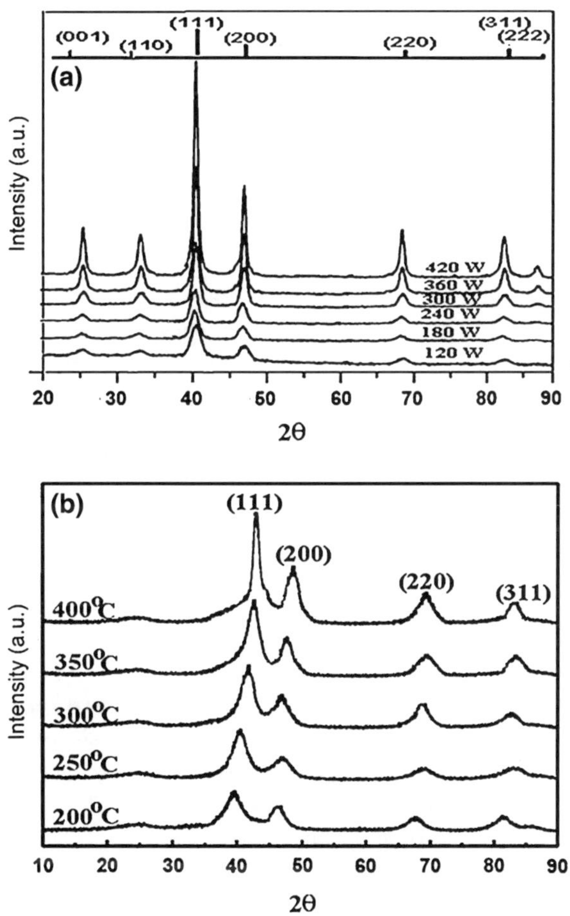

Fig. 3 XRD of Fe–Pt NPs **a** at different power level for MH **b** at different temperature for CH samples

Table 1 Particles size from XRD and TEM of as-synthesized MH- and CH-Fe–Pt nanoparticles at different preparative parameters

Microwave route			Conventional route		
Microwave power (W)	Fe–Pt particle size (nm) from		Temperature (°C)	Fe–Pt particle size (nm) from	
	XRD	TEM		XRD	TEM
120	7		200	20	
180	8		250	21	
240	10	7	300	23	25
300	13	10	350	25	27
360	15	15	400	27	33
420	17	20	450	30	35

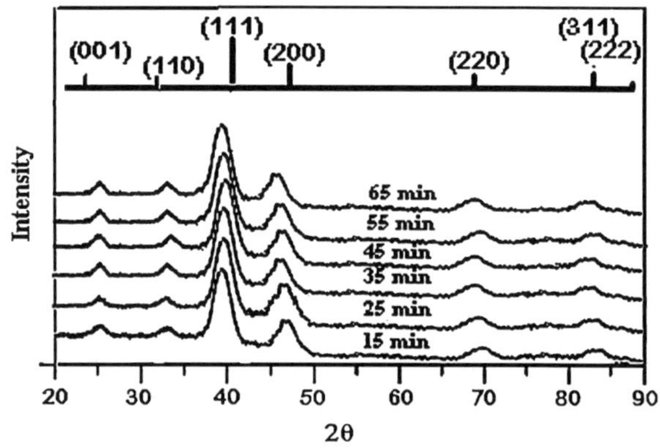

Fig. 4 XRD of MH-Fe–Pt NPs synthesized at different reaction time

formation chemically ordered *fct* structure of Fe–Pt, without any post-annealing treatment. The CH-Fe–Pt have to be annealed at 550 °C (optimized temperature) to get fct-phase.

All samples were synthesized in the form of stable colloidal solution for all synthesis conditions. The width of the XRD peaks provides information about the X-ray coherence length which is close to the average size of the single crystalline domain inside the nanocrystal. Using Scherrer formula average nanocrystallites sizes were calculated for all samples from the width of (111) peak of XRD and listed in Table 1. Systematic analysis of the XRD data reveals that the variations of preparative parameters such as microwave power for MH and temperature for CH during synthesis strongly influences the size of crystallites. With the increase of microwave power from 120 to 420 W and temperature from 200 to 450 °C, average size of crystallites are found to be raised in range of 7–17 nm and 20–30 nm, for MH and CH-samples respectively. On the other hand, reaction time has not any significant influence for both routes. It can also be viewed from XRD (Fig. 4). TEM images for MH and CH samples synthesized at different conditions are shown in Fig. 5a, b, respectively. The increase in size of crystallite with

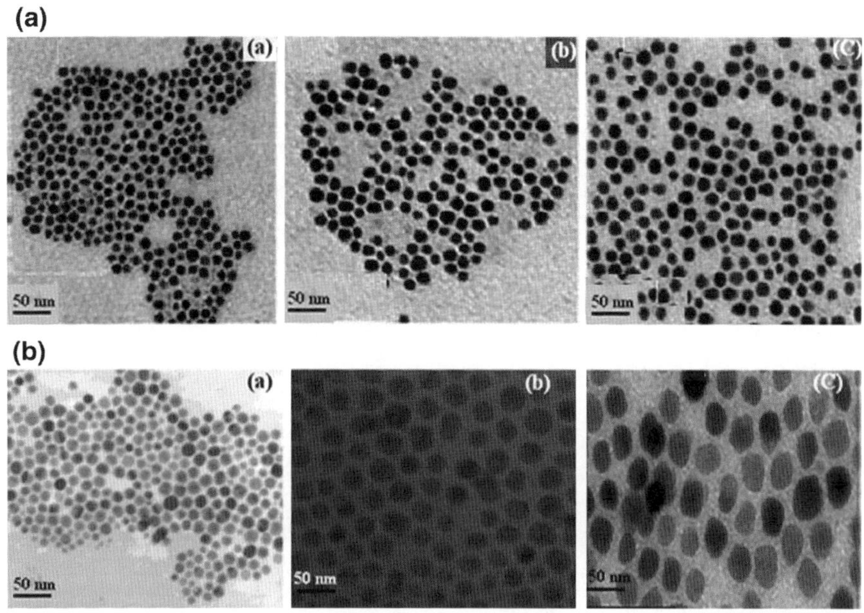

Fig. 5 TEM images **a** MH-sample synthesized at different power level, **b** CH-sample synthesized at different temperature

microwave power in MH and temperature in CH can be clearly observed from Fig. 5a, b, respectively. TEM image of the CH-Fe–Pt before and after annealing are shown in Fig. 6a, b. Pre-annealing images shows well separated highly monodispersed particles of Fe–Pt. Post-annealing images indicate agglomeration and coagulations morphology of the particles. Average crystallites size of Fe–Pt NPs, calculated by XRD and observed from TEM for all samples are listed in Table 1. Compositions of Fe:Pt NPs were determined by EDX technique. The EDX measurements indicate that all compositions of MH-Fe–Pt and CH-Fe–Pt are very close to the ratio 55:45.

3.1 Understanding of Role of Microwave for Direct Formation of fct-Fe–Pt Phase

Microwave heating is the result of absorption of microwave energy by a material exposed to the electromagnetic field distribution within a reflective cavity. It is based on the power absorbed per unit volume [Eq. (1)]

$$P = \sigma|E|^2 = 2\pi f \varepsilon_0 \varepsilon''_{eff}|E|^2 = 2\pi f \varepsilon_0 \varepsilon'_r \tan\delta|E|^2 \quad (1)$$

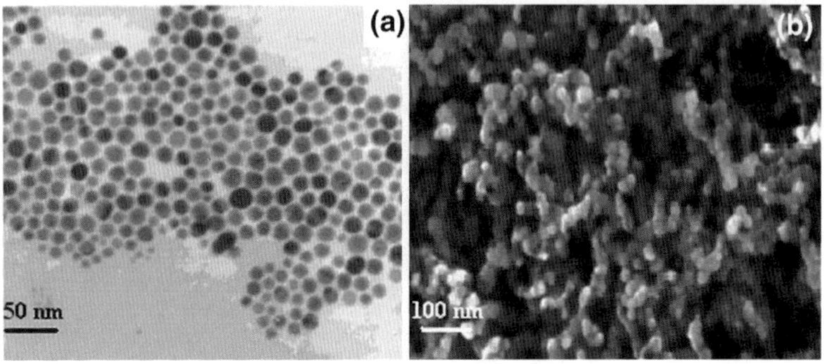

Fig. 6 TEM images of CH sample before and after annealing

where E—is the magnitude of the internal electric field, ε''_{eff}—is the relative effective dielectric factor, ε_0—permittivity of free space, *f*-microwave frequency, σ—total effective conductivity, ε'_r—relative dielectric constant, and $\tan\delta$—energy loss required to store a given quantity of energy. As can be seen from this equation, the dielectric properties (ε'_r, ε''_{eff} and $\tan\delta$) assume a significant role in the extent of power absorbed by a material. The majority of the absorbed microwave power is converted into heat within the materials, as shown in Eq. (2).

$$\frac{\Delta T}{\Delta t} = \frac{2\pi f \varepsilon_o \varepsilon''_{eff} |E|^2}{PC_p} \tag{2}$$

where T is the temperature, t is time, ρ is the density, and Cp is the heat capacity, notice that there are no structural parameters (atomic, micro structural or otherwise) in the equation structural features are assumed to be accounted for by changes in the dielectric properties (ε'_r, ε''_{eff} and $\tan\delta$).

The dielectric properties are important parameters in determining the depth to which the microwaves will penetrate into the materials. As can be seen by Eq. (3), the higher the values of ε'_r and $\tan\delta$, the smaller the depth of penetration for a specific wavelength.

$$D = \frac{3\lambda_o}{8.686\pi \tan\delta \left(\frac{\varepsilon'_r}{\varepsilon_o}\right)^{1/2}} \tag{3}$$

where, D is the depth of penetration at which the incident power is reduced by one half, λ_o is the incident wavelength.

The depth of penetration is important, since it will determine the uniformity of heating, curing etc. throughout the material. High frequencies and large values of the dielectric properties will result in surface heating, while low frequencies and small values of dielectric properties will result in more volumetric heating. Therefore, when a mixture of two materials, A and B of different dielectric constants. ε'_A and ε'_B is exposed to microwave field, the material with higher values of

ε' absorbs energy preferentially and gets heated rapidly compared to the other. A precursor material can be chosen such that it has a high value of ε' and decomposes by preferential microwave absorption to yield the desired materials. Thus if a metal organic or an inorganic complex salt, of reasonable high value of dielectric constant is dissolved in a suitable liquid medium of low dielectric constant and irradiated with microwaves, one can expect the fast reduction and decomposition of metal salts into metals. The liquid chosen should be one, which does not decompose before its normal boiling temperature. Also the resulting product should not be soluble in it. The insoluble product, however, should be prevented from growing into large particles. This is further accomplished by addition of suitable capping agents to the initial solution. It is also necessary to control the microwave power so that the solvent does not boil violently.

In the present work, benzyl ether is used as a solvent and $Pt(acac)_2$ and $Fe(CO)_5$ metal complex salts as precursor for Fe and Pt, respectively. Benzyl ether has substantially lower dielectric constant (3.86). Both the metal salts are very recognized as good microwave absorber During synthesis of Fe–Pt, solution containing $Pt(acac)_2$ and $Fe(CO)_5$ exposed to microwave, the salts absorbs microwave energy preferentially and gets heated rapidly compared to the solvent. It causes the reduction of $Pt(acac)_2$ with formation of Pt-rich nuclei and decomposition of $Fe(CO)_5$ to form Fe. The Fe ions diffuse towards the Pt-nuclei to form Fe–Pt. The rate of diffusion of Fe towards Pt decides the rate of formation and ordering of Fe–Pt. The solvent of the reaction play very important role in deciding the rate of diffusion. In the present case, solvent (benzyl ether) has low dielectric constant and thus low microwave absorption capability, the rate of heating and temperature of solvent is less than the reactants. It reduces speed of the diffusion of Fe and slows down the reaction kinetics. This gives favorable time for chemically ordering of Fe–Pt nuclei and help to form directly the ordered fct-Fe–Pt phase.

3.2 *Growth Process of Fe–Pt NPs by Microwave-Assisted Chemical Route*

Generally, in nanomaterials synthesis by chemical route, the particles size can be tuned by the rate of nucleation. Fast nucleation during the reaction leads to formation of high concentration of seeds. Consequently, low concentrations of monomer are available in the reactant for growth of seeds yielding small crystallites. However, slow nucleation provides low concentration of seeds, consuming the same amount of monomer fasten the growth of crystallites. Hence control over nucleation rate during the reaction is a very important criterion to control the size of crystallites on nanoscale by avoiding Ostwald ripening. Conditions of synthesis play a very important role in the manipulation of the rate of nucleation. In the present work, microwave power and synthesis temperature was found to regulate the size of crystallite of Fe–Pt. It implies that the rate of nucleation is controlled by

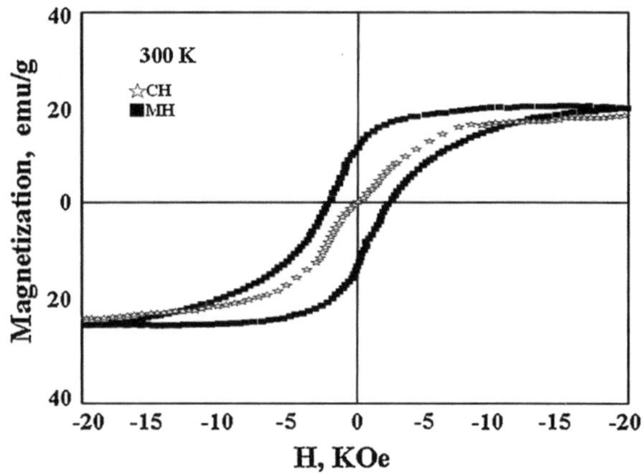

Fig. 7 Hysteresis loop of Fe–Pt NPs **a** MH and **b** CH sample

the power levels of microwave and temperature by conventional route. The role of power levels on the rate of nucleation should be understood.

During microwave processing of chemical reaction, temperature of reaction depends upon the power dissipated in the solvent. The dissipation of power per unit volume (P) is described by $P = \sigma(E)^2$, where σ is the conductivity and E is the electric field related to applied microwave power. It means that the reaction temperature is manipulated by the power level of microwave.

In chemical synthesis route, rate of heating and the temperature of the reaction play dominant role in controlling the growth of nucleation. In microwave-assisted synthesis, in contrast to convective heating; dielectric heating raises the temperature of the total volume of the reactants by transforming energy selectively to microwave absorbing materials. The larger the microwave cross-section for a particular constituent, the more dramatic the heating process is. The intrinsic temperature localized around the ion is significantly higher than that of the bulk solution. It can therefore be understood that localized heating drives the rate of nucleation and controls the Ostwald ripening process. The microwave power level decides the localized heating rate. On the other hand, exposure of reactant in microwave at constant power as function of time has not any significant change in intrinsic temperature. This behavior is attributed to constant rate of nucleation and it may avoid Ostwald ripening processes.

The as-synthesize MH-Fe–Pt NPs are in chemically ordered *fct*-phase, while CH-Fe–Pt NPs in chemically disordered fcc-phase are confirmed from XRD analysis. The magnetic behavior of the MH and CH are shown in hysteresis curves Fig. 7. The hysteresis measurements of MH-Fe–Pt NPs and CH-Fe–Pt NPs observed at 300 K. It exhibits superparamagnetic behavior of a CH-Fe–Pt and ferromagnetic of MH-Fe–Pt NPs. Figure 8 shows a set of measurements performed for characterizing the size- dependent magnetic properties of as-synthesized MH NPs. From the hysteresis loops, it can be concluded that all the MH-Fe–Pt samples are found to exhibit ferromagnetic while CH-Fe–Pt are found to show supermaramagnetic behavior at room temperature. For MH samples results reveal that

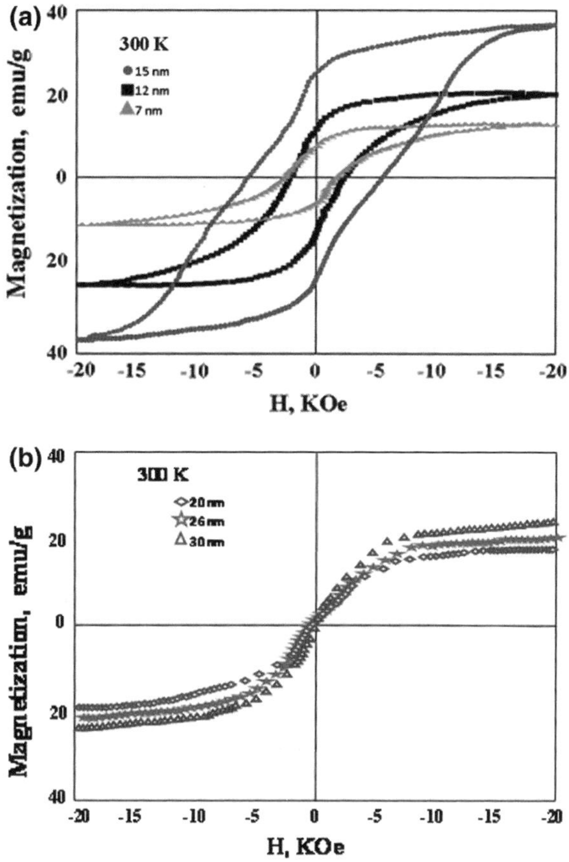

Fig. 8 Size dependent magnetic properties of MH-Fe–Pt

coercivity decreases with decreasing particle size. This is due to easy reorientation of the magnetic moment for small particles. The drop in the saturation magnetization with decreasing NPs size can be attributed to an increase of the surface-to-volume ratio.

4 Conclusion

In the present work, two different mode of heating; microwave and conventional heating route have been used for the synthesis of Fe–Pt NPs. The effects of conditions of heating (microwave power, irradiation time, temperature) on the growth process of Fe–Pt NPs are systematically investigated. We have succeeded in the direct synthesis of $L1_0$ structure of Fe–Pt without any post synthesis treatments by microwave- assisted route, while conventional route gives rise to disordered face centered cubic (fcc) phase. The particle size and magnetic properties of the as-prepared Fe–Pt are found to be very sensitive to the preparative parameters such as microwave power and conventional temperature. The size of the particles was tuned in the range of 7–17 nm by varying power level by 120–420 W and 20–30 nm by varying temperature 200–450 °C. The hysteresis measurements were performed at 300 K to study magnetic properties of as-synthesized Fe–Pt as a function of mode of heating and crystallite size. Conventionally synthesized Fe–Pt NPs display super paramagnetic behavior, whereas microwave-assisted sample exhibits ferromagnetic behavior. Size dependent magnetic behavior of microwave-assisted synthesized samples shows that coercivity and saturation magnetization is observed to be decreasing with diminishing particle size. For 15, 12, 7 nm particles coercivity are found to be 5, 4 and 3.5 KOe, respectively. Saturation magnetization are observed to be 42, 20 and 11 emu/g for the 15, 12, 7 nm particles of Fe–Pt, respectively.

Acknowledgments The authors also thank Dr. S. K. Kulkarni, Professor, IISER, Pune for magnetic characterization and Dr. R. M. Singru, Professor, IIT Kanpur for fruitful discussion and suggestion. This work is part of SAP programme, UGC, New Delhi.

References

1. S. Weller, A. Moser, Thermal effect limits in ultrahigh density magnetic recording. IEEE Trans. Magn. **35**, 4423 (1999)
2. J.A. Christodoulides, Y. Huang, Y. Zhang, G.C. Hadjipanayis, I. Panagiotopoulos, D. Niarchs, CoPt and FePt thin film for high density recording media. J. Appl. Phys. **87**, 6938 (2000)
3. D.L. Peng, T. Hihara, K. Sumiyama, Formation and magnetic properties of Fe-Pt alloys clusters by Plasma-gas condensation. Appl. Phys. Lett. **82**, 350 (2003)
4. S. Sun, Recent advances in chemical synthesis, self-assembly, and applications of FePt nanoparticles. Adv. Mater. **18**, 393 (2006)
5. S. Sun, C.B. Murray, D. Weller, L. Folks, A. Moser, Monodisperse FePt nanoparticles and ferromagnetic FePt nanocrystal superlattices. Science **287**, 1989 (2000)
6. H. Zeng, J. Li, Z.L. Wang, J.P. Liu, S. Sun, Interparticle interactions in annealed FePt nanoparticle assemblies. IEEE Trans. Magn. **38**, 2598 (2002)
7. H. Zeng, J. Li, J.P. Liu, Z.L. Wang, S.H. Sun, Exchange—coupled nanocomposite magnets by nanoparticles self- assembly. Nature. **420**, 395 (2002)
8. H. Kodama, S. Monose, T. Sugimoto, T. Uzumaki, A. Tanaka, Chemically synthesized FePt nanoparticles material for ultrahigh density recording. IEEE Trans. Magn. **41**, 665 (2005)
9. B. Jeyadevan, A. Hobo, K. Urakawa, C.N. Chinnasamy, K. Shinoda, K. Tohji, Towards direct synthesis of fct-FePt nanoparticles by chemical route. J App. Phys. **93**, 7574 (2001)
10. B. Jeyadevan, Direct synthesis of fct-FePt nanoparticles by chemical route. Japan J. Appl. Phys **42**, L350–L352 (2003)
11. K. Sata, B. Jeyadevan, K. Tojhi, Preparation and properties of ferromagnetic FePt dispersion. J. Magn. Magn. Mater. **266**, 227 (2003)
12. S. Kang, Z. Jia, S. Shi, D.E. Nikles, J.W. Harrell, Easy axis alignment of chemically partially ordered FePt nanoparticles. Appl. Phys. Lett. **86**, 62503 (2005)
13. M. Howard, H.L. Nguyen, S.R. Giblin, B.K. Tanner, I. Terry, A.K. Hughes, O.S. Evans, A synthetic routes to size-controlled fcc and fct FePt nanoparticles. J. Am. Chem. Soc. **127**, 10140 (2005)

14. R. Minamia, Y. Kitamoto, T. Chikata, S. Kato, Direct synthesis of L10 type Fe–Pt nanoparticles using microwave-polyol method. Electrochim. Acta **51**, 864 (2005)
15. L. Colak, G.C. Hadjipanayis, Chemically synthesized FePt nanoparticles with controlled particle size, shape and composition. Nanotechnology **20**, 485602 (2009)
16. C. Rong, D. Li, J.P. Liu, V. Nandwana, N. Poudyal, Y. Ding, Z.L. Wang, H. Zeng, Size dependent chemical and magnetic ordering in LI_0-FePt. Adv. Mater. **18**, 2984 (2006)
17. V. Nandana, K.E. Elkins, N. Poudyal, G.S Chanbey, K. Yano, J.P. Liu, Size and shape control of monodisperse FePt nanoparticles. J. Phys. Chem. C. **111**, 4185 (2007)
18. M. Chen, J.P. Liu, S. Sun, One step synthesis of FePt nanoparticles with tunable size. J. Am. Chem. Soc. **126**, 8394 (2004)
19. H. Hu, H. Yang, P. Huang, D. Cui, Y. Peng, J. Zhang, F. Lu, J. Lian, D. Shi, Unique role of ionic liquid in microwave-assisted synthesis of monodisperse magnetic nanoparticles. Chem. Commun. **46**, 3866 (2010)
20. Y. Qing, W. Zhou, S. Jia, F. Juo, D. Zhu, Effect of heat treatment on the microwave electromagnetic properties of carbonyl iron/epoxy-silicone resine coatings. J. Mater. Sci. Technol. **26**(11), 1011 (2010)

Effect of Coordination Agent on the Structure and Morphology of ZnS Nanostructure

S. A. Acharya, S. Darulkar and P. Choudhary

Abstract The high temperature structure of ZnS, (Wurtzite over 1,020 °C), is successfully prepared at low temperature (170 °C) by co-precipitation and hydrothermal route in presence of ethylenediamine (EN) as coordination agent. The effect of reaction route, and EN concentration on structure and morphology of ZnS nanocrystals are systematically investigated. We have demonstrated the structure transformation process from zinc blend to wurtzite with change in concentration of EN. Besides the structure evolution, morphological transformation from particles to rods and sheets, are also observed. To understand the low temperature structure transformation process of ZnS in presence of EN, chemical bonding approach is used. The bonding between Zn and S has been found to be admixture of covalent and ionic bond and transition from one bond state to other give rise to the phase transformation. The EN assists for formation of intermediate complex Zn-EN, which is responsible to modify the nature of bond of Zn–S, it initiates structure transformation. The geometry of structure of Zn–EN complex decides the growth direction of ZnS and thus morphology.

Keywords ZnS · Phase transition · Ethylenediamine · Nanosheet · Nanorods

1 Introduction

The crystalline structure and morphology of semiconductor nanocrystallites are important factors for determining their photophysical properties [1, 2]. Generally, the difference in the crystalline structure of the solid leads to considerable change

S. A. Acharya (✉) · S. Darulkar · P. Choudhary
Department of Physics, RTM Nagpur University, Nagpur, Maharashtra 440033, India
e-mail: saha275@yahoo.com

P. K. Giri et al. (eds.), *Advanced Nanomaterials and Nanotechnology*,
Springer Proceedings in Physics 143, DOI: 10.1007/978-3-642-34216-5_6,

in the effective masses of electrons and holes in their electronic bands. While the surface morphology plays an important role in determining the thermodynamics of the system especially at nanoscale due to large variation of surface-to-volume ratio. The simultaneous controlling of structures and morphology of semiconductors nanocrystallites provides opportunities to tune and explore their optical properties. Currently, morphology control has raised significant concern in the fabrication of semiconductor nanocrystals [3–11].

ZnS, as one of the important II–VI group semiconductor materials has been of great research interest as a result of its unique applications in optoelectronic and luminescent devices [12–14]. ZnS has been observed in variety of polytypes, which are basically related to two structures: the cubic zinc-blende structure and the hexagonal wurtzite structure [15]. The most stable form of zinc sulfide is the cubic structure and in the bulk it transforms to wurtzite structure at 1,020 °C, wurtzite is thermodynamically metastable state [16]. The bulk zinc sulfide melts at a temperature of 1,650 °C. The nanocrystalline cubic ZnS has been found to transform to wurtzite at a temperature 400 °C in a vacuum [17]. The hexagonal structure has been found to be transformed spontaneously to the cubic phase by contact with some organic molecules at ambient temperature [18]. Therefore, it is great challenge to directly synthesize wurtzite ZnS and stabilize it at room temperature by controlling its transformation to the cubic phase.

Morphology control of ZnS nanocrystallites is another important factor in the manipulation of optical properties. The research efforts in the field of morphology control of ZnS have been mostly focused on fabricating 1D nanowires or nanorods and 2D nanosheet [19–24]. In the present work, we report synthesis of ZnS in 2D sheet-like and 1D rods-like morphology by using coprecipitation and hydrothermal route, respectively in presence of ethylenediamine (EN) as a bident ligand. ZnS is synthesized in cubic phase without using EN and in presence of EN developed wurtzite phase of ZnS.

2 Experimental Procedure

The starting precursors were zinc acetate dihydrate ($Zn(CH_3COO)_2 . 2(H_2O)$), and thiourea (CH_4N_2S) procured from Merck, India. All reagents were used as received. Equal moles (0.8 M) of zinc acetate and thiourea were mixed with EN solution. The concentrations of EN were varied from 12, 40, 67, 90 and 100 % (to the total volume of solution, 100 ml), respectively, as a volume ratio in x EN to $(100 - x)$ H_2O. The hydrothermal and co-precipitation routes are used for synthesis. The samples synthesized by hydrothermal route are designated as **HY–ZnS** and samples by co-precipitation route by **CO–ZnS**. For hydrothermal synthesis each solution was introduced into Teflon lined autoclave and sealed within a stainless steel container. For co-precipitation method the solutions were introduced into 1,000 ml round bottom flask and proceed on heating plate with magnetic stirrer. For both the methods, processing temperature and time were optimized.

Best possible results are obtained at 170 °C for 8 h by co-precipitation route and for hydrothermal routes 12 h processing at the same temperature gave the results. The final products are filtered and dried in vacuum.

Crystalline information was obtained using X-ray diffractometry PANanalytical X'pert PRO from Phillips Holland equipped with copper target ($\lambda_{K\alpha l} = 1.5405$ Å) at the scanning rate 1° min^{-1}. The morphology and energy dispersive x-ray (EDAX) analysis were carried out by using a JEOL, JSM, 6380 Analytical SEM. The transmission electron microscopy (TEM) was carried out using a Technai 20 G^2 operated at an accelerating voltage of 200 keV using carbon coated formvar grids and the sample was prepared dispersing the nanostructures in N,N dimethylformamide (DMF) using ultrasonication. The absorption spectra of the samples were measured using a Perkin-Elmer Lambda-950 UV-Vis spectrophotometer (double distilled water is used as a reference).

3 Results and Discussion

3.1 XRD Study of ZnS Nanocrystallites

The XRD pattern of samples prepared by hydrothermal and co-precipitation routes using different concentration of EN are shown in Fig. 1a and b, respectively. The diffraction peaks are observed at 29.01, 48.96 and 57.16° for HY–ZnS-0 % EN and at 28.91, 48.03, 58.11° for CO–ZnS-0 % EN. Both XRD are well matched to the (111), (220) and (311) diffraction peaks of sphalerite or zinc blende structures of ZnS (JCPDS no. 5–566). It indicates that both the samples are in the cubic zinc

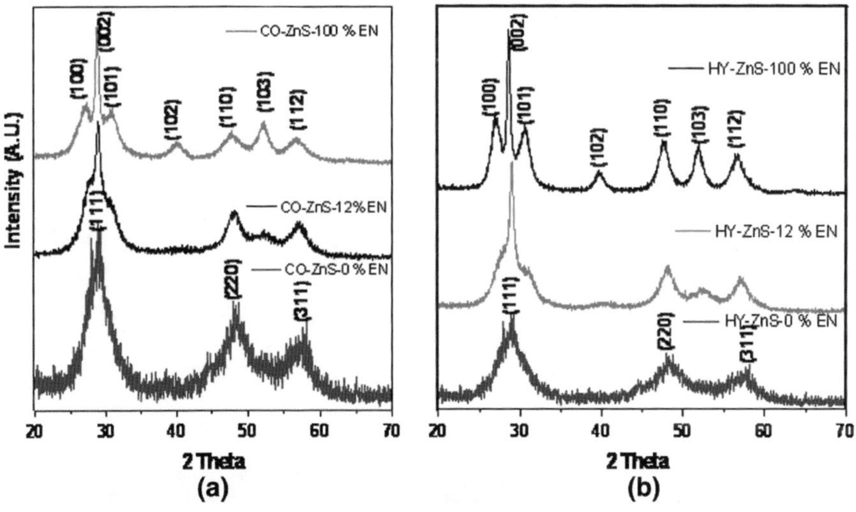

Fig. 1 XRD of ZnS synthesis by **a** co-precipitation **b** hydrothermal method

blende structure, synthesis route has no effect on phase formation. With the addition of EN (12 %) into an aqueous solution, the diffraction peaks are detected at 28.96, 48.07, 52.26 and 57.02° for CO–ZnS, whereas at 28.61, 47.73, 52.16 and 56.86° by HY–ZnS. The XRD peaks are matched with (002), (110), (103) and (112) planes of wurtzite phase of ZnS (some peaks of the phase are missed).

The XRD of ZnS-100 % EN shows peaks at 27.04, 28.96, 30.82, 39.95, 48.07, 52.26 and 57.02 for CO–ZnS and 27.10, 28.61, 30.54, 39.70, 47.73, 51.92, 56.64 for HY–ZnS; all peaks are well related to wurtzite phase of ZnS. The XRD data is the clear evidence for the phase transformation from of zinc-blende-to-wurtzite with addition of 12 % EN. The wurtizite ZnS are found to be stabilized for a longer time more than 1 month.

3.2 Surface Morphology of ZnS Nanocrystallites by SEM and TEM

The surface morphology of the ZnS nanocrystallites prepared under different conditions is investigated and the results are shown in Figs. 2, 3 and 4. SEM image of CO–ZnS prepared without EN shows (Fig. 2) the highly agglomerated spherical particles having size varying in the range of 30–60 nm. However, CO–ZnS with 40 % EN (Fig. 3) demonstrates sheet-like morphology with rectangular lateral dimensions. The results demonstrate (Fig. 3b) that a few sheets are stacked parallel; while some of the sheets are detected to be randomly oriented (Fig. 3a). The lateral dimensions of the sheet are observed to be in micrometer, whereas thickness of the sheet is in the range of 100–150 nm.

SEM image of HY–ZnS (Fig. 4) depicts rods like morphology with diameter in range of 90–100 nm and the average length is more than ~1 μm. The transmission electron microscopy (TEM) (Fig. 5) confirms the rod-like morphology of HY–ZnS at nano-scale. However, it was found very difficult to obtain TEM images of sheet-like morphology of the ZnS.

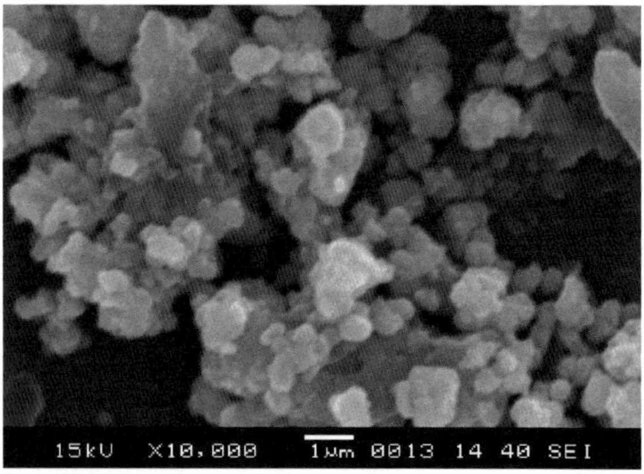

Fig. 2 SEM of CO–ZnS-0 % EN

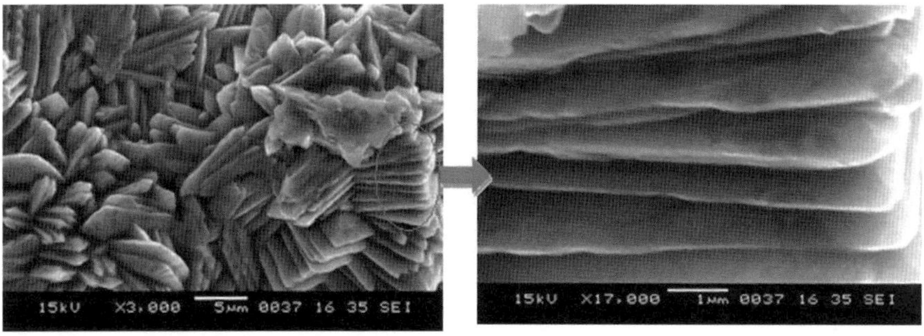

Fig. 3 SEM of as-synthesized CO–ZnS-67 % EN

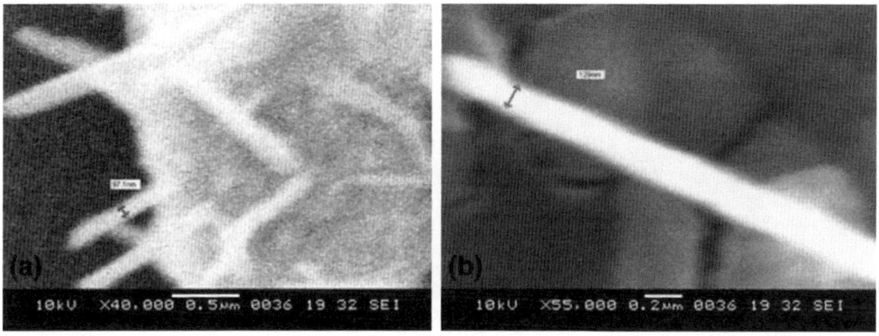

Fig. 4 SEM of as-synthesized HY–ZnS-67 % EN

Fig. 5 TEM of as-synthesized HY–ZnS-67 % EN

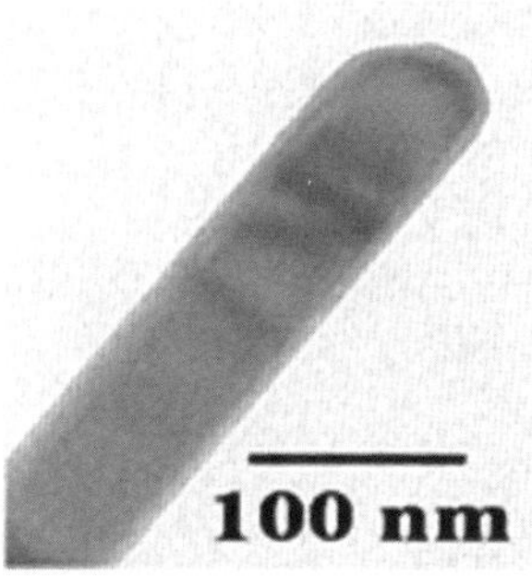

Yu et al. [25] also reported the nanosheet of ZnS in wurtzite structure obtained by solvothermal methods in inert atmosphere at temperature above 350 °C. Sheet-like morphology of ZnS is obtained by co-precipitation route in the present work, which is totally different than the earlier reported work. We have got the morphology at atmospheric conditions and much low temperature (170 °C).

The EDAX pattern of CO–ZnS-40 %EN is presented in Fig. 6. It can be noticed that the peaks due to Zn, S, Si, C and Pt are present. The strongest peak is due to

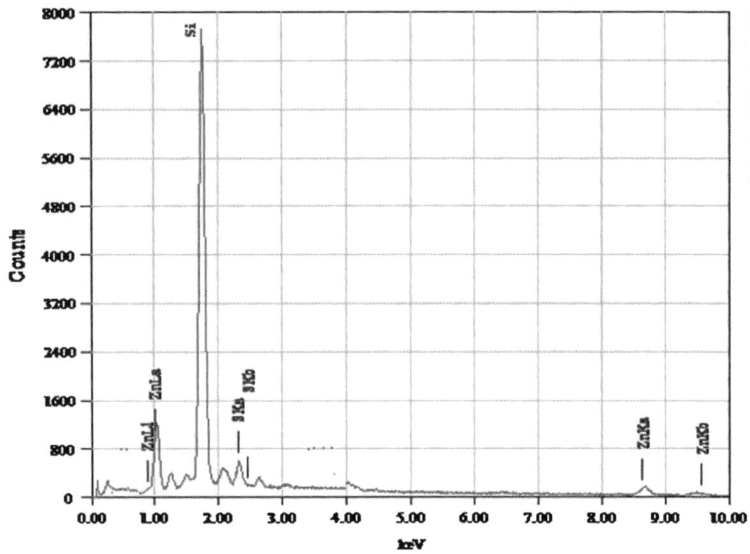

Fig. 6 EDAX pattern of CO–ZnS-67 % EN

the silicon substrate. Pt is due to the coating used for avoiding the charging and carbon can be due to the fragment remaining from the EN used in the sample synthesis. Some contamination by carbon can also be due to the presence of unintentional carbon in the vacuum chamber of the SEM. The average atomic percentage ratio of Zn–S is 56:44. That is, the sample is rich in Zinc.

3.3 UV-Absorption Spectral Characteristics of ZnS

It is well known that, as a consequence of quantum confinement of the photogenerated electron–hole pair, the UV-Vis absorption spectra of semiconductor nanoparticles are size dependent. In particular, the wavelength at the band maximum decreases as the nanoparticles size decreases. The effect of EN concentration on the size of as-synthesized ZnS nanocrystallites are investigated here. UV-Vis absorption spectra of the as-prepared ZnS rods and sheets-like morphology are shown in Fig. 7a and b. From Figs it can be noticed that the absorption peak has blue shift by about 25–40 nm as compared to the value ($\sim$345 nm) expected for the bulk ZnS materials. The blue shift is attributed to the quantum confinement effects in rod-like and sheet-like morphology. However, change in EN concentrations has not shown any noticeable change in blue shift. This indicates that EN is not playing any significant role in the tuning of nano-dimensions of the rod-like and sheet-like morphology of ZnS.

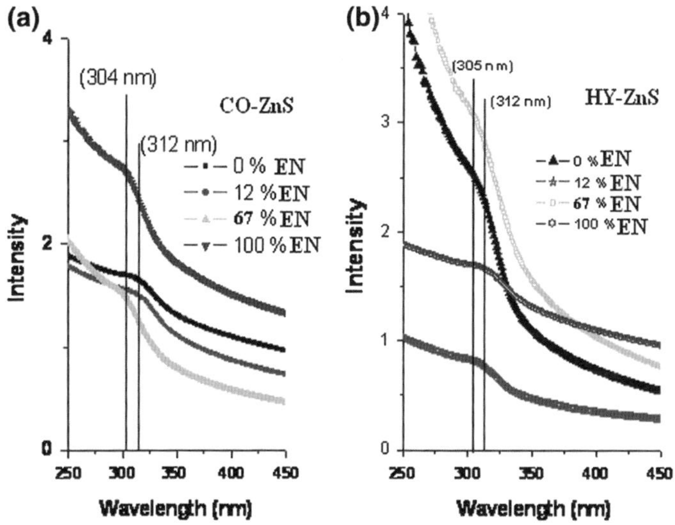

Fig. 7 UV-Vis absorption spectra of **a** CO–ZnS **b** HY–ZnS with different % of EN

3.4 Understanding the Role of EN in Phase and Morphology Formation of ZnS

The concentration of EN is found to be controlling both the structures and morphologies of the as-prepared ZnS products. In order to recognize the role of EN in phase deformation of ZnS, chemical bonding approach is used. The nature of bond in more stable phase of Zn (sphalerite) [space group F43 m (no 216)] is understood as, the S^{2-} anions form a face-centered cubic arrangement with Zn^{2+}cations occupying half the tetrahedral holes (or vice versa). The sulfide ions adopt a cubic-closed packed structure. There are covalent interaction between Zn^{2+} and S^{2-}. The sp^3 hybridization of atomic orbitals of cation and anions is responsible for covalent interaction in the tetrahedral surrounding. These are formed by the available s and p valence orbitals of both zinc and sulphur. The valence orbitals of Zn and S contribute 8 valence electrons (6e− from S and 2e− from Zn) which are equally distributed in the hybridized orbitals and the effective formal charges to Zn^{2-} and S^{2+}. But there is questionability of occurrence of Zn^{2-}.

Pauli's Equation for quantitative prediction of nature of bonding on basis of electronegativity difference is given as:

$$P_{AB} = 1 - \exp\left[-\frac{1}{4}(X_A - X_B)\right]$$

where, P_{AB} is ionic bond strength, x is average electronegativity, subscript A and B are cations and anions, respectively. For ZnS having electronegativity difference 0.93 indicates 74 % covalency and 26 % ionicity of Zn–S bond. That is 74 % out of total valence electrons of Zn and S (i.e. 8 electrons) are shared between Zn and S and 26 % charge are shifted from Zn to S. Both these evidences point toward the nature of bonding in ZnS, which is intermediate between the extremes of pure covalent or pure ionic. The Zn–S bond is admixture of covalent and ionic. The tuning of transition from covalent to ionic bond state or vice versa is possible by using external agencies like temperature, pressure or presence template or surfactant during synthesis. This may be responsible for existence of ZnS in different polytypes.

In our present study, synthesis of ZnS without using of EN has developed sphalerite structure, which is identified as more symmetric and stable phase of ZnS. Since it is less energetic structure; the natural tendency to form this phase easily, when Zn and S come in contact through their precursors in normal synthesis conditions. On the other hand the ZnS synthesis reaction with presence of EN proceeds with formation of intermediate complex due to covalent interaction of EN and Zinc from zinc acetate. As EN is strong coordination agent responsible to produce strong electrostatic field by its electron pair bond with Zn. This results out strong covalent interaction between Zn and EN, which reduces the covalent interactions between Zn and S or increase of ionic interaction between Zn and S. On other hand, the Zn–EN complex modified the interaction between Zn and S. This tune the covalent interaction between Zn and S, which is calculated 74 % in stable phase. The decrease of covalent bond strength increases the interatomic spacing, which forced the phase transformation.

The morphology formation mechanism of ZnS nanorods and nanosheets with EN is also very interesting. In absence of EN, due to high symmetry of the sphalerite structure, cubic ZnS nanocrystals are formed in spherical shape. EN plays a critical role in the formation of ZnS nanostructures. It acts as a molecular soft template to change the direction of growth of ZnS crystallites. Deng et al. [26], Chen et al. [27] and Yue et al. [28] discussed the very important role that EN took part in the formation of ZnS nanorods and nanosheets but they didn't discuss how is interaction between electron donating ligand EN with Zn and its impact on specific crystal faces which decides the growth direction of ZnS on their resultant geometry. This will be discussed in the latter part of this paper by considering role of EN as a structure-directing and surface-passivating coordination agent. The growth geometry of the Zn–EN complex in formation of sheets and rods-like morphologies are displayed in Figs. 8 and 9, respectively

As discussed earlier, synthesis of ZnS with EN developed Intermediate complex of Zn–EN. According to Werner complex theory, the most common structure of the complex is tetrahedral, square planar, and octahedral respectively. For Zn^{2+} having $3d^{10}$ configurations, tetrahedral or square planar structure of Zn–EN complex are more probable. The coordination of Zn with EN to form the complex is the function of synthesis conditions. In general on the basis of geometry of the complex structure, the atoms at the corner of the same plane have the strongest bonding force as compared to the atoms at the corner of perpendicular plane. For Zn–EN complex with tetrahedral geometry the crystal face directed in the plane of tetrahedral is easy direction for bonding of S^{2-} and thus growth. The EN directed

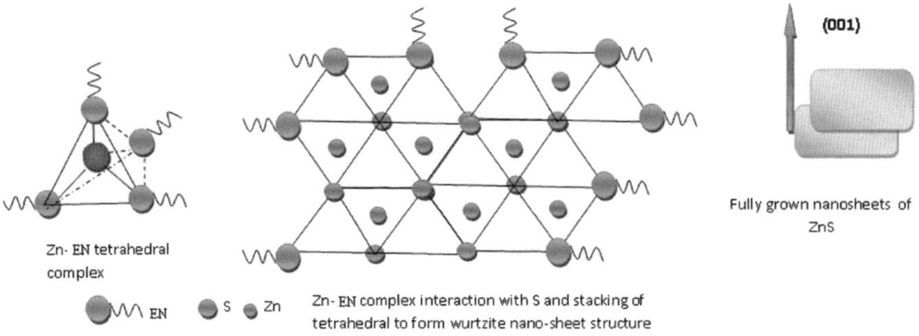

Fig. 8 Geometry of stacking of Zn–S tetrahedra from Zn–EN complex to form nano-sheet like morphology

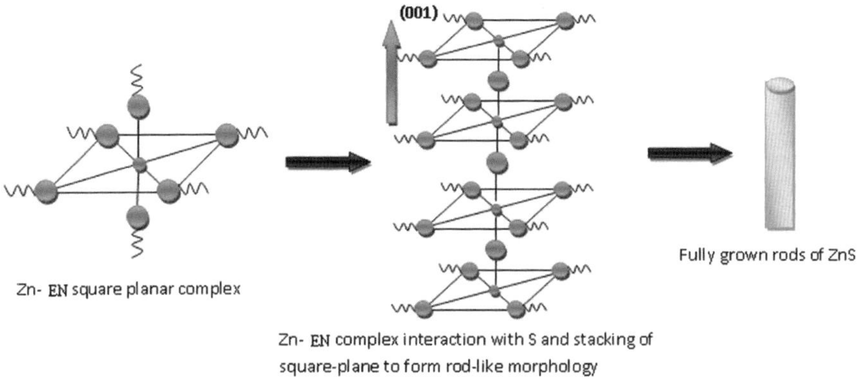

Fig. 9 Geometry of stacking of Zn–S from square-planar Zn–EDA complex to form nano-rods like morphology

the faster growth of ZnS along that direction while passivating other directions. However, in square planar geometry, there is only one easy direction for S^{2-} to line up while other directions are passivated by EN bonding. The growth of ZnS crystallite is faster along this direction and consequently the product obtained displayed a rod-like morphology.

4 Conclusion

Both sphalerite and wurtzite ZnS nanocrystals are successfully synthesized by hydrothermal and co-precipitation method using EN as structure-directing and surface-passivating coordination agent. The phases of the ZnS nanocrystallites are tuned by varying EN concentration. While, shapes of the ZnS are controlled by synthesis route in presence of EN. The rods-like morphology is obtained for ZnS synthesized by hydrothermal route and sheets-like morphology is obtained by co-precipitation route. The phase transformation from sphalerite to wurtzite is observed during synthesis in presence of EN. The wurtzite ZnS sheets have a lateral dimension up to hundreds of microns and thickness is ranging in 100–150 nm. At the same time, wurtzite ZnS rods have diameter 90–100 nm and the length in microns. The SEM and TEM images confirm the morphology and fine size at nanoscale of both the samples. The EDAX pattern of ZnS verify average atomic percentage ratio of Zn–S 56:44. That is, the sample is rich in Zinc. Blue shift in UV-Vis spectra shows Quantum confinement effect in the developed structure of ZnS.

References

1. A. P. Alivisatos, Semiconductor clusters, nanocrystals, and quantum dots. Science, **271**, 933 (1996)
2. P.C. Kuo, H.W. Wang, S.Y. Chen, Synthesis and photoluminescent properties of wurtzite ZnS nanorods by hydrothermal and co-precipitation methods. J. Ceramic Soc. Japan **114**, 918 (2006)
3. J. Wang, H. Luo, T. Chen, Z. Yuan, Facile shape-selective growthof ZnO nanotips and graded nanowires from its oriented nanorods in a saturated ZnS solution. Nanotechnology **21**, 505603 (2010)
4. C. Liangyuan, L. Zhiyong, B. Shouli, Z. Kewei, L. Dianqing, C. Aifan, C.C. Liu, Synthesis of 1-dimensional ZnO and its sensing property for CO. Sens. Actuators B **143**, 620 (2010)
5. X. Jian, Y. Liu, Y. Gao, X. Zhang, L. Shi, Preparation of one-dimensional nanostructured ZnO. Particuology **8**, 383 (2010)
6. J.Y. Zhu, J.X. Zhang, H.F. Zhou, W.Q. Qin, L.T. Chai, Y.H. Hu, Microwave-assisted synthesis and characterization of ZnO nanorods arrays. Trans. Non Ferrous Met. Soc. China, **19**, 1578 (2009)
7. J. Zhao, Z.G. Jin, T. Li, X.X. Liu, Nucleation and Growth of ZnO nanorods on the ZnO coated seed surface by solution chemical method. J. Eur. Ceram. Soc. **26**, 2769 (2006)
8. Z. Deng, J. Qi, Y. Zhang, Q. Liao, Y. Huang, J. Cao, Synthesis, structure and growth mechanism of ZnS nanowire with high Aspect Ratio. Acta Phys.-Chim., Sin. **24**(2), 193 (2008)
9. F. Li, Y. Jiang, L. Hu, L. Liu, Z. Li, X. Huang, Structural and luminescent properties of ZnO nanorods and ZnO/ZnS nanocomposites. J. Alloys Comp. **474**, 531 (2009)
10. H. Chen, D. Shi, J. Qi, J. Jia, B. Wang, The stability and electronic properties of wurtzite and Zinc-blende znS nanowires. Phys. Lett. A **373**, 371 (2009)
11. N. Uekawa, T. Matsumoto, T. Kojima, F. Shiba, K. Kakeguwa, Synthesis of stable sol of ZnS nanoparticles by heating the mixtures of ZnS precipitate and ethylene glycol"l. Colloids Surf. A: Physicochem. Eng. Aspects, **361**, 132 (2010)
12. P. Calander, M. Gofferdi, V.T. Liveri, Study of the growth of ZnS nanoparticles in water/AOT/n-heptane microemulsions by UV-absorption spectroscopy. Colloids Surf. A: Phys. Eng. Aspects, **160**, 9 (1999)
13. T.V. Prevenslik, Acoustoluminescence and sonoluminescence. J. Lumin. **87** 1210 (2000)
14. T. Trindade, P. O'Brien, N.L Pickett, Nanocrystalline semiconductors: synthesis, properties, and perspectives. Chem. Mater. **13** 3843 (2001)
15. J. Baars, G. Brandt, Structural phase transitions in ZnS. J Phys. Chem. Solids **34**, 905 (1973)
16. L. Brus, Zero-dimensional "excitons" in semiconductor clusters. IEEE J. Quantum Electron. **22**, 1909 (1986)

17. S.B. Qadri, E.F. Skelton, D. Hsu, A.D. Dinsmore, J. Yang, H.F. Gray, B.R. Ratan, J. Yang, H.F. Gray, B.R. Ratan, Size-induced transition-temperature reduction in nanoparticles of ZnS. Phys. Rev. B **60**, 9191 (1999)
18. K. Murakoshi, H. Hosokawa, N. Tanaka, M. Saito, Y. Wada, T. Sakata, H. Morib, S. Yanagida, Phase transition of ZnS nanocrystallites induced by surface modification at ambient temperature and pressure confirmed by electron diffraction. Chem. Commun. **3** 321 (1998)
19. R.F. Zhuo, H.T. Feng, D. Yan, J.T. Chen, J.J. Feng, J.Z. Liu, P.X. Yan, Rapid growth and photoluminescence properties of doped ZnS one-dimensional nanostructures. J. Cryst. Growth **310**, 3240 (2008)
20. T. Thongtem, C. Pilapong, S. Thongtem, Silica gel-assisted solvothermal production of CdS, CuxS ($x = 1, 2$) and ZnS with different morphologies. Trans. Nonferrous Met. Soc. China **19**, 105 (2009)
21. L. Dong, Y. Chu, Y. Zhang, Microemulsion-mediated solvothermal synthesis of ZnS nanowires. Mater. Lett. **61**, 4651 (2007)
22. M.S Niasari, F. Davar, M.R.L Estarki, Controllable synthesis of thioglycolic acid capped $ZnS(Pn)_{0.5}$ nanotubes via simple aqueous solution route at low temperatures and conversion to wurtzite ZnS nanorods via thermal decompose of precursor. J. Alloys Comp. **494** 199 (2010)
23. L. Li, Y. Tang, Y. Zhang, J. Yang, B. Du, Synthesis of ZnS hollow nanospheres with holes using different amine templates. Front. Chem. China **3**, 76 (2008)
24. C. Wang, Y. Ao, P. Wang, S. Zhang, J. Qian, J. Hou, A simple method for large-scale preparation of ZnS nanoribbon film and its photocatalytic activity for dye degradation. Appl. Surf. Sci. **256**, 4125 (2010)
25. S.H. Yu, M. Yoshimura, Shape and phase control of ZnS nanocrystals: Template fabrication of wurtzite ZnS single-crystal nanosheets and ZnO flake-like dendrites from a lamellar molecular precursor ZnS. $(NH_2CH_2CH_2NH_2)_{0.5}$. Adv. Mater. **14**, 296 (2002)
26. Z.X. Deng, W. Chen, X.M. Sun, Y.D. Li, Structure-directing coordination template effect of ethylenediamine in formations of ZnS and ZnSe nanocrystallites via solvothermal route. Inorg. Chem. **41**, 869 (2002)
27. X.J. Chen, H.F. Xu, N.S. Xu, F.H. Zhao, W.J. Lin, G. Lin, Y.L. Fu, Z.L. Huang, H.Z. Wang, M.M. Wu, Kinetically controlled synthesis of wurtzite ZnS nanorods through mild thermolysis of a covalent organic-Inorganic Network. Inog Chem, **42**, 3100 (2003)
28. G.H. Yue, P.X. Yan, D. Yan, J.Z. Liu, D.M. Qu, Q. Yang, X.Y. Fan, Synthesis of two-dimensional micron-sized single-crystalline ZnS thin nanosheets and their photoluminescence properties. J. Crys. Growth **293**, 428 (2006)

Silver Nanoparticles: Novel Synthesis Technique and Characterizations

Shweta Rajawat and M. S. Qureshi

Abstract In the present work synthesis of nanostructures of silver nanoparticles is done using X-ray films. Exposed X-ray films consisting silver nanoparticles after photo reduction, are cut into small pieces of size 1 × 1 cm. The pieces of photo graphic films are heated in distilled water at 70 °C temperature. These nanoparticles get separated from heated films which are simultaneously collected through electrolytic deposition using copper and carbon rods as electrodes. The carbon rod is wrapped over by LDPE sheet for easy extraction of silver nanoparticles depositing over it. This process was carried in two different environments (1) in broad daylight and (2) on a cloudy day. Characterization of the as synthesized silver nanoparticles was done using XRD, TEM, and UV-vis spectroscopy. XRD of the particles gave peaks well in accordance with JCPDS file 04-0783. TEM revealed that the interaction of silver nanoparticles with sunlight gave chain like structures whereas in the absence of interaction with sunlight, cloudy day, Nano flowers were formed. Nano structures were more prominent for bigger particles because with the increase in size cohesive forces become prominent bringing particles closer. UV-Visible spectroscopy of the sample was also done which confirms formation of silver nanoparticles.

Keywords Silver nanoparticles · X-Ray · Nano-chain · Nano-flowers

S. Rajawat (✉) · M. S. Qureshi
Department of Physics, Maulana Azad National Institute of Technology
M.A.N.I.T, Bhopal, M.P, India
e-mail: srajawat73@gmail.com

M. S. Qureshi
e-mail: shumsqureshi@rediffmail.com

P. K. Giri et al. (eds.), *Advanced Nanomaterials and Nanotechnology*,
Springer Proceedings in Physics 143, DOI: 10.1007/978-3-642-34216-5_7,

1 Introduction

Metal nanoparticles are of increasing interest due to their use in a large number of application fields, such as bio-sensing, plasmonics, biology, biomedical, optics, microelectronics, solar energy conversion and catalysis. The fabrication of nano structures has aroused the interest of many researchers and great varieties of synthetic methods have been developed [1]. Silver nanoparticles are found to have wide applications in various areas like optical receptors [2], bio-labelling [3], sensors [4], bio active materials [5], signal enhancers in SERS [6] based enzyme Immunoassay. Silver nanoparticles are known for their antimicrobial activity. Silver is an effective antibacterial agent [7] with low toxicity. Given its broad spectrum activity, silver nanoparticles have been the focus of increasing interest and are being used as an excellent candidate for therapeutic purposes [8].

Most prevalent method for synthesizing silver nanoparticles is chemical reduction method using reducing agents. Chemical methods for metal nanoparticle fabrication usually involve toxic chemicals, which are usually expensive and potentially dangerous for the environment. The methods applied to recover silver from photographic waste are important in reducing cost and time, and have a positive effect on environmental pollution. Compared to other films, waste X-ray photographic film contains a great deal of silver to enhance its sensitivity to transmitted X-rays and 1.5–2.0 % silver by weight survives in the emulsion on the polyester film base, even after the developing and fixing processes. These waste films are a good source of silver that is reused for a variety of purposes including light-sensitive materials [9]. Various studies have been carried out over a long period of time to recover the silver from these wastes. The silver recovery methods from waste films can be classified as (a) burning the films directly (b) oxidation of the metallic silver following electrolysis, and (c) stripping the gelatine-silver layer using different solutions. Generally second and third methods are been used more for the recovery of silver than the first method. In general, the methods reported for the recovery of silver from films involve two steps; the first is the separation of the silver from the film base, and the second the recovery of the silver by smelting or electrolysis. In this paper second method is used for the extracting silver. Different nanostructures and separated silver nanoparticles of size less than 150 nm are obtained by varying weather conditions.

2 Methodology

Experimental set up for the synthesis of silver nanoparticles is as shown in Fig. 1. The experimental set up required is very simple and easy to install. It consists of a beaker, filled with water, placed over magnetic stirrer with hot plate at 70 °C. Copper and carbon rods were taken as anode and cathode respectively. Carbon rod is wrapped with low density polyethylene (LDPE) for easy collection of AgNPs. Small

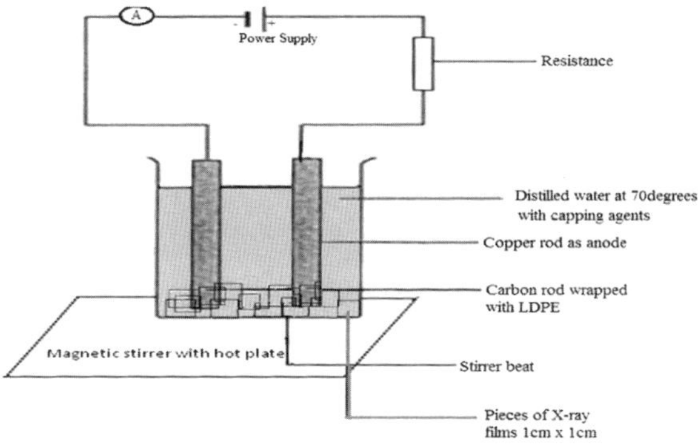

Fig. 1 Experimental set up

pieces of X-ray films were cut and placed in beaker containing water at 70 °C. The temperature of the water was kept 70 °C throughout the experiment. Electrodes, placed in beaker, were connected to a Daniel cell of 2 V output. Contents of beaker are continuously stirred using magnetic stirrer in order to avoid agglomeration. Experiments were carried out without capping agents in different conditions: a bright sunny day of summer season and a cloudy day of winter season keeping rest all parameters: current, temperature, stirring speed and electrodes same.

3 Results and Discussions

The as-synthesized silver nanoparticles were characterized using XRD, TEM and UV-vis spectroscopy. Silver nanoparticles obtained were highly pure and the XRD pattern as shown in Figs. 2 and 3 were exactly in accordance with the standard JCPDS file 04-0783.

The morphology and the crystal structure of synthesized silver nanoparticles were examined using HR-TEM. The results of HR-TEM confirm the shape of some of the silver nanoparticles to be spherical and some oval with varied sizes. Silver nanoparticles were spherical in shape for small size particles and oval for bigger particles. The spherical and oval shape of the particle is due to the fact that when a particle is formed, it in its initial state tries to acquire a shape that corresponds to minimum potential energy. The spherical and oval shapes correspond to state of minimum potential energy. The size of the as-synthesized silver particles is 10–150 nm in both the cases. Silver nanoparticles are formed when the X-ray films containing silver halides are exposed to the sunlight. At the time of exposure the size of the particle is dependent on the size of the aperture and the time of exposure. The difference in the structures obtained is due to presence of less and more sunlight. When the X-ray film containing reduced silver is heated with water at 70 °C, silver nanoparticles are recovered. The recovered silver nanoparticles are collected by electrolysis process.

Fig. 2 XRD pattern of as-synthesized silver nanoparticles forming chain structure

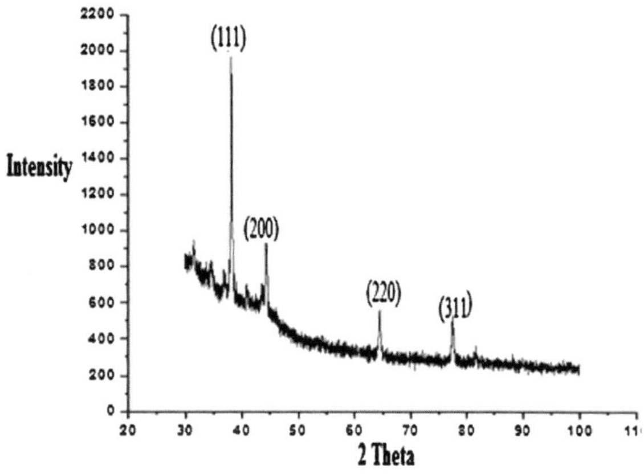

Fig. 3 XRD pattern of silver nanoparticles forming flower structure

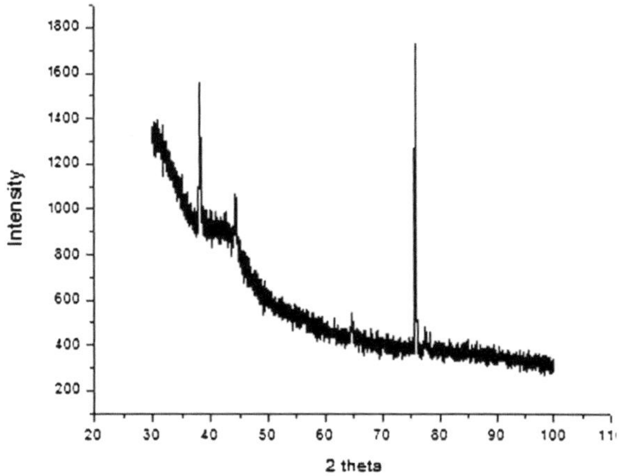

The as-synthesized silver nanoparticles, on a very hot sunny day of summer season, formed chain like structures whereas silver nanoparticles, on a cloudy day of winter season rendered flower like structures. The chain like structures on a bright sunny day is due to the property of silver nanoparticles interacting highly with sunlight generating surface Plasmon. It is also observed that when the particle size is smaller they align in a chain fashion but do not exactly give a chain structure but as the size increases, they have higher tendency of forming chain structures due to higher cohesive forces as obvious from the Fig. 4a, b, c, d.

In case of cloudy conditions, at the time of winters, i.e. very small amount of sunlight is available hence the interaction of silver nanoparticles with the sunlight is very less and then the shape of structures of silver nanoparticles thus obtained is determined by the cohesive forces between these particles. The structures thus obtained are, for smaller size particles in flower like fashion and for bigger particles almost flower like structures i.e. structure with least potential energy. In Fig. 5a the silver nanoparticles are very small ($\sim$5 nm). These can be seen

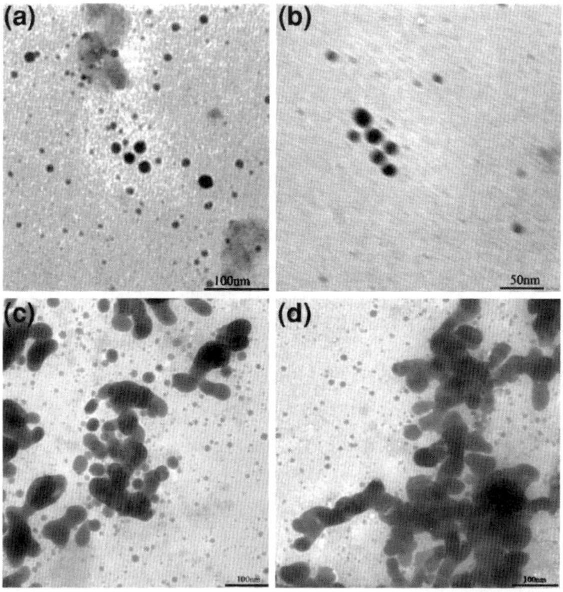

Fig. 4 **a** TEM picture of silver nanoparticles of size ~20 nm. **b** TEM picture of silver nanoparticles of size ~20 nm aligning in chain fashion. **c** TEM picture of silver nanoparticles of size ~25–100 nm forming chain structures. **d** TEM picture of silver nanoparticles of size less than 100 nm confirming forming of chain structures

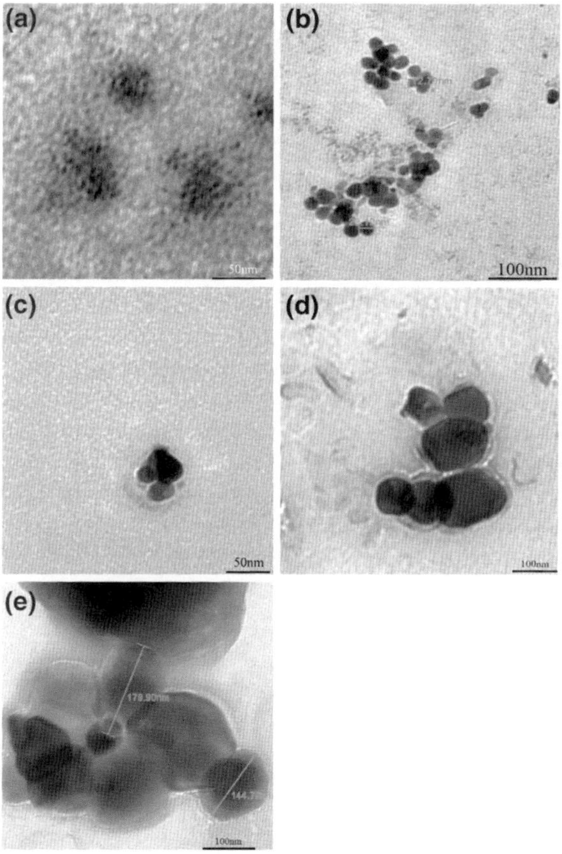

Fig. 5 **a** TEM pictures of silver nanoparticles of size ~5 nm showing tendency of nanoparticles arranging in almost flower like structure. **b** TEM pictures of silver nanoparticles of size ~10–20 nm showing formation of flower like structures. **c** TEM pictures of silver nanoparticles of size ~25–50 nm showing tendency of forming flower like structures. **d** TEM pictures of silver nanoparticles of size ~60–120 nm showing formation of flower like structures. **e** TEM pictures of silver nanoparticles of size ~100 nm showing flower like structures

Fig. 6 Uv-vis graph for the silver nanoparticles

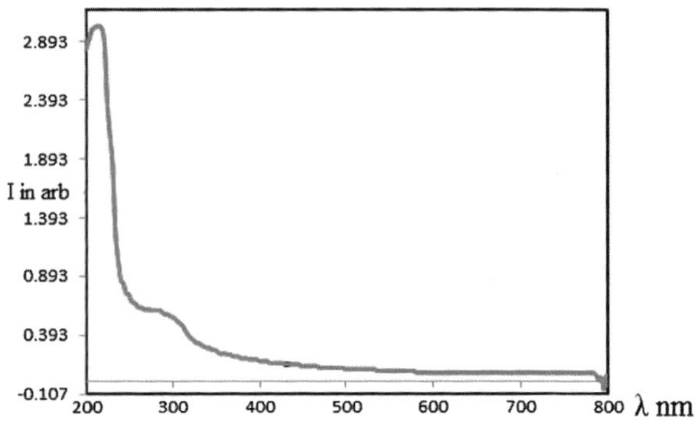

grouping in circular fashion, a state of minimum potential energy. In case of smaller particles flower structure is not clearly visible. When the particle size increases flower like structures can be clearly seen in Fig. 5e.

UV-vis spectroscopy of the as-synthesized silver nanoparticles was done. Graph obtained by UV-vis spectrometry, shown in Fig. 6, indicates formation of nanoparticles.

4 Conclusions

Silver nanoparticles can be synthesized using X-ray films. The method used is simple, easy, cost effective and environment friendly. The size of silver nanoparticle obtained on X-ray film depends on time of exposure of X-ray film and aperture of the opening for film. Different structures of silver nanoparticle can be obtained by carrying the extraction process of silver nanoparticles in different weather conditions. If the process is carried in broad daylight of sunny day of summer season then interaction of the nanoparticles with sunlight dominates silver nanoparticle structure formation and results in chain like structures whereas in case of absence of sunlight or dim light on a cloudy day of winter season, the concentration and cohesive forces between silver nanoparticles dominates the structure formation and as a result flower like structures are formed. These structures find different applications in solar cells for example chain like structure or arrays of silver nanoparticles can be coated upon the conventional solar cell and due to the property of silver nanoparticles of generating surface Plasmon on interacting with sunlight, efficiency of solar cells can be increased. Metal nanostructures show great applications in chemical sensing, biomedical detection, optical-thermal therapy, and optical communications because of their electromagnetic field enhancement properties. Such strong optical field enhancement induced by the localized surface plasmon resonance is dependent on the configurations and the sizes of the metal nanoparticles. The field enhancement factor in welded silver nanostructures is much larger than in separated silver nanoparticles [10].

Acknowledgments Author is thankful to the Director, M.A.N.I.T., Bhopal, India for the financial support in the form of scholarship and the facilities at the institute, Dr. M.S. Qureshi, M.A.N.I.T., Bhopal for the valuable guidance, IUC Indore for XRD facility, HSADL, Bhopal for the TEM facility, Civil Engineering department M.A.N.I.T., Bhopal for providing UV vis-spectrophotometer facility. I am also thankful to all my colleagues for the valuable help and suggestions.

References

1. S. Jradi, L. Balan, X.H. Zeng, J. Plain, D.J. Lougnot, P. Royer, R. Bachelot, S. Akil, O. Soppera, L. Vidal, Spatially controlled synthesis of silver nanoparticles and nanowires by photosensitized reduction. Nanotechnology 2010, vol. 21(6) (2010). doi:10.1088/0957-4484/21/9/095605
2. S. Schultz, D.R. Smith, J.J. Mock, D.A. Schultz, Single-target molecule detection with nonbleaching multicolor optical immunolabels. Proc. Natl. Acad. Sci. **97**, 996–1001 (2000)
3. M.A. Hayat, *Colloidal Gold: Principles, Methods and Applications*, vol. 1 (Academic Press, San Diego, CA., USA, 1989)
4. S.T. Dubas, V. Pimpan, Green synthesis of silver nanoparticles for ammonia sensing. Talanta **76**, 29–33 (2008)
5. J.J. Blaker, S.N. Nazhat, A.R. Boccaccini, Development and characterisation ofsilver-doped bioactive glass-coated sutures for tissue engineering and wound healing applications. Biomaterials **25**, 1319–1329 (2004)
6. X. Chen, E.S. Simon, Y. Xiang, M. Kachman, P.C. Andrews, Y. Wang, Quantitative Proteomics Analysis of cell cycle-regulated golgi disassembly and reassembly. J. Biol. Chem. **285**, 7197–7207 (2009)
7. J.R. Morones, J.L. Elechiguerra, A. Camacho, K. Holt, J.B. Kouri, J.T. Ramírez, M.J. Yacaman, The bactericidal effect of silver nanoparticles, IOP Science. Nanotechnology **16**(10) (2005)
8. M. Ramgopal, C.H. Saisushma, I. Hamad Attitalla, A.M. Alhasin, A facile green synthesis of silver nanoparticles using soap nuts. Res. J. Microbio. **6**(5), 438 (2011)
9. N. Nakiboglu, D. Toscali, G. Nisli, A novel silver recovery method from waste photographic films with NaOH stripping. Turk. J. Chem. **27**, 127–133 (2003)
10. T. Zhang, X.-Y. Zhang, X.-J. Xue, X.-F. Wu, C. Li, A. Hu, Plasmonic properties of welded metal nanoparticles. Open Surf. Sci. J. **3**, 76–81 (2011)

Synthesis of Core–Shell Quantum Dots and Their Potential Application

Udit Soni, Vikas Arora, Gurpal Singh, Manzoor Hussain and Sameer Sapra

Abstract We report the synthesis of core–shell CdSe/ZnS, CdSe/CdS, CdSe/CdS/ZnS nanoparticles. Fluorescent properties of these core/shell nanoparticles depends on the type of ions present on the surface of nanoparticles; a cation-rich surface enhances the emission intensity, while an anion-rich surface leads to deterioration of the fluorescence signal. We used this phenomenon for the synthesis of fluorescent and non-fluorescent quantum dots. We synthesized CdSe/S in which the fluorescence emission is completely quenched, and CdSe/S/Cd and CdSe/S/Zn type of core–shell nanoparticles, which exhibit high fluorescence emission. Although various fluorescence sensors for metal cations based on nanoparticles were previously developed, to the best of our knowledge we first explored the potential application of non-fluorescent quantum dots for the detection of cadmium and zinc in organic medium.

Keywords Nanoparticles · Core–shell · CdSe/S · Metal detection

1 Introduction

Semiconductor nanoparticles are very interesting research topic in both theoretical and applied areas [1–3] due to their size and shape dependent optical properties. CdSe is an important candidate of luminescent II–VI family having bright fluorescence in the broad visible range of optical spectra. This material has shown

U. Soni · V. Arora · M. Hussain · S. Sapra (✉)
Department of Chemistry, IIT Delhi, New Delhi, India
e-mail: sapra@chemistry.iitd.ac.in

G. Singh
Centre for Biomedical Engineering, IIT Delhi, New Delhi, India

P. K. Giri et al. (eds.), *Advanced Nanomaterials and Nanotechnology*, Springer Proceedings in Physics 143, DOI: 10.1007/978-3-642-34216-5_8,

significant potential to be used in light emitting diodes [4, 5], solar cells [6], biological labels [7–9], and fluorescent sensors [10]. The fluorescence quantum yield of the as synthesized nanoparticles is further improved by the growth of a proper epitaxial shell of a higher band gap semiconductor material as CdS and ZnS. Change in nature of the surface of the nanoparticles can subsequently change the fluorescent and optical properties of the as synthesized nanoparticles. Tuning the nature of the surface of the nanoparticles to change the optical and fluorescent properties is relatively intact field. Most of the groups worked with synthesis of core–shell nanoparticles and less number of studies are available depicting systematic study in the change of optical properties due to surface of core–shell nanoparticles. In this paper, synthesis of colloidal CdSe/S, CdSe/S/Zn, and CdSe/S/Cd core–shell nanoparticles from CdSe containing different ions on the surface of nanoparticles is reported [11]. Fluorescence properties of the core–shell are studied systematically as a function of ions present on the surface of nanoparticles. Recently, we have demonstrated that a modified SILAR (successive ion layer adsorption and reaction) technology can be extended to the growth of core/shell semiconductor nanoparticles having cation or anion rich surfaces. The key feature of the modified SILAR method [12] is that the precursors of the anionic (Sulphur) and cationic (Cadmium or Zinc) components of the shell material are introduced in an alternating fashion. Fluorescent properties of these core–shell nanoparticles depends on the type of ions present on the surface of nanoparticles, a cation-rich surface enhances the emission intensity, while an anion-rich surface leads to weakening of the fluorescence signal [11]. We used this phenomenon for the synthesis of CdSe/S in which the fluorescence emission is completely quenched, and CdSe/S/Cd and CdSe/S/Zn type of core–shell nanoparticles, which exhibit high fluorescence emission. The unique fluorescence properties of colloidal semiconductor nanoparticles with anion rich surface have potential as local nanosensors due to their high sensitivity to local environment. For example, CdSe/S nanoparticles placed in metal ion solution demonstrate significant changes in fluorescence intensity. This phenomenon can potentially be used for optical probing of metal ions at room temperature. In this study, we demonstrate high sensitivity of fluorescence of colloidal nanoparticles to the metal ions. In a series of measurements, we investigate the fluorescence properties of CdSe/S core/shell colloidal nanoparticles dissolved in toluene in the presence of metal ion solution at room temperature.

2 Experimental

2.1 Chemicals

Cadmium oxide, (CdO, 99.5 %), Trioctylphosphine oxide (TOPO, 99 %), Trioctylphosphine (TOP, 90 %), 1-octadecene (ODE, 90 %), oleic acid (90 %), Zinc oxide powder (99.99 %), sulfur powder (99.98 %), and Selenium powder mesh (99.99 %) were purchased from Sigma-Aldrich. Octadecylamine (ODA, 97 %)

was purchased from Merck Chemicals. All the solvents were anhydrous and were purchased from Merck Chemicals.

2.2 Solution for Injection

The Cd injection solution (0.2 M) was prepared by dissolving 0.3 mmol of CdO and 2.4 mmol of OA in ODE at 270 °C, and the Zn solution (0.2 M) was prepared by dissolving 0.3 mmol of ZnO and 2.4 mmol of OA in ODE at 300 °C. The Sulfur injection solution (0.2 M) and Selenium injection solution (0.2 M) was prepared by dissolving in ODE at 100 °C. All the above solutions were prepared in Ar atmosphere and were heated at 60 °C prior to injections. Metal–acid complex were prepared by dissolving metal acetates in acid and ODE and heating the solution to 280 °C until a clear solution is obtained. For our experiments we used for four acids, Octenoic acid, Oleic acid, Stearic acid and Lauric acid.

2.3 Core CdSe Nanoparticles

CdSe nanoparticles were synthesized by previously reported method [12–15]. A 0.1 mmol of Se was dissolved in 3.5 mL of tri-n-octylphosphine (TOP) in a glovebox. A 0.1 mmol of CdO and 0.4 mmol of oleic acid (OA) were added to 10 mL of 1-octadecene (ODE) and heated to 270 °C with continuous stirring under an Ar atmosphere until a colourless solution was obtained. The temperature was lowered to room temperature, and 3 g of octadecylamine (ODA) and 2 g of tri-n-octylphosphine (TOPO) were added to the solution and heated to 280 °C under an Ar atmosphere with continuous stirring. After attaining a constant temperature, TOP-Se solution was injected swiftly in the Cd solution. The temperature fell down to 260 °C and the reaction was allowed to proceed until the desired size of the nanoparticles was achieved. The solution was cooled down and washed twice in a 1:1 mixture of hexane-methanol, to remove unreacted precursor. The nanoparticles were precipitated by adding methanol and redissolved in toluene. The concentration of the nanoparticles was determined by the method described in the literature [16].

2.4 Core–Shell Nanoparticles

For synthesis of CdSe/S core–shell structure a 2.5×10^{-7} mol of core CdSe nanoparticles were taken in 5 ml of octadecene (ODE) along with 2.5 g octadecylamine (ODA) and a calculated amount of sulfur stock solution was added according to the number of atoms required for the next complete monoloyaer [17]

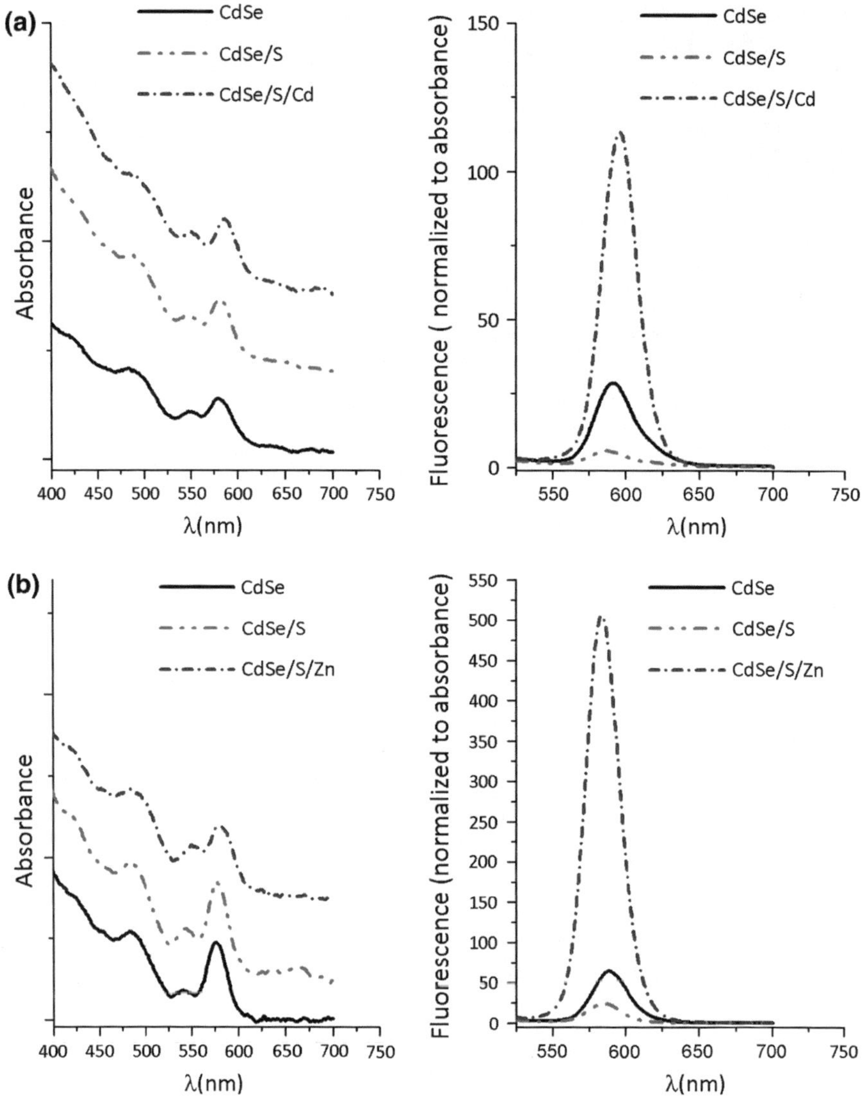

Fig. 1 Uv-vis and fluorescence spectra of **a** CdSe, CdSe/S and CdSe/S/Cd and **b** CdSe, CdSe/S and CdSe/S/Zn

and heated at 200 °C for 30 min. All the reactions were carried out in Ar atmosphere. The fluorescence intensity drops down to a very low value for the sulfur coated CdSe nanoparticles as seen from Fig. 1a, b. For the synthesis of CdSe/S/Zn and CdSe/S/Cd, after the addition of S solution Zn-oleate and Cd-oleate solution were added respectively and heated for 30 min.

2.5 *Characterization*

UV-vis absorption spectra were acquired on Shimadzu UV-2450 spectrophotometer. Dilute solutions of semiconductor nanoparticles in toluene were placed in four window 1 cm quartz cuvettes, and their absorption and corresponding fluorescence were measured. The fluorescence spectra were taken on Perkin-Elmer LS-45 spectrometer. The excitation wavelength was kept at 380 nm for all of the measurements reported here, and the same sample was used for absorption and fluorescence measurements.

Transmission electron microscopic (TEM) images were recorded on a Technai G^2 20 electron microscope operated at an accelerating voltage of 200 kV. Samples were prepared on 300-mesh carbon coated Cu grids by dropping nanoparticles solutions dissolved in toluene.

3 Results and Discussion

Figure 1 represents how the photoluminescence intensity of semiconductor nanocrystals changes. Here we coat the CdSe nanoparticles with S and Cd or Zn ions alternatively. The spectra show that the first addition of the S solution reduces photoluminescence intensity of the nanoparticles, but the second addition i.e. the addition of Cd or Zn solution increases the photoluminescence intensity of the nanoparticles. It is also seen that core/shell nanocrystals with cationic surfaces exhibit higher fluorescence intensity compared to CdSe/S surfaces as shown in Fig. 1a, b, something which has been previously observed for core nanoparticles with Cd and Zn terminated outer layers [11]. TEM images of CdSe, CdSe/S, CdSe/S/Cd and CdSe/S/Zn are shown in Fig. 2. The images indicate increase in size of nanoparticles after shell formation thus confirming that the uniform shell growth is taking place.

The solutions were then cooled down to room temperature, and calculated amounts of the metal–acid complex were added to the CdSe/S nanoparticles. The reaction mixture was kept for 6 h at room temp. After that, aliquots were taken and mixed with hexane for recording optical absorption and fluorescence spectra. In Fig. 3a, b, the sequential changes in the luminescence band and the intensity of the luminescence band are shown. Addition of the metal-carboxylates shows a red shift of the fluorescence bands in confirmation with the reduction in the quantum size effects due to an increased size and a large increase in the intensity of the luminescence. We interpret the large increase in the fluorescence intensity as a confirmation of the good surface passivation with a thickness that is sufficient to confine the exciton to the CdSe core with only partial leakage into the shell. The fluorescence of ion-free CdSe/S nanoparticles were used as control solution (seen from Fig. 3a, b).

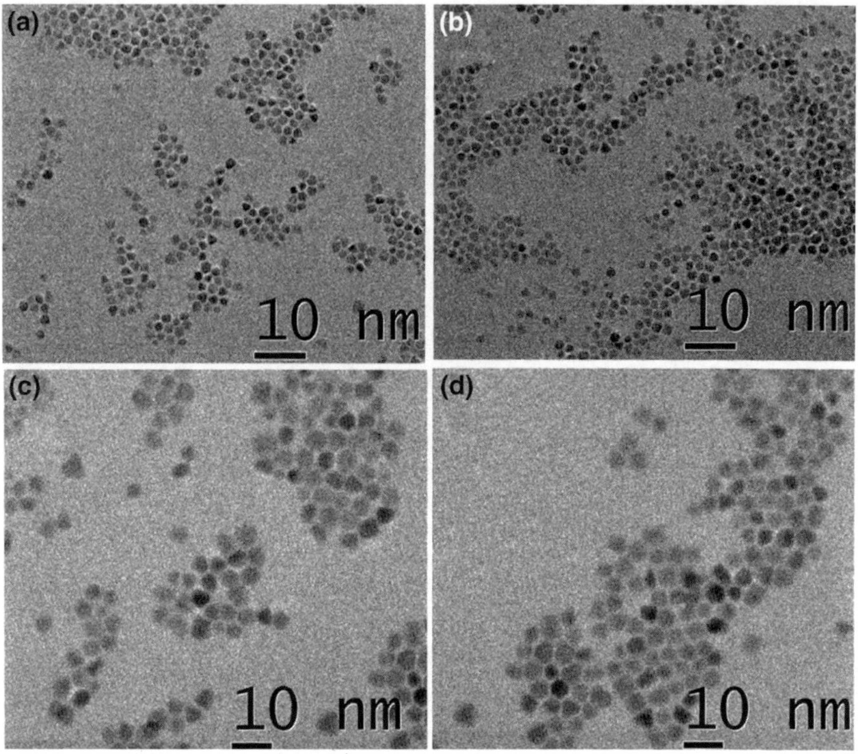

Fig. 2 TEM images of **a** CdSe, **b** CdSe/S, **c** CdSe/S/Cd and **d** CdSe/S/Zn

Not only is the fluorescence enhanced upon addition of the metal-carboxylate complexes, the extent to which this happens should also be noted. The selectivity of the CdSe/S was investigated by testing the fluorescence efficiency in the presence of different acid, including octenoic acid, oleic acid, stearic acid and lauric acid. At first, the fluorescence efficiency was measured upon addition of a single type of metal ions with different acids to one nanomole of CdSe/S (Fig. 3a, b). The result showed excellent increase in the fluorescence intensity for metal-octenoate and metal-oleate complex, but the other complexes, namely the stearate and the laure-atehad hardly any significant increase in the intensity. As a control, the experiment was also conducted by addition of only the neat acid without the metal. This resulted in fluorescence spectra that were practically unchanged from that of CdSe/S. This shows conclusively that the increase in fluorescence intensity is achieved only by the addition of metal ion solutions as seen in Fig. 4.

In order to evaluate the effect of Zn and Cd, we then tested the effect of different cations on the fluorescence intensity of CdSe/S by the addition of Mn, Fe, Co, Cu and Pb as their oleates (Fig. 5). Substantial fluorescence intensity was observed for cadmium and zinc which was significantly greater than for any of the other cation tested. Only Co and Pb produced fluorescence intensity but of a much lesser

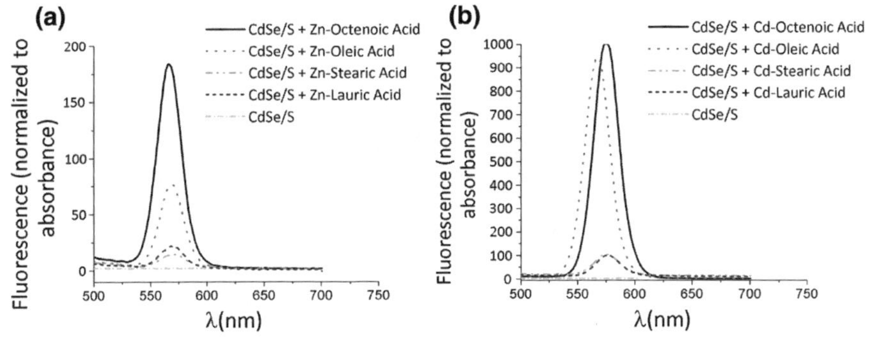

Fig. 3 Fluorescence spectra of CdSe/S in presence of **a** Zn-acid complex and **b** Cd-acid complex

magnitude than for Zn and Cd, highlighting that CdSe/S has good selectivity over other cations. The most likely reason for the increase of fluorescence upon addition of Cd/Zn is a better surface passivation of CdSe/S [11].

In order to obtain quantitative information on the varying amount of metal-carboxylate, fluorescence intensities were plotted as a function of amount of the carboxylate using the metal-oleate complex as the model for our study. 1 nmol CdSe/S was mixed with 2 ml of metal-oleate, and the emission spectrum is illustrated in Fig. 6. Addition of the metal-oleate to the CdSe/S resulted in significant increase of the fluorescence emission. It can be seen that the fluorescence intensity increases with increasing metal-oleate concentrations. The concentration dependence of the fluorescence intensity follows the binding of metal-carboxylate to the surface of the CdSe/S nanoparticles and is well defined by surface passivation of nanoparticles. Here we reported change in fluorescence intensity up to micromolar concentrations of metal-oleates. The result indicates the high sensitivity of the CdSe/S system, and suggests the possibility of determination of low concentrations of Cd and Zn metals in organic solvents.

Fig. 4 Fluorescence spectra of CdSe/S in the presence of acid only

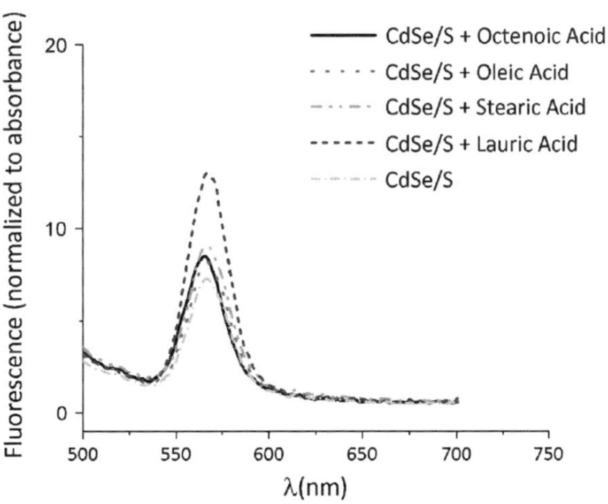

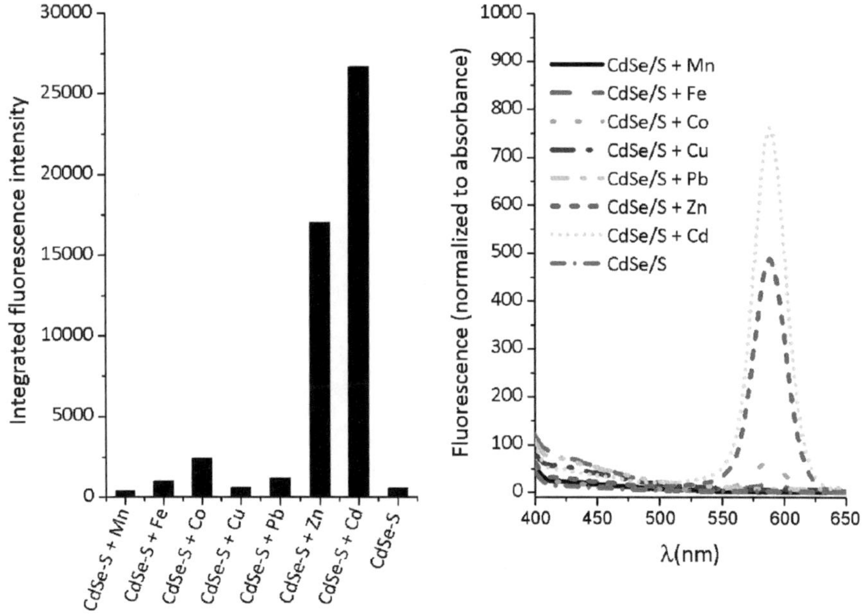

Fig. 5 Integrated fluorescence and fluorescence spectra of CdSe/S after addition of metal-oleate complex

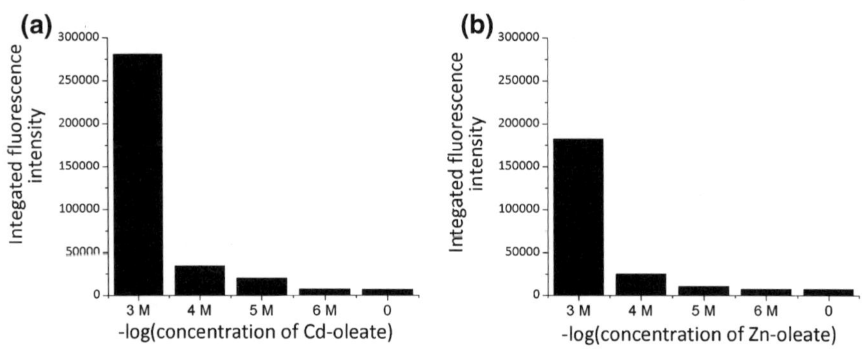

Fig. 6 Integrated fluorescence emission of CdSe/S after addition of **a** Cd-oleate and **b** Zn-oleate

4 Conclusion

Hydrophobic CdSe/S nanoparticles have been synthesized and used to investigate the effect on their fluorescence responses to divalent metal ions, such as Zinc, and Cadmium ions. The changing trends of fluorescence are nearly similar for both Zn and Cd. The enrichment effect on the fluorescence response of CdSe/S nanoparticles to calcium and zinc ions could be explained by better surface passivation.

It was the first time the qualitative and quantitative application of non-fluorescent CdSe/S nanoparticles in ion analysis in organic medium has been used.

Acknowledgments We are thankful to Dr. Pramit Chowdhury for help with the absorption measurements. SS acknowledges DST for the SERC-FAST Track scheme, and US and VA are thankful to CSIR for their SRF fellowship. We are also thankful to Nano Science Unit of NSTI, DST for TEM facility in IIT Delhi.

References

1. V.I. Klimov, A.A. Mikhailovsky, S. Xu, A. Malko, J.A. Leatherdale, C.A. Hollingsworth, H.J. Eisler, M.G. Bawendi, Optical gain and stimulated emission in nanocrystal quantum dots. Science **290**, 314 (2000)
2. H. Mattousi, J.M. Mauro, E.R. Goldman, G.P. Anderson, V.C. Sundar, F.V. Mikulec, M.G. Bawendi, Self-assembly of CdSe-ZnS quantum dot bioconjugates using an engineered recombinant protein. J. Am. Chem. Soc. **122**, 12142 (2000)
3. M. Nirmal, L. Brus, Architectonic quantum dot solids. L. Acc. Chem. Res. **32**, 407 (1999)
4. P.O. Anikeeva, J.E. Halpert, M.G. Bawendi, V. Bulovic, Quantum dot light-emitting devices with electroluminescence tunable over the entire visible spectrum. Nano Lett. **9**, 2532 (2009)
5. S. Nizamoglu, E. Mutlugun, O. Akyuz, N.K. Perkgoz, H.V. Demir, L. Liebscher, S. Sapra, N. Gaponik, A. Eychműller, White emitting CdS quantum dot nanoluminophores hybridized on near-ultraviolet LEDs for high-quality white light generation and tuning. New J. Phys. **10**, 023026 (2008)
6. P.V. Kamat, Quantum dot solar cells. Semiconductor nanocrystals as light harvesters. J. Phys. Chem. C **101**, 18737 (2008)
7. A.H. Fu, W.W. Gu, B. Boussert, K. Koski, D. Gerion, L. Manna, M.L. Gros, C.A. Larabell, A.P. Alivisatos, Semiconductor quantum rods as single molecule fluorescent biological labels. Nano Lett. **7**, 179 (2007)
8. O.T. Bruns et al., Real-time magnetic resonance imaging and quantification of lipoprotein metabolism in vivo using nanocrystals. Nat. Nanotechnol. **4**, 193 (2009)
9. G. Singh, Z.H. Nelam, U. Soni, M. Gautam, R. Jackeray, H. Singh, S. Sapra, Detection of bioconjugated quantum dots passivated with diffrent ligands for bio-applications. J. Nanosci. Nanotechnol. **11**, 1 (2011)
10. L. Dyadyusha, H. Yin, S. Jaiswal, T. Brown, J.J. Baumberg, F.P. Booy, T. Melvin, Quenching of CdSe quantum dot emission, a new approach for biosensing. Chem. Commun. **25**, 3201 (2005)
11. U. Soni, S. Sapra, The importance of surface in core-shell semiconductor nanocrystals. J. Phys. Chem. C **114**, 22514 (2010)
12. J.J. Li, Y.A. Wang, W. Guo, J.C. Keay, T.D. Mishima, M.B. Johnson, X. Peng, Large-scale synthesis of nearly monodisperse CdSe/CdS core/shell nanocrystals using air-stable reagents via successive ion layer adsorption and reaction. J. Am. Chem. Soc. **125**, 12567 (2003)
13. R. Xie, U. Kolb, J. Li, T. Basche, A. Mews, Synthesis and characterization of highly luminescent CdSe-core CdS/Zn0.5Cd0.5S/ZnS multishell nanocrystals. J. Am. Chem. Soc. **127**, 7480 (2005)
14. D. Battaglia, B. Blackman, X. Peng, Coupled and Decoupled Dual Quantum Systems in One Semiconductor Nanocrystal. J. Am. Chem. Soc. **127**, 10889 (2005)
15. S. Sapra, S. Mayilo, T.A. Klar, A.L. Rogach, J. Feldmann, Bright white-light emission from semiconductor nanocrystals: By chance and by design. Adv. Mater. **19**, 569 (2007)
16. W.W. Yu, L. Qu, W. Guo, X. Peng, Experimental determination of the extinction coefficient of CdTe, CdSe, and CdS nanocrystals. Chem. Mater. **15**, 2854 (2003)
17. S. Sapra, D.D. Sarma, Evolution of the electronic structure with size in II–VI semiconductor nanocrystals. Phys. Rev. B **69**, 125304 (2004)

Water Dispersible Semiconductor Nanorod Assemblies Via a Facile Phase Transfer and Their Application as Fluorescent Biomarkers

Ambarish Sanyal and Kevin M. Ryan

Abstract We demonstrate the formation of water dispersed nanorod assemblies by phase transfer of semiconductor (CdS, CdSe, CdTe) nanorods from the organic to the aqueous using pluronic triblock copolymers. On phase transfer, the randomly dispersed nanorods in the organic medium close pack in the form of discs encapsulated in the hydrophobic core of water dispersible micelles. The assemblies showed excellent cellular uptake exhibiting membrane and cell specific fluorescence at low light intensity under confocal microscopy.

Keywords Nanorod assemblies · Copolymer · Nanorod supercrystals · Micelle · Biomarker

1 Introduction

Close packed semiconductor nanorod assemblies where each one dimensional nanocrystal adopts a side by side ordering and uniform axial orientation are a significant recent discovery. While discrete nanorods show polarised light emission, length dependent photon absorption and high fluorescence cross-sections [1, 2], their collective assembly can allow scalable applications in next generation LEDs, photovoltaics and field emission devices [3–10]. Nanorods have been

A. Sanyal (✉)
Planning and Performance Division, CSIR Hqrs, New Delhi 110001, India
e-mail: ambarish@csir.res.in

A. Sanyal · K. M. Ryan
Materials and Surface Science Institute, University of Limerick, Limerick, Ireland
e-mail: kevin.m.ryan@ul.ie

P. K. Giri et al. (eds.), *Advanced Nanomaterials and Nanotechnology*,
Springer Proceedings in Physics 143, DOI: 10.1007/978-3-642-34216-5_9,

directed into perpendicularly ordered assemblies either by using external electric fields or by evaporation control at solid–liquid or liquid–air interfaces [11–14]. The aspect ratio and polydispersity of the rods in addition to solvent evaporation dynamics play a key role in achieving vertical alignment and close packing from solution. In effect, the gradual packing of randomly dispersed nanorods into a reducing solvent volume allows attractive dipole–dipole and van der Waals forces between ligand capped nanorods to dominate the assembly behaviour at a substrate [15]. Assemblies of nanorods dispersed in solution are also interesting in particular where individual properties such as linearly polarised emission can be collectively harnessed for example as novel fluorescent biomarkers [16, 17]. In non-polar solvents, ligand capped nanorods are randomly dispersed in dilute solutions but can form nematic liquid crystal (LC) phases in more concentrated dispersions [18]. The LC behaviour is dynamic and is destroyed outside of local temperature and concentration variations limiting discrete application. More robust and potentially biocompatible nanorod assemblies can be achieved when nanorods are stabilised in the aqueous phase as micelles. Nie et al. functionalised gold nanorods with hydrophilic CTAB ligands grafted along the rod sides with hydrophobic polystyrene on the rod ends which assembled into either discrete chains or spherical bundles depending on the water concentration in mixed DMF or THF solvent [19]. More recently Zhuang et al. achieved water dispersible disc shaped CdS–CdSe nanorod assemblies in a two stage process by further co-ordinating individual alkyl capped nanorods with a dual interaction ligand (dithiol-functionalised Tween-20, Tween-SH) followed by annealing in ethylene glycolat 80 °C [20].

In this report, we demonstrate a facile single-step, room temperature route to the formation of water dispersible nanorod assemblies of CdS, CdSe and CdTe using triblock copolymer surfactants of the form $PEO_xPPO_yPEO_x$ (PEO: polyethylene oxide, PPO: polypropylene oxide). These types of copolymers are known to self-assemble to form morphological variations of spherical, vesicular or mesophasic aggregates in aqueous solutions and are widely used in the pharmaceutical industry to solubilise hydrophobic drugs [21, 22]. Here, we discovered that the facile stirring of an organic solution of alkyl capped nanorods mixed with an aqueous solution of the triblock copolymer resulted in the formation of very stable aligned nanorod assemblies in the aqueous phase. The phase transfer occurs where by discrete alkyl capped nanorods in the organic phase are transferred to the aqueous phase via encapsulation in hydrophobic polypropylene oxide core of a triblock micelle. The resultant close packed disk shaped assemblies consists of rods tethered only by the interdigitation of the original long chain alkyl ligands suggesting the triblock copolymer does not co-ordinate to each individual nanorod but to the rod ensemble. In vitro analysis showed an excellent uptake of the water dispersed nanorod assemblies by human vascular smooth muscle cell and the occurrence of various spectral properties via this interaction (membranous and intracellular) demonstrates their potential as fluorescent labels.

2 Experimental

Materials: Cadmium oxide (>99 %) was purchased from Fluka, trioctylphosphine (TOP, 90 %), tri-n-octylphosphine oxide (TOPO, 99 %), sulfur (99.98 %), selenium (99.98 %) and tellurium (99.98 %) were purchased from Aldrich. n-Octadecylphosphonic acid (ODPA), n-tetradecylphosphonic acid (TDPA), n-hexylphosphonic acid (HPA), and n-octylphosphonic acid (OPA) were obtained from PolyCarbon Industries, Inc. (PCI). The block coploymers F88, F127, P123, and P65 were obtained from BASF, Germany. All the chemicals were used as received. Filter paper of pore sizes 2.5 micron and 0.45 micron was purchased from Whatman and Millipore Corporation, respectively, for size selective separation of the nanorods assemblies through filtration.

2.1 Synthesis of CdS Nanorods

Method: The CdS nanorods utilised in this study were synthesised essentially through the pyrolysis method [23], redispersed in toluene and kept under N_2 environment inside a glove box. The reaction was quenched after 30 min by adding a suitable quantity of toluene. A stoichiometric amount of acetone was added to precipitate the nanorods (average length and diameter around 25–30 and 6–8 nm, respectively), following which they were centrifuged and redispersed in toluene. A drop of CdS nanorods solution was then deposited and allowed to evaporate onto carbon-coated Cu grids for transmission electron microscopy (TEM) characterisation. Evaporation of the resulting solution led to the formation of a 2D array of CdS super lattice with the rods aligned in a perpendicular direction.

2.2 Synthesis of CdSe Nanorods

Method: CdSe nanorods were synthesised with slight modifications to the published procedure [4]. The general approach for the synthesis of nanorods involves the stepwise addition of cadmium precursor and followed by chalcogenide precursor at a high temperature. The nanorods growth was terminated by removal of the heating mantle, and at 80 °C 2–4 ml anhydrous toluene were added to the mixture to quench the reaction. The nanorods were purified by dissolution in toluene and precipitation from anhydrous isopropanol. They were cleaned thrice with a toluene and isopropanol mixture and redispersed in toluene for further measurements.

Synthesis of CdTe nanorods/dipods/tetrapods: CdTe was also synthesised in a procedure similar to that of CdSe. It eventually turned out that the concentration of

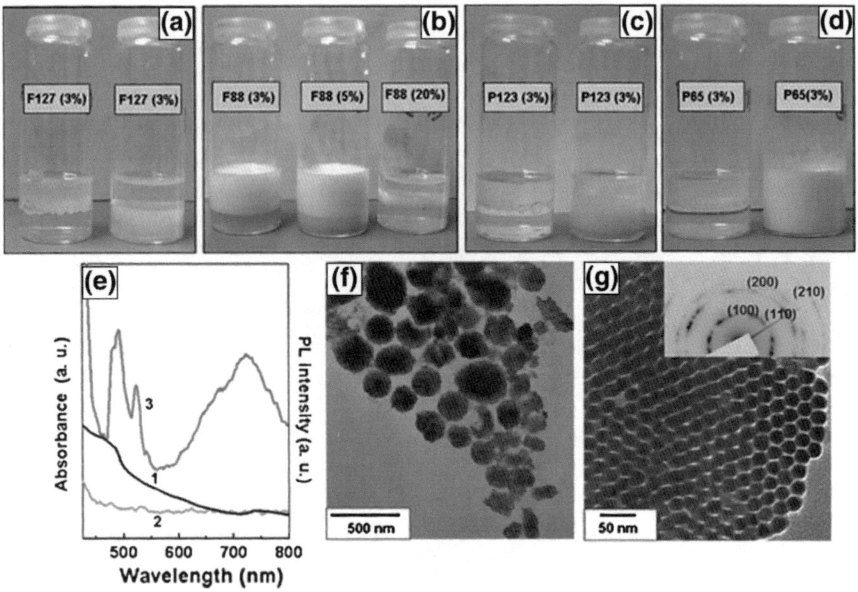

Fig. 1 Macroscopic, optical and TEM imaging of phase transferred CdS nanorods. **a** Photographs of sample vials showing the phase transfer of CdS nanorods from toluene with 3 % F127, **b** 3, 5 and 20 % F88, **c** 3 % P123 and **d** 3 % P65 into water. In parts **a**, **c** and **d** CdS nanorods are in toluene in the bottle in the *left-hand side* (*upper layer*), whereas they are transferred to the aqueous phase (*bottom layer*) in the *right-hand side* bottle. **e** UV–Vis and photoluminescence spectra of CdS nanorods phase transferred with F127 (*curves 1* and *3*, respectively) in the aqueous phase. *Curve 2* is the emission spectra of pristine F127. **f** Representative TEM images of the phase transferred CdS nanorods in water by F127 revealing NRSCs and **g** perpendicular alignment of the nanorods magnified from one of the assemblies shown in **f**. The inset in **g** shows inverse selected area electron diffraction (SAED) pattern taken from the corresponding assembly of CdS nanorods

CdTe tetrapods and dipods was more than the nanorods in the resultant product. They were then cleaned thrice with a toluene and methanol mixture and redispersed in toluene for further measurements.

2.3 Phase Transfer of CdS, CdSe and CdTenanorods/tetrapods

Method: 5 ml of 3 % block copolymers (Pluronic F88, F127, P123 and P65, obtained from BASF, Germany) were mixed with 5 ml of the diluted CdS nanorods in toluene and allowed to stir intensively for 12 h until there was a clear observation of an appearance of colour in the aqueous phase (see photographs in Fig. 1). The top organic layer was carefully separated and later discarded. The remaining aqueous layer was centrifuged and washed a couple of times with deionised water in order to remove excess unbound surfactants. Aliquots from these resulting solutions were taken and used for further characterisation. Phase transfer was accomplished with 5 and 20 % F88 in the same manner. In order to have an idea of the flexibility of the process, CdSe nanorods and CdTedipods/tetrapods dispersed in toluene were also phase transferred with 3 % F127 and treated in a similar manner before characterisation.

2.4 Cell Culture and Cellular Uptake of CdS Nanorods

Method: Vascular smooth muscle cells (VSMCs) (PromoCell, Germany) were grown in monolayers at 37 °C in a humidified atmosphere at 5 % CO_2 in 5 % fetal bovine serum—VSMC media supplemented with penicillin/streptomycin. At 85 % confluence, cells were detached from a 75 cm^2 flask surface by treatment with 0.25 % trypsin and a 0.02 % EDTA in Ca^{2+}- and Mg^{2+}-free PBS (pH 7.3). Cells were plated for 24 h before the experiment on poly-L-lysine coated glass cover slips at 5×10^{-3} cell cm^2. Mouse fibroblasts, another cell line, were grown in a medium containing 71 % Dulbecco's modified eagle's medium (DMEM) (Sigma, UK), 17.5 % Medium 199 (Sigma, UK), 5 % fetal calf serum (FCS) (Life Technologies Ltd., UK), 1.6 % 200 mM L-glutamine (Life Technologies Ltd., UK), and 0.9 % 100 mM sodium pyruvate (Life Technologies Ltd., UK). Human vascular endothelial cells were grown in a commercially supplied growth factor rich media (PromoCell). After incubation for 12 h, 5 and 7 days with CdS nanorods/VSMC media at 37 °C in a humidified 5 % CO_2 incubator, cells were washed three times with PBS in order to remove any loosely attached or non-attached nanorods and then fixed in a 4 % formaldehyde solution in PBS at room temperature for 10 min. The cells were then washed twice with PBS and incubated with 0.1 % Triton X-100 (Sigma-Aldrich) plus 1 % bovine serumalbumin (BSA, Sigma-Aldrich) in PBS at room temperature for 15 min followed by 5 mg ml^{-1} 4, 6-diamidino-2-phenylindole (DAPI, Molecular Probes) staining in PBS for 3 min at room temperature. The slides were washed twice with PBS, mounted with glycerol on the glass slide for microscopical observation.

2.5 Characterisation

Transmission electron microscopy (TEM) was performed using a JEOL 2011 TEM operating at an accelerating voltage of 200 kV. Selected area electron diffraction (SAED) was also observed from the regions of perpendicularly aligned CdS nanorods. UV–Vis spectroscopy and fluorescence measurements of the phase transferred solutions were carried out on a Cary-300 Bio UV–Vis spectrophotometer operated at a resolution of 1 nm and a Varian Cary Eclipse Fluorescence

Spectrophotometer, respectively. X-Ray photoelectron spectroscopy (XPS) measurement of the phase transferred nanorods was carried out in a Kratos Axis 165 spectrometer using monochromated Al Kα radiation (hν = 1,486.58 eV). High resolution spectra were taken for the following photoelectron transitions: Cd3d, S2p, O1 s and C1s at a fixed analyser pass energy of 20 eV. The core level spectra were background corrected using the Shirley algorithm and the chemically distinct species were resolved using a mixed Gaussian–Lorentzian fitting procedure. The core level binding energies (BEs) were calibrated using C1s at 284.8 eV. In order to decipher the plausible mechanism of the phase transfer process, a drop of the phase transferred CdS nanorods by F127 was observed on a clean glass slide under an optical microscope(Model: Axio Imager MAT, Carl Zeiss AG Light microscopy).

Confocal microscopy was performed on "Carl-Zeiss Meta 710" instrument excited with a 405 nm laser. Scanning electron microscopy (SEM) of the phase transferred samples, drop-caston a Si (111) substrate was performed by a Hitachi S-4800 machine. Localisation of nanoparticles was performed using a spectral laser scanning microscope Meta 710 based on the Axio Observer II (Carl Zeiss, Germany) equipped with a Zeiss Neofluor 20 X objective. Excitation wavelengths were 405 and 488 nm. All confocal images were acquired with the same settings with respect to laser intensity and detector gain. Optical sectioning was performed to obtain the pattern of spatial distribution nanorods (cell surface, intracellular compartments).

3 Results and Discussion

3.1 Structural Evolution and Characterisation of Water Dispersed Nanorod Assemblies

The phase transfer of the nanorods was conducted by vigorous stirring of the nanorod–toluene solution with aqueous solution containing triblock pluronic copolymers. During this process, the transfer of CdS nanorods from the organic to the aqueous phase could be followed visually (Fig. 1a) where the bright yellow colour of the CdS nanorods in toluene was transferred to the aqueous layer containing 5 ml of 3 % F127. Aqueous solutions containing 3 and 5 % F88 resulted in an incomplete transfer with the characteristic yellow colour present in both the organic and the aqueous medium, Fig. 1b. A clear organic layer was only obtained when the concentration of F88 was raised to 20 %, however, the aqueous layer became extremely viscous, due to excess surfactants, and could not be separated, Fig. 1b. P123 was also found to be quite effective at phase transfer, achieving 100 % phase transfer with a 3 % surfactant concentration, Fig. 1c. However, in this case, CdS nanorods from the aqueous phase were found to form a separate layer when further diluted with water. Probably, P123 was able to drag the rods into aqueous layer but not able to provide an effective hydrophilic environment for solubilising in a polar solvent thoroughly. P65 was found to be least effective, regardless of percentage weight in solution, and a vigorous stirring of organic and aqueous layer generated an emulsion inseparable (Fig. 1d) even after 3 days. Visual observation of the yellow colour from the aqueous part of F127 solution was in accord with the UV–Vis and photoluminescence spectra (Fig. 1e, curves 1 and 3) which show characteristic features similar to CdS synthesised in the organic medium.

The emission peak for F127 phase transferred CdS nanorods (curve 3 in Fig. 1e) centred at $\sim$490 nm (band gap $\approx$2.53 eV) is almost 30 nm blue-shifted as compared to the emission spectra of bulk CdS (ca. 510 nm). This is probably due to an enhanced quantum confinement within the nanorods. The presence of a satellite peak ($\sim$523 nm, green shifted) instead of a single sharp peak is possibly from the group of nanorods, a phenomenon which is generally observed when the nanorods are not isolated and are coupled to other nanocrystals [24]. The other broad absorption band ranging from 650 to 800 nm is generally attributed to the emission caused due to non-radiative decay of surface trap states of the semiconductor materials. In contrast, pristine F127 did not produce any emission in the visible region when excited at 350 nm (curve 2, Fig. 1e). Figure 1f shows the transmission electron micrograph of F127 protected CdS *nanorod super crystals* (*NRSCs*), ranging from 100 to 600 nm, that have resulted after a phase transfer to the aqueous medium. A higher magnification image, Fig. 1g, shows that each consists of an ensemble of axially aligned and close packed CdS nanorods. Interestingly, the inter nanorod distance ($\sim$3.5 nm) is comparable to the distance obtained when the same nanorods are assembled on a substrate from an organic solution under highly oriented pyrolytic graphite (HOPG) as reported previously [12]. The absence of enhancement in the inter-rod spacing in the NRSCs when compared to rods assembled under HOPG strongly suggests that the rods are still tethered only by the long chain alkyl ligands in the close packed assembly. The triblock copolymer therefore does not interact individually with each alkyl capped rod but encloses a group of rods as an ensemble. The selected area electron diffraction (SAED) pattern obtained from F127 capped CdS nanorods, inset of Fig. 1g, is indexed to the reflections of wurtzite CdS (a, b = 4.1409 Å, c = 6.7198 Å, PCPDF file no.: 41-1049). Similar results were obtained in the case of phase transfer of CdS nanorods by F88 where NRSCs ranging from 100 to 400 nm were obtained with the individual rods being perpendicularly aligned within these structures. Although the UV–Vis spectra of as synthesised CdS nanorods in toluene were similar to that of the assembled nanorods in aqueous medium, yet there was a conspicuous difference in the emission spectra from the assembled nanorods which suggests a compound effect from the corralled nanorods in aqueous medium.

Phase transfer with CdSe nanorods and CdTe dipods/tetrapods which were previously prepared through a similar pyrolytic route [4] having ODPA/OPA/HPA/TOPO as the surface capping agent was also studied. The phase transfer of monodisperse CdSe nanorods resulted in the formation of individual NRSCs

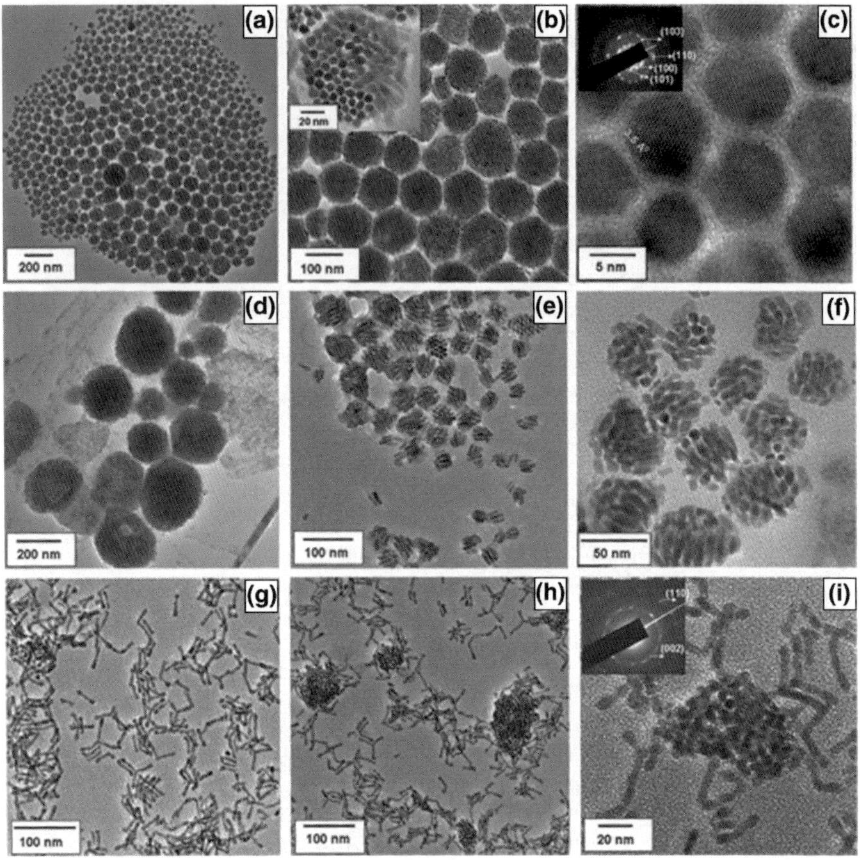

Fig. 2 **a** Representative TEM image of the formation of F127 mediated CdSe nanorod assemblies extending from 20 to 300 nm. Closed-packed pseudohexagonal assembly of the nanorods is revealed in **b** and its corresponding inset shows the hexagonal stacking of the individual nanorods inside the micelle like structure. **c** High resolution TEM image of the individual nanorods within the NRSCs showing the crystallographic planes with the corresponding SAED pattern shown in the inset. Stepwise size selective centrifugation of NRSCs resulting in the larger [>200 nm, TEM image (**d**)] and consequently smaller assemblies (20–80 nm, TEM image (**e**)]. Part **f** shows the TEM image of the smallest disc (40–50 nm) resulted from the alignment and packing of the individual nanorods. **g** Representative TEM image of the phase transferred CdTe dipods/tetrapods. **h** and **i** are TEM image snapshots of the CdTe solution showing formation of few assemblies as compared to that of CdSe. SAED pattern (inset of **i**) confirm the crystalline nature of the phase transferred CdTe NRSCs

ranging from 20 to 300 nm, Fig. 2a. A further magnified image, Fig. 2b, shows that NRSCs of similar size can further organise on a larger scale into a pseudo-hexagonal arrangement that mimics the close packing of individual rods within each super crystal (inset, Fig. 2b).

A high resolution image, Fig. 2c, of nanorods within an NRSC shows the lattice parameter (0.35 nm) from the highly oriented crystallographic planes (002) of the individual nanorods. The CdSe NRSCs were more monodisperse than equivalent CdS NRSCs and showed remarkably high water dispersibility requiring high centrifugation speeds for precipitation. Judicious size selection of the NRSCs was achieved by sequentially increasing the speed with the largest (>200 nm) precipitating at 5,500 rpm (Fig. 2d) and the smallest (20–80 nm) at 13,000 rpm (Fig. 2e). The smallest NRSCs which form a discernible disc from the axial alignment and close packing of nanorods were typically 40–50 nm in diameter, Fig. 2f.

TEM images (Fig. 2g–i) of phase transferred CdTe dipods/tetrapods are somewhat different, though assembly can be found in some regions of the sample (Fig. 2g and h). The lack of NRSC formation tendency in tetrapods can be attributed to the fact that dipods/tetrapods cannot be compressed easily like nanorods in the form of compact close packed structures. In spite of this fact, CdTe tetrapod-assemblies are somewhat formed and transported to the aqueous phase which clearly establishes the efficacy of F127 for phase transfer of semiconductor nanocrystals. To prove the crystalline nature of phase transferred CdSe and CdTe, SAED patterns are given as inset in Fig. 2c and i, respectively, which correspond to the wurtzite structures (CdSe: a, b = 4.299 Å, c = 7.010 Å, PCPDF file no.: 08-0459; CdTe: a, b = 4.58 Å, c = 7.5 Å, PCPDF file no.: 19-0193). X-Ray photoelectron spectroscopy (XPS) of the CdS nanorods phase transferred by F127 and F88 was also performed. The C1s spectra in Fig. 3a and b clearly show the presence of two component peaks. The lower binding energy peak at 284.8 eV is assigned to carbon bonded to carbon or hydrogen. The appearance of a second component peak at 286.3 eV (deconvoluted curve 2 in Fig. 3a and b) can be assigned to C–O bonding present in both F127 and F88.

The O1s spectra, Fig. 3c and d show the related oxygen from the C–O bond appearing at 532.5 eV. The ratio of C:O for the C–O bonding is 2.1 and 1.9 for F127 and F88 transferred CdS nanorods, respectively. This is in good agreement with the C:O ratio from the molecular formula of the ether linkages in F127 ($PEO_{106}PPO_{69}PEO_{106}$) and F88 ($PEO_{102}PPO_{39}PEO_{102}$). In contrast, C1s peak of only ODPA/TOPO capped CdS nanorods in the organic medium did not show any such high energy C1s peak corresponding to C–O. The C–O containing species in the phase transferred samples constitute 50–60 % of the surface composition as determined by XPS. In addition, the intensity of the alkyl peak (284.8 eV) corresponding to surfactant species decreases by half relative to that of CdS nanorods in the absence of the polymer, while the ratio of alkyl C:Cd of the nanorods remains comparable. This can be viewed as resulting from the screening effect imposed by the polymer, which supports the postulation that the polymer encapsulates the nanorods ensemble by interacting with the capping layer. The proposition of encapsulation of a group of nanorods by F127 was further supported by scanning electron microscope (SEM) images.

Discrete NRSCs could be resolved by SEM and at low resolution the structures appear as a mixture of spherical and anisotropic particles, Fig. 4a. HRSEM, Fig. 4b (scale bar = 100 nm) and further magnified image in Fig. 4c (scale bar = 50 nm), revealed the individual nanorod close packing within the structures

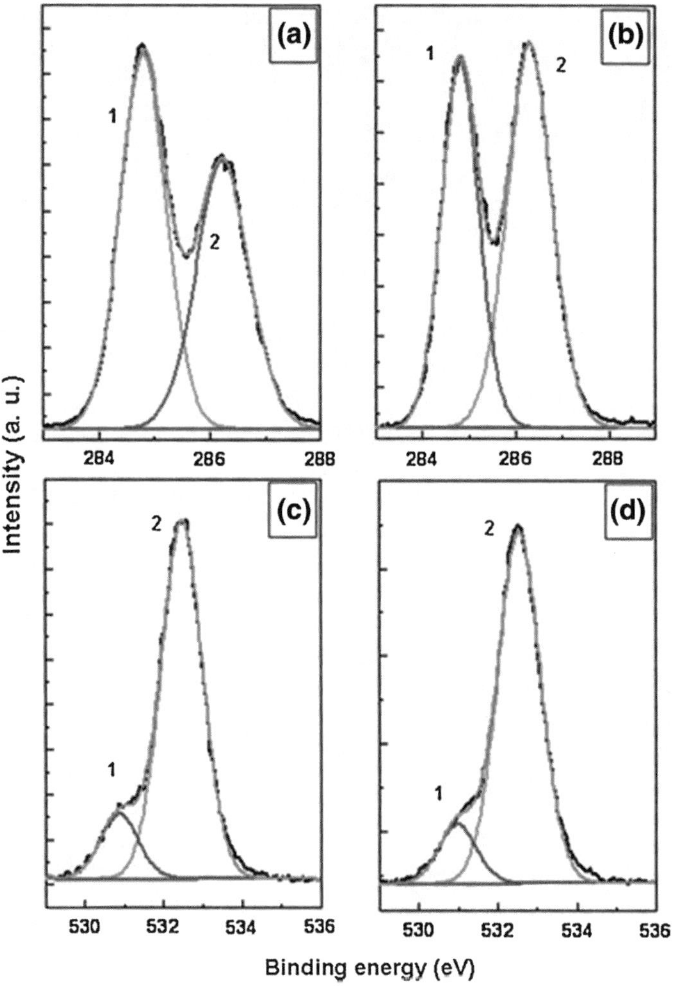

Fig. 3 C1s and O1s core level spectra recorded from a drop-cast film on a glass substrate from the CdS nanorods in aqueous phase that have been phase transferred with F127 (**a**) and (**c**) and with F88 (**b**) and (**d**), respectively. The chemically distinct deconvoluted components have been shown as *curves 1* and *2* in parts **a–d**

which forms a perfect hexagonal framework. The HRSEM image further confirms that all the NRSCs have a disc morphology predominantly consisting of a single monolayer of axially aligned rods. In the NRSC labelled 1 (Fig. 4b) the disc has come to rest flat on the substrate whereas in two the disc is on its edge. Some stray rods lie in a parallel orientation on the 2D assemblies but the predominant rod orientation is close packed within the disc. An estimate on a single NRSC as shown in the magnified image of Fig. 4c revealed that as many as 800 nanorods are bound together in this hexagonal closed-packed assembly. The occurrence of

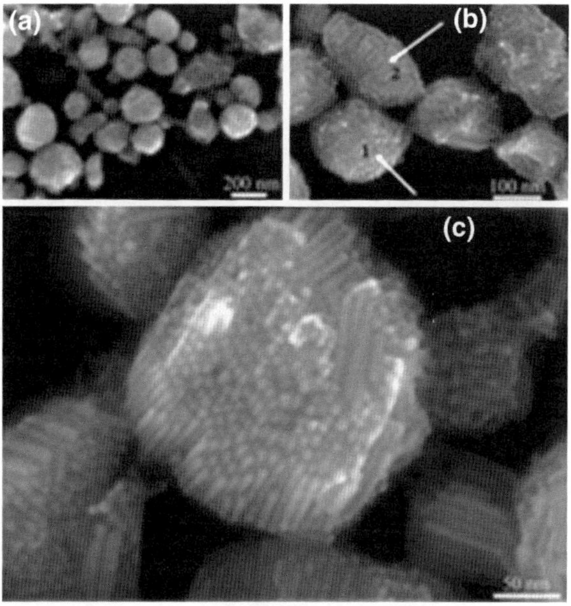

Fig. 4 **a** SEM image of CdS nanorods phase transferred by F127 in the aqueous phase showing different sized anisotropic nanorods assemblies. **b** HRSEM image where one of the NRSCs are lying flat (*point 1*) or are on its edge (*point 2*). **c** Further magnified image of one of the NRSCs revealing the multiple number of individual CdS nanorods stacked into a single structure

2D assembly from nanorods in solution suggests a significant driving force for side by side packing of the rods. This effect parallels that observed from nanorods assembled on a substrate where super-crystallisation was evident with the nanorods finding their preferred place on the growing assembly before locking in thus allowing the lowest energy consideration to be reached [11, 12, 15]. Clearly, the similar formation of highly ordered nanorod discs from randomly dispersed nanorods in solution suggests a similar phenomenon is occurring on phase transfer. Particle size analysis, optical microscopy and the ability to selectively precipitate by mass (centrifuge) or by size done separately confirmed that the NRSCs were real features present in solution and do not occur due to solvent evaporation.

3.2 *Proposed Mechanism of Formation*

A plausible mechanism of formation is the phase transfer occurs by the aggregation and encapsulation of individual alkyl capped nanorods as a hydrophobic core most likely co-ordinated to a PPO corona and PEO shell of a triblock copolymer micelle. It is anticipated that there are two factors responsible for the phase transfer of CdS nanorods from toluene to water namely (1) hydrophobic–hydrophobic interaction between hydrophobic head groups of the copolymer and the long chain hydrocarbon of ODPA capping ligands on the CdS nanorods and (2) the hydrophilic tail of the copolymer should be sufficient to provide a good hydrophilic environment to the compact micelle like structure. Essentially it was seen that mere presence of a large hydrophobic head or a large hydrophilic tail is not sufficient for proper phase transfer. Among the set of four different triblock copolymers used in the experiment, F127 has the longest head and tail. So both factors are applicable in this case, enabling a good protection to the hydrophobic core as well as providing a good hydrophilic environment, thus F127 worked best for the phase transfer. F88 has equivalent long hydrophilic tails like F127 but the head group is relatively small. Thus factor (1) is relatively weak and is not enough for an effective phase transfer. Therefore a lower concentration of F88 is not sufficient for 100 % phase transfer, but on increasing the F88 concentration, the number of molecules interacting with CdS increases and as a result phase transfer capability was enhanced. The tail of the copolymer F88 was long and thus able to make the NRSC water dispersible without any difficulty. P123 has a long hydrophobic head, so it can also successfully phase transfer CdS to water and satisfy factor (1), but the short hydrophilic tail is responsible for its poor water dispersibility. For similar reasons phase transfer by P65 does not occur due to the combination of short head (PPO) as well as short tail (PEO).

3.3 Biological Application

Preliminary investigation of biological application was carried out by in vitro analysis in a human vascular smooth muscle (VSM) cell line which is a primary candidate for biomedical treatment for many cardiovascular pathologies [25]. Excitation ($\lambda = 405$ nm) of the phase transferred CdS NRSCs by CLSM (confocal laser scanning microscope, Fig. 5a, yielded a characteristic fluorescence intensity in the blue and further confirmed the ease of dispersibility and homogeneous size distribution of the disc shaped assemblies (0.2–0.6 mm range). The VSM cell line was incubated with the NRSCs and their internalisation, cellular uptake and intracellular distribution were further monitored. Uptake occurred without any transfection agents into the adherent VSM cell line and was efficient, fast and led to high loading into the cells, Fig. 5b and e. Rapid and irreversible internalisation starts during the first 20 min of incubation with no excretion or further aggregation even after 24 h. With application of optical sectioning we were able to show two types of interaction between the cells and NRSCs: adsorption of the nanostructures on the cell membrane giving membranous fluorescence in the blue spectral region, Fig. 5c and internalisation of the NRSCs inside the cell giving intracellular fluorescence in the red spectral region, Fig. 5d. Overall, there is a significant fluorescence intensity signal from NRSC treated cells (Fig. 5b–e) which allowed visualisation of all minor structural details of cell shape, cell surface morphology and adhesion to substrate. Strong fluorescence, cellular uptake and prolonged persistence in the human cells (up to 7 days in this study) make these ensembles potentially useful as fluorescent probes in a range of biological applications including for example flow cytometry techniques, cell–scaffold interaction studies and DNA markers [26, 27]. Similar results were obtained with several other cell

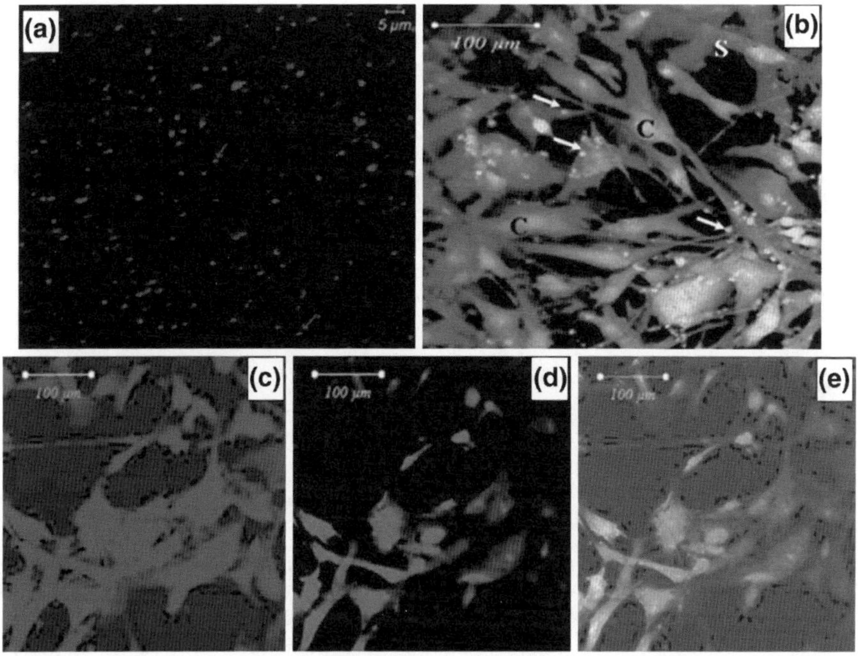

Fig. 5 Preliminary investigation of cellular biocompatibility of CdS assemblies, cellular uptake and maintenance of the fluorescent properties after internalisation into cytosol. **a** LSCM image of phase transferred CdS NRSCs (excitation laser $\lambda = 405$ nm; emission detection were collected in the range $\lambda = 420$–480 nm) and **b** cellular uptake of CdS nanorods by VSM cells. LSCM images of CdS NRSCs adsorbed onto VSM cell outer membrane after 3 h incubation and internalisation of CdS NRSCs into the cell cytoplasm of VSMC via endocytosis can be observed on the cell surface membrane. After internalisation spectral properties of the NRSCs are shifted to red spectra. Arrows shows the location of event of CdS uptake by the cells, **c** LSCM image of CdS NRSCs localised onto VSMC cell outer membrane after 24 h incubation and they are shown to produce only a fluorescence signal in the blue spectral region (excitation laser $\lambda = 405$ nm; emission detection were collected in the range $\lambda = 420$–480 nm), **d** internalised CdS NRSCs into the cell cytoplasm of VSMC are shown to have an emission in the red spectral region $\lambda = 660$–690 nm, and **e** total fluorescent signal from the CdS nanorods treated cell shown on a merged image of the two spectral detection channels

lines like mouse fibroblasts "Primo Cells" and human vascular endothelial cells (data not shown to avoid repetition). However, details of the uptake mechanism, cytotoxicity, and photostability, will clearly need to be addressed.

Water soluble NRSCs may have advantages over discrete nanorods in that the total intensity of emission is proportional to the number of rods and their size can potentially be tuned to specifically match different biological entities for example a protein (5–50 nm), a virus (20–500 nm), or a gene (2 nm wide and 10–100 nm long). As an example, larger 300–400 nm NRSCs, while difficult to translocate to cell compartments, would still be small enough to transport in the blood vessel and capillary system [28]. NRSCs in the 100–300 nm range may have promising applications in imaging techniques of cardiovascular pathology, biosensors, liquid chromatography and bioseparation methodology [29–33]. Since the NRSCs in this size range would be larger than the targeted species, they most likely would participate in binding without significant perturbation of the binding equilibrium and provide an efficient detection of a low amount of a target by microscopy or spectroscopical tools [34, 35]. Assemblies in the 30–100 nm range will be similar in size or smaller than the target species and may find use in applications such as FRET analysis, [36] hybridisation or nanoscale analysis of low abundant proteins and peptides (nanoproteomics) [37].

4 Conclusion

In conclusion, we have demonstrated a facile route for the formation of water soluble NRSCs of CdS, CdSe and CdTe. The phase transfer occurs at room temperature with the encapsulation and close packing of a group of nanorods in the hydrophobic core of a triblock copolymer micelle. The method of transfer is generally applicable to any monodisperse solution of ligand capped nanorods. The order and axial alignment of the assemblies allows discrete nanorod properties such as fluorescence to be enhanced by collective emission. In this study, uptake of the water dispersible nanorod ensembles in a cell line showed strong membrane and cell specific fluorescence at low laser intensity. The range of sizes (50–600 nm) obtainable with water dispersible nanorod assemblies and the biocompatibility of their polymer shells have the potential to greatly extend both the range and diversity of biological applications possible with semiconductor nanocrystals.

Acknowledgments The work was principally supported by Science Foundation Ireland (SFI) under the Principal Investigator Programme Contract No. 06/IN.1/I85. Institutional funding from INSPIRE is further acknowledged. A. Sanyal acknowledges the Irish Research Council for Science, Engineering and Technology (IRCSET) for a research postdoctoral fellowship. The authors thank Professor Noel Buckley for access to Hitachi S-4800 HRSEM. Dr Gordon Armstrong is also acknowledged for help in optical measurements.

References

1. J. Hu, L.-S. Li, W. Yang, L. Manna, L.-W. Wang, A.P. Alivisatos, Linearly polarized emission from colloidal semiconductor quantum rods. Sci. **292**, 2060 (2001)
2. S.A. Empedocles, R. Neuhauser, K. Shimizu, M.G. Bawendi, Photoluminescence from single semiconductor nanostructures. Adv. Mater. **11**, 1243 (1999)
3. A.P. Alivisatos, Perspectives on the physical chemistry of semiconductor nanocrystals. J. Phys. Chem. **100**, 13226 (1996)
4. I. Gur, N.A. Fromer, M.L. Geier, A.P. Alivisatos, Air-stable all-inorganic nanocrystal solar cells processed from solution. Sci. **310**, 462 (2005)

5. X.F. Duan, Y. Huang, R. Agarwal, C.M. Lieber, Single-nanowire electrically driven lasers. Nat. **421**, 241–245 (2003)
6. N.C. Greenham, X.G. Peng, A.P. Alivisatos, Charge separation and transport in conjugated-polymer/semiconductor-nanocrystal composites studied by photoluminescence quenching and photoconductivity. Phys. Rev. B: Condens. Matter **54**, 17628 (1996)
7. J.S. Steckel, J.P. Zimmer, S. Coe-Sullivan, N.E. Stott, V. Bulovic, M.G. Bawendi, Blue luminesce from (CdS) ZnS core shell nanocrystals. Angew. Chem. Int. Ed. **43**, 2154 (2004)
8. S. Coe, W.K. Woo, M. Bawendi, V. Bulovic, Electroluminescence from single monolayers of nanocrystals in molecular organic devices. Nat. **420**, 800 (2002)
9. R. Xie, D. Battaglia, X. Peng, Colloidal InP nanocrystals as efficient emitters covering blue to near-infrared. J. Am. Chem. Soc. **129**, 15432 (2007)
10. T. Zhai, X. Fang, Y. Bando, Q. Liao, X. Xu, H. Zeng, Y. Ma, J. Yao, D. Golberg, Morphology-dependent stimulated emission and field emission of ordered CdS nanostructure arrays. ACS. Nano. **3**, 949 (2009)
11. K.M. Ryan, A. Mastroianni, K.A. Stancil, H. Liu, A.P. Alivisatos, Electric-field-assisted assembly of perpendicularly oriented nanorod superlattices. Nano. Lett. **6**, 1479 (2006)
12. S. Ahmed, K.M. Ryan, Self-assembly of vertically aligned nanorod supercrystals using highly oriented pyrolytic graphite. Nano. Lett. **7**, 2480 (2007)
13. C. O'Sullivan, S. Ahmed, K.M. Ryan, Gold tip formation on perpendicularly aligned semiconductor nanorod assemblies. J. Mater. Chem. **18**, 5218 (2008)
14. C. Kang, C.W. Lai, H.C. Peng, J.J. Shyue, P.T. Chou, 2D self-bundled CdS nanorods with micrometer dimension in the absence of an external directing process. ACS. Nano. **2**, 750 (2008)
15. Y. Min, M. Akbulut, K. Kristiansen, Y. Golan, J. Israelachvili, The role of inter-particle and external forces in nanoparticle assembly. Nat. Mater. **7**, 527 (2008)
16. J.K. Jaiswal, H. Mattoussi, M.J. Mauro, S.M. Simon, Long-term multiple color imaging of live cells using quantum dot bioconjugates. Nat. Biotechnol. **21**, 47 (2003)
17. W.J. Parak, D. Gerion, T. Pellegrino, D. Zanchet, C. Micheel, S.C. Williams, R. Boudreau, M.A.L. Gros, C.A. Larabell, A.P. Alivisatos, Biological applications of colloidal nanocrystals. Nanotechnol. **14**, R15–R27 (2003)
18. L.S. Li, J. Walda, L. Manna, A.P. Alivisatos, Semiconductor nanorod liquid crystals. Nano. Lett. **2**, 557 (2002)
19. Z. Nie, D. Fava, E. Kumacheva, S. Zou, G.C. Walker, M. Rubinstein, Self-assembly of metal-polymer analogues of amphiphilic triblock copolymers. Nat. Mater. **6**, 609 (2007)
20. J. Zhuang, A.D. Shaller, J. Lynch, H. Wu, O. Chen, A.D.Q. Li, C.Y. Cao, Cylindrical superparticles from semiconductor nanorods. J. Am. Chem. Soc. **131**, 6084 (2009)
21. K.M. Ryan, N.R.B. Coleman, D.M. Lyons, J.P. Hanrahan, T.R. Spalding, M.A. Morris, D.C. Steytler, R.K. Heenan, J.D. Holmes, Control of pore morphology in mesoporous silicas synthesized from triblock copolymer templates. Langmuir **18**, 4996 (2002)
22. A.V. Kabanov, E.V. Batrakova, V.Y. Alakhov, Pluronic block copolymers as novel polymer therapeutics for drug and gene delivery. J. Controlled Release **82**, 189 (2002)
23. H. Liu, J.S. Owen, A.P. Alivisatos, Mechanistic study of precursor evolution in colloidal group II–VI semiconductor nanocrystal synthesis. Am. Chem. Soc. **129**, 305 (2007)
24. H. Htoon, J.A. Hollingworth, A.V. Malko, R. Dickerson, V.I. Klimov, Light amplification in semiconductor nanocrystals: Quantum rods versus quantum dots. Appl. Phys. Lett. **82**, 4776 (2003)
25. U. Westedt, L. Barbu-Tudoran, A.K. Schaper, M. Kalinowski, H. Alfke, T. Kissel, Deposition of nanoparticles in the arterial vessel by porous balloon catheters: Localization by confocal laser scanning microscopy and transmission electron microscopy. AAPS PharmSciTech. **4**, 206 (2002)
26. J.W.M. Bulte, M.M.J. Modo (eds.), *Nanoparticles in biomedical imaging emerging technologies and applications in fundamental biomedical technologies*, vol 3 (Springer, New York, 2008)
27. A. Merkoci, Electrochemical biosensing with nanoparticles. FEBS J **274**, 310 (2007)

28. F. Gentile, M. Ferrari, P. Decuzzi, The transport of nanoparticles in blood vessels: The effect of vessel permeability and blood rheology. Ann. Biomed. Eng. **36**, 254 (2008)
29. T. Vo-Dinh (ed.), *Biofunctionalization of fluorescent nanoparticles, nanotechnology in biology and medicine—methods, devices and applications* (Taylor & Francis Group, Boca Raton, 2007)
30. J.L. Swift, D.T. Cramb, Nanoparticles as fluorescence labels: Is size all that matters. Biophys. J. **95**, 865 (2008)
31. D.P. Kalogianni, T. Koraki, T.K. Christopoulos, P.C. Ioannou, Nanoparticle-based DNA biosensor for visual detection of genetically modified organisms. Biosens. Bioelectron. **2006**, 21 (1069)
32. W.-J. Chen, P.-J. Tsai, Y.-C. Chen, Functional nanoparticle-based proteomic strategies for characterization of pathogenic bacteria. Anal. Chem. **81**, 1722 (2009)
33. H.M.E. Azzazy, M.M.H. Mansour, S.C. Kazmierczak, Nanodiagnostics: A new frontier for clinical laboratory medicine. Clin. Chem. (Washington, D. C., 2006), 52, 1238
34. M. Brehm, T. Taubner, R. Hillenbrand, F. Keilmann, Infrared spectroscopic mapping of single nanoparticles and viruses at nanoscale resolution. Nano. Lett. **6**, 1307 (2006)
35. M. Breunig, S. Bauer, A. Goepferich, Polymers and nanoparticles: intelligent tools for intra cellular targeting. Eur. J. Pharm. Biopharm. **68**, 112 (2008)
36. W. Jiang, B.Y.S. Kim, J.T. Rutka, W.C.W. Chan, Nanoparticle-mediated cellular response is size-dependent. Nat. Nanotech. **3**, 145 (2008)
37. D.A. Egas, M.J. Wirth, Fundamentals of protein separations: 50 years of nanotechnology, and growing. Annu. Rev. Anal. Chem. **1**, 833 (2008)

Introduction of $BiFeO_3$ Nanoparticles in 5CB Nematic Liquid Crystal Host: Effect on Texture and the Non Responsive Behaviour of the Sample in a Twisted Nematic Configuration

Subhojyoti Sinha, Ayan Mukherjee, Sanat Kumar Chatterjee, Ajit Kumar Meikap and Soumen Basu

Abstract Here we report the introduction of multiferroic $BiFeO_3$ nanoparticles in a 5CB nematic liquid crystal host having positive dielectric anisotropy. The $BiFeO_3$ nanoparticles had an average diameter of ~ 50 nm and were prepared by modified Pechini method. Its effect on the Schlieren Texture and the formation of the isolated stripe domains was observed under optical polarizing microscope. When the sample was poured into a 90° twisted nematic cell and observed under an optical polarizing microscope in cross polarizer configuration it yielded normally black state in contradiction to the expected normally white and remains unchanged even with the application of external voltage. The polarizing microscope photograph of the cell observed under parallel polarizer configuration suggests the accommodation of the nanoparticles along the cell wall grooves.

Keywords: Nematic liquid crystal · Bismuth ferrite nanoparticles · Schlieren texture · Isolated stripe domains · TN cell

1 Introduction

Extensive research in the past and in recent years [1, 2] have been done on the composite systems prepared by doping particles in liquid crystal. When the particles are dispersed in these anisotropic fluids there exists a lot of interactions or forces [3–8] between these particles. The stability of the dispersions depends upon the balance between these forces. If the particles are dispersed in a nematic liquid

S. Sinha (✉) · A. Mukherjee · S. K. Chatterjee · A. K. Meikap · S. Basu
Department of Physics, National Institute of Technology, Durgapur, India
e-mail: sinhasubhojyoti@gmail.com

P. K. Giri et al. (eds.), *Advanced Nanomaterials and Nanotechnology*,
Springer Proceedings in Physics 143, DOI: 10.1007/978-3-642-34216-5_10,

crystal elastic deformation of the director field happened and which results in dipolar or quadrapolar type forces depending upon the symmetry of the director configuration around the particle [9–12]. The director field configuration around a spherical particle in nematic liquid crystal was reported by Stark [13]. Assuming that the molecules prefer homeotropic orientation at the surface of the particle, he investigated the three possible configurations, a dipole, a saturn ring, and a surface ring. This was done by numerically minimizing the Frank free energy, supplemented by a magnetic field and surface energy term. Long ago, before these observations were done, Brochard and de Gennes proposed doping a suitable liquid crystal with ferromagnetic grains [14]. The sample prepared by adding magnetic nanoparticle in the nematic liquid crystal is called ferronematics. With the introduction of a magnetic nanoparticle a considerable increase (or decrease) in the magnetic (or electric) threshold field was observed by Koneracka et al. [14, 15]. The macroscopic collective behavior and the new texture in magnetically doped liquid crystal was reported by Chen et al. [16]. Here we report the introduction of the $BiFeO_3$ nanoparticles in the 5CB nematic liquid crystal. Multiferroics are the materials that simultaneously exhibit more than one of the four basic ferroic order parameters- ferromagnetism, ferroelectricity, ferroelasticity and ferrotoroidicity. As the multiferroics possess both ferromagnetism and ferroelectricity the introduction of multiferroic may have potential application in reducing both the magnetic and electric Fredericksz transition fields. An optimization of the particle loading to obtain the desired effect is of utmost importance. However in this report we just focused our attention to the effect on the Schlieren texture and the response behavior when the mixture was taken in a twisted nematic liquid crystal cell.

2 Experimental Procedure

The $BiFeO_3$ nanoparticles were prepared by modified Pechini method [17]. In the first step, the precursor were prepared by mixing $Bi(NO_3)_3$. $5H_20$ and $FeNO_3.9H_2O$ in the molar ratio 1:1 in distilled water with the addition of HNO_3 (to adjust PH 1–2) by ultrasonication for 2 h. In the second step maleic acid (0.03 mol) was dissolved in distilled water (30 ml) in a separate beaker. Metal nitrate precursor solution was added to the maleic acid under constant ultrasoniation. Polyethylene Glycol (PEG) in a molar ratio of 1:1 to respective maleic acid was finally added as a polymerizing agent. The solutions were evaporated into dry polyester precursors in a hot air oven. Finally the grinded precursor powders were calcined in air with a heating rate of 400° Ch^{-1} and a soaking time of 2 h at temperatures of 600 °C. The TEM image of the nanoparticles is shown in Fig. 1.

Then 0.122 gm of these nanoparticles was mixed in 0.1 ml 5CB and the mixture was ultrasonicated for 2 h. The sample was then then observed under cross and parallel polarizer configuration under an OPM.

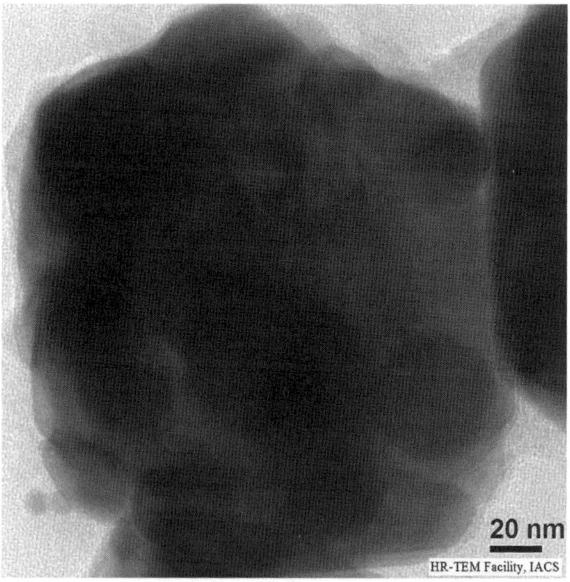

Fig. 1 TEM image of the $BiFeO_3$ nanoparticles

3 Result and Discussion

When the sample was observed under an Optical polarizing microscope (OPM) in cross polarizer configuration it was observed that in the Schlieren there was aggregation of the nanoparticles at the point of singularities resulting in the occurrence of birefringence at those points as evident in Figs. 2, 3 and 4.

As other regions of the samples were observed some isolated stripe domains [18] were found as shown in Fig. 5. When the sample was taken in a homogeneous 90° twisted nematic liquid crystal cell of cell gap ~6.8 μm under optical

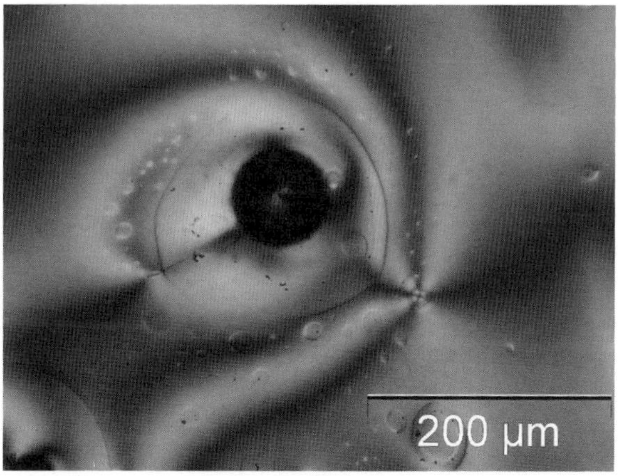

Fig. 2 Schlieren texture in the $BiFeO_3$ dispersed 5CB liquid crystal. Observed in region 1

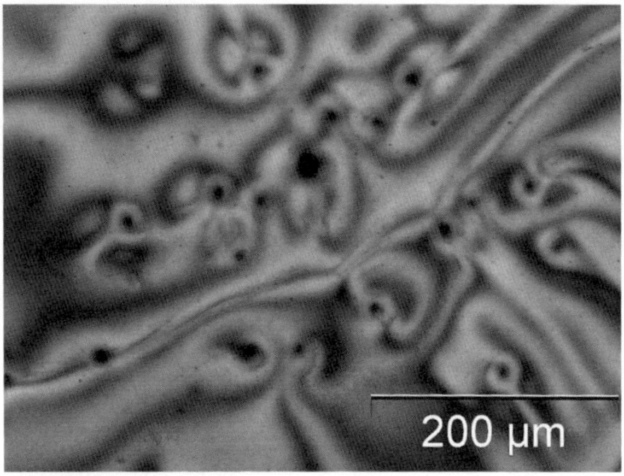

Fig. 3 Schlieren texture in the $BiFeO_3$ dispersed 5CB liquid crystal. Observed in region 2

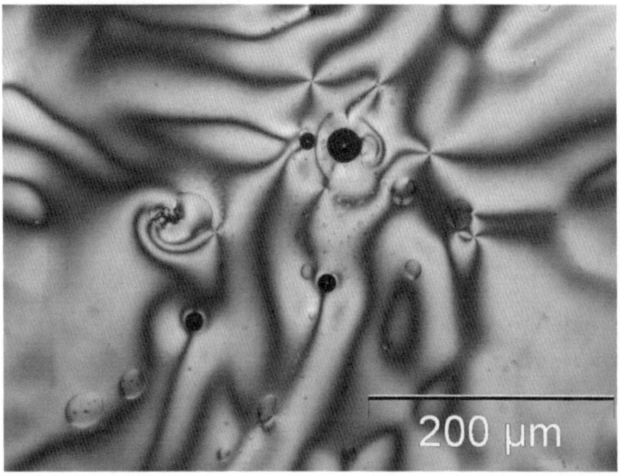

Fig. 4 Schlieren texture in the $BiFeO_3$ dispersed 5CB liquid crystal. Observed in region 3

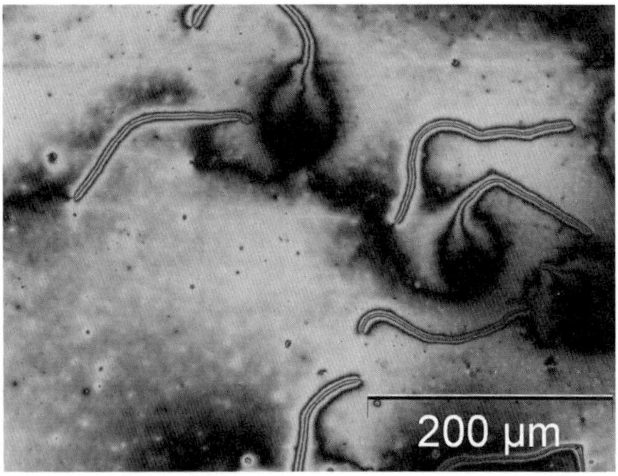

Fig. 5 Formation of the isolated stripe domains in the sample

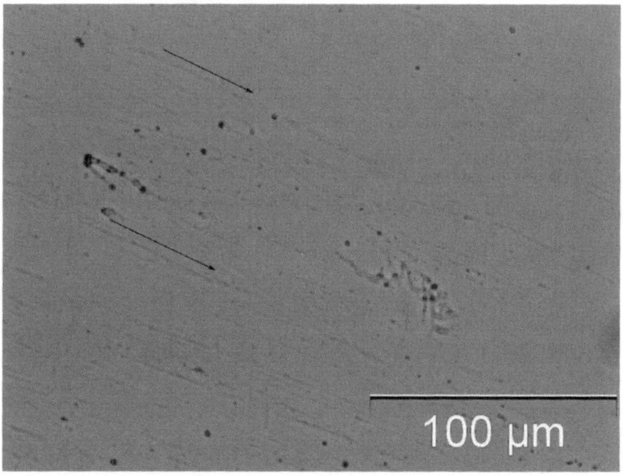

Fig. 6 The arrangement of the $BiFeO_3$ nanoparticles along the grooves in a TN liquid crystal cell observed under parallel polarizer configuration of the OPM. The *parallel arrows* are showing the direction of rubbing on the polyimide layer on ITO glass plates

polarizing microscope in cross polarizer configuration it yielded a normally black state in place of the normally white. Additionally with the application of ac and dc voltages there was no change in transmittance. When the cell was investigated under OPM in parallel polarizer configuration in 20 X magnification it was observed that the $BiFeO_3$ nanoparticles had arranged themselves along the aligning grooves on the polyamide alignment layer on the Indium Tin Oxide (ITO) coated electrodes which was actually meant for the arrangement of the LC molecules as shown in Fig. 6.

4 Conclusion

Thus we may conclude that the $BiFeO_3$ nanoparticles have arranged themselves in the microgrooves due to larger interaction energy with the cell walls. Due this unwanted self-accommodation of the $BiFeO_3$ nanoparticles along the rubbed grooves the liquid crystal molecules could not undergo 90° twists between the two glass plates and it also blocks the effect of external field on the molecules.

Acknowledgments The authors are grateful to MHRD, Government of India for the financial support. S. K. Chatterjee acknowledges with thanks the University Grants Commission (New Delhi) for supporting the work with Emeritus fellowship. S. Basu and A. Mukherjee thank DST (SR/FTP/PS-66/2008) for financial support.

References

1. J.C. Loudet, P. Barios, P. Poulin, Colloidal ordering from phase separation in a liquid-crystalline continuous phase. Nature **407**, 611–613 (2000)
2. C.D. Cruz, O. Sandre, V. Cabuil, Phase behavior of nanoparticles in a thermotropic liquid crystal. J. Phys. Chem. B **109**, 14292–14299 (2005)

3. W.B. Russel, D.A. Saville, W.R. Schowalter, *Colloidal Dispersions* (Cambridge University Press, Cambridge, 1995)
4. M. Krech, *The Casimir Effect in Critical Systems* (World Scientific, Singapore, 1994)
5. V.M. Mostepanenko, N.N. Trunov, *The Casimir Effect and its Application* (Clarendon Press, Oxford, 1997)
6. A.D. Dinsmore, A.G. Yodh, D.J. Pine, Entropic control of particle motion using passive surface microstructures. Nature **383**, 239–242 (1996)
7. A.D. Dinsmore, D.T. Wong, P. Nelson, A.G. Yodh, Hard spheres in vesicles: curvature-induced forces and particle-induced curvature. Phys. Rev. Lett. **80**, 409–412 (1998)
8. D. Rudhardt, C. Bechinger, P. Leiderer, Direct measurement of depletion potentials in mixtures of colloids and nonionic polymers. Phys. Rev. Lett. **81**, 1330–1333 (1998)
9. S. Ramaswamy, R. Nityananda, V.A. Raghunathan, J. Prost, Power-law forces between particles in a nematic. Mol. Cryst. Liq. Cryst. **288**, 175–180 (1996)
10. R.W. Ruhwandl, E.M. Terentjev, Long-range forces and aggregation of colloid particles in a nematic liquid crystal. Phys. Rev. E **55**, 2958–2961 (1997)
11. P. Poulin, H. Stark, T.C. Lubensky, D.A. Weitz, Novel colloidal interactions in anisotropic fluids. Science **275**, 1770–1773 (1997)
12. T.C. Lubensky, D. Pettey, N. Currier, H. Stark, Topological defects and interactions in nematic emulsions. Phys. Rev. E **57**, 610–625 (1998)
13. H. Stark, Director field configurations around a spherical particle in a nematic liquid crystal. Eur. Phys. J. B **10**, 311–321 (1999)
14. M. Koneracka, V. Zavisova, P. Kopcansky, J. Jadzyn, G. Czehowski, B. Zywucki, Study of the magnetic fredericksz transition in ferronematics. J. Magn. Magn. Mater. **157/158**, 589–590 (1996)
15. M. Koneracka, V. Zavisova, P. Kopcansky, T. Kuczynski, Study of magnetic fredericksz transition in ferronematic. J. Magn. Magn. Mater. **140–144**, 1455–1456 (1995)
16. S.H. Chen, N.M. Amer, Observation of macroscopic collective behavior and new texture in magnetically doped liquid crystals. Phys. Rev. Lett. **51**, 2298–2301 (1983)
17. M.P. Pechini, US Patent No.3.330.697, July 1 (1967)
18. T. Hegmann, H. Qi, V.M. Marx, Nanoparticles in liquid crystals: synthesis, self– assembly, defect formation and potential applications. J. Inorg. org. Polym. **17**, 483–508 (2007)

Magnetic Properties in Multiferroic $Bi_{1-x}Fe_{1+x}O_3$ ($x \le 0.5$) of Nanoparticles

T. Karan, S. Ram and R. K. Kotnala

Abstract Magnetic properties have been studied in the $Bi_{1-x}Fe_{1+x}O_3$ ($x \leq 0.5$) series in order to establish how a partial $Bi^{3+} \rightarrow Fe^{3+}$ substitution can be explored in tuning the functional magnetic and multiferroic properties. A self-propagating combustion of a solid precursor comprising the nitrates in the presence of camphor (as a fuel) has been used to synthesize this compound $Bi_{1-x}Fe_{1+x}O_3$ as nanoparticles at low temperature such as 550 °C in open air. An almost single phase compound has formed easily during burning the precursor in air. The 550 °C temperature is an optimized value to get ideal single magnetic domain particles without undergoing a significant disintegration. The X-ray diffraction patterns reveal a rhombohederally distorted perovskite with a space group R3c. The observed lattice parameters are $a = 0.5586$ nm and $c = 1.3903$ nm with cell density $\rho = 7.2$ g/cm^3, assuming a crystal lattice of $z = 6$ formula units. An average crystallite size was found using the Debye–Scherrer formula, which varies from 36 to 42 nm for the different compositions. Different samples studied by varying the x-value in small steps of 0, 0.1, 0.2, 0.5 and 0.8 reveal a maximum magnetization 8.23 emu/g in the bismuth ferrite phase on an optimal $x = 0.5$ substitution. A maximum coercivity of 172 Oe is achieved at lower $x = 0.2$. The sintered pellet (at 550 °C for 4 h) of sample $Bi_{0.5}Fe_{1.5}O_3$ ($x = 0.5$), measures a wide ferroelectric hysteresis loop at room temperature, showing a polarization of 0.07 μC/cm^2, remnant polarization of 0.06 μC/cm^2, and coercivity 39.2 kV/cm for an applied electric field up to 60 kV/cm.

T. Karan (✉) · S. Ram (✉)
Materials Science Centre, Indian Institute of Technology, Kharagpur 721 302, India
e-mail: karantrisha@gmail.com

S. Ram
e-mail: sram@iitkgp.ernet.in

R. K. Kotnala
National Physical Laboratory, New Delhi 110 012, India

P. K. Giri et al. (eds.), *Advanced Nanomaterials and Nanotechnology*,
Springer Proceedings in Physics 143, DOI: 10.1007/978-3-642-34216-5_11,

Keywords Magnetic properties • Multiferroic • Self-combustion • Nanoparticles • Ferroelectric hysteresis loop

1 Introduction

Multiferroics are a specific class of smart materials which offer exchange coupled ferroelectric, ferromagnetic, and ferroelastic properties simultaneously in a single system [1–3]. Co-operative interactions in these three physical properties give rise to multiferroic behavior which is useful for several applications including ferroelectric memory devices, spintronics, storage devices, and sensors [4, 5]. $BiFeO_3$ is one of the ABO_3 type single phase multiferroic perovskite oxide, where A and B site cations are sources of ferroelectricity and magnetism, respectively [6, 7]. Ferroelectricity in $BiFeO_3$ is driven by the stereochemical activity of $6s^2$ lone pair of Bi^{3+} ions, whereas the Fe^{3+} ions order antiferromagnetically. In other perovskites, viz; $BaTiO_3$ and $PbTiO_3$, coexistence of the two properties in a single material is not possible as ferroelectricity requires transition metal ions (B) with d^0 electronic configuration, whereas ferromagnetism requires B with an odd number of d electrons. $BiFeO_3$ is the only multiferroic single compound known so far in which the magnetic and ferroelectric orders co-exist well above the room temperature with the ferroelectric Curie temperature T_c ~836 °C and the Neel temperature T_N ~370 °C [8, 9]. It is a G-type antiferromagnet with a long-range cycloidal spin arrangement of wavelength 62 nm incommensurate with the lattice [10, 11]. Ferro- or ferrimagnetism instead of antiferromagnetism along with an enhanced ferroelectric polarization is highly required for low-field applications. Nevertheless, use of the bismuth ferrite in the applications is limited because it suffers from structural defects and non-stoichiometry which cause low electrical resistivity with unwantedly large current leakage. These limitations can be overcome by an isovalent substitution on the A-site leading to improve phase stability of a single phase compound $BiFeO_3$ [12]. Enhancements in the magnetization and the ferroelectric polarization were reported in $Bi_{1-x}Nd_xFeO_3$ and an A-site doped system of $Bi_{0.9-x}Tb_xLa_{0.1}FeO_3$ where Tb and La are isovalent to Bi^{3+} ions. Therefore, substitution on A-site by Fe^{3+} ion may enhance magnetic properties with improved phase stability which is the driving factor for the work reported here. In this paper, we have reported the phase analysis and the magnetic properties of $Bi_{1-x}Fe_{1+x}O_3$ ($x \leq 0.5$) nanoparticles synthesized at moderate temperatures (500–800 °C) by a self-propagating combustion method.

2 Experimental Details

A series of $Bi_{1-x}Fe_{1+x}O_3$ ($x \leq 0.5$) of nanoparticles has been synthesized using a self-propagating combustion method upon post annealing at moderate temperatures 500–800 °C in open air. $Fe(NO_3)_3{\cdot}9H_2O$ and $Bi(NO_3)_3{\cdot}5H_2O$ were used as precursors with camphor ($C_{10}H_{19}O$) as a fuel that provides the required energy to form the oxides from the nitrates in preparing this series. A prerequisite amount of the precursors were mixed well in a mortar using a pestle and then the mixture was ignited with camphor in obtaining the ferrites. The samples were annealed at two different temperatures 550 and 800 °C in tailoring the microstructures and the functional properties for possible applications.

The phase formation in the ferrites $Bi_{1-x}Fe_{1+x}O_3$ ($x \leq 0.5$) was analysed in terms of X-ray diffraction (XRD). The XRD patterns were recorded on a Philips X-ray diffractometer (model PW 1710) with filtered 0.15405 nm CuK_a radiation. The data were collected at a scanning rate 0.05°/s over the diffraction angle 2θ in the 20–80° range. An average crystallite size (D) in the bismuth ferrite and the partially substituted derivatives were calculated from the full width at half maximum values in the XRD peaks as per the Debye–Scherrer relation. The size and morphology in the different $Bi_{1-x}Fe_{1+x}O_3$ compositions were measured using a high-resolution scanning electron microscope (SEM) (Oxford model Leo 1550), with an accelerated voltage 2–10 kV. The magnetic properties were measured at room temperature with help of a vibrating sample magnetometer (DMS-1600). The P-E hysteresis loop of a poled sintered pellet was measured at room temperature using a Radiant ferroelectric loop tracer (M/S Technology Inc, USA).

3 Results and Discussion

In order to identify the crystal structure and the various phases present in the samples $Bi_{1-x}Fe_{1+x}O_3$ ($x \leq 0.5$), we analyzed the XRD patterns as presented in Fig. 1 for the selective compositions (x = 0, 0.1, 0.2, and 0.5) upon thermal annealing (at 500–800 °C for 1–2 h in ambient air). The XRD patterns reveal a well-defined rhombohederally distorted perovskite crystal structure with a space group R3c (JCPDS, File No. 86-1518), which has been observed also in a bulk $BiFeO_3$. The cell density $\rho = 7.2$ g/cm^3 and the average lattice parameters $a = 0.5586$ nm and $c = 1.3903$ nm were calculated from the XRD patterns using a hexagonal frame of reference, assuming a crystal lattice of z = 6 formula units.

The D-value, which varies from 36 to 42 nm for different compositions, is found to be minimum ~36 nm at x = 0.5. A regular shift in the peaks (110) and (104), which is a characteristic of change in the cell parameters retaining the original crystal structure, incurs on the $Fe^{3+} \rightarrow Bi^{3+}$ substitution. The peaks shift towards a higher d_{hkl}-value (interplanar spacing) in the samples with x = 0.5 (a = 0.560 nm and c = 1.393 nm), with respect to the value at x = 0. In contrast, the shift is towards a lower d_{hkl}-value in an intermediate substituted sample with x = 0.2 (a = 0.556 nm and c = 1.388 nm) and x = 0.1 (a = 0.557 nm and c = 1.389 nm).

The FESEM images from the $Bi_{1-x}Fe_{1+x}O_3$, x = 0, 0.1, 0.2, 0.5, in Fig. 2, portray a marked variation in size (80–350 nm) and morphology of the nano-crystallites and aggregates. The $Bi_{0.5}Fe_{1.5}O_3$ sample (x = 0.5) consists of rod-like

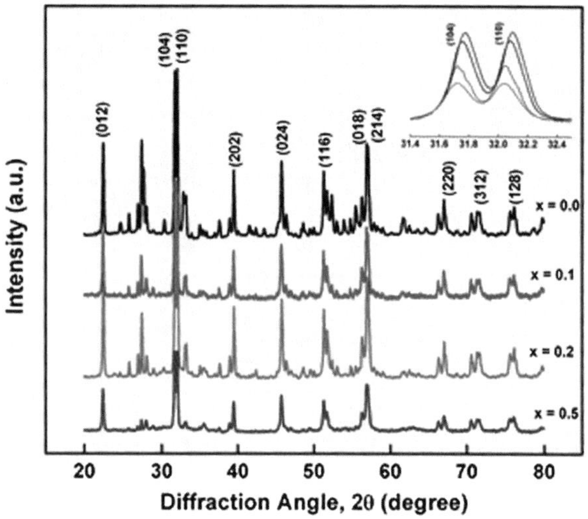

Fig. 1 XRD patterns in $Bi_{1-x}Fe_{1+x}O_3$ ($x \leq 0.5$) ferrites annealed at 550 °C for 1 h in ambient air atmosphere, showing a rhombohederally distorted perovskite structure. A shift in the (110) and (104) peaks on a $Fe^{3+} \rightarrow Bi^{3+}$ substitution is shown in the inset

nanoparticles with a small size (~80 nm). The rod-like shape might be originated from an anisotropic behavior of the subsumed bismuth ferrite. The sample $Bi_{0.8}Fe_{1.2}O_3$ ($x = 0.2$) displays square nanoplates with particle size (~0.35 μm) whereas a plate-like structure develops in $BiFeO_3$ ($x = 0$) with particle size (~0.3 μm). The sample $Bi_{0.9}Fe_{1.1}O_3$ ($x = 0.1$) has almost same morphology as the parent ferrite ($x = 0$) and with particle size ~100 nm. The smaller size of nanoparticles at $x = 0.5$ as compared to the value in the other compositions is possibly due to a lower diffusion of Fe^{3+} species in comparison to that in the Bi^{3+} species. A change in the morphology in the $Fe^{3+} \rightarrow Bi^{3+}$ substituted $Bi_{1-x}Fe_{1+x}O_3$ ferrites indicates a change in the kinetics of the grain growth. Uniform particle size distribution and a homogenous microstructure are evident in all the compositions studied over here.

In order to comprehend how the D-value and the morphology affect the magnetic properties in $Bi_{1-x}Fe_{1+x}O_3$ ($x \leq 0.5$), we studied the magnetic hysteresis loops at room temperature with applied magnetic fields up to ±5 kOe as displayed in Fig. 3. The sample with $x = 0.5$ having a small D-value (36 nm) exhibits negligibly small values of hysteresis loss, coercivity ~123 Oe and remanence ~1.07 emu/g, with a large magnetization ~8.23 emu/g which does not saturate up to 5 kOe. These are the characteristics of ideal single magnetic domains in this specific sample. The sample ($x = 0.2$) having larger D-value (42 nm) attains the values of coercivity and remanence enhanced by 40 and 17.7 %, respectively, with the magnetization reduced by 46.5 %. The magnetic properties observed in the different samples are summarized in Table 1.

The nanoparticles in $Bi_{1-x}Fe_{1+x}O_3$ ($x \leq 0.5$) with a D-value less than the incommensurate wavelength, i.e., 62 nm, show an enhanced magnetic property relative to its bulk counterpart due to the breaking of spiral spin structure. In a bulk bismuth ferrite, magnetic structure is a canted two-sublattice antiferromagnet with a

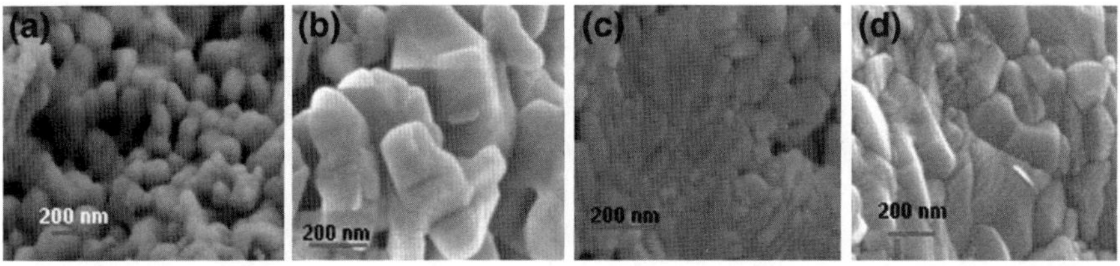

Fig. 2 FESEM images of $Bi_{1-x}Fe_{1+x}O_3$ powders annealed at 550 °C for 1 h in ambient air; **a** x = 0.5, **b** x = 0.2, **c** x = 0.1, and **d** x = 0

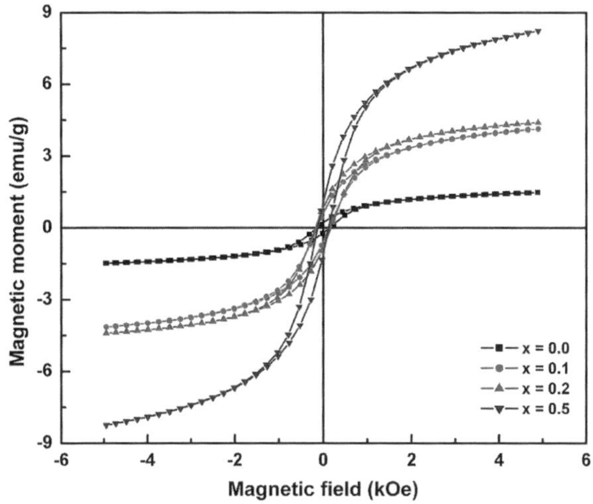

Fig. 3 Magnetic hysteresis loops for $Bi_{1-x}Fe_{1+x}O_3$, $x \leq 0.5$, at room temperature

Table 1 Magnetic properties observed in $Bi_{1-x}Fe_{1+x}O_3$, $x \leq 0.5$, ferrites at room temperature

Size (*d*, nm)	x-value	*Magnetization* M_S at 5 kOe		*Coercivity* H_C (Oe)	M_r/M_S
		(emu/g)	(μ_B/Fe)		
40	0.0	1.47	0.09	142.47	0.15
38	0.1	4.14	0.21	137.63	0.14
42	0.2	4.40	0.22	172.33	0.20
36	0.5	8.23	0.94	123.04	0.13
Bulk[a]		0.34	0.02		

[a] Park et al. [14]

spiral of wavelength 62 nm [13]. An enhanced ferromagnetic property with a high value of magnetization (M_S) at a given field value was observed for the modified bismuth ferrites, $Bi_{1-x}Fe_{1+x}O_3$, $x \leq 0.5$ as compared to the value in the bulk counterpart. In sample $Bi_{0.5}Fe_{1.5}O_3$, the magnetization measures as large as 8.23 emu/g and it does not saturate up to the field 5 kOe. The magnetic moment in the $Bi_{-x}Fe_{1+x}O_3$ ($x \leq 0.5$) samples was calculated using the relation, $M_s = N \times 9.7 \times 10^{-21} \times \delta$, where *N* is the number of atoms per gram and δ is the Bohr's magneton (μ_B) per atom.

At a specific x = 0.5, the δ-value is found to be $\sim$0.94 μ_B/Fe which is considerably larger than the value in a bulk bismuth ferrite (0.02 μ_B/Fe).

The large increase in magnetization at x = 0.5 is a consequence of a distortion in crystal structure on Fe^{3+} substitutions that helped in releasing the potential magnetization locked in the spiral spin structure. Moreover, the decrease in particle dimensions ($\sim$80 nm) in order to possess ideal single magnetic domains introduced coordination distortion and lattice disorder by surface strain that permeate throughout the entire particle, as opposed to being confined only at the surface, with the net result of enhanced magnetization in $Bi_{1-x}Fe_{1+x}O_3$ [14]. Uncompensated magnetic spins associated with the surface atoms, which lead to a frustrated spin system, become significant in these nanostructures describing an enhanced magnetization at room temperature.

To establish to the correlation of large M_S observed in $Bi_{1-x}Fe_{1+x}O_3$ ($x \leq 0.5$) with the ferroelectric properties, we studied ferroelectric hysteresis loops in the specific sample with x = 0.5, with different electric fields E_a (260–780 V) at room temperature as illustrated in Fig. 4.

The final shape and size of the hysteresis loops vary with the E_a – value, showing E_a- field dependent hysteresis loops. At E_a – value 780 V the sintered $Bi_{0.5}Fe_{1.5}O_3$ pellet (at 550 °C for 4 h) exhibits a wide hysteresis loop at room temperature. The sample measures coercivity 39.2 kV/cm with saturation polarization and remnant polarization $\sim$0.07 and 0.06 μC/cm^2 respectively at an applied electric field up to 60 kV/cm. Both the magnetic and ferroelectric loops have maximum coercivity in ideal single domains of a randomly oriented assembly. When an external field is applied it grows favourably along it at the expense of less favourably oriented domains. With increasing a bias electric field, a P-E loop thus grows larger and wider as per the magnetoelectric feature [15]. A discontinuity is observed during loading and unloading applied electric field and it increases with higher E_a – value. This discontinuity is possibly due to electric-field induced reorientation of suitable domains.

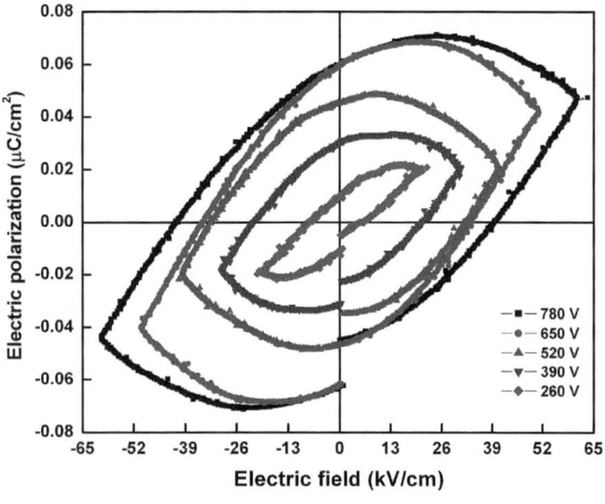

Fig. 4 Ferroelectric hysteresis loops measured on a sintered $Bi_{0.5}Fe_{1.5}O_3$ pellet, at different applied voltages (260–780 V) after poling at room temperature

4 Conclusion

In summary, the effect of Fe^{3+} ion substitution was studied in terms of structural and magnetic properties in nanocrystalline multiferroic $BiFeO_3$, synthesized by self-propagating combustion of a solid precursor comprising the nitrates in the presence of camphor (as a fuel) at low temperature such as 550 °C in open air. XRD patterns indicate that all these materials have rhombohederally distorted perovskite structure. A dramatic increase, nearly two orders of magnitude in magnetic moment ($\sim$0.94 μ_B/Fe) was observed in the substituted bismuth ferrite with $x = 0.5$. The reasons for the enhanced magnetic properties in $Bi_{1-x}Fe_{1+x}O_3$ ($x \leq 0.5$) is expected to be a combined effect of local ferrimagnetization by A-site substitution and the breakdown of cycloid spin structure due to smaller particle size. The ferrite $Bi_{1-x}Fe_{1+x}O_3$ ($x = 0.5$), show a ferroelectric behavior at room temperature with a maximum polarization value $\sim$0.07 $\mu C/cm^2$ at an applied electric field of 13.47 kV/cm.

Acknowledgments This work has been supported by Council of Scientific and Industrial Research.

References

1. H. Schmid, Mutiferroic magnetoelectrics. Ferroelectrics **162**, 665 (1994)
2. T. Kimura, T. Goto, H. Shintani, K. Ishizaka, T. Arima, Y. Tokura, Magnetic control of ferroelectric polarization. Lett. Nat. Nat. **426**, 55 (2003)
3. W. Eerenstein, N.D. Mathur, J.F. Scott, Multiferroic and magnetoelectric materials. Nature **442**, 759 (2006)
4. J.R. Cheng, N. Li, L.E. Cross, Structural and dielectric properties of Ga-modified $BiFeO_3$–$PbTiO_3$ crystalline solutions. J. Appl. Phys. **94**, 5153 (2003)
5. J.T. Han, Y.H. Huang, X.J. Wu, C.L. Wu, W. Wei, B. Peng, W. Huang, J.B. Goodenough, Tunable synthesis of bismuth ferrites with various morphologies. Adv. Mater. **18**, 2145 (2006)
6. G. Catalan, J.F. Scott, Physics and application of bismuth ferrite. Adv. Mater. **21**, 2463 (2009)
7. N.A. Hill, Why are there so few magnetic ferroelectrics? J. Phys. Chem. B **104**, 6694 (2000)
8. M.K. Singh, R. Katiyar, W. Preller, J.F. Scott, The Almeida-Thouless line in $BiFeO_3$: is bismuth ferrite a mean field spin glass? J. Phys.: Condens. Matter. **21**, 042202 (2009)
9. A.K. Pradhan et al., Magnetic and electrical properties of single phase multiferroic $BiFeO_3$. J. Appl. Phys. **97**, 093903 (2005)
10. I. Sosonowska, M. Lowenhaupt, W.I.F. David, R.M. Ibberson, Investigation of the unusual magnetic spiral arrangement in $BiFeO_3$. Phys. B: Condens. Matter **180 & 181**, 117 (1992)
11. S. Basu, S.K.M. Hussain, D. Chakravorty, M. Pal, Enhanced magnetic properties in hydrothermally synthesized Mn-doped $BiFeO_3$ nanoparticles. Curr. Appl. Phys. **11**, 976 (2011)
12. S.M. Selbach, M.A. Einarsrud, T. Grande, On the thermodynamic stability of $BiFeO_3$. Chem. Mater. **21**, 169 (2009)

13. I. Sosnowska, W. Schäfer, W. Kockelmann, K.H. Andersen, I.O. Troyanchuk, Crystal structure and spiral magnetic ordering of $BiFeO_3$ doped with manganese. Appl. Phys. A Mater. Sci. Process. **74**, S1040 (2002)
14. T.J. Park, G.C. Papaefthymiou, A.J. Viescas, A.R. Moodenbaugh, S.S. Wong, Size-dependent magnetic properties of single crystalline multiferroic $BiFeO_3$ nanoparticles. Nano Lett. **7**, 766 (2007)
15. K.K. Bhargav, S. Ram, S.B. Majumder, A liquid-precursor synthesis of single-phase magnetoelectric $LaFeO_3$ nanocrystallites. Mater. Express **1**, 210 (2011)

Study of Photocurrent and Dark Current in ZnS Nano Particles Prepared by Precipitation Method

Nitin Pandey, Rajneesh Kumar Srivastava and S. G. Prakash

Abstract In present work an attempt has been made to study photoconductivity and dark conductivity in ZnS nano particles synthesized by precipitation method. Voltage dependence of photocurrent and dark current has been measured at room temperature under visible illumination. Rise and decay curve exhibits anomalous behavior of photocurrent, where in the photocurrent decreases even during steady illumination. Structural study has been performed using XRD and SEM.

Keywords Photoconductivity · ZnS · Nanoparticles · Precipitation method

1 Introduction

The ZnS is an important optoelectronic device material for its use in the violet and blue regions owing to its wide band gap (−3.7 eV). It is transparent in visible region having exciton binding energy of 40 meV. Induced sub band gap transitions in ZnS occur at energies in the visible range that allows the optical detection of traps, relative recombination centers and surface states. The small size and high optical activity of ZnS nanoparticles (NPs) make them interesting for various optoelectronic applications [1–4].

There are several methods to synthesize the ZnS nanoparticles (NPs) such as sol–gel method, co-precipitation method, solid state reaction method, auto-combustion method etc. In the present work ZnS nanoparticles (NPs) have been

N. Pandey · R. K. Srivastava (✉) · S. G. Prakash
Department of Electronics and Communication, University of Allahabad,
Allahabad 211002, India
e-mail: rkumarsau@gmail.com

P. K. Giri et al. (eds.), *Advanced Nanomaterials and Nanotechnology*,
Springer Proceedings in Physics 143, DOI: 10.1007/978-3-642-34216-5_12,

synthesized by precipitation method. Precipitation method is a simple, inexpensive and high yield method.

Photoconductivity is a useful tool to study the properties of semiconductor. The photoconductivity arises due to generation of electron hole pairs because of the interaction of the photons with bound electrons of lattice atoms. The conductivity of the materials depends upon the carrier density and complex process of generation, trapping, and recombination of carriers [5, 6]. Photoconductivity is also a function of temperature, applied field, intensity of light and energy of radiation. In the present work, photoconductivity and dark conductivity properties of ZnS nanoparticles are studied. Structural characterizations are done by X-ray diffraction (XRD) and scanning electron microscopy (SEM).

2 Experimental Section

2.1 Chemicals

Zinc acetate (Zn $(CH_3COOH)_2{\cdot}2H_20$), and Sodium sulphide (Na_2S) were purchased from E. Merk Ltd., Mumbai. These chemicals were directly used without any special treatment.

2.2 Sample Preparation

10 ml of 1 M zinc acetate was mixed in 80 ml of double distilled water. In the above solution 10 ml of 1 M of Na_2S was added drop wise under vigorous stirring for 24 h. A white precipitate was obtained, which was separated by centrifugation. The precipitate was washed several times with double distilled water and ethanol. The precipitate was dried under vacuum at 80 °C overnight to get the powder sample.

2.3 Instrumentation

Crystal structure of ZnS nanoparticles were characterized by X-Ray diffraction using rigaku D/MAX -2200H/PC with Cu Kα radiation ($\lambda = 1.54178$ Å). For measuring photoconductivity and dark conductivity, a cell is formed by spreading a thick layer of powdered sample in between two Cu electrodes etched on a Cu plate (PCB), having a spacing of 1 mm (Fig. 1a). The powdered layer was pressed with transparent glass plate. This glass plate has slit for providing illumination area of 0.25 cm^2. In this cell type device, the direction of illumination is normal to the

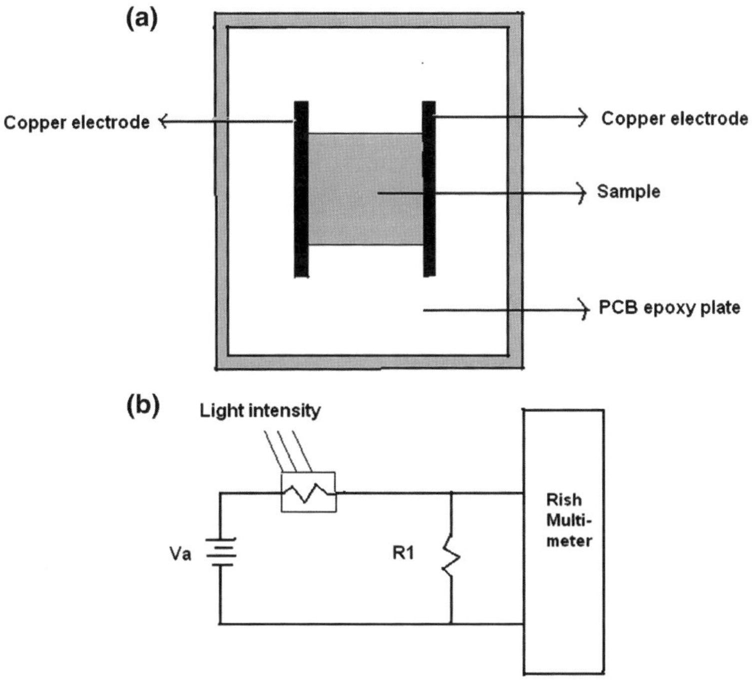

Fig. 1 **a** Photoconductive cell. **b** Measurement setup used for observation of photoconductivity

field across the electrodes. The cell was mounted in a dark chamber with a slit where from the light is allowed to fall over the cell. The visible photo response was measured using commercial bulb of 200 W as a photo excitation source. A stabilized dc field (50–500 V/cm) was applied across the cell to which a digital dc nano meter for measurement of current and Rish multi 18 S with adapter were connected in series (Fig. 1b). The light intensity over the surface was changed by varying the distance between slit and the light source. The cell is kept in dark till it attains equilibrium before measuring the photoconductivity.

3 Results and Discussion

3.1 *Structural Study*

Figure 2 shows the XRD pattern of ZnS nano particles (NPs) synthesized by precipitation method at room temperature. The obtained peaks are well matched with that of standard cubic ZnS. The peak broadening in XRD pattern indicates that nano particles have been formed. The XRD patterns show peaks at 2θ values

Fig. 2 XRD pattern of ZnS NPs at room temperature

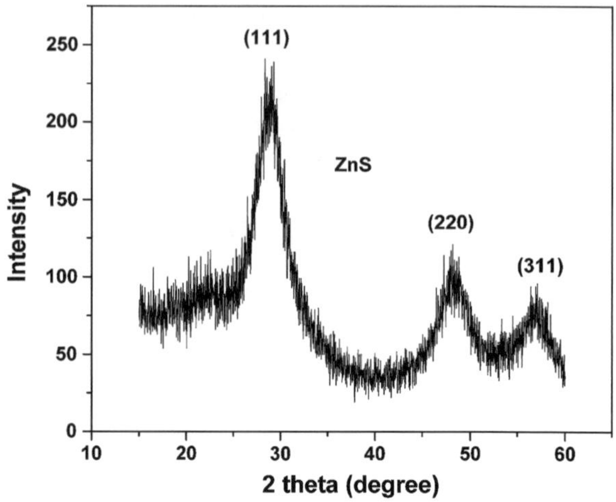

of 29.16, 48.67 and 56.92 referring to diffraction from (111), (220), and (311) planes. The average crystallite size is calculated from the Scherer formula [7, 8]

$$D = \frac{0.9\,\lambda}{\beta\,\cos\theta} \tag{1}$$

where D is the crystallite size, λ is the wavelength of radiation used, θ is the Bragg angle and β is the full-width at half-maximum (FWHM) measured in radian. The average crystallite size of sample lies in the range of 28.40 to 31.94 nm.

3.2 *Morphology Study*

Figure 3 shows the SEM of ZnS NPs. The NPs are spherical in nature and are highly agglomerated. The particle size lies in the range of 38–45 nm.

3.3 *Photoconductivity Study*

3.3.1 Voltage Dependence of Photocurrent

Figure 4 shows variation of photocurrent (I_{pc}) with applied voltage for ZnS nano particles (NPs) on ln–ln scale. The *ln* (*Ipc*) versus *ln* (*V*) curve is straight line with single segment. The variation of photocurrent may be expressed by power law, i.e. $Ipc \propto V^r$, where *r* is slope of straight line and is 1.39. The ZnS nanoparticles (NPs) exhibit super linear variation at low voltage as well as high voltage. The super linear behavior suggests that some charge carriers are injected into the sample from one of the electrodes [9].

Fig. 3 SEM image of the ZnS NPs

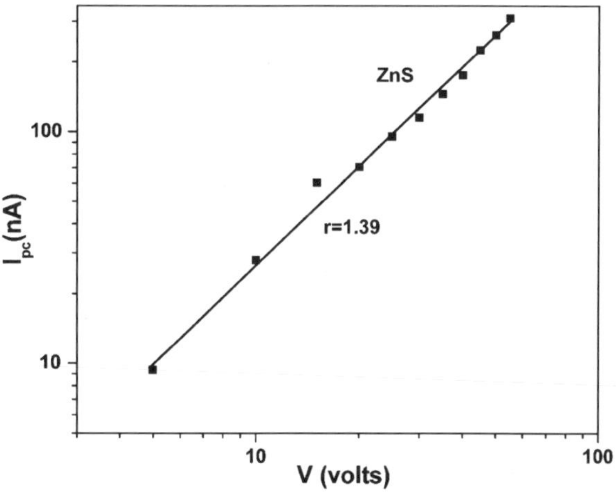

Fig. 4 Variation of photocurrent with applied voltage on ln–ln scale

3.3.2 Voltage Dependence of Dark Current

Figure 5 shows variation of dark current (I_{dc}) with applied voltage (V) for ZnS (NPs) on ln–ln scale. The ln (I_{dc}) versus ln (V) for the dark current curve is straight line with two segments with different slopes. The variation of dark current may be expressed by powder law, i.e. $I \propto V^r$, where r is slope of straight line. At low voltage (up to 20 V), the dark current varies super linearly as r = 1.41. This may be due to injection of additional charge carriers into the sample from one of the electrodes [9]. At high voltage, r is 2.07 that suggest space-charge limited trap limited variation of dark current with voltage.

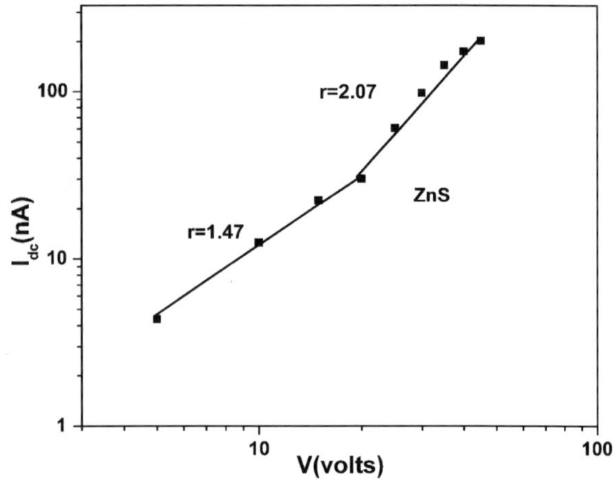

Fig. 5 Variation of dark current with applied voltage on ln–ln scale

Ratio of photocurrent to dark current at low voltage is approximately 2.33 up to 20 V, after which this reduces and becomes approximately 1.2. The reduction in photocurrent with respect to dark current at high voltage as compared to that at low voltage may be due to higher threshold voltage for occurrence of space charge limited photo current during illumination as compared to that in dark.

3.3.3 Rise and Decay of Photocurrent

Figure 6 shows the time—resolved rise and decay photocurrent spectra of ZnS nanoparticles (NPs). ZnS nanoparticles are found to exhibit anomalous behavior of photo current. When the light is switched on, the photocurrent increases, reaches to a maximum value and then it starts decreasing even during steady visible illumination, which may be explained on the basis of adsorption/desorption process of O_2 molecules. When light is switched on, the absorbed photons generate electron–hole pairs from defect levels and the increased concentration cause the photocurrent to increase. In the dark, O_2 molecules adsorbed on the surface of NPs are converted to O_2^- by capturing electrons. Light induced desorption of O_2 molecules releases these captured electrons into conduction band, which also increase photocurrent when light is switched on. During steady illumination, the reduction in photoconductivity may be attributed to dominance of the process of readsorption of oxygen molecules [10, 11]. Similar anomalous behavior has also been reported by Ahn et al., Nelson et al., Bera et al. in ZnO nanowires, TiO_2 nanocrystalline films and ZnO nanowires [12–15] respectively. When the light is terminated, the electrons recombine with holes and are captured by reabsorbed O_2 molecules causing an exponential decay in the current [10, 15]. The rise and decay curves are also governed by trap levels and recombination centers lying in forbidden region of photoconductor, so these curves are used to know the nature and distribution of traps and recombination centers also [16].

Fig. 6 Rise and Decay of photocurrent in ZnS nanoparticles (NPs)

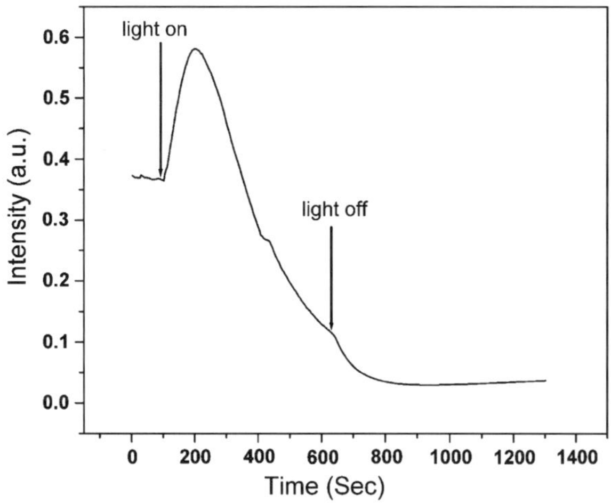

4 Conclusion

ZnS NPs are synthesized by precipitation method. XRD pattern indicates formation of cubical ZnS. Peak broadening is due to nano dimensions. Particle size calculated by XRD peaks lies in the range 28.40–31.94 nm. SEM image indicates ZnS NPs are spherical and highly agglomerated. Variation of photocurrent with voltage is superlinear. Dark current varies with applied voltage super linearly at low as well high voltage. At high voltage the dark current is space charge limited trap limited current. When the light is switched on, the photo current initially increases, reaches a maximum value and then it reduces even during steady illumination. This anomalous behavior of photocurrent may be attributed to dominance of the process of readsorption of O_2 molecules.

Acknowledgments The author Rajneesh K. Srivastava is thankful to UGC for its financial assistance in form of a project 37–395 (2009) (SR). The authors are also thankful to N. C. E. M. P., University of Allahabad for providing XRD and SEM facilities.

References

1. A.P. Alivisatos, Science **271**, 933 (1996)
2. H.S. Yang, P.H. Holloway, B.B. Ratna, J. Appl. Phys. **93**, 586 (2003)
3. J.P. Borah, K.C. Sarma, Acta. Phys. Polon. A **114**, 713 (2008)
4. C. Wang, Q. Li, B. Hu, Chin. Opt. Lett. **7**, 432 (2009)
5. N.V. Joshi, *Photoconductivity: Art, Science, and Technology* (Marcel Dekker, New York, 1990)
6. A. Rose, *Concepts in Photoconductivity and Allied Problems* (Krieger, New York, 1978)
7. B.D. Cullity, S.R. Stock, *Elementary of X-ray Diffraction, Englewood Cliffs*, 3rd edn. (Prentice-Hall, New Jersey, 2001)

8. L. Guo, S. Yang, C. Yang, P. Yu, J. Wang, W. Ge, G.K.L. Wong, Appl. Phys. Lett. **76**, 2901 (2000)
9. R.K. Srivastava, S.G. Prakash, Nat. Acad. Sci. Lett. **30**, 1114 (2007)
10. A. Bera, D. Basak, Appl. Phys. Lett. **93**, 053102 (2008)
11. Q.H. Li, T. Gao, Y.G. Wang, T.H. Wang, Appl. Phys. Lett. **86**, 123117 (2005)
12. S.E. Ahn, H.J. Ji, K. Kim, G.T. Kim, C.H. Bae, S.M. Park, Y.K. Kim, J.S. Ha, Appl. Phys. Lett. **90**, 153106 (2007)
13. S.J. Nelson, A.M. Eppler, I.M. Ballard, J. Photochem. Photobiol. **A1**(48), 2328 (2002)
14. J. Carry, H. Carrere, M.L. Kahn, B. Chaudret, X. Marie, M. Respaud, Semicond. Sci. Technol. **23**, 025003 (2008)
15. A. Bera, D. Basak, Appl. Phys. Lett. **94**, 163119 (2009)
16. R.K. Srivastava, S.G. Prakash, Natl. Acad. Sci. Lett. **30**, 393 (2007)

Synthesis and Photoluminescence Studies of Ni Substituted ZnO Nanoparticles

S. Bobby Singh, K. Nomita Devi, Ibetombi Soibam and H. Nandakumar Sarma

Abstract Nickel (Ni) substituted ZnO nanopowders were synthesized by simple and efficient co-precipitation method with zinc nitrate and nickel nitrate as the starting materials. Effect of Ni substitution on structural and photoluminescence properties of ZnO nanopowders has been studied. Structural characterization of the as prepared powders was performed with X-ray diffraction (XRD) studies. It was confirmed that all the powders were of zinc oxide having polycrystalline nature and possessing typical hexagonal wurtzite structure. Both lattice parameters *a* and *c* decreases from 3.257 to 3.214 Å and 5.217 to 5.156 Å, respectively with increase in Ni content from 0 to 10 wt % while the crystallite size calculated from the XRD data decreases from 29.8 to 14.2 nm, respectively. The effect of Ni substitution on the photoluminescent properties of ZnO nanopowders was also studied in order to exploit their possible application in nanoscale optoelectronic devices. The synthesized ZnO nanopowder samples showed a broad UV photoluminescence (PL) emission peak at 3.14 eV.

Keywords Chemical synthesis · Photoluminescence · Nanoparticles · ZnO

S. Bobby Singh · H. Nandakumar Sarma
Department of Physics, Manipur University, Canchipur 795003, India

K. Nomita Devi (✉)
Department of Physics, North Eastern Regional Institute of Science and Technology (NERIST), Nirjuli 791109, Arunachal Pradesh, India
e-mail: nomita_k@rediffmail.com

I. Soibam
National Institute of Technology, Takyel, Imphal 795001 Manipur, India

P. K. Giri et al. (eds.), *Advanced Nanomaterials and Nanotechnology*,
Springer Proceedings in Physics 143, DOI: 10.1007/978-3-642-34216-5_13,

1 Introduction

Synthesis and characterization of zinc oxide (ZnO), a versatile semiconductor material, has been attracting attention because of the commercial demand for optoelectronic devices operating at blue and ultraviolet regions. Although in the past, GaN and GaN-based materials have dominated this wavelength range, ZnO enters the arena with several advantages most important properties being wide band gap of 3.3 eV and large exciton binding energy of 60 meV at room temperature[1–4]. Advantage associated with large band gap includes higher breakdown voltages, ability to sustain large electric field, lower electronic noise and high temperature and high power application. ZnO has many other unique properties such as good optical and mechanical properties, piezoelectric properties and strong room temperature luminescence etc. [5, 6]. Owing to these properties, ZnO is used in numerous applications ranging from antireflection coatings, transparent electrodes in solar cells, thin film gas sensors, varistors, spintronic devices, photodetectors, surface acoustic wave devices and light emitting diodes to nanolasers [7–14].

In order to fabricate ZnO nanoparticles different kinds of synthesis methods have been used. Most of them include traditional high temperature solid state reaction having disadvantage in the particle properties. It is difficult to achieve size selective synthesis of such small particles, by using the traditional approach. As such various chemical method have been studied which include solvothermal, hydrothermal, chemical vapour deposition, radio frequency (RF) sputter, molecular beam epitaxy, metal organic chemical vapour deposition, chemical co-precipitation etc. [15–17]. Among them chemical co precipitation method offers certain advantages such as simplicity, versatility, low cost of the route and low processing temperature. Also the resulting powder was observed to have a finer and more uniform particle size distribution. On the one hand ZnO doped with nickel show enhancement in the optical absorption which indicate that it can be used as an efficient catalyst. Therefore a study of Ni substitution in ZnO leading to various modifications in the properties will be interesting.

In the present work undoped and Ni doped ZnO nanoparticles were synthesized using chemical co-precipitation method and the structural and photoluminescence studies carried out.

2 Experimental

2.1 Synthesis

Nanocrystalline $Zn_{1-x}Ni_xO$ (x = 0.00, 0.02, 0.04, 0.06, 0.08, and 0.1) powder samples were synthesized by a simple, low cost and low temperature chemical co-precipitation method. All the chemicals used were of analytical grade purity.

In a typical synthesis of $Zn_{1-x}Ni_xO$ samples, the appropriate proportion of zinc nitrate $Zn(NO_3)_2{\cdot}6H_2O$) and nickel nitrate ($Ni(NO3)_{2.}6H_2O$) were completely dissolved in distilled water. The aqueous solution so obtained was kept under constant stirring using magnetic stirrer for about half an hour to mix the solution uniformly. After the precursor was dissolved completely, NaOH solution was added drop by drop under constant stirring condition. pH factor of the final solution was maintained at constant value of 11. After these precipitate starts forming. The reaction was further allowed to proceed for about 1 h at a constant temperature of 70 °C. The precipitate was then filtered and washed several times with distilled water. It was dried for 24 h in an oven at 40 °C to obtain the nanopowders. The resulting powders of undoped ZnO were white in colour while nickel doped samples were greenish in colour which became darker with increasing Ni content.

2.2 Characterization

The crystalline structure was determined by X-ray diffraction using X'pert PRO (PANalytical) diffractometer with Cu K_α radiation ($\lambda = 0.15405$ nm) and employing a scanning rate of 5° min^{-1}. Photoluminescence (PL) spectrum of ZnO powders was investigated by Perkin Elmer (LS55) Fluorescence spectrometer. All the measurements were performed at room temperature.

3 Result and Discussion

3.1 Structural Characterization

The crystal structure and phase purity of as prepared undoped ZnO and different composition of Ni-doped ZnO nanopowders were characterized using X-ray diffraction. Figure 1 shows a typical XRD spectrum of pure ZnO and $Zn_{1-x}Ni_xO$ (x = 0.02, 0.04, 0.06, 0.08, 0.1) nanoparticles. Using X'Pert High Score software, the observed XRD pattern can be indexed as hexagonal wurtzite structure of ZnO which are in good agreement with the ICDD Reference Pattern: zinc oxide, 01-070-8070. No phase corresponding to nickel or other nickel compounds was observed in the XRD patterns which implies that Ni atoms replace zinc in the hexagonal lattice. The most prominent peak observed in the undoped film corresponds to (101). Other peaks corresponding to (100), (002), (102), (110), (103), (200), (112) and (201) were observed with varying intensities with (200) being the lowest. The crystallite size *D* is calculated using the Scherrer's formula [18]

$$D = \frac{0.9\lambda}{\beta \cos \theta} \quad (1)$$

Fig. 1 XRD pattern of undoped and Ni doped ZnO nanopowders

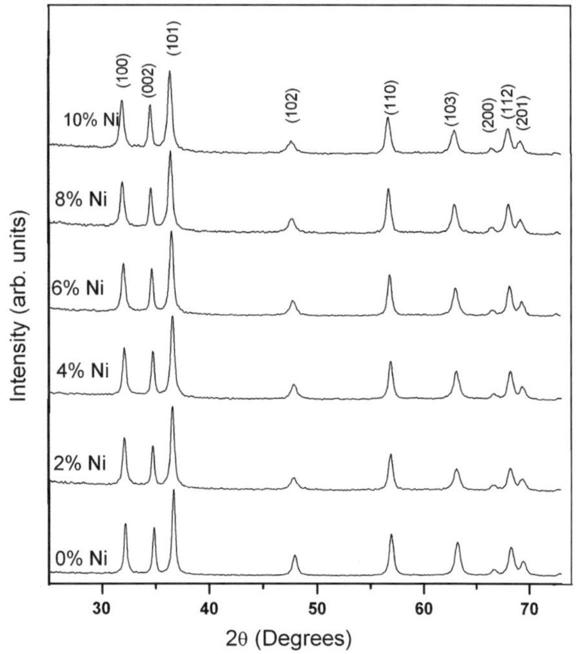

where λ is the wavelength of the incident X-ray ($\lambda = 1.5418$ Å), β is the corrected FWHM for instrumental broadening of the maximum intensity peak and θ is the angle at which the maximum peak occurs. The lattice parameter and crystallite size of ZnO nanoparticles calculated for most prominent peak with Ni dopant concentration is shown in Table 1. It is observed that compared to undoped ZnO, the lattice parameters in the Ni doped nanoparticles decreases. As compared to undoped ZnO, the crystallite size in the Ni-doped samples decreases with increasing Ni content. This is attributed to the replacement of relatively bigger Zn^{2+} ions by the relatively smaller Ni^{2+} ions during the formation of the Ni doped ZnO nanoparticles. As the ionic radius of Ni^{2+} ion (0.069 nm) is smaller than that of Zn^{2+} ion (0.074 nm), it leads to a decrease in the lattice constants, which in turn is responsible for the changes in the crystallite size.

3.2 Photoluminescence Studies

In many cases ZnO shows, under proper excitation conditions, a green photoluminescence band, centered at about 550 nm, and a UV band at about 370 nm. The green band comes from deep defect levels attributed to oxygen vacancies or

Table 1 Structural properties of undoped and Ni doped ZnO nanopowders

Sample	Crystallite size (nm) calculated for the most prominent peak	Lattice parameters (Å)	
		a	c
Undoped (0 wt % Ni) ZnO	29.8	3.257	5.217
2 wt %Ni-doped ZnO	27.9	3.232	5.181
4 wt %Ni-doped ZnO	16.9	3.229	5.179
6 wt %Ni-doped ZnO	16.1	3.228	5.176
8 wt % Ni-doped ZnO	15.5	3.226	5.175
10 wt %Ni-doped ZnO	14.2	3.213	5.156

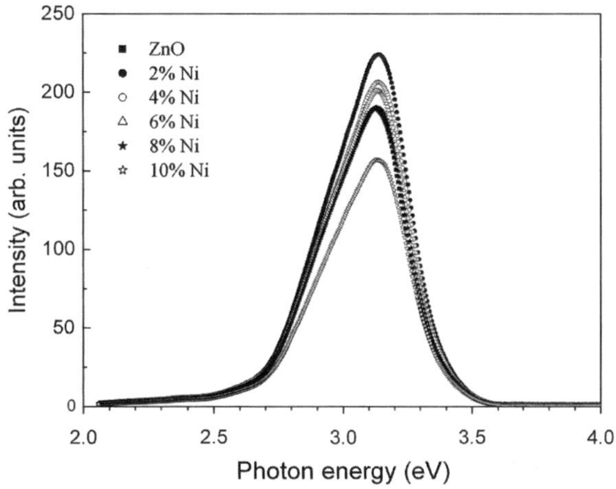

Fig. 2 Room temperature PL spectra of undoped and Ni doped ZnO nanopowders

interstitial zinc ions, whereas the UV band has excitonic nature [19, 20]. The relationship between the two luminescence bands depends strongly on the preparation method and post-preparation treatment.

Figure 2 shows the PL spectra obtained at room temperature for ZnO samples with various Ni doping concentrations over wavelength range 300–700 nm on irradiating at wavelength $\lambda_{ex} = 250$ nm by a xenon lamp. The excitation energy was chosen from photoluminescence excitation (PLE) spectra of the specimens which show an intense peak at 250 nm. The PL spectra of all the samples exhibit one prominent emission peak (−3.14 eV) in the UV position. The PL peak intensity increased from 0 to 2 wt % and then decreases gradually with increase in Ni concentration and the position of UV emission peak is slightly blue shifted (3.12 to 3.14 eV) with increase of Ni doping. It is to be noted that, the band emission is not related to the defects or impurities in the samples and is due to excitonic emission indicating the high crystal quality of the material. No luminescence was found in the visible region. In particular the typical broad green emission appearing for wavelengths larger than 500 nm is completely absent.

4 Conclusion

Pure and Ni doped ZnO nanopowders were synthesized by chemical co precipitation method. XRD studies confirmed the hexagonal wurtzite structure of the samples. The lattice parameters and crystallite size decreases with the increase in Ni concentration. Prominent UV peak ascribed to the free exciton emission was observed in the PL spectra of all the samples. No visible emission was observed.

References

1. D.C. Look, Mat. Sci. Eng. B **80**, 383 (2001)
2. Z.K. Yang, P. Yu, G.K.L. Wong, M. Kawasaki, A. Ohtomo, H. Koinuma, Y. Segawa, Solid State Comm. **103**, 459 (1997)
3. D.M. Bagnall, Y.F. Chen, Z. Zhu, T. Yao, S. Koyama, M.Y. Shen, T. Goto, Appl. Phys. Lett. **70**, 2230 (1997)
4. A. Ohtomo, M. Kawasaki, I. Ohkubo, H. Koinuma, Z.K. Tang, P. Yu, G.K.L. Wong, B.P. Zhang, Y. Segawa, Sci. Eng. B **56**, 239 (1998)
5. N. Kumar, R. Kaur, R.M. Mehra, J. Lumin. **12692**, 784 (2007)
6. W.J. Liang, L. Kuan, Q. Xiu-juan, S. Guang-jie, J. Harbin, Inst. Tech. **36**(2), 226 (2004)
7. O. Kluth, G.G. Schope, J. Hupkes, C. Agashe, J. Muller, B. Rech, Thin Solid Films **442**, 80 (2003)
8. O. Lupan, S. Shishiyanu, L. Chow, T. Shishiyanu, Thin Solid Films **516**, 3338 (2008)
9. E. Suvaci, I.O. Ozer, J. Eur. Ceram. Soc. **25**, 1663 (2005)
10. Y. Zheng, J.C. Boulliard, D. Demaille, Y. Bernard, J.F. Petroff, J. Cryst. Growth **274**, 156 (2005)
11. S. Liang, H. Sheng, Y. Liu, Z. Huo, Y. Lu, H. Shen, J. Cryst. Growth **225**, 110 (2001)
12. W.C. Shin, M.S. Wu, J. Cryst. Growth **137**, 319 (1994)
13. X. Xu, W. Sun, Appl. Phys. Lett. **83**, 3806 (2003)
14. H.M. Huang, S. Mao, H. Feik, H. Yan, Y. Wu, H. Kind, E. Weber, R. Russo, P.D. Yang, Science **292**, 1987 (2001)
15. Z. Hui, Y. Deren, M. Xiangyang, J. Yujie, X. Xin, Nanotechnology **15**, 622 (2004)
16. J. Zhang, L.D. Sun, J.L. Yin, Chem. Mater. **14**, 4172 (2002)
17. W.J. Li, E.W. Shi, Z.W. Yin, J. Mater. Sci. Lett. **20**, 1381 (2001)
18. B.D. Cullity, *Elements of X-ray Diffraction*, 2nd edn. (Addison Wisley, Reading, 1978), p. 1866
19. A. Van Dijken, E.A. Meulenkamp, D. Vanmaekelberg, A. Meijerink, J. Lumin. **87–89**, 454 (2000)
20. A. Van Dijken, E.A. Meulenkamp, D. Vanmaekelberg, A. Meijerink, J. Lumin. **90**, 23 (2000)

Synthesis and Characterization of Linoleic Acid Capped Palladium Nanoparticles

Ratan Das, Siddhartha S. Nath and Ramendu Bhattacharjee

Abstract Many chemical reduction methods have been used to synthesize palladium nanoparticles from palladium salts. Here palladium nanoparticles have been synthesized through the chemical reduction of palladium ions by ethanol using linoleic acid as a protecting agent, which is then dispersed in chloroform to form homogeneous colloidal solution. Transmission electron microscope (TEM) reveal that the particles are mostly spherical in shape with an average size of nearly 6 nm. Fourier Transform infrared spectroscopy confirms the capping of these nanoparticles by linoleic acid. The UV/Vis absorption spectra show absorption peak at 390 nm which is due to surface plasmon resonance (SPR).

Keywords Linoleic acid · Surface plasmon resonance · Absorption spectra

1 Introduction

Nanoparticles (1–100 nm) have attracted great interest The unusual physical and chemical properties exhibited by nanomaterials are quite different from those of their bulk counterparts [1–5]. Noble metal nanoparticle synthesis and study of their size and properties are of technological importance because optical, electronic, and catalytic properties depend on their size, shape and chemical surroundings [1–5]. There is presently intense interest in the synthesis of stable gold nanoparticles and in their emission and absorption spectra. Here we have prepared palladium

R. Das (✉)
Department of Physics, Tripura University, Suryamaninagar, Tripura 799022, India
e-mail: dasratan22@gmail.com

S. S. Nath · R. Bhattacharjee
Department of Physics, Assam University, Silchar 788011 Assam, India

P. K. Giri et al. (eds.), *Advanced Nanomaterials and Nanotechnology*,
Springer Proceedings in Physics 143, DOI: 10.1007/978-3-642-34216-5_14,

nanoparticles by reduction method [6] which is a simple and convenient process. In preparing these nanoparticles, we have employed linoleic acid as the protective agent to control the formation of the palladium colloids and to stabilize them. The prepared palladium nanoparticles have been dispersed in chloroform and then have been examined using Transmission Electron Microscopy (TEM), Fourier transform infrared (FTIR) spectroscopy, and UV/Vis absorption spectroscopy (UV–Vis).

2 Synthesis Process

Palladium nanoparticles can be obtained through the reduction of palladium ions by ethanol at temperature between 70 and 100 °C in presence of linoleic acid and sodium linoleate [6]. In this synthesis process, 20 ml of aqueous solution containing palladium chloride (0.4 g of $PdCl_2$), 1.5 g sodium linoleate (C_{18} $H_{32}ONa$), 10 ml ethanol (C_2H_5OH) and 2 ml linoleic acid (C_{18} H_{32} O_2) are added in a capped tube under constant stirring. The system is kept at the temperatures between 70 and 100 °C for 2 h. In the aqueous solution of palladium chloride, sodium linoleate and the mixture of linoleic acid and ethanol are added in order. The ethanol in the liquid and solution phases reduced palladium ions into palladium nanoparticles. The linoleic acid caps the palladium nanoparticles along with the reduction process. The product, collected at the bottom of vessel after cooling to room temperature, is dispersed in chloroform to form a homogenous colloidal solution. Palladium nanoparticles dispersed in chloroform show dark brown colour.

3 Results and Discussion

3.1 TEM Image Analysis

A drop of palladium nanoparticles colloidal solution is taken on a copper grid and when evaporation takes place, transmission electron microscope (TEM) image is recorded. TEM image of the prepared colloidal solution of palladium nanoparticles is shown in the Fig. 1. The Pd nanoparticles (black portion) are spherical in shape with a smooth surface morphology. The average diameter (size) of the nano particles is found to be approximately 6 nm.

3.2 Analysis of FTIR Spectroscopy

Capping of Pd nanoparticles by linoleic acid has been examined by FTIR spectroscopy. The FT-IR absorption spectra of the samples are shown in Fig. 2 with resolution of 4 cm^{-1}, which was performed in Spectrum BX series. The peak

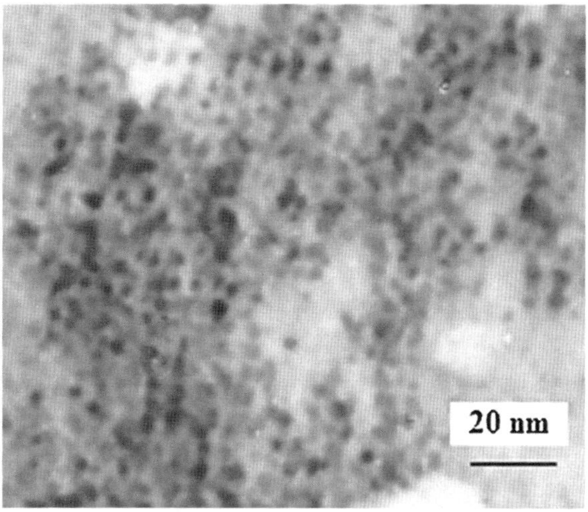

Fig. 1 TEM image of palladium nanoparticles

around 3,014 cm^{-1} is due to C = C stretching mode. The lack of broad peak due to OH stretching of the free ligand in the range 3,000 cm^{-1}–3,100 cm^{-1} is due to the chemisorptions of linoleic acid on the Pd nanoparticles.

3.3 UV/V is Spectroscopy of Pd Nanoparticles

The UV/Vis absorption spectra of the palladium nanoparticles dispersed in chloroform show an absorption peak at 390 nm as shown in the Fig. 3. This strong absorption band in the visible range is due to surface plasmon resonance (SPR).

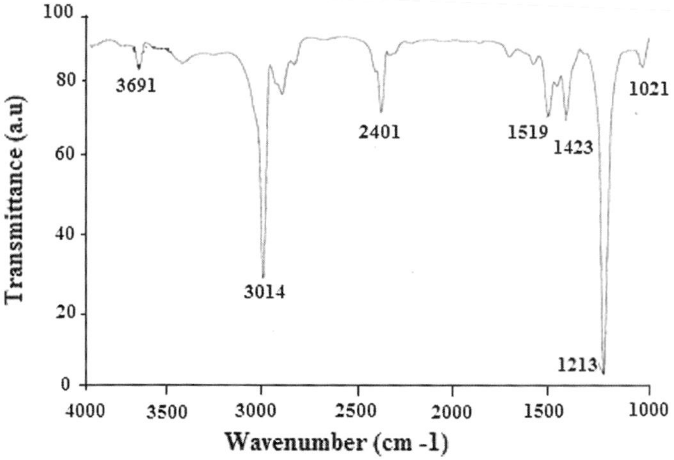

Fig. 2 FT-IR spectra of linoleic acid capped Pd nanoparticles dispersed in chloroform

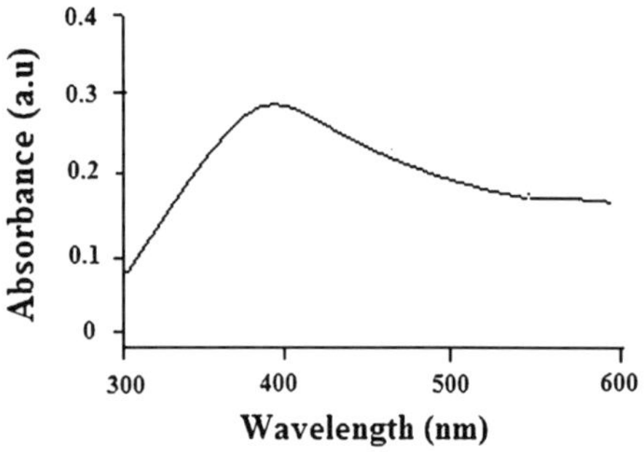

Fig. 3 UV- Vis spectra of Pd nanoparticles

Surface plasmon resonance is the coherent motion of free electrons in the conduction band caused by interaction with electromagnetic wave [1–5, 7, 8]. The electric field of an incoming light wave induces a polarization of electrons with respect to the heavier ionic core of the nanoparticles, which in turn induces a dipolar oscillation of all the free electrons with the same phase. When the frequency of the light wave become resonant with the electron motion, a strong absorption occurs which is the origin of observed colour of colloids. Here Pd nanoparticles dispersed in chloroform show dark brown colour.

4 Conclusion

Palladium nanoparticles have been prepared through the reduction method. Here linoleic acid acts as a stabilizer. UV/Vis spectra, and TEM image reveal that the prepared nanoparticles are spherical in shape with average size of 6 nm. SPR peak is observed at 390 nm.

Acknowledgments Dr. B. DKHR (S.O.) NEHU, Shillong, India, and Dr. Sudip Choudhury, Lecturer, G.C. College, Silchar, Assam, India for their suggestions and assistance during the work.

References

1. C.P. Poole, F.J. Owens (eds.), *In Introduction to Nanotechnology* (Wiley, New Jersey, 2003)
2. G. Cao (ed.), *In Nanostructures and Nanomaterials* (Imperial College Press, London, 2004)
3. P.M. Ajayan, L.S. Schadler, P.V. Braun (eds.), *Nanocomposite Science and Technology* (Wiley, Weinheim, 2003)
4. M. Dragoman, D. Dragoman (eds.), *Nanoelectronics Principle and Devices* (Artech House, London, 2005)
5. C. Sonnichsen, T. Franzl, T. Wilk, G. von Plessen, J. Feldmann, New J. Phys. **4**, 93.1–93.8 (2002)
6. X. Wang, J. Zhuang, Q. Peng, Y. Li, Nature **431**, 03968 (2005)
7. S.K. Ghosh, T. Pal, Chem. Rev. **107**, 4797–4862 (2007)
8. S. Link, M.A. El-Sayed, Int. Rev. Phys. Chem. **19**(3), 409–453 (2000)

Preparation and Gas Sensing Properties of Nanostructured $ZnSnO_3$ Thin Films

I. G. Pathan, D. N. Suryawanshi, A. R. Bari, D. S. Rane and L. A. Patil

Abstract Transparent conducting $ZnO:SnO_2$ and perovskite nanostructured $ZnSnO_3$ thin films have been deposited onto preheated glass substrate by a spray pyrolysis technique using water soluble Zinc chloride and Stannic chloride as precursors. A fine spray of the source solution using air as a carrier gas has grown highly textured thin films. The concentration of Zinc chloride and Stannic chloride precursors are kept fixed and their ratio were varied as: 50:50 (wt %), 40:60 (wt %) and 30:70 (wt %) respectively. The thin film samples with Zinc chloride and Stannic chloride with compositions of: 50:50 (wt %), 40:60 (wt %) and 30:70 (wt %)were referred as samples S1, S2 and S3 respectively. Thin films S1 and S2 were found to be the composite of $ZnO:SnO_2$. S3 films were found to be the perovskite nanostructured thin films of $ZnSnO_3$. The crystallinity of the prepared samples were analyzed by X-ray diffraction spectroscopy. The X-ray diffractogram of thin film S3 clearly indicated the stoichiometric and perovskite nanostructured nature of $ZnSnO_3$. The band gap energy of S1, S2 and S3 samples was observed to be 4.00, 4.06 and 4.13 eV respectively. We tested gas sensing performance of the samples S1, S2 and S3. H_2 gas response of $ZnSnO_3$ S3 sample is presented. The perovskite oxides were used as potential gas sensing material for their stability in thermal and chemical atmospheres.

Keywords Transparent conducting oxides · perovskite thin films · Spray pyrolysis · H_2 sensor

---

I. G. Pathan
Arts, Commerce and Science College, Navapur 425418, India

D. N. Suryawanshi
Rani Laxmibai College, Parola 425111, India

A. R. Bari · D. S. Rane · L. A. Patil (✉)
Nanomaterials Research Lab, Pratap College, Amalner 425001, India
e-mail: Plalchand_phy_aml@yahoo.co.in

P. K. Giri et al. (eds.), *Advanced Nanomaterials and Nanotechnology*,
Springer Proceedings in Physics 143, DOI: 10.1007/978-3-642-34216-5_15,

1 Introduction

Due to finite small size and the high surface-to volume ratio, nanostructred materials often exhibit novel and sometimes unusual properties [1–6]. The optical, electrical, thermal and chemical properties depend to a large extent on the particle size and shape of these materials [4–6]. The large number of surface and edge atoms provides active sites for catalyzing surface reactions. In the area of chemical and gas sensitive semiconductor devices and nanoelectronics, ZnO, SnO_2 and their composites are finding great interest and attention in recent years [4–10]. Zinc oxide (ZnO) is an n-type wide band gap (3.20–3.43 eV at 298 K) semiconductor with a large exciton energy of 60 meV and a wurtzite structure [10, 11]. Tin dioxide (SnO_2) is an n-type semi conductor of rutile crystal structure with a wide band gap (3.6–3.97 eV) and high power gas detecting/sensing capabilities at low operating temperatures [12, 13]. Zinc and tin oxides have recently attracted considerable attention because they exhibit interesting technological properties. Transparent conducting oxides such as zinc stannate (ZTO) in the phase space SnO_2–ZnO, whereas $ZnSnO_3$ is a perovskite-type oxide material. The reported band gap energy of this compound oxide is 3.4–3.6 eV. The band gap values depend on the type of material, single crystal or bulk, method of preparation, size and shape. The data on the synthesis of $ZnSnO_3$ are ambiguous and contradictory [14], and among the large studies of materials, the details on ternary oxide systems with spinel or perovskite structure have been rarely published [13, 14].

Perovskite materials display a wide range of properties that make them attractive for a variety of electrical ceramic applications. The structure is named after the naturally occurring mineral perovskite ($CaTiO_3$) [15, 16] and is usually depicted in pseudocubic form. Figure 1 shows crystal structure of $ZnSnO_3$ perovskite. The structure contains two cation sites in the crystal lattice: the larger cation (Sn) resides on the corners of the unit cell, and the smaller cation (Zn) is in the center of the unit cell. The oxygen ions (O) are on the centers of the faces, and the structure is formed via a network of corner linked oxygen octahedra, with the larger cation (Sn) filling the dodecahedral holes and the smaller cation (Zn) filling the octahedral sites. Interestingly, and of technological importance, a variety of compositions crystallizes in the perovskite (ABO_3) structure.

Recently, one-dimensional (1D) nano-/microstructures of semiconductive oxides have been extensively studied owing to their practical prospects and theoretical significance. Many methods, including ultrasonic spray pyrolysis [17], thermal evaporation deposition [18–20], chemical vapor deposition [21], hydrothermal synthesis [22–25] and others [26, 27], have been used to produce ZnO, SnO_2 nanoparticles. In this study, ZnO–SnO_2 nano composite and nanostuctured $ZnSnO_3$ thin films were synthesized by a simple Spray pyrolysis method. Mixing a water soluble Zinc Chloride with Stannic chloride in various proportion were tends to composite of ZnO–SnO_2 and perovskite of $ZnSnO_3$ thin films. The gas sensing properties of metal oxides are related to the material surface and the gases are always adsorbed and react with the surface [28]. Therefore, in comparison with

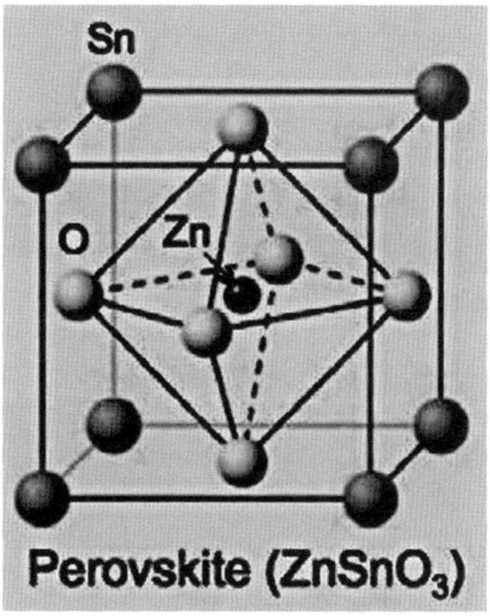

Fig. 1 Crystal structure for the $ZnSnO_3$ perovskite

conventional sintered bulk gas sensors, thin film gas sensing materials have good sensitivity and selectivity [29, 30]. Furthermore, development of gas sensors to monitor combustible gases is imperative due to the concern for safety requirements in homes and for industries, for detection of hazardous gases.

2 Experimental

Transparent conducting stoichiometric nanocrystalline $ZnSnO_3$ thin films and non stoichiometric nanocrystalline composite ZnO–SnO_2 thin films were prepared from aqueous solution of zinc chloride ($ZnCl_4 \cdot 5H_2O$) and stannic chloride ($SnCl_4 \cdot 5H_2O$) (Loba chem extra pure) dissolved in double distilled water to a concentration of 0.1 M for the preparation of thin films. The stock solution was delivered to nozzle with constant and uniform flow rate of 5 ml/minute using air as a carrier gas. The spray (mist) produced by nozzle was sprayed onto the glass substrates heated at 400 $\pm$ 5 °C. Various parameters such as nozzle-to-substrate distance, deposition time, flow rate of solution, deposition temperature and concentration of source solution were optimized to obtain highly textured thin films of good quality. The concentration of Zinc chloride and Stannic chloride precursors are kept fixed and the ratio of Zinc chloride and Stannic chloride solution were varied as: 50:50, 40:60 and 30:70 respectively. The thin film samples with compositions of: 50:50, 40:60 and 30:70 were referred as S1, S2 and S3 respectively. The samples were annealed at 500 °C for 1 h. Thin films S1 and S2 were found to be the composite of $ZnO:SnO_2$. Thin films S3 were found to be the perovskite nanostructured thin films of $ZnSnO_3$.

3 Material Characterizations

As prepared films were characterized by X-ray diffractometer (Philips PW 1730) using CuKα ($\lambda = 1.5418$ Å) radiation. The surface morphology of the thin films was studied using scanning electron microscope (model JEOL JSM –6,360 Å). The quantitative elemental analysis of the films was estimated by computer controlled energy dispersive X-ray analyzer attached to the scanning electron microscope. Absorption spectroscopy is used to determine the band gap energy of the samples. Electrical and gas sensing properties were measured using a static gas sensing system. The sensor performance on exposure of LPG, carbon dioxide, hydrogen, ammonia, ethanol and chlorine was tested.

3.1 Crystal Structure using X-Ray Diffraction

Figure 2 shows the X-ray diffractogram of as grown transparent composite thin film samples S1, S2 and perovskite sample S3. The diffraction peaks from various planes and d-values are matching well with standard ASTM data for ZnO (JCPD#41-1445), SnO_2 (JCPD#05-0664) and $ZnSnO_3$ (JCPD#28-1486). It reveals from XRD that the films are polycrystalline in nature. The $ZnSnO_3$ particles had a (100) face-exposed hexahedron structure and the composite ZnO–SnO_2 particles had a (092) face-exposed octahedron structure. The average particle sizes for the ZnO–SnO_2 composite and $ZnSnO_3$ particles were 23 and 20 nm respectively.

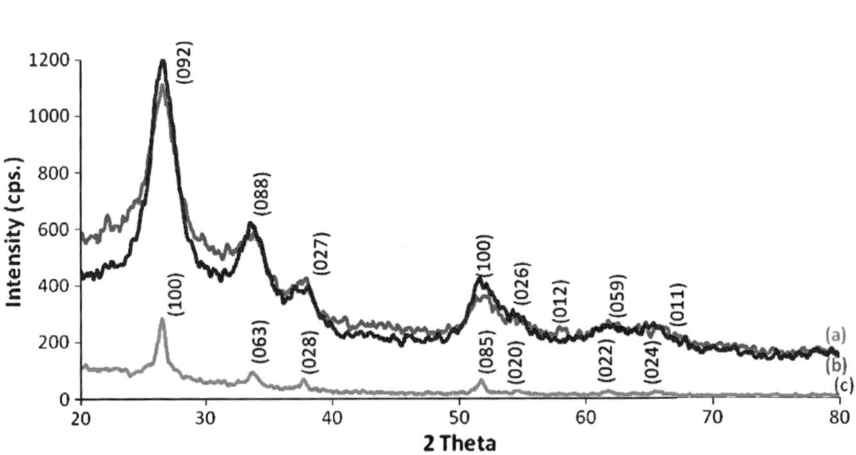

Fig. 2 X-ray diffractogram of nanocrystalline ZnO thin films: **a** sample S1, **b** sample S2 and **c** sample S3

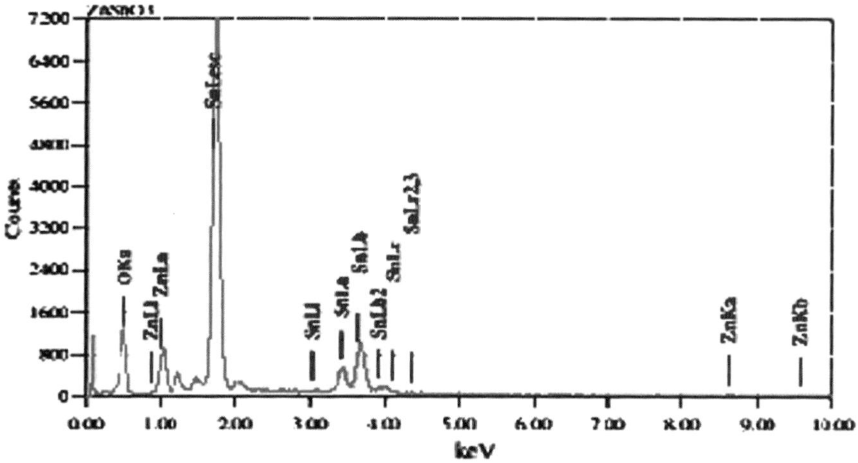

Fig. 3 EDAX Analysis of Perovskite $ZnSnO_3$ thin films sample S3

3.2 Elemental Composition

It is clear from the table that the composite thin films are zinc as well as tin rich and oxygen deficient and hence are non stoichiometric in nature. For the samples S2 and S3, it is found that with increase of $ZnCl_4$, the wt. % of Sn goes on decreasing. The percentage of Sn is larger while percentage of Zn is smaller in sprayed S1, S2 and S3 samples. The EDAX analysis for $ZnSnO_3$ nanostructured thin films S3 obtained by JEOL-2300 is shown in Fig. 3 (Table 1).

3.3 Surface Morphology

Figure 4(a) and (b) depict SEM images of composite nanostructured S1 and S2 thin films samples. Figure 4(c) represents SEM image of nanostructured perovskite $ZnSnO_3$ thin films sample S3. Figure 4(a) and (b) shows randomly distributed flat plate-like grains with shape distribution. Figure 4(c) depicts the microstructure

Table 1 The composition of composite and perovskite thin films S1, S2 and S3

Sample No.	$SnCl_4{\cdot}5H_2O$ (ml)	Addition of $ZnCl_4{\cdot}5H_2O$ (ml)	Mass %					
			SnO_2	Sn	O	O	Zn	ZnO
–	0	100	0	0	0	48.30	51.70	100
–	100	0	100	51.70	48.30	0	0	0
S1	50	50	51.70	26.25	25.45	22.4	25.9	48.30
S2	60	40	61.70	31.53	30.37	17.75	20.55	38.30
S3	70	30	–	19.68	61.76		18.56	–

of sample S3 consisting of smaller spherical grains of $ZnSnO_3$. The spherical grains may be due to large number of Zn-nucleation centers. The sample S3 is expected to be more porous, giving larger effective area (for the gas to react and in turn higher sensitivity may be expected). Image in Fig. 4(b) shows relatively larger spherical and elliptical shaped grains as compared to Fig. 4(a). Grain size is becoming more and more spherical, and numbers of grains are increasing with the increase of ZnO percentage in the composition as evidenced from Fig. 4(c). Atomic radius of Zn^{+2} is smaller than Sn^{+2}. Larger the number of Zn^{+2} ions, larger would be the nucleation centers. The feed would therefore, be divided and this would result into larger number of grains with smaller size and spherical shape.

3.4 Absorbance Spectra

Figure 5 shows the absorbance spectra of nanostructured thin films for various Sn and Zn compositions and perovskite $ZnSnO_3$ S3 thin films synthesized on glass substrates. The absorbance of perovskite $ZnSnO_3$ thin film sample (S3) is larger as compared to the composites samples (S2 and S3).The band gap energy calculated form absorbance spectra are 4.00, 4.06 and 4.13 eV of samples S1, S2 and S3 respectively. Reported band gap of $ZnSnO_3$ is 3.6 eV. The band gap of $ZnSnO_3$ was observed to be enhanced (4.13 eV) as compared to the reported band gap (3.6 eV). It may be due to nanocrystalline nature of the film.

3.5 Gas Sensing Characteristics

Gas response, selectivity, response time and recovery time are the important characteristics in gas sensing. Gas response is defined as the ratio of change in conductance of a sample on exposure to a test gas to the conductance in air. The gas response can be written as

$$S = \frac{Gg - Ga}{Gg} \quad \frac{\Delta G}{Gg}$$

where Gg and Ga are the conductance in the presence of test gas and in air respectively. Selectivity can be defined as the ability of a sensor to respond to a certain gas in the presence of other gases. Response time is defined as the time needed for a sensor to attain 80 % of maximum change in conductance upon the exposure to a test gas, while recovery time as the time taken by a sensor to get back 80 % of the original conductance in air [31].

Fig. 4 SEM images of samples: **a** S1, **b** S2 and **c** S3

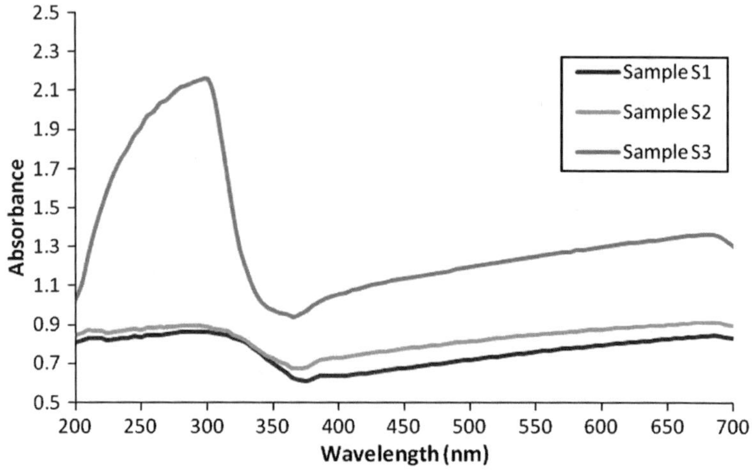

Fig. 5 Absorbance spectra of thin film samples S1, S2 and S3

3.5.1 Gas Response with Operating Temperature

Composite Films (Sample S1 and S2)

Figures 6 and 7 depict the variation of gas response with operating temperature (°C) of S1 and S2 samples for LPG, H_2, CO_2, C_2H_5OH, NH_3 and Cl_2 gases. It is clear from figure that the nature of gas responses is similar for all gases. The response of C_2H_5OH gas goes on increasing with increasing the operating temperature reaches to maximum at 350 °C and then decreases with increasing the operating temperature. More the oxygen ions adsorbed, more electrons would be

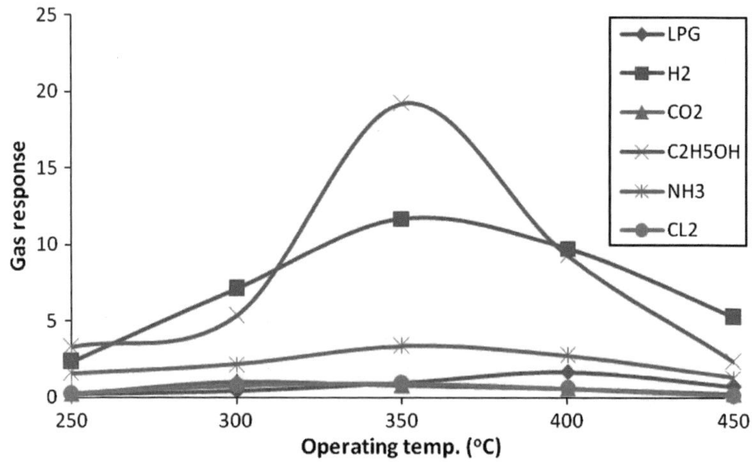

Fig. 6 Gas response of composite thin film S1 sensor

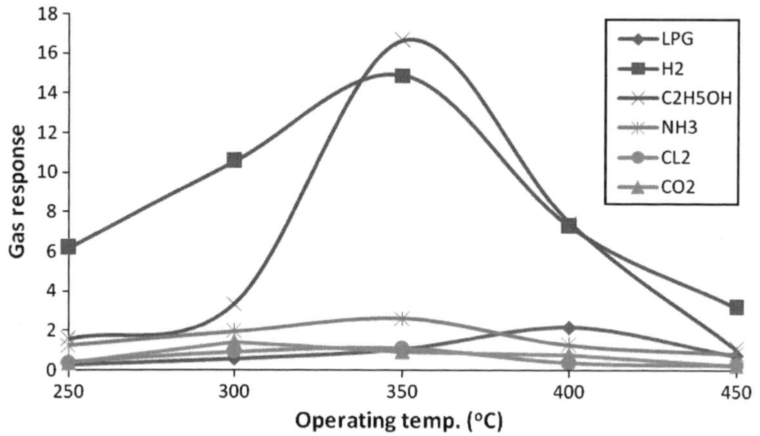

Fig. 7 Gas response of composite thin film S2 sensor

removed from the surface. This would make the film more resistive before exposure of C_2H_5OH gas. On exposure of C_2H_5OH gas, it would be oxidized and all the abstracted electrons would be returned back to material giving high conductivity. Smaller conductivity before exposure and larger conductivity after exposure of C_2H_5OH gas would give higher gas response. The sensitivity (gas response) of sensor was observed largest to ethanol gas for sample S1 (19.23) and sample S2 (16.67) against the other gases. The sensor also showed remarkable response to H_2 for sample S1 (11.71) and for sample S2 (14.87).

Perovskite ZnSnO3 Thin Film (Sample S3)

Figure 8 depict the variation of gas response with operating temperature of most sensitive nanostructured perovskite $ZnSnO_3$, base sensor for all the gases under test. It is clear from figure that the nature of gas responses is similar for all gases. The response of H_2 gas goes on increasing with increasing operating temperature, reaches to maximum (112.93) at 350 °C and decreases with the further increasing the operating temperature. Nanostructures materials are attractive as sensing elements owing to their large surface area to volume ratio allowing higher adsorption of gas molecules. H_2 is reducing gas which decreases the device resistance. It is unanimous that H_2 molecules react with O^- adsorbates on the semiconductor crystals. It is assumed that H_2 molecules collide directly with O^- ions (Eley-Rideal mechanism) and H_2O molecules formed are desorbed instantly from the surface. It produces a reversible change in the resistivity with the exchange of charges between H_2 and the $ZnSnO_3$ surface leading to changes in the depletion length [32, 33]. Initially the molecular oxygen are adsorbed on the surface of $ZnSnO_3$ nanostructure and electrons are consumed thus increases resistance of the film. When hydrogen is exposed on nanostructured $ZnSnO_3$, its atoms react with these

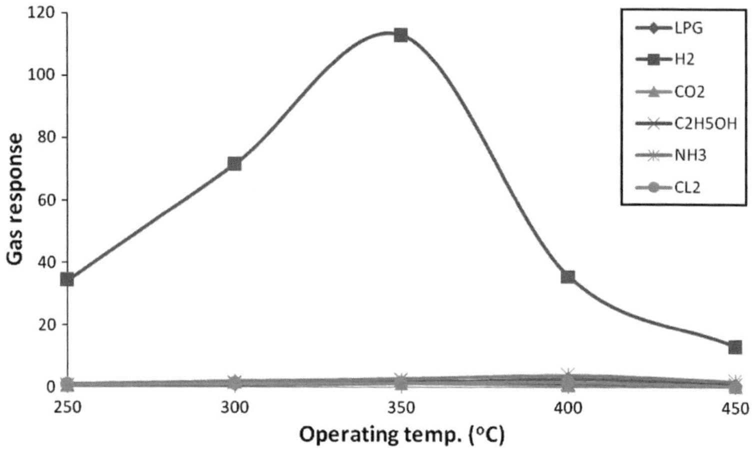

Fig. 8 Response of sensor S3 (Perovskite $ZnSnO_3$)

chemisorbed oxygen ions and produce H_2O molecules consuming chemisorbed oxygen from the surface by releasing electrons. As a result electrons will be released back to the conduction band and will contribute to current increase through the $ZnSnO_3$ thin film. This also results in a reduction of surface depletion region and increase conductivity.

Ethanol Response to Sample S1, S2 and S3

Figure 9 represents the superimposition of sensitivity values of C_2H_5OH gas of the sensor as a function of temperature. Comparing the sensitivity values of the sensors S1, S2 and S3, it is clear that sensor S1 shows the maximum gas response to ethanol and S3 shows the minimum gas response, where as the gas response value of sensor S2 lies between S1 and S3 at 350 °C. The response shows that maximum sensitivity was given by sample S1, on the other hand, the response falls as the percentage of Zn was increased in Sn in the S1, S2 and S3 samples.

Hydrogen Response to S1, S2 and S3

Response of sensors S1, S2 and S3 for Hydrogen gas is given in Fig. 10. The response goes on increasing with increasing operating temperature, reaches to maximum (112.93) at 350 °C and decreases with the further increase of operating temperature. It is clear from the figure that max. response to H_2 was given by perovskite $ZnSnO_3$. The Sensitivity was found to be increase for S1 and S2 and it was maximum for ZnSnO3.

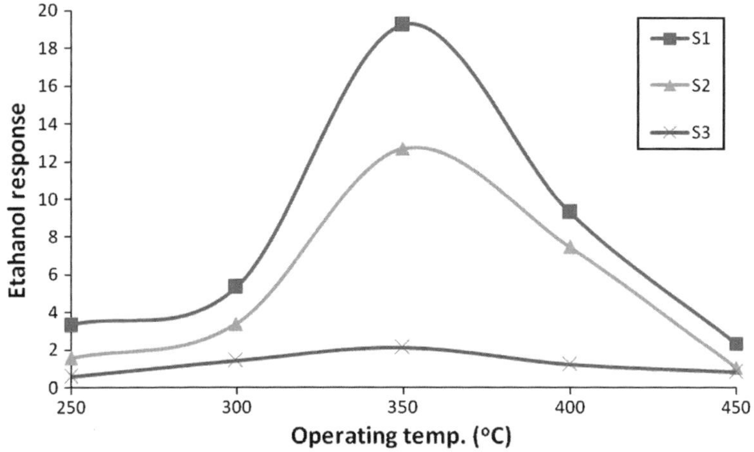

Fig. 9 Sensitivity of ZnO doped SnO_2 thin films to C_2H_5OH as a function of temperature

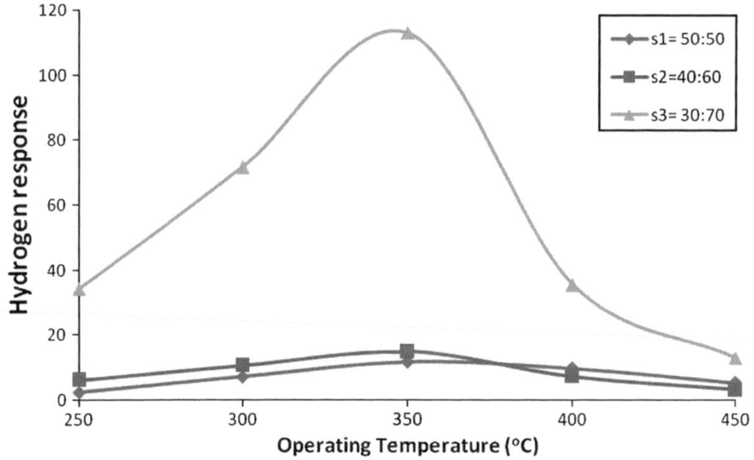

Fig. 10 Sensitivity of Sensors S1, S2 and S3 to H_2 as a function of temperature

3.5.2 Selectivity

Figure 11 depict the selectivity of H_2 gas in comparison with all other gases tested at 350 °C. It is clear from histogram that gas response to H_2 gas goes on increasing with increase in ZnO doping level in SnO_2.

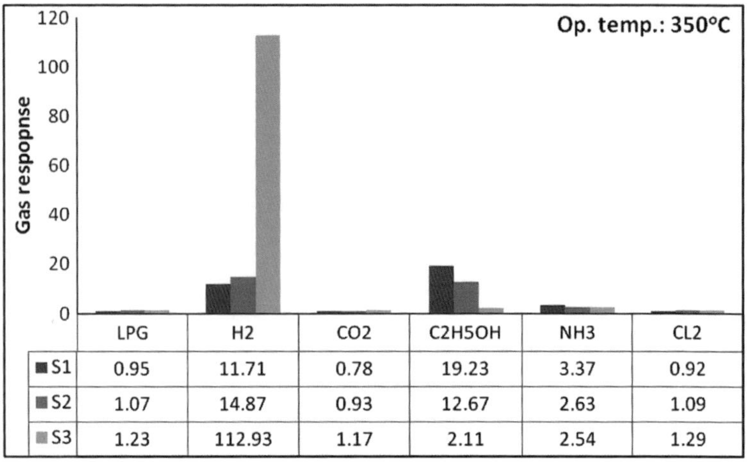

Fig. 11 Histogram comparing sensitivities of various gases

3.6 Discussion

The preferable operating temperature for semiconductor gas sensors ranges between 200 °C and 500 °C. At temperatures below 150 °C, oxygen is adsorbed in molecular form (neutral or charged). When the temperature is increased, the adsorbed oxygen molecule dissociates into atomic oxygen (neutral, singly or doubly charged), and when the temperature is too high, oxygen atoms desorbs from the sensor. Exposure to reducing gases changes the density of ionosorbed oxygen that alters the sensor conductance. The conductance of n-type oxides increases in the presence of reducing gases oxidizing gases and decreases in the presence of oxidizing gases reducing gases, and the opposite behavior is observed for p-type metal oxides.

Sunandan Baruah et al. [34] reported that gas sensors based on n-type semiconductors like ZnO and SnO_2 have been studied for the detection of inflammable or toxic gases such as H_2, CH_4, C_2H_5OH, CO or LPG [35–42]. H_2 is reducing gas which decreases the device resistance, its atoms react with these chemisorbed oxygen ions and produce H_2O molecules consuming chemisorbed oxygen from the surface by releasing electrons. The sensing mechanism for H_2 at room temperature can be represented by the following relation:

$$O^- + H_2 = H_2O + e$$

As a result electrons will be released back to the conduction band and will contribute to current increase through the $ZnSnO_3$ thin film. This also results in a reduction of surface depletion region and increase conductivity. The reaction is exothermic in nature [43] and the molecular water desorbs quickly from the surface. Due to the fact that water forming reaction is sophisticated and mechanism is not well defined, therefore the H_2 sensing mechanism still needs more understanding.

The sensing mechanism of ethanol is closely related to its decomposition and/or oxidation reaction. Jinkawa et al. [44] reported that the decomposition of ethanol at elevated temperatures depended on the acid–base properties of the oxide catalyst used.

$$C_2H_5OH(g) \rightarrow CH_3CHO(g) + H_2(g) \quad \text{(basic oxide)}$$

$$C_2H_5OH(g) \rightarrow C_2H_4(g) + H_2O(g) \quad \text{(acidic oxide)}$$

The higher gas response of the basic-oxide-added SnO_2 shows that the oxidation of CH_3CHO and H_2 with the negatively charged surface oxygen (Osurf −) induces a larger increase in conductance. The electro negativity of Zn^{2+} (1.7) is slightly smaller than that of Sn^{4+} (2.0). However, the addition of ZnO to SnO_2 decreased the response to ethanol, and it was clear from figure [10]. This suggests that in the gas sensing, the decrease in surface area with increasing the ZnO content has a larger effect than the change in the acid–base properties with increasing the temperature to 350 °C of the sensor can be explained by the rapid kinetics of ethanol oxidation at elevated temperatures. However, the detection of ethanol in the composite of SnO2 and ZnO at 350 °C cannot be explained by the change in the acid–base properties or by the similar intermediate phase. Although more study will be needed to determine the detailed sensing mechanism, the hetero junction between SnO_2 and ZnO might be a reason.

4 Conclusions

1. Nanostructured composite thin films of ZnO: SnO_2 and perovskite $ZnSnO_3$ thin films were successfully prepared by simple spray pyrolysis technique.
2. The structural and morphological properties of samples showed the nanostructures in S1, S2 and S3 samples.
3. The band gap energy of S1, S2 and S3 samples was observed to be 4.00, 4.06 and 4.13 eV. The band gap of $ZnSnO_3$ was observed to be enhanced (4.13 eV) as compared to the reported band gap (3.6 eV). It may be due to nanocrystalline nature of the film.
4. The composite samples S1 and S2 showed higher gas response to ethanol gas.
5. The ethanol response of S1, S2 and S3 samples was found to decrease with increase of Zn wt % in the film composition.
6. The perovskite S3 samples showed higher gas response to H_2 gas as compared to other gases.
7. The H_2 response of S1, S2 and S3 samples was found to increase with increase of Zn wt % in the film composition.

Acknowledgement One of the authors IGP is very much thankful to UGC (Western Regional Office, Pune) for providing financial support through scheme no. 47-659/08 (WRO). The Principal, Pratap College, Amalner, and Principal, Arts, Commerce and Science College, Navapur (Maharashtra, India) are acknowledged for providing laboratory facilities.

References

1. L.M. Liz-Marzan, P.V. Kamat, *Nanoscale Materials* (Springer, Berlin, 2003). (Kluwer Academic Publishers, Boston)
2. A.S. Edelstein, R.C. Cammarata (eds.), *Nanomaterials: Synthesis, Properties and Applications* (Institute of Physics, Bristol, 1996)
3. G.M. Chow, K.E. Gonsalves (eds.), *Nanotechnology*: *Molecularly Designed Materials*. ACS Symposium Series 622, Washington (1996)
4. G.C. Hadijipanyis, R.W. Siegel, *Nanophase Materials: Synthesis, Properties, Applications* (Kluwer Academic Publications, London, 1994)
5. Z.L. Wang, Nanowires and Nanobelts—Materials, Properties and Devices, vol. 1. Metal and Semiconductor Nanowires. (Kluwer Academic Publisher, Dordrecht, 2003)
6. Z.L. Wang, Nanowires and Nanobelts—Materials, Properties and Devices, vol. 2. Metal and Semiconductor Nanowires. (Kluwer Academic Publisher, Dordrecht, 2003)
7. J.H. Sung, Y.S. Lee, J.W. Lim, Y.H. Hong, D.D. Lee, Sens. Actuators B **66**, 149 (2000)
8. H. Giefers, F. Porsch, G. Wortmann, Solid State Ionics **167**, 199 (2005)
9. O.A. Fouad, J. Nanosci. Nanotechnol. **6**, 2090 (2006)
10. J. Jie, G. Wang, X. Han, J.G. Hou, J. Phys. Chem. B **108**, 17027 (2004)
11. L. Sangaletti, L.E. Depro, A. Dieguez, G. Marca, J.R. Morante, A.R. Rodringuez, G. Sberveglieri, Sens. Actuator B **44**, 268 (1997)
12. C. Ristoscu, L. Cultera, A. Dima, A. Perrone, R. Cutting, H.L. Du, A. Busiakiewicz, Z. Klusek, P.K. Datta, S.R. Rose, Appl. Surf. Sci. **247**, 95 (2005)
13. H. Zhu, D. Yang, G. Yu, H. Zhang, D. Jin, K. Yao, J. Phys. Chem. B **110**, 7631 (2006)
14. J. Xu, X. Jia, X. Lou, J. Shen, Solid-State Electron. **50**, 504 (2006)
15. B. Jaffe, W.R. Cook, H. Jaffe, *Piezoelectric Ceramics* (Academic, New York, 1971)
16. F.S. Galasso, *Structure, Properties and Preparation of Perovskite- type Compounds* (Pergamon Press, Oxford, 1969)
17. U. Alver, T. Kilinc, E. Bacaksiz, T. Kuecuekoemeroglu, S. Nezir, I.H. Mutlu, F. Aslan, Synthesis and characterization of spray pyrolysis zinc oxide microrods. Thin Solid Films **515**, 3448–3451 (2007)
18. H.B. Cheng, J.P. Cheng, Y.J. Zhang, Q.M. Wang, Large-scale fabrication of ZnO micro- and nano-structures by microwave thermal evaporation deposition. J. Cryst. Growth **299**, 34–40 (2007)
19. P. Klason, K. Magnusson, O. Nur, Q.X. Zhao, Q.U. Wahab, M. Willander, Synthesis and structural and optical properties of ZnO micro- and nanostructures grown by the vapour–liquid–solid method. Phys. Scr. **126**, 53–56 (2006)
20. L. Khomenkova, P. Fernandez, J. Piqueras, ZnO nanostructured microspheres and elongated structures grown by thermal treatment of ZnS powder. Cryst. Growth Des. **7**, 836–839 (2007)
21. F.H. Su, W.J. Wang, K. Ding, G.H. Li, Y.F. Liu, A.G. Joly, W. Chen, Pressure dependence of the near-band-edge photoluminescence from ZnO microrods at low temperature. J. Phys. Chem. Solids **67**(11), 2376–2381 (2006)
22. O. Lupan, L. Chow, G.Y. Chai, B. Roldan, A. Naitabdi, A. Schulte, H. Heinrich, Nano fabrication and characterization of ZnO nanorod arrays and branched microrods by aqueous solution route and rapid thermal processing. Mater. Sci. Eng. B **145**, 57–66 (2007)
23. Z.Y. Huang, C.F. Chai, B.Q. Cao, Temperature-dependent emission shifts of Peanut like ZnO microrods synthesized by a hydrothermal method. Cryst. Growth Des. **7**, 1686–1689 (2007)

24. R.B. Kale, S.Y. Lu, Structural, morphological, and optical properties of double ended needle-like ultra-long ZnO micro/nanorods. J. Phys. Condens. Matter **19**, 096209/1–096209/7 (2007)
25. Y.J. Kim, C.H. Lee, Y.J. Hong, G.C. Yi, S.S. Kim, H. Cheong, Controlled selective growth of ZnO nanorod and microrod arrays on Si substrates by awet chemical method. Appl. Phys. Lett. **89**, 63128/1–163128/3 (2006)
26. C.L. Kuo, T.J. Kuo, M.H. Huang, Hydrothermal synthesis of ZnO microspheres and hexa gonal microrods with sheetlike and platelike nanostructures. J. Phys. Chem. B **109**, 20115–20121 (2005)
27. J.M. Du, Z.M. Liu, Y. Huang, Y.N. Gao, B.X. Han, W.J. Li, G.Y. Yang, Control of ZnO morphologies via surfactants assisted route in the subcritical water. J. Cryst. Growth **280**, 126–134 (2005)
28. X.Q. Liu, S.W. Tao, Y.S. Shen, Preparation and characterization of nanocrystalline _Fe2O3 by a sol–gel process. Sens. Actuators B **40**, 161–165 (1997)
29. Y. Zhu, H. Lu, Y. Lu, X. Pan, Characterization of SnO_2 films deposited by d.c. gas discharge activating reaction evaporation onto amorphous and crystalline substrates. Thin Solid Films **224**, 82–86 (1993)
30. C.C. Chai, J. Peng, B.P. Yan, Preparation and gas-sensing properties of _$-Fe_2O_3$ thin films. J. Electron. Mater. **24**, 799–804 (1995)
31. T. Ishihara, K. Kometani, M. Hashida, Y. Takita, J. Electrochem. Soc. **138**, 173–177 (1991)
32. H.L. Hartnagel, A.L. Dawar, A.K. Jain, C. Jagadish, *Semiconducting Transparent Thin Films* (IOP, Bristol, 1995)
33. G. Heiland, D. Kohl, in *Chemical Sensor Technology*, vol. 1, ed. by T. Seiyama, (Kodansha, Tokyo, 1983), p. 113
34. S. Baruah, J. Dutta, Sci. Technol. Adv. Mater. **12**, 013004 (18 pp), 14 (2011)
35. O. Lupan, L. Chow, S. Shishiyanu, E. Monaico, T. Shishiyanu, V. Sontea, C.B. Roldan, A. Naitabdi, S. Park, A. Schulte, Mater. Res. Bull. **44**, 63 (2009)
36. M.C. Carotta et al, Sens. Actuators B **137**, 164 (2009)
37. S.M. Chou, L.G. Teoh, W.H. Lai, Y.H. Su, M.H. Hon, Sensors **6**, 1420 (2006)
38. E. Comini, G. Faglia, G. Sberveglieri, Z. Pan, Z.L. Wang, Appl. Phys. Lett. **81**, 1869 (2002)
39. N. Koshizaki, T. Oyama, Sens. Actuators B **66**, 119 (2000)
40. Y.I. Lee, K.J. Lee, D.H. Lee, Y.K. Jeong, Y.H. Choa, Curr. Appl. Phys. **9**, S79 (2009)
41. P. Mitra, A.K. Mukhopadhyay, Bull. Polish Acad. Sci. **55**, 281 (2007)
42. M.W. Ahn, K.S. Park, J.H. Heo, D.W. Kim, K.J. Choi, J.G. Park, Sens. Actuators B **138**, 168 (2009)
43. S. Saito, M. Miyayama, K. Kuomoto, H. Yanagida, J. Am. Ceram. Soc. **68**, 40 (1985)
44. T. Jinkawa, G. Sakai, J. Tamaki, N. Miura, N. Yamazoe, Relationship between ethanol gas sensitivity and surface catalytic property of tin oxide sensors modified with acidic or basic oxides. J. Mol. Catal. A Chem. **155**, 193–200 (2000)

Synthesis and Characterization of Single and Few Layer Graphene for Field Effect Transistor

Pawan Kumar Srivastava and Subhasis Ghosh

Abstract Single and few-layer graphene sheets with sizes up to 100 micron were synthesized using physico-chemical method by dispersing highly ordered pyrolytic graphite (HOPG) flakes into *N, N*-dimethylformamide (DMF) solution. The layer structure and thickness of graphene sheets were characterized with transmission electron microscopy (TEM), scanning electron microscopy (SEM), atomic force microscopy (AFM) and Raman spectroscopy. Synthesized graphene sheets were transferred to solid substrates for further processing. Field-effect transistor with individual graphene sheet was fabricated. In ambient conditions, graphene sheets were found to exhibit p-type behavior.

Keywords Graphene · TEM · SAED · FET · AFM

1 Introduction

Graphene, a monolayer of graphite has recently attracted considerable interest due to its outstanding properties [1]. This two dimensional configuration consists of sp^2 bonded carbon atoms are organized into a honeycomb lattice in graphene. The maximum excitement over graphene is due to its exceptional electronic properties. It has an amazingly good electrical conductivity. Unlike other semiconducting materials, it shows extremely high conductivity with a mobility $\sim$10,000 cm^2/Vs, which is much higher than most conventional semiconductor materials [2]. It is envisaged that on the basis of its excellent electrical and mechanical properties graphene can replace well established silicon technology.

P. K. Srivastava (✉) · S. Ghosh
School of Physical Sciences, Jawaharlal Nehru University, New Delhi 110067, India
e-mail: pksri08@gmail.com

P. K. Giri et al. (eds.), *Advanced Nanomaterials and Nanotechnology*,
Springer Proceedings in Physics 143, DOI: 10.1007/978-3-642-34216-5_16,

Now a days various methods have been developed to produce graphene. Exfoliation of thin graphene sheets from bulk graphite using scotch tape has been used widely with good reproducibility. CVD and chemical methods have also been used frequently to produce large area graphene sheets. All these methods are time consuming and requires wide variety of resources. Few years ago Dai et al. [3] synthesized fine quality graphene sheets using some organic solvents, but their method consists of several steps and few days to intercalate solvents into graphitic layers.

2 Experimental Details

In our work, we present physico-chemical method to produce single and few layer graphene using Highly Ordered Pyrolitic Graphite (HOPG) dispersed in pure *N*, *N*-dimethylformamide (DMF) solution which consists of following three steps:

(1) Ultrasonication of HOPG flakes dispersed in DMF solution.
(2) Centrifugation of ultrasonicated solution of HOPG and DMF.
(3) Transferring thin graphene layers suspended in dispersed solution onto the SiO_2 or heavily doped Si substrates.

Highly ordered pyrolytic graphite contains large number of graphene flakes stacked together by weak van der Walls forces. Sonicating HOPG flakes into DMF solution causes its molecules to intercalate between the flakes which results in weakening of van der Waals forces. Alternatively, large sheets and aggregates may be fractured into uniform dispersions of smaller single sheets by repeated ultrasonication. Dispersions of single layer sheets are created by precipitating out higher density multilayers by means of centrifugation. Good quality and large area graphene sheets can be synthesized by optimizing the quantity of HOPG into DMF solution, ultrasonication and centrifugation time. After transferring Graphene nanosheets onto the substrates, topographical measurements were carried out. Transmission electron microscopy (TEM) was done by drop casting the dispersed solution directly onto the TEM-grid. AFM and Raman spectroscopy were carried out in order to probe the morphology and layered structure of graphene sheets respectively. After morphological characterizations, we fabricated graphene field effect transistors to investigate the transport properties of graphene sheet. This was done by drop casting the graphene containing solution on to the SiO_2 substrates and copper source-drain electrodes (35 nm) were deposited by thermal evaporation technique. Electrode separation was 50 μm, patterned through conventional shadow masking technique. Gate voltage was applied to the heavily doped silicon substrate in back gate geometry.

3 Results and Discussions

Figure 1a shows TEM image of one of our samples. Image shows the folded graphene sheet where edges tend to scroll. Black spots in the TEM image show some tiny HOPG flakes. In order to identify the layer structure of graphene, selected area electron diffraction (SAED) patterns were observed experimentally (Fig. 1b). By analyzing experimentally observed diffraction patterns, we could distinguish single layer graphene from multilayer graphene. The centre of the hexagon appears as one unit in the image, whereas six carbon atoms should be located at each corner of the hexagon [3]. The key for the identification of monolayer graphene is that its reciprocal space has only the zero order Laue zone, therefore diffraction intensities of all the points should remain same [3, 4]. Where as for multilayer graphene, intensities of diffraction points vary strongly.

Although TEM images can be used to find single sheets of graphene, the images are representative of electronic structures, not topography. Hence AFM is needed to establish the thickness and surface roughness of single and few layer graphene sheets. Figure 2a, b represents the non-contact AFM topographical image ($5 \times 5\ \mu m^2$) of graphene sheet and its height profile respectively. As observed by AFM measurement, thickness of sheet is around 1.4 nm, showing a bilayer graphene sheet [5]. Raman spectroscopy is one of the most powerful characterization technique for carbon materials. It has been shown that Raman scattering can be used as a hall mark for identification of single and few layer graphene. Graphite has three most intense features at $\sim$1,580 cm^{-1} (G band), $\sim$1,350 cm^{-1} (D band), and $\sim$2,700 cm^{-1} (2D band) [6]. For bulk HOPG, G/2D ratio is $\sim$3.2, whereas for single layer graphene it is $\sim$0.5 [6]. Along with the intensity ratio of I(G)/I(2D), 2D band shape and position are good finger prints of single and bilayer graphene to distinguish them from multilayers. Figure 3 shows the Raman spectra for bulk HOPG and graphene layer having G band $\sim$1,630 cm^{-1} and 2D band

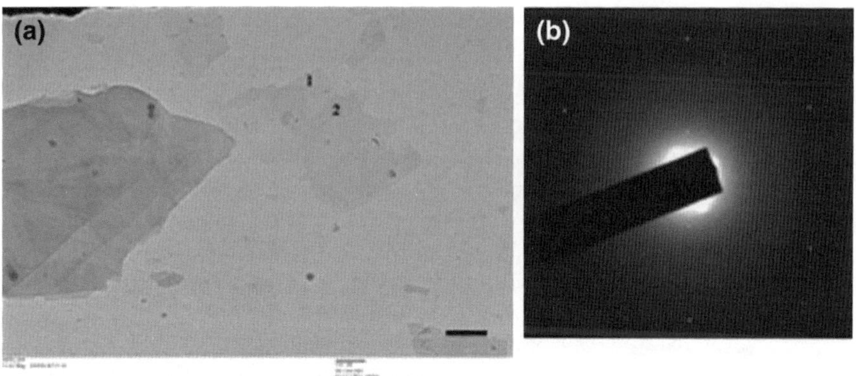

Fig. 1 **a** TEM image of single (*point 1*) and few layer graphene (*point 2*), **b** shows SAED pattern of monolayer graphene. Scale bar is 500 nm

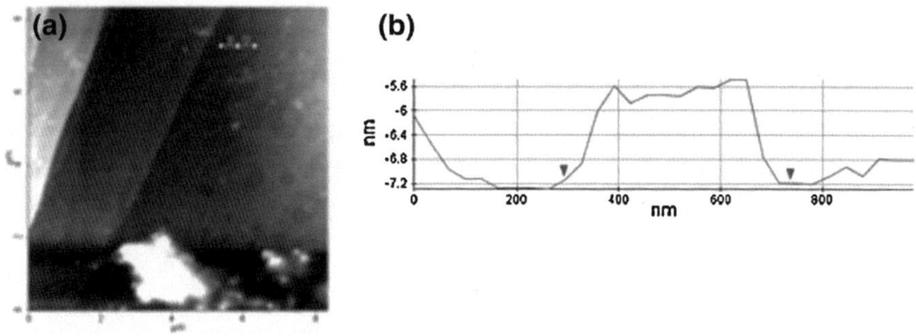

Fig. 2 **a** AFM topographic image of bilayer graphene sheet with scan size 5 × 5 μm^2, **b** height profile corresponding to cursers shown in figure (**a**)

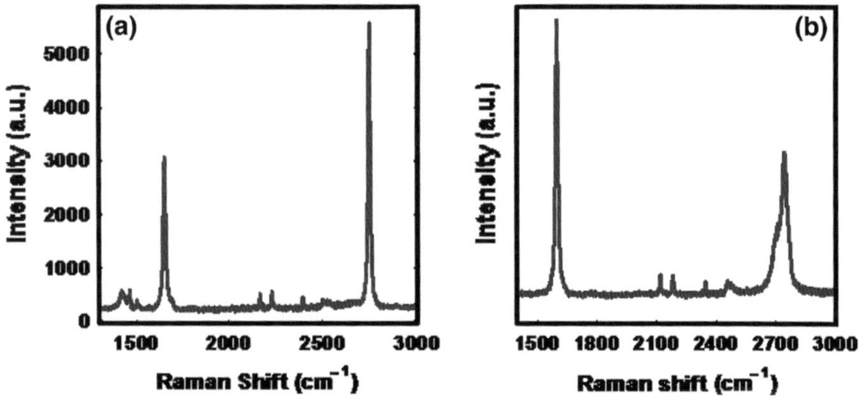

Fig. 3 **a** Raman spectrum of G and 2D band of graphene, ratio of the G/2D shows characteristics of monolayer graphene sheet, **b** Raman spectrum of bulk HOPG

$\sim$2,750 cm^{-1} with I(G)/I(2D) $\approx$ 0.8, indicating single layer graphene. We observed 50 and 9 cm^{-1} blue shift in comparison to HOPG raman spectra for G and 2D band respectively. This blue shift in G and 2D peak position may be due to the unintentional doping in graphene by the charge impurities present on the Si/SiO2 substrate [7]. For 2D band, FWHM was found to be 25 and 50 cm^{-1} for graphene sheet and bulk HOPG respectively, which is in agreement with the data reported earlier [6].

Figure 4 shows V_{DS} (source-drain voltage) versus I_{DS} (source-drain current) with gate bias ranging from 0–50 V with step of 10 V. Linear I_{DS} vs V_{DS} curve indicates good ohmic conduction between graphene and Cu electrodes. Inset shows I_{DS} vs V_G curve of same device taken at $V_{DS} = 0.1$ V. This transfer characteristics indicates typical p-type behavior. Mobility of the charge carriers was calculated by the equation

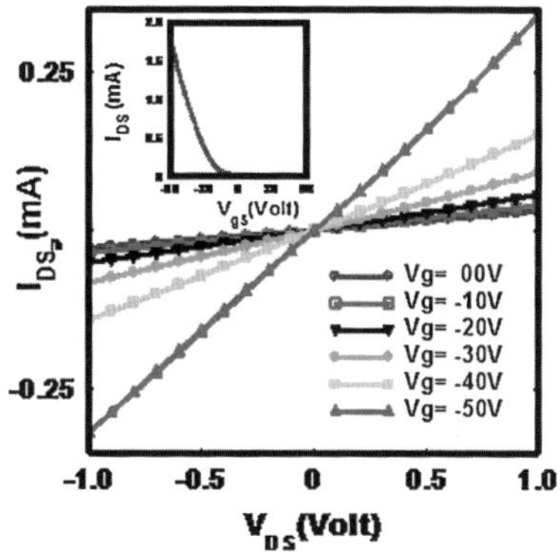

Fig. 4 Output characteristics of graphene field effect transistor. Inset shows transfer characteristics of the same device at $V_{DS} = 0.1$ V

$$I_{DS} = \frac{W}{L}\mu C_i(V_G - V_T)V_{DS} - \frac{W}{2L}\mu C_i V_{DS}^2$$

where V_G is the gate voltage, V_T is the threshold voltage, $\frac{W}{L}$ is the width to length ratio of the channel and C_i is the intrinsic capacitance of the dielectric layer (SiO_2). By putting these geometrical constants and experimentally observed quantities, mobility was extracted to be 3,300 cm^2/V-s indicating graphene sheets exhibits excellent quality for electronic applications.

4 Conclusion

In summary, we demonstrated growth of single and few layer graphene using physico-chemical method. Unlike the conventional mechanical cleavage method, adhesive tapes are not needed so graphene made via this process is free from contamination. Additionally our method is simple, yielding $\sim$100 micron graphene sheets. Characterizations were done using various imaging techniques. Raman spectroscopy and electron diffraction measurement were carried out in order to confirm the layer structure of graphene. Graphene based field effect transistor was fabricated by transferring graphene sheets on Si/SiO_2 substrates.

Acknowledgements The authors thank Dr. Gajendra, AIRF, JNU and Dr. Fouran Singh for their help in TEM measurements and Raman spectroscopic measurements respectively. One of the author PKS thanks council of scientific and industrial research for financial assistance through fellowship.

References

1. A.K. Geim, K.S. Novoslov, Rise of graphene. Nat. Mater. **6**(183) (2007)
2. X. Li, G. Zhang, X. Bai, X. Sun, X. Wang, E. Wang, H. Dai, Highly conducting graphene sheets and langmuir–Blodgett films. Nat. Nanotechnol. **3**(538) (2008)

3. Y.J. Suh, S.Y. Park, M.J. Kim, High resolution TEM and electron diffraction study of graphene layers. Microsc. Microanal. **15**(2), 1168 (2009)
4. J.C. Meyer, A.K. Geim, M.I. Katsnelson, K.S. Novoselov, T.J. Booth, S. Roth, The structure of suspended graphene sheets. Nature **446**, 60 (2007)
5. S. Park, R.S. Rouf, Chemical methods for the productions of graphenes. Nat. Nanotechnol. **4**, 217 (2009)
6. A.C. Ferrari, J.C. Meyer, V. Scardaci, C. Casiraghi, M. Lazzeri, F. Mauri, S. Piscanec, D. Jiang, K.S. Novoselov, S. Roth, A.K. Geim, Raman spectrum of graphene and graphene layers. Phys. Rev. Lett. **97**, 187401 (2006)
7. A. Das, C. Bishwanath, A.K. Sood, Raman spectroscopy of graphene on different substrates and influence of defects. Bull. Mater. Sci. **31**, 579 (2008)

Oxygen Partial Pressure Dependent Properties of Nanocrystalline Nickel Oxide Films

A. Mallikarjuna Reddy, Y. Ashok Kumar Reddy, Ch. Seshendra Reddy, A. Sivasankar Reddy and P. Sreedhara Reddy

Abstract Nickel oxide thin films have been successfully deposited by dc reactive magnetron sputtering technique on glass substrates at different oxygen partial pressures. X-ray diffraction analysis revealed that the preferred orientation changed from (200) to (220) with increasing oxygen partial pressure. Fine grains with RMS roughness of 9.4 nm were observed at an oxygen partial pressure of 6×10^{-4} mbar.

Keywords Oxide coating · Sputtering · Oxygen partial pressure · Surface morphology

1 Introduction

Nickel oxide is a semi-transparent, wide band gap, and p-type transparent conducting oxide. In the fabrication process of electronic and optoelectronic oxide semiconductor devices, it is necessary to combine a p-type and an n-type semiconductor to form p-n junction. However, most of the oxide semiconductor materials show n-type properties because cation interstitials formed much more easily than anion interstitials. As a candidate of p-type oxide semiconductor, nickel oxide thin films with NaCl-type structure have a variety of applications due to excellent chemical stability,

A. Mallikarjuna Reddy (✉) · Y. Ashok Kumar Reddy · Ch. Seshendra Reddy
P. Sreedhara Reddy
Department of Physics, Sri Venkateswara University, Tirupati 517502, India
e-mail: malliphysics@gmail.com

A. Sivasankar Reddy
Division of Advanced Materials Engineering, Kongju National University, Budaedong, Cheonan, South Korea

P. K. Giri et al. (eds.), *Advanced Nanomaterials and Nanotechnology*,
Springer Proceedings in Physics 143, DOI: 10.1007/978-3-642-34216-5_17,

as well as optical, electrical, and magnetic properties [1]. Several techniques like spray pyrolysis [2], sputtering [3], vacuum evaporation [4], electron beam evaporation [5], chemical deposition [6], sol–gel [7], pulsed laser deposition [8] and plasma-enhanced chemical vapor deposition [9] have been employed for the deposition of NiO thin films. Among these methods, reactive sputtering is considered to be the most widely useful technique having high deposition rates, uniformity over large areas of the substrates and easy control over the composition of the deposited films. In the present study, NiO thin films were deposited using dc reactive magnetron sputtering technique and studied the effect of oxygen partial pressure on the structural and morphological properties.

2 Experimental Details

Nickel oxide thin films were deposited on Corning 7,059 glass substrates by dc reactive magnetron sputtering at various oxygen partial pressures. The sputtering conditions maintained during the growth of NiO films were as follows: Sputtering target is pure Nickel (99.98 %), target to substrate distance was maintained as 70 mm, ultimate pressure in the chamber was 5×10^{-6} mbar. Nickel oxide films were deposited in the oxygen partial pressure range of 4×10^{-4}–8×10^{-4} mbar at a sputtering power of 150 W, substrate temperature of 523 K and sputtering pressure of 4×10^{-2} mbar.

3 Results and Discussion

All the deposited films were polycrystalline in nature with cubic structure. At the oxygen partial pressure of 4×10^{-4} mbar, the films showed (220) orientation along with (200), on further increasing the oxygen partial pressure to 6×10^{-4} mbar the intensity of the (220) peak was increased and becomes sharper, and (200) peak was disappeared. Beyond this oxygen partial pressure, the intensity of the (220) peak was decreased (Fig. 1). The grain size of the films was calculated from the (220) orientation by using Scherrer formula [10]. The grain size of the films increased from 28 to 29 nm with increase of oxygen partial pressure from 4×10^{-4} to 6×10^{-4} mbar, thereafter it was decreased to 22 nm at higher oxygen partial pressure. The decreasing of the grain size was attributed to the segregation of oxygen into the grain boundaries, which causes limited grain boundary mobility, and therefore a decrease in grain size [11]. Figure 2 shows the AFM images of NiO films at different oxygen partial pressures (Fig. 2). The surface roughness of the films was reduced from 11.2 to 9.4 nm with increasing of the oxygen partial pressure from 4×10^{-4} to 6×10^{-4} mbar. Beyond this oxygen partial pressure the films getting rough with RMS roughness of 10 nm. From the histograms (Fig. 3), the grain size of the films was 29, 41 and 25 nm for oxygen

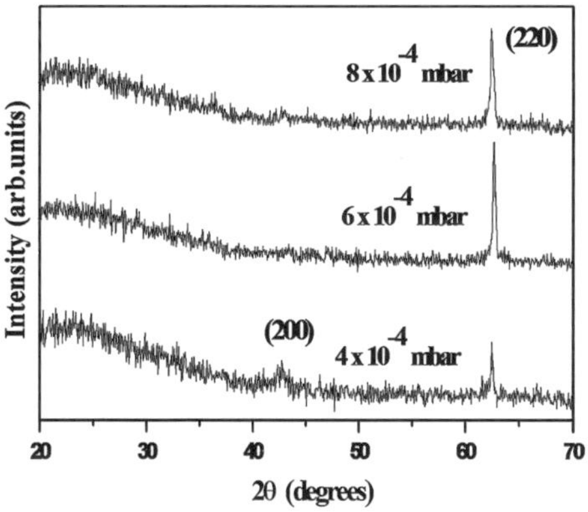

Fig. 1 X-ray diffraction profiles of NiO films at various oxygen partial pressures

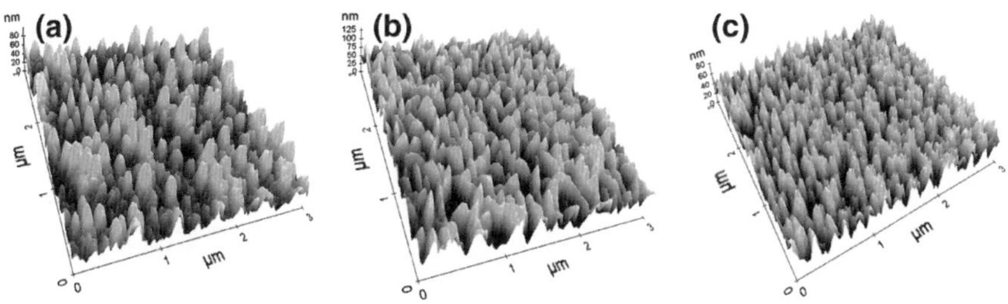

Fig. 2 AFM images of NiO films at different oxygen partial pressures **a** 4×10^{-4} mbar **b** 6×10^{-4} mbar **c** 8×10^{-4} mbar

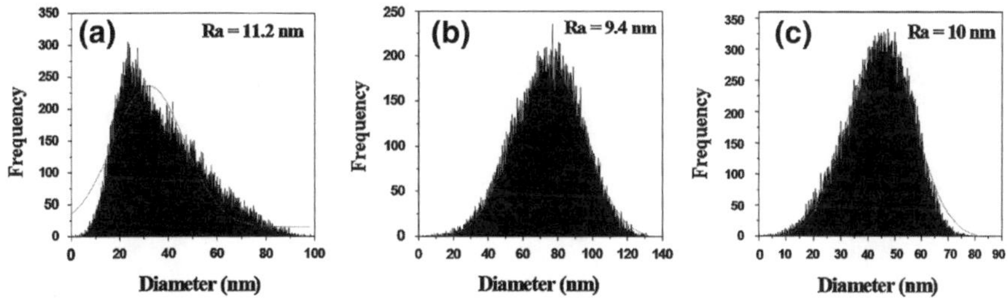

Fig. 3 Histograms of NiO films at different oxygen partial pressures **a** 4×10^{-4} mbar **b** 6×10^{-4} mbar **c** 8×10^{-4} mbar

partial pressure of 4×10^{-4}, 6×10^{-4} and 8×10^{-4} mbar, respectively. The variation in the grain size values were in accordance to the XRD results, but the grain sizes determined by AFM histograms were greater than that measured by XRD. This is caused by the different grain size criteria, underlying the different methods. From the AFM and XRD results, the films growth towards the (220) orientation exhibited smooth surface with bigger grains.

4 Conclusion

The deposited films as a function of oxygen partial pressure exhibited (220) as preferred orientation. Fine and uniform grains with RMS Roughness of 9.4 nm were observed at oxygen partial pressure of 6×10^{-4} mbar.

References

1. I. Hotovy, J. Huran, J. Janik, A.P. Kobzev, Deposition and properties of nickel oxide films produced by DC reactive magnetron sputtering. Vacuum **51**, 157 (1998)
2. J.D. Desai, S.K. Min, K.D. Jung, O.S. Joo, Spray pyrolytic synthesis of large area NiO_x thin films from aqueous nickel acetate solutions. Appl. Surf. Sci. **253**, 1781 (2006)
3. H.Y. Ryu, G.P. Choi, W.S. Lee, J.S. Park, Preferred orientations of NiO thin films prepared by RF magnetron sputtering. J. Mater. Sci. Lett. **39**, 4375 (2004)
4. J. Scarminio, W. Estrada, A. Andersson, A. Gorenstein, F. Decker, H insertion and electrochromism in NiO_x thin films. J. Electrochem. Soc. **139**, 236 (1992)
5. A. Agarwal, H.R. Habibi, R.K. Agarwal, J.P. Cronin, D.M. Roberts, C.P.R. Sue, C.M. Lampert, Effect of deposition pressure on the microstructure and electrochromic properties of electron beam-evaporated nickel oxide films. Thin Solid Films **221**, 239 (1992)
6. M.A.V. Hurtado, A.M. Galvan, Optical and structural characterization of nickel oxide-based thin films obtained by chemical bath deposition. Mater. Chem. Phys. **107**, 33 (2008)
7. R.C. Korosec, P. Bukovec, Solgel prepared NiO thin films for electro.chromic applications. Acta Chim. Slov. **53**, 136 (2006)
8. M. Tanaka, M. Mukai, Y. Fujimori, M. Kondoh, Y. Tasaka, H. Baba, S. Usami, Transition metal oxide films prepared by pulsed laser deposition for atomic beam detection. Thin Solid Films **281**, 453 (1996)
9. E. Fujii, A. Tomozawa, S. Fujii, H. Torii, M. Hattori, and R. Takayama, NaCl-type oxide films prepared by plasma-enhanced metalorganic chemical vapor deposition. Jpn. J. Appl. Phys. **32**, L1448 (1993)
10. B.D. Cullity, *Elements of X-ray diffraction*, 2nd edn. (Addison Wesley, London, 1978)
11. K. Ellmer, Magnetron sputtering of transparent conductive zinc oxide: relation between the sputtering parameters and the electronic properties. J. Phys. D Appl. Phys. **33**, R17 (2000)

Observation of Nonlinear Optical Properties of Chemically Synthesized Cu^{2+} Doped ZnS Nanoparticles

A. K. Kole and P. Kumbhakar

Abstract ZnS and Cu^{2+} (0.50 and 1.00 %) doped ZnS (i.e., ZnS:Cu) nanoparticles (NPs) are synthesized by chemical co-precipitation method at room temperature. X-ray diffraction (XRD) studies and the analysis of the selected area electron diffraction pattern (SAED) obtained from transmission electron microscopy (TEM) confirmed the formation of zinc blende structure of all the synthesized samples. Irrespective of the samples, the average particle sizes, as obtained from the XRD and TEM is about 2.5 nm. The room temperature photoluminescence (PL) emission measurements revealed the presence of green emission band in all the ZnS:Cu samples which is attributed to Cu^{2+} incorporation in ZnS. It is found by Gaussian deconvolution of the measured PL spectra of the synthesized samples that two peaks appeared at 405 and 445 nm in undoped ZnS NPs. In addition to the above two peaks, a third peak is noted in green region for ZnS:Cu NPs, the second and third peak shows a red shift with increasing Cu^{2+} concentration. Three photon absorption (3PA) and nonlinear refraction (NLR) coefficients are also measured at 532 nm, the second harmonic of a Q-switched Nd:YAG laser radiation, by using the z-scan technique, viz., in 0.5 % Cu^{2+} doped ZnS:Cu sample. From the analysis of open aperture (OA) z-scan data it is found that three photon absorptions (3PA) is the dominant mechanism for appearance of nonlinear absorption (NLA) in the sample and the extracted value of the intrinsic 3PA coefficient is $\sim 10^9$ times larger than that of bulk ZnS.

Keywords II–VI semiconductor · Nanostructures · Optical properties · Photoluminescence · Nonlinear optical properties · Z-scan

---

A. K. Kole · P. Kumbhakar (✉)
Nanoscience Laboratory, Department of Physics, National Institute of Technology, Durgapur 713209, West Bengal, India
e-mail: pathik.kumbhakar@phy.nitdgp.ac.in; nitdgpkumbhakar@yahoo.com

P. K. Giri et al. (eds.), *Advanced Nanomaterials and Nanotechnology*, Springer Proceedings in Physics 143, DOI: 10.1007/978-3-642-34216-5_18,

1 Introduction

Recently, research studies on doped ZnS nanomaterials have drawn considerable attention because of their potential applications in optoelectronic devices, blue/green light emitting diodes (LEDs), electro luminescence devices (ELDs), UV light sensitive photo detectors, solar cells etc., [1–10]. Semiconductor nanocrystallites (NCs) have different optical and electronic properties than bulk materials, arising from 'quantum confinement effects' [7] and their band gap energy can be controlled by controlling the particle sizes. As a result of 'quantum confinement effect', a blue shift in the absorption spectrum with decreasing particle sizes can be observed. Amongst all other II–VI semiconductor nanocrystals, ZnS has a well known wide direct band gap (3.67 eV at room temperature) with a large exciton binding energy ($\sim$40 meV) [7–10]. It has a high refractive index and well known for its photoluminescence property. Luminescence properties of impurity added ZnS NPs are widely different from those of bulk ZnS because impurities can easily influence the electronic structure and transition probabilities between various states. Nanocrystallites doped with optically active colour center provide new opportunities for luminescent study and applications [9–17]. The chemical properties of transition metal ions such as Mn^{2+}, Co^{2+} and Cu^{2+} etc., are very similar to those of Cd^{2+} or Zn^{2+} and thus Mn^{2+}, Cu^{2+} etc., can be easily doped with II–VI ZnS semiconductor host. The first report on doping of nanocrystals was on ZnS:Mn system [8]. Followed by that report, extensive research work is going on doped II–VI semiconductor nanomaterials to enhance their emission activities and thereby making them good candidate for optoelectronic applications. By using various synthesis mechanisms such as chemical precipitation method, solvothermal method, hydrothermal method, reverse micelle method, sol–gel method etc., doping of transition metal ions and rare earth ions can be done in ZnS host semiconductor [4]. The chemical co-precipitation method is most popular synthesis mechanism due to its various advantages, such as it is more economical than the other synthesis methods, doping of foreign atoms are easily possible, less instrumentation is required as well as a large quantity of the synthesized materials can be obtained. The size of the synthesized ZnS:Cu NPs are $\sim$2.5 nm. No such significant change in the particle sizes are observed because the concentration of Cu^{2+} ions incorporated in ZnS is small and the synthesis environment is almost same for all the samples. A blue shift is noted in the UV–Visible optical absorption spectra of the prepared ZnS:Cu NPs with respect to the bulk ZnS and this has occurred as a result of 'quantum confinement effect' [7]. The effect of Cu^{2+} doping on the photoluminescence property of chemically synthesized ZnS NPs is studied in the present work. Different groups have reported various types of PL emissions bands (blue and green) from ZnS:Cu NPs [11, 17–24]. However, in case of ZnS:Cu nanoparticles, the luminescence mechanism is still controversial. In this report, it is found by Gaussian fittings of the measured PL spectra of our samples that two peaks appeared at 405 and 443 nm for undoped ZnS. For ZnS:Cu NPs, in addition to the above two, a third peak is noted at $\sim$500 nm.

There exists variety of techniques for measurements of nonlinear refraction and absorption of materials. Recently a technique called the z-scan has been widely employed to characterize the nonlinear optical (NLO) properties of semiconductors. Z-scan technique is very sensitive and also by employing this technique it is possible to determine the sign as well as the magnitude of the NLR coefficient of a NLO material [31]. In the recent past, various researchers have studied NLO properties of ZnS nanocolloids and thin films doped with CdS and ZnS nanoparticles [25–30]. There are only few reports on the observation of multiphoton absorption (MPA) either in nano or in bulk ZnS [25–30]. So far there is no report on the observation of three-photon absorption (3PA) in ZnS:Cu NPs. Here, the 3PA nonlinear absorption (NLA) and nonlinear refraction (NLR) coefficients of the ZnS:Cu (0.50 %) NPs dispersed in methanol are reported which are measured by employing z-scan technique with a Q-switched Nd:YAG laser radiation at 532 nm wavelength and with pulse duration of 10 ns, repetition rate of 10 Hz.

2 Experiment

ZnS and ZnS:Cu NPs are synthesized at room temperature in methanol by chemical co-precipitation method [32], which is described below briefly. All the chemicals used are of AR grade (Merck and SD fine chemicals) and they are used without further purification. At first calculated molar amounts $Zn(CH_3COO)_2$ and $Cu(II)(CH_3COO)_2$ are dissolved separately in 50 ml methanol. Both the solutions are continuously stirred in a magnetic stirrer for 20 min. Then Na_2S solution prepared in 50 ml methanol and $Cu(II)(CH_3COO)_2$ solution is added drop wise to the $Zn(CH_3COO)_2$ solution under continuous stirring. The mixture is continuously stirred in a magnetic stirrer for another 1 h. After 1 h of continuous stirring, a greenish-white precipitate is deposited at the bottom of the flask. The precipitate is separated from the reaction mixture by centrifugation (Eltek Refrigerated Centrifuge RC4100D) for 10 min at 5,000 rpm and washed several times with distilled water and methanol to remove the remaining sodium particles, if any. The wet precipitate is then dried for further measurements. Three different samples those are synthesized; henceforth will be called as S1, S2, and S3 corresponding to undoped ZnS, ZnS:Cu (0.50 %) and ZnS:Cu (1.00 %).

The UV–visible optical absorption spectra of the prepared ZnS:Cu NPs dispersed in methanol are recorded using a UV–visible spectrophotometer (Hitachi, U-3010). The formation of ZnS:Cu NPs are confirmed by transmission electron microscopy (TEM, JEOL 2000 FX-II). The XRD patterns are recorded using an X-ray diffractometer (PANALYTICAL) with Cu K_α radiation of wavelength $\lambda = 0.15406$ nm in the scan range $2\theta = 20$–$70°$. The photoluminescence (PL) spectra of the ZnS:Cu NPs dispersed in methanol are recorded using a spectrofluorimeter (Perkin Elmer LS-55).

For the measurement of NLO properties, a standard z-scan set up is used [31]. The experimental z-scan transmission traces are measured at room temperature with the second harmonic radiation of a Q-switched Nd:YAG laser delivering 532 nm radiation with 10 ns pulse duration and 10 Hz repetition rate. The spatial transverse distribution of the laser pulses is Gaussian after passing through a spatial filter. The pulses are divided into two beams using a beam splitter. To monitor the fluctuation of the energy of the incident laser beam, the energy of the reflected beam representing the incident beam is measured simultaneously with that of the transmitted beam. The transmitted laser beam is focused using a lens of focal length 20 cm. The beam waist (w_0) and the confocal parameter (z_0) at the focus are thus, respectively, 24 μm and 3.6 mm. The synthesized nanopowders are dispersed in methanol and kept in a quartz cuvette of path length (L) 2 mm so that the "thin sample" approximation (i.e., $L < z_0$) is satisfied. The cuvette is mounted in a translation stage which can translate the sample in the pre-focal and post-focal directions along the direction of propagation (z-axis) of the transmitted laser beam.

3 Results and Discussions

Figure 1 shows the UV–Visible absorption characteristics of all the synthesized samples. The linear optical absorption characteristics of the prepared samples are measured by at first dispersing the dried powder samples in methanol and then taken in quartz cuvette of 10 mm path length (d). For obtaining the absorption characteristics of all the samples, at first transmittance (T) at different wavelengths (λ) are measured and then absorbance (α) at the corresponding wavelengths are calculated by using Beer-Lambert relation:

$$\alpha = \frac{1}{d}\ln\left(\frac{1}{T}\right), \tag{1}$$

where d is the path length. As shown in Fig. 1, a sharp excitonic peak is clearly observed at about 305 nm for all the samples and is fairly blue shifted from the absorption band edge of the bulk ZnS (334 nm). Such blue shift in the band edge takes place due to 'quantum confinement effect' [7]. The absorption coefficient α and band gap E_{gn} of the synthesized samples are related by,

$$(\alpha h\nu)^m = A\,(h\nu - E_{gn}) \tag{2}$$

where A is a constant and the band gap of the studied material is denoted as E_{gn}. Exponent "m" depends upon the type of the transition: m may have values 2, 1/2, 2/3 and 1/3 corresponding to the allowed direct, allowed indirect, forbidden direct and forbidden indirect transitions, respectively [33]. For ZnS, the value of m is always 2 i.e., the fundamental absorption corresponds to allowed direct transition.

To calculate the band gap values of the prepared samples, $(\alpha h\nu)^2$ versus $h\nu$ is plotted and is shown in the inset of Fig. 1. The exact value of the band gap is determined by extrapolating the straight line portion of $(\alpha h\nu)^2$ versus $h\nu$ graph to the $h\nu$ axis. It is found that the calculated values of band gap of ZnS and ZnS:Cu

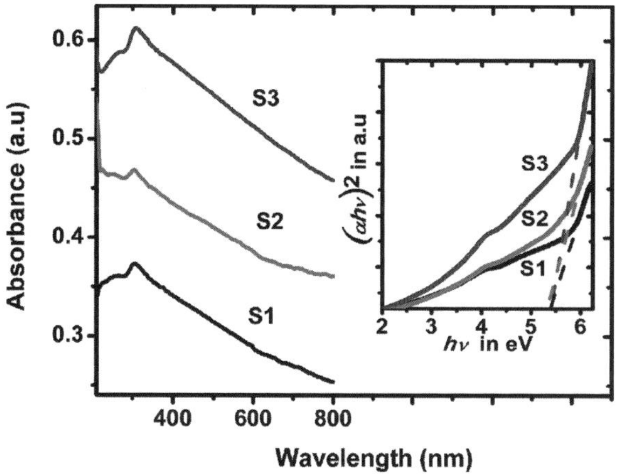

Fig. 1 UV-visible absorption spectra of the ZnS and ZnS:Cu nanoparticles dispersed in methanol. All the graphs are rescaled to plot together. Inset shows the plot of $(\alpha h\nu)^2$ versus $h\nu$ of ZnS and ZnS:Cu nanoparticles. Here, *S*1, *S*2, and *S*3 corresponding to the ZnS, ZnS:Cu (0.50 %) and ZnS:Cu (1.00 %), respectively

NPs are higher than that of bulk ZnS. The shift of the band gap of the synthesized samples from the bulk value of 3.71 eV (334 nm) of ZnS is observed due to quantum confinement effect. By using the calculated values of band gap and using Brus' equation, we can also calculate the particle sizes with radius *r* (nm) as [30],

$$E_{gn} = E_{gb} + 2.7442/r^2 - 0.2963/r. \tag{3}$$

The calculated values from optical absorption studies matches well with those obtained from transmission electron microscope (TEM) micrograph analyses as described later.

Figure 2 shows the XRD patterns of all the samples. Three diffraction peaks at 2θ values 28.15°, 47.72° and 56.21° corresponding to the (111), (220) and (311) planes of cubic (zinc blende) structure appeared in the XRD pattern of all the samples, and those matches well with the standard card (JCPDS no. 80–020). The broadening of the XRD peaks indicates the nanocrystalline nature of the samples.

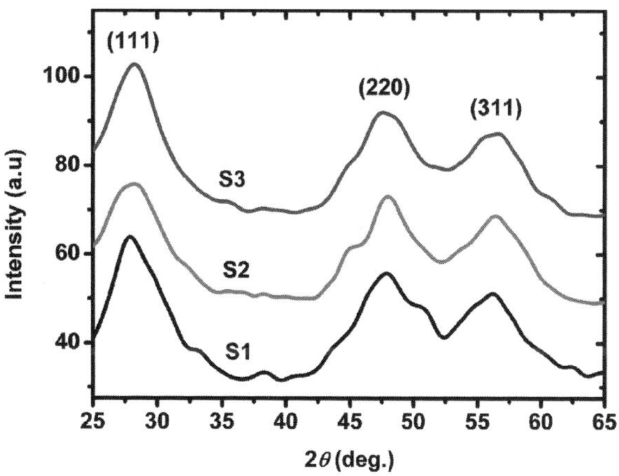

Fig. 2 X-ray diffraction (XRD) pattern of ZnS and ZnS:Cu samples. All the curves are rescaled to plot together. Here, *S*1, *S*2, and *S*3 corresponding to the ZnS, ZnS:Cu (0.50 %) and ZnS:Cu (1.00 %), respectively

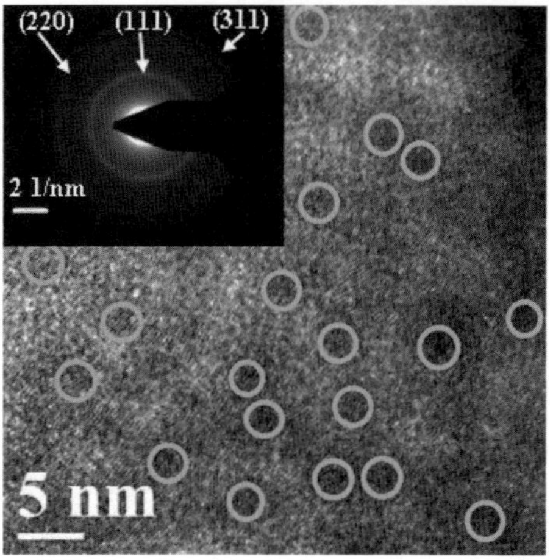

Fig. 3 TEM image of 0.50 % Cu^{2+} doped ZnS (S2) and its inset shows the SAED pattern which confirms the diffraction from *(111)*, *(220)*, and *(311)* planes of cubic ZnS

In addition, no impurity related diffraction peaks are detected. From the value of full width at half maximum (FWHM) of the most prominent peak in XRD pattern, the mean crystallite sizes are calculated using Debye-Scherer's formula [34],

$$D = 0.89\lambda/(\beta\cos\theta). \tag{4}$$

where D is the average crystallite size, λ is the X-ray wavelength, θ is the diffraction angle and β is the FWHM of the most prominent diffraction peak. The calculated average crystallite sizes from XRD data, viz., for S2 sample is about 2.5 nm.

Direct information about particle size and morphology of the prepared NPs can be obtained by analyzing TEM micrographs of the samples. However, a typical TEM image of S2 sample, viz., is shown in Fig. 3 and its inset shows the corresponding selected area electron diffraction (SAED) pattern. From the Fig. 3 it is seen that particles in various sizes are present in the prepared samples. The calculated average particle size from TEM is about 2.5 nm which matches well with the value of the same as obtained from optical studies and XRD analysis as described earlier. The lattice fringes are clearly observed which indicates the crystalline nature of the synthesized samples. Three circular rings appeared in SAED pattern are assigned to (111), (220) and (311) crystalline planes of cubic ZnS.

In Mn^{2+} doped ZnS NPs (ZnS:Mn), the dopant Mn^{2+} ion has a d^5 configuration and an yellow orange photoluminescence is well known due to the ${}^4T_1$–${}^6A_1$ transition of Mn^{2+} ions excited via energy transfer from the host ZnS [9]. In ZnS:Eu NPs, PL emission (colour range from blue to yellow) due to $4f_7$–$4f_6\ 5d_1$ transition was also observed [35]. However, in case of ZnS:Cu NPs, the luminescence mechanism is still controversial. Figure 4 shows the room temperature PL emission graph of all the samples measured at an excitation wavelength of 305 nm. Different groups have reported various types of PL emissions bands (blue and green) from ZnS:Cu nanocrystals [17–24]. Xu et al. [24] have observed PL

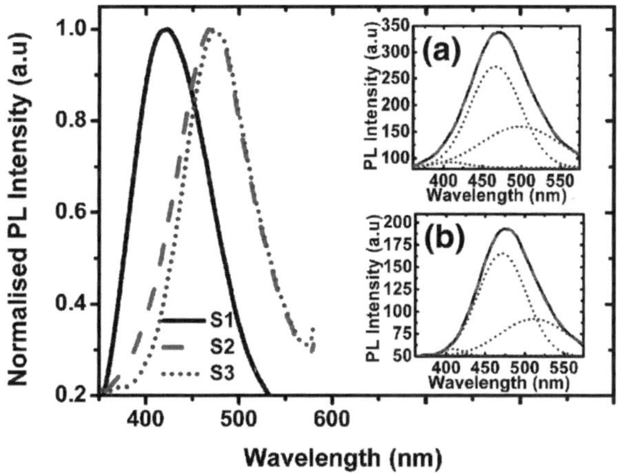

Fig. 4 Room temperature photoluminescence (PL) spectra of ZnS and ZnS:Cu nanoparticles at 305 nm excitation wavelength. Inset (**a**) and (**b**) shows the Gaussian fittings of individual PL spectra of *S*1 and *S*3 samples, respectively. In the insets *dashed curves* (*red* colour) shows the theoretical Gaussian fitting of experimentally obtained curves (*solid black* colour) and the *dotted* curves (*blue* colour) show the individual *Gaussian curves*

spectrum of Cu-doped ZnS consists of two emission bands; one is at 460 nm while another is at 507 nm. Lee et al. [21] has reported PL emissions at 420 and 520 nm from ZnS:Cu NPs. Khosravi et al. [23] have observed single green (480 nm) emission band and Huang et al. [22] have observed a single blue emission band (415 nm) from nanocrystalline ZnS:Cu. Bol et al. [18] has also reported a red emission from nanocrystalline ZnS:Cu. Very recently, Zhang et al. [17] has reported the PL emission centered at 500–505 nm originated from substitutional Cu centers (Cu^{2+}) in ZnS:Cu NPs. ZnS can host large radii ions of Cu^{2+} (atomic radius 0.71 Å) where Zn^{2+} ion can be easily substituted by Cu^{2+} ion because Zn is more active in chemical reactions than Cu. When Cu^{2+} ions are doped in host ZnS, the tetrahedral crystal field of the four S^{2-} ligands the $3d^9$ ground state splits into higher lying t_2 levels and lower lying e levels [11, 18]. Whenever an excited electron makes a transition from shallow (delocalized) donor level to the t_2 level of Cu^{2+}, a green emission takes place. The origin of blue emission is not clear and possibly, it is related to the colour centre formed by the spatial association of a substitutional Cu and interstitial Cu [11]. By Gaussian fitting of the measured PL spectra of S1, it is found that there are two emission peaks centered at 405 and 443 nm and is shown in the inset (a) of Fig. 4. The PL peak at 405 nm is well known due to the recombination between the sulfur-vacancy related donor and valance band [11].The peak at 443 nm is attributed to the trap states emission of ZnS, related with the zinc vacancy. The PL spectra of the ZnS:Cu NPs (i.e., S2 and S3) are also fitted with Gaussian fitting. The inset (b) of Fig. 4 shows a typical example of the Gaussian de-convolution of S3 sample. For Cu^{2+} doped ZnS samples in addition to the above two, a third peak are noted in green region. As the third peak is absent in undoped ZnS (i.e., S1), we can conclude that this green emission is due to Cu^{2+} ion incorporation in ZnS host. According to Peng et al. [11], the green emission arises from the recombination between the shallow donor level (sulfur vacancy) and the t_2 level of Cu^{2+}. A red shift (from 495 to 511 nm) in third peak is noted with increasing Cu^{2+} content. Such red shift is noted as the t_2 energy level of Cu^{2+} ions is shifted further from the valance band with increasing Cu^{2+} concentration [18].

The experiment for measurement of NLO properties is carried out in two parts. For OA z-scan measurement, the transmission detector was kept fully open so that the total surface area of the sensing detector is exposed. For closed aperture (CA) z-scan measurement, an aperture having a transmittance of $S = 0.07$ was inserted in front of the transmission detector. The OA z-scan scheme allowed measuring the nonlinear absorption coefficient (β), whereas the CA z-scan scheme allowed determining the sign and magnitude of the nonlinear index (n_2) [31].

The intensity dependent nonlinear absorption coefficient (β) in the presence of 3PA can be written as [30],

$$\alpha = \alpha_0 + \beta_{3\mathrm{PA}} I_0^2 \tag{5}$$

and the intensity dependent nonlinear refraction coefficient (γ) is written as,

$$n = n_0 + \gamma I_0 \tag{6}$$

where α_0 is the linear absorption coefficient at the concerned wavelength in cm^{-1} and two-photon absorption (2PA) (β_{2PA}) and three photon absorption (3PA) (β_{3PA}) coefficients are expressed in cm/GW and cm^3/GW^2 units, respectively. The input peak intensity I_0 is expressed in W/cm^2 and NLR coefficient (γ) is in cm^2/W. Here n_0 is the linear refractive index and the nonlinear refractive index, n_2 (esu) $= 10^{-4}(cn_0/40\pi)\ \gamma\ (cm^2/W)$, c being the velocity of light in vacuum in cm/sec. In the case of far field OA z scan with S = 1, the normalized transmittance $T_{m\mathrm{PA}}\ (z)$ in the presence of multiphoton absorption (mPA) ($m = 2$, 3, and 4 for two-, three-, and four-photon absorption cases, respectively) can be written as [36],

$$T_{m\mathrm{PA}}(z) = \left[1 + m\beta_{m\mathrm{PA}} L_{eff}^{(m-1)} \left(I_0/\left(1 + (z/z_0)^2\right)\right)\right]^{-1/m} \tag{7}$$

where $L_{eff}^{(m-1)} = [1\text{-}\exp\{-(m-1)\ \alpha_0 L\}]/(m-1)$, is the effective thickness of the sample taking into account of mPA, and L is the actual thickness of the sample. Figure 5 show the OA z-scan data for sample S2. The symbols denote the experimental points; and the solid lines are the theoretically fitted curves by considering 3PA processes, respectively. The bandgap (E_{gn}) of the NPs are >5.1 eV, while the incident photon energy $E_p = 2.33$ eV. Therefore the condition $2E_p < E_{gn} < 3E_p$ is satisfied for all the samples. It clearly eliminates the possibility of occurrence of 2PA and shows that 3PA is the operative and dominant mechanism for the present experimental observations [28]. We have also verified such dominance of 3PA by another simple method proposed by He et al. [37] and the result is shown in the inset of (a) of Fig. 5, viz., for sample S2. By using linear

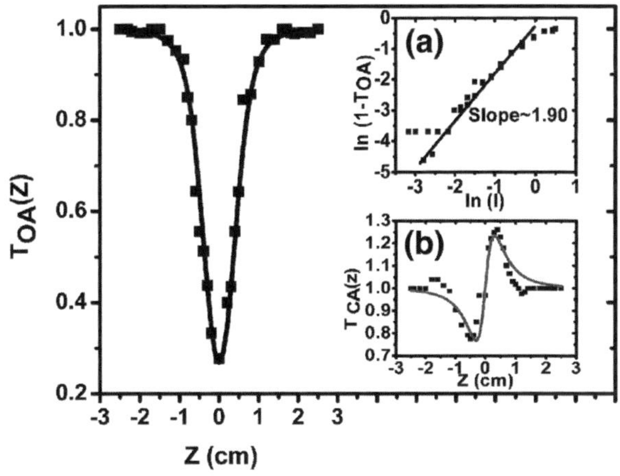

Fig. 5 Open aperture z-scan transmission trace of 0.50 % Cu doped ZnS (S2) sample. The *solid curve* is theoretically fitted one considering 3PA in the sample. Symbols are experimental points in every case. The inset (**a**) shows the variation of *ln*(1-*TOA*) with *ln*(*I*). The *solid line* represents the linear fit to the data which gives the slope ~2.0, indicating 3PA process. The inset (**b**) shows the closed aperture z-scan transmission traces at an intensity of I0 = 1.64 GW/cm^2. Symbols are experimental points and *solid curves* are theoretically fitted ones as described in the text

fit to the plot of $\ln(1-T_{OA})$ versus $\ln(I)$, where $I = I_0/(1 + z^2/z_o^2)$, it is found that the slope of the curve is ~2, which indicates the presence of 3PA process.

Using the value of β_{3PA} of NP solution obtained from theoretical fittings, the value of intrinsic 3PA coefficient β_{Int3PA} for the sample have been calculated using $\beta_{Int3PA} = \beta_{3PA} n_{0s}^3 \Big/ (n_{0QD}^3 |f|^6 f_v)$ where f is the local field correction that depends on the dielectric constants of the solvent methanol and the NPs and f_v is the volume fraction of the NPs in the solution [27]. The calculated value of β_{Int3PA} for S2 (0.50 % ZnS:Cu) at a peak intensity of 1.64 GW/cm^2 is 3.77 × 10^6 cm^3/GW^2 which is 10^9 times larger than that of bulk ZnS measured at 760 nm [37] and 10^2 times larger than that of ZnS and ZnS:Mn NPs [15].

The CA z-scan traces are also taken at 532 nm and are shown in the inset (b) of Fig. 5. Though NLR can be obtained from CA z-scan traces, but due to absorption effect the pure or absolute nonlinear refraction coefficients cannot be obtained directly from CA z-scan traces. In order to separate the pure NLR from the experimental CA z-scan traces, various methods have been proposed [31, 38, 39]. In the Seikh-Bahae method [31], the experimental normalized CA transmittance ($T_{\rm CA}^{\rm exp}$) is divided by the experimental normalized OA transmittance ($T_{\rm OA}^{\rm exp}$) to obtain roughly the pure nonlinear refraction, i.e., the normalized transmittance for pure NLR is given as, $T_{\rm CA}^{\rm Pure\ NLR} = T_{\rm CA}^{\rm exp}/T_{\rm OA}^{\rm exp}$. In Chappel method [38], the normalized transmittance for pure NLR is given as, $T_{\rm CA}^{\rm pure\ NLR} = T_{\rm CA}^{\rm exp}/[T_{CA}^{\rm exp}(z/1.25)]^{S+0.67(1-S)^2}$ where S = 1− $exp(-2r_{\rm a}^2/w_{\rm a}^2)$, which is the linear aperture transmittance, $r_{\rm a}$ is the aperture radius and $w_{\rm a}$ is the beam radius at the aperture. In the present experiment, the CA z-scan data for pure NLR are extracted following another method, known as Guo et al. [39] method because this method is more accurate than other methods for extracting NLR by using the z-scan technique with Gaussian beam [15]. According to this method, the expression for the normalized transmittances T_{CA} in the presence of both NLA and NLR with finite aperture size S can be written as,

$$T_{CA} = 1 + \left[(1-S)^y \sin\varphi / S\left(1+x^2\right)\right]\Delta\varphi + \left[\{(1-S)^y \cos\varphi - 1\}/S\left(1+x^2\right)\right]\Delta\psi \tag{8}$$

With $y = 2(3+x^2)/(9+x^2)$, $\phi = -4\,x\ln(1-\mathrm{S})/(9+x^2)$, $\Delta\phi = k_0 L_{\mathrm{eff}}$, and $\Delta\psi = \beta I_0 L_{\mathrm{eff}}/2$. Where, $x = z/z_0$, indicates the relative position of the sample (z_0 is the Rayleigh length); At low intensity of incident radiation, $\Delta\phi$ is the on-axis nonlinear phase-shifts caused by nonlinear refraction at the focus and $\Delta\psi$ is the proportion of irradiance absorbed on–axis at the focus. The theoretical normalized transmittance T_{CA} from the contribution of pure NLR can be obtained by putting $\Delta\psi = 0$ in the above equation and can be written as,

$$T_{CA} = 1 + \left[(1-S)^y \sin\varphi / S\left(1+x^2\right)\right]\Delta\phi \tag{9}$$

The contribution of the pure NLR from the experimentally obtained z-scan traces can be obtained by using the following analytical relationship,

$$T_{\mathrm{CA}}^{\mathrm{NLR}}(x) = T_{\mathrm{CA}}(x) + [1 - T_{\mathrm{OA}}(x)]A/S \tag{10}$$

Here $A = [1-(1-S)^{2/3}]$. The above equation is more reasonable compared with the other methods because the normalized transmittance of the CA z-scan arises from the contribution of superposition of both NLR and NLA but is not multiplying or dividing of both. By defining the difference between the normalized peak and valley transmittances in $T_{\mathrm{CA}}^{\mathrm{NLR}}$ (x) as ΔT_{pv}, the relation between ΔT_{pv} and $\Delta\varphi$ for any given S can be written as,

$$\begin{aligned}\Delta T_{\mathrm{pv}} \cong 0.406\big[1 - 0.4070 - 0.1461A^2 - 0.4841A^3 \\ +0.3826A^4 + 0.0423A^5 - 0.3696A^6\big]\,\Delta\phi\end{aligned} \tag{11}$$

The NLR coefficient (γ) of the samples is calculated by following the procedure as described below. At first the experimental CA z-scan trace is collected using an aperture with S = 0.07 and then $T_{\mathrm{CA}}^{\mathrm{NLR}}$ (x) at different values of x are collected by using Eq. (10). After that, the value of ΔT_{pv} is obtained and with that value of ΔT_{pv} and Eq. (11), the value of $\Delta\varphi$ is found out. In Fig. 5, the solid lines represent the theoretical curves drawn by calculated values of $\Delta\varphi$ and Eq. (9). The valley-peak shape of the CA z-scan traces indicates that S2 sample exhibit positive (i.e., self-focusing) nonlinearity. The analysis of Kramers–Kronig (K–K) relations predicts that a positive nonlinearity will occur if $E_p/E_g < 0.69$. For the used samples, $E_p/E_g < 0.46$, hence we expect self focusing in all the samples. The calculated values of γ and hence n_2 for sample S2 are 3.47×10^{-18} m^2/W and 21.53×10^{-12} esu, respectively.

4 Conclusion

ZnS:Cu NPs of average sizes ~2.5 nm have been successfully synthesized by chemical precipitation methods at room temperature by using methanol as solvent. From the Gaussian fittings of the measured PL spectra it is found that three PL peaks are present in the PL spectra of ZnS:Cu samples. A red shift is found with increasing Cu^{2+} concentration. Such red shift is noted as the t_2 energy level of Cu^{2+} ions is moved away from the valance band with increasing Cu^{2+} concentration. We have also reported for the first time, self focusing NLR and 3PA absorption in ZnS:Cu NPs. The efficient visible emission as well as large NLA and NLR of the synthesized samples showed that they can be used as good materials in optoelectronic and photonic devices.

Acknowledgments Authors are grateful to Department of Science and Technology (SR/FTP/PS-67/2008), Government of India, for the financial support. They express their sincere thanks to Dr. U. Chatterjee of Physics Dept, Burdwan University and Dr. R. Sarkar, NIT Durgapur for their technical help during experiments. AKK is grateful to the Ministry of Human Resource Development (MHRD), National Institute of Technology Durgapur, India for the maintenance scholarship.

References

1. T. Nakada, K. Furumi, A. Kunioka, High-efficiency cadmium-free Cu(In, Ga)Se2 thin-film solar cells with chemically deposited ZnS buffer layers. IEEE Trans. Electron Devices **46**, 2093 (1999)
2. F. Akihito, W. Hideo, S. Ken-Ichiro, N. Shigeru, H. Masato, Diamond-ZnS composite infrared window. Proc. SPIE **4375**, 206 (2000)
3. D.V. Talapin, J. Lee, M.V. Kovalenko, E.V. Shevchenko, Prospects of colloidal nanocrystals for electronic and optoelectronic applications. Chem. Rev. **110**, 389 (2010)
4. X. Fang, Y. Bando, U.K. Gautam, T. Zhai, H. Zeng, X. Xu, M. Liao, D. Golberg, ZnO and ZnS nanostructures: ultraviolet-light emitters, lasers, and sensors. Crit. Rev. Solid State Mater. Sci. **34**, 190 (2009)
5. B. Liu, Y. Bando, X. Jiang, C. Li, X. Fang, H. Zeng, T. Terao, C. Tang, M. Mitome, D. Golberg, Self-assembled ZnS nanowire arrays: synthesis, in situ Cu doping and field emission. Nanotechnol. **21**, 375601 (2010)
6. T. Tohda, Y. Fujita, T. Matsuoka, A. Abe, New efficient phosphor material ZnS:Sm, P for red electroluminescent devices. Appl. Phys. Lett. **48**, 95 (1986)
7. L. Brus, Electronic wave functions in semiconductor clusters: experiment and theory. J. Phys. Chem. **90**, 2555 (1986)
8. R.N. Bhargava, D. Gallagher, X. Hong, A. Nurmikko, Optical properties of manganese-doped nanocrystals of ZnS. Phys. Rev. Lett. **72**, 416 (1994)
9. R.N. Bhargava, Doped nanocrystalline materials -physics and applications. J. Lumin. **70**, 85 (1996)
10. S. Sapra, A. Prakash, A. Ghangrekar, N. Periasamy, D.D. Sharma, Emission properties of manganese-doped ZnS nanocrystals. J. Phys. Chem. B **109**, 1663 (2005)
11. W.Q. Peng, G.W. Cong, S.C. Qu, Z.G. Wang, Synthesis and photoluminescence of ZnS:Cu nanoparticles. Opt. Mater. **29**, 313 (2006)

12. W. Chen, A.G. Joly, J. Malm, J. Bovin, Upconversion luminescence of Eu^{3+} and Mn^{2+} in ZnS: Mn^{2+}, Eu^{3+} codoped nanoparticles. J. Appl. Phys. **95**, 667 (2004)
13. C.S. Tiwari, P. Kumbhakar, A.K. Mondal, A.K. Mitra, Synthesis and enhanced green photoluminescence emission from BCT ZnS nanocrystals. Phys. Status Solidi A **207**, 1874 (2010)
14. S.S. Nath, D. Chakdar, G. Gope, J. Kakati, B. Kalita, A. Talukdar, D.K. Avasthi, Green luminescence of ZnS and ZnS:Cu quantum dots embedded in zeolite matrix. J. Appl. Phys. **105**, 094305 (2009)
15. M. Chattopadhyay, P. Kumbhakar, R. Sarkar, A.K. Mitra, Enhanced three-photon absorption and nonlinear refraction in ZnS and Mn^{2+} doped ZnS quantum dots. Appl. Phys. Lett. **95**, 163115 (2009)
16. A.K. Kole, P. Kumbhakar, Effect of manganese doping on the photoluminescence characteristics of chemically synthesized zinc sulfide nanoparticles. Appl. Nanosci. **2**, 15 (2012)
17. W. Zhang, H. Lee, Synthesis and optical property of water-soluble ZnS:Cu quantum dots by use of thioglycolic acid. Appl. Opt. **49**, 2566 (2010)
18. A.A. Bol, J. Ferwerda, J.A. Bergwerff, A. Meijerink, Luminescence of nanocrystalline ZnS:Cu^{2+}. J. Lumin. **99**, 325 (2002)
19. M. Kuppayee, G.K.V. Nachiyar, V. Ramasamy, Synthesis and characterization of Cu^{2+} doped ZnS nanoparticles using TOPO and SHMP as capping agents. Appl. Surf. Sci. **257**, 6779 (2011)
20. P. Yang, M. Lü, D. Xü, D. Yuan, G. Zhou, Synthesis and photoluminescence characteristics of doped ZnS nanoparticles. Appl. Phys. A **73**, 455 (2001)
21. S. Lee, D. Song, D. Kim, J. Lee, S. Kim, I.Y. Park, Y.D. Choi, Effects of synthesis temperature on particle size/shape and photoluminescence characteristics of ZnS:Cu nanocrystals. Mater. Lett. **58**, 342 (2004)
22. J. Huang, Y. Yang, S. Xue, B. Yang, S. Liu, J. Shen, Photoluminescence and electroluminescence of ZnS:Cu nanocrystals in polymeric networks. Appl. Phys. Lett. **70**, 2335 (1997)
23. A.A. Khosravi, M. Kundu, L. Jatwa, S.K. Deshpande, U.A. Bhagwat, M. Sastry, S.K. Kulkarni, Green luminescence from copper doped zinc sulphide quantum particles. Appl. Phys. Lett. **67**, 2702 (1995)
24. S.J. Xu, S.J. Chua, B. Liu, L.M. Gan, C.H. Chew, G.Q. Xu, Luminescence characteristics of impurities-activated ZnS nanocrystals prepared in microemulsion with hydrothermal treatment. Appl. Phys. Lett. **73**, 478 (1994)
25. R.A. Ganeev, M. Baba, M. Morita, D. Rau, H. Fujii, A.I. Ryasnyansky, N. Ishizawa, M. Suzuki, H. Kuroda, Nonlinear optical properties of CdS and ZnS nanoparticles doped into zirconium oxide films. J. Opt. A:Pure Appl. Opt. **6**, 447 (2004)
26. X.B. Feng, G.C. Xing, W. Ji, Two-photon-enhanced three-photon absorption in transition-metal-doped semiconductor quantum dots. J. Opt. A: Pure Appl. Opt. **11**, 024004 (2009)
27. J. He, W. Ji, J. Mi, Y. Zheng, J.Y. Ying, Three-photon absorption in water-soluble ZnS nanocrystals. Appl. Phys. Lett. **88**, 181114 (2006)
28. G. Xing, W. Ji, Y. Zheng, J.Y. Ying, Two- and three-photon absorption of semiconductor quantum dots in the vicinity of half of lowest exciton energy. Appl. Phys. Lett. **93**, 241114 (2008)
29. B. Gu, J. Wang, J. Chen, Y. Fan, J. Ding, H. Wang, Z-scan theory for material with two- and three-photon absorption. Opt. Express **13**, 9230 (2005)
30. M. Chattopadhyay, P. Kumbhakar, C.S. Tiwary, R. Sarkar, A.K. Mitra, U. Chatterjee, Multiphoton absorption and refraction in Mn^{2+} doped ZnS quantum dots. J. Appl. Phys. **105**, 024313 (2009)
31. M. Sheik-Bahae, A.A. Said, T.H. Wei, D.J. Hagan, E.W. Van Stryland, Sensitive measurement of optical nonlinearities using a single beam. IEEE J. Quantum Electron **26**, 760 (1990)

32. R. Sarkar, C.S. Tiwary, P. Kumbhakar, S. Basu, A.K. Mitra, Yellow-orange light emission from Mn^{2+} doped ZnS nanoparticles. Phys. E **40**, 3115 (2008)
33. J.I. Pankove, *Optical Processes in Semiconductor* (Prentice-Hall, New Jersey, 1971)
34. B.D. Cullity, *Elements of X-ray Diffraction*, 2nd edn. (Addison Wesley Company, USA, 1978)
35. W. Chen, J.-O. Malm, V. Zwiller, Y. Huang, S. Liu, R. Wallenberg, J.-O. Bovin, L. Samuelson, Energy structure and fluorescence of Eu^{2+} in ZnS:Eu nanoparticles. Phys. Rev. B **61**, 11021 (2000)
36. P. Kumbhakar, M. Chattopadhyay, A.K. Mitra, Nonlinear optical properties of doped ZnS quantum dots. Int. J. Nanosci. **10**, 177 (2011)
37. J. He, Y.L. Qu, H.P. Li, J. Mi, W. Ji, Three-photon absorption in ZnO and ZnS crystals. Opt. Express **13**, 9235 (2005)
38. P.B. Chapple, J. Staromlynska, J.A. Hermann, T.J. Mckay, R.G. Mcduff, Single-beam z-scan: measurement techniques and analysis. J. Nonlinear Opt. Phys. Mater. **6**, 251 (1997)
39. S. Guo, K. Xu, J. Yun, H. Wang, X. You, N.B. Ming, An accurate method for extracting nonlinear refraction by z-scan technique in the presence of nonlinear absorption. J. Nonlinear Opt. Phys. Mater. **12**, 307 (2003)

Structural and Optical Study of ZnS Nanoparticles Doped with Different Concentration of Co

Harish Lakhotiya, Gagandeep, Chetan Saini, Ankit Goyal, K. V. R. Rao and S. L. Kothari

Abstract Zinc Sulfide is an II–VI semiconductor material which subjected to study due to their size and shape dependent optical characteristics. Zinc Sulfide particles which size is in nanometer range follow the quantum confinement phenomenon. The whole II–VI semiconducting family is important but among the family ZnS, CdS, ZnO, CdTe are foremost candidate because of optical and electronic properties. Synthesis of ZnS nanoparticles pristine and doped with different concentration (wt%) of Co has been done using chemical co-precipitation method. Novel luminescence characteristics, such as, stable and visible-light emission with different colors were observed from doped ZnS nanocrystals at room temperature. Co is interest of working as a dopant due to showing a novel luminescence characteristics. XRD and TEM analysis has been done for structural studies. Peak broadening in XRD analysis shows grain size is in nanometer range. Average particle size has been calculated by Scherer Equation is in nanometer range. TEM analysis confirms nanometer size of particles. A slight shifting in photoluminescence emission peak was observed with the Co concentration. UV-visible spectrophotometer measurements has been done for band gap determination. Increase in band gap is observed as with increase in dopant concentration due to quantum confinement. Enhance optical properties of ZnS can be modulated and used in LEDs and LCDs.

Keywords Optical properties · ZnS · Nanoparticles · Structural analysis · XRD

H. Lakhotiya (✉) · Gagandeep · C. Saini · A. Goyal · K. V. R. Rao · S. L. Kothari
Centre for Converging Technologies, University of Rajasthan, Jaipur 302055, India
e-mail: harish.cct@gmail.com; ankitcct@gmail.com

K. V. R. Rao
Department of Physics, University of Rajasthan, Jaipur 302055, India

S. L. Kothari
Department of Botany, University of Rajasthan, Jaipur 302055, India

P. K. Giri et al. (eds.), *Advanced Nanomaterials and Nanotechnology*,
Springer Proceedings in Physics 143, DOI: 10.1007/978-3-642-34216-5_19,

1 Introduction

The synthesis and characterization of semiconductor nanoparticles are great interest of area due to presence of outstanding potential applications, owing to their nonlinear optical and luminescence properties, quantum size effect and other important physical and chemical properties [1]. Noncrystalline material that are semiconducting in nature play a major role in nanophotonics and nanoelectronics [2]. Semiconducting nanoparticles are mostly use in optical switching, single charge memories, single electron transistors, etc. [3]. Many of method are present for preparation of nanoparticle, one of the important over other methods is chemical routes. The advantages of chemical routes over other synthesis methods are many (1) relatively cheap (2) easier control of the oxidation states (3) ability to make nanostructures of different sizes and shapes [4]. Zinc Sulfide that is II-VI semiconducting nanoparticle [5] is more studies due to their size and shape dependent optical characteristics. Particle that size in nanometer range follow the quantum confinement phenomenon [2]. The whole II–VI semiconducting family is important but among the family ZnS, CdS, ZnO, CdTe are foremost candidate because of optical and electronic properties. ZnS is more commercially important due to presence of large band gap over other [5] 3d transition-metal impurities (Ti, V, Cr, Mn, Fe, Co, Ni, Cu) in semiconductors form deep levels within the band gaps of the host materials. They are technologically important as luminescence centers and charge compensations as well as unwanted traps. Novel luminescence characteristics, such as, stable and visible-light emission with different colors were observed from doped ZnS nanocrystals at room temperature [6]. Co is interest of working as a dopant due to showing a noval luminescence characteristics [7]. In this paper the structural and luminescence properties of ZnS nanoparticles alone and with different concentration of Co^{+2} ion as a dopant studies. For this all XRD, TEM, UV-Vis, PL characterization has been done and their results is basis of this paper.

2 Experimental Details

ZnS nanoparticle have been synthesized using chemical co precipitation method. Four samples of ZnS nanoparticle doped with Co^{+2} ion concentration by wt% 0, 2, 4, 6 % have been prepared, for simplification we have named as sample 1, sample 2, sample 3, sample 4 respectively. The zinc nitrate and cobalt sulfate were used as a source of Zn and Co respectively.

$$Zn(No_3)_2 + Na_2S \xrightarrow{\text{Roomtemp}} ZnS + 2\ NaNo_3.$$

Color of sample darken with increase in cobalt concentration. X-Ray diffraction pattern was recorded with Bruker AXSD-8 deffractometer and TEM image obtained by the HRTEM (FEI Tecnai-20). The optical studies has been done with help of absorption spectra and photoluminescence emission spectra by the Shimadzu UV Probe (ver. 2.32) and Shimadzu spectrofluorophotometer (RF-5301PC) respectively. Sample that were dry and in powder form, were dissolved into distal water in case of absorption spectra measurement and for emission spectra nanoparticles dispersed in methanol solvent.

3 Results and Discussion

3.1 Structural Measurement

Structural studies has been done using XRD and TEM analysis. XRD pattern of all sample are shown in Fig. 1, the peak broadening in pattern confirmed particle in nanometer range. Crystallite size calculated by Scherer's equation is in nanometer range which is also confirmed by TEM analysis (~3–4 nm particle size) as shown in Fig. 2. The crystallite size was estimated from the full width at half maximum of the major XRD peak using the Debye–Scherrer equation (Eq. 1)

$$D = 0.9\lambda / \beta\cos\theta \text{ (nm)}$$

with D being the crystallite size, λ the X-ray wavelength β the full peak width at half-maximum (FWHM), and θ representing the Bragg angle [9]. Figure 1 spectrum shows cubic zinc blende structure of ZnS with the three lattice planes (111), (220) and (311). Doping of Co does not make any significant change in structure and phase of pure form. As increase the concentration of dopant, highest intense peak slightly shift toward right angle. Figure 3 reveals the continuous increase in grain size and decrease in full width at half maximum (FWHM) as increase in dopant concentration.

3.2 Optical Measurement

The optical absorption spectra of all samples were taken in the wavelength range 250–700 nm. The relation between the absorption coefficient, α, and the incident photon energy, hν, can be written as:

$$\alpha h\nu = A\left(h\nu - E_g\right)^n$$

where α is the absorption coefficient, h is Planck's constant, ν is the photon frequency and E_g is the optical band gap, A is a constant depending on the transition probability and *n* is an index that characterizes the optical absorption process and is theoretically equal to n = 1/2 and 3/2 for direct allowed and direct

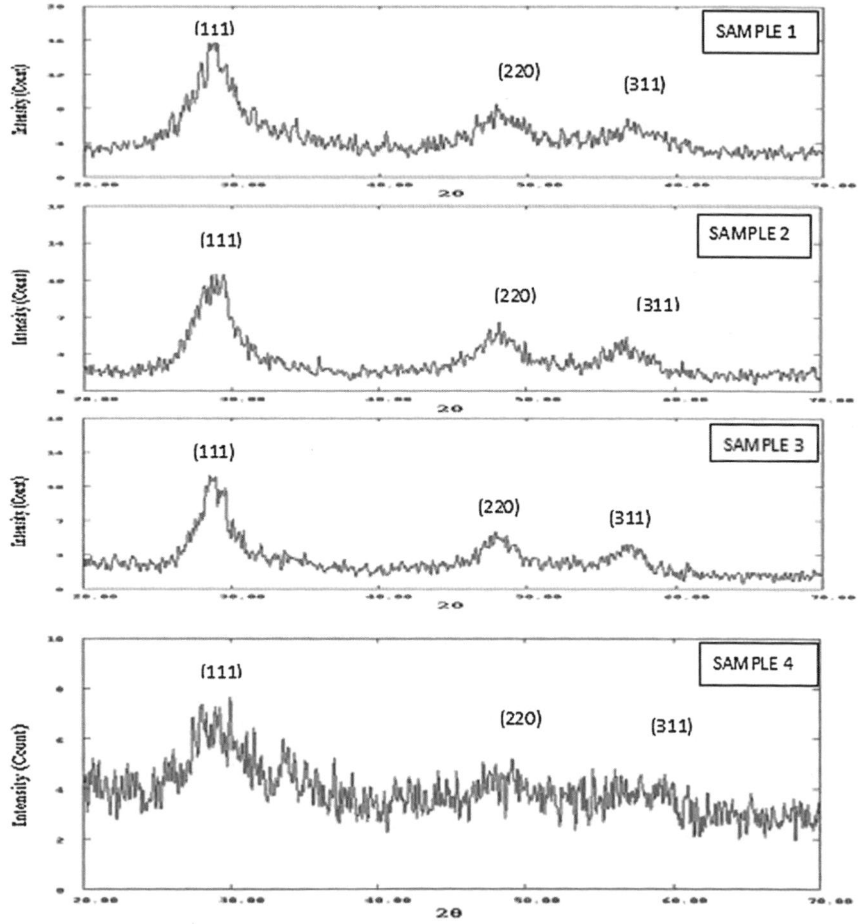

Fig. 1 XRD pattern of all samples

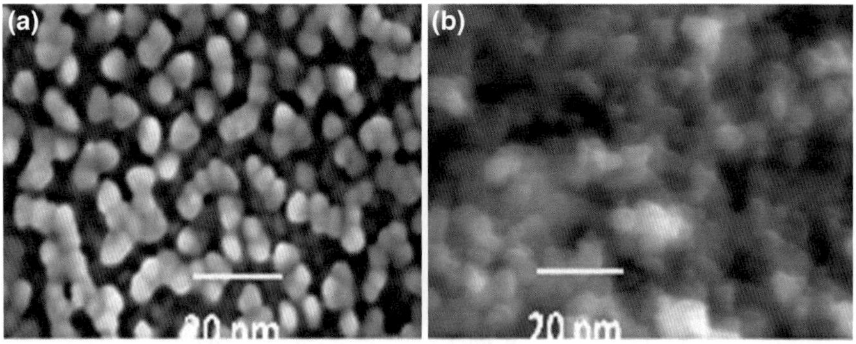

Fig. 2 TEM image of sample 1 and sample 2

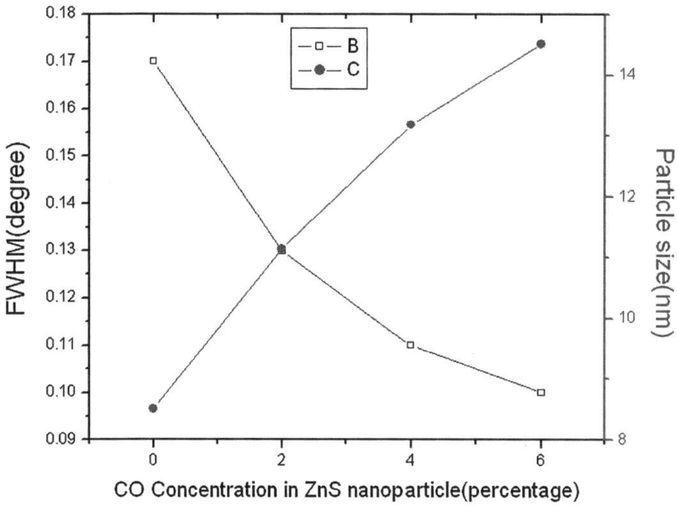

Fig. 3 Particle size and FWHM variation with dopant concentration

forbidden transition, respectively, while n = 2 and 3 for indirect allowed and indirect forbidden transition, respectively. The direct band gap for allowed transition of all samples were calculated by plotting $(\alpha h\nu)^2$ v/s hν [8]. An extrapolation of the linear region of a plot of $(\alpha h\nu)^2$ on the y-axis versus photon energy (hν) on the x-axis as shown in Fig. 4 gives the value of the optical band gap E_g. Band gap of pure ZnS is also more than bulk ZnS which can be related to particle size of samples. Band gap of ZnS sample (3.95 eV) is higher than bulk ZnS (3.6 eV) due to the quantum confinement in ZnS nanoparticle and continuous increase with dopant concentration that shows blue shift in nature. Doping of Co affects the grain boundaries of ZnS, which increases the band gap Fig. 5 shows photoluminescence spectra of all samples taken at same excitation wavelength (280 nm). Doping of Co reveal slightly blue shift in emission spectra than pure form which justify the band gap increment as it happens in UV-Vis results. Dopant concentration basically tuned the intensity of the pure ZnS samples and such revealed results are significantly different from that reported by Vatankhah et al. [9] due to more electron-hole pairs form with the cobalt concentration in the samples that enhanced luminescence intensity. ZnS shows increase in band gap and material can be optimized to fabricate tunable optical device.

Increases in intensity have been observed in PL results which is follows the LASER phenomenon. This can be due to presence of sub energy levels induced due to doping.

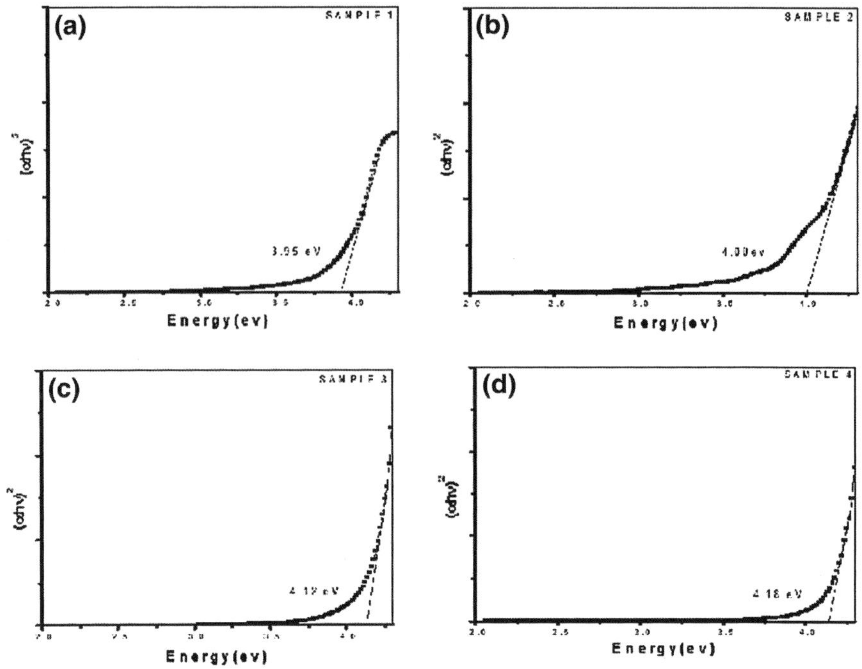

Fig. 4 Plot of $(\alpha hv)^2$ versus (hv) for all sample

Fig. 5 Photoluminescence spectra of all sample

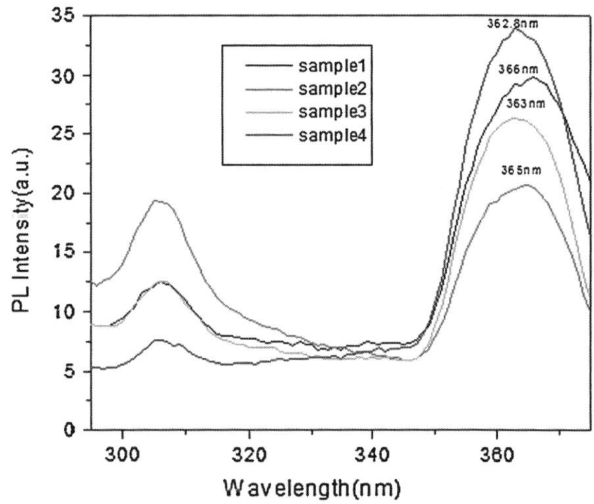

4 Conclusions

A systematic study reveals the following conclusion:

1. There is no change in Phase which reveals that basic properties are not altered by the experiment.
2. TEM confirms uniform particle size with uniform distribution.
3. Blue shift in Band gap shows possibility of Tunable optical device. This phenomenon can be used in LEDs and LCDs.
4. Increased Intensity in PL spectra is like LASER phenomenon and opens new possibilities of research on this material.

Acknowledgments Authors are thankful to Mr. Praveen Jain, CCT, UOR for their guidance and TEM characterization.

References

1. W. Wang, I. Germanenko, M.S. El-Shal, Room temperature synthesis and characterization of nanocrystalline CdS, ZnS and $Cd_xZn_{1-x}S$. Chem. Mater. **14**, 3028–3033 (2002)
2. E.K. Athanassiou, R.N. Grass, W.J. Stark, One-step large scale gas phase synthesis of Mn^{+2} doped ZnS nanoparticles in reducing flames. Nanotechnology **21**, 215603 (2010)
3. A. Goyal, V. Sharma, A. Sharma, R. Agarwal, K.B. Sharma, S.L. Kothari, Synthesis, structural and optical study of Cds nanoparticles doped with different concentration of Cu. J. Nano. Electron. Phys. **3,** 254 (2011)
4. S. Singhal, A.K. Chawla, H.O. Gupta, R. Chandra, Influence of cobalt doping on the physical properties of Zn0.9Cd0.1S nanoparticles. Nanoscale Res. Lett. **5**(2), 323–331 (2010)
5. J.P. Borah, J. Barman, K.C. Sarma, Structural and optical properties of Zns nanoparticles. Chalcogenide **5**(9), 201–208 (2008)
6. P. Yang, M. Lu, D. Xu, D. Yuan, C. Song, S. Liu, X. Cheng, Luminescence characteristics of ZnS nanoparticles co-doped with Ni^{2+} and Mn^{2+}. Opt. Mater. **24**(3), 497–502 (2003)
7. P. Yang, M. Lu, G. Zhou, D. Yuan, D. Xu, Synthesis and photoluminescence characteristics of doped Zns nanoparticles. Appl. Phys. A **73**(4), 455–458 (2001)
8. B.S. Rema Devi, R. Raveendaran, A.V. Vaidyan, Synthesis and characterization of Mn2+ -doped ZnS nanoparticles. PARANAMA J. Sci. **68**(4), 679–687 (2007)
9. C. Vatankhah, M.H. Yuosefi, A.A. Khosravi, M. Savarian, The study of optical properties of ZnS, ZnS:Co2+ nanoparticles. Eur. Phys. J. Appl. Phys. **48**, 20601 (2009)

Trapping Parameters of Thermally Stimulated Luminescence Glow Curves in Y_2O_3: Tb^{3+} Nanophosphor

S. Som, S. K. Sharma and S. P. Lochab

Abstract This paper reports the structural and thermoluminescence properties of terbium doped yttrium oxide nanophosphor. This terbium (Tb^{3+}) activated yttrium oxide (Y_2O_3) compound was prepared by combustion synthesis method. The prepared compound was characterized by X-ray diffraction (XRD) and Thermoluminescence (TL) techniques. The sharp and single peaks in XRD pattern indicate the existence of single phase in this compound. The crystallite size calculated by Debye–Scherrer formula was found in the range 10–20 nm. The Y_2O_3 nanophosphor exhibits the cubic (body centered-type) structure having unit cell dimension 10.60 Å with the space group Ia3. The structural parameters such as particle density (D_x), dislocation density (δ) and microstrain (ε) were also calculated from XRD. The phosphor was irradiated with gamma rays dose ranging from 250 Gy to 10 KGy for TL studies. This results in the TL glow curves showing a single peak at 325 °C. The intensity of this glow peak increases linearly with gamma dose up to 1.5 KGy. The composite TL glow peak was fitted by GlowFit program in order to obtain the isolated peaks. The trapping parameters associated with the isolated peaks were then calculated by using different glow curve analysis methods.

Keywords Nanophosphor · XRD · Thermoluminescence · Trapping parameter

S. Som (✉) · S. K. Sharma
Department of Applied Physics, Indian School of Mines, Dhanbad 826004, India
e-mail: sudipta.som@gmail.com

S. P. Lochab
Inter University Accelerator Centre, New Delhi 11006, India

P. K. Giri et al. (eds.), *Advanced Nanomaterials and Nanotechnology*,
Springer Proceedings in Physics 143, DOI: 10.1007/978-3-642-34216-5_20,

1 Introduction

The phosphor materials play an important role in modern life with tremendous applications ranging from color display, scintillators, fluorescent lamps, intensifying screens, dosimeter and many more [1, 2]. The incorporation of rare earth impurities to phosphor material changes its themoluminescence properties as well as dose response behavior with ion beam or any ionizing radiations like X-rays or γ-rays [1]. This variation is due to the change in chemical composition, defects, degree of structural disorder and by the presence of dopants/impurities.

Thermoluminescence (TL) technique plays a vital role in getting the information about the physical parameters such as order of kinetics, trap depth (activation energy) and frequency factor of the traps present in the materials. These parameters quantitatively describe the trapping and emitting centers and can be calculated by various methods. GlowFit computer program [3] was used for isolation of TL glow peaks from the composite TL glow curve. The trapping parameters of isolated TL glow peaks were then obtained through this program. Other thermoluminescence methods such as glow curve shape (Chen, Grosweiner, Kivitis and Hagebeuk) and different heating rate (Hoogenstraaten and Opanowicz) were also used to determine these parameters.

The present studies aims on the calculation of trapping parameters associated with the isolated TL glow peaks in order to get information about the mechanism of trapping and recombination of charge carriers within the traps. The knowledge of the mechanism of trapping and recombination of charge carriers taking place through localized states formed by traps is essential for the understanding of optoelectronic properties of materials. Some researchers had reported the luminescence properties of rare earth doped Y_2O_3 phosphors prepared by different methods such as Pechini, Co-precipitation and Wet chemical [4, 5]. To the best of our knowledge, so far no work has been reported till date on the calculation of trapping parameters of this doped phosphor. Keeping this in view, the present studies were carried out.

2 Experimental

Tb^{3+} doped Y_2O_3 nanophosphor was prepared by combustion synthesis method using terbium oxide (Tb_4O_7), yttrium oxide (Y_2O_3), nitric acid (HNO_3) and urea ($CO(NH_2)_2$) as starting raw materials. The stock solutions of $Y(NO_3)_3$ and $Tb(NO_3)_3$ were prepared by dissolving Y_2O_3 and Tb_4O_7 in nitric acid and diluting with deionized water. $Y(NO_3)_3$ and $Tb(NO_3)_3$ solutions were mixed according to the formula $(Y_{0.95}Tb_{0.05})_2O_3$ in a beaker and then a suitable amount of urea was added to it keeping urea to metal nitrate molar ratio as 2.5 [6]. The mixture was then dissolved properly to achieve a uniform solution and dried by heating at 80 °C. Finally the solid residue was transferred to crucible and heated at 600 °C in a furnace for an hour [6]. The synthesis reaction is [6]

$$(2-2x)Y(NO_3)_3 + 2xRE(NO_3)_3 + 5(NH_2)_2CO \rightarrow (Y_{1-x}RE_x)_2O_3 + 5CO_2 + 8N_2 + 10H_2O.$$

X-Ray diffractogram of the prepared compound was recorded in a wide range of Bragg angle 2θ ($20° \leq 2\theta \leq 90°$) using Philips Holland TW 1710 XRD measuring instrument with Co target radiation ($\lambda = 0.179$ nm) at IIT Kharagpur. The sample was irradiated with ^{60}Co gamma rays and the thermoluminescence studies were carried out using HARSHAW thermoluminescence analyzer at Inter University Accelerator Center, New Delhi. The calculation of different trapping parameters was done using GlowFit program and other TL analysis methods.

3 Results and Discussions

3.1 XRD Studies

The X-ray diffractogram of Tb^{3+} doped Y_2O_3 compound is shown in Fig. 1. The sharp and single diffraction peaks of the XRD pattern confirmed the formation of single phase compound. The doped Y_2O_3 compound exhibited the Cubic (body centered-type) structure with the space group Ia3. The (h k l) values of most prominent peaks and the corresponding d-values are shown in the XRD pattern. The lattice parameters of the unit cell were $a = b = c = 10.60$ Å with $\alpha = \beta = \gamma = 90°$ which was in correspondence with the JCPDS database of pdf number 83-0927 as shown in the Fig. 1.

For estimating the crystallite size (D) of prepared compound, Scherrer formula [3] was used:

$$D = 0.9\lambda / \beta\cos\theta \quad (1)$$

Here λ is the wavelength of the X-rays, β is the full-width at half maximum (FWHM) and θ is the angle of diffraction. The estimated crystallite size of Tb^{3+} doped Y_2O_3 was found to be in the range of 10–20 nm. The various structural parameters such as particle density (D_x), dislocation density (δ) and microstrain (ε) were also calculated from the formulae [7, 8] using XRD data:

$$D_x = 16M / \left(Na^3\right) \quad (2)$$

$$\delta = 1/D^2 \quad (3)$$

$$\varepsilon = \beta\cos\theta/4 \quad (4)$$

where M is the molecular mass, N is the Avogadro's number and a is the lattice parameter. The values of above structural parameters are summarized in Table 1.

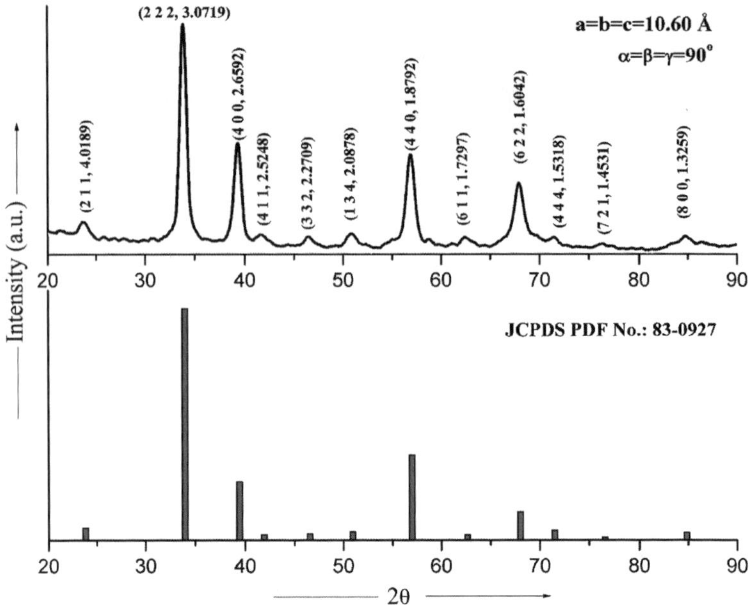

Fig. 1 XRD pattern of Tb doped Y_2O_3

Table 1 Structural parameters of Y_2O_3: Tb^{3+} phosphor

Maximum Peak (2θ) (deg.)	FWHM (β) (deg.)	Lattice parameter a (Å)	Crystallite size D (nm)	Particle density (D_x) gm/cm^3	Dislocation density (δ) ($\times 10^{15}/m^2$)	Microstrain ε ($\times 10^{-3}$)
33.84	0.81	10.60	13	5.0365	5.9	3.4

3.2 *Thermoluminescence Studies*

Thermoluminescence glow curves of Tb^{3+} doped Y_2O_3 phosphor were recorded after the irradiation by gamma rays. Before any TL measurements, the phosphor was annealed at 300 °C for 1 h to erase out the previous radiation memory from the phosphor material. The composite TL glow curves for different gamma doses at a heating rate 5 °C/s are shown in Fig. 2.

The TL glow curves show a broad peak with maximum intensity peak at around 325 °C. The samples were irradiated with different gamma doses from 250 Gy to 10 KGy to see the dose dependence effect on the glow curves. For all the doses, there is no significant change in the position of peak temperatures and they are within the experimental error ±3 °C.

The TL intensity of glow peaks increases linearly up to the dose of 1.5 KGy as shown in Fig. 3, after which it saturates and sub linearity was observed. The linear range up to 1.5 KGy indicates that this nanophosphor can be used as a dosimeter for the dose estimation of Gamma rays in clinical dosimetry applications.

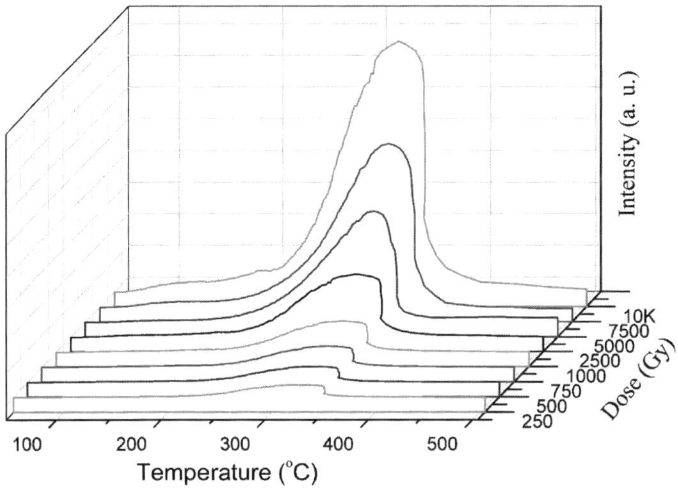

Fig. 2 Thermoluminescence glow curves at different gamma doses

Fig. 3 Thermoluminescence dose response curve

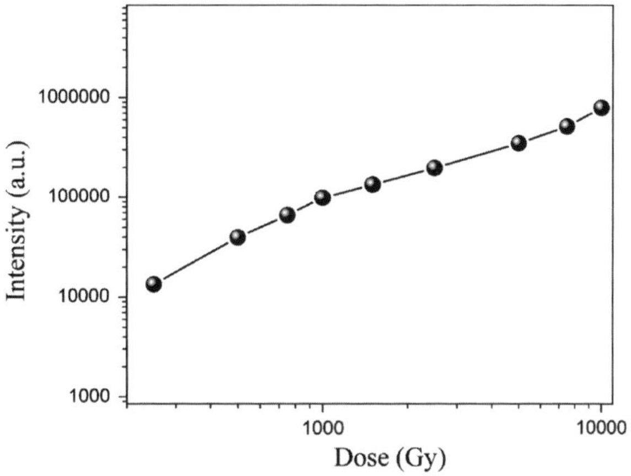

3.3 Determination of Trapping Parameters

The TL glow curve is related to the trap levels lying at different depths in the band gap between the conduction and the valence bands of a solid. These trap levels are characterized by different trapping parameters such as trap depth, order of kinetics and frequency factor [3]. The loss of dosimetry information stored in the materials after irradiation is strongly dependent on the position of trapping levels within the forbidden gap which is known as trap depth or activation energy (E). The mechanism of recombination of detrapped charge carriers with their counterparts is known as the order of kinetics (b). The frequency factors (s) represents the product of the number of times an electron hits the wall and the wall reflection coefficient, treating the trap as a potential well. Thus, a reliable dosimetry study of thermoluminescent material is based on its trapping parameters. Hence, trapping parameters of trap levels are investigated by different methods.

3.3.1 Order of Kinetics by Randall and Wilkins Model

The order of kinetics can be estimated by a model proposed by Randall and Wilkins [9, 10]. According to this model, the position of maximum intensity for first order kinetics can be estimated by the following equation:

$$\beta = (\frac{sK}{T})T_m^2 \exp(-\frac{E}{KT_m}) \tag{5}$$

where β is the linear heating rate and T_m is the temperature corresponding to the maximum intensity I_m of glow peak. Increasing the heating rate β on the left-hand side of Eq. 5 would give an increase in the right-hand side. Since $T_m^2 \exp(-E/kT_m)$ is an always increasing function of T_m, a growth in the value of the function necessarily implies an increase in argument T_m. This clearly indicates that with increasing heating rate both the parameters i.e. T_m and I_m of the glow peak increases.

According to above theory, the peak temperature of glow peaks are expected to change only with heating rate for first order of kinetics ($b = 1$). Figure 4 shows the glow peaks at different heating rates (3, 5, 7 °C/s). The shift in peak temperature of the glow peaks was observed around 20 °C with heating rates and the peak shifts towards higher temperature. This indicates that these glow peaks are of first order kinetics.

Moreover, for a constant heating rate, the peak temperature should not be affected by other experimental parameters and should be fairly constant within the limit of experimental errors [11]. Figure 5 demonstrate this fact, in which the peak temperature is almost constant within the limit of experimental errors for the different weights (2.5, 5, 7.5 mg) of the sample taken at a constant heating rate of 5 °C/s.

The above two facts are in favour that the observed TL glow peak at 325 °C follows first order kinetics.

3.3.2 Order of Kinetics by Chen Method

Chen [12] developed a method for finding the order of kinetics based on the shape of glow curve. This method utilizes the following shape parameters to determine the order of kinetics: the total half intensity width ($\omega = T_2 - T_1$), the high temperature half width ($\delta = T_2 - T_m$) and the low temperature half width ($\tau = T_m - T_1$). Here T_m is the peak temperature, T_1 and T_2 are the temperatures on either side of T_m corresponding to half peak intensity. Chen introduced a symmetry factor (μ_g) to differentiate between first and second order TL glow peaks:

$$\mu_g = \delta/\omega = (T_2 - T_m)/(T_2 - T_1) \tag{6}$$

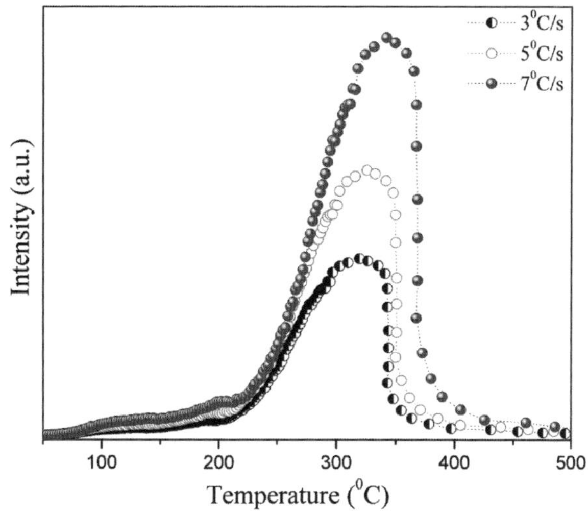

Fig. 4 Glow peak at different heating rates

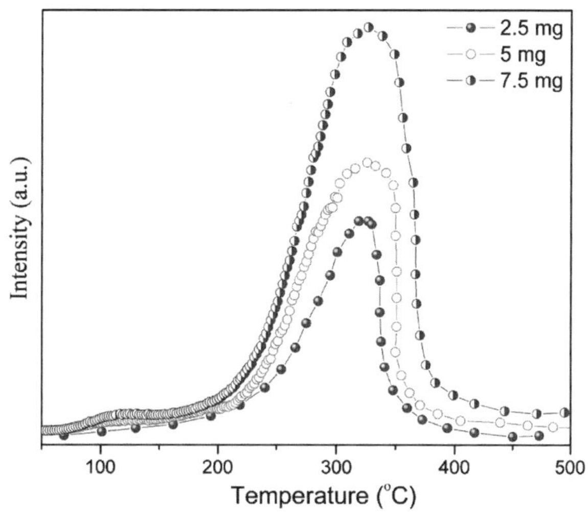

Fig. 5 TL glow curves of the sample by varying amount (at constant heating rate)

$$\mu_g = 0.42 \text{ for first order kinetics}$$
$$\mu_g = 0.52 \text{ for second order kinetics}$$

In the present case, the value of μ_g for the composite peak at three different heating rates was found to be around 0.42 which confirms it to be of first order. The values of shape parameters are given in Table 2.

3.3.3 Trap Depth and Frequency Factor by GlowFit Program

GlowFit program is applicable for finding the parameters of trap states when the glow peak is of first order. In the present case, the GlowFit program was employed

Table 2 Shape parameters of the composite glow peak

Heating rate (°C/s)	$\tau = T_m - T_1$ (°C)	$\delta = T_2 - T_m$ (°C)	$\omega = T_2 - T_1$ (°C)	μ_g
3	57	35	82	0.43
5	36	26	62	0.42
7	48	35	83	0.42

to obtain the trapping parameters. The recorded composite glow curves for gamma dose 2,500 Gy at different heating rates 3, 5 and 7 °C/s were first fitted by this computer program in order to obtain the isolated peaks. The parameter describing the quality of fitting, called Figure of Merit (FOM) was calculated for each fitting according to equation [13]:

$$FOM[\%] = \frac{\sum_i |y_i - y(x_i)|}{\sum_i y_i} \times 100\% \tag{7}$$

where y_i is content of the channel i, $y(x_i)$ represents the value of fitting function in the centre of channel i. Glow curves with FOM values greater than 5 % are subjected to further investigation to determine the reasons for the poor fit. Hence, in the present case minimum three peaks are required to get the best fit. The fitted TL glow curves at the different heating rates are shown in the Fig. 6. In each case the isolated peaks are designated as peak 1, peak 2 and peak 3. The trapping parameters of these isolated peaks at heating rates 3, 5 and 7 °C/s were calculated by GlowFit program and are shown in Table 3.

The trap depths calculated by GlowFit program are also verified using other glow curve analysis methods like various shape methods and different heating rate methods.

3.3.4 Trap Depth by Glow Curve Shape Methods

Trap depth (E) was calculated by using the Chen's equations [12], which gives the trap depth in terms of τ, δ and ω. A general formula for E was given by:

$$E_\gamma = c_\gamma(k\,T_m^2/\gamma) - b_\gamma(2\,k\,T_m) \tag{8}$$

where γ is τ, δ or ω. The constants c_γ and b_γ for the three equations (τ, δ and ω) and for first order kinetics are $c_\tau = 1.51$, $b_\tau = 1.58$, $c_\delta = 0.976$, $b_\delta = 0$, $c_\omega = 2.52$, $b_\omega = 1.0$.

Grosweiner [14] obtained a formula for the trap depth (E) in case of first-order kinetics:

$$E = G\frac{kT_mT_1}{\tau} \tag{9}$$

where $G = 1.51$, $\tau = T_m - T_1$, T_m is the temperature of the TL peak-maximum intensity (I_m) and $T_1 < T_m$ is the temperature at the TL intensity $I = 0.5\ I_m$.

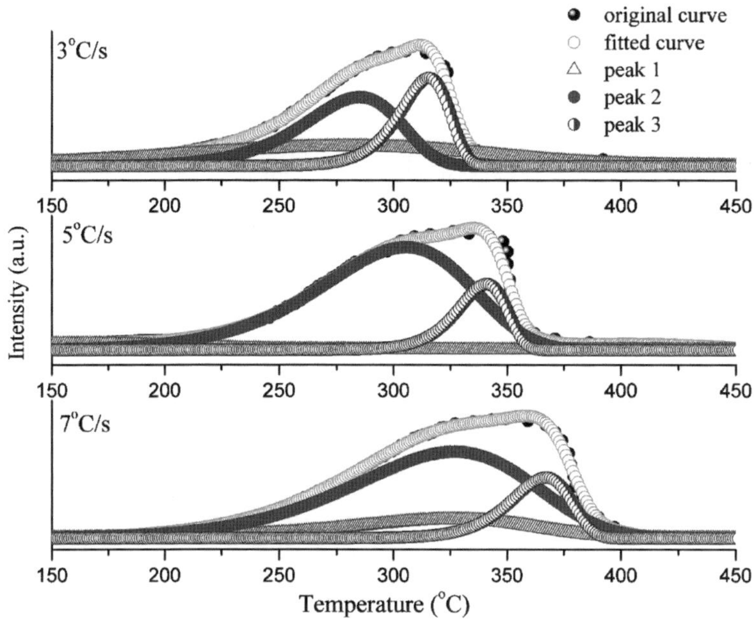

Fig. 6 Fitting of TL glow curves at heating rates 3, 5 and 7 °C/s

Table 3 Trap depth (E) and frequency factor (s) of isolated peaks by GlowFit program

Heating rate (°C/s)	Peak	T_m (°C)	E (eV)	s (s^{-1})	FOM (%)
3	Peak 1	189	0.26	97.4	1.87
	Peak 2	289	1.01	4.03×10^{8}	
	Peak 3	320	2.99	3.12×10^{25}	
5	Peak 1	192	0.25	22.15	2.04
	Peak 2	301	1.15	4.6×10^{9}	
	Peak 3	332	3.15	1.6×10^{26}	
7	Peak 1	270	0.23	39.85	1.97
	Peak 2	313	0.97	7.2×10^{7}	
	Peak 3	344	2.97	1.43×10^{24}	

Kivitis and Hagebeuk [15] found a formula for trap depth:

$$E = C_{KH} \frac{kT_m T_1}{\tau} \tag{10}$$

where $C_{KH} = \ln(2 + 3\tau/2\delta)$, $\delta = T_2 - T_m$, T_2 is the temperature at $I = 0.5\ I_m$.

Table 4 Trap depth (E) of isolated peaks by glow curve shape methods

Methods (°C/s)	Peak 1	Peak 2	Peak 3
Chen method			
3	0.23	1.01	3.05
5	0.20	1.15	3.18
7	0.23	0.95	3.04
Grosweiner et al. Method			
3	0.24	0.99	2.98
5	0.20	1.14	3.10
7	0.23	0.95	3.01
Kivitis and hagebeuk Method			
3	0.24	0.99	2.98
5	0.20	1.14	3.11
7	0.23	0.94	2.98

The values of trap depth calculated by the above formulae for isolated peaks at heating rates 3, 5 and 7 °C/s are given in Table 4.

3.3.5 Trap Depth by Different Heating Rate Methods

In the Arrhenius/Hoogenstraaten methods, the slope of $\ln(T_m^2/\beta)$ versus $1/kT_m$ plot gives the trap depth according to the equation [16]:

$$\ln(T_m^2/\beta) = E/kT_m + \ln(E/ks) \tag{11}$$

where β is the linear heating rate and s (s^{-1}) indicates frequency factor.

Opanowicz [3] derived relation between the half-width of the TL peak ($T_m - T_1$) and the sample heating rate β:

$$\frac{\beta}{|T_m - T_i|} = D_i \exp\left(-\frac{E}{kT_m}\right) \tag{12}$$

where D_i is a constant dependent on the trap parameters, i = 1, 2; T_i is the temperature at $I = 0.5\,I_m$, $T_1 < T_m$, $T_2 > T_m$. So plotting $\ln(\beta/|T_m - T_i|)$ versus $1/T_m$ (T_m and T_i for different values of β) one obtains a straight line with a slope –E/k.

Table 5 Trap depth (E) of isolated peaks by different heating rate methods

Methods	Peak 1	Peak 2	Peak 3
Arrhenius/hoogenstraaten method	0.19	1.09	1.21
Opanowicz method ($T_i = T_1$ or T_2)	0.18	1.14	1.20

Table 5 shows the values of trap depth calculated by the above two methods, this show a close agreement between the trap depth calculated by different thermoluminescence methods.

4 Conclusion

XRD studies reveal the cubic structure of the prepared compound with crystallite size in the range 10-20 nm. The composite TL glow curves of Y_2O_3: Tb^{3+} nanophosphor in the gamma dose range 250 Gy–10 KGy are described as a superposition of three glow peaks which indicates three different sets of traps are being activated having its own trapping parameters. The trap depth found by GlowFit method was in good agreement with that calculated by Glow curve shape and different heating rate methods. This indicates the reliability of the methods used and accuracy of the measurements. The linear TL dose response up to 1.5 KGy indicates that prepared nanophosphor may be suitable for the application as gamma-rays thermoluminescence dosimeter (TLD) in that range. The existence of TL glow peak at high temperature indicates that Y_2O_3: Tb^{3+} nanophosphor can be used as dosimeter for measuring radiation levels at high temperature areas like reactors and space ships.

Acknowledgments The authors are thankful to University Grants Commission, New Delhi, Government of India for funding this work [Project F. No. 37-200/2009 (SR)].

References

1. J.M. Sung, S.E. Lin, W.C.J. Wei, Synthesis and reaction kinetics for monodispersive Y2O3:Tb^{3+} spherical phosphor particles. J. Europian Ceram. Soc. **27**, 2605–2611 (2007)
2. K.S. Kang, H.Y. Park, D.S. Bae, Synthesis and characterization of Eu doped Y_2O_3 particles by a glycothermal processing. J. Electroceram. **23**, 492–496 (2009)
3. A. Choubey, S.K. Sharma, S.P. Lochab, D. Kanjilal, Argon ions induced thermoluminescence properties of Ba0.12Sr0.88SO4 phosphor. J. Phys. Chem. Solids **72**, 136–143 (2011)
4. V. Singh, V.K. Rai, I.L. Rak, S. Watanabe, T.K.G. Rao, J.F.D. Chubaci, L. Badie, F. Pelle, S. Ivanova, NIR to visible up-conversion, infrared luminescence, thermoluminescence and defect centres in Y_2O_3: Er phosphor. J. Phys. D Appl. Phys. **42**, 065104–065112 (2009)
5. F. Vetrone, J.C. Boyer, J.A. Capobianco, A. Speghini, M. Bettinelli, J. Mater. Res. **19**, 3398–3406 (2004)
6. N. Vu, T.K. Anh, G. Yi, W. Strek, Photoluminescence and cathodoluminescence properties of Y2O3:Eu nanophosphors prepared by combustion synthesis. J. Lumin. **122–123**, 776–779 (2007)
7. Fu Yen-Pei, Preparation and characterization of Y_2O_3:Eu phosphors by combustion process. J. Mater. Sci. **42**, 5165–5169 (2007)
8. A. Vij, R. Kumar, A.K. Chawla, S.P. Lochab, R. Chandra, N. Singh, Swift heavy ion induced synthesis and enhanced photoluminescence of SrS:Ce nanoparticles. Opt. Mater. **33**, 58–62 (2010)

9. S.W.S. McKeever, *Thermoluminescence of Solids* (Cambridge University Press, Cambridge, 1988)
10. R. Chen, S.W.S. McKeever, *Theory of Thermoluminescence and Related Phenomena* (World Scientific Publishing co Pte. Ltd., Singapore, 1997)
11. A.N. Yazici, M. Topaksu, The analysis of thermoluminescence glow peaks of unannealed synthetic quartz. J. Phys. D Appl. Phys. **36**, 620–627 (2003)
12. X. Chen, E. Ma, G. Liu, Energy levels and optical spectroscopy of Er^{3+} in Gd2O3 nanocrystals. J. Phys. Chem. C **111**, 10404–10411 (2007)
13. M. Puchalska, P. Bilski, GlowFit—a new tool for thermoluminescence glow-curve deconvolution. Radiat. Meas. **41**, 659–664 (2006)
14. L.I. Grossweiner, A note on the analysis of first order glow curves. J. Appl. Phys. **24**, 1306–1309 (1953)
15. P. Kivits, H.J.L. Hagebeuk, Evaluation of the model for thermally stimulated luminescence and conductivity; reliability of trap depth determinations. J. Lum. **15**, 1–11 (1977)
16. R. Chen, Y. Kirsh, *Analysis of Thermally Stimulated Processes* (Pergamon Press, oxford, 1981)

Statistical Characterization of Flicker Noise Fluctuation of a Nano-Scale NMOS Transistor

Sarmista Sengupta and Soumya Pandit

Abstract The statistical variability of flicker noise for a nano-scale NMOS transistor due to intra-die variations in channel length, oxide thickness and mobility has been characterized. This has been done through theoretical estimations as well Monte Carlo-HSPICE simulation technique. BSIM-SPICE process parameters have been chosen for characterization purpose. The present work explicitly depicts the dependence of flicker noise variability on various causes of variability and device design parameters. 45 nm CMOS process technology has been considered in the present work. The theoretical model can be effectively used for statistical characterization which is essential for robust analog and RF circuit design.

Keywords Flicker noise · Intra-die process variability · Channel length · Oxide thickness · Mobility · Mean · Standard deviation

1 Introduction

With the downscaling of CMOS transistors in the sub-90 nm regime, the impact of process parameter variations on the performances of CMOS analog and digital circuit is becoming increasingly important [1]. The increasing amount of intra-die variability leads to critical challenges in the conventional design procedure of nano-scale integrated circuits. A circuit optimized with the conventional design methodology is more susceptible to random performance variability. For robust

S. Sengupta (✉) · S. Pandit
Institute of Radio Physics and Electronics, University of Calcutta, Kolkata, India
e-mail: sarmista.sengupta@gmail.com

S. Pandit
e-mail: soumya_pandit@ieee.org

P. K. Giri et al. (eds.), *Advanced Nanomaterials and Nanotechnology*,
Springer Proceedings in Physics 143, DOI: 10.1007/978-3-642-34216-5_21,

design of integrated circuits, it becomes essential to accurately characterize and model the amount of uncertainty of circuit performances due to the process variations [1, 2].

The intra-die variation consists of the parametric fluctuations between identically designed MOS transistors within a short distance. It introduces systematic variability in patterning and random variability in patterns. The major sources of random intra-die variability are random discrete dopants, channel length variation, oxide thickness variation, mobility variation etc., [3]. The variations in the channel length of the transistor have significant impact on the various electrical performances of a MOS transistor [4]. In advanced CMOS technology with oxide thickness of about 1 nm, the oxide thickness variation due to the comparable thickness of interface roughness affects many electrical performances of a MOS transistor [5]. Mobility variations arise from several complex physical mechanisms such as fluctuations in effective fields, fixed oxide charges, doping, surface roughness etc. The variations of mobility have also significant impact on the various electrical performances of a MOS transistor [6].

Flicker noise is an important parameter that needs to be critically considered for the design of analog circuits as well as RF circuits. However, this has not received enough attention by the researchers except recently in [7]. In [7], a flicker noise variability model has been developed considering the intra-die variations of traps. However, the issues of channel length variation, oxide thickness variation and mobility variation have not been taken care of. This paper for the first time presents an approach for statistical characterization of the variability of flicker noise power spectral density of an NMOS transistor due to intra-die process variations considering channel length variation, oxide thickness variation and mobility variation. The BSIM parameters corresponding to these physical parameters have been chosen for characterization. The statistical study has been performed theoretically as well as through Monte Carlo-HSPICE simulation technique. The theoretical estimates have been compared with simulation results and good accuracy has been obtained. The theoretical study gives an insight about the dependence of the coefficient of variation on various design parameters such as channel length and oxide thickness.

The rest of the paper is organized as follows. Section 2 presents the theoretical technique for characterization. Results and discussion are provided in Sect. 3 and finally conclusion is drawn in Sect. 4.

2 Statistical Characterization

2.1 Flicker Noise Model

The various mechanisms responsible for flicker noise and their physical explanations were reported much earlier [8]. There are two general concepts explaining the source of flicker noise in MOS transistors. In the carrier number fluctuation theory, the source of flicker noise is explained by the random fluctuation in the number of carriers in the channel and fluctuations in the surface potential. These are caused by trapping and releasing of carriers by traps located near the silicon–silicon dioxide interface. The mobility fluctuation theory attributes flicker noise to mobility fluctuations due to carrier interactions with lattice fluctuations. The recent approach is to unify the two theories [9, 10]. According to the unified theory the total drain current noise power is [9]

$$i_n^2(f) = \frac{kTq^2 I_D \mu_s}{a\gamma f L_{eff}^2 C'_{ox}} \left[A \ln \frac{N_0}{N_L} + B(N_0 - N_L) + \frac{1}{2} C\left(N_0^2 - N_L^2\right) \right] \tag{1}$$

Where A, B and C are technology dependent parameters, N_0 and N_L are the carrier densities at the source and drain ends of the channel, respectively. I_D is the drain current, μ_s is the surface carrier mobility and L_{eff} is the effective channel length. $C'_{ox} = \varepsilon_{ox}/t_{ox}$ is the oxide capacitance per unit area, t_{ox} is the oxide thickness. The other constants bear their significances as described in [9]. The surface carrier mobility is defined as

$$\mu_s = \frac{\mu_0}{1 + \theta_1(V_{GS} - V_T) + \theta_2(V_{GS} - V_T)^2} \tag{2}$$

μ_0 is the low field carrier mobility. θ_1 and θ_2 are the mobility degradation coefficients due to the vertical channel field. The values of these parameters depend upon the oxide thickness. The effective channel length L_{eff} is defined as

$$L_{eff} = L_{drawn} + XL - 2dL \tag{3}$$

where L_{drawn} is the drawn gate length, XL is the channel length offset due to mask/etch effect and dL is the offset due to source/drain lateral diffusion.

Since it is difficult to handle (1) analytically for statistical characterization purpose, we write (1) in the form

$$i_n^2 = \frac{KF I_D}{C'_{ox} L_{eff}^2 f} \Delta f \tag{4}$$

where KF is the Flicker noise coefficient. The value of the flicker noise coefficient depends upon the operating voltages, surface carrier mobility and effective channel length. These are approximated in the present model through the following three parameter functions.

$$KF(V_{GS}) = A1 + A2.V_{GS} + A3.V_{GS}^2 \tag{5}$$

$$KF\left(L_{eff}\right) = B1 + B2.L_{eff} + B3.L_{eff}^2 \tag{6}$$

$$KF(\mu_s) = C1 + C2.\mu_s + C3.\mu_s^2 \tag{7}$$

Here the values of the various parameters are to be determined based upon the chosen process technology.

The equivalent input-mean-square noise voltage at the gate of the MOS transistor is given as

$$e_n^2 = \frac{i_n^2}{g_m^2} = \frac{KF\left(V_{GS}, \mu_s, L_{eff}\right)}{2fC_{ox}'^2 W_{eff} L_{eff} \mu_{eff}} \Delta f \tag{8}$$

where W_{eff} is the effective channel width.

In presence of the lateral field, the surface mobility of the carriers is modified and the effective carrier mobility is given as, $\mu_{eff} = \mu_s.f\left(L_{eff}\right)$. This functional form is written as,

$$f\left(L_{eff}\right) = D1 - D2\exp\left(-\frac{L_{eff}}{D3}\right) \tag{9}$$

Here $D1$, $D2$ and $D3$ are three unknown parameters whose values are to be determined based upon the chosen process technology.

2.1.1 Flicker Noise Coefficient Extraction

In order to match with HSPICE simulation results corresponding to 45 nm CMOS process technology, the exact values of the flicker noise coefficient have to be extracted. This depends upon the applied gate-source voltage. For extraction purpose, it is assumed that the noise voltage is measured at 1 Hz bandwidth so that the term Δf is unity. With this (8) can be written as

$$\log\left[e_n^2\right] = \log\left[\frac{KF}{2\mu_{eff} W_{eff} L_{eff} C_{ox}'^2}\right] - \log[f] \tag{10}$$

The intercept of the log [*f*] vs. $\log\left[e_n^2\right]$ is given by

$$\log\left[\frac{KF}{2\mu_{eff} W_{eff} L_{eff} C_{ox}'^2}\right] \tag{11}$$

The input noise spectral density is simulated through HPSICE using 45 nm process technology and the necessary plot is obtained. From (8), the value of the coefficient *KF* can be extracted by substituting the values of the necessary process and geometry parameters. However, the value of effective mobility μ_{eff} needs to be extracted from the HSPICE simulation of the $I_D - V_{GS}$ curve, which is a standard procedure.

2.2 Theoretical Formulation

Let the process parameter set be $\mathbf{p} = \left[p_1, p_2, \ldots, p_{N_p}\right]$ and the mean values of the process parameters be $\boldsymbol{m_p}$. Expanding the performance parameter e_1 around the mean and assuming that the elements of the set $\mathbf{p}$ are independent and normally distributed gives

$$e_n(\mathbf{p}) \approx e_n(\boldsymbol{m_p}) + \sum_{i=1}^{N_p} \left.\frac{\partial e_1}{\partial p_i}\right|_{\mathbf{p}=m_\mathbf{p}} (p_i - m_{p_i}) \tag{12}$$

The mean and variance of e_1 are defined as follows

$$\boldsymbol{m_{e_n}} = \boldsymbol{e_n}(\boldsymbol{m_p}) \tag{13}$$

$$\sigma_{e_n}^2 = \sum_{i=1}^{N_p} \left(\left.\frac{\partial e_n}{\partial p_i}\right|_{p=m_p} \right)^2 \sigma_{p_i}^2 \tag{14}$$

In (13) and (15), $\left.\frac{\partial e_n}{\partial p_i}\right|_{\mathbf{p}=m_\mathbf{p}}$ represents the sensitivity of the performance parameter with respect to the chosen process parameters. An underlying assumption behind this is that the variations in the electrical performances are approximated as linear functions of the process parameters. However, this approximation is often found to be valid and sufficient to model all of the observed electrical performances.

2.2.1 BSIM Process Parameter Variability

The BSIM model parameters can be divided into two groups: process parameters and fitting parameters. The process parameters are directly related to the chosen process technology. In the present work, the BSIM process parameters that have been chosen are (1), *XL* (2) *TOXE* which is the electrical gate equivalent oxide thickness and (3) *U*0 which is the low field carrier mobility.

The variations of *XL* represent the intra-die variations of the channel length variability. The variations of *TOXE* represent the intra-die oxide thickness variations and the variations of *U*0 represent the intra-die variations of low field carrier mobility. The variations of these process parameters are assumed to be Gaussian. The 3σ variations are assumed to be $\pm$ 10 % of the technology specified nominal values. These are based upon ITRS specifications and that available in literature [2].

The sensitivity of the flicker noise with respect to the parameters *XL*, *TOXE* and *U*0 are defined from (5)–(14) as follows.

$$\sigma^2_{en(XL)} = \left(\frac{\partial e_n}{\partial L_{eff}} \cdot \frac{\partial L_{eff}}{\partial XL}\right)^2 . \sigma^2_{XL} \tag{15}$$

$$\frac{\partial e_n}{\partial L_{eff}} = \frac{C}{2}\sqrt{\frac{B1 + B2L_{eff} + B3L^2_{eff}}{L_{eff}\left\{D1 - D2\exp\left(-\frac{L_{eff}}{D3}\right)\right\}}} \left[\frac{B2 + 2B3L_{eff}}{B1 + B2L_{eff} + B3L^2_{eff}} - \frac{1}{L_{eff}} - \frac{\frac{D2}{D3}\exp\left(-\frac{L_{eff}}{D3}\right)}{D1 - D2\exp\left(-\frac{L_{eff}}{D3}\right)}\right] \tag{16}$$

From (3) we have, $\frac{\partial L_{eff}}{\partial XL} = 1$. In (16), $C = \frac{t_{ox}}{\varepsilon_{ox}} \frac{1}{\sqrt{2fW_{eff}\mu_s}}$.

$$\sigma^2_{en(U0)} = \left(\frac{\partial e_n}{\partial \mu_s} \cdot \frac{\partial \mu_s}{\partial U0}\right)^2 . \sigma^2_{U0} \tag{17}$$

$$\frac{\partial e_n}{\partial \mu_s} = \frac{A}{2}\sqrt{\frac{C1 + C2\mu_s + C3\mu_s^2}{\mu_s}}\left[\frac{C2 + 2C3\mu_s}{C1 + C2\mu_s + C3\mu_s^2} - \frac{1}{\mu_s}\right] \tag{18}$$

$$\frac{\partial \mu_s}{\partial U0} = \frac{1}{1 + \theta_1(V_{GS} - V_{TH}) + \theta_2(V_{GS} - V_{TH})^2} \tag{19}$$

where, $A = \frac{t_{ox}}{\varepsilon_{ox}\sqrt{2fL_{eff}W_{eff}f(L_{eff})}}$

$$\sigma^2_{en(TOXE)} = \left(\frac{\partial e_n}{\partial TOXE}\right)^2 . \sigma^2_{TOXE} \tag{20}$$

$$\frac{\partial e_n}{\partial TOXE} = B.\sqrt{1 + \frac{E1}{TOXE} + \frac{E2}{TOXE^2}} + \frac{B}{2} TOXE \frac{\left(-\frac{E1}{TOXE^2} - 2\frac{E2}{TOXE^3}\right)}{\sqrt{1 + \frac{E1}{TOXE} + \frac{E2}{TOXE^2}}} \tag{21}$$

where, $B = \frac{1}{\varepsilon_{ox}}\sqrt{\frac{KF}{2fW_{eff}L_{eff}U0f(L_{eff})}}$.

3 Results and Discussion

The theoretical as well as the numerical HSPICE simulation results that have been provided in this work are based upon 45 nm CMOS Predictive Technology Model [11].

In order to extract the values of the flicker noise coefficient *KF* as function of the gate voltage V_{GS} we assume $L_{eff} = 22.5$ nm, $W_{eff} = 90$ nm and $t_{ox} = 1.75$ nm.

The necessary spectral density plot for this as obtained from HSPICE simulation results is shown in Fig. 1. Using the extracted values of *KF*, we compare the results obtained from HSPICE simulation and analytical model (8). The coincidence of the simulated and model predicted noise spectral density proves accuracy of the model. *KF* is extracted for five different V_{GS} values. It is observed that with increase of V_{GS}, the value of *KF* increases which means that the noise also increases. The flicker noise spectral density depends upon channel length and carrier mobility. The variations of *KF* with effective length L_{eff} and surface carrier mobility μ_s are obtained from HSPICE simulation. We observe that as the effective length is reduced, the noise increases so as the value of *KF*. These two variations are justified from (1). The functional form of $f(L_{eff})$ as described in (9) is shown in Fig. 2. The values of the various unknown parameters of (5)–(7) and (9) are extracted by curve fitting technique and are summarized in Table 1.

The statistical variability of flicker noise spectral density for an NMOS transistor due to intra-die variations of channel length, oxide thickness and mobility variations are studied through Monte Carlo simulations using HSPICE.

The corresponding BSIM parameters, i.e., *XL*, *TOXE* and *U*0 are varied by $\pm$ 10 % of the technology specified mean values. We observe that the coefficient of variations $\left({}^{\sigma}/_{\mu}\right)$ for channel length variation, oxide thickness variation and mobility variation are 4.21, 2.84 and 0.81 % respectively for fifty Monte Carlo runs. Considering single MOS transistor, these figures are alarming about the impact of intra-die process variations on flicker noise variability. This demonstrates the need of accurate characterization of the various causes of flicker noise variability. It is also observed that channel length variation is the primary source of variation regarding flicker noise variability among the three sources of variations considered in the present work.

In order to validate the theoretical estimations of flicker noise variability with actual HSPICE-Monte Carlo simulation results, we compare the two. These are shown in Figs. 3, 4 and 5. It is performed for $V_{GS} = 0.7$ V. Here the process input

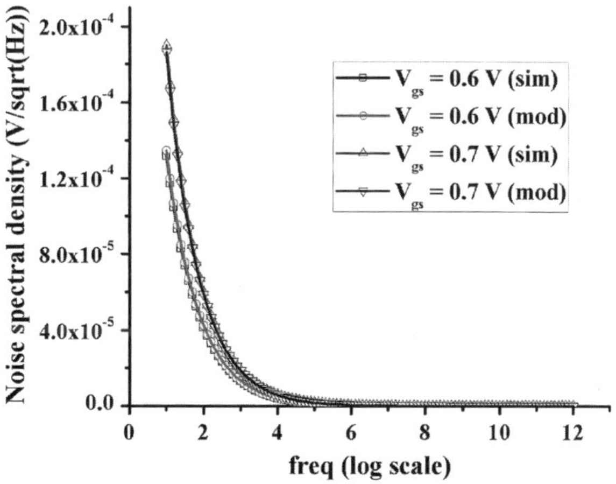

Fig. 1 Flicker noise spectral density plot used to extract *KF*

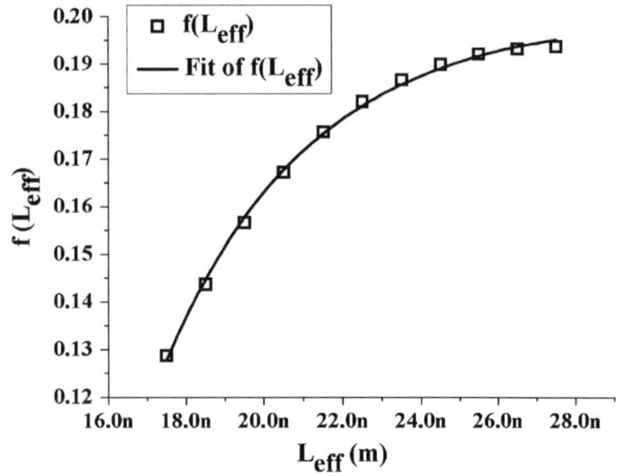

Fig. 2 Functional form of $f(L_{eff})$

Table 1 Values of the various unknown parameters as used in the present work

Parameters	Values	Parameters	Values
$A1$	5.69e–27 F-A	$C1$	−5.98e–28 F-A
$A2$	−2.80e–26 F-A/V	$C2$	8.40e–26 C^2/m^2
$A3$	3.66e-26 F-A/V^2	$C3$	8.91e–25 C^2V sec/m^4
$B1$	1.05e-26 F-A	$D1$	0.2
$B2$	-3.76e-19 F-A/m	$D2$	7.94
$B3$	3.73e-12 F-A/m^2	$D3$	3.73e–9

parameters are assumed to follow a Gaussian distribution, and resulting performance variations are also found to follow the same. One thousand Monte Carlo runs are performed for the simulation purpose. The values of the mean and standard deviation as estimated from theoretical model and obtained through

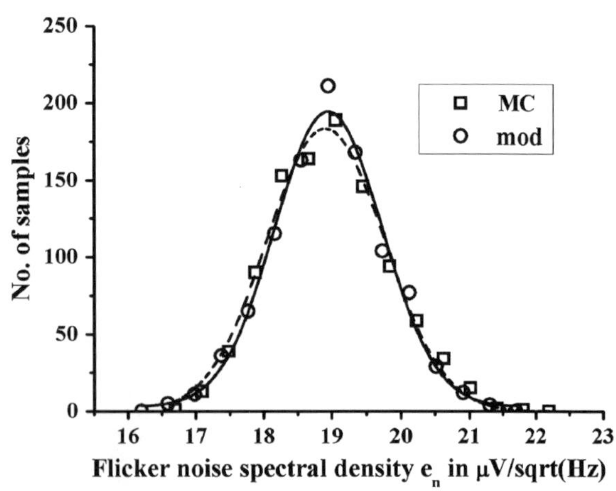

Fig. 3 Comparison of distribution curves (theoretical model and Monte Carlo-HSPICE simulation) of flicker noise variability with channel length variations. $V_{GS} = 0.7$ V

Fig. 4 Comparison of distribution curves (theoretical model and Monte Carlo-HSPICE simulation) of flicker noise variability with oxide thickness variations. $V_{GS} = 0.7$ V

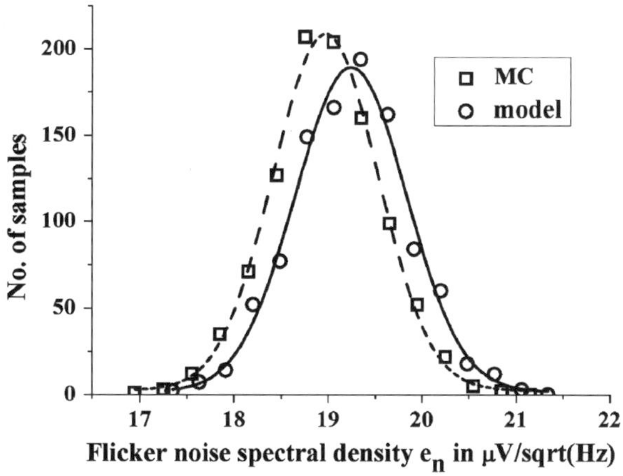

Fig. 5 Comparison of distribution curves (theoretical model and Monte Carlo-HSPICE simulation) of flicker noise variability with mobility variations. $V_{GS} = 0.7$ V

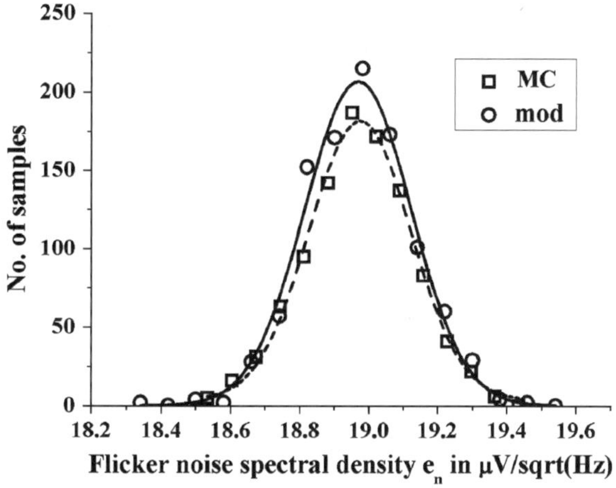

Table 2 Comparison between mean and standard deviation of flicker noise as obtained from theoretical model and Monte Carlo-HSPICE simulation results

V_{GS} (V)	Parameters	Mean m (model) $(V/\sqrt{Hz})$	Mean m (Monte carlo) $(V/\sqrt{Hz})$	% error	Standard deviation σ (model) $(V/\sqrt{Hz})$	Standard deviation σ (Monte carlo) $(V/\sqrt{Hz})$	% error
0.7	*XL*	18.90e–6	19.00e–6	0.53	794.97e–9	835e–9	4.79
	TOXE	19.21e–6	19.00e–6	1.11	586.72e–9	586e–9	0.12
	U0	18.95e–6	19.00e–6	0.26	153.43e–9	155e–9	1.01

Monte Carlo simulation results are summarized in Table 2. It is observed from all the comparisons that the theoretical estimations are reasonably accurate, as it is evident from the % errors reported which has a maximum value of 4.79 % in one of the cases and in most of the cases it is around 1 % or even less.

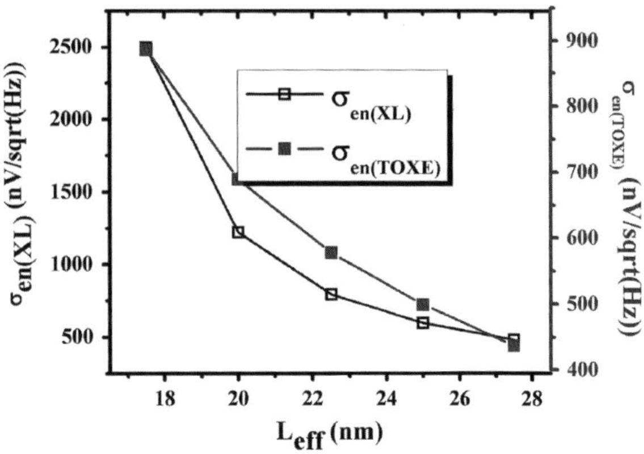

Fig. 6 Variation of flicker noise variability due to *XL* and *TOXE* as functions of effective channel length

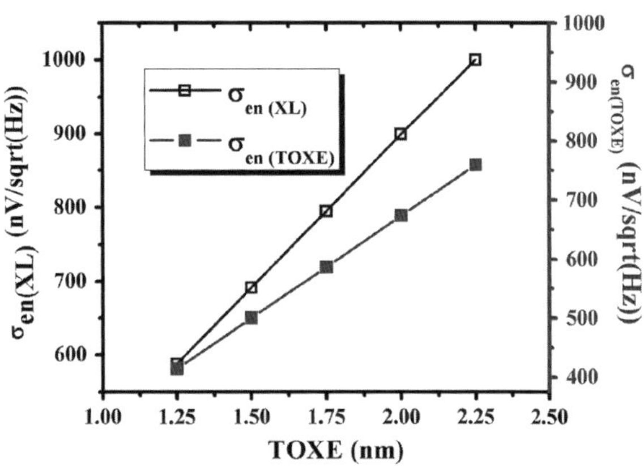

Fig. 7 Variation of flicker noise variability due to *XL TOXE* as functions of oxide thickness

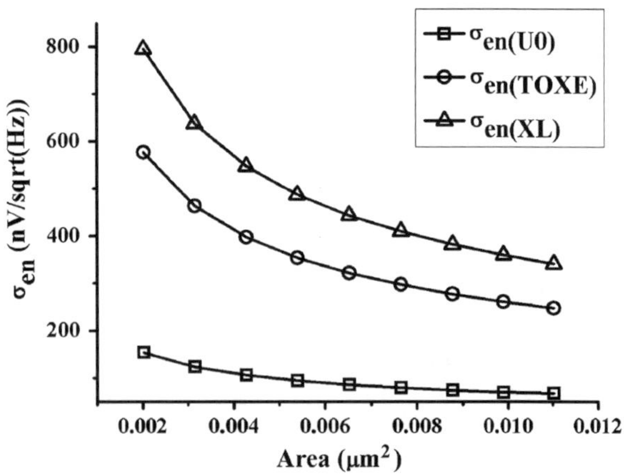

Fig. 8 Variation of flicker noise variability due to *XL*, *TOXE* and *U*0 as a function of gate area

Figure 6 shows the variations of flicker noise variability due to channel length variation and oxide thickness variation as function of the channel length. It is observed from Fig. 6 that higher effective channel length is beneficial so that the variability is less for both the variations.

Figure 7 shows the flicker noise variability due to *XL* and *TOXE* as function of oxide thickness. It is observed that the flicker noise variability is reduced for lower oxide thickness in both the cases. This is significant as far as device design is considered. Thus we can conclude from these graphs that flicker noise variability due to channel length variation can be minimized by choosing some optimum combination of *TOXE* and L_{eff}.

Figure 8 explicitly shows the area dependence of the flicker noise variability. It is observed that the variability increases with decrease in device sizes. The flicker noise variability model we derive correctly describes this fact.

4 Conclusion

This paper presents statistical study and characterization of flicker noise variability for nano-scale NMOS transistor with intra-die channel length variations, oxide thickness variations and mobility variations. The study has been made theoretically as well as through HSPICE-based Monte Carlo simulation results. BSIM-SPICE process parameters have been chosen for characterization purpose. The theoretical estimates have been compared with MC simulation and good accuracy has been achieved. The theoretical model thus can be used for accurate characterization of flicker noise variability, which is essential for robust design of analog and RF circuits in the nano-scale regime.

Acknowledgments The authors thank the Centre for Research in Nanoscience and Nanotechnology, University of Calcutta and Department of Science and Technology, Govt. of India (under Fast Track Young Scientist Scheme) for supporting the research work financially.

References

1. S.K. Saha, Managing process variability in scaled CMOS technology. IEEE Des. Test Comput. **27**(2), 8–16 (2010)
2. S.S. Sapatnekar, Overcoming variations in nanometer-scale technologies. IEEE J. Emerg. Sel. Top. Circuits Syst. **1**(1), 5–18 (2011)
3. K. Kuhn et al., Managing process variations in intel's 45 nm CMOS technology. Intel Technol. J. **12**(2), 92–110 (2008)
4. D. Sylvestar, K. Agarwal, S. Shah, Variability in nanometer CMOS: impact, analysis, and minimization. Integr. VLSI J. **41**, 319–339 (2008)
5. A. Asenov, S. Kaya, J.H. Davis, Intrinsic threshold voltage fluctuations in decanano MOSFETs due to local oxide thickness variations. IEEE Trans. Electron Devices **49**(1), 112–119 (2002)

6. S.K. Saha, Process variability modeling for VLSI circuit simulation. Workshop Compact Model. NSTI-Nanotechnol. **2**, 751–755 (2011)
7. T.H. Morshed, M.V. Dunga, J. Zhang, D.D. Lu, A.M. Niknejad, C. Hu, Compact modeling of Flicker noise variability in small size MOSFETs', IEDM, pp. 30.6.1–30.6.4 (2009)
8. M.B. Weissman, 1/f noise and other slow, nonexponential kinetics in condensed matter, Rev. Mod. Phys., 60(2) (1988)
9. K.K. Hung, P.K. Ko, C. Hu, Y.C. Cheng, A unified model for the Flicker noise in metal-oxide-semiconductor field-effect transistors. IEEE Trans. Electron Devices **37**(3), 654–665 (1990)
10. Y. Nemirovsky, I. Brouk, C.G. Jakobson, 1/f noise in CMOS transistors for analog applications. IEEE Trans. Electron Devices **48**(5), 921–927 (2001)
11. W. Zhao, Y. Cao, New generation of predictive technology model for sub-45 nm early design exploration. IEEE Trans. Electron Devices **53**(11), 2816–2823 (2006)

Structural, Dielectric and Magnetic Properties of Nano-Crystalline Ni-Mg Ferrites Prepared by Citrate-Gel Auto Combustion Method

Mohd. Hashim, Alimuddin, M. Wasi Khan, Shalendra Kumar, Sagar E. Shirsath and Ravi Kumar

Abstract Nano-particles of $Ni_{0.5}Mg_{0.5}Fe_{2\text{-}x}Cr_xO_4$ $(0 \leq x \leq 1.0)$ have been synthesized by citrate-gel auto combustion method. X-ray diffraction (XRD) and infrared spectroscopy (IR) were used to study the structural changes taking place with varying Cr concentration. Morphological analysis was made by scanning electron microscopy. The dielectric and magnetic properties were investigated by ac impedance spectroscopy and vibrating sample magnetometer (VSM), respectively. X-ray diffraction analysis assured the formation of single phase cubic spinel structure. Infrared spectroscopic studies revealed two main absorption bands in the range 400–800 cm^{-1} arising due to tetrahedral (A) and octahedral (B) site vibrations. The dielectric properties have been carried out as a function of frequency (42 Hz–5 MHz) at room temperature. The dielectric behavior is explained on the basis of interface polarization, arising from the heterogeneous nature of ferrites structure. In general, substitution of chromium ion plays an important role tailoring the structural, dielectric and magnetic properties of the ferrite material.

Keywords Ferrite · Nanoparticles · Dielectric constant · Doping additives

Mohd. Hashim (✉) · Alimuddin · M. Wasi Khan
Department of Applied Physics, Aligarh Muslim University, Aligarh 202002, India
e-mail: md.hashim09@gmail.com

S. Kumar
School of Nano and Materials Engineering, Changwon National University, Uichang-gu, South Korea

S. E. Shirsath
Department of Physics, Vivekanand College, Aurangabad 431001 Maharashtra, India

R. Kumar
Centre for Material Science Engineering, National Institute of Technology, Hamirpur 177005 Himachal Pradesh, India

P. K. Giri et al. (eds.), *Advanced Nanomaterials and Nanotechnology*,
Springer Proceedings in Physics 143, DOI: 10.1007/978-3-642-34216-5_22,

1 Introduction

During the last few decades all we are interested in the nanoscale materials because it is at this scale the properties of materials are entirely different from those at a large scale. Two principal factors cause the properties of nanomaterials to differ significantly from other materials: increased relative surface area and quantum effects. These factors can change or enhance the properties such as reactivity, strength and electrical characteristics. Ferrites are significant because of the rapidly varying characteristics with composition and are useful for systems that do not obey the reciprocity principle. The high frequency radiation can be propagated through such devices containing oxide ferrimagnetic materials like ferrites with very little attenuation. Microwave devices such as isolators, circulators and phase shifters are made by using magnetic ferrite materials having a very low conductivity and loss tangents [1]. Ni–Mg ferrites are among the most widely used soft magnetic materials because of high frequency applications as they possess high electrical resistivity and low eddy current losses [2–4]. The substitution of trivalent ions in pure ferrites leads to the modification of the structural, electrical and magnetic properties [5]. The interesting features are that their properties can be tailored by controlling the preparation methodology and type and amount of substituents. Khalid et. al. have reported the finite size effect and influence of temperature on electrical properties of nanocrystalline Ni–Cd ferrites [6]. Kharabe et al. have studied the dielectric and magnetic properties of Cd substituted Li–Ni ferrites [7]. Soibam et al. have reported the high value of magnetization for Ni–Li–Zn ferrites [8]. The electrical properties of Li–Ni–Eu and Li–Mg–Ti ferrites have been reported by Al-Hilli et al. [9] and by Bellad and Chougule [10], respectively. The electrical properties of spinel ferrites are determined by the different factors, of which synthesis method, chemical composition and grain size have the most important role. In this work, we present the influence of Cr^{3+} ions substitution on dielectric and magnetic properties of Ni–Mg ferrites synthesized by citrate–gel auto combustion process. All measurements were carried out at room temperature and the results are presented in this paper.

2 Experimental Procedure

Nanoparticles of basic composition $Ni_{0.5}Mg_{0.5}Fe_{2-x}Cr_xO_4$ (x ranging from 0 to 1 in steps of 0.2) have been prepared by using citrate-gel auto combustion method using analytical (AR) grade chemicals Ni $(NO_3)_2.6H_20$, Mg $((NO_3)_2.6H_2O$, Cr $(NO_3)_3.9H_2O$, Fe $(NO_3)_3.9H_2O$ as starting raw materials. Metal nitrates have been taken in the required stoichiometric ratio and dissolved in a minimum amount of distilled water with the ratio 1:1 of nitrate to citrate. The Ph of starting solutions was adjusted in the range $\sim$8 by adding NH_4OH. The resulting solution was continuously heated on the magnetic stirrer at 70 °C until transformed into gel.

When the dried gel was ignited at any point, the combustion rapidly propagated forward unit all the gel was burnt completely to form loose powder. Finally, a brown product fluffy precursor was yielded. The obtained gel was annealed at 200 °C for 24 h and then grinding for half an hour till the fine powder is obtained. The powder formed was heated for 36 h at 600 °C to remove any organic material present in the system and then grounded for half an hour [11]. The dried mixtures were mixed with PVA (binder) and ground into smooth powders in an agate jar, pressed into pellets. The X-ray powder diffraction (XRD) studies using a Rikagu Miniflex (II) with Cu Kα radiation ($\lambda = 2.2897$ Å) at a sweep rate of 2°/min was used to study the single phase nature and nano-phase formation of the pure and doped Ni–Mg–Cr ferrite nanoparticles at room temperature. The infrared spectra of the powders (as pellets in KBr) were recorded by Fourier transform infrared spectrometry [FT–IR] (PerkinElmer Instruments, Waltham, MA, USA) in the range of 200–2,200 cm^{-1} with a resolution of 1 cm^{-1}. The capacitance, dielectric and impedance spectroscopy measurements were carried out in the frequency range 42 Hz–5 MHz using LCR HI-Tester (HIOKI 3532–50). For pellet formation, a load of 5 t for 5 min was applied. The surface of all the samples were polished well to be parallel and smooth, and then coated with the silver paste as contact material for the dielectric measurements. The value of real dielectric constant ($\acute{\varepsilon}$) has been calculated using the following relation:

$$\varepsilon' = \frac{C_p \times t}{\varepsilon_o A} \tag{1}$$

where $\varepsilon_0 = 8.85 \times 10^{-12}$ F/m, known as permittivity of the free space t is the thickness of pellets, A is the area of cross sectional area of the flat surface of the pellets and Cp is the capacitance of the pellet in (F). The imaginary dielectric constant has been calculated by the relation:

$$\varepsilon'' = \varepsilon' \tan\delta \tag{2}$$

where tan δ is the dielectric loss tangent which is proportional to the 'loss' of energy from the applied field into the sample (in fact this energy is dissipated into heat) and therefore denoted as dielectric loss.

The loss tangent has been calculated from the following relation:

$$\mathbf{tan}\delta = \frac{1}{2\pi f \varepsilon_0 \varepsilon' \rho} \tag{3}$$

where f is the frequency of the applied field and ρ is the resistivity.

Magnetic measurements were carried out using vibrating sample magnetometer (Lake Shore 7,305) for all samples.

3 Results and Discussion

Figure 1 highlights the experimental XRD pattern, which clearly confirms that the prepared samples are of exclusively single phase cubic spinel structure. No additional phase corresponding to any structure in starting and doped samples was detected.The lattice constants were calculated for the cubic phase from the observed d-values and by using the following relation;

$$\mathbf{a} = \frac{d_{\mathrm{hkl}}}{\sqrt{(h^2 + k^{2+} l^2)}} \tag{4}$$

where (h, k, l), are the miller indices and d_{hkl} is the inter-planer spacing obtained from the XRD pattern. The average crystallite sizes of all the samples were calculated from the X-ray broadening of the diffraction peaks by using the Scherrer formula [12].

$$D = \frac{\lambda.K}{\beta.\cos\theta} \tag{5}$$

where D is the apparent volume-weighted crystallite size, λ is the wavelength of X-rays (1.54060 Å, in this case), K = 0.89 (the Scherrer constant), θ is the angle of Bragg diffraction, β = B−b, B is the full width, at half maximum and b represents the instrumental line broadening. Lattice constants values are found to be decreased with increasing the doping ion concentration and are shown in Fig. 2. This behavior can be explained based on the difference in ionic radii of Fe^{3+} and Cr^{3+}. In the present series $Ni_{0.5}Mg_{0.5}Fe_{2-x}Cr_xO_4$ $(0 \leq x \leq 1.0)$, larger Fe^{3+} (0.073 nm) ions are replaced by smaller (0.067 nm) ones; therefore a decrease in lattice constant is expected. The broad diffraction lines exhibited by the ferrite

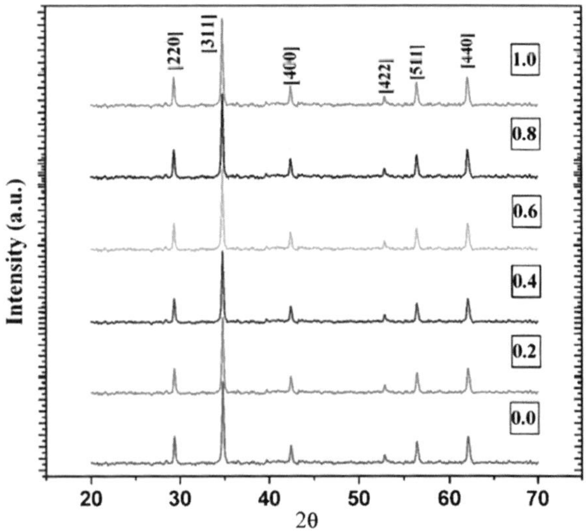

Fig. 1 X-ray diffractograms for $Ni_{0.5}Mg_{0.5}Fe_{2-x}Cr_xO_4$ ferrite Nps

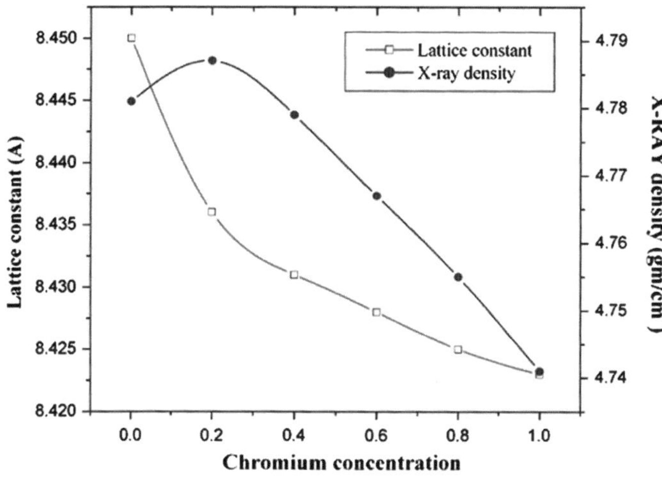

Fig. 2 Variation of lattice constant and X-ray density with chromium ion concentration

powder are indicative of the fine particle nature of the ferrite material. The X-ray density (theoretical) of the prepared samples was calculated by using the following relation:

$$D_{\text{hkl}} = \frac{8M}{Na^3} \tag{6}$$

where M is molecular weight of the sample, N the Avogadro's number, a lattice constant and 8 (in ferrites) represents the number of molecules in a unit cell of spinel ferrite. The X-ray density depends on the lattice constant and the molecular weight of the sample while the apparent density. The X-ray densities as a function of Cr contents are shown in Fig. 2. The X-ray density decreases with Cr substitution, which may be due to the fact that the density and atomic weight of Cr are 8.902 gm/cm^3 and 58.71, respectively, which are greater than that of Fe ions (7.87 gm/cm^3, 58.93). In order to understand the morphology, and shape of the grown nanoparticles, SEM measurements were carried out. The SEM micrographs were taken at 1,000 magnifications by selecting different parts of the samples. The SEM images of pure samples are shown in Fig. 3. The chemical composition of samples was estimated by EDAX technique. The EDAX pattern confirms homogeneous mixing of Ni, Mg, Fe, and O atoms in pure samples. The observed composition is almost equal to that of the samples produced by stoichiometric calculations while taking oxygen as balanced.

3.1 IR Spectroscopy

The infrared spectra gives us information about the chemical and molecular structure change in the ferrite due to changes in the Fe^{3+}–O^{2-} bond during heat

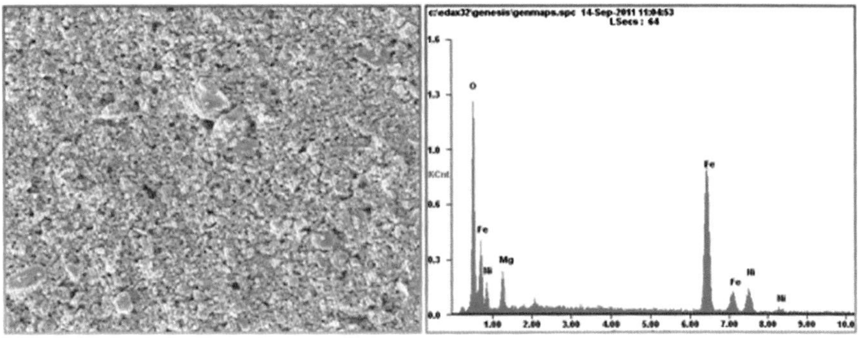

Fig. 3 Representative SEM images and corresponding EDAX spectra of Co–Ni Nps grown by citrate-gel auto combustion (for x = 0.0)

treatment or when some foreign atom is introduced in the parent ferrite compound. The IR spectra of chromium-substituted nickel-magnesium ferrite were recorded at room temperature in the frequency range 200–2,200 cm^{-1} and the obtained results are shown in Fig. 4. For ferrites, generally two assigned absorption bands appear around 600 cm^{-1}: ν_1, which is attributed to stretching vibration of tetrahedral groups Fe^{3+}—O^{2-} and around 400 cm^{-1}: ν_2, which is attributed to the octahedral groups complex Fe^{3+}—O^{2-}. In our IR measurement, the higher frequency band (ν_1) (584–670 cm^{-1}) and lower frequency band (ν_2) (412–422 cm^{-1}) are assigned to the tetrahedral and octahedral sites respectively [13, 14]. It explains that the normal mode of vibration of tetrahedral cluster is higher than that of octahedral cluster. It should be attributed to the shorter bond length of tetrahedral cluster and longer bond length of octahedral cluster [15]. It is also observed from Fig. 4 that the intensity of the bands decreases while broadening increases.

3.2 Dielectric Properties

Figures 5 and 6 shows the variation of ε'' and tanδ, with composition at selected frequencies. The dielectric constant decreases with the increase in applied frequency which is similar to the normal behavior of ferrite material. The dispersion in dielectric constant with frequency can be explained on the basis of Maxwell–Wagner two layer model and Koop's phenomenological theory [16, 17]. In this model a dielectric medium has been assumed to be made up of well conducting grains which are separated by the poorly conducting grain boundaries. The electron reaches the grain boundary by hopping and the resistance of the grain boundary is relative more. The electron piles up at grain boundaries where they produce the polarization. When the frequency of applied field is increased, the electrons reverse their direction of motion more often which decreases probability of electrons reaching the grain boundary and as a result polarization decreases,

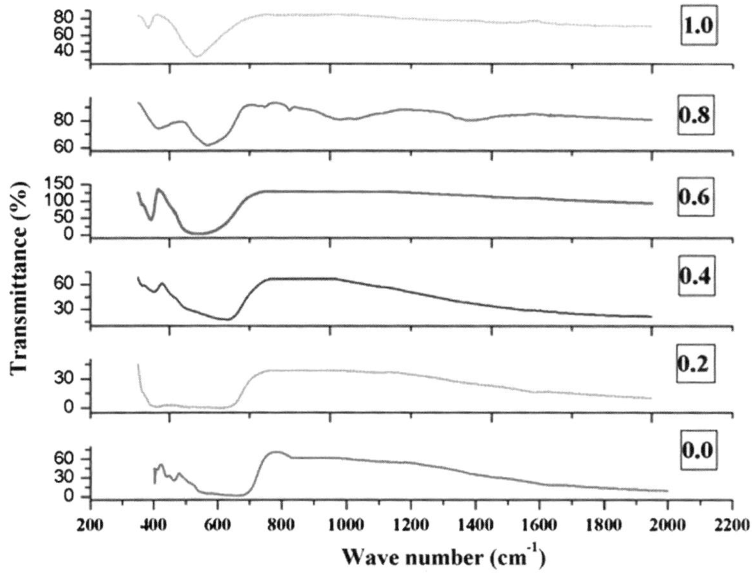

Fig. 4 FT-IR spectra of $Ni_{0.5}Mg_{0.5}Fe_{2-x}Cr_xO_4$ ferrite nanoparticles

therefore, decreasing the dielectric constant and dielectric loss of the developed materials with the increase in frequency. It has been found that the loss tangent shows similar normal behavior like dielectric constant. The samples show an anomalous behavior when relaxation peak appear at higher frequency region. This behavior can be explained in the light of Rezlescu Model [18]. According to this model the peaking nature occurs when the hopping frequency matches exactly with the applied frequency of the field.

Fig. 5 Variation of imaginary part of dielectric constant with composition

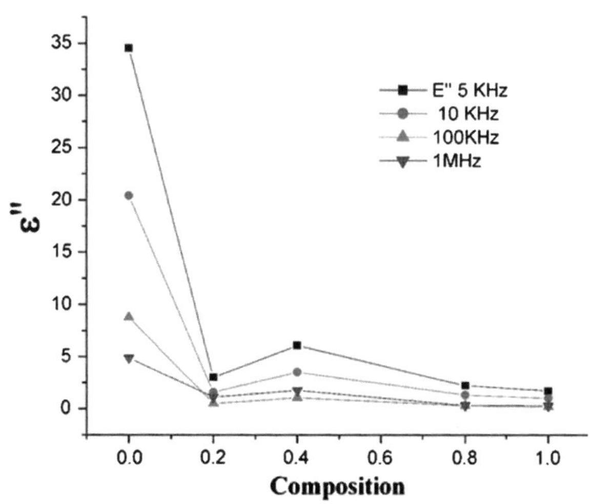

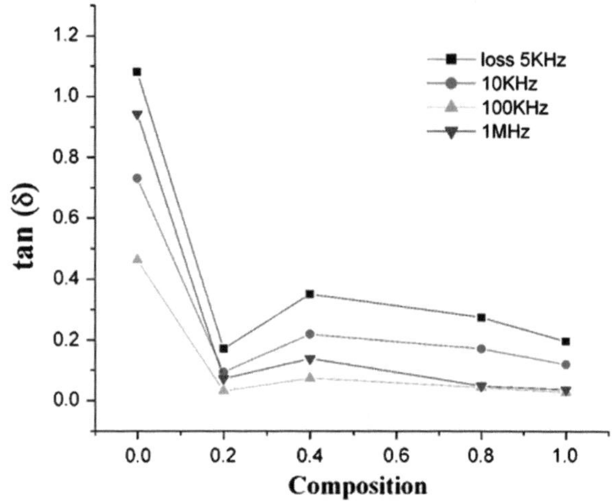

Fig. 6 Variation of dielectric loss (tan δ) with composition

3.3 M–H Curve

The Magnetic hysteresis loops for pure and doped samples were obtained by using vibrating sample magnetometer at room temperature and an applied maximum field of 6 koe. M–H curve helps in understanding the magnetic response of synthesized material and provides the information regarding the magnetic parameters

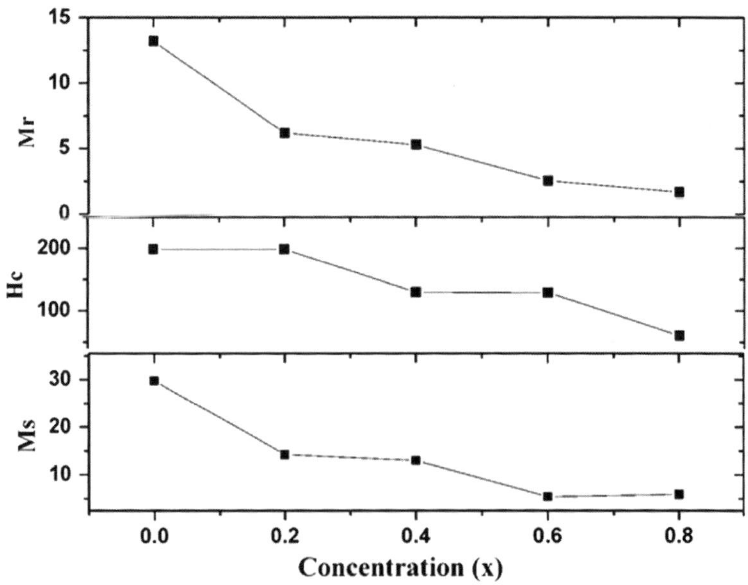

Fig. 7 Saturation magnetization (M_S), coercive field (H_C) and remnant magnetization (M_r) as a function of Cr doping

such as saturation magnetization (Ms), coercivity (Hc) and remanence magnetization (Mr). It is evident that the saturation magnetization Ms (magnetic moment per unit mass) decreased as the percentage of Cr ion increases (Fig. 7). It is found that the saturation magnetization decreases slowly with increase in the Cr concentration up to $x = 0.4$, but after this the sample shows an abrupt decrease. The decrease in the magnetization value may also be attributed to the presence of Fe^{2+} ion in the octahedral site [19].

4 Conclusion

We have successfully synthesized the single phase Cr doped Ni-Mg ferrites with an average crystallite size of 30–35 nm. Structural analysis with XRD reveals that the system confirms the formation of Ni-Mg ferrite with lattice constant decreases proportionally to the chromium concentration. FT-IR spectra show the presence of two vibrational modes corresponding to the tetrahedral and octahedral complexes. Dielectric constant shows normal behavior with respect to frequency which has been explained on the basis of Maxwell–Wagner two layer model.

Acknowledgments Grateful thanks are due to U.G.C. New Delhi, India for financial support through grant MANF-MUS-UTT-1092. Thanks are also due to Center of excellence in Materials Science (Nanomaterials), Department of Applied Physics, Aligarh Muslim University, Aligarh, to sensible help in characterization of as-prepared specimens.

References

1. C.B. Murray, C.R. Kagan, M.G. Bawendi, Synthesis and characterization of monodisperse nanocrystals and close-packed nanocrystal assemblies. Annu. Rev. Mater. Sci. **30**, 545 (2000)
2. X. He, G. Song, Zhu, Non-stoichiometric Ni-Zn ferrite by sol-gel processing. J. Mater. Lett. **59**, 1941 (2005)
3. Y. Matsuo, M. Inagaki, T. Tomozawa, F. Nakao, High performance Ni-Zn ferrite. IEEE Trans. Magn. **37**, 2359 (2001)
4. P.S. Anil Kumar, J.J. Shrotri, S.D. Kulkarni, C.E. Deshpande, S.K. Date, Low temperature synthesis of Ni0.8Zn0.2Fe2O4 powder and its characterization. Mater. Lett. **27**, 293 (1996)
5. A.M. El-Sayed, Influence of zinc content on some properties of Ni-Zn ferrites. Ceram. Int. **28**, 363 (2002)
6. K.M. Batoo, S. Kumar, C.G. Lee, Alimuddin, Finite size effect and influence of temperature on electrical properties of nanocrystalline Ni-Cd ferrites. Curr. Appl. Phys. **9,** 1072 (2009)
7. R.G. Kharabe, R.S. Devan, B.K. Chougale, Structural and electrical properties of Cd-substituted Li-Ni ferrites. J. Alloy. Compd. **463**, 67 (2008)
8. I. Soibam, S. Phanjoubam, C. Prakash, Magnetic and Mossbauer studies of Ni substituted Li-Zn ferrite. J. Magn. Mater. **321**, 2779 (2009)
9. M.F. Al-Hilli, S. Li, K.S. Kassim, Microstructure, electrical properties and Hall coefficient of europium-doped Li-Ni ferrites. Mater. Sci. Eng. **158**, 1 (2009)
10. S.S. Bellad, B.K. Chougule, Composition and frequency dependent dielectric properties of Li-Mg-Ti ferrites. Mater. Chem. Phys. **66**, 58 (2000)

11. M. Hashim, Alimuddin, S. Kumar, S. Ali, B.H. Koo, H. Chung, R. Kumar, Structural, magnetic and electrical properties of Al3+ substituted Ni-Zn ferrite nanoparticles. J. Alloy. Compd. **511**, 107 (2012)
12. M. George, S.S. Nair, A.M. John, P.A. Joy, M.R. Anantharaman, Structural, magnetic and electrical properties of the sol-gel prepared Li0.5Fe2.5O4 fine particles. J. Phys. D Appl. Phys. **39**, 900 (2006)
13. P. Priyadharsini, A. Pradeep, P.S. Rao, G. Chandrasekaran, Structural, spectroscopic and magnetic study of nanocrystalline Ni-Zn ferrites. Mater. Chem. Phys. **11**, 207 (2009)
14. M. Kaiser, Effect of nickel substitutions on some properties of Cu-Zn ferrites. J. Alloy. Compd. **468**, 15 (2009)
15. R. Laishram, S. Phanjoubam, H.N.K. Sarma, C. Prakash, Electrical and magnetic studies of the spinel system Li0.5+t CrxSbt Fe2.5$-x-2t$O4. J. Phys. D Appl. Phys. **32**, 2151 (1999)
16. J.C. Maxwell, *Electricity and magnetism*, vol. 1 (Oxford University Press, Oxford, 1929). (Section 328)
17. K.W. Wagner, Zur theorie der unvollkommenen dielektrika. Ann. Phys. **40**, 817 (1913)
18. N. Rezlescu, E. Rezlescu, Dielectric properties of copper containing ferrites. Phys. Status Solidi A **23**, 575 (1974)
19. A.M. Abo, E. Ata, M.K. E. Nimra, S.M. Attia, D. E. Kony, and A.H. Al-Hammadi, Studies of AC electrical conductivity and initial magnetic permeability of rare-earth-substituted Li-Co ferrites. J. Magn. Magn. Mater. **297**, 33 (2006)

Acoustical Studies of the Intermolecular Interaction in Copper Oxide-Ethylene Glycol Nanofluid

M. Nabeel Rashin and J. Hemalatha

Abstract Stable Copper oxide-Ethylene glycol nanofluids of different concentrations have been synthesized. The structural, morphological and particle–fluid interaction studies have been made through X-ray diffraction, HRSEM and ultrasonic technique respectively. The intermolecular interactions between nanosized copper oxide and ethylene glycol have been discussed elaborately for different concentrations and temperatures.

Keywords Ultrasonic · Nanofluid · Compressibility · Suspension · Intermolecular interaction

1 Introduction

Nanofluids are novel heat transfer fluids engineered by suspending nano-solid particle in base fluids. There is great attraction towards nanofluids due to their enhanced thermal conductivity and their wide spread applications in transportation, power generation, micro-manufacturing, engines, thermal therapy, heating, cooling, ventilation, air conditioning loud speakers, bearings, MEMS, optoelectronic devices etc. [1, 2].

In this paper the authors have reported the synthesis of stable Copper Oxide-Ethylene glycol nanofluids of different concentrations through ultrasonically assisted two step method. PVP (Poly vinyl pyrrolidone) is used as the surfactant. Also, the structural analysis, morphological and intermolecular interaction studies are included.

M. Nabeel Rashin · J. Hemalatha (✉)
Advanced Materials Lab, Department of Physics, National Institute of Technology, Tiruchirappalli, Tamilnadu 620015, India
e-mail: hemalatha@nitt.edu

P. K. Giri et al. (eds.), *Advanced Nanomaterials and Nanotechnology*,
Springer Proceedings in Physics 143, DOI: 10.1007/978-3-642-34216-5_23,

2 Experimental Details

The chemicals, Copper Nitrate (Cu $(NO_3).3H_2O$), Ethylene glycol (CH_2OH–CH_2OH), Sodium Hydroxide (NaOH), Acetone ($(CH_3)_2$ CO) were purchased from Merck. The chemicals were used as purchased without further treatment. The CuO nanopowders were synthesized by co-precipitation [3]. This involves reacting aqueous solution of copper nitrate with sodium hydroxide at pH 9.5 and a post annealing of 200 °C for 3 h.

The crystalline structure, phase composition and crystallite size of copper oxide were identified from XRD patterns obtained using Cu Kα radiation ($\lambda = 1.541$ Å) for 2θ value ranging from 10° to 80° in X-ray diffractometer (Model Rigaku Ultima III). The morphology of the nano-copper oxide sample was obtained using High Resolution Scanning Electron Microscope-F E I Quanta FEG 200.

Appropriate amount of synthesized nano CuO powder samples were mixed in ethylene glycol using PVP as surfactant. The mixture was sonicated for 30 min using ultrasonic bath. The homogenous suspension was obtained without any phase separation (Fig. 1b). It is evident that the ultrasonic treatment to the fluids increases the stability of the suspension. In the similar way nanofluids of various concentrations (0.5–1.5 wt %) were prepared for further studies.

The synthesized various concentrations of nanofluids were subjected to ultrasonic studies at different temperatures (35, 45 and 55 °C). The required temperatures were maintained by circulating hot water through the outer jacket of the ultrasonic cell by using thermostatically controlled water bath. The velocity values of ultrasonic wave propagation through nanofluid samples were measured using single frequency continuous wave ultrasonic interferometer (Model F81, Mittal Enterprises, and New Delhi) with an accuracy of ± 0.05 % at frequency of 2 MHz. The density of the fluid was determined using specific gravity bottle (5 cc).

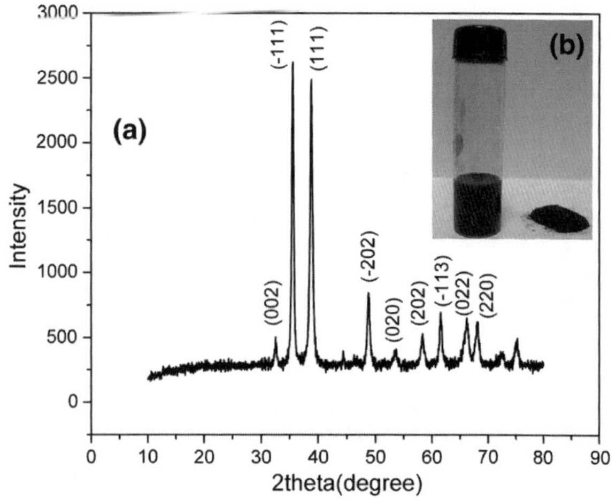

Fig. 1 **a** XRD of nano-CuO powder, **b** CuO-EG nanofluid

Fig. 2 HRSEM image of nano-copper oxide

3 Results and Discussion

Structural determination of CuO was done with the help of XRD (Fig. 1a. From XRD pattern it is clear that, the synthesized nano CuO is of pure phase. All the peaks match well with the standard JCPDS 89-2529. Using Debye–Scherrer equation the average grain size is found to be 20 nm and the crystallites possess monoclinic end-centered structure with lattice parameter a = 4.688, b = 3.422, c = 5.131 Å.

The High Resolution Scanning Electron Microscopic image of nano-copper oxide powder is shown in the Fig. 2.

The nano-copper oxide particles are agglomerated into uniform shuttle like structures. This is because of the small nanocrystals posses large surface energy, which leads the nanocrystals to aggregate in order to lower their surface energy during crystal growth [4].

The acoustical parameters like adiabatic compressibility (β) and acoustic impedance (Z) were calculated using the velocity (c) and density (ρ) data obtained through experiments. The relations used for the calculation are given below [5–8]

$$\beta = \left(\frac{1}{\rho c^2}\right) \tag{1}$$

$$Z = \rho \cdot c. \tag{2}$$

The variation of velocity with temperature for various concentrations is shown in Fig. 3. It is observed that the velocity of ultrasonic waves in the nanofluids decreases with increase of temperature and also with increase of nanoparticle concentration.

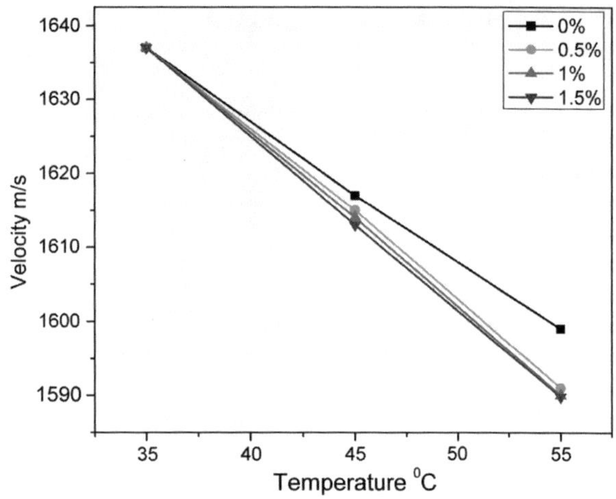

Fig. 3 Velocity verses temperature for nanofluids of different concentrations

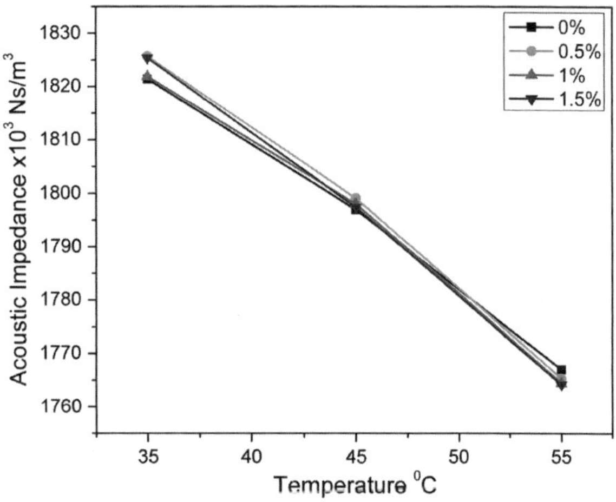

Fig. 4 Acoustic impedance verses temperature for nanofluids of different concentrations

The reduction in the ultrasonic velocity with the increase of concentration is ascribed to the variation in density and compressibility. It reveals the intermolecular interaction in fluid. The decreasing velocity with the increasing temperature can be attributed to the well known behavior of non aqueous liquids. The variation in acoustic impedance (Fig. 4) also supports the above result.

Increase of concentration leads to the rise in density due to comparative increase in mass and there by a reduction in velocity. With the increase of temperature, even though there is a chance for decrease in density, due to the enhancement of volume and there by an increase in velocity, it is overridden by the increase in compressibility. The fluid becomes more compressible (Fig. 5) at high temperature and elevated concentration, may be due to reduction in intramolecular interaction.

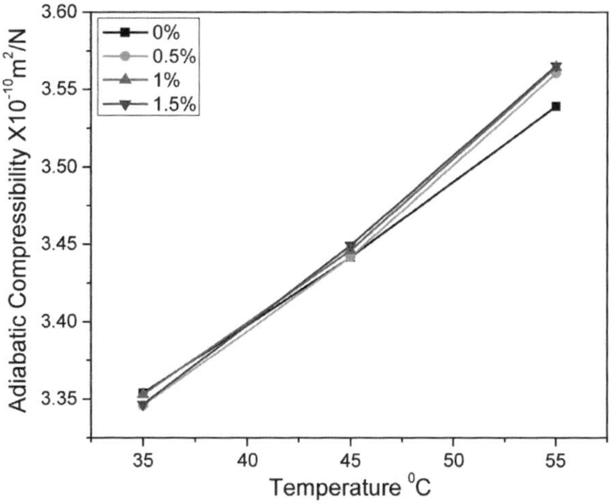

Fig. 5 Adiabatic compressibility verses temperature for nanofluids of different concentrations

4 Conclusion

Stable Copper oxide–Ethylene glycol nanofluid has been prepared. The structural and morphological investigations are done with the help of X-ray diffraction technique and HRSEM respectively which reveal the monoclinic end-centered structure of crystallite and shuttle like morphology of agglomerates. The acoustical studies performed using ultrasonic wave propagation indicates the strong dependence of particle–fluid interaction on temperature and concentration. Also, it is observed that the fluid becomes more compressible at elevated temperatures and higher concentrations.

References

1. D. Erickson, Towards numerical prototyping of labs-on-chip: modeling for integrated microfluidic devices. Microfluid. Nanofluid. **1**, 301 (2005)
2. N.T. Nguyen, A. Beyzavi, K.M. Ng, X. Huang, Kinematics and deformation of ferrofluid droplets under magnetic actuation. Microfluid. Nanofluid. **3**, 571 (2007)
3. A. Punnoose, H. Magnone, M.S. Seehra, Bulk to nanoscale magnetism and exchange bias in CuO nanoparticles. Phys Rev B **64**, 174420 (2001)
4. G.N. Rao, Y.D. Yao, J.W. Chen, Superparamagnetic behavior of antiferromagnetic CuO nanoparticles. IEEE Trans. Magn. **41**, 3409 (2005)
5. A.J. Matheson, *Molecular acoustics* (Wiley, Newyork, 1971)
6. J. Hemalatha, T. Prabhakaran, R.P. Nalini, A comparative study on particle–fluid interactions in micro and nanofluids of aluminium oxide. Microfluid. Nanofluid. **10**, 263 (2011)
7. J.S. Rowlinson, F.L. Swinton, *Liquid and Liquid Mixtures*, 3rd edn. (Butterworths, London, 1982), pp 16–17
8. M.J.W. Povey, *Ultrasonic Techniques for Fluids Characterization*, (Academic Press, USA, 1997), p 25

Fabrication and Characterization of Aluminum Nanostructures Using Femtosecond Ablation Technique

G. Krishna Podagatlapalli, Syed Hamad, S. Sreedhar, Surya P. Tewari and S. Venugopal Rao

Abstract In this paper we report the fabrication of nanostructures through the interaction of ultrashort laser pulses ($\sim$40 fs) with bulk Aluminum immersed in various liquid media of different polarity (chloroform which is polar, carbon tetrachloride which is non-polar, and water which is polar) using femtosecond laser ablation. Except water medium other two media showed yellow coloration after ablation indicating the formation of nanoparticles in the solution. The coloration of the laser exposed portion in the Al substrate was golden yellow and its closer view depicted micro-grating and nano-ripple (period 330 nm) formation. The investigation of polarization dependence on the ablation was performed for the liquid media. Depending on the focal position we observed sub-micron structures and nano-ripples on the surface. As the rate of ablation depends on the position of the focus on the Al substrate and beam waist and other laser parameters, we have studied the liquid level dependence of ablation with different water levels on the Al substrate and we compared these patterns when focus was beyond the sample and on the sample. Size distribution of the generated nanoparticles in the liquid in which ablation took place and surface morphology of the substrates were investigated using SEM technique.

Keywords Femtosecond · Laser ablation · Nanoparticles · Nano-ripples

G. K. Podagatlapalli · S. Sreedhar · SuryaP. Tewari · S. V. Rao (✉)
Advanced Center of Research in Higher Energy Materials (ACRHEM),
University of Hyderabad, Hyderabad 500046, India
e-mail: svrsp@uohyd.ernet.in

S. Hamad ·
SuryaP. Tewari
School of Physics, University of Hyderabad, Hyderabad 500046, India

P. K. Giri et al. (eds.), *Advanced Nanomaterials and Nanotechnology*,
Springer Proceedings in Physics 143, DOI: 10.1007/978-3-642-34216-5_24,

1 Introduction

Metal nanostructures are known to exhibit distinct optical characteristics, which differ from those observed in the bulk and form the basis for a variety of spectroscopic techniques such as surface enhanced Raman scattering (SERS) and second harmonic generation (SHG) [1–5]. Nanoparticles synthesis, nanostructure formation and their characterizations are of great interest in the present day advancement of science because of the physics associated with them and are described by the intermediate regime between quantum physics and classical physics [6]. The electronic properties of the nano-sized systems change dramatically since the density of states and spatial length scale of electronic motion are reduced with decreasing size. For these nano-entities, Eigen states are determined by the systems boundaries and hence the surface effects become very important [7]. Femtosecond laser ablation (FLA) of metals immersed in liquid environment is the best method among all other methods to generate impurity free nanoparticles of narrower size distribution with reduced porosity [8] and nanostructures of different sizes. Furthermore, as generated nanoparticles remains in the liquid in which ablation takes place, it can reduce the contamination of the surrounding air medium from the expulsion of generated nanoparticles into it. This technique provides the possibility of generating a large variety of NPs those are free of both surface-active substances and counter ions [9, 10]. If the input laser energy density is high enough than the damage threshold of the metal used, then the deposition of high energy density through an ablation of a femtosecond laser pulse on a metal immersed in liquid results in rapid heating of the surface to high temperatures since the peak intensity of fs laser pulse is so high, and hence eventually it leads to plasma formation [11]. Under suitable high laser fluences, surface of the target melts, and the melt is subsequently dispersed in the surrounding liquid and formation of nanostructures take place on the substrate. This structure formation was because of the recoil pressure of the vapor of the liquid which surrounds the laser beam waist at the focus on the surface of the target. In this process along with nanoparticles, free atoms, ions and target fragments will come from the plume. When a metal passes through a melt state due to ablation, stimulation of the physicochemical processes of interaction of metal with liquid medium takes place causing the formation of more complex structures instead of usual spherical particles. The growth of nanoparticles and aggregation takes place under the influence of surrounding liquid environment and aggregation depends upon the permittivity and the polarity of the molecules of the surrounding liquid medium [12]. At fluence close to the melting threshold of the material, short laser pulses with duration of less than 1 ns melt only the micro protrusions on the target surface, giving rise to efficient formation of nanostructures. Under these conditions, the viscous interaction of the vapor of the liquid on the target surface with molten target layer may be responsible for several instabilities like Kelvin–Helmholtz or Rayleigh–Taylor instabilities. These instabilities are tentatively assigned to the observed nanostructure formation on the substrate [13]. One more condition to fabricate nanostructures on the metal surface is the existence of an initial roughness of the target surface since no nanostructures are formed on the optically polished metals [14].

Al is the cheapest metal among the plasmonic metals such as Au, Ag, and it has low melting point (660 °C) in comparison with other metals, Al nanoparticles can be easily generated. Furthermore Al nanoparticles have large surface area to volume ratio and these are the important fuel in high energetic material applications like propellants, munitions and pyrotechnics and can be used as additives for plastics, and powder metallurgy [15]. Al NPs particles provide enhanced heat release during their exothermal oxidation. The rate of reactivity increases as the size of Al NP decreases i.e., bare Al is highly reactive. Al is highly reactive to ambient oxygen, pure Al nanoparticles fabrication is restricted by the unwanted oxidation effects which in turn reduce the burn rate and hence the velocity of detonation. Thus the reactivity of the Al nanoparticles diminishes with thin oxide cladding present on it. This effect becomes more pronounced as particle size decreases, since the oxide layer represents a significant fraction of its mass. In this context, femtosecond laser pulses are helpful in quenching of nanostructured Al and thus preserve its metallic nature as it is due to the shorter time of interaction with liquid environment. In order to avoid the instantaneous oxidation caused by the dissolved oxygen in liquids, we have chosen CCl_4 and $CHCl_3$, since one of the barriers for bare Al NPs formation is the dissolved oxygen in the medium in which ablation takes place. In this paper we report the structuring on bulk Al in liquids of carbon tetrachloride, chloroform and water under the action of $\sim$40 fs pulses.

2 Experiment

Pure Al targets were washed with acetone after sonication to remove organic dopants from the surface. This study (Fig. 1) has been done by a chirped pulse amplified (CPA) of Ti:sapphire laser system (LEGEND, Coherent) delivering nearly bandwidth limited laser pulses ($\sim$40 fs, 1 kHz repetition rate) at 800 nm as an excitation source. Figure 1 shows the schematic of the experimental setup. The amplifier was seeded with $\sim$15 fs (55–60 nm FWHM) pulses from an oscillator (MICRA, Coherent, 1 W, 80 MHz, 800 nm). The average and peak powers after the amplifier were 2.5 W and 67 GW, respectively. The target was placed into a Pyrex cell and covered by a layer of absolute liquids—water, CCl_4 and $CHCl_3$. After ensuring the sample was perfectly parallel to the optical bench, laser pulses were allowed to focus onto the Al sample using a plano-convex lens of focal length 8 cm. Initial beam polarization was taken as S-polarization. To get the control over the polarization and the energy of the laser beam, we used the combination of quarter wave plate and Brewster window. It is very difficult to adjust the focus exactly on the surface of the Al substrate immersed in liquid because of the refractive index of the medium the focal position increases which fabricate the nanoparticles at a poor rate. To get the good rate of fabrication of nanoparticles

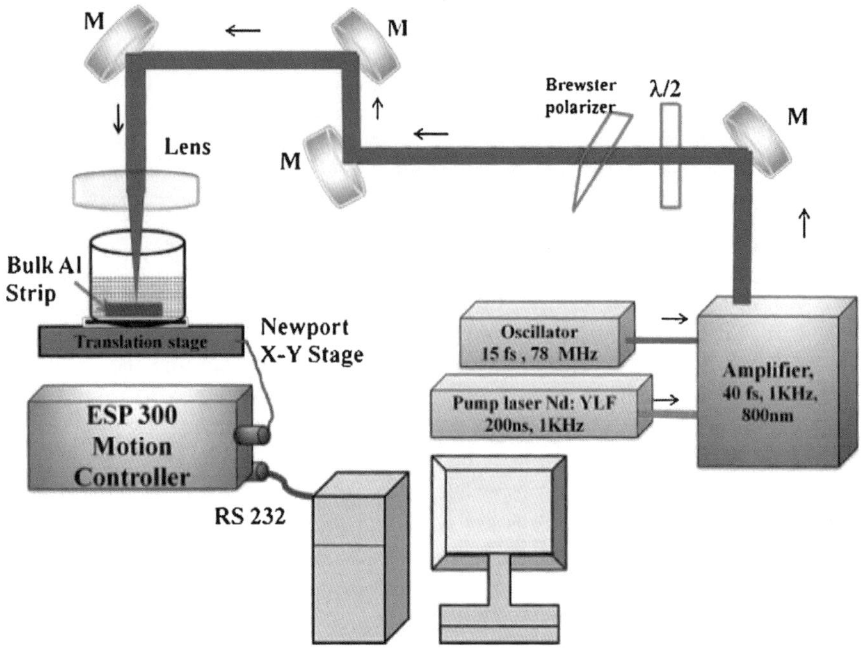

Fig. 1 Experimental scematic of laser ablation of bulk Al immersed in liquid media

the focus should be placed on the surface of the Al substrate. Not only the nanoparticles but the structure formation was also depends on the position of the focus whether it was on the substrate or beyond the substrate. The beam waist estimated at the focus was ~15 μm. Initially the level of liquid was ~2 to 3 mm above the Al sample. Al targets were placed normal to the laser beam on a motorized X–Y stage, which can be operated through the motion controller. Typical pulse energies used were ~250 μJ. The scanning speeds of the X–Y stages were 0.2 and 0.4 mm/sec. Both the stages were controlled by Newport ESP 300 motion controller. The motorized stages (Newport) were moved in such a way to write periodic line pattern on the Al sample, at a separation of ~150 μm. Time of exposure was about 5 min and each scan resulted in 20 periodic lines. At first, we performed the ablation in water, CCl_4 and $CHCl_3$ for a fixed polarization (S-polarization). Secondly, ablation was carried out in water, CCl_4 and $CHCl_3$ for S, P and circular polarizations. Finally, ablation was performed in water only for different levels of the water above the Al substrate. We compared the structure formed on the Al substrate when the focus was beyond the sample and focus exactly remained on the sample, for different water levels. The position of the focus was approximated to lie at the point where plasma was generated when it focused in air. Taking this as the confirmation we could manually change the position of the focus beyond the sample and exactly on the surface of the sample using a translational stage (along Z-direction) on which sample was fixed. We observed that the CCl_4 and $CHCl_3$ solutions changed to a gold-yellow color due to the dispersion of ejected Al nanoparticles into the liquid [16] irrespective of the polarization. But water did not suffer any change in coloration in all of the three studies mentioned above probably due to aggregation effects. The coloration of the NPs colloidal solution of Chloroform was thicker (dark yellow) than CCl_4 colloidal solution. FESEM (Ultra 55 from Carl ZEISS) analysis and the UV absorption spectra and Energy Dispersive Spectrometer (OX-Ford Instruments) of these colloidal solutions proved the presence of well dispersed nanoparticles. We observed that the coloration of the laser exposed portion on the metal changed to yellow color. FESEM images of the laser exposed portions in the Al substrate were showing different structure formations as previously reported mushroom structures by Stratakis et al. [16, 17].

3 Results and Discussion

Femtosecond laser ablation in air medium did not show any structure formation on the substrate other than removal of surface matter (Fig. 2). When the ablation was performed in water (Fig. 3a), CCl_4 (Fig. 3b) and $CHCl_3$ (Fig. 3c) for same input fluence (2–3 J/cm^2) and for the same liquid level and polarization (S), well dispersed nanoparticles were observed in CCl_4 and $CHCl_3$ along with different grating formation on the substrates whereas no nanoparticles was observed in water but structures on surface were observed. These observations were confirmed by FESEM. Microstructure formation was observed only when the focus of the laser beam was exactly on the Al substrate, which provided large amount fluence than the threshold of the metal. In general, when we place a liquid on the Al substrate, the refractive index of the liquid medium caused the focus to go beyond the sample. So the displacement of the focus above the sample may be the combined effect of the manual positioning of the sample and the refractive index of the liquid. Laser

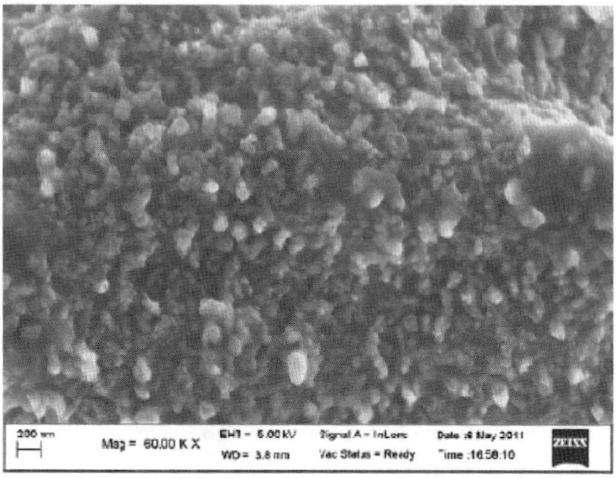

Fig. 2 FESEM image of laser exposed portion of Al substrate in airwhich did not show any structure formation

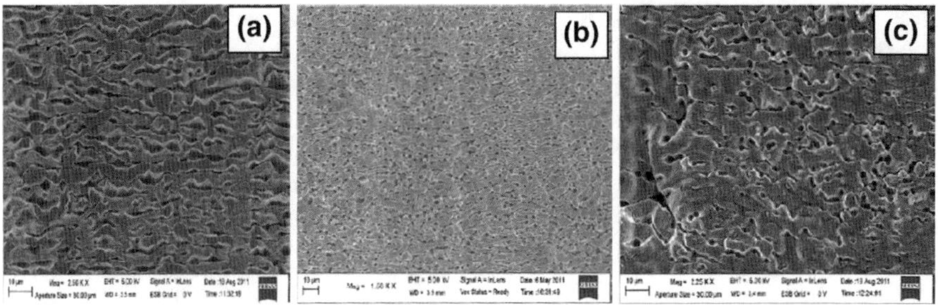

Fig. 3 FESEM images of the laser exposed portions of Al for same input beam parameters (S-Pol, 2-3 J/cm^2 in (**a**) water (**b**) carbon tetrachloride and (**c**) chloroform depicting microgratings.

exposed region of Al target acquired yellowish color, being visible at angles close to the normal incidence. The coloration of the Al strip was more pronounced when ablation was carried out in water at a fluence of $\sim$2 to 3 J/cm^2 and at scanning speed of 0.2 mm/sec. This coloration can be assigned to structuring of Al surface and similar nanostructures were previously reported on both Ag and Au [14, 18]. Secondly, we performed the ablation for different polarizations of the input beam without changing the liquid level above the Al substrate and focusing conditions. In this case also we observed the same micro-grating formation but from the FESEM images it was confirmed that the way of ablation was different for different polarizations (Fig. 4). FESEM images of the laser exposed portions of Al substrate in CCl_4 (Fig. 4a–c), in water (Fig. 4d–f) for S, P, Circular polarizations, exhibited a different way of ablation for different polarizations. Finally, water level dependence of ablation above the Al substrate also demonstrated a different structure than mentioned above. We performed this in two ways (1) Ablation of Al for different

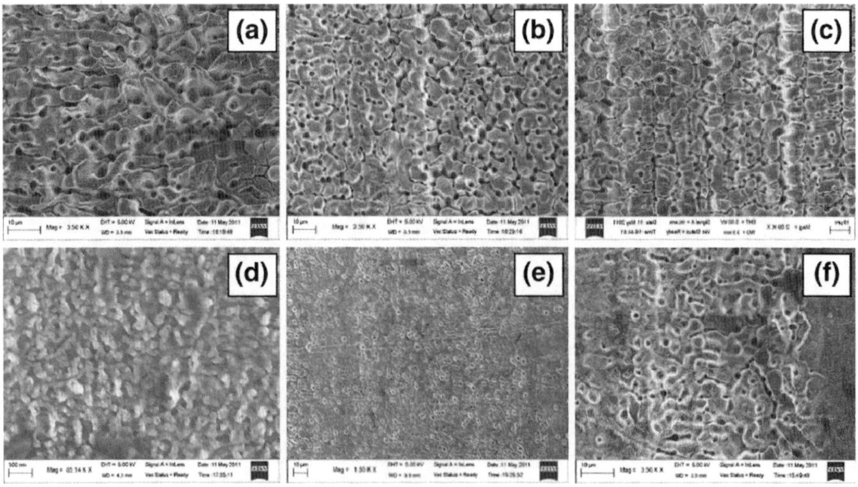

Fig. 4 FESEM imaging of laser exposed portions for the same liquid level in (1) water (*top*) (**b**) carbon tetrachloride (*bottom*). (**a**, **d**: S-pol, **b**,**e**: P-pol, **c**,**f**: Circular polarization)

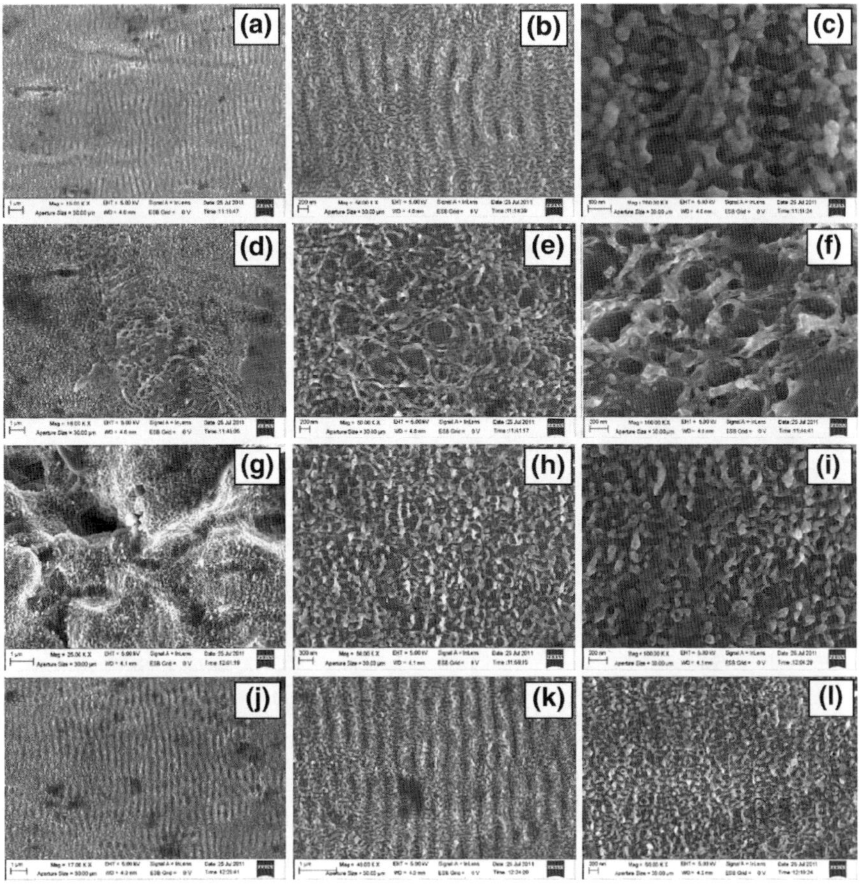

Fig. 5 FESEM images (**a**, **b** and **c**) of ablated Al surface at a level of 1.5 mm above the substrate showing nanoripples with a period 330 nm at different magnifications; (**d**, **e** and **f**), (**g**, **h** and **i**) and (**j**, **k** and **l**) are at 2.5, 3.5 and 5.5 mm water levels showing mushroom structures.

levels of water by setting the focus position exactly (some extent) on the sample. (2) Ablation for different levels of water by setting the focus manually beyond the sample. Particularly, in the first case we observed micro grating with a more or less periodicity and in the second case we observed somewhat nanoripples. FESEM images (Fig. 5) showed the of nano-ripples (with a period ~330 nm) which were observed on the Al substrate immersed in water when the water level is 1.5 mm above the sample (Fig. 5a–c). The interference of the incident electromagnetic wave with the surface electromagnetic wave (SEW) generated on the Al substrate could be the reason for these nanoripples. The closer view of these ripples demonstrated mushroom like nanostructures, tentatively assigned to Rayleigh–Taylor like instabilities which probably occurred between vapor of the liquid and the metal melt. Under the above mentioned instabilities redistribution of the melt might take

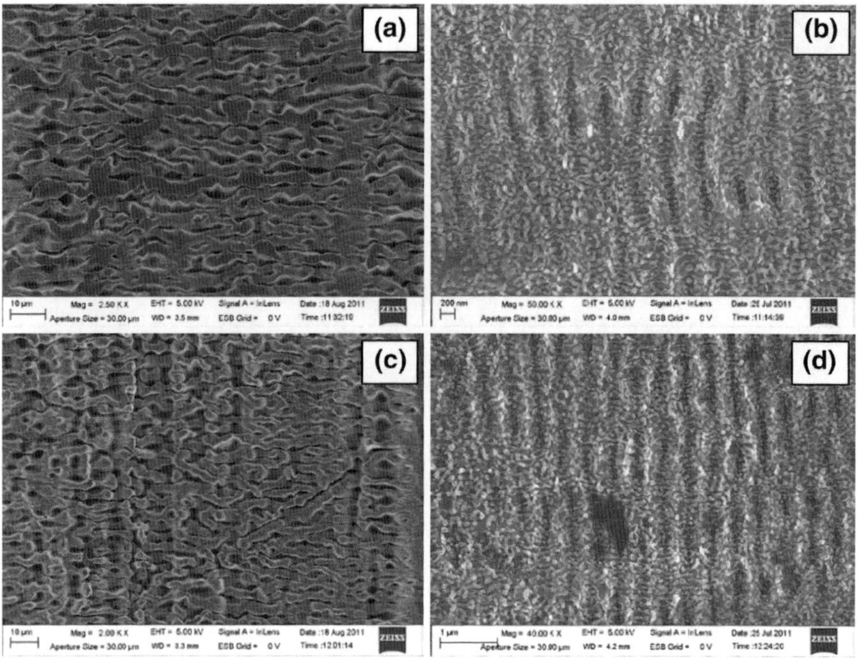

Fig. 6 FESEM images of (**a**, **c**) micrograting on the Al substrate for the water levels 1.5 and 4 mm when the focus was beyond the sample, (**b**, **d**) nanograting for the same levels when the focus was exactly on the substrate.

place and hence mushroom like structures were observed [13]. Increasing the water level in steps of 0.5 ml demonstrated the way ablation changed to a great extent. These are shown in (Fig. 5d–i). At a level 4 mm again we observed periodic nanoripples of period greater than 400 nm (Fig. 5j–l). We observed a difference between the ripples formed at a level 1.5 mm and 4 mm, in the former one distiguihability of the ripples at closer view (200 nm scales) was possible but in the latter this was missing at higher resolutions. Micro-grating and nano-ripples formed for different level of liquid are shown in Fig. 6. Our aim is to produce a pattern on the Al substrate to get the enhanced Raman signal from it. Because the electric field of the signal depends on the substrate's ability to generate surface plasmon resonances, much research has focused on different methods for positioning nanoparticles in ways conducive to generating strong signals. Creating patterned surface structures with a fs laser has yielded high electric fields when used in conjunction with methods such as lithography or chemical plating or laser machining in silver nitrate solutions or on Ag-doped materials has also produced exceptional electric fields due to the nanostructures formed by the laser processing [19]. A systematic study is being undertaken at present for the understanding of micro-/nano-structures, nanoparticle formation taking into account (a) surface roughness effect (b) pulse duration effects (experiments are being carried out with 2 ps pulses) (c) laser parameters (fluence, beam waist, pulse shaping etc.).

4 Conclusions

In summary, FLA of bulk Al sample in oxygen free liquid media like CCl_4 and $CHCl_3$ in an open air environment led to Al nanoparticle generation and structure formation of both micro and nano scales depend on the focusing conditions. These structures can be useful in enhancing the Raman signal since these nano structures are involved with surface plasmons which are the collective Vibrational modes of conduction electrons in the surface.

Acknowledgments This work is supported by DRDO, India.

References

1. C.K. Chen, A.R.B. de Castro, Y.R. Shen, Surface enhanced second harmonic generation. Phys. Rev. Lett. **46**, 145 (1981)
2. A. Wokaun, J.G. Bergman, J.P. Heritage, A.M. Glass, P.F. Liao and D.H. Oslon, "surface second harmonic generation from metal island films and micro lithographic structures", Phys. Rev. B **24,** 849 (1981)
3. F.J. Garcia-Vidal and J.B. Pendry, "collective theory of surface enhanced Raman scattering", Phys. Rev. Lett. **77**, 1163 (1996)
4. Y.W.C. Cao, R.C. Jin and C.A. Mirkin, "Nanoparticles with Raman spectroscopic fingerprints for DNA and RNA detection", Science, **297**, 1536 (2002)
5. W. Fritzsche and T.A. Taton, "Metal nanoparticles as labels for heterogeneous, chirp based DNA detection", Nanotechnology, **14**, R63 (2003)
6. S. Eliezer, N. Eliaz, E. Grossman, "Nanoparticles and nanotubes induced by femtosecond lasers", Laser and particle beams, **23**, 15 (2005)
7. S. Link, M.A. El- Syed, "Optical properties and ultrafast dynamics of nanocrystals", Annu. Rev. Phys. Chem, **54,** 331 (2003)
8. S. Eliezer, N. Eliaz, E. Grossman, "synthesis of nanoparticles with femtosecond laser pulses", Phys. Rev. B**, 69,** 144119 (2004)
9. G.A. Shafeev, *Formation of Nanoparticles Under Laser Ablation of Solids in Liquids* (Nova Science Publishers Inc, New York, 2008), p. 1
10. E. Stratakis, V. Zorba, M. Barberoglou, C. Fotakis, G.A. Shafeev, Laser writing of nanostructures on bulk Al via its ablation in liquids. Nanotechnology **20**, 10530 (2009)
11. J. Radziemski and D.A.Creamers, Laser Induced Plasma and Applications (Marcel Dekker, New York, 1989
12. R.M. Tilaki, A. Irajizad, S.M. Mahdavi, "stability, size and optical properties of silver nanoparticles prepared by laser ablation in different carrier media", Appl. Phys A, **84** 215 (2006)
13. P.V. Kazakevich, A.V. Simakin, G.A. Shafeev, "Formation of periodic structures upon laser ablation of metal targets in liquids", Quantum Electronics **35**, 831 (2005)
14. E.V. Zavedeev, A.V. Petrovskaya, A.V. Simakin, G.A. Shafeev, "Generation of surface nanostructures on nickel by liquid phase laser ablation and their surface enhanced Raman scattering activity", Quant. Electron. **36**, 978 (2006)
15. J. Suna, S.L. Simon, "The melting behavior of Aluminum nanoparticles", Thermochima Acta, **463,** 32 (2007)

16. E. Stratakis, V. Zorba, M. Barberoglou, C. Fotakis, G.A. Shafeev, "Femtosecond laser writing of nanostructures on bulk Al via its ablation in air and liquids", Appl. Surf. Sci, **255,** 5346 (2009)
17. E. Stratakis, M. Barberoglou, C. Fotakis, G. Viau, C. Garcia, G.A. Shafeev, "Generation of Aluminum nanoparticles via ablation of bulk Al in liquids with short laser pulses", Opt. Express **17**, 12650 (2009)
18. S. Lau Truong, G. Levi, F. Bozon-Verduraz, A.V. Petrovskaya, A.V. Simakin, G.A. Shafeev, "Generation of Ag nano spikes via laser ablation in liquid environment and their activity in SERS of organic molecules", Appl. Phys. A, **89,** 373 (2007)
19. A. Hamdorf, M. Olson, C.H. Lin, L. Jiang, J. Zhou, H. Xiao, H.L. Tsai, "Femtosecond andnanosecond laser fabricated substrates for surface enhanced Raman scattering", Opt. Lett, **36**, 3353 (2011)

Magnetic Resonance Studies of Nano Phase Conducting Polyaniline

B. Kavitha, N. Narsimlu, D. Srinivasu, CH. Srinivasm and K. Siva Kumar

Abstract Polyaniline (PANI) was synthesized in Emaraldine form by in situ chemical oxidation method by using Ammonium perdisulphate as oxidizing agent. The XRD pattern indicates that PANI is a semi crystalline solid with d-spacing 4.801 and 4.358 Å. The SEM results show that the particle size lies as an average of 80 nm and length 550 nm. Similarly the UV-Visible spectra of this polymer indicate two absorption bands at around 320 nm and 640 nm. These absorption bands are attributed to the transitions of $\pi \rightarrow \pi^*$ and Benzenoid to Quinoid respectively. The FT-IR spectrum of PANI shows strong bands at 3,442, 2,925, 1,598, 1,494, 1,453, 1,176, 1,112 and 744 cm^{-1} respectively. ^1H NMR spectrum shows 6 peaks with chemical shifts $\delta = 7.452$, 7.468, 7.505, 7.525, 7.604, 7.638 ppm. The lower three peaks at ($\delta = 7.452$, 7.468, 7.505 ppm) are attributed to the protons related to the ^{14}N nucleus. These spectral lines intensity ratio lies as 1:1:1. Another three peaks at ($\delta = 7.525$, 7.604, 7.638 ppm) are attributed to the three protons, which are present at CH Benzenoid, NH, CH at Quinoid. The intensity ratio of these spectral lines is 1:2:1. The room temperature ESR spectrum of PANI Emeraldine salt shows an unresolved peak with lande's g factor 2.010 due to polarons.

Keywords PANI · XRD · UV-visible · NMR · ESR

B. Kavitha · N. Narsimlu · D. Srinivasu · CH.Srinivasm · K. Siva Kumar (✉)
Department of Physics, Osmania University, Hyderabad 500007, India
e-mail: sivakumarou@gmail.com

P. K. Giri et al. (eds.), *Advanced Nanomaterials and Nanotechnology*,
Springer Proceedings in Physics 143, DOI: 10.1007/978-3-642-34216-5_25,

1 Introduction

Over 30 years ago conjugated polymers were as futuristic new materials that would lead to the next generation of electronic and optical devices. The development of "plastic electronics" devices based on conjugated polymers has evolved the states of these materials from academic curiosity to the rapidly growing new electronic industry. Among these conjugated polymers polyaniline possess excellent thermal and environmental stability. Polyaniline have been used in variety applications like diodes, Field effect transistors, solar cells, memory devices, corrosion resistance and in gas sensors [1]. The polyaniline can be synthesized into different sizes and shapes like nanofibers, nanorods, nanotubes and into thin films. Not only tunable morphology, polyaniline can be synthesized with tunable properties like electrical conductivity and optical band gap. The presence of imine and amine nitrogen functional groups in the polyaniline is the main reason for the tunable properties of polyaniline as shown in Fig. 1. These two peculiar properties have given a lot of importance to polyaniline in nano-electronic and optical devices technology. Morphological and spectroscopic study of these polymers is an important subject to understand the transport phenomena of these conjugated polymers to apply them in electronic industry [2, 3]. Hence we attempted to synthesis the polyaniline nanofibers, and characterization of these synthesized nano-polymers with help of different spectroscopic techniques.

2 Experimental Section

2.1 *Synthesis of Polyaniline*

Polyaniline was synthesized by *In-Situ* chemical polymerization of aniline in the presence of hydrochloric acid as a catalyst and ammonium peroxydisulphate as an oxidant. For the synthesis, we took 50 ml, 1 M HCl, and 2 ml of aniline were added into a 250 ml beaker equipped with a Teflon coated magnetic stirrer at about 0 °C temperature. Then 5 g of ammoniumperdisulphate ($(NH_4)_2S_2O_8$) aqueous solution in 50 ml 1 M HCl was drop wise added into the above solution. The polymerization temperature 0 °C was maintained for 10 h to complete the reaction. Then the precipitate obtained was filtered. The product was washed successively by 1 M HCl followed by double distilled water until the wash solution turned colorless. The product PANI was dried at 60 °C for 24 h to get powder form PANI. This powder was pelletized with help of hydraulic machine for characterization.

Fig. 1 Polyaniline

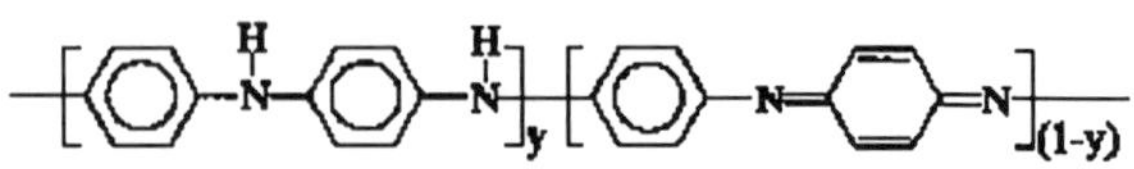

2.2 Characterization

The X-Ray diffraction studies of polyaniline were performed on Phillips XPERT diffractometer with Cu K_α X-ray ($\lambda = 1.54$ Å). Scanning electron microscope (SEM) experiments are performed by employing Hitachi SU 1,510 microscope. The Fourier transform infrared (FTIR) spectra were obtained using a Bio-rad FT-IR (model FT-155) spectrophotometer. The samples were mixed with KBr and pressed into a pellet. The room temperature optical absorption spectrum of these materials was recorded by employing the Varian spectrometer (model Cary 5E) in the 190–1,000 nm wavelength region. The ESR spectra were recorded by using JEOL spectrometer and NMR studies are performed in CFRD Osmania University by employing the Bruker spectrometer.

3 Results and Discussion

The X-ray diffraction pattern of polyaniline emeraldine salt form shows two sharp peaks at $2\theta = 21.10$ and 23.59 as shown in Fig. 2. The inter-planar distances are calculated as 42.10 and 37.70 nm respectively. The average crystallite size is calculated by Debye Scherer equation $D = [0.89\ \lambda]/[\beta \text{ Cos } \theta$ where β is full width of half maxima (FWHM). Generally polymers are considered to be amorphous, but here the synthesized polymer is showing crystalline structure due to their fiber nature and planar nature of Benzenoid and Quinoid functional groups [3]. The SEM micrographs reveal that they are actually agglomeration of nanofibers Fig. 3. The diameter and length of nanofibers are measured as 80 and 550 nm respectively. Agglomeration is taking place due to the conventional filtration of polymer powder in the synthesis process. The Fourier transform infrared spectra of Polyaniline nanofibers are also consistent with the existing literature with strong bands

Fig. 2 XRD of polyaniline

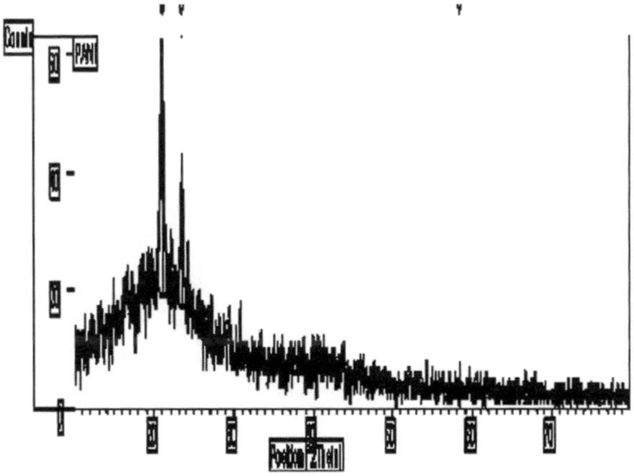

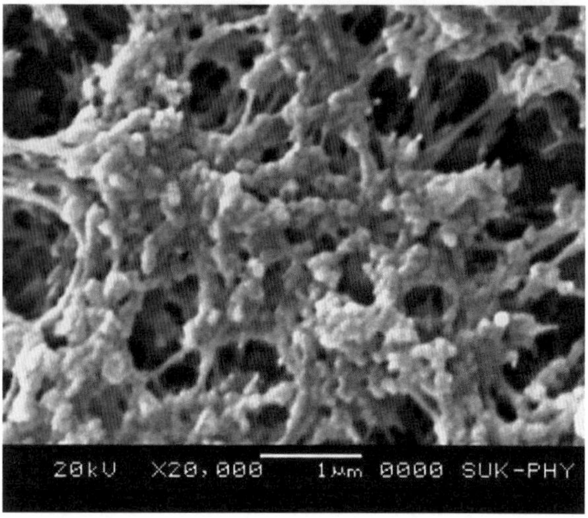

Fig. 3 SEM image of polyaniline

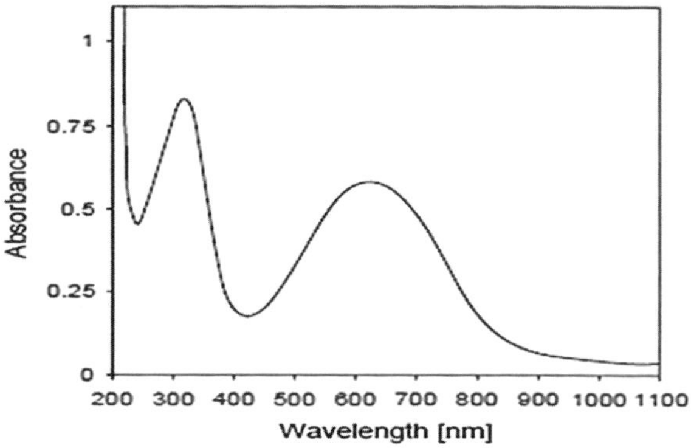

Fig. 4 UV-visible spectra of polyaniline emeraldine base (*EB*) form

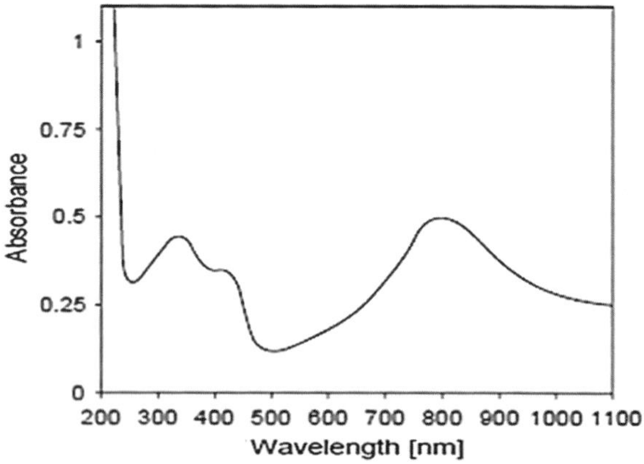

Fig. 5 UV-visible spectra of polyaniline emeraldine salt (*ES*) form

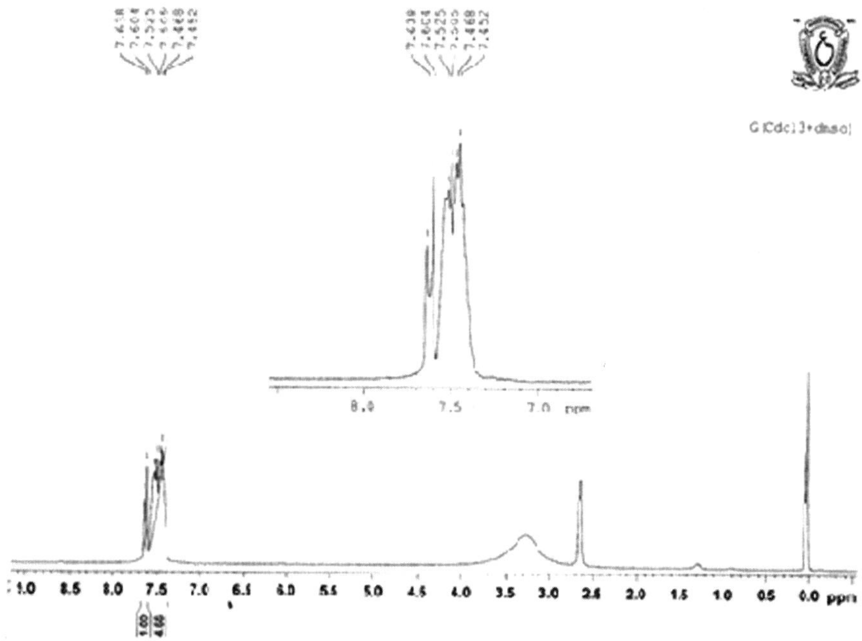

Fig. 6 NMR spectra of polyaniline (*ES*)

Fig. 7 ESR spectra of polyaniline (*ES*)

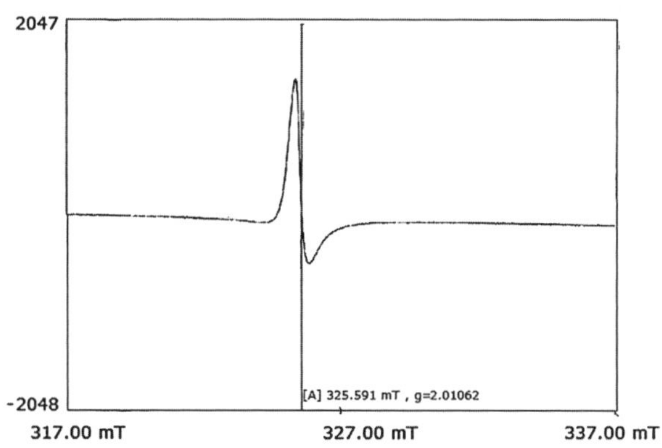

at around 2,925, 1,558 and 14,690 cm^{-1} corresponds to N–H bending, Benzenoid and Quinoid rings respectively [4]. The UV-Visible spectrum of Polyaniline Emeraldine base (EB) form shows two absorption peaks at 325 and 625 nm (Fig. 4). These peaks are assigned to the π-π* transition and polarans band transitions respectively. Similarly, the UV-Visible spectrum of HCl doped Polyaniline Emeraldine salt (ES) shows three absorption bands at 340, 440 and 800 nm (Fig. 5). They are the π-π* transition located 340 nm, the polaron-π* transition at ~440 nm and polaron-π transition around 800 nm. This indicates that the synthesized polyaniline have been effectively doped protonic acid. These two features are fingerprint of electronically conducting Polyaniline Emeraldine salt form [5, 6].

The room temperature ^{1}H NMR spectrum of polyaniline shows 6 peaks with chemical shifts $\delta = 7.452, 7.468, 7.505, 7.525, 7.604, 7.638$ ppm (Fig. 6). The lower three peaks at ($\delta = 7.452, 7.468, 7.505$ ppm) are attributed to the protons related to the ^{14}N nucleus. These spectral lines intensity ratio lies as 1:1:1. Another three peaks at ($\delta = 7.525, 7.604, 7.638$ ppm) are attributed to the three protons, which are present at CH Benzenoid, NH, CH at Quinoid [7–9]. The intensity ratio of these spectral lines is 1:2:1. The ESR spectrum of PANI Emeraldine salt form at room temperature shows an unresolved peak with lande's g factor 2.010 ($h\nu = g\beta H$) (Fig. 7). This may be attributed to the half spin polarans which is created in the PANI back bone due to doping of HCl acid. This result is quite supporting the UV-Visible spectra of PANI Emeraldine salt form. As polyaniline is doped, the resonance line amplitude of the positive lobe (I+) is not equal to the amplitude of the negative lobe, (I−). This change in the resonance line shape is due to the skin effect in polyaniline [10, 11].

4 Conclusions

We synthesized polyaniline nano-fibers by chemical oxidative polymerization method. The morphological studies (XRD, SEM) indicate the radius of polymer fiber is around 80 nm and length around 550 nm respectively. The Emeraldine salt form of polyaniline is confirmed with FTIR, UV/Visible and ESR spectra. The proton NMR studies reveals that peak at a triplet is caused by −NH free radical.

Acknowledgments The authors are thankful to the Head Department of Physics, Osmania University, Hyderabad, India for providing laboratory facilities.

References

1. A.J. Heegar, *Nobel lecture*, 8 Dec 2000
2. C.-H. Chen, J. Appl. Polym. Sci. **89**, 2142–2148 (2003)
3. J. Ouyang, C.-W. Chu, R.J. Seng, A. Prakash, Y. Yang. Hand book of conducting polymers. Proceesing Appl. 3rd edn, pp 8–1 (2006)
4. R.H. Goncalves, W.H. Schreiner, E.R. Leite, Langmuir article. **26**(14), 11657–11662 (2010)
5. J. Huang, R.B. Kaner, Handbook of conducting polymers, in *Conjugated Polymers: Theory, Synthesis, Properties, and Characterization*, ed. by T.A. Skotheim, J.R. Reynolds, 3rd edn. (CRC Press Taylor & Francis Group, 1998), pp 7–16
6. J. Huang, R.B. Kaner, *Hand Book of Conjugated Polymers*, 3rd edn. 7–14 (1998)
7. S. Mu, Y. Yang, J. Phys. Chem. B **112**, 11558–11563 (2008)
8. K. Gupta, P.C. Jana, A.K. Melkap J. Phys. Sci. **12**, 233–238 (2008)
9. D.B. Dupare, M.D. Shirsat, A.S. Aswar Yhe Pacific, J. Sci. Technol. **10**, 1 may 2009

10. D. Stauffer, A. Aharony, *Introduction to percolation theory* (Taylor & Francis, London, 1992)
11. M. Chipara,G.H. Aldicaa, D. Huib, M. Dimoniec, K.T. Laud, L. Georgescue, I. Munteanue, H. Marascoiuc, J. Optoelectron. Adv. Mater. **6**(1), 297 (2004)
12. S.K. Shukla, A. Bhradvaja, A. Tiwari, S. Pilla, G.K. Parashar, G.C. Dubey, Adv. Mater. Lett. **1**(2), 129–134 (2010)

Influence of Co Doping on the Structural and Optical Properties of ZnO Nanostructures

S. Chakraborty, A. K. Kole and P. Kumbhakar

Abstract Undoped and 2.5 % Co doped ZnO nanostructures have been synthesized by simple chemical method at the temperature of 65 °C. The prepared samples are characterized by X-ray diffraction (XRD) analysis, UV–visible absorption (UV–Vis) and photoluminescence (PL) emission spectroscopy. The XRD results indicate that the synthesized ZnO powders are a pure single phase wurtzite structure and also it is found that the lattice parameter of the synthesized ZnO nanostructure varies due to Co doping. It is also found that PL emission and UV–Vis absorption peaks are red shifted due to Co incorporation in ZnO. The results confirmed that Co is incorporated as Co^{2+} in the core of these nanocrystals, occupying the Zn^{2+} sites in the wurtzite structure of ZnO. Localized selected area electron diffraction (SAED) patterns shows that the nanorods are single crystals. The UV PL emission band of doped ZnO nanorods exhibits a red shift from 382 to 386 nm, indicating a band-edge bending due to the presence of doping material in ZnO nanostructure. The defect-related PL emission band centered at 600 nm observed in undoped ZnO sample is suppressed considerably in doped nanorods, revealing the quenching of surface defects present in the synthesized materials.

Keywords Zinc oxide · Optical properties · Photoluminescence

S. Chakraborty · A. K. Kole · P. Kumbhakar (✉)
Nanoscience Laboratory, Department of Physics, National Institute of Technology, Durgapur 713209, West Bengal, India
e-mail: pathik.kumbhakar@phy.nitdgp.ac.in; nitdgpkumbhakar@yahoo.com

P. K. Giri et al. (eds.), *Advanced Nanomaterials and Nanotechnology*, Springer Proceedings in Physics 143, DOI: 10.1007/978-3-642-34216-5_26,

1 Introduction

ZnO has received considerable attention in the field of semiconductor materials due to its direct wide band gap (3.37 eV) at room temperature and transparency in the visible range [1]. It is a promising candidate for transparent electronics as well as in ultraviolet (UV) optoelectronics, including for development of laser because of its high exciton binding energy (60 meV) at room temperature, hardness, chemical stability, optical transparency, and piezoelectric properties [1–3]. Methods for introducing new magnetic, optical, electronic, photophysical or photochemical properties to semiconductor nano-crystals are attracting intense interest as it has different applications in nanotechnology. An effective method for manipulating the physical properties of semiconductors involves impurity doping. A great deal of attention has been paid to dope bulk semiconductors with magnetic ions. Diluted magnetic semiconductors (DMSs) have attracted extensive scientific interests because of their potential applications in spintronics and optoelectronics [4–7]. Most of these systems are derived from various wide-band-gap semiconductors doped with transition-metal (TM) ions. Among the reported DMSs, the wide-band-gap ZnO is particularly interesting because of its optical properties, controllable magnetic properties, and Curie temperature above the room temperature (RT). The coexistence of magnetic, electric, and optical properties increases the potential of TM-doped ZnO (ZnO:TM) to be a multifunctional material.

Recently, 3d transition-metal elements (Co, Ni, Mn and Cu) have been alloyed with ZnO and their properties have been investigated. Cobalt due to its unique chemical stability is also recognized as one of the most efficient doping element to improve and tune the optical, electrical and magnetic properties of ZnO nano materials. Thus, the study of optical and magnetic properties of Co-doped ZnO is important from both fundamental and applied points of view. Usually, doped ZnO nanowires [8] and nanorods [9] are synthesized by thermal evaporation, vapor phase transport and condensation methods, all of which require high-temperature conditions. It has been noted that such high-temperature synthesis strongly influence the properties of ZnO nanomaterials. However, wet chemical synthetic approach provides a promising option for low-temperature in situ doping during the synthesis of ZnO nanostructures. This method is easily reproducible and applicable to large scale industrial production.

Hoang et al. presented a simple microwave-assisted synthesis of $Zn_{1-x}Co_xO$ nano powders in which the surfactant Triethanolamine (TEA) plays an important role on the morphology of $Zn_{1-x}Co_xO$ nanoparticles [10]. Hexagonal ring like superstructures composed of hexagonal single-crystal Co-doped ZnO nanorods were synthesized via a combined hydrothermal and electrochemical process by Li et al. [11]. A detailed investigation of the influence of the surface modification, namely, polymer encapsulation and inorganic capping, on the magnetic properties of Co-doped ZnO colloidal has been carried out by Pereira et al. [12]. It is known that knowledge of the electronic structure of the dopants may enhance the understanding of the mechanisms inducing high-temperature ferromagnetism; therefore, optical properties such as linear optical absorption and photoluminescence (PL) emission characteristics have been of interest to many researchers. However, in the existing literature, most of the papers deal with the absorption properties of Co-doped ZnO; meanwhile, there are few papers dealing with the PL properties of this material here, we report the synthesis of Co doped ZnO prepared by the chemical precipitation method at 65 °C in distilled water medium without using any capping agent or surfactant. The effect of Co doping in structural and optical characteristics of ZnO nanostructures is studied. The three most intense peaks appeared in the XRD pattern clearly show the shifting of the diffraction peaks slightly towards lower angles by Co substitution indicating an expansion of unit cell by an amount of 1.5 %. The samples are also characterized by Transmission electron microscopy (TEM), Scanning electron microscope (SEM) energy-dispersive X-ray analysis (EDAX), and selected area electron diffraction (SAED). SAED pattern shows ring patterns without any additional spots and rings of secondary phases revealing a highly crystalline ZnO wurtzite structure. The EDAX results indicate the presence of Zn, Co and O in the doped sample. SEM analyses of the ZnO sample show the presence of transparent sheets like nano structures. TEM analyses of the ZnO show the presence of both plates like and sheet like 2D nanostructures. SEM of the doped sample show the formation of thin transparent sheet like structure mixed with some tube like formation, which is further confirmed by TEM. Both the PL emission and UV–Vis absorption peaks are red shifted due to Co incorporation. The defect-related PL emission band centered at 600 nm observed in undoped ZnO sample is suppressed considerably in doped nanorods, revealing the quenching of surface defects present in the nanostructures

2 Experimental Details

Co-doped ZnO nanostructures are synthesized at 65 °C temperature in distilled water medium by chemical co-precipitation method, which is described below. All the chemicals used are of AR grade (Merck and SD fine chemicals) and they are used without further purification. Co doped (2.5 wt %) ZnO nanostructures (ZnO:Co) are prepared by mixing calculated amounts of zinc acetate solution in distilled water and cobalt nitrate solution in distilled water under continuous stirring followed by drop-wise addition of sodium hydroxide solution in distilled water up to pH–8. The mixture is continuously stirred in a magnetic stirrer for another 1 h. After 1 h of continuous stirring, a blue-white precipitate is deposited at the bottom of the flask. The precipitate is separated from the reaction mixture by centrifugation (Eltek Refrigerated Centrifuge RC4100D) for 10 min at 5,000 rpm and washed several times with distilled water and methanol to remove the remaining sodium particles, if any. The wet precipitate is then dried for further measurements.

The optical transmission spectra of the prepared ZnO:Co nanostructures dispersed in water are recorded using a UV–Vis spectrophotometer (Hitachi, U-3010).

The formation of ZnO:Co nanostructures have been confirmed by transmission electron microscopy (TEM, JEOL 2000 FX-II). The XRD pattern is recorded by using an X-ray diffractometer (PANLYTICAL) with Cu K_α radiation of wavelength $\lambda = 0.15406$ nm in the scan range $2\theta = 20$–$80°$. The photoluminescence (PL) spectra of the ZnO:Co nanostructures dispersed in water are recorded using a spectrofluorimeter (Perkin Elmer LS-55).

3 Results and Discussion

Figure 1 shows the XRD pattern of both the undoped and doped ZnO samples. From Fig. 1 it is seen that the synthesized powders of either pure or doped ZnO present a wurtzite structure (hexagonal) with a P63mc space group and no impurity related peak is present in the XRD patterns(JCPDS Card No. 80-0075). This indicates also that no correlated phases like CoO, Co_3O_4, etc. are formed in the samples. For all samples, all of the diffraction peaks can be indexed as the pure hexagonal phase of ZnO with a typical wurtzite structure having lattice constants (Table 1) $a = 3.25$ Å and $c = 5.21$ Å. All the peaks of the XRD pattern clearly show a shifting of the diffraction peaks slightly towards lower angles due to Co substitution. The effective ionic radius of Co^{2+} in the tetrahedral configuration (0.58 Å) is slightly smaller than that of Zn^{2+} (0.60 Å); the ionic radius of Co^{2+} in the 12-coordinated metal configuration (1.25 Å) is smaller than that of Zn^{2+} (1.39 Å). Therefore, when Zn^{2+} ions are substituted by Co^{2+} ions, the lattice is shrunk so that the lattice constants are decreased, which is contradictory to our result. The blue shift of XRD peaks can be explained as follows, when Zn atoms are replaced by Co atoms, it leads to an increase of out plane lattice constant may be due to its large Possion's ratio [13].

As can be seen in Fig. 1 that the full width half maximum (FWHM) of (002) peak is narrowest among all diffraction peaks. It exhibits the evidence of the *c*-axis texture [14]. This observation accords with the SAED results. There is no obvious characteristic diffraction peaks associated to secondary phases or clusters in any of our samples. The $d_{(002)}$ values were calculated from the corresponding XRD peaks and is observed to increase linearly with increasing Co content, consistent with other reported results on this type of systems [15]. A linear increase of the lattice spacing thus indicates, in accordance with Vegard's law [16], that at least up to x = 2.5 % doping, Co ions are substituted in ZnO without changing the wurtzite structure. The interplanner spacing (d_{hkl}) as calculated from XRD and JCPDS data card (80-0075) and corresponding (*hkl*) values are summarized in Table 2, for some major XRD peaks. Qiu et al. [9] has proposed a linear relationship between red shift and square of lattice volume, which clearly demonstrate that the band gap energies of direct wide bandgap semiconductors can be tuned by lattice volume.

The morphological structures studied from FESEM and TEM are shown in Figs. 2 and 3, respectively. From Figs. 2a and 3a it is seen that the undoped sample seems to be made up with different ZnO nanostructures with varying shape and sizes. From Fig. 2a it is seen that the there are very thin nanosheets, which are

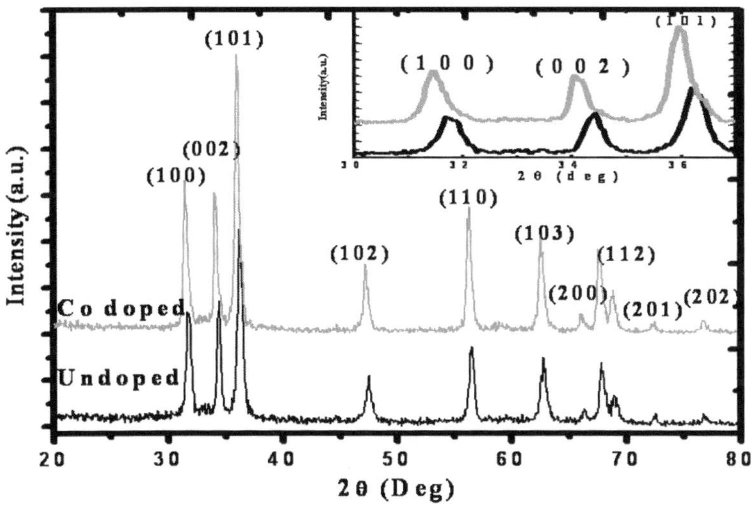

Fig. 1 XRD patterns of undoped ZnO and ZnO:Co (2.5 %) samples. The *inset* shows the XRD pattern of the three most intense peaks [(100), (002), and (101)] of the Co-doped ZnO sample in expanded form

Table 1 Lattice constants of the prepared samples

Sample	*a* (Å)	*c* (Å)	*c/a*	Unit cell volume $(Å)^3$
ZnO	3.250	5.21	1.6	47.65
ZnO:Co (2.5 %)	3.276	5.25	1.6	48.80

folded and trying to take a tubular shape. These sheets are interconnected with each other. If sheets are flattened length of them will be ~2–5 μm. The nanosheets are very thin in width and are also transparent. In undoped ZnO, formation of 2D plate like nanostructure with bended contours is also observed as can be seen from the TEM image as shown in Fig. 3a. Such two-dimensional plate like structure has been reported recently in metal nanoparticles by Viswanath et al. [17]. Multisheet structures are obtained in the doped ZnO samples as can be seen from the FESEM image as shown in Fig. 2b. The micro sheet of undoped sample seems to be divided into several small nanosheets. For the doped sample there is also presence of few nanorods as can be seen from the TEM image which is shown in Fig. 3b.

The formation of nanorod structure of Co doped ZnO has been reported earlier by several authors [9, 11, 12]. TEM image of Co doped ZnO shown in of Fig. 3b reveals that these rods are interconnected through some ribbon like transparent lamellar path and the average diameter of the nanorods are ~40 nm and lengths are lying within 200–250 nm, having aspect ratio varies between ~5 and 6. Some nano sheets in the outer surface are rolled into nanorods. Such nanorods are seems to be hollow i.e. they are very much transparent to the electron beam, so it would be justified to say them nanotubes rather than nanorods.

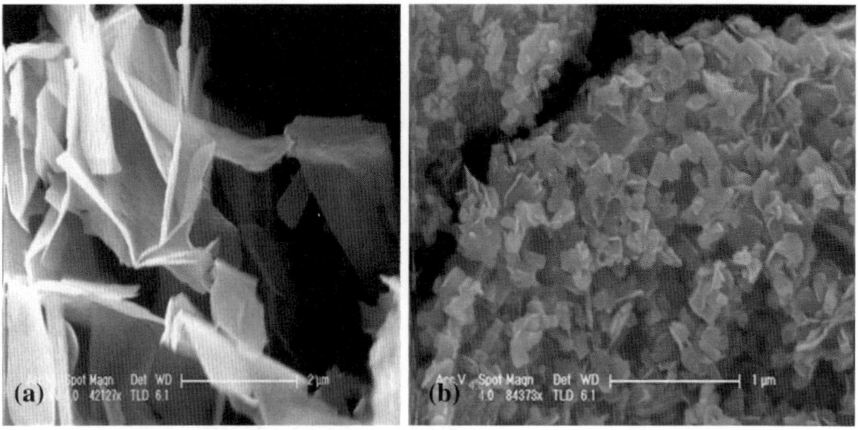

Fig. 2 Field emission scanning electron microscope (FESEM) images of the undoped and ZnO:Co samples are shown in (**a**) and (**b**), respectively

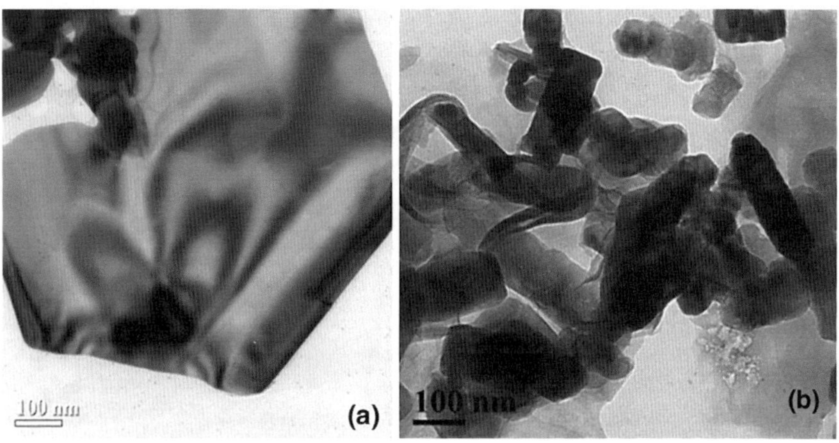

Fig. 3 Transmission electron microscope (TEM) images of the undoped and ZnO:Co samples are shown in (**a**) and (**b**), respectively

The EDAX results as shown in Fig. 4 indicate that Co^{2+} substitute into ZnO lattice. The small Cu peak appearing in the spectra is due to the coating of the sample surface required for the measurements. The SAED pattern as shown in the inset of Fig. 4 shows ring patterns without any additional spots and rings of secondary phases revealing a highly crystalline ZnO wurtzite structure. Several fringe patterns are observed with plane distances of 2.81, 2.68, 2.43 and 1.81, 1.61, 1.28, and 1.20 Å, in the electron diffraction pattern which corresponds to (100), (002), (101), (102), (110), (103) and (112) planes, respectively, of pure wurtzite hexagonal structure of ZnO.

Table 2 Interplanner spacing (d_{hkl}) from XRD and JCPDS data card and corresponding (*hkl*) values

d_{XRD} (Å) (ZnO)	d_{XRD} (Å) (Co doped ZnO)	d_{XRD} (Å) (JCPDS)	(*hkl*)
2.81	2.83	2.82	(100)
2.60	2.62	2.60	(002)
2.475	2.49	2.48	(101)
1.91	1.92	1.91	(102)
1.62	1.63	1.63	(110)
1.477	1.48	1.48	(103)
1.378	1.38	1.38	(112)

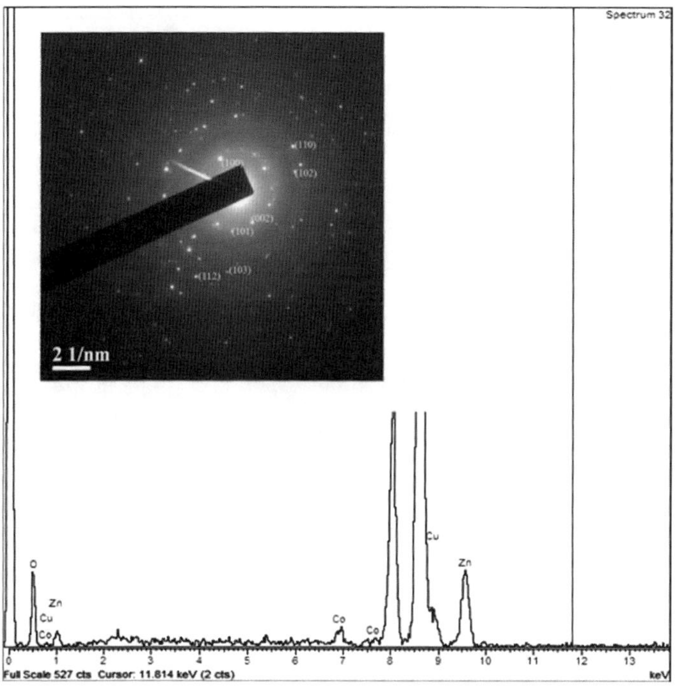

Fig. 4 EDAX spectrum of ZnO:Co sample. The *inset* shows the selected area electron diffraction pattern (SAED) of ZnO:Co sample

A simple reaction mechanism is proposed for the formation of ZnO:Co nanostructures is described below by following report of Pannigrahy et al. [18]. Zinc cation (Zn^{2+}) from zinc acetate, Co^{2+} from cobalt nitrate-hexa hydrate, and a hydroxide ion (OH^{-}) from water react in aqueous solution to form a Zinc-Co hydroxide quasi-precursor. Which decomposes by heating at temperatures of 65 °C to form ZnO:Co nuclei. After nucleation they grow in a specific pattern to form such ZnO:Co nanostructures.

Figure 5 shows the UV–Vis absorption characteristics of all the synthesized samples. To obtain the absorption characteristics of all the samples, at first

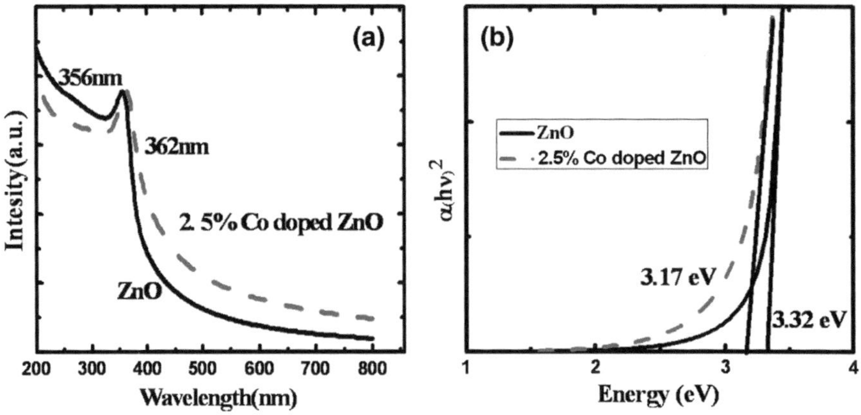

Fig. 5 UV-Vis absorption characteristic of the prepared samples dispersed in water are shown in **a** and **b** shows the calculation of band gap of the samples

transmittance (T) at different wavelengths (λ) are measured and then absorbance (α) at the corresponding wavelengths are calculated by using Beer–Lambert relation: $\alpha = 1/d \ \ln(1/T)$, where d is the path length. The linear optical transmission characteristics of the prepared samples are measured by dispersing the powdered samples in water and then it is taken in a quartz cuvette of 10 mm path length (d). From the absorption onset of corresponding UV–Vis spectra, as shown in Fig. 5, the band gap energy can be estimated.

From Fig. 5 it is seen that the band edge for the undoped ZnO sample appears at 356 nm, while the band edge for the doped sample is shifted to lower energy at 362 nm. The absorption coefficient α and band gap E_g of a semiconductor material can be related as, $\alpha h\nu^m = A\ (h\nu - E_{gn})$. Here A is a constant and the band gap of the studied material is denoted as E_{gn}. Exponent "m" depends upon the type of the transition; m may have values 2, 1/2, 2/3 and 1/3 corresponding to the allowed direct, allowed indirect, forbidden direct and forbidden indirect transitions, respectively [19]. For ZnO, the value of m is always 2, i.e., the fundamental absorption corresponds to allowed direct transition. To calculate the band gap values of the prepared samples, $(\alpha h\nu)^2$ versus $h\nu$ has been plotted and it is shown in Fig. 2. The value of the band gap is determined by extrapolating the straight line portion of ($\alpha h\nu$) is 3.34 and 3.18 eV for pure and doped sample. This red shift of band gap E_g with incorporation of Co into ZnO has already been reported by several researchers [6, 12, 20]. Since the exciton Bohr radius of ZnO is much smaller than the diameter of D the synthesized nanorods there is no indication of quantum confinement. This red shift in E_g is mainly due to sp-d exchange interactions between the band electrons and the localized d electrons of Co ions substituting Zn ions [6]. The s-d and p-d exchange interactions lead to a negative and positive correction to the conduction band and the valence band edges, resulting a band gap narrowing. The interaction leads to corrections in the energy

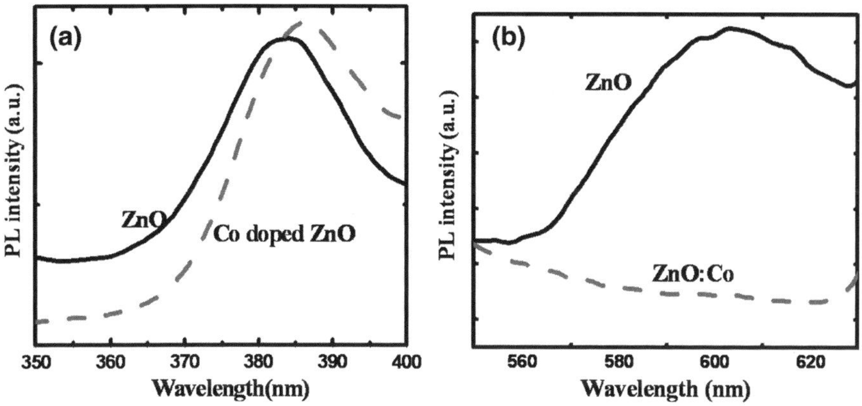

Fig. 6 Photoluminescence (PL) emission spectra of the samples. UV band is shown in (**a**) and yellow–orange band is shown in (**b**)

bands the conduction band is lowered and the valence band is raised causing the bandgap to shrink [9, 19].

Photoluminescence (PL) emission studies for pure ZnO and ZnO:Co nanostructures at room temperature reveal the presence of a UV PL emission peak centered at around 382 nm and a broad yellow-orange emission band ranging from 550 to 730 nm, as shown in Figs. 6a, b. The UV emission is attributed to the near band edge emission i.e. due to radiative recombination of a hole in the valence band and an electron in the conduction band. This peak is found to be shifted from 382 to 386 nm, i.e. toward lower energy due to the substitution of Co ions in ZnO nanostructure. The shift of the UV emission peak to the longer wavelength side is due to the lowering of the band gap of the ZnO and which arises due to strong exchange sp-d interactions between the doping ion Co^{2+} and of the host band ZnO [18, 21, 22]. For the doped ZnO samples, the broad emission peaks become very weak as compared with the pure ZnO, indicating the presence of Co^{2+}in the sample as the intensity reduces considerably after doping. The quenching of the luminescence of the visible region in 2 and 5 % Co-doped ZnO samples had also been reported earlier by Antony et al. [7]. The doping of the transition metal ions in pure ZnO minimizes the concentration of defects that was present in the undoped sample. The cobalt cations provide competitive pathways for recombination, which results in quenching of the broad emission.

4 Conclusions

Unopded and Co doped ZnO nanostructures of varying shapes and sizes are successfully synthesized via direct precipitation route using wet chemical reaction method at room temperature. By analyzing different experimental data the substitution of Co into ZnO lattice is confirmed. The analysis of XRD pattern showed the presence of pure wurtzite crystal structure only in the samples. FESEM and TEM analysis confirmed the formation of different nanostructures such as nanosheet, nanoplate and nanorod. Both the undoped and doped samples exhibit PL emissions which consist of two band; one in the UV region at ~382 nm, which arises due to near band edge emission and another appeared in the yellow-orange region ranging from 550 to 730 nm. The UV PL emission peak has been found to be red shifted from 382 to 386 nm due to the smaller value of the band gap of the ZnO:Co in comparison that of undoped ZnO and this arises due to strong exchange interactions between the *d* electrons of the doping ion and the *s* and *p* electrons of the host lattice. It is found that the PL emission in the visible region is quenched in the doped sample, which indicates high crystallinity of the samples and passivation of surface defects in the doped sample.

Acknowledgments Authors are grateful to Department of Science and Technology (SR/FTP/PS-67/2008), Government of India, for the financial support. They express their sincere thanks to Sri C. S. Tiwary for his help in FESEM analysis.

References

1. U. Ozgur, Y.I. Alivov, C. Liu, A. Teke, M.A. Reshchikov, S. Dogan, V. Avrutin, S.J. Cho, H. Morkoc, A comprehensive review of ZnO materials and devices. J. Appl. Phys. **98**, 41301 (2005)
2. A. Janotti, C.G.V. de Wale, Fundamentals of zinc oxide as a semiconductor. Rep. Prog. Phys. **72**, 126501 (2009)
3. L.S. Mende, J.L.M. Driscoll, ZnO-nanostructures, defects, and devices. Mater. Today **10**, 40 (2007)
4. G. Glaspell, P. Dutta, A. Manivannan, A room-temperature and microwave synthesis of M-doped ZnO (M = Co, Cr, Fe, Mn & Ni). J. Cluster Sci. **16**, 523–536 (2005)
5. A. Singhal, S.N. Achary, J. Manjanna, S. Chatterjee, P. Ayyub, A.K. Tyagi, Chemical synthesis and structural and magnetic properties of dispersible cobalt- and nickel-doped ZnO nanocrystals. J. Phys. Chem. C **114**, 3422–3430 (2010)
6. S. Maensiri, P. Laokul, S. Phokha, A simple synthesis and magnetic behavior of nanocrystalline $Zn_{0.9}Co_{0.1}O$ powders by using Zn and Co acetates and polyvinyl pyrrolidone as precursors. J. Magn. Magn. Mater. **305**, 381–387 (2006)
7. J. Antony, S. Pendyala, A. Sharma, X.B. Chen, J. Morrison, L. Bergman, Y. Qiang, Room temperature ferromagnetic and ultraviolet optical properties of Co-doped ZnO nanocluster films. J. Appl. Phys. **97**, 97 (2005)
8. W. Liang, D.Y. Benjamin, D. Yuhas, P. Yang, Magnetotransport in Co-doped ZnO nanowires. Nano Letter **9**, 892 (2009)
9. X. Qiu, L. Li, G. Li, Nature of the abnormal bandgap narrowing of highly crystalline $Zn_{1-x}Co_xO$ nanorods. Appl. Phys. Lett. **88**, 114103 (2006)
10. L.H. Hoang, P.V. Hai, N.H. Hai, P.V. Vinh, X.B. Chen, H.P Van, I.S. Yang, The microwave-assisted synthesis and characterization of $Zn_{1-x}Co_xO$ nanopowders. Mater. Lett. **64**, 962 (2010)
11. Y.J. Li, C.Y. Wang, M.Y. Lu, K.M. Li, L.J. Chen, Electrodeposited hexagonal ringlike superstructures composed of hexagonal Co-doped ZnO nanorods with optical tuning and high-temperature ferromagnetic properties. Cryst. Growth Des. **8**, 2598 (2008)

12. A.S. Pereira, A.O. Ankiewicz, W. Gehlhoff, A. Hoffmann, S. Pereira, T. Trindade, M. Jrundmann, M.C. Carmo, N.A. Sobolev, Surface modification of Co-doped ZnO nanocrystals and its effects on the magnetic properties. J. Appl. Phys. 103 07D140 (2008)
13. Y.Z. Yoo, T. Fukumura, Z. Jin, K. Hasegawa, M. Kawasaki, P. Ahmet, T. Chikyow, H. Koinuma, ZnO-CoO solid solution thin films. J. Appl. Phys. **90**, 4246 (2001)
14. S. Ilican, Y. Caglar, M. Caglar, Preparation and characterization of ZnO thin films deposited by sol-gel spin coating method. J. Optoelectron Adv Mater **10**, 2578 (2008)
15. S. Guo, X. Zhang, Y. Huang, Y. Li, Z. Du, Investigation on electronic structures and nature of charge-transfer transition of ZnO:Co with variation of Co content. Chem. Phys. Lett. **459**, 82 (2008)
16. M. Naeem, S.K. Hasanain, M. Kobayashi, Y. Ishida, A. Fujimori, S. Buzby, S. Ismat Shah, Effect of reducing atmosphere on the magnetism of $Zn_{1-x}Co_xO(0 \leq x \leq 0.10)$ nanoparticles. Nanotechnology **17**, 2675 (2006)
17. B. Viswanath, P. Kundu, B. Mukherjee, N. Ravishankar, Predicting the growth of two-dimensional nanostructures. Nanotechnology **19**, 195603 (2008)
18. B. Panigrahy, M. Aslam, D. Bahadur, Aqueous synthesis of Mn- and Co-doped ZnO nanorods. J. Phys. Chem. C **114**, 11758 (2010)
19. J.I. Pankove, *Optical Processes in Semiconductor* (Prentice-Hall, New Jersey, 1971)
20. M. Ivill, S.J. Pearton, S. Rawal, L. Leu, P. Sadik, R. Das, A.F. Hebard, M. Chisholm, J.D. Budai, D.P. Norton, Structure and magnetism of cobalt-doped ZnO thin films. New J. Phys. **10**, 065002 (2008)
21. I. Ozerov, F. Chabre, W. Marine, Incorporation of cobalt into ZnO nanoclusters. Mater. Sci. Eng. C **25**, 614 (2005)
22. P. Lommens, P.F. Smet, C.M. Donega, A. Meijerink, L. Piraux, S. Michotte, S.M. Tempfli, D. Poelman, Z. Hens, Photoluminescence properties of Co^{2+}-doped ZnO nanocrystals. J. Lumin. **118**, 245 (2006)

Isoelectronic Centers in Quantum Dots and Photoluminescence Decay

Kiran G. Sonawane, Ch. Rajesh and Shailaja Mahamuni

Abstract It is hypothesized that Te forms an isoelectronic trap in ZnSe. These isoelectronic centers show blue and green band luminescence at low temperature. Quantum confinement effects reveal isoelectronic trap related luminescence at room temperature in contrast to bulk $ZnSe_{1-}$ yTey. To find the effect of these isoelectronic center on Mn^{2+} *d–d* transition luminescence, Mn doped $ZnSe_{0.99}$ $Te_{0.01}$ QDs are synthesized. Mn doped ZnSe shows dominating orange emission related to Mn^{2+} *d–d* transitions. This Mn emission increases at the cost of band edge emission. Addition of Te as small as 1 % in ZnSe strongly quenches photoluminescence of Mn-doped ZnSe QDs showing predominance of Te-isoelectronic centers. Orange emission and near band edge luminescence in Mn doped $ZnSe_{0.99}Te_{0.01}$ are not correlated as they are in case of Mn-doped ZnSe QDs. Time resolved photoluminescence and photoluminescence excitation study revealed these isoelectronic center changes the recombination path ways. The changes in relaxation path ways are responsible for distinct emission behavior of $ZnSe_{0.99}Te_{0.01}$QDs.

Keywords Quantum dots · Isoelectronic doping · Luminescence · Time resolved photoluminescence

K. G. Sonawane · Ch. Rajesh · S. Mahamuni (✉)
Department of Physics, University of Pune, Ganeshkhind Road, Pune 411007, India
e-mail: shailajamahamuni@yahoo.co.in

P. K. Giri et al. (eds.), *Advanced Nanomaterials and Nanotechnology*,
Springer Proceedings in Physics 143, DOI: 10.1007/978-3-642-34216-5_27,

1 Introduction

In nano regime band gap can be tuned by varying the particle size [1, 2]. Band gap tailoring is also offered by alloying of semiconductor [3–8]. Effect of Te incorporation on optical properties of ZnSe is well studied in bulk [9–11]. For lower concentration (1–5 %) of Tellurium (Te) in $ZnSe_{1-y}Te_y$, Te acts as an isoelectronic impurity and traps the electron from the exciton [9–11]. Because of the large electro-negativity difference (Se is more electronegative than Te) valance band electron originally belonging to Te tends to localize near Se. Consequently, Te atom lacks an electron and acts as a trap centre for the exciton [4]. When Te is isoelectronically doped in ZnSe, two photoluminescence (PL) emission bands are observed. Blue band (between 2.60 and 2.75 eV) [12] is attributed to $Te_{n=2}$ (exciton bound to two Te atoms) and green band (between 2.35 and 2.55 eV) [12] is attributed to $Te_{n>2}$ (exciton bound to more than two Te atoms) isoelectronic centers [11, 12] Few reports indicate that green band arise either from ZnTe/ZnSe type II quantum structure or $Te_{n>2}$ isoelectronic centre [3, 11–15].

In the present work, colloidal $ZnSe_{1-y}Te_y$ quantum dots (QDs) are prepared using high temperature wet chemical method. Amount of Te is limited up to 4 % in order to achieve only isoelectronic doping. We focus our attention on the behavior of isoelectronic luminescence as a function of composition. Due to quantum confinement effect, isoelectronic luminescence can be observed at room temperature. Notably, in earlier reports the luminescence for isoelectronic centre and/or type II quantum structure has been observed only at low temperature [3, 9–12]. In order to check effect of isoelectronic center on Mn emission in ZnSe, Mn doped $ZnSe_{0.99}Te_{0.01}$ QDs were prepared. For Mn doped ZnSe QDs appearance of *d–d* transition emission at the cost of near band edge emission is reported [16]. Te isoelectronic centers quenches Mn emission in Mn doped ZnSe QDs [17]. The quenching of Mn emission and insensitivity of near band edge luminescence to Mn concentration suggest change in exciton decay mechanism. Time resolved photoluminescence (TRPL) and photoluminescence excitation (PLE) measurements are performed to understand the emission behavior of $ZnSe_{1-y}Te_y$ and Mn- doped $ZnSe_{1-y}Te_y$ QDs.

2 Experimental Details

$ZnSe_{0.99}Te_{0.01}$ QDs were synthesized by modifying the reported method [18] for ZnSe QDs. The organometallic precursor containing diethyl zinc (0.8 mM) and a mixture of selenium (Se) and tellurium (Te) powder (1 M) in trioctylphosphine (TOP), was rapidly injected into the hot octadecylamine (ODA) (10 g) at high temperature. The amount of Se and Te was varied to attain different Te doping levels. The particles were isolated [19] and a fine powder of $ZnSe_{1-y}Te_y$ QDs was obtained. Undoped ZnSe QDs were also synthesized by the same method for comparison. Mn doped $ZnSe_{0.99}Te_{0.01}$ were prepared as reported earlier [17]. All concentration mentioned in the manuscript are in solution unless and other wise specifically mentioned.

Phase determination and average size of $ZnSe_{1-y}Te_y$, Mn doped ZnSe and Mn doped $ZnSe_{0.99}Te_{0.01}$ QDs were determined by using x-ray diffraction (XRD). XRD measurements were carried out on Bruker D8 advance powder X-ray diffractometer, using Cu K_α ($\lambda = 1.5402$ Å) as an incident radiation. Optical absorption studies were performed using Perkin Elmer Lambda 950 spectrophotometer to estimate the gap between the highest occupied molecular orbital (HOMO) and lowest unoccupied molecular orbital (LUMO). The PL measurements were carried out on Perkin Elmer LS55 PL spectrometer. For TRPL measurements Ti–Sapphire picosecond laser and a time-correlated single photon counting setup, coupled to a micro-channel plate photomultiplier was employed. Pulses of 1 ps duration of 777 nm radiation from the Ti–Sapphire laser were frequency doubled to 388 nm, using a frequency doubler. For all the TRPL measurements, samples were excited at 388 nm, at a pulse repetition rate of 4 MHz. The instrument response function (IRF) was obtained at 388 nm, using a very dilute colloidal suspension of dried nondairy coffee whitener. The width (full width at half maximum) of the IRF was ~40 ps. The decay was deconvoluted with respect to the IRF and analyzed using a sum of discrete exponentials.

3 Results and Discussion

Figure 1 shows the XRD pattern of Mn doped ZnSe, $ZnSe_{1-y}Te_y$ and Mn doped $ZnSe_{0.99}Te_{0.01}$ QDs. X-ray reflections of bulk ZnSe, and bulk ZnTe are given for comparison. The XRD patterns of Mn doped ZnSe, $ZnSe_{1-y}Te_y$ and Mn doped $ZnSe_{1-y}Te_y$ QDs show broad peaks corresponding to (1 1 1), (2 2 0), and (3 1 1) planes of cubic zinc-blende phase of ZnSe. No secondary phase formation was observed even for sample with Te as high as 4 % and Mn as high as 10 %. Scherrer formula was employed to calculate the average particle diameter. When Te is incorporated in the lattice no appreciable change in the diameter is observed. For $ZnSe_{1-y}Te_y$ QDs, size, as estimated from Scherrer formula is 3.2 ± 0.6 nm. Mn incorporation causes a slight change in diameter. Such a behavior is also observed in ZnSe QDs and attributed to change in reaction kinematics due to Mn precursors [16]. For Mn doped $ZnSe_{1-y}Te_y$ QDs, size estimated from Scherrer formula is 3.0 ± 0.9 nm for 10 % Mn doping level.

Figure 2 shows absorption and PL spectra for undoped ZnSe, Mn doped ZnSe, $ZnSe_{1-y}Te_y$, and Mn doped $ZnSe_{1-y}Te_y$ QDs. With incorporation of Te band gap decreases, indicating successful incorporation of Te in ZnSe QDs. Whereas, with incorporation of Mn the band gap of Mn doped $ZnSe_{1-y}Te_y$ QDs increases from that of $ZnSe_{1-y}Te_y$ by virtue of decrease in size. PL spectra of ZnSe QDs show only band edge luminescence. Even with incorporation of Te as small as 1 %, near band edge emission shift to 440 nm. This shift in emission is accompanied by

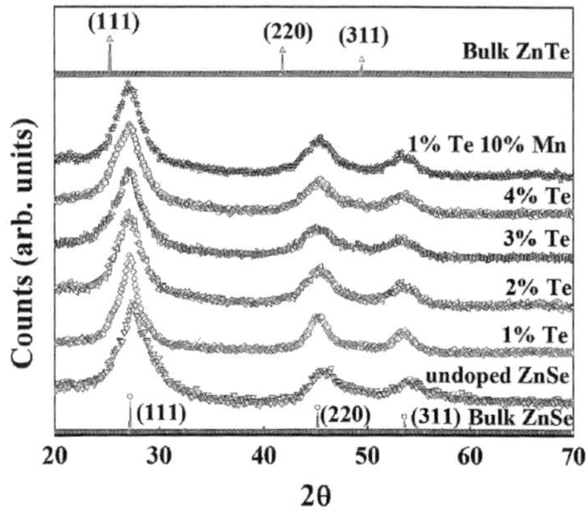

Fig. 1 XRD spectra for $ZnSe_{1-y}Te_y$, Mn doped ZnSe, and Mn doped $ZnSe_{0.99}Te_{0.01}$ QDs. Bulk ZnSe and ZnTe cubic zinc blende spectra were given for reference

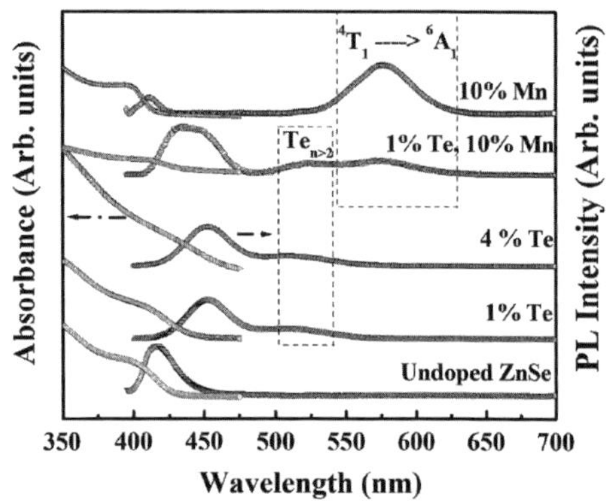

Fig. 2 Optical absorption and PL spectra (excited at 375 nm) for undoped ZnSe, $ZnSe_{1-y}Te_y$, Mn doped $ZnSe_{0.99}Te_{0.01}$ and Mn doped ZnSe QDs

appearance of a hump like feature at 522 nm. These features are assigned to originate from $Te_{n=2}$ (blue band) and $Te_{n>2}$ (green band) isoelectronic centers. It is worthwhile to mention that the isoelectronic center related emission is observed at room temperature in QDs. This is in contrast to bulk and epilayers [3, 9–12]. Te being less electronegative than Se, forms an isoelectronic trap centers. These isoelectronic centers trap electron from the exciton. Thermal energy can ionize exciton bound to Te isoelectronic center resulting in formation of free exciton. The free exciton can recombine and give band edge emission [14]. In case of QDs, excitons are confined in proximity of trap center. The ionized electron from isoelectronic center can again easily get trapped by isoelectronic center. In other words quantum confinement avoids thermal ionization of trap centers leading to observation of isoelectronic luminescence at room temperature.

In the literature, green band emission is either attributed to formation of ZnTe/ZnSe type II core/shell structure or to the Te-related isoelectronic centers [3, 10, 14]. In the present work, amount of incorporated Te is way lower than the amount of precursor needed to form even one monolayer thick shell of ZnTe. Thereby, it can be inferred that under the present experimental conditions, it is not possible to form core/shell structure of ZnSe/ZnTe or ZnTe/ZnSe. Thus hump at 522 nm can

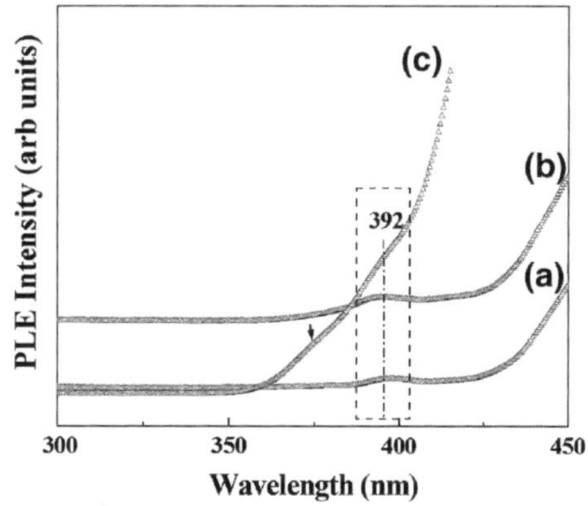

Fig. 3 PLE spectra for $ZnSe_{0.99}Te_{0.01}$ QDs recorded at emission wavelength (**a**) 528 nm, (**b**) 522 nm, and (**c**) 440 nm

be assigned to originate from $Te_{n>2}$ isoelectronic centers rather than type II structure. To confirm that the emission is only from isoelectronic center photoluminescence excitation (PLE) measurements were carried out.

Emission from type II structure comes from band to band energy transfer, whereas, emission from the isoelectronic center requires energy transfer from host ZnSe levels. As a result, PLE spectra for type II emission and isoelectronic emission is distinctly different. It is conjectured [10] that green band emission has two different origins. The high energy region of green band emission arises from isoelectronic center. Type II structure contributes to low energy region of green band. Moreover, PLE spectra of blue band and high energy region of green band are excited from free excitons from ZnSe. Also it is observed that, PLE spectra of type II structures from green band and blue band are distinctly different. PLE measurements of chemically grown $ZnSe_{0.99}Te_{0.01}$ QDs are depicted in Fig. 3a, b,c [with fixed emission wavelength at 528, 522, and 440 nm respectively]. PLE feature appears at 392 nm in all cases and is attributed [16] to ZnSe QDs. In all the cases emission comes from the same energy level. No band to band transition for type II structure is observed. From the measurement it is clear that, green band and blue band emission originate from excitation through host ZnSe and not from inter band excitation. These measurements confirm green band originate from isoelectronic centers only. It is interesting to study effect of such isoelectronic center on *d–d* transion of Mn doped ZnSe QDs. Mn doped ZnSe QDs shows band edge emission at 414 nm along with a feature at 575 nm (Fig. 2). Orange (575 nm) emission arises [16] from $^4T_1 \rightarrow {}^6A_1$ Mn *d–d* transition. In case of Mn doped ZnSe, orange emission is the most dominating emission. Most interestingly, reduction of Mn impurity emission is observed in case of Mn doped $ZnSe_{1-y}Te_y$ QDs (Fig. 2). In order to further probe the origin of Mn emission quenching PLE measurements were carried out on Mn emission.

Figure 4a shows PLE spectra recorded at Mn emission from Mn doped ZnSe and Mn doped $ZnSe_{0.99}Te_{0,01}$ QDs. PLE spectra for Mn doped ZnSe and Mn doped $ZnSe_{0.99}Te_{0.01}$ QDs are distinctly different. PLE spectra of Mn doped ZnSe shows most dominating contribution to Mn emission comes from ZnSe band gap states.

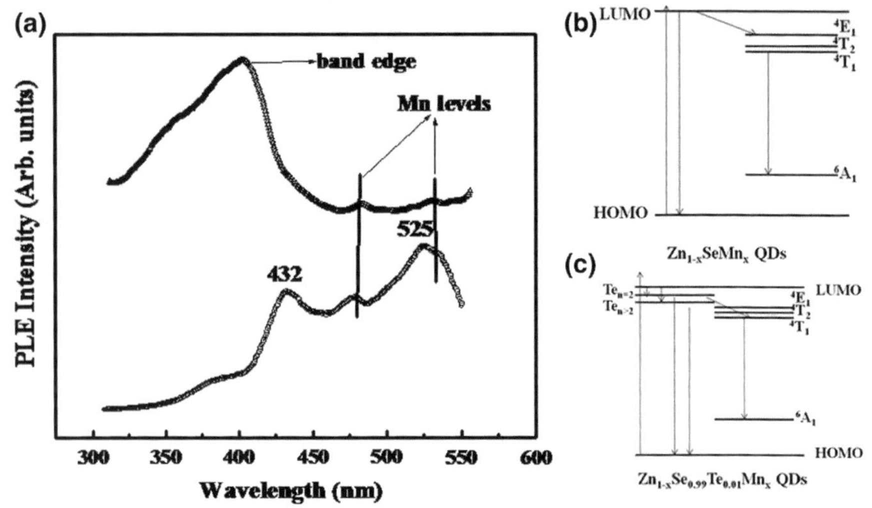

Fig. 4 **a** PLe spectra of Mn doped ZnSe and Mn doped $ZnSe_{0.99}Te_{0.01}$ recorded at Mn emission. Schematic of energy level decay pathways in **b** Mn doped ZnSe and **c** Mn doped $ZnSe_{0.99}Te_{0,01}$ QDs

Fig. 5 TRPL spectra of Mn doped ZnSe and Mn doped $ZnSe_{0.99}Te_{0.01}$ QDs recorded at near band edge emission

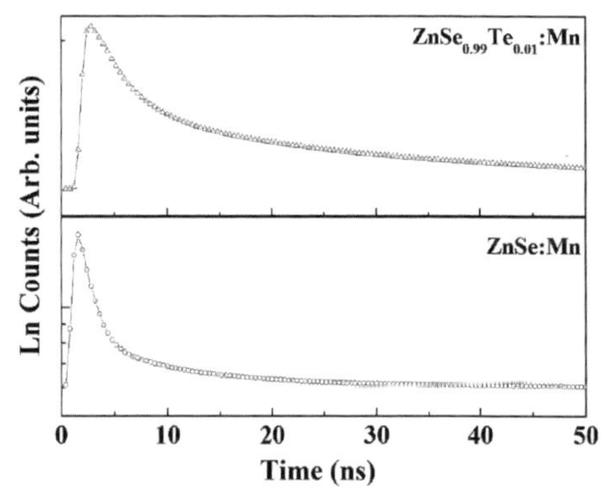

Where as in case of Mn doped $ZnSe_{0.99}Te_{0,01}$ QDs, most dominating contribution is from level around 432 and 525 nm these levels are the isoelectronic centers seen in the PL spectra. PLE measurements indicate that, Te isoelectronic centers changes decay path in Mn doped ZnSe. In case of Mn doped ZnSe exciton can decay by transferring energy to Mn site or via band edge emission. Schematic of the process is shown in Fig. 4b. In case of Mn doped $ZnSe_{0.99}Te_{0,01}$ QDs isoelectronic center dominate the emission and influence the decay paths. The exciton produced is readily captured by isoelectronic center and then can decay through energy transfer to Mn or isoelectronic luminescence. Schematic of mechanism is shown in Fig. 4c.

Table 1 Values of τ_1, τ_2, and τ_3 for Mn doped ZnSe and Mn doped $ZnSe_{0.99}Te_{0.01}$ QDs

Sample	τ_1 (ns)	τ_2 (ns)	τ_3 (ns)
Mn doped ZnSe	0.091	0.762	3.48
Mn doped $ZnSe_{0.99}Te_{0.01}$	0.262	1.637	9.17

Energy transfer mechanism is probed by TRPL measurements. Decay time of Mn-related emission cannot be probed as it is too long (μs). Intensity of green emission being rather weak, TRPL spectra of this band could not be recorded. TRPL spectra of blue emission from QDs can be fitted with three exponentials. TRPL spectra for Mn doped ZnSe, and Mn doped $ZnSe_{0.99}Te_{0.01}$ QDs are shown in Fig. 5. Time constant associated with first exponential (τ_1) is attributed to scattering loss and hence is not considered further. Time constant associated with second exponential (τ_2) and that with third exponential (τ_3) are attributed to transition from dark and bright excitonic states emerging due to the exchange interaction [20, 21]. Life time of Mn doped $ZnSe_{0.99}Te_{0,01}$ QDs is higher than that of Mn doped ZnSe. (Table 1).

PL decay rate depend on radiative, non radiative and charge transfer rate as

$$\frac{1}{\tau_{PL}} = \frac{1}{\tau_R} + \frac{1}{\tau_{NR}} + \frac{1}{\tau_{CT}} \tag{1}$$

where, τ_{PL} is life time of the sample under consideration, τ_R is radiative life time, τ_{NR}, non-radiative life time τ_{CT} charge transfer life time.

Incorporation of Te in Mn doped ZnSe increases the life time drastically. Increase in life time indicate that, charge transfer rate ($1/\tau_{CT}$) is drastically reduced in Mn doped $ZnSe_{0.99}Te_{0,01}$ QDs by presence of isoelectronic center. Indicating reduced overlap of Mn and isoelectronic center wave functions. In other word isoelectronic center prefer to decay via direct recombination of exciton rather than transferring the energy to Mn levels.

PLE and TRPL measurement indicate that, isoelectronic centers in QDs play a dominating role in exciton relaxation. As soon as exciton is created it is captured by isoelectronic cite. Wave function overlap of isoelectronic center and Mn ion is very small (evident from PL, PLE, and TRPL measurements) than the LUMO and Mn level, resulting in Mn emission quenching.

4 Conclusion

Undoped ZnSe, Mn doped ZnSe, $ZnSe_{0.99}Te_{0.01}$, Mn doped $ZnSe_{0.99}Te_{0.01}$ were synthesized chemically. The observation of isoelectronic center at room temperature is understood by confinement of charge carriers. PLE measurement confirm that the green band in these QDs originate from isoelectronic centers rather than type II core/shell structure. TRPL measurements on these samples show, reduced overlap of isoelectronic center and Mn ion wavefunction. The reduced wave-function overlap is responsible for Mn emission quenching by isoelectronic center.

Acknowledgments KGS and CR are thankful to BRNS and ISRO-UOP for the financial support respectively. Authors are extremely thankful to Prof. N. Periasamy of TIFR for his interest in the work and several suggestions. We are also thankful to Ms. Madhuri of TIFR for help in recording the TRPL spectra. Authors are also thankful to D. S. T. Unit on Nanoscience, CNQS and D. S. T. FIST program for various characterization facilities.

References

1. A.P. Alivisatos, Semiconductor clusters, nanocrystals, and quantum dots. Science **271**, 933 (1996)
2. B.O. Dabbousi, J. Rodriguez-Viejo, F.V. Mikulec, J.R. Heine, H. Mattoussi, R. Ober, K.F. Jensen, M.G. Bawendi, (CdSe)ZnS core—shell quantum dots: synthesis and characterization of a size series of highly luminescent nanocrystallites. J. Phys. Chem. B. **101**, 9463 (1997)
3. Y.C. Lin, W.C. Chou, W.C. Fan, J.T. Ku, F.K. Ke, W.J. Wang, S.L. Yang, W.K. Chen, W.H. Chang, C.H. Chi, Time-resolved photoluminescence of isoelectronic traps in $ZnSe_{1-x}Te_x$ semiconductor alloys. Appl. Phys. Lett. **93**, 241909 (2008)
4. R. Hill, Energy-gap variations in semiconductor alloy. J. Phys. C: Solid State Phys. **7**, 521 (1974)
5. J.E. Bernard, A. Zunger, Optical bowing in zinc chalcogenide semiconductor alloys. Phys. Rev. B. **34**, 5992 (1986)
6. R.E. Bailey, S. Nie, Alloyed semiconductor quantum dots: tuning the optical properties without changing the particle size. J. Am. Chem. Soc. **125**, 7100 (2003)
7. W. Jiang, A. Singhal, J. Zheng, C. Wang, C.W. Warren, Optimizing the synthesis of red- to near-IR-emitting CdS-capped $CdTe_xSe_{1-x}$ alloyed quantum dots for biomedical imaging. Chem. Mater. **18**, 4845 (2006)
8. F.C. Liu, T.L. Cheng, C.C. Shen, W.L. Tseng, M.Y. Chiang, Synthesis of cysteine-capped $Zn_xCd_{1-x}Se$ alloyed quantum dots emitting in the blue–green spectral range. Langmuir **24**, 2162 (2008)
9. T. Yao, M. Kato, J.J. Davis, H. Tanino, Photoluminescence of excitons bound at Te isoelectronic traps in ZnSe. J. Crystal Growth **86**, 552 (1988)
10. Y. Gu, I.L. Kuskovsky, M. Van der Voort, G.F. Neumark, X. Zhou, M.C. Tamargo, Zn-Se-Te Type-II quantum structures and isoelectronic centers multilayers with submonolayer quantities of Te. Phys. Rev. B. **71**, 045340 (2005)
11. I.L. Kuskovsky, C. Tian, G.F. Neumark, J.E. Spanier, I.P. Herman, W.C. Lin, S.P. Guo, M.C. Tamargo, Optical properties of δ-doped ZnSe:Te grown by molecular beam epitaxy: The role of tellurium. Phys. Rev. B. **63**, 155205 (2001)
12. L.W. Chang, J.H. Cheng, C.H. Hsu, H.Y. Chao, W. Li, Y.H. Chang, K.Y. Chen, Y.F. Chen, C.T. Laing, Isoelectronic centers and type-II quantum dots: Mechanisms for the green band emission in ZnSeTe alloy. J. Appl. Phys. **105**, 113511 (2009)
13. C.S. Yang, D.Y. Hong, C.Y. Lin, W.C. Chou, C.S. Ro, W.Y. Uen, W.H. Lan, S.L. Tu, Optical properties of the $ZnSe_{1-x}Te_x$ epilayers grown by molecular beam epitaxy. J. Appl. Phys. **83**, 2555 (1998)
14. Y. Gong, W. MacDonald, G.F. Neumark, M.C. Tamargo, I.L. Kuskovsky, Optical properties and growth mechanism of multiple type-II ZnTe/ZnSe quantum dots grown by migration-enhanced epitaxy. Phys. Rev. B. **77**, 155314 (2008)

15. M.C.-K. Cheung, A.N. Cartwright, I.R. Sellers, B.D. McCombe, I.L. Kuskovsky, Time-resolved photoluminescence of type-II quantum dots and isoelectronic centers in Zn–Se–Te superlattice structures. Appl. Phys. Lett. **92**, 032106 (2008)
16. S. Mahamuni, A.D. Lad, S. Patole, Photoluminescence properties of manganese doped ZnSe quantum dots. J. Phys. Chem. C. **112**, 2271 (2008)
17. K.G. Sonawane, CH. Rajesh, M. Temgire, S. Mahamuni, A case study: Te in ZnSe and Mn doped ZnSe quantum dots, Nanotechnology 22, 305702_1–305702_7 (2011)
18. V.V. Nikesh, A.D. Lad, S. Kimura, S. Nozaki, S. Mahamuni, Electron energy levels in ZnSe quantum dots. J. Appl. Phys. **100**, 113520 (2006)
19. M.A. Hines, P. Guyot-Sionnest, Bright UV-blue luminescent colloidal ZnSe nanocrystals. J. Phys. Chem. B. **102**, 3655 (1998)
20. M. Nirmal, D.J. Norris, M. Kuno, M.G. Bawendi, A.L. Efros, M. Rosen, Observation of the "Dark exciton" in CdSe quantum dots. Phys. Rev. Lett. **75**, 3728 (1995)
21. Ch. Rajesh, A.D. Lad, A. Ghangrekar, S. Mahamuni, Exciton recombination dynamics in zinc selenide quantum dots. Solid State Commun. **148**, 435 (2008)

Structural and Optical Evaluation of Pristine and Partially Iodized Ag Thin Films

D. Rajesh and C. S. Sunandana

Abstract Pristine and partially iodized noble metal thin films are of interest in plasmonic, 'plexitonic' (involving Plasmon-exciton interaction and transition) and nanophotonic applications as well as potential substrates for SERS. It is crucial to basically characterize such films for optimization and implementation of structures for specific practical applications. Iodization offers an attractive, simple and effective experimental procedure to modify the surface of noble metal films and also to grow semiconductor nanostructures. Interestingly one can observe the development of excitonic and fundamental electronic band structure of the semiconductor which is AgI in the present case. In this paper we focus on structural and optical characterization of (a) <30 nm thick Ag films obtained by thermal evaporation and (b) the same Ag films surface-modified by partial iodization. The Structural properties, the crystal structure, Surface morphology and chemical composition of the deposited Ag films were studied by X-Ray diffraction (XRD), Field Emission Scanning electron microscopy (FESEM). XRD Pattern showed that the films had preferred orientation along (111) of the face centered cubic (fcc) structure. The mechanical properties (The intrinsic stress, Dislocation density, grain size) are also calculated. The optical properties pristine and partially iodized Ag films of were investigated by measuring the absorbance from 300 to 800 nm.

Keywords Ag thin films · Partial iodization · Plasmons · Excitons

D. Rajesh · C. S. Sunandana (✉)
School of Physics, University of Hyderabad, Hyderabad, Andra Pradesh, India
e-mail: sunandana@gmail.com

D. Rajesh
e-mail: rajesh.desapogu@gmail.com

P. K. Giri et al. (eds.), *Advanced Nanomaterials and Nanotechnology*, Springer Proceedings in Physics 143, DOI: 10.1007/978-3-642-34216-5_28,

1 Introduction

Metal nanostructures have attracted considerable attention both fundamentally and technologically because of their unique physical and chemical properties and functionalities compared to their bulk counterparts. One of the most fascinating aspects is their optical properties. Pristine and partially iodized noble metal thin films are of interest in plasmonic, 'plexitonic' (involving Plasmon-exciton interaction and transition) and nanophotonic applications as well as potential substrates for SERS [1–3]. On the nano scale many metals, e.g., silver and gold exhibit strong absorption in the visible region of the spectrum. The origin of this absorption is attributed to collective conduction band electron oscillation in response to the electrical field of the electromagnetic radiation of light. This optical absorption is termed surface plasmons, partly because net charges are displaced transiently on the particle surface during electron oscillation. Surface plasmon absorption band are somewhat dependent on the size of spherical nanostructures, they strongly depend on the shape of the nanostructures. For a metal such as silver or gold, almost any color or absorption in any part of the visible spectrum can be produced by controlling the shape or structure of the nano material. Examples of various non spherical metal nanostructures reported in the literature include aggregates, nano rods, nano wires, nano cages, nano spheres, nano prisms, and nano plates or sheets [4–12]. The controllable and tunable optical properties of metal nanostructures are highly desired for many applications that rely on light absorption of metal, including surface Plasmon resonance (SPR), surface-enhanced Raman scattering (SERS), sensing, imaging, therapy, and photo catalysis. Therefore, the optical properties of metal nano materials can be altered at will by manipulating the details of the structure in a controllable manner. The modification of the surfaces for an enhancement of optical absorption, hence, provides a good method to improve efficiency of an optoelectronic device involving photon absorption [13].

In this paper, the control of the aggregation of spherical silver nano particles were examined by reaction of iodine vapors with Ag film at normal conditions. It has been proved that the partial iodization modifies the Ag film surface and produces reshaped silver nano particles on the surface of the Ag film.

2 Experimental Part

Ag films were deposited on borosilicate glass substrates by thermal evaporation using a molybdenum filament. The material was highly pure Ag wire (Aldrich, 99.999) placed in the filament which was thermally evaporated on to the cleaned borosilicate glass substrates kept at an ambient temperature. The borosilicate substrates were cleaned as follows: glass substrates were first immersed 10 % soap solution with 90 % water, rubbed with cotton 10–15 min with single distilled water to remove weathering, kept in chromic acid up to boiling point for removing organic contaminates, washed in cold water to remove surface contaminants followed by ultra-sonification in iso-propyl alcohol for 10 min duration, and, finally these substrates were dried in air before loading into the system for deposition. The glass substrates were kept at a distance of 20 cm from the source for deposition [14]. Ag film thickness measured as 20 nm with 0.1 nm/s rate of deposition by using thermal evaporator attached thickness monitor (Model CTM-200) during the experiment, and also the thickness was measured using Stylus profiler XP-1, Ambios Technology, USA. The deposition rate was controlled by adjusting the current through the filament. Vacuum of the order of 4.5×10^{-6} mbar was maintained in the chamber thought the deposition. The thermal evaporated films were iodized for selected durations up to 8 min using a set up described earlier [2].

Crystallographic study was carried out using a PHILIPS X-Ray diffractometer with Co-Kα ($\lambda = 1.78897$ Å) in the 2θ range from 20^0 to 100^0. The microstructure of the Ag thin films on the glass substrate was studied by using a Field emission scanning electron microscope (FESEM) (Model ULTRA-55) attached with an Energy Dispersive Scattering (EDS) analyzer to measure quantitatively the Ag film chemical stoichiometry. In order to study optical properties, optical absorption spectra were recorded in the wavelength range of 300–800 nm using JASCO V-570 UV–VIS-NIR double beam spectrometer in visible range from 300 nm to 800 nm with scanning rate 4 nm/s.

3 Results and Discussion

X-ray diffraction (XRD) is a popular and powerful technique for determining crystal structure of crystalline materials. By examining the diffraction pattern, one can identify the crystalline phase of the material. Small angle scattering is useful for evaluating the average inter particle distance while wide angle diffraction is useful for refining the atomic structure of nano particle [15]. The width of the diffraction lines are closely related to strain and defect size and distribution in nano crystals. As the size of the nano crystals decrease, the line width is broadened due to loss of long-range order relative to the bulk. The presence of peaks at 2θ values 44.6^0, 52^0, 76.6^0, 93.2^0, 98.7^0 correspond to (111), (200), (220), (311) Planes of silver respectively. Thus the XRD pattern confirmed the completely crystalline nature of silver thin film. These data agree with the standard JCPDS data card [16]. The lattice constant of silver deduced from the pattern is a = 4.085Å compares well to the standard data of silver. This XRD line width can be used to estimate the size of the particle by using the Debye–Scherrer formula (Fig. 1).

$$D = \frac{0.9\lambda}{W\cos\theta} \tag{1}$$

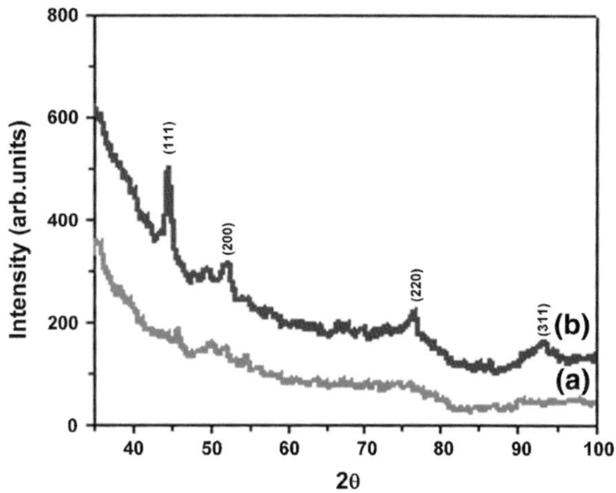

Fig. 1 XRD pattern of the 20 nm Ag film before and after iodization time. **a** 0 min, **b** 8 min

where D is the nano crystal diameter, λ is the wavelength of light, W is the full width half-max (FWHM) of the peak in radians, and θ is the Bragg angle. The crystallite size of silver Nanoparticles has been found to be 14 nm. In order to distinguish the effect of crystallite size induced broadening and strain induced broadening at FWHM of XRD profile,

$$W\cos\theta = C\lambda/D + 2\varepsilon\sin\theta$$

Where W is FWHM in radians, t is the grain size in nm, ε is the strain in nm, l is X-ray wavelength in nanometers, and C is a correction factor taken as 0.9. The grain size and strain of the sample have been found to be 14 nm and 88.35×10^{-2}, respectively. The dislocation density (δ) in the nanoparticles has been determined using expression [17]

$$\delta = 15W\cos\theta/4\text{aD}$$

The X-ray line profile analysis has been was used to determine the dislocation density of silver nano particles and found to be as $4.024 \times 10^{-4}\ \text{m}^{-2}$ respectively.

3.1 Optical Properties of Ag Films

When the Ag film was exposed to iodine vapors at room temperature, the film surface will modify according to the time of iodine expose, before iodization a broad absorption band was observed at 522 nm. The gradual changes of the peak position and FWHM observed in 0–8 min of iodization seem to be responsible for reshaped silver nano particles. If increases the iodization time the absorption band full width half maxima (FWHM) also increased, at 2 min iodization time the broad

absorption band was absorbed at 517 nm. Within this duration iodine vapors will try to modify the film surface and at the 4, 6, 8 min iodization time silver film the surface particles will stabilized and form a reshaped spherical nano particles in Fig. 2b, c. The variation of absorption spectrum with iodization time is given in Fig. 2. In the absorption spectrum a small shoulder at 380 nm this peak is assigned to the silver clusters less than 1 nm in size (Loginov et al. 2002). When iodization is continued, the small peak at 380 nm gets reduced in intensity, and the absorbance of the long wavelength tail is increased. Equally significantly, the minimum in absorbance preceding the absorption maximum which is related to the reflectivity rises upon progressive iodization. These observations indicate the growth and aggregation of silver.
Absorption co-efficient ($\Delta\alpha$) versus iodization time for the SPR minimum, maximum is shown in the Fig. 2b. It shows two roughly linear segments with a break at 4 min. But $\Delta\alpha$ versus iodization time plot for the SPR maximum is nonlinear saturating up to 8 min, while SPR peak shows a linear increase (Fig. 2c). These features are quantitatively similar to the $\Delta\alpha$ versus iodization time for UV minimum. The linear increases in SPR absorbance with increasing iodization time offers an energy calibration which could be used for applications.

3.2 FESEM

Field emission scanning electron micrographs (FE-SEM) of the 20 nm Ag films before and after 2, 8 min iodization and corresponding electron dispersive scattering (EDS) spectrum are shown in Fig. 3. Before iodization, the film surface looked rough and silver cracks were observed on the whole surface in Fig. 3a. Within two minutes iodization, the silver particles (size 10–20 nm) are growing because iodine vapors are modifying the Ag film surface (Fig. 3b). When the iodization was continued up to 8 min, the surface was covered with silver nano particles of relatively uniform size $\sim$40 nm and density of silver nano particles was increased and the particles grow in size. Most of the particles were in spherical shape and the distribution of silver nano particles at the film surface was almost homogeneous particles [18, 19].

The elemental analysis of Ag film has been performed using EDS spectroscopy (Fig. 3d). The peaks observed at 2.6–3.2 keV correspond to the binding energies of Ag. The remaining peaks came due to carbon coated tape; glass substrate and gold paint, no peaks of any other impurity have been detected. Therefore, the EDX profile of Ag film indicates silver nanoparticles only in which case we are using iodine for surface modification purpose only. Within the 8 min iodization time AgI has not yet formed [20].

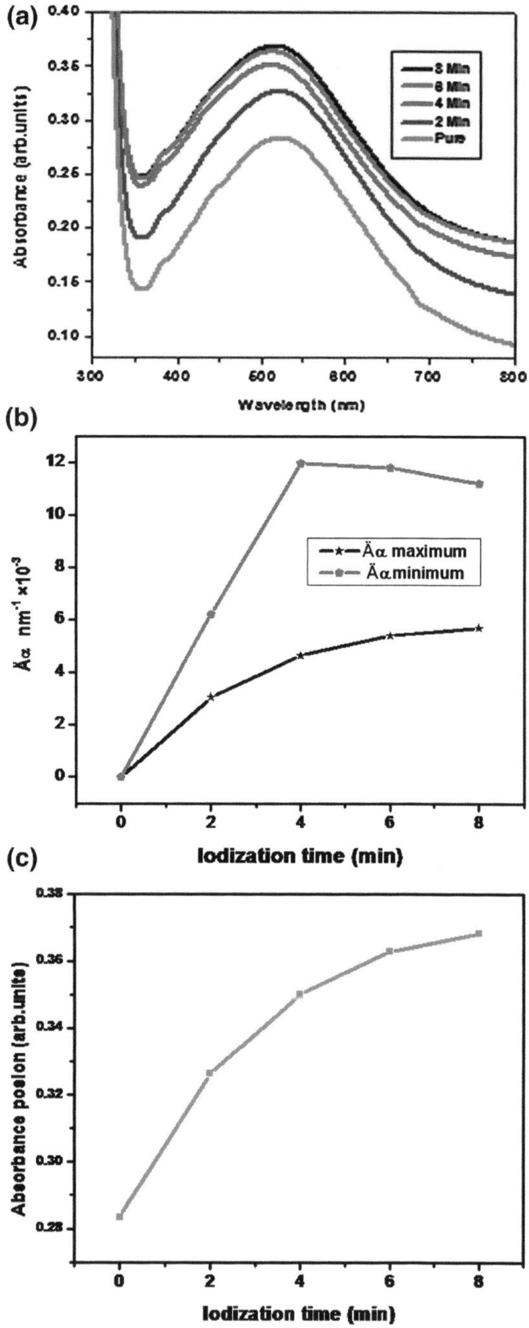

Fig. 2 **a** UV-Vis absorption spectrum of 20 nm Ag film pure and different time duration of iodized Ag film. **b** Iodization time versus absorption co-efficient minimum, maximum. **c** Iodization time versus SPR position

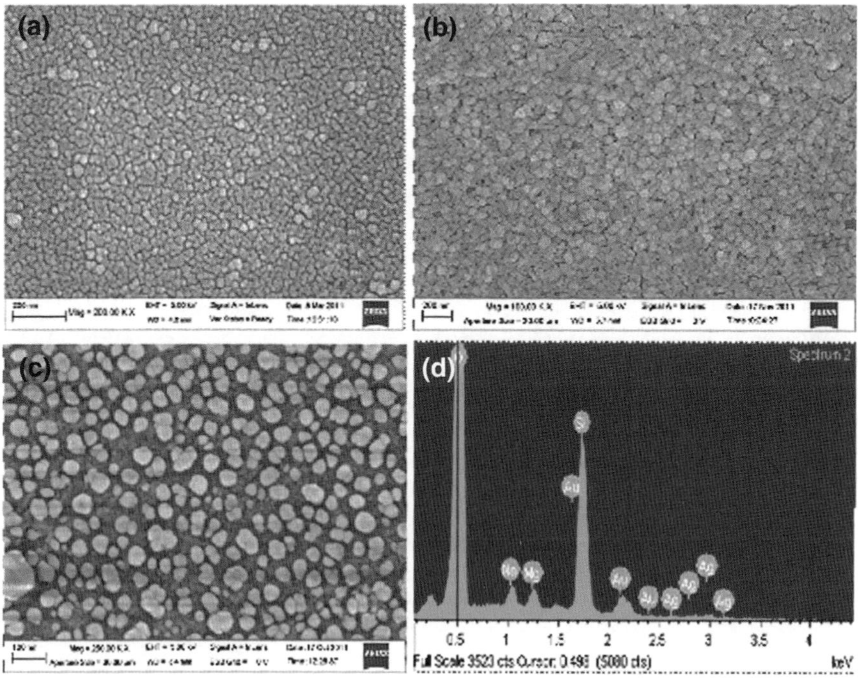

Fig. 3 Field emission scanning electron microscopy (FE-SEM) graphs of the 20nm Ag film before and after iodization time: (**a**) 0 min, (**b**) 2 min, (**c**) 8 min iodization shows that silver nano particles are spherical in shape having smooth surface and are well dispersed (**d**) EDS spectrum of reshaped Ag nano particles

4 Conclusion

Formation and aggregation of reshaped silver nanoparticles at the surface of silver film of have been investigated by partial iodization. The Structural properties, the crystal structure, Surface morphology and chemical composition of the deposited Ag films were studied by X-Ray diffraction (XRD), FE-SEM observations have been carried out to characterize silver nanoparticles at the film surface. Silver nano particles of $\sim$40 nm size were formed on the surface. According to the FE-SEM observation, the growth of the particle diameter and aggregation of Nanoparticles took place after partial iodization.

References

1. C.S. Sunandana, M. Gnanavel, in *Photonics 2010*: X International conference on fiber optics and photonics, ed. by S Khijwania, A K Sharma. (Viva Books, New Delhi, 2010) p. 127

2. M. Gnanavel, C.S Sunandana, in *SPIE Proceedings, Plasmonic: Metllic Nanostrutures and their Optical Properties* VII, ed. by M.I. Stockman. 7394, 73942D (2009)
3. C. Yang, Y.T. Xie, M.M.F. Yuen, X. Xiong, C.P. Wong, A facile chemical approach for preparing a SERS active silver substrates. Phys. Chem. Chem. Phys. **12**, 4459 (2010)
4. J.Z. Zhang, C. Noguez, Plasmonic optical properties and applications of metal nanostructures. Plasmonics **3**, 127 (2008)
5. C.J. Murphy, T.K. Sau, A.M. Gole, C.J. Orendorff, J. Gao, L. Gou, S.E. Hunyadi, T. Li, Anisotropic metal nanoparticles: Synthesis, assembly, and optical applications. J. Phys. Chem. B **109**, 13857 (2005)
6. L. Gou, C.J. Murphy, Fine-tuning the shape of gold nanorods. Chem. Mater. **17**, 3668 (2005)
7. C.J. Murphy, A.M. Gole, S.E. Hunyadi, C.J. Orendorff, One-dimensional colloidal gold and silver nanostructures. Inorg. Chem. **45**, 7544 (2006)
8. B. Nikoobakht, M.A. El-Sayed, Preparation and growth mechanism of gold nanorods (nrs) using seed-mediated growth method. Chem. Matter **15**, 1957 (2003)
9. Y. Sun, B.T. Mayers, Y. Xia, Template-engaged replacement reaction: a one-step approach to the large-scale synthesis of metal nanostructures with hollow interiors. Nano Lett. **2**, 481 (2002)
10. Y. Sun, B. Mayers, Y. Xia, Transformation of silver nanospheres into nanobelts and triangular nanoplates through a thermal process. Nano Lett. **3**, 675 (2003)
11. J. Chen, J.M. McLellan, A. Siekkinen, Y. Xiong, Z.Y. Li, Y. Xia, Facile synthesis of gold-silver nanocages with controllable pores on the surface. J. Am. Chem. Soc. **128**(46), 14776 (2006)
12. A.M. Schwartzberg, J.Z. Zhang, Novel optical properties and emerging applications of metal nanostructures. J. Phys. Chem C **128**(28), 10323 (2008)
13. M. Ihara, K. Tanaka, K. Sakaki, I. Honma, K. Yamada, Enhancement of the absorption coefficient of *cis*-$(NCS)_2$ Bis(2,2‘-bipyridyl-4,4‘-dicarboxylate)ruthenium(II) Dye in Dye-Sensitized solar cells by a silver island film. J. Phys. Chem. B. **101**, 5153 (1997)
14. E.N. Economou, surface plasmons in thin films. Phys. Rev. **182**, 539 (1969)
15. A.P. Alivisatos, Semiconductor clusters, nanocrystals, and quantum dots. Science **271**, 933 (1996)
16. J. Spreadborough, J.W. Christian, J. Sci. Instru. **36**, 116 (1959)
17. Y.P. Venkata Subbaiah, P. Prathap, K.T. Ramakrishna Reddy, Structural, electrical and optical properties of ZnS films deposited by close-spaced evaporation. Appl. Surf. Sci. **253**, 2409 (2006)
18. D. Rajesh, C.S. Sunandana, Briefly brominated Ag thin films: XRD, FESEM, and optical studies of surface modification. Appl. Sur. Sci. **259**, 276 (2012)
19. D. Rajesh, C.S. Sunandana, XRD, optical and AFM studies on pristine and partially iodized Ag thin film. Results Phys. **2**, 22, (2012)
20. D. Rajesh, C.S. Sunandana, Controlled structural and optical properties of nano structured Ag films reshaped by brief iodization. Thin Solid Films **524**, 316 (2012)

Ground- and Excited-State Dipole Moments of 1, 2-Diazine Nanomaterials Using Solvatochromic Method

S. R. Manohara, V. Udaya Kumar, Shivakumaraiah and L. Gerward

Abstract The ground-state (μ_g) and excited-state (μ_e) dipole moments of three 1, 2-diazine nanomaterials (pyrrolo-pyridazine derivatives) were determined using the solvatochromic shift methods, i.e. Lippert-Mataga, Bakhshiev, Kawski-Chamma-Viallet and Reichardt equations. All these equations are based on the variation of Stokes shift with solvent's dielectric constant and refractive index. Theoretical μ_g values were also evaluated by quantum chemical calculations using the DFT method by adopting B3LYP/6-31G* level of theory (Gaussian 03). It was observed that all the three 1, 2-diazine nanomaterials possess higher dipole moment values in the excited singlet-state than in the ground-state. This confirms that the excited state of these nanomaterials is more polar than the ground state. Therefore, the solvent–solute interactions should be stronger in the excited singlet-state than in the ground-state, demonstrating an important redistribution of charge densities between both electronic states.

Keywords Dipole moment · Diazine nanomaterials · Solvatochromic shift method · Fluorescence · Stokes shift · Ab-initio calculations · Gaussian 03

S. R. Manohara (✉)
Department of Physics, Siddaganga Institute of Technology, Tumkur 572103, Karnataka, India
e-mail: sr.manohara@yahoo.com

V. U. Kumar · Shivakumaraiah
Department of Chemistry, Siddaganga Institute of Technology, Tumkur 572103, Karnataka, India

L. Gerward
Department of Physics, Technical University of Denmark, DK–2800 Lyngby, Denmark

P. K. Giri et al. (eds.), *Advanced Nanomaterials and Nanotechnology*,
Springer Proceedings in Physics 143, DOI: 10.1007/978-3-642-34216-5_29,

1 Introduction

Dipole moment is a direct measure of the electron distribution in a molecule of known geometry. Excitation of a molecule by a photon causes a redistribution of charges leading to conformational changes in the excited-state. This can result in an increase or decrease of the dipole moment of an excited-state as compared to the ground-state. The excited-state dipole moment of a molecule is an important parameter, which reveals information on the electronic and geometrical structures of the molecule in the short-lived excited-state. The knowledge of dipole moments of first electronically excited-singlet (S_1) state of the molecules is quite useful in designing materials with nonlinear optical properties [1], and elucidating the nature of the excited states. Further, the experimental data on dipole moments are useful in parameterization in quantum chemical methods [2]. The excited-state dipole moments of fluorescent molecules reported in this article also determine the tunability range of the emission energy as a function of the polarity of the medium.

The dipole moment is a physical constant, which can be obtained by the experimental and theoretical methods. A number of excellent texts, monographs, and reviews devoted to dipole moments are available and contain a detailed discussion on the theory of dipole moments [3–8]. Experimental methods for the determination of singlet excited–state dipole moments are based on experimental ground–state dipole moments and a change of the position of a spectral band (in an electronic spectrum) caused by an electric field which can be either external (electrochromism) or internal (solvatochromism). The electro-optic methods (electrochromism) such as electronic polarization of fluorescence [9], electric-dichroism [8], microwave conductivity [10] and Stark splitting [11] are generally considered to be very accurate, but their use is limited because they are considered as equipment sensitive and the studies have been restricted to relatively very simple molecules. The solvatochromic method is experimentally much simpler and widely accepted, as it does not use any external field [12, 13]. This method is based on a linear correlation between the wave numbers of the ultraviolet–visible absorption and fluorescence maxima, and a solvent polarity function which involves both dielectric constant, ε, and refractive index, n, of the medium [14–20]. Under suitable conditions, the solvatochromic method yields fairly satisfactory results [20].

Of late, the 1, 2-diazine derivatives have proved to be invaluable materials in the field of medicine, optoelectronics and agriculture due to their applications in areas from biomedical to material sciences and electronics [21]. 1, 2-diazines were found to play an important role due to their pharmaceutical applications, which includes a wide range of biological activities, and also their derivatives exhibit liquid crystal properties. In the field of agriculture, diazines are useful as herbicides and also serve as grow up factors for plants.

Mangalagiu et al. [22] and Zbancioc et al. [23] have reported the synthesis, and studied the absorption and fluorescence spectra of three 1, 2-diazine nanomaterials (Fig. 1) in different organic solvents. In the present work, we report the estimation

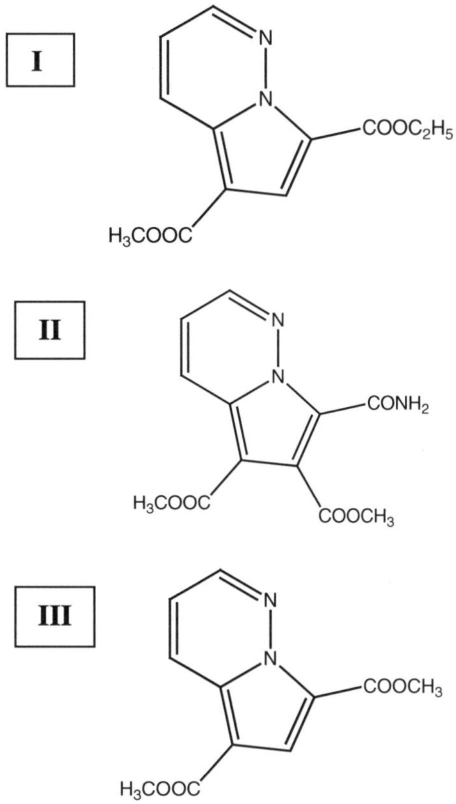

Fig. 1 Molecular structures of 1, 2-diazine nanomaterials (I, II, and III)

of ground- (μ_g) and excited-state (μ_e) dipole moments of three diazine nanomaterials. To the best of our knowledge, there are almost no reports on the determination of μ_g and μ_e values for these 1, 2-diazine nanomaterials, and hence there is a dearth of information on their dipole moments. This prompted us to carry out the present investigation. In this article, we have also compared the experimental μ_g values with theoretical μ_g values evaluated by quantum chemical calculations using Gaussian 03 program [24].

2 Methods

2.1 Theoretical Calculations of Ground-State Dipole Moments

Ground-state dipole moments (μ_g) of the three 1, 2-diazine nanomaterials were calculated theoretically by quantum chemical calculations, i.e. Gaussian 03 program [24], using DFT method by adopting B3LYP/6-31 + G* basis set with appropriate scale factors. All these computations were carried out with a Intel Core i5 PC on Windows 7 operating system.

2.2 Experimental Excited-State Dipole Moments

Experimental excited singlet-state dipole moments were determined by the solvatochromic method, using the equations proposed by Lipert-Megata Eq. (1) [14, 15], Bakhshiev Eq. (2) [16], Kawski-Chamma-Viallet Eq. (3) [17–19], and Reichardt Eq. (13) [13, 25, 26].

Lipert-Megata equation [14, 15]:

$$\bar{\nu}_a - \bar{\nu}_f = m\,F(\varepsilon, n) + \text{constant}. \tag{1}$$

Bakhshiev equation [16]:

$$\bar{\nu}_a - \bar{\nu}_f = m_1\,F_1(\varepsilon, n) + \text{constant}. \tag{2}$$

Kawski-Chamma-Viallet equation [17–19]:

$$\frac{\bar{\nu}_a + \bar{\nu}_f}{2} = m_2\,F_2(\varepsilon, n) + \text{constant} \tag{3}$$

where $\bar{\nu}_a$ and $\bar{\nu}_f$ are absorption and fluorescence maxima wave numbers in cm^{-1}. $F(\varepsilon, n)$, F_1 (ε, n), and F_2 (ε, n) are Lipert-Megata, Bakhshiev, Kawski-Chamma-Viallet polarity functions, respectively, and are given by,

$$F(\varepsilon, n) = \frac{\varepsilon - 1}{2\varepsilon + 1} - \frac{n^2 - 1}{2n^2 + 1} \tag{4}$$

$$F_1(\varepsilon, n) = \frac{2n^2 + 1}{n^2 + 2}\left(\frac{\varepsilon - 1}{\varepsilon + 2} - \frac{n^2 - 1}{n^2 + 2}\right) \tag{5}$$

$$F_2(\varepsilon, n) = \frac{2n^2 + 1}{2(n^2 + 2)}\left(\frac{\varepsilon - 1}{\varepsilon + 2} - \frac{n^2 - 1}{n^2 + 2}\right) + \frac{3(n^4 - 1)}{2(n^2 + 2)^2} \tag{6}$$

where ε and n are the dielectric constant and the refractive index of the solvent.

From Eqs. (1)–(3) it follows that, $\bar{\nu}_a - \bar{\nu}_f$ versus $F(\varepsilon, n)$, $\bar{\nu}_a - \bar{\nu}_f$ versus $F_1(\varepsilon, n)$, and ½ $(\bar{\nu}_a + \bar{\nu}_f)$ versus $F_2(\varepsilon, n)$ should give linear graphs with slopes m, m_1, and m_2 respectively, and are given by,

$$m = \frac{2(\mu_e - \mu_g)^2}{h\,c\,a^3} \tag{7}$$

$$m_1 = \frac{2(\mu_e - \mu_g)^2}{h\,c\,a^3} \tag{8}$$

and

$$m_2 = \frac{2\left(\mu_e^2 - \mu_g^2\right)}{h\,c\,a^3} \tag{9}$$

where μ_g and μ_e are the ground- and excited-state dipole moments of the solute molecule. The symbols h and c are Planck's constant and velocity of light in vacuum. 'a' is the Onsager cavity radius of the solute molecule and its value can be evaluated by atomic increment method [27].

Assuming that the symmetry of the investigated solute molecule remains unchanged upon electronic transition, and the ground- and excited-state dipole moments are parallel, the following relations can be arrived from Eq. (8) and (9).

$$\mu_g = \frac{|m_2 - m_1|}{2}\left(\frac{h\,c\,a^3}{2\,m_1}\right)^{1/2} \tag{10}$$

$$\mu_e = \frac{|m_2 + m_1|}{2}\left(\frac{h\,c\,a^3}{2\,m_1}\right)^{1/2} \tag{11}$$

By dividing (10) by (11):

$$\frac{\mu_e}{\mu_g} = \frac{|m_2 + m_1|}{|m_2 - m_1|} \quad (\text{For}\, m_2 > m_1) \tag{12}$$

2.3 Molecular-Microscopic Solvent Polarity Parameter, E_T^N

The *empirical molecular-microscopic solvent polarity parameter*, E_T^N, correlates with the solvatochromic data better than the traditionally used bulk solvent polarity functions. The theoretical basis for the correlation of the spectral band shift with E_T^N was proposed by Reichardt [13, 25] and developed by Ravi et al. [26].

Reichardt equation [13, 25]:

$$\bar{\nu}_a - \bar{\nu}_f = m_3\,E_T^N + \text{constant}. \tag{13}$$

The meaning of $\bar{\nu}_a$ and $\bar{\nu}_f$ is same as in earlier equations. The normalized solvent polarity of Reichardt, E_T^N, is a solvatochromic parameter based on the absorption wave number of a standard betaine dye in the corresponding solvent and is given by,

$$E_{\mathrm{T}}^{\mathrm{N}} = \frac{E_{\mathrm{T}}(30)_{\text{solvent}} - E_{\mathrm{T}}(30)_{\text{TMS}}}{E_{\mathrm{T}}(30)_{\text{Water}} - E_{\mathrm{T}}(30)_{\text{TMS}}} = \frac{E_{\mathrm{T}}(30)_{\text{solvent}} - 30.7}{32.4} \quad (14)$$

where TMS represents tetramethylsilane, known as a non-polar solvent. $E_{\mathrm{T}}(30)_{\text{solvent}} = hcN\bar{\nu}_{\mathrm{ai}} = 2.8591 \times 10^{-3}\bar{\nu}_{\mathrm{ai}}$ (in kcal mol^{-1}). $\bar{\nu}_{\mathrm{ai}}$ is the absorption maxima wave number of the standard betaine dye in *i*th solvent, in cm^{-1}.

From Eq. (13) it follows that, $\bar{\nu}_a - \bar{\nu}_f$ versus $E_{\mathrm{T}}^{\mathrm{N}}$ should give linear graph with slope m_3, and is given by

$$m_3 = 11307.6\left[\left(\frac{\Delta\mu}{\Delta\mu_{\mathrm{B}}}\right)^2\left(\frac{a_{\mathrm{B}}}{a}\right)^2\right] \quad (15)$$

where $\Delta\mu_{\mathrm{B}}$ [=9 Debye units (D)] and a_{B} [=6.2 Å] are the dipole moment change on excitation and the Onsager cavity radius of a betaine dye, and $\Delta\mu = (\mu_{\mathrm{e}} - \mu_{\mathrm{g}})$ and a are the corresponding quantities for the solute molecule of interest. Eq. (15) can be written as,

$$\Delta\mu = \mu_e - \mu_g = \sqrt{\frac{m_3 \times 81}{11307.6\,(6.2/a)^3}} \quad (16)$$

3 Results and Discussion

The spectral properties (electronic absorption and fluorescence emission maxima wavelengths, Stokes shift and arithmetic mean of Stokes shift values) of three 1, 2-diazine nanomaterials, mentioned earlier, in the organic solvents of varying polarities (protic polar: ethanol, aprotic polar: chloroform, and non-polar: cyclohexane) are summarized in Table 1. Absorption and fluorescence maxima wave numbers of these nanomaterials were calculated from the wavelength maxima values given in Ref. [22, 23]. The values of solvent polarity functions (F, F_1, and F_2) and molecular-microscopic solvent polarity parameter ($E_{\mathrm{T}}^{\mathrm{N}}$) of organic solvents, used under the present study, were calculated using Eqs. (4)–(6) and (14), respectively, and are given in Table 2. The dielectric constant, ε, the refractive index, n, and $E_{\mathrm{T}}(30)$ values of solvents were taken from literature [28, 29].

The dipole moments of the three 1, 2-diazine nanomaterials were evaluated applying Eqs. (1)–(3) and (13), i.e. Lippert-Mataga, Bakhshiev, Kawski-Chamma-Viallet, and Reichardt equations. The first step involves plotting the graphs of $\bar{\nu}_a - \bar{\nu}_f$ versus F, $\bar{\nu}_a - \bar{\nu}_f$ versus F_1, ½ $(\bar{\nu}_a + \bar{\nu}_f)$ versus F_2, and $\bar{\nu}_a - \bar{\nu}_f$ versus $E_{\mathrm{T}}^{\mathrm{N}}$ by employing Origin 6.1 program. In the second step, a linear regression and fitting the data to a straight line were carried out, and the slopes m, m_1, m_2, and m_3 were evaluated using the same program. In the third step, the dipole moment values were calculated using the Eqs. (7)–(11) and (16). The results of the

Table 1 Electronic absorption and fluorescence spectral data of the 1, 2-diazine nanomaterials (I, II, and III) in different organic solvents [22, 23]

Nanomaterial	Solvent[a]	$\lambda_a{}^b$	$\lambda_f{}^c$ (quantum yield %)	$\bar{\nu}_a{}^d - \bar{\nu}_f{}^e$	$(\bar{\nu}_a + \bar{\nu}_f)/2$
I	Ethanol	349	429 (66)	5,343	25,982
	Chloroform	358	422 (91)	4,236	25,815
	Cyclohexane	361	416 (85)	3,662	25,870
II	Ethanol	356	429 (63)	4,780	25,700
	Chloroform	362	431 (82)	4,422	25,413
	Cyclohexane	366	434 (76)	4,281	25,182
III	Ethanol	349	430 (66)	5,397	25,955
	Chloroform	357	423 (91)	4,371	25,826
	Cyclohexane	361	416 (85)	3,662	25,870

[a] Solvents are listed in the order of decreasing dielectric constant
[b] Electronic absorption maxima wavelength, in nm
[c] Fluorescence emission maxima wavelength, in nm
[d] Absorption maxima wave number, in cm^{-1}
[e] Fluorescence maxima wave number, in cm^{-1}

Table 2 Solvent polarity functions, F, F_1, F_2, and the molecular-microscopic solvent function, E_T^N, used in the Lippert-Mataga, Bakhshiev, Kawski-Chamma-Viallet, and Reichardt equations, respectively

Solvent	n^a	ε^a	$E_T(30)^b$	F^c	F_1^d	F_2^e	E_T^{Nf}
Ethanol	1.3614	24.55	51.9	0.2888	0.8129	0.6525	0.6543
Chloroform	1.4429	4.81	39.1	0.1492	0.3725	0.4864	0.2593
Cyclohexane	1.426	2.02	31.2	−0.0016	−0.0031	0.2875	0.0154

[a] ε and n are dielectric constant and refractive index values of organic solvents from Ref. [28]
[b] $E_T(30)$ values taken from Ref. [29]
[c] Values of Lippert-Mataga solvent polarity function calculated from Eq. (4)
[d] Values of Bakhshiev solvent polarity function calculated from Eq. (5)
[e] Values of Kawski-Chamma-Viallet solvent polarity function calculated from Eq. (6)
[g] Values of molecular-microscopic solvent polarity parameter calculated from Eq. (14)

statistical treatment of the Lippert-Mataga, Bakhshiev, Kawski-Chamma-Viallet, and Reichardt correlations, namely the slopes, intercepts and correlation coefficients are given in Table 3.

The ground-state (μ_g) and first excited-singlet-state (μ_e) dipole moments of the three 1, 2-diazine nanomaterials are evaluated using Eqs. (10) and (11). Figs. 2, 3 show the plot of $\bar{\nu}_a - \bar{\nu}_f$ versus F_1 and ½ $(\bar{\nu}_a + \bar{\nu}_f)$ versus F_2 from which the slopes m_1 and m_2 are obtained and then μ_g, μ_e and (μ_e/μ_g) are evaluated. *Example*: Consider II and the slopes were found to be $m_1 = 617\ cm^{-1}$ and $m_2 = 1{,}411\ cm^{-1}$. Using Eqs. (10) and (11), we get $\mu_g = 1.18$ D and $\mu_e = 3.01$ D and change in dipole moments is $\Delta\mu = (\mu_e - \mu_g) = 1.83$ D. Similarly μ_g, μ_e and $\Delta\mu$ were calculated for other diazine nanomaterials, namely I and III.

Table 3 Statistical treatment of the correlations of solvent spectral shifts of the 1, 2-diazine nanomaterials (I, II, and III)

Nanomaterial	Slope (cm^{-1})	Intercept (cm^{-1})	Correlation coefficient
Lipert-Megata correlation			
I	5,763	3,576	0.98
II	1,708	4,246	0.96
III	5,959	3,610	0.99
Bakhshiev correlation			
I	2,073	3,597	0.99
II	617	4,251	0.98
III	2,132	3,637	1.00
Kawski-Chamma-Viallet correlation			
I	288	25,752	0.62
II	1,411	24,761	0.99
III	218	25,780	0.61
Reichardt correlation			
I	2,647	3,594	1.00
II	793	4,249	0.99
III	2,705	3,639	1.00

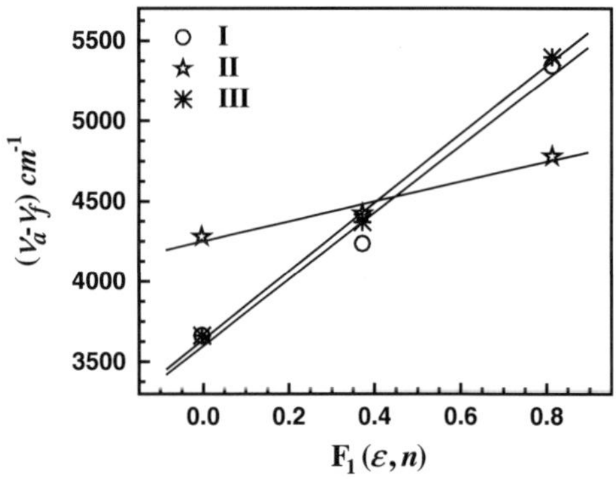

Fig. 2 The variation of Stokes shift with F_1 using Bakhshiev equation in different solvents

The slopes m and m_3 are obtained from the graph of $\bar{\nu}_a - \bar{\nu}_f$ versus F, and $\bar{\nu}_a - \bar{\nu}_f$ versus E_T^N. The linear E_T^N dependence on Stokes shift (Fig. 4) indicates the presence of a general type of solute–solvent interactions as well as *H*-bonding interaction in the solvents. Using Eqs. (7), (9) and (16), $\Delta\mu$ [=($\mu_e - \mu_g$)] values were evaluated.

The μ_e, μ_g and the ratio (μ_e/μ_g) of the studied 1, 2-diazine nanomaterials estimated from Eqs. (10), (11), and (12), respectively, are presented in Table 4. The change in dipole moment values, $\Delta\mu$ [=($\mu_e - \mu_g$)], estimated using the Lippert-Mataga, Bakhshiev, and molecular–microscopic solvent polarity relations

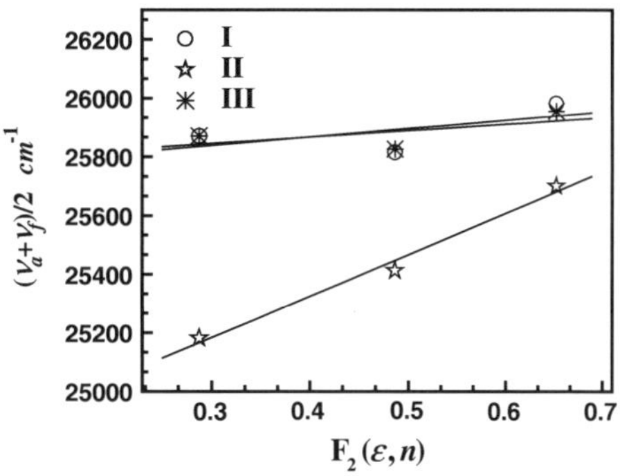

Fig. 3 The variation of arithmetic mean of Stokes shift with F_2 using Kawski-Chamma-Viallet equation in different solvents

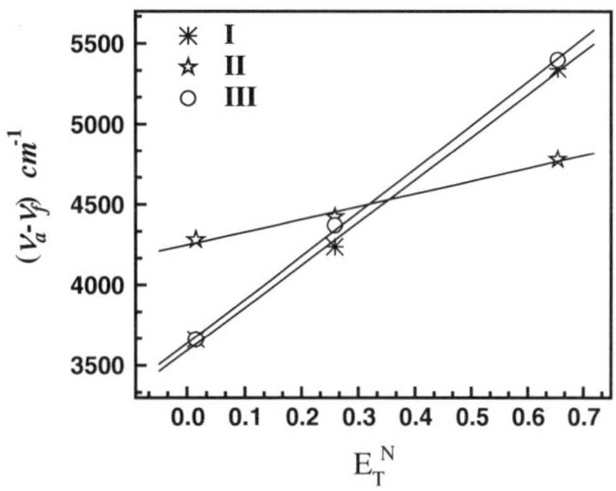

Fig. 4 The variation of Stokes shift with E_T^N using Reichardt equations in different solvents

are also given in Table 4. It is observed that, the $\Delta\mu$ values obtained for these nanomaterials from the Lippert-Mataga method are large compared to values obtained by all other methods. This is due to the fact that, this method does not consider polarizability of the solute. The $\Delta\mu$ values obtained by all the methods are moderately high, which suggest that, the fluorescence emission in these nanomaterials originates from a state, which is more polarized than the ground-state. There is a discrepancy between the estimated $\Delta\mu$ values, which may be partially due to the various assumptions and simplifications made in the use of different solvatochromic methods [30]. We also observed that (Tables 3, 4), diazine nanomaterial having large slope values has large $\Delta\mu$ values, and the values of μ_e and μ_g depend not only on slopes but also on the Onsager cavity radius, a, as calculated by Edward's method [27].

The ground-state dipole moments of the three diazine nanomaterials were computed in the gas phase by ab initio calculations using Gaussian 03 program [24]. The values of theoretical ground-state dipole moments are given in Table 4. There is a discrepancy between the values of μ_g obtained by quantum chemical calculations (theoretical) and solvatochromic methods (experimental). One of the deficiencies of the quantum chemical methods is that they tend to exaggerate uneven distribution of electrons in a molecule and thus make it more polar (with a

Table 4 Onsager cavity radii, a, (in Å), Ground-state (μ_g) and excited singlet-state (μ_e) dipole moments (in Debye units*, D) of 1, 2-diazine nanomaterials (I, II, and III)

Nanomaterial	a	$\mu_g{}^a$	$\mu_g{}^b$	$\mu_e{}^c$	$\Delta\mu^d$	$\Delta\mu^e$	$\Delta\mu^f$	$(\mu_e/\mu_g)^g$
I	3.65	3.65	1.36	1.80	5.28	3.17	2.07	1.32
II	3.80	1.51	1.18	3.01	3.05	1.83	1.14	2.55
III	3.58	3.87	1.40	1.72	5.22	3.12	1.93	1.23

* 1 Debye (D) = 10^{-18} esu = 3.33564×10^{-30} Coulomb meter (C m)
[a] Theoretical μ_g values obtained by employing Gaussian 03 program [24]
[b] Experimental μ_g values calculated from the Eq. (10)
[c] Experimental μ_g values calculated from the Eq. (11)
[d] $\Delta\mu$ values calculated from Lipert-Megata model
[e] $\Delta\mu$ values calculated from Bakhshiev model
[f] $\Delta\mu$ values calculated from molecular-microscopic solvent polarity parameter, E_T^N
[g] Calculated from Eq. (12)

higher dipole moment) than it actually is. Also, the experimental methods take solvent and environmental effects into account (solute–solvent interactions), where as ab initio calculations give μ_g values only for a free molecule. Optimized molecular geometry obtained from ab initio calculations using Gaussian 03 program is shown in Fig. 5 for the diazine nanomaterial, namely I.

Table 4 shows that, for each 1, 2-diazine nanomaterials, the excited singlet-state dipole moment value is higher than ground-state counterpart, which indicates that these nanomaterials are significantly more polar in their excited singlet-state than in their ground-state. This is also an indication of the fact that, the excited-state geometry could be different from that of the ground-state. Therefore,

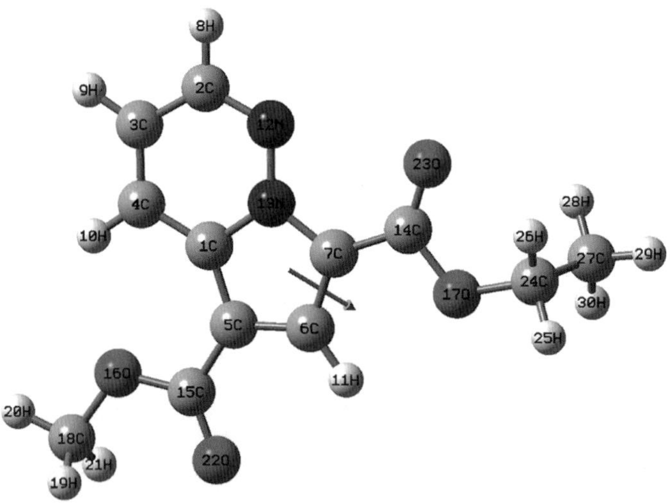

Fig. 5 Optimized molecular geometry and direction of ground-state dipole moment of 1, 2-diazine nanomaterial (I) obtained from ab initio calculations using Gaussian 03 program [24]

the solvent–solute interactions should be stronger in the excited singlet-state than in the ground-state.

4 Conclusions

In this article, the dipole moments of electronic ground- and excited–states of three 1, 2-diazine nanomaterials are calculated and compared by experimental and theoretical methods. It is noticed that, all the three 1, 2-diazine nanomaterials possess higher dipole moment values in the excited singlet-state than in the ground-state. For the studied 1, 2-diazine nanomaterials, the increase in the dipole moment in the excited singlet-state is in the range $1 - 6$ D. This confirms that the excited singlet-state of diazine nanomaterials is more polarized than the ground-state, and, therefore, is more sensitive to solvent effects.

The reason for the discrepancy in the results of $\Delta\mu$ values is that, in the solvatochromic methods, the dipole moment of the solute in solution is related to the dipole moment of isolated solute molecule, its static polarizability and internal electric field. This discrepancy may be partially due to the various assumptions and simplifications made in the different solvatochromic methods.

The Eq. (12) can be used to estimate the value of the excited-state dipole moment by the pre-knowledge of the value of ground-state dipole moment, without the necessity of knowing the Onsager radius of the solute.

References

1. D.S. Chemla, J. Zyss, *Non-linear optical properties of organic molecules and crystals* (Academic Press, New York, 1987)
2. S. Kumar, V.C. Rao, R.C. Rastogi, Excited-state dipole moments of some hydroxycoumarin dyes using an efficient solvatochromic method based on the solvent polarity parameter, E_T^N. Spectrochim. Acta A **57**, 41–47 (2001)
3. R.J.W. Le Fèvre, *Dipole moments* (Methuen, London, 1953)
4. J.W. Smith, *Electric dipole moments* (Butterworths, London, 1955)
5. V.I. Minkin, O.A. Osipov, Y.A. Zhdanov, *Dipole moments in organic chemistry* (Plenum Press, New York, 1970)
6. G.J. Moody, J.D.R. Thomas, *Dipole Moments in Inorganic Chemistry*, ed. by E. Arnold (London, 1971)
7. O. Exner, *Dipole Moments in Organic Chemistry*, ed. by G. Stuttgart (Thieme, 1975)
8. W. Liptay, *Excited States*, vol. 1, ed. by E.C. Lim (Academic, New York, 1974)
9. A. Kawski, J.F. Rabek, (eds.), In *Progress in Photochemistry and Photophysics*, vol. 5 (CRC Press, Boca Raton, 1992)
10. J. Czekella, Dektrische fluoreszenpolarisation: die bestimmung von dipolmomenten angeregter moleküle aus dem polarisationsgrad der fluoreszenz in starken elektrischen feldern. Z. Elektrochem. **64**, 1221–1228 (1960)
11. J. Czekella, Zwei elektro-optische methoden zur bestmmung von dipolmomenten angeregter molküle. Chimica. **15**, 26–31 (1961)

12. C. Párkányi and J.J. Aaron, Dipole moments of aromatic heterocycles, theoretical and computational chemistry, vol. 5 (Elsevier Science, New York 1998)
13. C. Reichardt, Solvatochromic dyes as solvent polarity indicators. Chem. Rev. **94**, 2319–2358 (1994)
14. E. Lippert, Spektroskopische bestimmung des dipolmomentes aromatischer verbindungen im ersten angeregten singulettzustand. Ber. Bunsenges. Phys. Chem. **61**, 962–975 (1957)
15. N. Mataga, Y. Kaifu, M. Koizumi, Solvent effects upon fluorescence spectra and the dipole moments of excited molecules. Bull. Chem. Soc. Japan **29**, 465–470 (1956)
16. N.G. Bakhshiev, Universal intermolecular interactions and their effect on the position of the electronic spectra of molecules in two -component solutions. Opt. Spektrosk. **16**, 821–832 (1964)
17. A. Kawski, Der wellenzahl von elecktronenbanden lumineszie- renden moleküle. Acta. Phys. Polon. **29**, 507–518 (1966)
18. L. Bilot, A. Kawski, Zur theorie des einflusses von liisungsmitteln auf die elektronenspektren der molecule, vol. 17a, ed. by Z. Naturforsch (1962), pp. 621–627
19. A. Chamma, P. Viallet, Détermination du moment dipolaire d'une molécule dans un état excité singulet: application á l'indole, au benzimidazole et á l'indazole, C. R. Acad. Sci. Paris Ser. C **270**, 1901–1904 (1970)
20. B. Koutek, Dipole moments in excited state. Collect. Czech. Chem. Commun. **43**, 2368–2386 (1978)
21. I.I. Mangalagiu, Recent achievements in the chemistry of 1, 2-diazines. Curr. Org. Chem. **15**, 730–752 (2011)
22. I.I. Mangalagiu, M. Florescu, G. Zbancioc, M. Caprosu, Synthesis of new 1, 2-Diazine nanomaterials through conventional and nonconventional methods. J. Phys: Conf. Series **61**, 484–486 (2007)
23. G. Zbancioc, T. Huhn, U. Groth, C. Deleanu, I.I. Mangalagiu, Pyrrolodazine derivatives as blue organic luminophores: Synthesis and properties Part 3. Tetrahedron **66**, 4298–4306 (2010)
24. M.J. Frisch et al., *Gaussian* 03 (*Revision D*.01), (Gaussian, Inc., Pittsburgh, 2003)
25. C. Reichardt, *Solvents and Solvent Effects in Organic Chemistry*, 3rd edn. (Wiley, Weinheim, 2003)
26. M. Ravi, T. Soujanya, A. Samanta, T.P. Radhakrishnan, Excited-state dipole moments of some coumarin dyes from a solvatochromic method using the solvent polarity parameter, E_T^N. J. Chem. Soc. Faraday Trans. **91**, 2739–2742 (1995)
27. J.T. Edward, Molecular volumes and the Stokes-Einstein equation. J. Chem. Educ. **47**, 261–270 (1970)
28. D.R. Lide, *CRC Handbook of Chemistry and Physics*, 80th edn. (CRC Press, New York, 1999)
29. C. Reichardt, *Molecular Interactions*, ed. by H. Ratajczak, W.J. Orville-Thomas, vol. 3, (Wiley, 1982)
30. J.J. Aaron, M. Buna, C. Párkányi, M.S. Antonious, A. Tine, L. Cissé, Quantitative treatment of the effect of solvent on the electronic absorption and fluorescence spectra of substituted coumarins: Evaluation of the first excited singlet-state dipole moment. J. Fluoresc. **5**, 337–347 (1995)

Effect of Process Parameters and Post Annealing Temperature on Structural and Optical Properties of $MgTiO_3$ Thin Films Deposited by RF Magnetron Sputtering

T. Santhosh Kumar, R. K. Bhuyan, A. Perumal and D. Pamu

Abstract $MgTiO_3$ (MTO) thin films have been deposited on to SiO_2 substrates at ambient temperature by RF magnetron sputtering at different oxygen mixing percentage (OMP). The MTO target has been synthesized by mechanochemical synthesis method and sintered at 1,350 °C for 3 h. The phase purity of the sputtering target was confirmed from X-ray diffraction pattern and it could be refined to $R\bar{3}$ space group with lattice parameters a = b = 5.0557(12) Å, c = 13.9003(9) Å. The effect of annealing on crystal structure, and optical properties of MTO thin films studied systematically. The as deposited films were X-ray amorphous irrespective of the processing parameters and exhibit refractive index in the range of 1.84–1.97 at 600 nm with an optical absorption edge value in between 3.72 and 3.78 eV. On annealing at 700 °C for 1 h, the films were polycrystalline accompanied by an increase in refractive index and band gap and decrease in transmittance. The increase in the refractive index and band gap on annealing can be attributed to the improvement in packing density and crystallinity decrease in porosity ratio.

Keywords $MgTiO_3$ thin films • RF sputtering • Annealing temperatures • Optical properties

1 Introduction

In recent years there has been substantial interest in nanostructured oxide thin films due to their potential in a number of applications, as the oxide layer of metal oxide semiconductor field effect transistors, optical coatings [1], buffer layers for

T. S. Kumar · R. K. Bhuyan · A. Perumal · D. Pamu (✉)
Department of Physics, Indian Institute of Technology Guwahati, Guwahati 781039, India
e-mail: pamu@iitg.ernet.in

P. K. Giri et al. (eds.), *Advanced Nanomaterials and Nanotechnology*,
Springer Proceedings in Physics 143, DOI: 10.1007/978-3-642-34216-5_30,

growing superconductors [2] and microwave integrated circuits [3, 4]. In general most of the optical devices employ multilayer structures for enhanced performance and efficiency. This would usually denote a stack of high index and low index films coated alternatively, the sequence being determined by the final device requirement. The breakdown of any such device determined mostly by the failure of any one of the layers and the degradation in its properties. Therefore a study of single layers of oxide films is as significant as the study of the multilayer.

$MgTiO_3$ (MTO) is one of the significant materials, which has good microwave dielectric properties such as dielectric constant (ε_r) $\sim$17, a high quality factor (Qxf_0) $\sim$1,60,000 GHz and a negative temperature coefficient of resonant frequency (τ_f) ~ -50 ppm/ °C [5] and good thermal stability. $MgTiO_3$ is known to form ilmenite structure with rhombohedral symmetry and $R\bar{3}$ space group. MTO in its bulk form has been widely used in microwave dielectric resonators and ceramic capacitor [6]. In addition to the good dielectric properties, MTO thin films also exhibit good optical properties to find applications in antireflection coatings, wave guides, high permittivity insulator coating in electronic devices and as protective coatings [7, 8]. Due to their excellent optical, electrical, thermal and dielectric properties, it is important to study the structural, microstrctural and optical properties of films processed under different experimental conditions. Optical constants of the deposited MTO thin films such as refractive index (n), packing density (p) and optical band gap (E_g) are important parameters, which determine the overall performance of the optical devices.

While there are few reports on preparation and characterization of MTO ceramics but those studies were mostly in bulk form [9, 10] and only few reports on deposition and characterization of MTO thin films [11, 12]. However, there is no report on systematic study on deposition, structural and optical characterization of MTO thin films. In the present study, MTO thin films were deposited by using RF magnetron sputtering at ambient temperature. The effect of processing parameters and post annealing on structural and optical properties of MTO thin films were studied systematically.

2 Experimental Procedure

The sputtering target of $MgTiO_3$ was synthesized by mechanochemical reaction method. Stoichiometric ratio of high purity oxide powders (>99.9 %) MgO and TiO_2 were weighed and milled for 30 h in planetary ball mill (Fritsch GmbH, Germany). After drying and sieving, the powder was uniaxially pressed into a cylindrical target of 56 mm in diameter and 4 mm thickness. The sputtering target was sintered at 1,350 °C for 3 h in air and cooled to room temperature at the rate of 2 °C per minute.

MTO thin films were deposited on SiO_2 substrates at ambient temperature using RF magnetron sputtering. A base pressure of 1.0×10^{-5} mbar was obtained by using diffusion pump. The processing gas was a mixture of high purity argon

(99.99 %) and oxygen (99.99 %) and was introduced into chamber through mass-flow controllers. All thin films were deposited at a fixed RF power of 80 W and for duration of 3 h. A sputtering pressure 6×10^{-2} mbar maintained constantly throughout the deposition with varying the oxygen mixing percentage (OMP) in the sputtering gas. The substrates were cleaned by rinsing in ultrasonic bath of methanol and the substrate to target distance of 5 cm was kept constant for all depositions. The target was pre-sputtered in an argon atmosphere for 5 min to remove surface impurities. Since the as deposited MTO thin films were amorphous to explore the properties of MTO thin films in crystalline form, films were annealed at 700 °C for 1 h.

The Phase purity of the sintered MTO sputtering target was analyzed by recording X-ray diffraction pattern using Bruker D8 X-ray Diffractometer with CuK_α ($\lambda = 1.5406$ Å) radiation and thin film XRDs were recording by using Rigaku high power X-ray Diffractometer (RINT 2,500 system TTRAX) with CuKα radiation ($\lambda = 1.5406$ Å). The XRD pattern of target material was refined by using Rietveld refinement technique and Fullprof program [13]. The transmittance spectra of thin films deposited on SiO_2 substrates were recorded using UV–VIS Spectrophotometer (UV 3101PC, SHIMADZU) in the wavelength range of 200–1,400 nm. The optical constants were estimated using envelope technique [14].

3 Results and Discussions

3.1 Structural Characterization of Target

Figure 1 shows the X-ray diffraction pattern along with the Rietveld refinement of MTO sputtering target sintered at 1,350 °C for 3 h. The Rietveld refinement was performed through Fullprof program [15]. The refinement was carried out by considering $R\bar{3}$ space group and by refining the lattice parameters, position of the Mg, Ti, and O atoms, occupancy, and thermal parameters.

The lattice parameters are found to be a = b = 5.0557(12) Å, c = 13.9003(9) Å. From the Rietveld refinement, it is confirmed that there is no secondary phase. The values of χ^2, R_{Brag} factor and R_f factor were found to be 1.09, 14.26 and 12.53 respectively. The above XRD analysis results are in agreement with those reported by Ferri et al. [16].

The XRD patterns of MTO films deposited at different OMP are shown in Fig. 2. All the as deposited films were amorphous, which crystallined to trigonal structure on annealing at 700 °C for 1 h in air. Along with the trigonal phase a secondary phase of Mg_2TiO_5 is present. It is well known that the oxide films naturally grow in the amorphous state unless the activation energy is provided to the process either in the form of temperature or ion bombardment. From the XRD patterns of the MTO thin films, it was observed that as the OMP increases, the intensities of the peaks decreased and the FWHM of the (104) peak increased. The

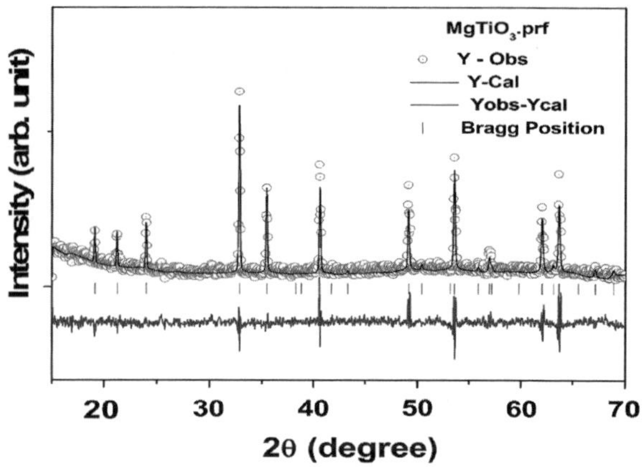

Fig. 1 XRD pattern along with rietveld refinement of $MgTiO_3$ target sintered at 1,350 °C. The *circles* and *solid line* represent the experimental and the Rietveld refined data, respectively. The dotted lines at the *bottom show* the difference between the experimental and the refined data. The *vertical bars* correspond to the allowed Bragg's peaks

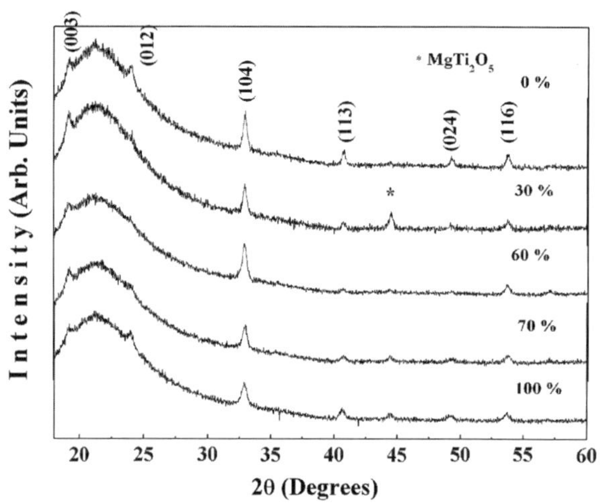

Fig. 2 XRD patterns of MTO thin films deposited on SiO_2 substrates at different OMP and annealed at 700 °C for 1 h

average crystallite size of the annealed films was estimated from the FWHM values using Scherrer's equation. The results indicated a decrease in average crystallite size with increase in OMP from 0 to 100 OMP and the crystallite size values were in between 29 to 21 nm. Figure 3 shows the variation in crystallite size as a function of OMP.

3.2 Optical Characterization

The optical transmittance spectra of MTO thin films as deposited and annealed at 700 °C deposited at 60 % OMP were shown in Fig. 4. The sharp fall in transmission and the disappearance of fringes at shorter wave lengths are due to the fundamental absorption of the films. All the transmittance spectra show

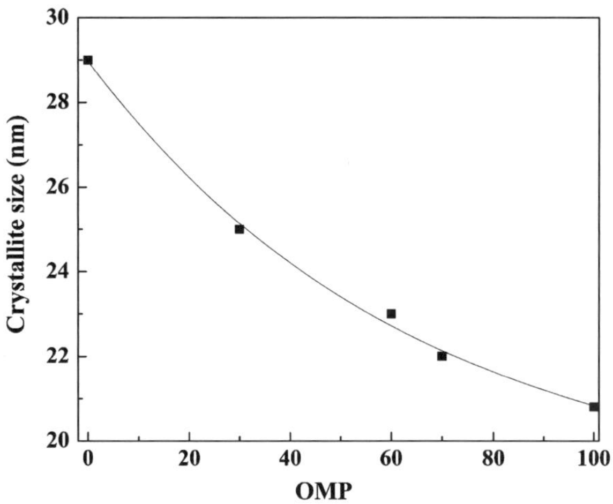

Fig. 3 Variation in crystallite size as a function of OMP

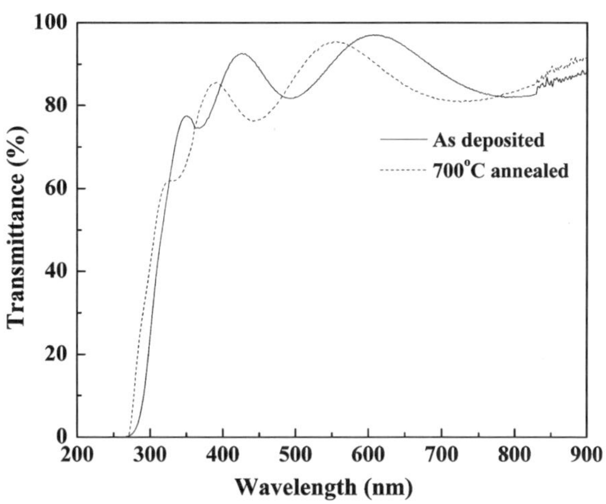

Fig. 4 Optical transmittance spectra of MTO thin films as deposited and annealed at 700 °C, deposited at 60 OMP

interference fringes, which originates due to the interference at the air and substrate-film interfaces. The thickness and optical constants of the deposited MTO thin films estimated by using envelop technique was found to be in the range of 350–450 nm. It was observed that the transmittance of the MTO thin films decreased on annealing and may be attributed to the scattering effects for its higher surface roughness and increase in film density. It was also observed that the absorption edge shifted to the lower wavelengths on annealing.

The optical packing densities (P) of the films were calculated using the relation given by Jacobson [17]

$$P = \left(\frac{{n_f}^2 - 1}{{n_f}^2 + 2}\right) \times \left(\frac{{n_b}^2 + 2}{{n_b}^2 - 1}\right) \tag{1}$$

The porosity ratio (the volume of pores per volume of film) of the films obtained using the following expression [18]

$$P = 1 - \left(\frac{n_f^2 - 1}{n_b^2 - 1}\right) \tag{2}$$

where n_f and n_b are the film and bulk refractive index of $MgTiO_3$ (single crystal refractive index =2.31) [19] respectively.

The variation in refractive index and packing density of as deposited and annealed MTO thin films as a function of OMP was shown in Fig. 5a and porosity ratio as a function of OMP as deposited and annealed films were shown in Fig. 5b. It was observed that, both the refractive index and packing density exhibited a similar behaviour as a function OMP and the increase in refractive index on annealing can be attributed to the decrease in porosity and improvement in crystallinity. The refractive index and packing densities for the as deposited films were in the range of 1.84–1.96 @ 600 nm and 0.75–0.82 respectively and increases to

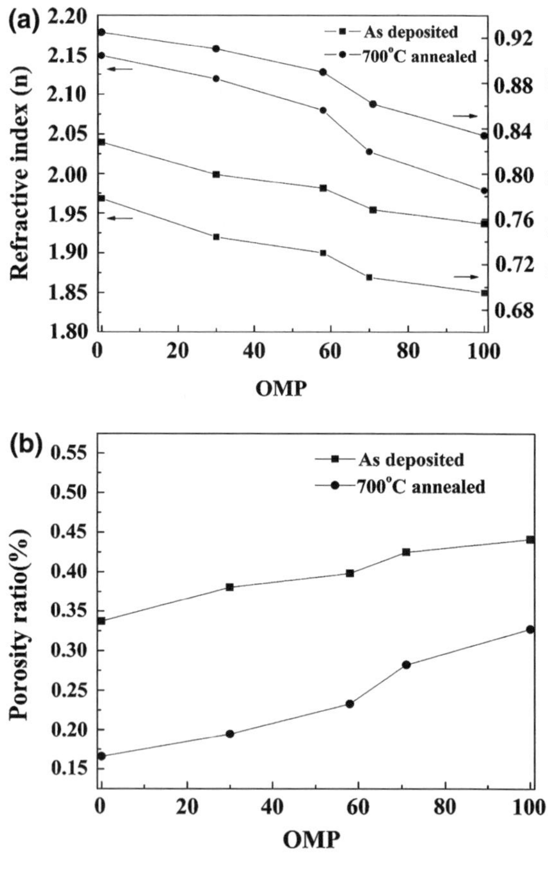

Fig. 5 Variation in **a** refractive index and packing density and **b** porosity ratio of as deposited and annealed films as a function of OMP respectively

1.98–2.14 and 0.83–0.92 respectively on annealing. For example for the as-deposited film at 30 % OMP, the refractive index was 1.92 (at 600 nm) which is increased to 2.12 at the same wavelength on annealing. It is well known that the refractive index of a transparent thin film, is directly proportional to its electronic polarization and the electronic polarization is in turn inversely proportional to the inter-atomic separation [20]. In the present study the as-deposited films are highly disordered due to their amorphous nature resulting in lower film density, which in turn results in the lowering of the refractive index. On annealing, there is a reduction in the inter-atomic spacing due to crystallization that leads to higher densification and hence an increase in the refractive index. The porosity ratios were found to decrease on annealing which complements the increase in refractive index and packing density of the films.

The optical band gap (E_g) for all the thin films was calculated using the Tauc equation [21], which was given by

$$(\alpha h\upsilon) = C\left(h\upsilon - E_g\right)^{\gamma} \qquad (3)$$

where C is a constant, α is an absorption coefficient, $h\upsilon$ is the incident photon energy and $\gamma = 0.5$, 1.5, 2 or 3 for allowed direct, forbidden direct, allowed indirect and forbidden indirect electronic transitions. In the present case the band gap energy has been estimated by assuming $r = 2$ according to the literature [22] the disordered titanates are characterized by an indirect allowed electronic transitions. The bandgap energy (E_g) of the films was obtained by extrapolating the linear portion of the plot of $(\alpha h\upsilon)^{1/2}$ against $h\upsilon$ to $(\alpha h\upsilon)^{1/2} = 0$. Figure 6a shows the variation in band gap as a function of OMP for the as-deposited and annealed MTO thin films. It was observed that the band gap values for MTO thin films found to increase on annealing and the similar behaviour was reported by Ferri et al. [16].

The increase in band gap values associated to decrease of intermediary energy levels within the optical band gap and was confirmed from optical transmittance spectra (Fig. 4), the absorption edge was shifted to lower wavelength which shows that on annealing the band gap of the films increases. Figure 6b shows $(\alpha h\upsilon)^{1/2}$ versus $h\upsilon$ for the MTO thin films deposited at 60 % OMP. The obtained band gap values were in the range 3.72–3.76 eV and 4.07–4.23 eV for as-deposited and annealed MTO thin films respectively. It is also observed that band gap values slightly increases with increase in OMP. Although the band gap energy is a constant for a material in the bulk form it is known to vary in thin films. The dependence of optical band gap energy on crystallite size in TiO_2 thin film deposited by reactive DC-sputtering has been demonstrated previously by one of the present authors [23]. The dependence of optical band gap on crystallite size is shown in Fig. 6c. It can be observed that the films with smaller crystallite exhibit larger band gap energy than those with larger crystallite size.

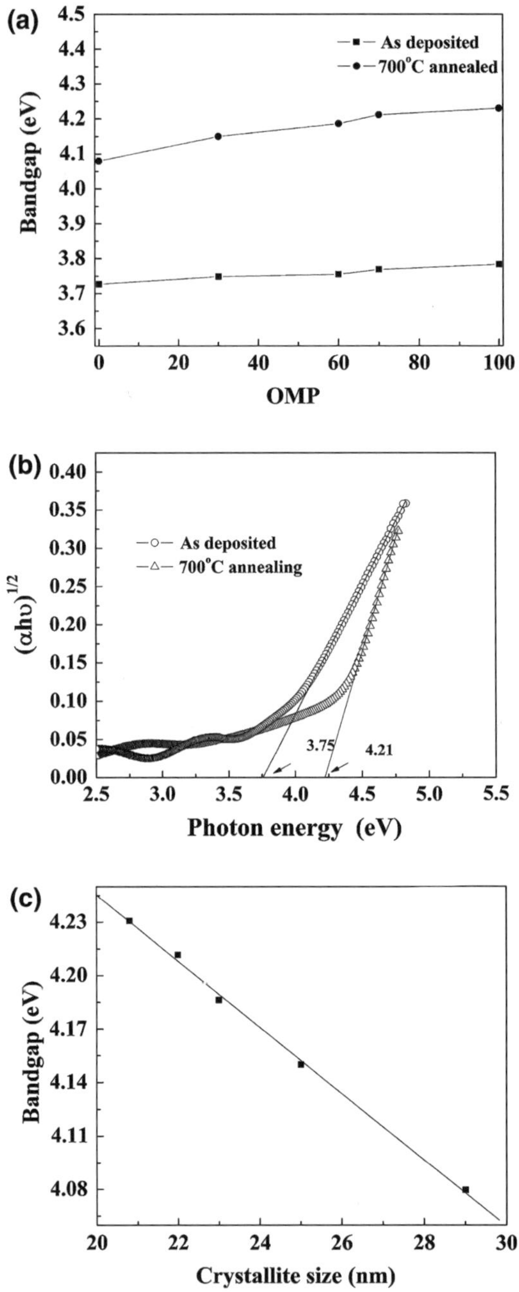

Fig. 6 **a** Variation in bandgap as a function of OMP for as deposited and annealed MTO thin films. **b** A plot of $(\alpha h\upsilon)^{1/2}$ versus $h\upsilon$ for the MTO thin films deposited at 60 % OMP 6. **c**. Variation in band gap as a function of crystallite size

4 Conclusion

MTO thin films have been deposited on to SiO_2 substrates at ambient temperature by RF magnetron sputtering at different OMPs. The MTO target has been synthesized by mechanochemical synthesis method. As deposited films found to be amorphous and crystallized on annealing at 700 °C for 1 h. The refractive indices and bandgap of MTO thin films are found to increase on annealing and is attributed to the improvement in the packing density and crystallinity. The refractive index and band gap values on annealing were in the range 1.98–2.14 and 4.07–4.23 eV respectively. It was observed that the processing parameters especially OMP and annealing temperature influenced the structural and optical properties of MTO thin films.

Acknowledgments The Authors acknowledges the financial support from DRDO and BRNS. Author acknowledges Dr. Biswanath Samantaray for Rietvied refinement analysis. XRD facility supported by DST, New Delhi, under FIST program (Ref. No: SR/FST/PSII-020/2009) is acknowledged.

References

1. T. Takada, S.F. Wang, S. Yoshikawa, S.J. Yang, R.E. Newnham, Effect of glass additives on BaO-TiO_2-WO_3 microwave ceramics. J. Am. Ceram. Soc. **77**, 1994 (1994)
2. R.J. Cava, W.F. Peck Jr, J.J. Krajewski, G.L. Reberts, B.P. Barber, Improvement of the dielectric properties of Ta_2O_5 through substitution with Al_2O_3. Appl. Phys. Lett. **70**, 1396 (1997)
3. V.M. Ferreira, J.L. Baptista, Role of niobium in magnesium titanate microwave dielectric ceramics. J. Am. Ceram. Soc. **79**, 1697 (1996)
4. S. Nishigaki, H. Kato, S. Yano, R. Kamimure, Microwave dielectric properties of (Ba, Sr)O-Sm_2O_3-TiO_2 ceramics. Am. Ceram. Soc. Bull. **66**, 1405 (1987)
5. V.M. Ferreira, J.L. Baptista, S. Kamaba, J. Petzdt, Dielectric spectroscopy of $MgTiO_3$-based ceramics in the 109–1014 Hz region. J. Mater. Sci. **28**, 5894 (1993)
6. N. Ichinose, H. Yamamoto, Effect of additives on microwave dielectric properties in low-temperature sintering (Mg, Ca)TiO_3 based ceramics. Ferroelectrics **201**, 255 (1997)
7. D. Pamu, K. Sudheendran, M. Ghanashyam krishna, K.C. James Raju, A.K. Bhatnagar, Ambient temperature stabilization of crystalline zirconia thin films deposited by dc magnetron sputtering. Thin Solid Films **517**, 1587 (2009)
8. M. Ferraris, E. Verne, P. Appendino, C. Moisescu, A. Krejewski, A. Ravaglioli, A. Piancastelli, Coatings on zirconia for medical applications. Biomaterials **21**, 765 (2000)
9. K. Suresh, J.K. Thomas, H. Sreemoolanadhan, C.N. George, A. John, S. Solomon, P.R.S. Wariar, J. Koshy, Synthesis of nanocrystalline magnesium titanate by an auto-igniting combustion technique and its structural, spectroscopic and dielectric properties. Mater. Res. Bull. **45**, 761 (2010)
10. J.H. Sohn, Y. Inaguma, S.O. Yoon, M. Itoh, T. Nakumura, S.J. Yoon, H.J. Kim, Microwave dielectric characteristics of ilmenute-type titanates with high Q values. Jpn. J. Appl. Phys. **33**, 5466 (1994)
11. B.D. Lee, H.R. Lee, K.H. Yoon, Y.S. Cho, Microwave dielectric properties of magnesium calcium titanate thin films. Ceram. Int. **31**, 143 (2005)

12. Y.B. Chen, C.L. Haung, Effects of O2/Ar mixing and annealing on the properties of $MgTiO_3$ films prepared by RF magnetron sputtering. Surf. Coat. Technol. **201**, 654 (2009)
13. R.A. Young, *The Rietveld Method*, International Union of Crystallography, (Oxford University Press, London, 1996)
14. R. Swanpoel, Determination of the thickness and optical constants of amorphous silicon. J. Phys. E: Sci. Instrum. **16**, 1214 (1983)
15. H.M. Rietveld, A profile refinement method for nuclear and magnetic structures. J. Appl. Crystallography **2**, 65 (1969)
16. E.A.V. Ferri, J.C. Sczancoski, L.S. Cavalcante, E.C. Paris, J.W.M. Espinosa, A.T. de Figueiredo, P.S. Pizani, V.R. Mastelaro, J.A. Varela, E. Longo, Photoluminescence behavior in $MgTiO_3$ powders with vacancy/distorted clusters and octahedral tilting. Mater. Chem. Phys. **117**, 192 (2009)
17. R. Jacobson, Inhomogeneous and co-evaporated homogeneous films for optical applications. Phys. Thin films **8**, 51 (1975)
18. Q. Ye, P.Y. Liu, Z.F. Tang, L. Zhai, Hydrophilic properties of nano-TiO_2 thin films deposited by RF magnetron sputtering. Vaccum **81**, 627 (2007)
19. J. Lee, C.W. Choi, sol- gel derived epitaxial $MgTiO_3$ thin films. Jpn. J. Appl. Phys. **38**, 3651 (1999)
20. P. Li, J.F. McDonald, T.M. Lu, Densification induced dielectric properties change in amorphous $BaTiO_3$ thin films. J. Appl. Phys. **71**, 5596 (1992)
21. J.C. Tauc, *Optical Properties of Solids*, (Amesterdam, North-Holland, 1972)
22. V.M. Longo, A.T. de Figueiredo, S. de Lázaro, M.F. Gurgel, M.G.S. Costa, C.O. Paiva-Santos, J.A. Varela, E. Longo, V.R. Mastelaro, F.S. de Vicente, A.C. Hernandes, R.W.A. Franco, Structural conditions that leads to photoluminescence emission in $SrTiO_3$: An experimental and theoretical approach. J. Appl. Phys. **104**, 023515 (2008)
23. D. Pamu, M. Ghanashyam Krishna, K.C. James Raju, A.K. Bhatnagar, Ambient temperature growth of nanocrystalline titanium dioxide thin films. Solid State Commun. **135**, 7 (2005)

Effect Sulfur and ZnO Nanoparticles on Stress Physiology and Plant (*Vigna radiata*) Nutrition

Prasun Patra, Samrat Roy Choudhury, Subhankar Mandal, Anirban Basu, Arunava Goswami, Robin Gogoi, Chitra Srivastava, Rajesh Kumar and Madhuban Gopal

Abstract Phytotoxic and agro beneficial properties of antimicrobially efficient Sulfur nanoparticles (SNPs) and ZnO nanoparticles (ZNPs) were evaluated on Mung (*Vigna radiata*). Mung seeds were germinated with an increasing concentration gradient of nanoparticles. The extent of phtotoxicity were assessed depending upon the effect of used nanoparticles on physical factors (relative root and shoot length, dry weight and area of leaves), photosynthetic pigments (chlorophyll, carotene and xanthophyll content) and mitochondrial stress indicator level (thiol), A simultaneous study was undertaken to understand the effect of SNPs and ZNPs on overall plant growth and nutrition. The nutritive values of nanoparticles

P. Patra (✉) · S. R. Choudhury (✉) · A. Goswami
Biological Sciences Division, Indian Statistical Institute, 203 B. T. Road, Kolkata 700108 West Bengal, India
e-mail: samratroychoudhury@gmail.com

A. Goswami
e-mail: agoswami@isical.ac.in

S. Mandal
Techno India, EM-4/1, Salt Lake City, Sector-V, Kolkata 700091, India

A. Basu
Himalayan Pharmacy Institute, Majhitar, East Sikkim 737136, India

R. Gogoi
Department of Plant Pathology, Indian Agricultural Research Institute, Pusa Campus, New Delhi, India

C. Srivastava
Department of Entomology, Indian Agricultural Research Institute, Pusa Campus, New Delhi, India

R. Kumar · M. Gopal
Department of Agricultural Chemicals, Indian Agricultural Research Institute, Pusa Campus, New Delhi, India

P. K. Giri et al. (eds.), *Advanced Nanomaterials and Nanotechnology*, Springer Proceedings in Physics 143, DOI: 10.1007/978-3-642-34216-5_31,

were determined in terms of total lipid and protein content. All the aforementioned biochemical assays were performed in triplicate for statistical viability.

Keywords Sulfur nanoparticles · ZnO nanoparticles · Stress physiology · Plant nutrition

1 Introduction

During the last few decades, there has been a rapid growth of interest in the areas of nanoscience and nanotechnology because of the realization that nanosized materials are effective in a multitude of agricultural and biomedical sectors [1]. The great potential of using nanoscale particles for different biological and biomedical applications include gene and drug delivery, bio sensing, diagnostic, and tissue engineering [2].

While most investigations are focused on studying the effects of different nanomaterials on the cellular morphology, behaviour, functions, and selective killing of pathogens or diseased cells, only few studies were reported regarding effectiveness of nanoparticles on plants of agricultural and horticultural importance.

To enlighten the aforementioned aspect, we have screened sulphur nanoparticles (SNPs) and zinc oxide nanoparticles (ZNPs) against Mung bean [Vigna radiata]. Phytophysiological parameters (both stress responsive and nutritive) of the tested plants were observed along a concentration gradient of SNPs and ZNPs.

2 Materials and Methods

2.1 Preparation, Characterization and Application Doses of NanoParticles

Orthorhombic and monoclinic SNPs were prepared via liquid phase precipitation method and water in oil micro emulsion technique respectively [3, 4]. ZNPs (N) were prepared following available standard method [5] by adding NaOH solution into zinc acetate solution. ZNPs (T) were prepared by a new method by adding TRIS solution into aqueous zinc acetate solution (Unpublished data). Nanoparticles thus prepared were characterized for their size distribution (with Dynamic Light Scattering), shape (with Scanning Electron Microscopy) and elemental composition (with Energy Dispersive X-ray Spectroscopy). All the nanoparticles in use were dissolved in Millipore water and applied at 4 different concentrations (500, 1,000, 2,000 and 4,000 ppm of the stock) along with the untreated replica.

2.2 Seed Germination Assay

Sterilized (with 10 % sodium hypochlorite for 10 min) seeds (variety: Samrat) were obtained from the Central Seed Research Institute, Baharampur, West Bengal. Nanoparticles solutions were sprayed (once per day for 10 days) to the seeds and allowed to germinate for 24 h (in dark) at 30 °C. Filter papers (100 × 15mm) soaked in different concentrations nanoparticles concentrations were then applied over the seed surface for prolonged supplementation.

2.3 Assessment of Root and Shoot Length, Leaf Area and Dry Weight

Relative length of root and shoot was measured on the 7th day of germination. Area of the first and second leaves for each of the concentrations was measured. All the derived data were compared with the untreated replica. Net increase in the dry weight of the treated plants was recorded to assess the nutritive value of the nanoparticles.

2.4 Increase in Proteins, Amino Acids and Lipids

The method of Lowry et al. [6] was used to measure the total protein content for the present experiments. Moreover, total amino acids from the plant samples were measured following the modified method of Moore et al. (1954) [7]. Total lipid content in the treated samples was measured (modified Bligh and Dyer [8]) to evaluate the effect of nanoparticles on the plant cell wall components.

2.5 Effect on Photosynthetic Pigments

Total chlorophyll, carotene and xanthophylls content form the leaves of treated and untreated plant samples were measured to understand the effect of the incorporated (in use) nanoparticles on the photosynthetic ability of plants.

2.6 Effect on Thiol Leakage

Total thiol assay was carried out to understand the effect of used nanoparticles on mitochondrial stress physiology. For total thiol assay, a modified [9] method was followed. 0.5 g of fresh plant sample was taken and ground using a cold mortar and pestle and extracted with 4 mL of 50 m Mice-cold phosphate buffer (pH 7.0). The homogenate was centrifuged at 2,000 g for 20 min at 4.1 °C and the supernatant was collected. Three millilitres of the supernatant was mixed with 2 mL phosphate buffer (pH 7.0) and 5 mL distilled water and they were mixed well to get a10 ml reaction mixture. 20 μl of 0.01 M DTNB solution was added to 3 mL of the reaction mixture, shaken well and absorbance was recorded at 412 nm. Thiol content was calculated using extinction coefficient 13,600 M^{-1} cm^{-1} for DTNB at 412 nm.

3 Results and Discussions

3.1 Physical Characters of Nanoparticles

DLS revealed that hydrodynamic diameters of ZNP (N) and ZNP (T) particles were ~95 and ~80 nm (Fig. 1). In comparison size distribution for orthorhombic and monoclinic SNPs were found to be 50 and 100 nm. Both the SNPs (Fig. 2) and ZNPs (Fig. 3) were nearly spherical in shape (as obtained from SEM micrographs with effective concentrations of sulphur and ZnO, as revealed from EDX spectra.

3.2 Rate of Seed Germination

The average germination rates of all the plant seeds were greater than 90 % as shown by a preliminary study. Although the rate of seed germinations were not much affected by the application of nanoparticles.

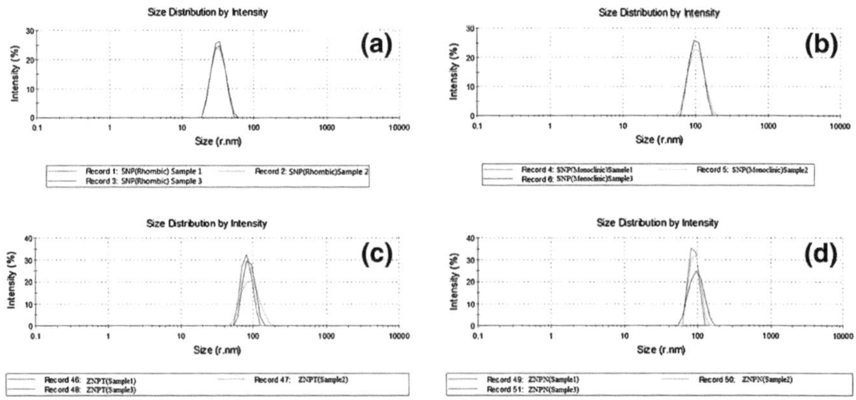

Fig. 1 Dynamic light scattering reveals hydrodynamic radius of orthorhombic SNPs (**a**), monoclinic SNPs (**b**), ZNPs (T) (**c**) and ZNPs (N) (**d**)

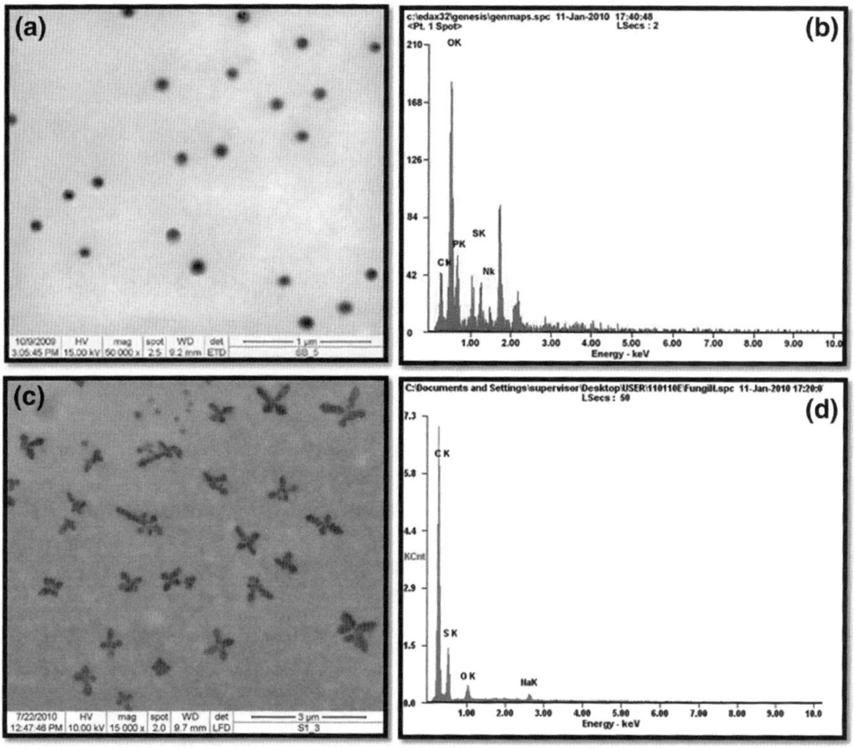

Fig. 2 Scanning electron micrographs and EDX spectra of orthorhombic (**a** and **b**) and monoclinic SNPs (**c** and **d**)

3.3 *Increase in Root and Shoot Length, Leaf Area and Dry Weight*

Total root and shoot length and area of the second leaf of the treated plants have been summarized (Fig. 4a). It was noted that for each of nanoparticles, dry weight (Fig. 4b) and leaf area were increased consistently.

3.4 *Change in Total Proteins, Amino Acids and Lipids*

Change in total lipids (Fig. 5b), proteins (Fig. 4c) and amino acids (Fig. 5a) were found to increase in comparison to the untreated replica. However, no significant trend of was observed for increasing concentrations of the nanoparticles.

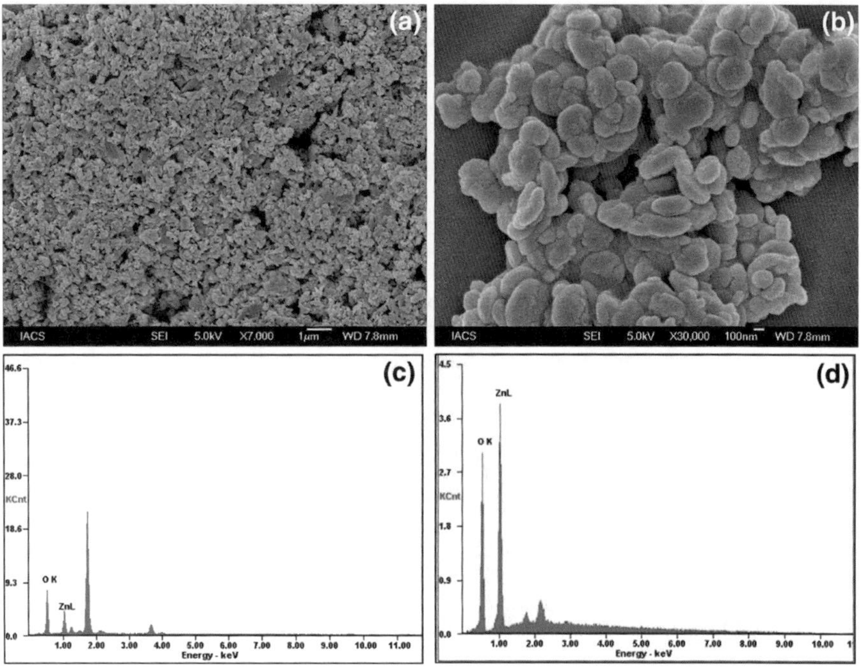

Fig. 3 Scanning electron micrographs and EDX spectra of ZNPs (T) (**a** & **b**) and ZNPs (N) (**c** and **d**)

3.5 Effect on Photosynthesis

Total chlorophyll content was increased than untreated replica for both the nanoparticles variety (Fig. 5c). However, no significant change was observed for carotene (Fig. 5d) and xanthophylls (Fig. 6a) content.

Significant increase in growth factors and nutritive values of the plant body were found for both the nanoparticles. However, SNPs were found to supplement more nutritive value to the plant bodies than that that of ZNPs. Noteworthy, changes in stress responsive elements for both the incorporated nanoparticles were negligible in comparison to the untreated replica.

The cumulative results strongly suggest safe use of SNPs and ZNPs in the agricultural fields with further suitable modifications.

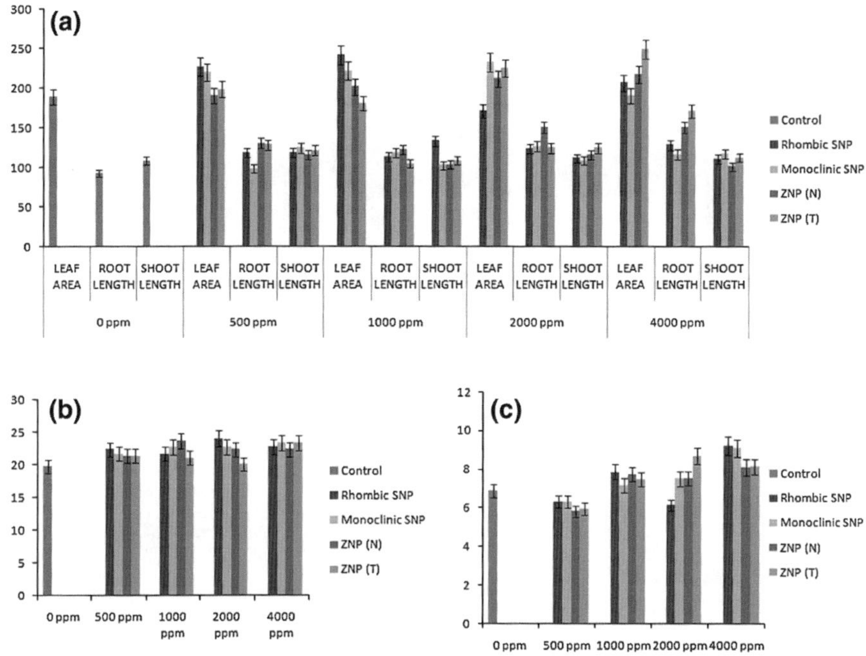

Fig. 4 Estimation of physical factors (**a** and **b**), total protein content (**c**) in treated plants

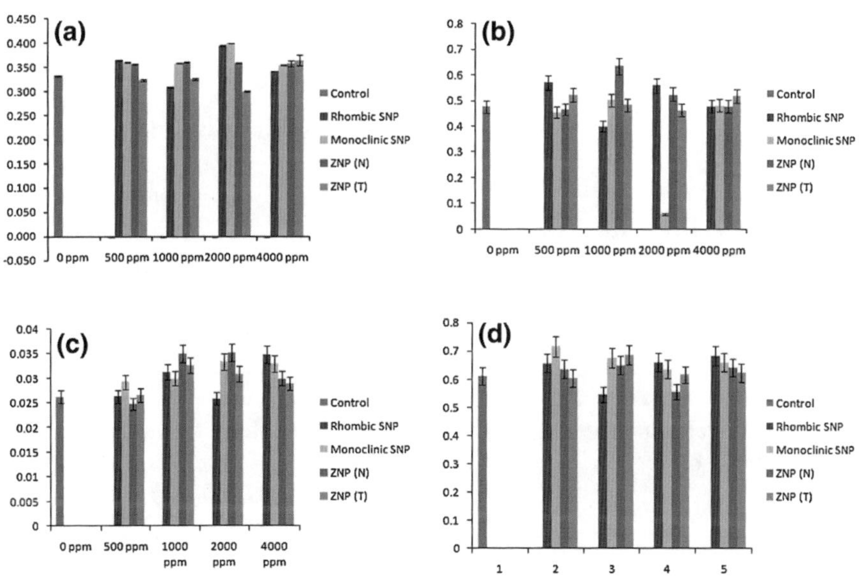

Fig. 5 Determination of total amino acids (**a**), total lipids (**b**), total chlorophyll (**c**) and carotene content (**d**) after treatment with nanoparticles

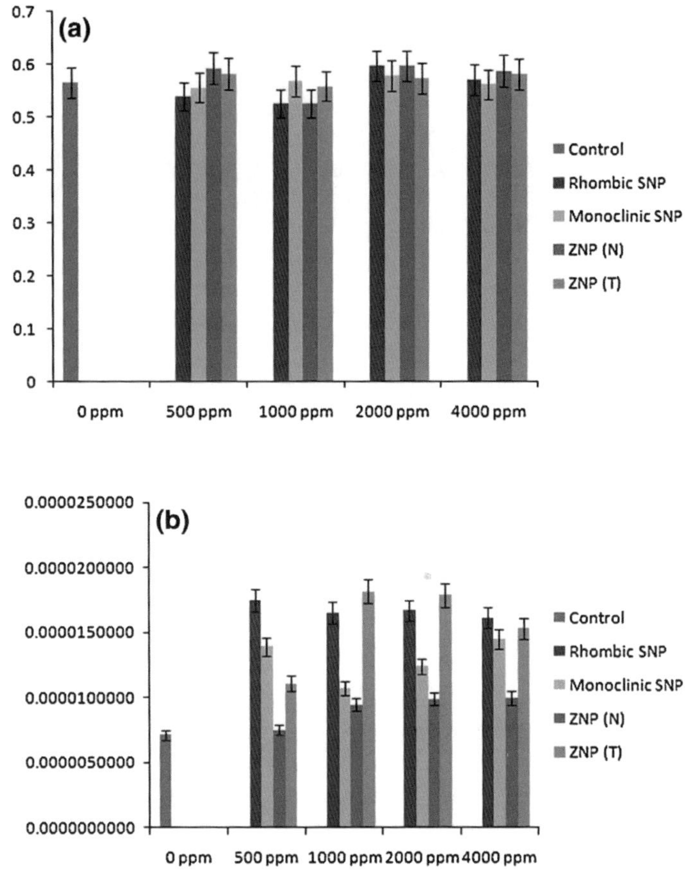

Fig. 6 Determination of total xanthophylls content (**a**) and thiol leakage (**b**) among treated plants

3.6 Change in Total Thiol Content

Highest leakage of thiol (−SH) content was observed for orthorhombic and ZNP (T) particles (Fig. 6b). Results indicate that the two aforementioned nanoparticles might have an adverse effect on mitochondrial integrity.

Cumulative data of this preliminary study strongly suggests the emergence of these two novel nanomaterials as a future generation fertilizer with additional antimicrobial efficacy.

References

1. R. Nair, S.H. Varghese, B.G. Nair, T. Maekawa, Y. Yoshida, D. Sakthi Kumar, Nanoparticulate material delivery to plants. Plant. Sci. **179**, 154–163 (2010)
2. K.S. Soppimath, T.M. Aminabhavi, A.R. Kulkarni, W.E. Rudzinski, Biodegradable polymeric nanoparticles as drug delivery devices. J. Control. Release. **70**(1–2), 1–20
3. S. Roy Choudhury, M. Ghosh, A. Mandal, D. Chakrovarty, M. Pal, S. Pradhan, A. Goswami, Surface modified sulfur nanoparticles: an effective antifungal agent against *Aspergillus niger* and *Fusarium oxysporum*. Appl. Microbiol. Biotechnol. **90**, 733–743 (2011)
4. Y. Guo, J. Zhao, S. Yang, K. Yu, Z. Wang, H. Zhang, Preparation and characterization of monoclinic sulfur nanoparticles by water-in-oil microemulsions technique. Powder Technol. **162**(2), 83–86 (2006)
5. C. Wu, X. Qiao, J. Chen, H. Wang, F. Tan, S. Li, A novel chemical route to prepare ZnO nanoparticles. Mater. Lett. **60**, 1828–1832 (2006)
6. O.H. Lowry, N.J. Rosebrough, A.L. Farr, R.J. Randall, Protein measurement with the Folin phenol reagent. J. Biol. Chem. **193**, 265–275 (1951)
7. S.W. Sun, Y.C. Lin, Y.M. Weng, M.J. Chen, Efficiency improvements on ninhydrin method for amino acid quantification. J. Food. Compos. Anal. **19**(2–3), 112–117 (2006)
8. E.G. Bligh, W.J. Dyer, Can. J. Biochem. Phys. **37**; 911–917 (1959)
9. G.L. Ellman, Tissue sulfhydryl groups. Arch. Biochem. Biophys. **82**, 70–77 (1959)

The Effects of Nano-Silver on Egg Quality Traits in Laying Japanese Quail

Amjad Farzinpour and Naser Karashi

Abstract Silver nanoparticles could prove to be a valuable alternative raw material for antibiotics and disinfectants as it is relatively free of adverse effects. Nano-silver is now been put to practical use in commonly used items, such as, clothes, electric home appliances, and electronic industry, but has not been widely applied in the medical or pharmacological fields. This study was designed to investigate the effects of nano-silver on egg quality traits in laying Japanese quail in completely randomized design with four treatments and six repetitions at 0, 4, 8 and 12 ppm of silver nanoparticle levels. Eggs collected daily and egg parameters, including egg weight, length, width; yolk weight and eggshell thickness were examined. The effect of the different silver nanoparticle levels was determined using the General Linear Model (GLM) of SAS procedure, whilst differences between the groups were determined using least significant difference test (LSD). Results indicated that silver nanoparticles at all levels caused significantly reduce of yolk weight and hen-day egg production for each week than the control treatment ($p < 0.05$), whilst silver nanoparticle had no significant effect on egg weight, egg length and width and eggshell thickness.

Keywords Nano-silver · Quail · Egg quality · Hen-day egg production

A. Farzinpour (✉) · N. Karashi
Department of Animal Science Faculty of Agriculture,
University of Kurdistan, Sanandaj, Iran
e-mail: amjadfarzinpour@gmail.com

P. K. Giri et al. (eds.), *Advanced Nanomaterials and Nanotechnology*,
Springer Proceedings in Physics 143, DOI: 10.1007/978-3-642-34216-5_32,

1 Introduction

Silver nanoparticles could prove to be a valuable alternative raw material for antibiotics and disinfectants. Nano-silver is now been put to practical use in commonly used items, such as, clothes, electric home appliances, and electronic industry, but has not been widely applied in the medical or pharmacological fields [1]. Silver nanoparticles show a broad band of antimicrobial effectiveness. The Pars Nano Nasb Company in Iran has been developing the technology of applying silver nano-particles to impose the efficiency of reducing bacterial contamination in the production of chicken and other animals. Therefore, the nanosilver is used today more as a high disinfectant in poultry production farms in Iran [2]. Although many studies have been done to evaluate the impact of nanosilver in birds, the effects exerted by silver nanoparticles administration on the overall reproductive performances of avian species has not been fully studied, especially when you know that the egg and meat of birds are the main protein sources for human. The present study investigated the effects of nanosilver on body weight (BW), egg production, egg quality parameters and mortality using a laboratory animal model, the Japanese quail.

2 Materials and Methods

Seventy two Japanese quail layers (n = 72) were used in the present study and divided into four groups or treatments at 8 weeks of age. Data were means of 6 replicate groups of three female Japanese quail each. Three female quails were assigned to each of 24 cages and were housed in a 45 × 16 × 15 cm–battery cage pen. To minimise the effects of body weight and physiological status in the egg-laying cycle, only females from the same hatch, of similar body weights (between 130 and 135 g) were used. The birds were reared in a light proof environmentally controlled house at 25 °C with a photoperiod of 14L:10D. The birds were supplied with commercial laying quail diets and tap drinking water ad libitum. Silver nanoparticles (Nanocid L2000, Colloidal Liquid, No. 357) were purchased from Nano Nasb Pars Company (Tehran, Iran). Physicochemical properties of the silver nanoparticles inclouded formula: TiO_2—Nano Ag, Concentration: TiO_2 ~99 % WT; Nano Ag ~1 % WT, Nano Ag Particles Size Distribution: 50 nm Max, Physical State: Solid—Powder, Appearance: White–Cream, Particle Diameter: 0.2 μm, Specific Area: 10.42 m^2/g, Temperature Endurance: UP to 2,000 °C. Four levels: 0, 4, 8 and 12 ppm of nanosilver were used in the drinking water during the experimental period. We determined the 0, 4, 8 and 12 ppm just by simply dilution from Nanocid. Every week the body weight (BW) of each bird was measured individually. The number of eggs produced and dead birds (if any) under each group were recorded every morning for 5 weeks. The collected eggs and the yolk weight were weighed. The hen-day egg production was recorded every week.

The hen-day egg production were then recorded daily at the same time and calculated as follows: total number of eggs collected divided by total number of live hens per day in each group. Egg shell quality was measured as deformation (abnormal eggs), shell weight, shell membrane weight, shell thickness and egg shape index. The yolks were carefully separated from the egg whites with forceps and blotting papers, and weighed. The shell membranes were completely removed from the egg shell using forceps after 30 min soaking in tap water. The membrane weight was taken after air drying in room temperature when the samples attained nearly constant weight. Eggshell thickness was measured with a digital micrometer (Mitutoyo, Japan). Length and breadth were measured using a vernier caliper and values were used to calculate the shape index (breadth $\times$ 100/length). The effect of the different silver nanoparticle levels data were evaluated using a $6 \times 4 \times 3$ (replicate $\times$ treatment $\times$ samples) factorial design. Data were subjected to the General Linear Models (GLM) procedures of SAS (SAS Institute, 1990). Means were separated using the least significant difference test. Significance was assumed at $P < 0.05$.

3 Results

The percentage of hen-day egg production in each week of the birds that received 4, 8 and 12 ppm of nanosilver were significantly ($P < 0.05$) lower than that of the control during 10–12 weeks (Fig. 1). Although neither the egg weight (Table 1) and eggshell thickness (Table 3), nor egg shape index (Table 4) were found to be markedly affected by nanosilver administration up to the end of the experimental period. The birds that received 12 ppm of nanosilver caused a significant decrease ($P < 0.05$) in yolk weight throughout the 12 week (Table 2) as compare with control group and birds that received 4 and 8 ppm of nanosilver and other weeks. The shell membrane weight weren't affected in treatment groups when were compared with control group. There wasn't any significant main effect of treatment groups on mortality (Table 5). The body weight was similar and there wasn't any differences between all groups during experimental period.

4 Discussion

The use of antibiotics in animal production has caused the development of antibiotic-resistant pathogen microorganisms. Recently, in many countries and especially in Iran, antibiotic-based growth promoters have been prohibited. Nevertheless, intensification of animal production and reduced immunity of highly productive animals require a search for new growth-promoting alternatives to antibiotics. Poultry species of very intensive growth kept in highly industrialised production systems are vulnerable to microbial infections affecting health and

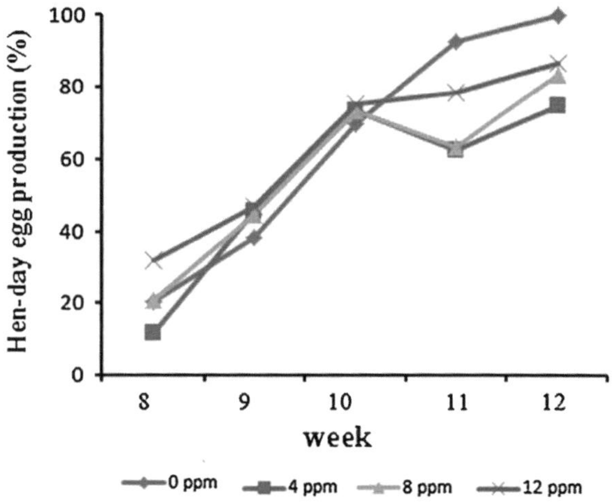

Fig. 1 Effect of nanosilver on hen-day egg production for each week in laying Japanese. The data of 10–12 weeks of the control are significantly different from those of the corresponding weeks of 4, 8 and 12 ppm of nanosilver ($P < 0.05$)

performance of animals. Numerous investigations have been carried out to develop feed additives (e.g. probiotics, herbs, preservatives), which would guarantee the profitability of Poultry production and which are not dangerous to human health as it happens in the case of growth promoters with antibiotic properties. Nanotechnological methods allow the production of nanoparticles of noble metals, which may exhibit unique biological properties. Noble metals such as silver and its ions have long been known to have unique anti-bacterial properties but even small doses of ionic silver may be toxic, limiting the use of silver as an anti-bacterial agent in animal production [3, 4]. However, the toxicity of Ag can be eliminated, when used in nanoparticle form [3]. Nanoparticles are structures from 1 to 100 nm. Due to their small size, the total surface area of the particles exposed in solution is maximised, resulting in the highest activity per unit of weight. Increased exposure to nanoparticles being produced in large scale industry facilities elicites concerns for the toxicity of certain classes of nanoparticles. In particular, the oral toxicity of silver nanoparticles is of particular concern to ensure public and consumer health. Due to the intensive commercial application of silver nanoparticles, health risk assessment of this nanoparticle is of great importance. New investigations demonstrated Ag-NPs caused toxicity in various cell-lines. Silver caused lethal damage to hepatocytes in rats and finally it lead to cell death [5]. Abnormal elevation of blood silver levels, Argyrialike symptoms and hepatotoxicity following the use of nanosilver coated dressings for burns in clinical application has been reported [6]. Silver does not only cause dermal and cosmetic toxic effects, but also it causes death in animals [7]. Soto et al. [8] have recently shown that silver nanoparticles can be nearly 50 % more toxic than chrysolite asbestos. Nanoparticles, such as silver nanoparticles, are showing severe toxic effects on the male reproductive system. The identified research suggests that nanoparticles cross the blood-testes barrier and are deposited in the testes [9]. A prolonged exposure to silver has shown toxic effects on CNS such as cerebral ataxia [7]. The available experimental animal studies have shown that silver is absorbed in and metabolized in soft tissues [7].

Table 1 Effects of nanosilver on egg weight in laying Japanese quail

Level (ppm)	Weeks 8	9	10	11	12
0	9.3 ± 0.12 (14)	10.1 ± 0.78 (24)	11.6 ± 0.10 (26)	11.7 ± 0.24 (26)	11.4 ± 0.39 (42)
4	9.1 ± 0.25 (8)	10.0 ± 0.79 (30)	11.2 ± 0.59 (30)	11.1 ± 0.63 (22)	10.9 ± 0.53 (30)
8	9.5 ± 0.11 (14)	10.2 ± 0.61 (24)	11.4 ± 0.31 (26)	11.3 ± 0.44 (22)	11.1 ± 0.32 (41)
12	9.7 ± 0.20 (20)	10.2 ± 0.54 (24)	11.2 ± 0.41 (28)	11.2 ± 0.52 (26)	11.4 ± 0.23 (42)

Values (g) are mean ± SD of egg weight in parentheses

Table 2 Effects of nanosilver on egg yolk weight in laying Japanese quail

Level (ppm)	Weeks 8	9	10	11	12
0	32.1 ± 1.20 (14)	29.4 ± 1.10 (12)	27.6 ± 2.32 (13)	29.9 ± 1.22 (13)	34.9 ± 2.23 (15)
4	33.0 ± 1.02 (8)	28.0 ± 1.52 (15)	29.6 ± 1.02 (15)	30.5 ± 1.17 (10)	34.0 ± 2.14 (15)
8	32.6 ± 0.12 (14)	27.6 ± 1.12 (12)	28.1 ± 1.14 (13)	31.0 ± 1.03 (10)	30.0 ± 2.51 (15)
12	29.5 ± 1.52 (10)	27.5 ± 1.62 (12)	30.6 ± 1.00(14)	29.4 ± 1.33 (13)	27.0 ± 2.12* (15)

Values (% to egg weight) are mean ± SD of the number of eggs in parentheses
* $P < 0.05$, when compared with the control in the same column

Table 3 Effects of nanosilver on eggshell thickness in laying Japanese quail

Level (ppm)	Weeks				
	8	9	10	11	12
0	0.198 ± 0.03 (7)	0.196 ± 0.06 (6)	0.197 ± 0.05 (6)	0.206 ± 0.08 (6)	0.215 ± 0.01 (7)
4	0.193 ± 0.07 (6)	0.199 ± 0.03 (7)	0.202 ± 0.01 (7)	0.211 ± 0.03 (5)	0.213 ± 0.03 (7)
8	0.200 ± 0.01 (7)	0.201 ± 0.01 (7)	0.201 ± 0.01 (6)	0.207 ± 0.06 (5)	0.211 ± 0.05 (7)
12	0.195 ± 0.11 (6)	0.197 ± 0.03 (7)	0.196 ± 0.08 (7)	0.209 ± 0.05 (7)	0.217 ± 0.01 (7)

Values (mm) are mean ± SD of the number of eggs in parentheses

Table 4 Effects of nanosilver on egg shape index in laying Japanese quail

Level (ppm)	Weeks				
	8	9	10	11	12
0	80.7 ± 0.31 (11)	80.9 ± 0.70 (34)	80.1 ± 0.04 (34)	79.5 ± 0.83 (28)	79.5 ± 0.41 (73)
4	77.6 ± 0.37 (7)	80.6 ± 0.37 (23)	79.6 ± 0.41 (46)	78.1 ± 0.03 (14)	77.4 ± 0.98 (53)
8	82.8 ± 0.84 (15)	79.2 ± 0.32 (32)	78.2 ± 0.42 (41)	79.2 ± 0.96 (17)	79.1 ± 0.84 (62)
12	80.1 ± 0.34 (28)	80.6 ± 0.83 (42)	79.2 ± 0.81 (44)	78.9 ± 0.02 (23)	79.2 ± 0.04 (63)

Values (%) are mean ± SD of the number of eggs in parenthesis

Table 5 Effects of nanosilver on mortality in laying Japanese quail

Level (ppm)	Weeks					Total
	8	9	10	11	12	8–12
0	1	1	0	4	0	6
4	1	1	2	1	0	5
8	1	1	3	1	0	6
12	1	1	2	1	0	5

Values (numbers) are the number of dead birds

According to Cheng et al. [10], Zhang and Sun [11], silver ions can also enter into the human body through the female genital tract. Silver nanoparticles have been shown to damage brain cells [12]; liver cells [13]. Silver and silver salts are distributed around the body and they accumulated and show some toxic effects in organs and tissues [7]. Silver ions show a high affinity for thiol in the liver [14]. Today silver nanoparticles can be applied to animal feeding as feed additive [15]. In an experiment with broiler chicks, dosage of metallic silver nanoparticles for 5 weeks was continued by 7 days of non-supplemented period. Silver retention was 0.035, 0.031 and 0.045 μg/g in muscular tissue and 0.113, 0.086 and 0.185 μg/g for the same treatments in liver tissue for 20, 30 and 40 silver ppm in diet, respectively [15]. Only 5 out of 10 animals given 20 and 30 ppm silver showed detectable concentration in muscles, while 6 and 7 out of 10 animals with the same treatments showed silver concentration in the liver [15]. We could consider the liver to be an integral part of that system because that is where the egg yolk lipid is formed. Laying birds require a normal liver to sustain maximum egg production over time. There is many factors that can adversely affect egg production. Egg production can be affected by such factors as feed consumption (quality and quantity), water intake, intensity and duration of light received parasite infestation, disease, toxins and numerous management and environmental factors. Although the silver as an antibiotic requirement of birds is relatively low, adequate levels are essential; and excessive amounts are toxic. The decreased hen-day egg production and yolk weight found in quails that received 12 ppm nanocide might be the result of hepatotoxic effects of silver. This idea coincided with the reports of some researchers who found silver nanoparticles to be restrictedly caused lethal damage to hepatocytes [13]. As a matter of fact, nanomaterials are widely used in many fields such as the chemical industry, biomedicine and poultry industry as a disinfectant and using of silver nanoparticles for a long time in layer or broiler breeder farms may have side effects on female birds and reproductive performances. Therefore, we hypothesised that nanosilver may affect the egg production in laying quail as a laboratory animal model. Our findings showed that administration of 12 ppm of nanosilver caused a significant decrease in yolk weight. The percentage of hen-day egg production for each week of the birds that received 4, 8 and 12 ppm of nanosilver were significantly lower than the control during 10–12 weeks. Our results indicate the importance of the length of silver nanoparticles exposure on egg production and yolk weight in laying quails during a long period of drinking water contained nanoparticles. The decreasing of yolk weight and egg production within the late week after nanocide administration indicated silver toxicity was a chronic effect. It seems that colloidal nanosilver has an effect on the egg production of the quail.

5 Conclusion

In conclusion, this investigation demonstrated for the first time that colloidal silver nanoparticles influenced egg production of quails; however, water containing 12 ppm silver nanoparticles significantly decreased the yolk weight. Furthermore, 4, 8 and 12 ppm nanocide decreased egg production. Finally, further research is required to elucidate the underlying mechanisms of silver nanoparticles on egg production, liver and reproductive performances. Determination of silver nanoparticle residual in meat and egg are also suggested in future studies.

References

1. S.H. Shin, M.K. Ye, H.S. Kim, H.S. Kang, The effects of nano-silver on the proliferation and cytokine expression by peripheral blood mononuclear cells. Int. Immunopharmacol. **7**, 1813–1818 (2007)
2. D. Shahbazzadeh, H. Ahari, N. Mohammad Rahimi, F. Dastmalchi, M. Soltani, M. Fotovat, J. Rahmannya, N. Khorasani, The effects of nanosilver (nanocid) on survival percentage of rainbow trout (oncorhynchus mykiss). Pak. J. Nutr. **8**, 1178–1179 (2009)
3. M. Grodzik, E. Sawosz, The influence of silver nanoparticles onchick enembryo development and bursa of fabricius morphology. J. Anim. Feed. Sci. **15**, 111–114 (2006)
4. E. Sawosz, M. Grodzik, M. Zielińska, P. Sysa, M. Szmidt, T. Niemiec, A. Chwalibog, Influence of hydrocolloidal silver nanoparticles on gastrointestinal microflora and morphology of enterocytes of quails. Arch. Anim. Nutr. **61**, 444–451 (2007)
5. C. Baldi, C. Minoia, A.D. Nuici, E. Capodaglio, L. Manzo, Effects of silver in isolated rat hepatocytes. Toxicology Lett. **41**, 261–268 (1988)
6. M. Trop, M. Novak, S. Rodl, B. Hellbom, W. Kroell, W. Goessler, Silver-coated dressing acticoat caused raised liver enzymes and argyria-like symptoms in burn patient. J. Trauma. **60**, 648–652 (2006)
7. N.R. Panyala, E.M. Peña-méndez, J. Havel, Silver or silver nanoparticles: a hazardous threat to the environment and human health. J. Appl. Biomedecine **6**, 117–129 (2008)
8. K.F. Soto, A. Carrasco, T.G. Powell, K.M. Garza, L.E. Murr, Comparative in vitro cytotoxicity assessment of some manufactured nanoparticulate materials characterized by transmission electron microscopy. J. Nanopart. Res. **7**, 145–169 (2005)
9. M.E. McAuliffe, M.J. Perry, Are nanoparticles potential male reproductive toxicants? a literature review. Nanotoxicology **1**, 204–210 (2007)
10. D. Cheng, J. Yang, Y. Zhao, Antibacterial materials of silver nanoparticles application in medical appliances and appliances for daily use. Chin. Med. Equip. J. **4**, 26–32 (2004)
11. Y. Zhang, J. Sun, A Study on the bio-safety for nano-silver as anti-bacterial materials. Chin. J. Med. Instrum. **31**, 35–38 (2007)

12. S.M. Hussain, M.K. Javorina, A.M. Schrand, H.M. Duhart, S.F. Ali, J.J. Schlager, The interaction of manganese nanoparticles with PC-12 cells induces dopamine depletion. Toxicol. Sci. **92**, 456–463 (2006)
13. S.M. Hussain, K.L. Hess, J.M. Gearhart, K.T. Geiss, J.J. Schlager, In vitro toxicity of nanoparticles in BRL 3A rat liver cells. Toxicol in Vitro **19**, 975–983 (2005)
14. P.L. Drake, K.J. Hazelwood, Exposure-related health effects of silver and silver compounds: a Review. Ann. Occup. Hyg. **49**, 575–585 (2005)
15. M. Fondevila, R. Herrer, M.C. Casallas, L. Abecia, J.J. Ducha, Silver nanoparticles as a potential antimicrobial additive for weaned pigs. Anim. Feed. Sci. Technol. **150**, 259–269 (2009)

Evaluation of Electromechanical, Damping and Dynamic Mechanical Properties of Silver Electrode IPMC Actuator

Dillip Kumar Biswal, Dibakar Bandopadhya and Santosha Kumar Dwivedy

Abstract Polymeric artificial muscle technologies are being developed that can produce similar strains and higher stresses like natural muscles using electro-static forces, electro-striation, ion interaction, and molecular conformational changes. Of specific interest, Ionic Polymer Metal Composite (IPMC), a class of polymer labeled as electroactive polymer (EAP) is being considered in application both as sensor and actuator. The proposed work is in line with the characterization of inexpensive IPMC of non-precious metal silver as the surface-electrode. At first, IPMCs are fabricated following chemical decomposition method. Several experiments are conducted for evaluation of surface resistance, damping properties, solvent loss, dynamic mechanical properties, tip displacement, and tip force to investigate the performance of the IPMC actuator. The experimental results suggest that the low cost Ag-IPMC is suitable for application as an active actuator.

Keywords Electro-active polymer · Ionic polymer metal-composite (IPMC) · Loss modulus · Surface-electrode resistance · Payload

1 Introduction

Polymeric artificial muscle technologies are being developed that can produce similar strains and higher stresses like natural muscles using electro-static forces, electro-striation, ion interaction, and molecular conformational changes. With the inception of electroactive polymers (EAPs), active polymeric materials have

D. K. Biswal · D. Bandopadhya (✉) · S. K. Dwivedy
Department of Mechanical Engineering, Indian Institute of Technology Guwahati, Guwahati, Assam 781039, India
e-mail: dibakarb@iitg.ernet.in

P. K. Giri et al. (eds.), *Advanced Nanomaterials and Nanotechnology*,
Springer Proceedings in Physics 143, DOI: 10.1007/978-3-642-34216-5_33,

received much attention nowadays for their potential applications in various research and practical fields during past 15 years. EAPs are usually classified into two groups i.e. ionic and electronic based on their operating principle. IPMCs belong to the ionic group of EAPs, are capable of exhibiting electromechanical and mechanoelectrical response when subjected to an applied electric potential and a mechanical deformation, respectively [1, 2].

Further, interesting properties such as i.e. lightness, flexibility, simple, workable under dry and wet environment, low power consumption etc. IPMCs are considered as a promising active material both for actuation and sensing. Researches also explored that IPMCs can also be effectively implemented in robotic as micro-grippers, dust wiper, biomimetic robots, fish, insects, artificial muscles, heart compression devices etc. [3]. Mostly noble materials such as gold or platinum are used as electroding material due to their corrosion resistance, high conductivity and chemical stability properties. However, the fabrication process incurs high manufacturing cost limiting its potential applications in the various fields.

To reduce the surface-electrode resistance, Shahinpoor and Kim [4] first introduced silver and copper as surface-electrode material by electroding a thin layer over a platinum electrode. It was observed that the surface resistance reduced considerably and hence the force output increases up to 20 %. Chung et al. [5] followed solution casting method to fabricate IPMC with silver nano-powders where an additional silver layer is deposited on the coated IPMC by chemical decomposition method to reduce the surface resistance. It was observed that low surface resistance enhances the actuation capability of IPMC. Solution casting technique is used further to fabricate IPMCs with nickel nano-powder while additionally gold layer is deposited over the membrane [6]. Park and Kim [7] followed ion-exchange method to fabricate IPMC with a nickel (II) sulfates hexahydrate solution. The fabricated IPMC shows high storage modulus which also shows improvement under the exposure of a magnetic field. Copper is also used as a surface-electrode [8]. To reduce production cost, non-precious metal such as silver was proposed for fabrication of IPMC for optimal performance and low cost. Experimentally deformation angle, surface resistance of IPMC were investigated [9].

In the present work, low cost IPMC of non-precious metal silver is fabricated with Nafion as the base polymer. The process considerably reduces the cost of IPMC while at the same time exhibits superior actuation performance compared to other electrode materials. The objectives of the present work include:

1. Evaluation of damping characteristics; surface resistance and solvent loss with time.
2. Evaluation of dynamic mechanical properties of the developed IPMC.

Chemical decomposition method is used to fabricate the proposed Ag-IPMC. Multi steps of trial-and-error method are carried out for optimizing the process parameters. The process significantly reduces the surface resistance and also eliminates oxidation problem, thus enhances the actuation performance and force generation capability [9].

The morphological analysis confirms that, silver particles dispersed well over the Nafion membrane surface uniformly. Experiments are carried out to measure the surface resistance, solvent loss, and damping characteristics of the actuator. Dynamic mechanical analysis (DMA) tests are also carried out to investigate the visco-elastic properties of the composite. The developed IPMC demonstrates superior performance with low power consumption. It also exhibits low surface resistance, high bending actuation, and larger tip force compared to other electrode material IPMCs.

2 Characterization of Ag-IPMC

2.1 Materials and Manufacturing of IPMC Actuator

The proposed work is carried out on a silver-electrode ionic polymer metal composite actuator. IPMCs are fabricated first using Nafion-117 as base polymer procured from Ion Power, Inc, New Castle, DE 19720 USA. The IPMCs are fabricated at Mechatronics Laboratory, Indian Institute of Technology Guwahati. Silver Nitrate ($AgNO_3$), dilute Ammonia solution (NH_3), Sodium Hydroxide (NaOH), Dextrose Anhydrous GR ($C_6H_{12}O_6$), Hydrochloric Acid (HCl), and lastly deionized/distilled water are purchased from Merck Specialities Private Limited, India.

Nafion is pretreated by roughening both the surfaces using metallography silicon carbide grinding paper of grain fineness P800 in order to increase the interfacial area between the base polymer and electrode. The Nafion membrane is then cleaned by boiling it in 2 N HCl solution for 30 min then rinsed in deionized water for 30 min to get rid of impurities and to swell the membrane. The membrane is kept immersed in a NaOH solution for diffusing Na^+ cations into it followed by $Ag(NH_3)_2OH$ solution to incorporate the $Ag(NH_3)_2^+$ at room temperature. A reducing agent ($C_6H_{12}O_6$) is added into Ag $(NH_3)_2OH$ solution drop by drop and stirred for a few minutes to reduce the silver salt into silver metal, which is deposited on the membrane surface. Finally, few microns thick layer of silver is developed on both faces of the Nafion membrane. Concentration of Ag $(NH_3)_2OH$, reducing agent, and immersing period are the main factors that governs the quality of the final IPMC. A detailed fabrication procedure is reported in the literature by Biswal et al. [10].

Figure 1a shows the photograph of the fabricated IPMC and Fig. 1b illustrates SEM based morphological structure of the Ag-IPMC surface. Figure 1b demonstrates that a uniform densely packed distribution of Ag particles is found over the Nafion surface. It is observed that the presence of lone micro-crack on the IPMC membrane may be due to the pretreatment flaws of Nafion surface. Further, morphological study reveals that the average diameter of the Ag-particle is around 0.5–0.6 μm [11].

2.2 Surface Resistance Measurement

Evaluation of surface resistance of the IPMC actuator is necessary as it influences the bending actuation and force generation capability of the actuator. It is found that low surface resistance of IPMC enhances the performance of the actuator [4] while surface resistance changes due to bending deformation of the actuator [12].

In this work, surface resistance of an IPMC actuator of size $40 \times 5 \times 0.2$ (mm^3) is evaluated in inactive mode in hydrated condition. Surface resistance is determined by using two-point AMPROBE 5XP-A multimeter between two points at a distance of 5 mm of the sample. Average value over five measurements is taken and plotted as shown in Fig. 2. Figure 2 also demonstrates the variation of surface resistance with time. It is observed that the mean value of surface resistance remains almost constant for first 3 days, while it increases with time for about 8 days. Further increase in time, surface resistance remains to be stable less than 2.5 Ohm. The results obtained by Chen et al. [9]; Fig. 7b, it is observed that the surface resistance of Ag-IPMC increases gradually with time/days and is observed to be less than 5 Ohm (after day one surface resistance is measured around 1.7 Ohm, after 14 days around 3.74 Ohm, and after 16 days it is around 4.29 Ohm), while in the present case, surface resistance is measured as 2.4 Ohm after 14 days and remains stable around 2.5 Ohm onwards. Hence, as evident, the low surface resistance implies better actuation and force generation capability of the actuator. Therefore, the proposed method and process parameters optimized for fabricating silver-electroded IPMC is proved to be efficient and reliable for fabricating low cost Ag-IPMC actuator.

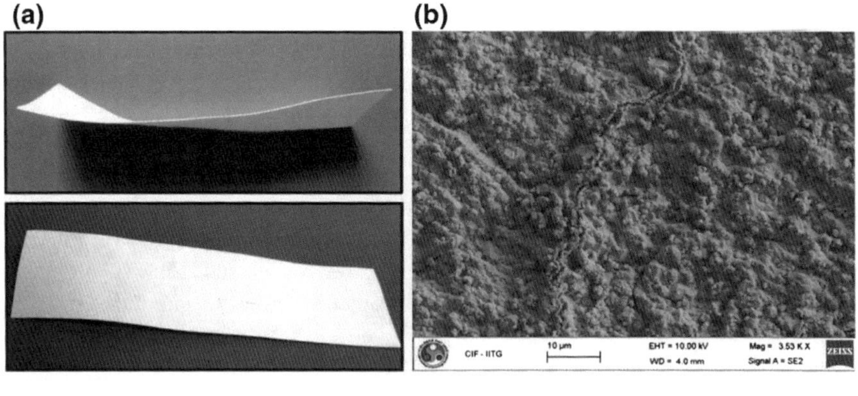

Fig. 1 **a** Photograph. **b** SEM micrographs of the surface of the developed IPMC

Fig. 2 Relations between surface resistance and time without application of input voltage

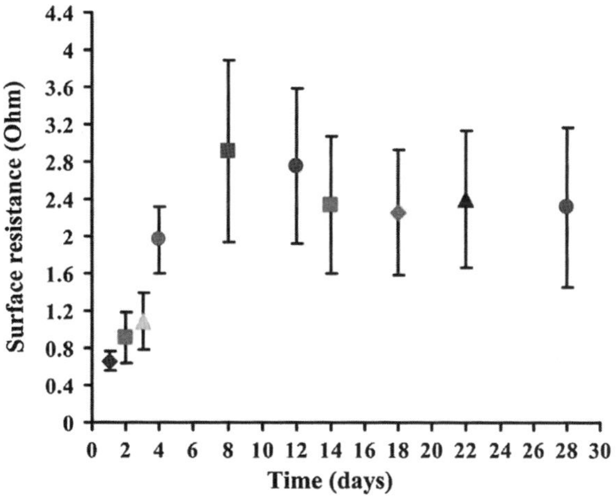

2.3 Damping Characteristics

The free vibration characteristic of an IPMC actuator of size $20 \times 5 \times 0.2$ (mm^3) is studied using Rotational Laser Vibrometer (RLV-5500, maker: Polytec, GmbH). The schematic diagram of the setup is shown in Fig. 3 while the photograph of the experimental setup is shown in Fig. 4. The actuator is allowed to vibrate under a small deflection at the tip for a time span before it comes to rest. The vibration response is then measured using NI PXI-1031 controller with NI PXI-4472 module and LABVIEW software, supplied by National Instruments.

The damping ratio and natural frequency of the actuator are investigated in fully hydrated (wet) condition. Figure 5a shows the free vibration response of the actuator. The logarithmic decrement (δ) and the damping ratio (ξ) can be expressed as [13]:

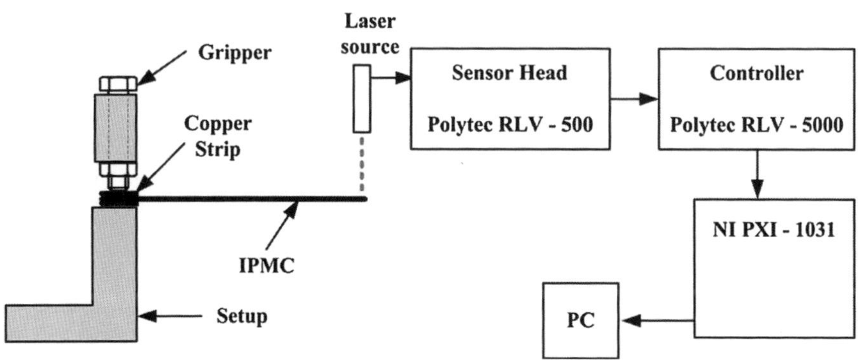

Fig. 3 Schematic diagram of the free vibration setup

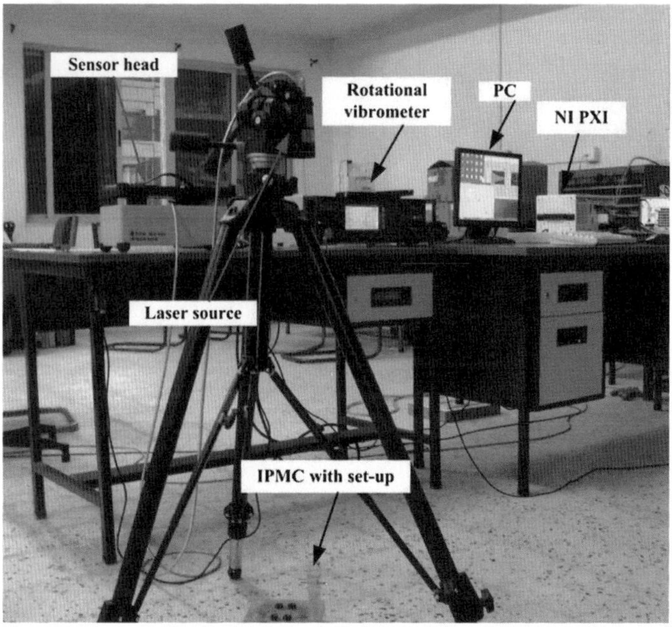

Fig. 4 Experimental setup for evaluation of vibration characteristics

$$\delta = \frac{1}{n} \ln\left(\frac{y_i}{y_{i+n}}\right) \tag{1}$$

$$\xi = \sqrt{\frac{\delta^2}{4\pi^2 + \delta^2}} \tag{2}$$

where, n is the number of cycles, y_i is the amplitude of displacement of ith cycle, y_{i+n} is the amplitude after n number of cycles. Taking ten successive cycles (n = 10), the damping ratio (ξ) of the actuator is determined to be 0.0179. Figure 5b shows the Fast Fourier transform (FFT) of the vibration response obtained. It is observed that the bandwidth of the vibration response remains within 2–20 Hz.

2.4 Solvent Loss Under Electric Potential

When an electrical potential is applied across the thickness of the IPMC actuator, the electromechanical transduction occurs due to transportation of cations with water molecules toward the cathode. This motion and accumulation of cations causes swelling near negative electrode (cathode) and shrinkage near the positive

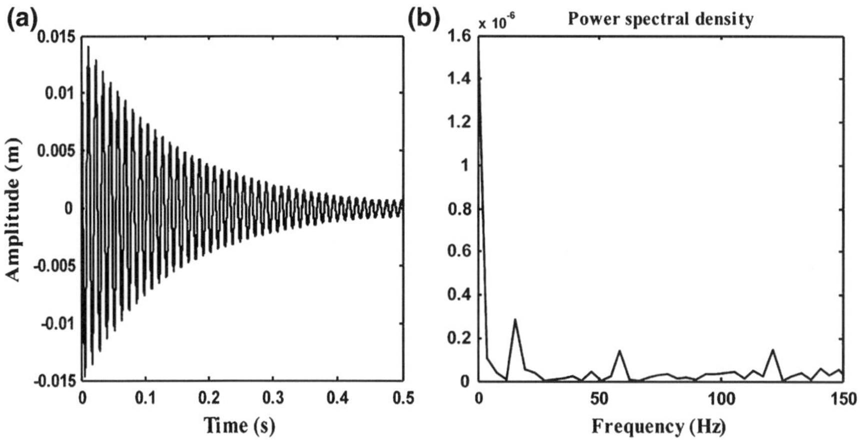

Fig. 5 **a** Vibration response taken at the tip of the actuator. **b** Corresponding FFT of the vibration response

electrode (anode) and initiates the bending toward the anode [14]. Hydration level is one of the key factors which influence the bending deformation of the actuator. For conventional IPMC, water is used as the polar solvent. It is observed that gradual dehydration of moisture contents and electrolysis beyond 1.23 V adversely affects the performance of the actuator [15]. In the present study, several experiments are carried out to find out the solvent loss under DC and AC input voltages. Before the experiment, the samples are immersed in the distilled water, boiled for 30 min and then kept for 24 h. The weight of the samples before and after full hydration condition is measured to calculate the water uptake. Water uptake can be defined as:

$$M_U = \frac{W_h - W_d}{W_d} \tag{3}$$

where, M_U is the water uptake, W_h and W_d is the weight of the fully hydrated and dried IPMC sample. Experimentally water uptake of a $20 \times 5 \times 0.2$ (mm^3) Ag-IPMC is found to be 23–24 %.

Solvent loss is calculated by experimenting a fully hydrated IPMC of size $20 \times 5 \times 0.2$ (mm^3). Both DC and AC (Sine wave, 1 Hz) input of 0.5 V is applied for different time periods. The weight of the sample before and after the experiment (example: after applying input voltage for 10, 20 s etc.,) is measured using a digital measuring balance (maker: Sartotius, model: BSA2245-CW).

The solvent loss (W_{loss}) is defined as [15].

$$W_{loss} = \frac{W_i - W_f}{M_U} \times 100\% \tag{4}$$

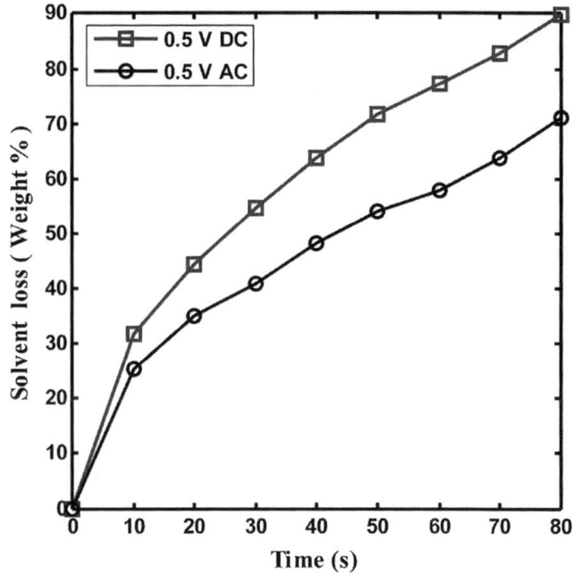

Fig. 6 Solvent (water) loss for both AC and DC input with time

where, W_i is the weight of the IPMC before the voltage is applied and W_f is the weight measured after the voltage is applied for a certain time period. Figure 6 shows the solvent loss from IPMC actuator for both AC and DC input voltage. It is observed that solvent loss is much less for AC input compared to DC input signal. Further, it is found that IPMC with 0.5 V DC input work for a time period of 80s suffered around 90 % solvent loss. On the other hand, with 0.5 V AC input, solvent loss is observed to be around 71 %. It is because of the fact that power consumption for DC input voltage gradually increases with time compared to the AC input voltage. The observation reveals that with AC input, the actuator can operate for a longer time period than same DC input voltage.

2.5 Dynamic Mechanical Analysis

Dynamic mechanical analysis (DMA) is carried out for the Ag-IPMC in dry condition for studying the visco-elastic properties. The test is conducted in open environment in tensile mode in an EXSTAR TMA/SS 6000 instrument. Applied stress frequency is taken as 0.01 Hz while the heating rate is kept constant at 5 °C/min.

Figure 7a demonstrates the change in storage modulus (E') of IPMC with temperature. The glass transition temperature (T_g) is obtained from the peak of tan δ and it is found to be 290.6 °C. At glass transition temperature (T_g), there is substantial change in the rigidity of the polymer for a short span of temperature. Figure 7b demonstrates the variation of dynamic loss modulus (E'') with temperature. It is anticipated that initial rise in the loss modulus is due to the chain

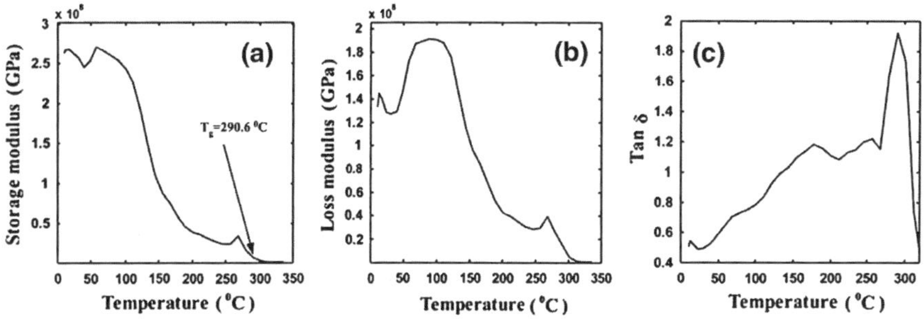

Fig. 7 **a** Storage modulus (E'). **b** Loss modulus (E''). **c** Tan δ as a function of temperature for Ag-IPMC at 0.01 Hz (heating rate 5 °C/min)

lengthening effect which causes the viscosity to rise with temperature. Figure 7c shows the tan δ curve which is obtained by taking the ratio E''/E', represents the damping characteristics of the sample. Tan δ indicates how efficiently a material losses energy due to molecular rearrangements. It is observed that Ag-IPMC shows simultaneously a high storage modulus and a high loss modulus value.

3 Performance of IPMC Actuator

Force developed at the tip of the IPMC due to input voltage is investigated on lifting capacity. An IPMC actuator of size 20 × 5 × 0.2 (mm^3) is tested at two different time intervals for finding the tip force/holding force of the actuator for an input of 1.0 V DC as shown in Fig. 8. Figure 8a shows the configuration of the same IPMC actuator with a payload of 2.1541 **g** while the weight of the IPMC

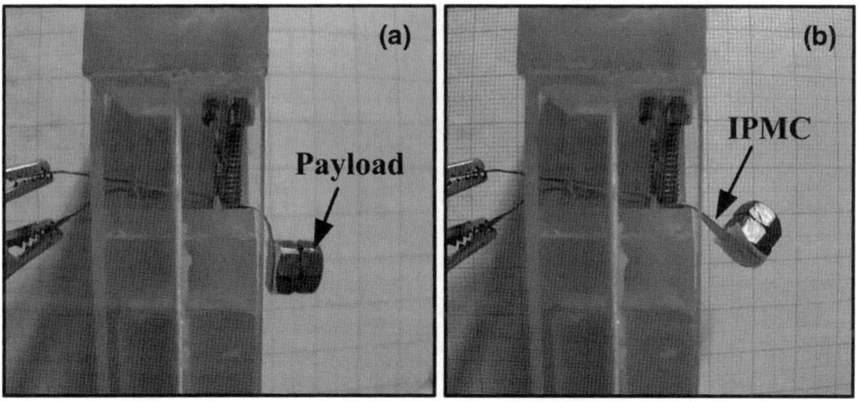

Fig. 8 An Ag-IPMC lifting a nut with an input of 1.0 V. **a** Time = 0 s. **b** Time = 3 s

actuator is 0.0425 g. Figure 8b demonstrates the position of the actuator at time t = 3 s. It is observed that the fabricated IPMC can lift a weight of almost 50 times more than its own weight just for an input of 1.0 V DC.

4 Conclusion

Non-precious metal silver is proved to be efficient and reliable for fabricating low cost IPMC. The experimental study validates the following observations:

- The low surface-electrode resistance of the developed IPMC proves better actuation and force generation capability of the actuator.
- Loss of solvent from IPMC is observed to be less when actuated under AC input potential than DC. This shows the operational life of the IPMC is more when actuated under AC than to DC input voltage. It is found that IPMC actuator of size 20 × 5 × 0.2 (mm^3) operate for a time period of 80 s experienced up to 90 % solvent loss for 0.5 V DC.
- Lifting payload capacity demonstrates the effectiveness of IPMC for actuation such as artificial muscle materials.

Acknowledgments The authors gratefully acknowledge the partial financial support of the Department of Science and Technology (DST), Government of India under SERC FAST Track Scheme (SR/FTP/ETA–076/2009). Authors also acknowledge the support from Central Instrument Facility (CIF), IIT Guwahati, for morphological analysis.

References

1. M. Shahinpoor, Y. Bar-Cohen, J.O. Simpson, J. Smith, Ionic polymer-metal composites (IPMCs) as biomimetic sensors, actuators and artificial muscles-a review. Smart. Mater. Struct. **7**, R15–R30 (1998)
2. Y. Bar-Cohen, S.P. Leary, A. Yavrouian, K. Oguro, S. Tadokoro, J.S. Harrison, J.G. Smith, J. Su, Challenges to the application of IPMC as actuators of planetary mechanisms in *Proceeding of Smart Structure and Materials*, SPIE 3987 (Newport Beach, CA March 06, 2000), pp 140–146
3. M. Shahinpoor, K.J. Kim, Ionic polymer-metal composites: IV. industrial and medical applications. Smart. Mater. Struct. **14**, 197–214 (2005)
4. M. Shahinpoor, K.J. Kim, The effect of surface-electrode resistance on the performance of ionic polymer-metal composite (IPMC) artificial muscles. Smart. Mater. Struct. **9**, 543–551 (2001)
5. C.K. Chung, P.K. Fung, Y.Z. Hong, M.S. Ju, C.C.K. Lin, T.C. Wu, A novel fabrication of ionic polymer-metal composites (IPMC) actuator with silver nano-powders. Sens. Actuators. B. **117**, 367–375 (2006)

6. B.-K. Fang, M.-S. Ju, C.-C.K. Lin, A new approach to develop ionic polymer-metal composites (IPMC) actuator: fabrication and control for active catheter systems. Sens. Actuators. A. **137**, 321–329 (2007)
7. Il-S. Park, K.J. Kim, Multi-field responsive ionic polymer-metal composite. Sens. Actuators. A. **135**, 220–228 (2007)
8. U. Johanson, U. Maeorg, V. Sammelselg, D. Brandell, A. Punning, M. Kruusmaa, A. Aabloo, Electrode reactions in Cu-Pt coated ionic polymer actuators. Sens. Actuators. B **131**, 340–346 (2008)
9. Q. Chen, K. Xiong, K. Bian, N. Jin, B. Wang, Preparation and performance of soft actuator based on IPMC with silver electrodes. Front. Mech. Engg. China **4**, 436–440 (2009)
10. D.K. Biswal, D. Bandopadhya, S.K. Dwivedy, Fabrication and thermo-mechanical analysis of pure silver-electrode ionic polymer-metal composite (IPMC) actuator. Appl. Mech. Mater. **110–116**, 1199–1206 (2012)
11. D.K. Biswal, D. Bandopadhya, S.K. Dwivedy, Electro-mechanical and thermal characterization of pure silver-electroded ionic polymer-metal composite (IPMC) actuator *in Proceeding of the Institute of Mechanical Engineers*, Part C: J. Mech. Eng. Sci. (2011). doi: 10.1177/0954406211424979
12. A. Punning, M. Kruusmaa, A. Aabloo, Surface resistance experiments with IPMC sensors and actuators. Sens. Actuators. A **133**, 200–209 (2007)
13. R.R. Craig, *Structural Dynamics an Introduction to Computer Methods* (Wiley, New York, 1981)
14. S. Nemat-Nasser, J. Li, Electromechanical response of ionic polymer-metal composites. J. Appl. Phys. **87**, 3321–3331 (2000)
15. S.-G. Lee, H.-C. Park, S.D. Pandita, Y. Yoo, Performance improvement of ionic polymer metal composites (IPMC) for a flapping actuator. Int. J. Control Autom. Syst. **4**, 748–755 (2006)

Stress Transfer Characterization at Fiber Break in Carbon Nanotube-Reinforced Composites

Sushen Kirtania and Debabrata Chakraborty

Abstract The present work deals with the study of stress distribution in the vicinity of a broken carbon nanotube (CNT) in a CNT-based nanocomposite. A square representative volume element (RVE) having nine uniformly spaced CNTs in the matrix has been considered for the analysis. Three dimensional finite elements (FE) analysis has been carried out for the square RVE and the effect of a broken CNT on the adjacent CNTs and matrix has been studied. CNTs and matrix in a RVE are considered linearly elastic, isotropic, and homogeneous materials and a perfect bonding is assumed between the CNT and matrix. The axial normal stress in the broken CNT, interfacial shear stress at the interface of the broken CNT and matrix, and axial normal stress in the adjacent CNTs has been calculated for different volume fractions and the effect of volume fraction on the stress redistribution has also been investigated. Results obtained from the present study conclude that the ineffective length of the broken CNT is dependent on the volume fraction of the composites as well as on the type of matrix materials. Stress concentration in the vicinity of the broken CNT is also influenced by the fiber volume fraction and the type of matrix materials.

Keywords Carbon nanotubes · CNT-based composites · Finite element analysis · Broken CNT · Stress concentration factor · Ineffective length

---

S. Kirtania (✉)
Mechanical Engineering Department, Tezpur University, Napaam,
Tezpur 784028, Assam, India
e-mail: sushen@tezu.ernert.in

D. Chakraborty
Mechanical Engineering Department, Indian Institute of Technology Guwahati,
IIT, Guwahati 781039, Assam, India
e-mail: chakra@iitg.ernet.in

P. K. Giri et al. (eds.), *Advanced Nanomaterials and Nanotechnology*,
Springer Proceedings in Physics 143, DOI: 10.1007/978-3-642-34216-5_34,

1 Introduction

Carbon nanotubes exhibit exceptionally high stiffness, strength and resilience, as well as superior electrical, thermal and mechanical properties. These nanotubes are also chemically inert and are able to sustain a high strain without breakage. These properties of CNTs are believed to be ideal for reinforcing high performance structural composites. Therefore, among many potential applications of nanotechnology, nanocomposites have been one of the important fields of research and hence number of works has already been reported in the direction of modeling and characterization of CNT-based nanocomposites. Due to their extremely small sizes, analytical models are difficult to be established, fabrication process and tests are extremely difficult and expensive to conduct. On the other hand, modeling and simulation can be advantageously used to analyze such nanocomposites. In spite of all the advantages sometimes the CNT-based nanocomposites may have one or more CNTs broken. This may be due to manufacturing defect or in-service damage. Therefore, the investigation of the stress redistribution in the broken fiber (i.e. CNT) and in the adjacent intact fibers is one of the important aspects which need to be studied. So, the present work is aimed at the calculation of the axial normal stress in the broken CNT, interfacial shear stress at the interface of the broken CNT and matrix, and axial normal stress in the adjacent CNTs for different volume fractions of CNT-based nanocomposites.

Carbon nanotubes have been the focus of researchers' interest since their discovery in 1991 [1]. There were earlier works [2–4] reported that the Young's modulus, and Poisson's ratio of CNTs are in the order of 1 TPa, and 0.28, respectively. A comprehensive reviews of manufacturing process, mechanical and electrical properties, and applications of CNT and CNT-based composites has been presented in two earlier works [5, 6]. A method has been proposed for developing structure–property relationships of nano-structural materials [7]. This method serves as a link between computational chemistry and solid mechanics by substituting discrete molecular structures with equivalent-continuum models. The use of multi-walled carbon nanotubes (MWCNTs) as intrinsic reinforcements for composite structures might not allow the maximum strength to be achieved due to non-uniform axial deformation inside the MWCNTs and hence the use of single-walled carbon nanotubes (SWCNTs) might be more beneficial for advanced composites structures [8]. The mechanical properties of CNT-based composites and load carrying capacities of CNTs in CNT-based composites have been investigated and represented in some earlier works [9, 10]. Effective mechanical properties of CNT-based composites have been evaluated using a three dimensional (3-D) nanoscale RVE based on the 3-D elasticity theory and solved by the FE method [9] and observed that with the addition of 3.6 % volume fraction of the CNTs in a matrix, axial Young's modulus of the composites increased by 33 % for the case of long CNT fibers. The experimental results of CNT/polystyrene (PS) composites [10] concluded that with only 1 % (by weight) adding of CNTs in polystyrene, elastic modulus and breaking stress were observed to increase by 36–42 % and −25 %, respectively, indicating significant load transfer across the nanotube-matrix interface. The FE analysis has been performed [11] for calculation of Young's modulus and coefficient of thermal expansion (CTE) of CNT/Epoxy composites for different orientation of CNTs in matrix, and concluded that CNTs could be more efficient as compared to conventional glass or carbon fibers. A molecular dynamics (MD) simulation have been performed [12] for two different polymer matrices with different volume fraction and determined the axial and transverse elastic moduli using constant-strain energy minimization technique and reported that interfacial bonding effect is important.

The performance of a composite materials system is critically controlled by the interfacial characteristics of the reinforcement and the matrix material. The interfacial characteristics of CNT-reinforced composite system have been reported in some earlier works [13–15] and suggested that the interfacial shear stress of the CNT-based composite was observed to be significantly higher than that of most carbon fiber reinforced composite systems. The shear stress transfer ability of MWCNT/polymer interfaces to be more than that of current advanced composites and is in the order of 500 MPa [15]. A pull out process in a SWNT/PS system have been done using MD dynamics simulation and reported that the interfacial bond strength could be up to 160 MPa [14], even without considering the chemical bond between CNT and matrix. From pullout experiments it has been found that the interfacial shear stress between a MWCNT and epoxy ranged from 35–375 MPa [13]. The effect of tube length and diameter on the distributions of tensile stress and interfacial shear stress of SWCNT in epoxy matrix has been investigated [16]. They also found that a CNT has greater stress transfer efficiency than a solid fiber, providing flexibility for toughness and tensile strength optimization. An analytical model has been developed to study the stress transfer in SWCNT reinforced polymer matrix composites [17]. A micromechanical model has been developed for assessing the interfacial shear stress transfer in CNT-reinforced polymer (NRP) composites [18]. There are two earlier works [19, 20] have been investigated the CNT-based sensors and their potential applications, and measurements on carbon nanotubes demonstrate the onset of nanotube damage in ambient conditions, with accelerated rate at higher bias voltages. Theses two papers have been explained the causes of the defects and damages of the CNTs.

There are many literatures available in the broad area of CNT-based composites and most of these studies were related to the thermoelastic properties of nanocomposites and investigated the load-carrying capacities of CNTs in a matrix. There are some literatures also available describing the manufacturing processes like resin transfer molding, powder metallurgy method and sonication techniques commonly used for manufacturing of CNT/Epoxy as well as CNT/Titanium composites [21, 22]. The above literatures report that the load carrying capacity of CNTs in matrix is significant and CNT-based composites have the potential to provide extremely strong lightweight new materials. There are also some works available on the analysis of CNT-based composites using FEM. However, there is no work available on the study of stress distribution in the vicinity of a broken CNT in the CNT-based composites. The stress concentration at a single fiber break for a conventional fiber (i.e. not CNT) reinforced composites have been analyzed using FE method [23] but in the present analysis the conventional fiber is replaced by a CNT in CNT-based composites. Therefore the present paper aims at studying the normal and shear stress distribution surrounding the broken CNT in a CNT-based composite and to study the effect of different important parameters on the failure of such nanocomposites.

CNTs are three types viz. zigzag, armchair, and chiral. Left side of Fig. 1 is shown the formation of different types of CNTs from a hexagonal graphene sheet and the right side of Fig. 1 are shown the side views of a zigzag SWCNT (14, 0) and a armchair SWCNT (10, 10). The atomic structure of CNTs can be described by the tube chiral vector and chiral angle (θ). The chiral angle determines the amount of twist in the tube. If $\theta = 0^{\circ}$ and 30° it will form zigzag and armchair CNTs, respectively and for $0^{\circ} < \theta < 30^{\circ}$, it will form a chiral CNT. The chiral vector also known as the roll-up vector and it can be describe by the following equation

$$\mathbf{C}_h = n\mathbf{a}_1 + m\mathbf{a}_2 \tag{1}$$

where the integers (n, m) are the number of steps along the zigzag carbon bonds and $\mathbf{a}_1$ and $\mathbf{a}_2$ are unit vectors as shown in Fig. 1.

The circumference of the CNTs also determine by the following equation

$$L = |\mathbf{C}_h| = a\sqrt{n^2 + m^2 + nm} \tag{2}$$

where a is the length of unit vector.

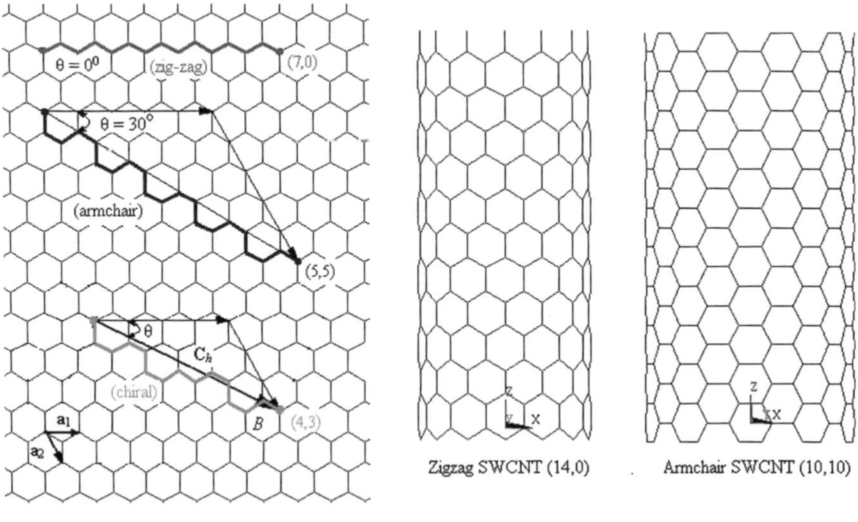

Fig. 1 Atomic structure of CNTs along with the side views of the FE meshes of a zigzag SWCNT (14, 0) and an armchair SWCNT (10, 10)

2 Modeling of CNT-Based Composites with a Broken CNT

2.1 Three Dimensional FE Model

The SWCNT is considered in the present work. It has been assumed that the CNTs and matrix in a RVE are linear, elastic, isotropic, and homogeneous materials. It has also been assumed that the CNTs and matrix are perfectively bonded with no slip at the interface in the RVE to be studied. The RVE has been modeled by taking two types of matrix materials (Epoxy and Titanium). In order to study the effect of a broken CNT on the adjacent CNTs as well as at the interface of the broken CNT and matrix, a square RVE with nine CNTs have been considered. The diameter of CNTs have been taken as 1.88 nm (which is equal to the diameter of zigzag (24, 0) CNTs), and the thickness (t) of CNT layer is considered 0.34 nm [4]. The volume fraction of the CNT in matrix of the square RVE is calculated by Chen and Liu [9]. Since the thickness of CNT is constant so the volume of the CNT is also constant but to get different volume fraction the volume of the matrix materials surrounding the CNT has been varied. In the present analysis for a 200 nm length of CNT is considered however, any other length of the CNT could also be taken in the FE model.

The broken fiber is considered to be placed at the centre of the RVE which is surrounded by the matrix materials. In the present work, it has been assumed that the orientation of the broken CNTs in the matrix is aligned and the corresponding RVE is used for the analysis. Since the RVE is modeled with 9-CNTs, so the broken fiber is surrounded by eight nearest fibers. SOLID45 elements embodied in ANSYS have been used for modeling the RVE. In fact mesh refinement has been done to capture the expected high stress gradient in the vicinity of the broken CNT. So a very fine mesh has been used near the broken CNT and as we moved away from the break along the length, coarse meshes were used for the FE analysis. As mentioned earlier, the volume fractions have been varied by varying the matrix volume surrounding a fixed size CNT; therefore while calculating the volume fractions based on the RVE in the range of 0.5 to 10.3, they were 0.5, 3.056 and 10.3. There was actually no basis for 3.056 %. It could have been any other in that range depending upon the size of RVE. The FE model of the square RVE with the front view and pictorial view are shown in Fig. 2a, b, respectively for a constant volume fraction 3.056 %.

2.2 Boundary Conditions

Figure 2a and b show a 3-D FE model for a square RVE with nine CNTs. The x–y plane is the transverse plane and the z-axis is the axial direction of the CNT. All the nodes at $z = 0$ are fully restrained except the nodes belonging to the

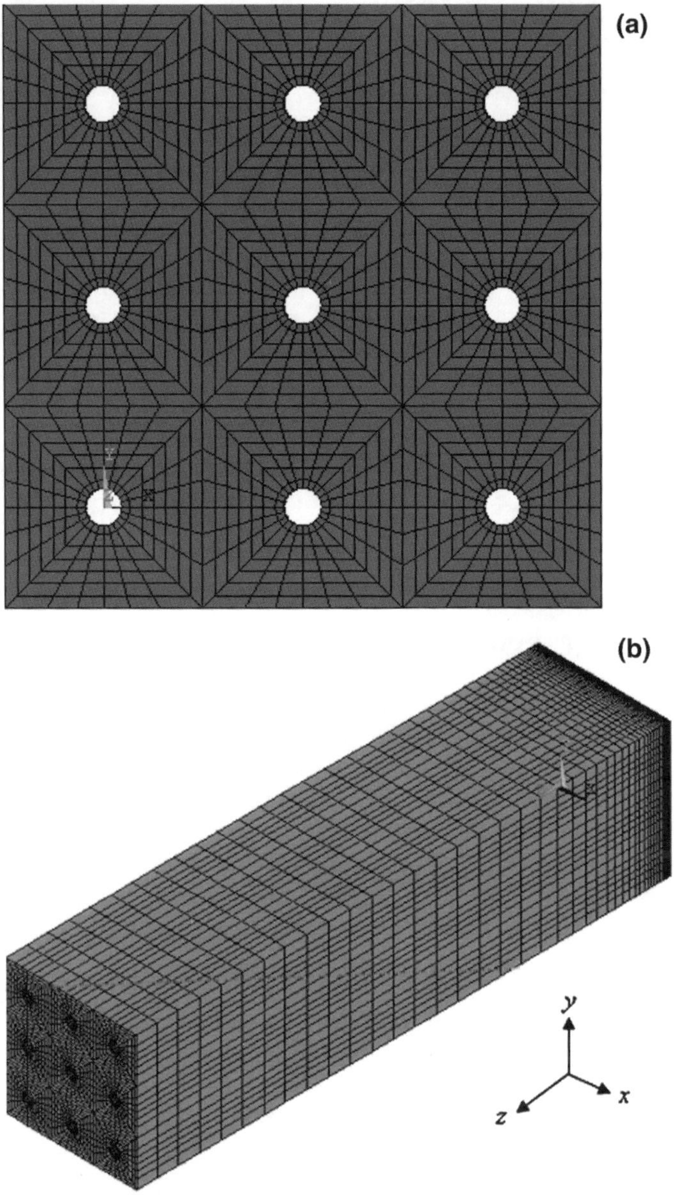

Fig. 2 A 3-D FE model for the square RVE with 9-CNTs **a** front view and **b** pictorial view

broken fiber. The axial mechanical loading in the nanocomposites is applied by applying a uniform axial displacement in the nodes at $z = L_a$. Here L_a is the axial length of the model.

2.3 Elastic and Mechanical Properties of the Constituents

In the present work, two different types of matrix materials (Epoxy and Titanium) have been chosen to study the effect of matrix materials on the CNT break in CNT-based composites. Properties of epoxy [24, 23] and the SWNT [3, 4] are as follows

SWNT $E_{nt} = 1{,}000$ GPa, $\nu_{nt} = 0.28$,
Epoxy $E_{me} = 3.89$ GPa, $\nu_{me} = 0.37$,
Titanium $E_{mt} = 116$ GPa, $\nu_{mt} = 0.32$

3 Results and Discussions

A Full 3-D finite element analysis has been performed to find out the stress distributions in the broken CNT, adjacent CNTs, and the interface of broken CNT and matrix. In the present study the axial normal stress in the broken CNT, interfacial shear stress at the interface of the broken CNT and matrix, and axial normal stress in the adjacent CNT of a square RVE with nine CNT have been studied at a constant volume fraction. The effect of volume fraction on the stress distribution has also been investigated. In the present analysis, stresses are calculated for the volume fractions of 0.5, 3.056, and 10.3 % of CNT in CNT-based composites.

In absence of any published results for a broken fiber CNT composites, qualitative trends of the stress distributions around the fiber break have been compared with those for a conventional graphite/epoxy composites by Nedele and Wisnom [23] and a good agreement has been observed.

3.1 Stress Distributions at a Constant Volume Fraction of 3.056 %

3.1.1 Axial Normal Stress in the Broken Fiber (i.e. CNT)

In the present analysis, a RVE with 9 CNTs has been considered with the central CNT broken at the mid span of its length. Therefore, in the FE analysis only half of the length has been modeled and the boundary conditions are put accordingly. Figure 3 shows the variation of the axial normal stress along the length of the broken fiber in the CNT-based composites and it could be observed that the broken fiber regains the nominal stress value after a length of 8 nm for CNT/Titanium and 50 nm for CNT/Epoxy composites. Therefore, it could be concluded that the ineffective length (where 95 % of the strength is regained) of the fiber in CNT/Titanium composites is 8 % but for CNT/Epoxy composites it is 50 % i.e. the

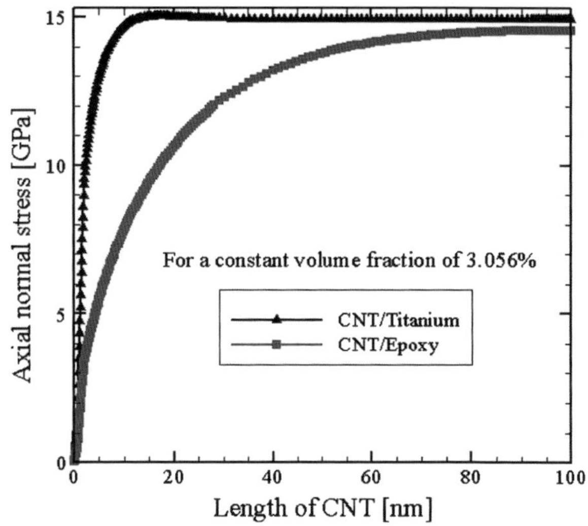

Fig. 3 Axial normal stress distribution in the broken fiber along the fiber length of CNT/Epoxy and CNT/Titanium nanocomposites for a constant volume fraction of 3.056 %

ineffective length of the CNT/Titanium composites is about 6 times less than the ineffective length of the CNT/Epoxy composites. The reason for this variation is the mismatch of the mechanical properties between the fiber and matrix.

3.1.2 Interfacial Shear Stress in the CNT-Matrix Interface

Figure 4 shows the interfacial shear stress distribution at the interface of the broken fiber and matrix for CNT/Epoxy and CNT/Titanium composites. Debonding/Delamination is cause by the interfacial shear stress and since near the broken CNT there has been a interfacial shear stress, therefore debonding may take place depending upon the magnitude of the interfacial shear stress and the corresponding strength. Magnitude of shear stress is higher in the case of CNT/Titanium composites compared to that in CNT/Epoxy composites.

3.1.3 Axial Normal Stress in the Adjacent CNTs

Figure 5 shows the variation of axial normal stress in the adjacent CNTs for CNT/Epoxy and CNT/Titanium nanocomposites. It could be observed that due to the presence of a broken CNT, the adjacent CNTs experience a stress concentration in the vicinity of the break. It could be seen from the Fig. 5 that the stress concentration factor (SCF) in the CNT/Titanium is 1.01 and that in CNT/Epoxy is 1.1. So, it could be concluded that there is a possibility that the adjacent CNTs may break as a consequence of over loading.

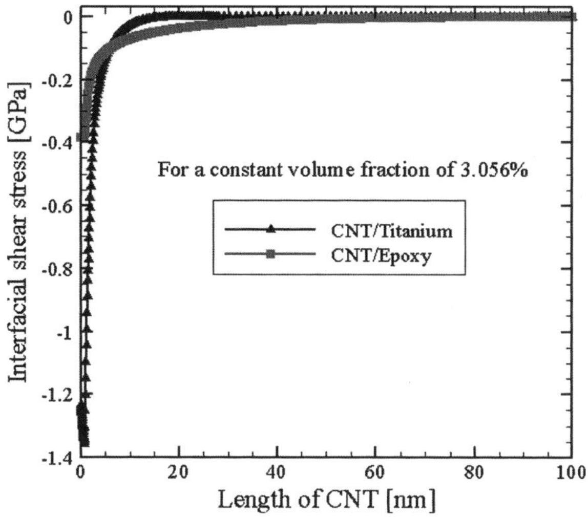

Fig. 4 Interfacial shear stress distribution at the interface of the broken fiber and matrix along the fiber length of CNT/Epoxy and CNT/Titanium nanocomposites for a constant volume fraction of 3.056 %

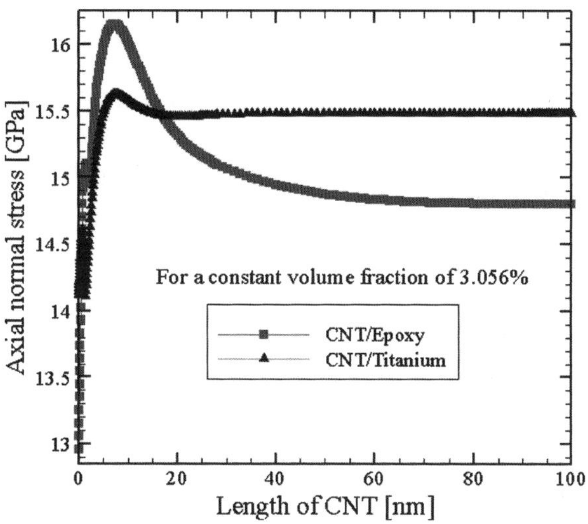

Fig. 5 Axial normal stress distribution in the adjacent fiber along the fiber length of CNT/Epoxy and CNT/Titanium nanocomposites for a constant volume fraction of 3.056 %

3.2 Effect of Volume Fraction

3.2.1 Variation of Axial Normal Stresses of the Broken CNT

Figure 6 shows the variation of the axial normal stress along the length of the broken fiber at different volume fractions for CNT/Epoxy and CNT/Titanium composites. It could be observed that the ineffective length of the broken fiber in CNT/Titanium composites is 8 nm for the volume fraction of 10.3 and 3.056 %. But 13 nm for the volume fraction of 0.5 %. However, there is not much

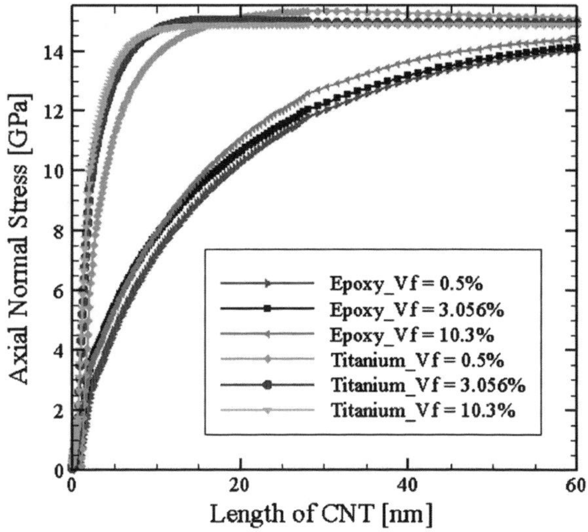

Fig. 6 Variation of axial normal stress distribution in the broken fiber along the fiber length of CNT/Epoxy and CNT/Titanium nanocomposites

difference in ineffective length for the volume fraction of 3.056 and 10.3 % of the CNT/Titanium composites. But in the case of CNT/Epoxy composites the ineffective length is 45 nm for the volume of 10.3 % but 50 nm for the volume fraction of 0.5 and 3.056 %. Therefore, in the case of CNT/Epoxy composites, for all the range of volume fractions, the broken CNT is ineffective and does not contribute to the load bearing of the composite. However, in the case of CNT/Titanium composites, a major length of the broken CNT contributes to the load bearing and the ineffective length of the broken CNT decreases with the increase in volume fraction.

3.2.2 Variation of Interfacial Shear Stress

Figure 7a, b show the variation of interfacial shear stress at the interface of the broken fiber and matrix for CNT/Epoxy and CNT/Titanium composites, respectively for different volume fractions. It could be seen that the magnitude of the interfacial shear stress increases with the increasing in volume fraction of CNT in CNT-based composites.

3.2.3 Variation of Axial Normal Stresses in the Adjacent CNTs

Figure 8a, b show the variation of axial normal stress for different volume fractions in the adjacent CNTs for CNT/Epoxy and CNT/Titanium composites, respectively. It could be observed that the axial normal stresses are different for different volume fractions. For CNT/Epoxy composites corresponding to the

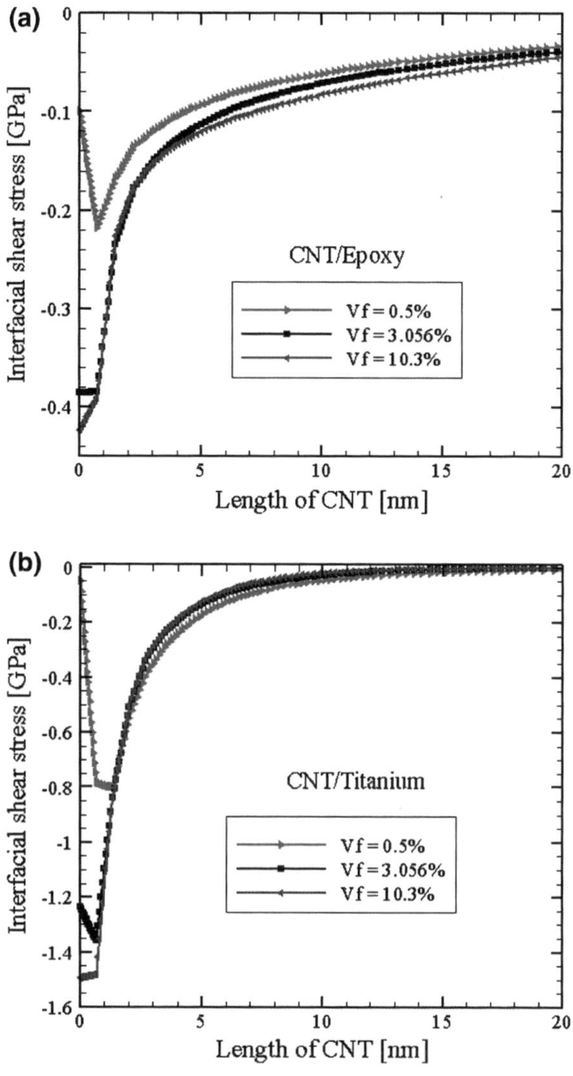

Fig. 7 Variation of interfacial shear stress distribution at the interface of the broken fiber and matrix along the fiber length **a** CNT/Epoxy and **b** CNT/Titanium nanocomposites

volume fractions of 0.5, 3.056, and 10.3 % the magnitudes of SCFs are 1.04, 1.09, and 1.11 respectively. But for CNT/Titanium composites corresponding to the volume fractions of 0.5, 3.056, and 10.3 the magnitude of SCFs are 1.003, 1.008, and 1.025 respectively. Above results conclude that the SCF increases with the increase in volume fraction for both the CNT/Epoxy and CNT/Titanium composites. It is also observed that for the same volume fraction, the SCF for CNT/Epoxy composites is greater than that the SCF for the CNT/Titanium composites.

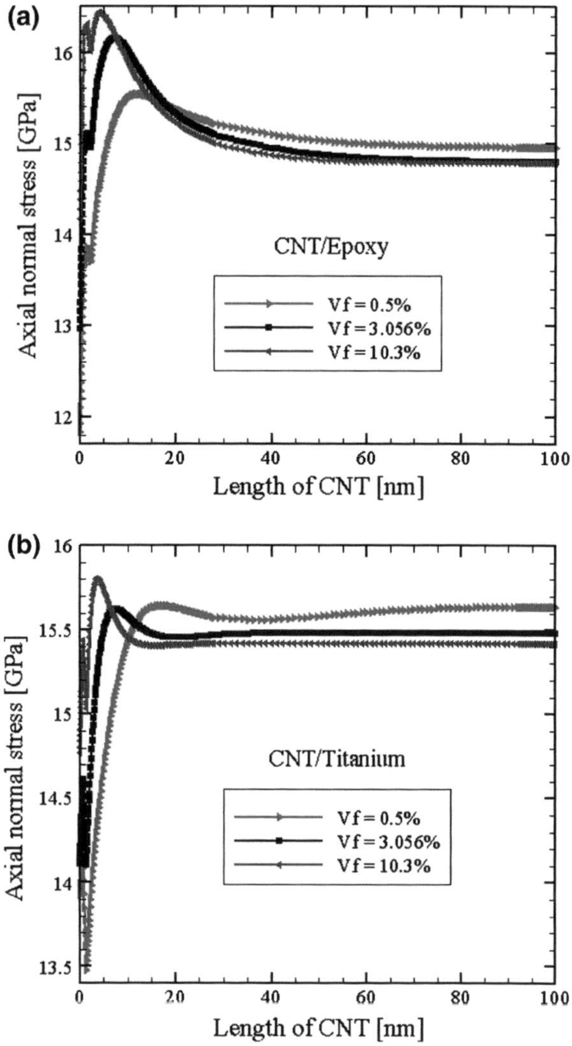

Fig. 8 Variation of axial normal stress distribution in the adjacent fiber along the fiber length **a** CNT/Epoxy and **b** CNT/Titanium nanocomposites

4 Conclusions

In the present work, the axial normal stress in the broken CNT, interfacial shear stress at the interface of the broken CNT and matrix, and axial normal stress in the adjacent CNTs for CNT-based nanocomposites has been determined using FE analysis and considering a square RVE with 9-CNTs. Two types of matrix materials have been chosen to investigate the effect of matrix material on the variations of axial normal stress as well as interfacial shear stress of the nanocomposites. The effect of volume fraction on the stress redistribution has also been investigated. Some of the important conclusions drawn from the above analysis for the CNT-based nanocomposites materials are.

- The ineffective length of the CNT in CNT/Titanium composites is 8 % but in CNT/Epoxy nanocomposites it is 50 % for a constant volume fraction of 3.056 %.
- Interface between the broken CNT and the matrix experience a high shear stress that might lead to debonding of the CNTs.
- Adjacent CNTs in CNT-based composites experience a SCF, which might lead to breaking of the adjacent CNTs, leading to successive failure of CNTs.
- With the increase in volume fraction the ineffective length decreases for both the CNT-based composites considered.
- The magnitude of the interfacial shear stress at the interface increases with the increasing in volume fraction of CNT in the CNT-based composites.
- The SCF in the adjacent CNTs increases with the increase in volume fraction for both the CNT-based composites considered.
- For the same volume fraction the SCF of CNT/Epoxy composites is greater than that in the CNT/Titanium composites.

References

1. S. Iijima, Helical microtubules of graphitic carbon. Nature **354**, 56 (1991)
2. S. Kirtania, D. Chakraborty, Finite element based characterization of carbon nanotubes. J. Reinf. Plast. Compos. **40**, 1557 (2007)
3. C. Li, T.W. Chou, A structural mechanics approach for the analysis of carbon nanotubes. Int. J. Solids Struct. **40**, 2487 (2003)
4. J.P. Lu, Elastic properties of carbon nanotubes and nanoropes. Phys. Rev. Lett. **79**, 1297 (1997)
5. K.T. Lau, D. Hui, The revolutionary creation of new advanced materials—carbon nanotube composites. Compos. Part B **33**, 263 (2002)
6. E.T. Thostenson, Z. Ren, T.W. Chou, Advance in the science and technology of carbon nanotubes and their composites: a review. Compos. Sci. Tech. **61**, 1899 (2001)
7. G.M. Odegard, T.S. Gates, L.M. Nicholson, K.E. Wise, Equivalent-continuum modeling of nano-structured materials. Compos. Sci. Tech. **62**, 1869 (2002)
8. K.T. Lau, D. Hui, Effectiveness of using carbon nanotubes as nano-reinforcements for advanced composites structures. Carbon **40**, 1605 (2002)
9. X.L. Chen, Y.J. Liu, Square representative volume elements for evaluating the effective material properties of carbon nanotube-based composites. Comput. Mater. Sci. **29**, 1 (2004)
10. D. Qian, E.C. Dickey, Load transfer and deformation mechanisms in carbon nanotube-polystyrene composites. Appl. Phys. Lett. **76**, 2868 (2000)
11. H.R. Lusti, A.A. Gusev, Finite element predictions for the thermoelastic properties of nanotube reinforced polymers. Model. Simul. Mater. Sci. Eng. **12**, S107 (2004)
12. Y. Han, J. Elliott, Molecular dynamics simulations of the elastic properties of polymer/carbon nanotube composites. Comput. Mater. Sci. **39**, 315 (2007)
13. C.A. Cooper, S.R. Cohen, A.H. Barber, H.D. Wagner, Detachment of nanotubes from a polymer matrix. Appl. Phys. Lett. **81**, 3873 (2002)

14. K. Liao, S. Li, Interfacial characteristic of a carbon nanotubes-polystyrene composites system. Appl. Phys. Lett. **79**, 4225 (2001)
15. H.D. Wagner, O. Lourie, Y. Feldman, R. Tenne, Stress-induced fragmentation of multiwall carbon nanotubes in a polymer matrix. Appl. Phys. Lett. **72**, 188 (1998)
16. K.Q. Xiao, L.C. Zhang, The stress transfer efficiency of a single-walled carbon nanotube in epoxy matrix. J. Mater. Sci. **39**, 4481 (2004)
17. A. Haque, A. Ramasetty, Theoretical study of stress transfer in carbon nanotube reinforced poly matrix composites. Compos. Struct. **71**, 68 (2005)
18. K. Li, S. Saigal, Micromechanical modeling of stress transfer in carbon nanotube-reinforced polymer composites. Mater. Sci. Eng. A **457**, 44 (2007)
19. K. Mølhave, S.B. Gudnason, A.T. Pedersen, C.H. Clausen, A. Horsewell, P. Bøggild, Transmission electron microscopy study of individual carbon nanotube breakdown caused by Joule Heating in Air. Nano Lett. **6**(8), 1663 (2006)
20. N. Sinha, J. Ma, J.T.W. Yeow, Carbon nanotube-based sensors. J. Nanosci. Nanotechnol. **6**, 573 (2006)
21. Q.F. Cheng, J.P. Wang, J.J. Wen, C.H. Liu, K.L. Jiang, Q.Q. Li, S.S. Fan, Carbon nanotube/epoxy composites fabricated by resin transfer molding. Carbon **48**, 260 (2010)
22. T. Kuzumaki, O. Ujiie, H. Ichinose, K. Ito, Mechanical characteristics and preparation of carbon nanotubes fiber-reinforced Ti composite. Adv. Engg. Mater. **2**(7), 416 (2000)
23. M.R. Nedele, M.R. Wisnom, Three-dimensional finite element analysis of the stress concentration at a single fiber breaks. Compos. Sci. Tech. **51**, 517 (1994)
24. M. Meyyappan, *Carbon Nanotubes Science and Applications* (CRC Press LLC, Boca Raton, 2005)

Microscopic Analysis of Mechanical Properties of Aligned Carbon Nanotube/ Epoxy Composite

S. Bal, J. P. Borah and C. Borgohain

Abstract This study presents a process of manufacturing and characterization of Aligned Multiwall Carbon Nanotubes (ACNTs) reinforced epoxy composites along with their microscopic analysis. Sonication method was used to prepare the composite plates with different conditions of curing along with various wt %. Mechanical characterization such as elastic modulus, stress–strain behaviour, hardness and microscopic characterization which include Scanning Electron Microscopy (SEM) and Raman spectroscopy were carried out. Nanoindentation method also had been adopted to obtain some mechanical parameters. The influence of aligned CNTs dispersion on the mechanical properties of nanocomposites have been evaluated through comparison and analysis of stress–strain curve of samples cured under room temperature and refrigerated conditions. With increase in wt % of CNTs, the modulus in latter case were increasing thus by making stronger samples. The aligned CNTs were observed to be more strongly bonded to the epoxy matrix in above samples along with bridging mechanism. Raman results also supplemented the above by showing increase in intensity and shifting of G band in case of refrigerated samples that confirms better reinforcement and stress transfer from matrix to nanotubes.

Keywords Nanocomposite • Epoxy matrix • Multiwall carbon nanotube • Young modulus • Raman spectroscopy

S. Bal (✉)
Deaprtment of Physics, NIT Silchar, Assam 788010, India
e-mail: smrutisikha_bal@yahoo.com

J. P. Borah
Department of Physics, NIT Nagaland, Nagaland, India

C. Borgohain
Deprtment of Instrumentation and USIC, Gauhati University, Assam, India

P. K. Giri et al. (eds.), *Advanced Nanomaterials and Nanotechnology*, Springer Proceedings in Physics 143, DOI: 10.1007/978-3-642-34216-5_35,

1 Introduction

Carbon nanotube (CNT) reinforced polymer composites have become attractive structural materials not only in the weight-sensitive aerospace industry, but also in the marine, armor, automobile, railway, civil engineering structures, and sporting goods industries because of their high specific strength and specific stiffness [1]. Considering excellent mechanical, electrical and thermal property of CNTs, different nanocomposites have been synthesized by incorporating them into various polymer matrices [2–4]. Epoxy resin is most often used as the polymer matrix with reinforcing nanotubes for advanced composite applications. The resins of this class have good stiffness, specific strength, dimensional stability and chemical resistance, and show considerable adhesion to the embedded fiber [5]. Because micro-scale fillers have successfully been synthesized with epoxy resin [6], nanoparticles, nanotubes, and nanofibers are now being tested as filler material to produce high performance composite structures with enhanced properties [7–9].

The reported studies on CNT/polymer nanocomposites [10–12] have shown that randomly oriented CNTs embedded in polymer matrices have failed to generate composites in which the full potential of superior properties of the CNTs can be exploited [13]. The final composite properties hinge on variables such as CNT dispersion, concentration, aspect ratio and orientation. A homogeneous dispersion of the CNTs in the polymer matrix is essential to obtain uniform properties and efficient load transfer during most applications. Good dispersion is usually hindered by the tendency of CNTs to aggregate as a result of Van der Waals attractions. The degree of alignment of the CNTs has a profound effect on the mechanical properties especially when the composite is loaded parallel or perpendicular to the CNT orientation direction. Many studies have been conducted on multiwalled carbon tube (MWNT) based composites earlier and recently on aligned carbon tube (ACNT) composites. MWNT composites have shown improved result in tensile modulus and yield strength [14]. However some studies reported decrease in flexural strength and pull out of CNTs in those composites [15]. In order to detect orientation and deformation of the CNTs in the nanocomposites, the tensile behaviour of both random and aligned MWNTs/Polystyrene nanocomposites was investigated by Thostenson and Chou [16]. They found the aligned CNTs composites showed more improved mechanical properties than random CNTs composites.

The present work aims to study the influence of reinforcing strategies of nanocomposites of different wt % (up to a maximum 0.5 %) of aligned CNTs in the epoxy matrix. To optimize reinforcement of these nanofillers in matrix, a new approach in the form of curing of composite specimens at refrigerated temperature was adopted. Moreover, Raman study supplementing the mechanical properties of aligned CNT/epoxy nanocomposites is yet to be reported in the literature. Macroscopic and microscopic properties of refrigerated nanocomposites have been analyzed and compared with room temperature cured samples. Investigating dispersion strategies and final properties of these nanocomposites will thus help to elucidate the mechanism favouring or hindering synthesis of superior novel materials.

2 Experimental Methods

2.1 Materials

Epoxy polymer matrix was prepared by mixing epoxy resin (Ciba-Geigy, araldite LY-556 based on Bisphenol A) and hardener HY-951 (aliphatic primary amine) in wt. ratio 100/12. Epoxy resin (5.3–5.4 equiv/kg) was of low processing viscosity and good overall mechanical properties. Aligned MultiwallNanotubes (ACNTs) used in this study for the preparation of nanocomposites were procured from Nanostructured & Amorphous Materials Inc. (NanoAmour), USA. They are 10–20 nm in diameter, 5–15 μm long and >95 % purity. TEM of nanotubes (Fig. 1a) reveal that they are partially aligned and entangled as well as multiwall in nature (Fig. 1b). Figure 2 presents Raman scattering spectra of ACNT where the peaks are observed in the regions of 1,000–4,000 cm^{-1}. The ratio (I_D/I_G) of the Raman intensities of the disordered band (D-band) at around 1,300 cm^{-1} to that of graphite band at around 1,600 cm cm^{-1} can be taken as a measure of the crystalline order in graphitic systems. Here, smaller ID/IG ratio suggests fewer defects, less amorphous carbon and higher graphitic order [17, 18].

2.2 ACNT/epoxy Composite Preparation

In the present work nanocomposites were prepared by taking very low loading of aligned CNTs (0.01–0.5 wt %) in the epoxy resin. The ACNTs were first dispersed in ethanol solution under manual stirring in order to optimise the dispersion. After complete evaporation of ethanol, the obtained ACNT powder was added to the epoxy resin and sonicated for 2 h at room temperature. Further, hardener was added to this epoxy/ACNT mixture and sonicated for 15 min. Then one part of the mixture was allowed to set in room temperature and another part was set under refrigeration condition. Refrigeration facilitates in preparing ductile samples by increasing settling time of the mixture in comparison to room temperature. All composite samples were cured for 5 h at 80 °C in an oven. The nomenclature of the composite samples is given in Table 1.

2.3 Mechanical Testing

2.3.1 Tensile Test

Five rectangular specimens from each sample were taken for three-point bend test as per ASTM D638. The tests were carried out using DEBEN MICROTEST 5KN tensile tester at ambient temperature keeping the cross-head speed 0.1 mm/min.

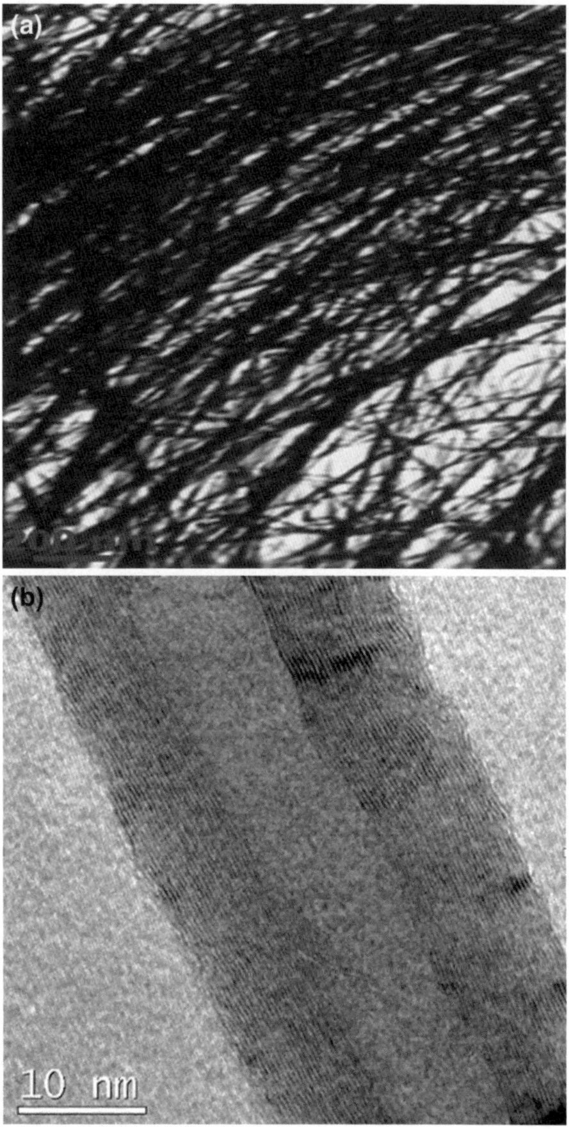

Fig. 1 TEM of Aligned CNT showing their **a** entanglement, **b** multiwall nature

Young modulus, tensile strength & ultimate strain of each sample were determined from the average value of five specimens.

2.3.2 Hardness Test

The hardness of all composite specimens was measured using a micro-hardness tester (Buehler 1600). A total of 10 points on each of the specimens were measured

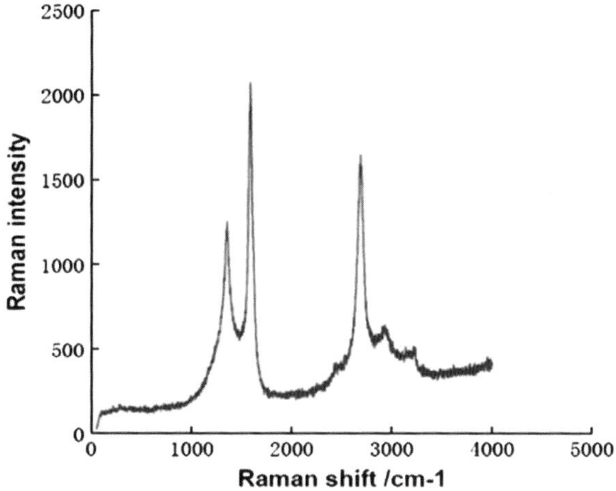

Fig. 2 Raman spectra of neat aligned carbon nanotube

Table 1 Identification of composite specimens

Sample description	Sample name (Cured at room temperature)	Sample name (Cured at refrigerated temperature)
Nanocomposite with 0.01 % of Aligned carbon nanotube	ACNT0.01	RACNT0.01
Nanocomposite with 0.1 % of Aligned carbon nanotube	ACNT0.1	RACNT0.1
Nanocomposite with 0.25 % of Aligned carbon nanotube	ACNT0.25	RACNT0.25
Nanocomposite with 0.5 % of Aligned carbon nanotube	ACNT0.5	RACNT0.5

in order to get average readings. The unit and magnitude of the hardness are defined by Vickers hardness and determined by measuring the average diagonal length of the indentation. The Vickers hardness test method [19] consists of indenting the test material with a diamond indenter as shown in Fig. 3. Vickers hardness (HV) was calculated using Eq. (1)

$$HV = \frac{2F\mathrm{Sin}\frac{136^0}{2}}{d^2} \text{ or } HV = 1.854\frac{F}{d^2}\,\text{(approximately)} \tag{1}$$

where F = load in kgf and d = arithmetic mean of the two diagonals, d1 and d2 in mm.

2.3.3 Nanoindentation

Because of the sub-nanometer (<20 nm) scale of the aligned CNTs, the mechanical characterization of the nanocomposites were employed using a

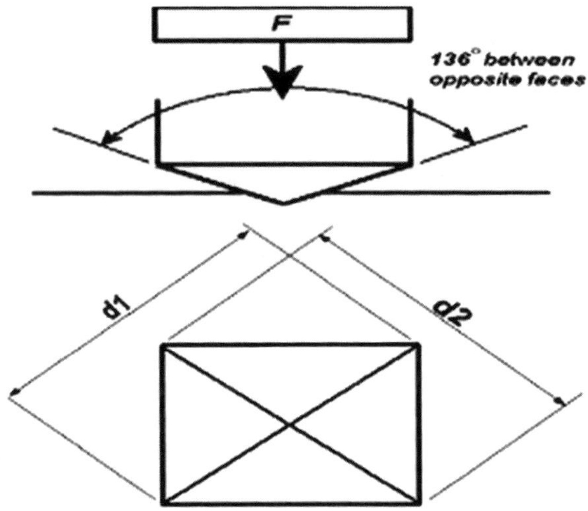

Fig. 3 Indentation geometry of hardness test

nanoindenter for providing high strain and displacement resolution. The nano-compression tests were performed using a CETR UNMT-1 nanoindenter. The nanoindenter monitors and records the load and displacement of the indenter with a force resolution of 10 mN and a displacement resolution of about 0.1 μm. By using a Berkovitch diamond tip the compression tests were performed inside the thermal and vibration insulated environmental chamber of the nanoindenter at room temperature (25 ± 0.5) °C and at a relative humidity of 45 ± 2 % on nanocomposite pillars.

2.4 Raman Spectroscopy

All Raman scattering spectra were recorded by using a Renishaw plc in via MicroRaman spectrometer equipped with 514 nm green laser having 2 cm^{-1} spectral resolution of Raman shift, X–Y step resolution of 0.1 lm and confocal resolution of 2.5 lm.

2.5 Electron Microscope (SEM and TEM) Study

Scanning electron microscope (Jeol JSM-6480 LV) was used to conduct the fracture surface topography characterization and dispersion behaviour. After mechanical test fracture surfaces were coated with a thin platinum layer. Cross section of neat ACNT sample was analyzed using Transmission Electron Microscope (FEI, TECHNAI G2).

3 Results and Discussion

3.1 Tensile Characterization

Figure 4 a, b shows the tensile stress–strain curves of room temperature (ACNT) and refrigerated nanocomposites (RACNT) with different carbon nanotube contents. All specimens fail immediately after the tensile stress reached the maximum value; however the stress–strain curves showed considerable non-linearity before reaching the maximum stress. The Young modulus and tensile strength of respective ACNT and RACNT samples with different wt % are presented in Tables 2, 3 respectively. It can be observed from both figure and table that the modulus of the nanophased epoxies increased continuously with increasing CNT content. It had already been reported recently that the mechanical properties of epoxy matrix increases after infusion of CNTs [20]. So in the present work we have mainly focused on comparison of properties between room temperature and refrigerated nanocomposites.

For the room temperature nanocomposites (Fig. 5a, Table 2), the Young modulus was almost same for ACNT0.01 (2867.5 MPa) and for ACNT0.1 (2879.5 MPa). With the increase of the ACNTs to 0.25 %, the tensile property was the best (4524.8 MPa, a rise scope of 58 %). However, it was observed that when the wt % was further increased to 0.5 %, there was a fall in the value of modulus (4371.7 MPa, a rise scope of 52 %) compared to the previous value. The cause of reduction was may be due to contraction of the matrix, which increased the clamping stress to the nanotube surface, and thus increased the frictional force between the nanotubes and the matrix. With the refrigerated nanocomposites (Fig. 5b, Table 3), with the content of ACNT to be 0.5 %, the Young modulus of the composite (RACNT0.01) increased significantly from 3107.9 to 7861.6 Mpa, a rise scope of 2.5 times. With the increase in CNTs loading, particle–particle interaction, particle–fluid interactions which includes orientation effect of CNTs also increase due to their cylindrical shape with very high aspect ratio. The relationship between tensile strength and CNT weight fraction was plotted in Fig. 6. As the CNTs loading increases, tensile strength of the refrigerated composites increases but that of room temperature composites shows variable behaviour. This indicates that the refrigerated nanocomposites are more efficient in transferring the applied load from matrix to aligned CNT in axial direction. Local stiffening due to nanotubes results in improved load transfer at the fibre/matrix interface [16]. It had been reported that the increase in elastic modulus between the random and aligned nanocomposite is a consequence of the nanotube orientation, not polymer chain orientation. However, the highest tensile strength for refrigerated samples was obtained with 0.25 wt % RACNT. At this loading, more than double enhancement in tensile strength (52.17 to 122.73 MPa) was observed. The tensile strength was decreased intensely in 0.5 wt % loading, although the gain in modulus was maintained. It may be due to the fact that the strain to failure was also reduced as the ACNT loading was increased from 0.25 to 0.5 wt %.

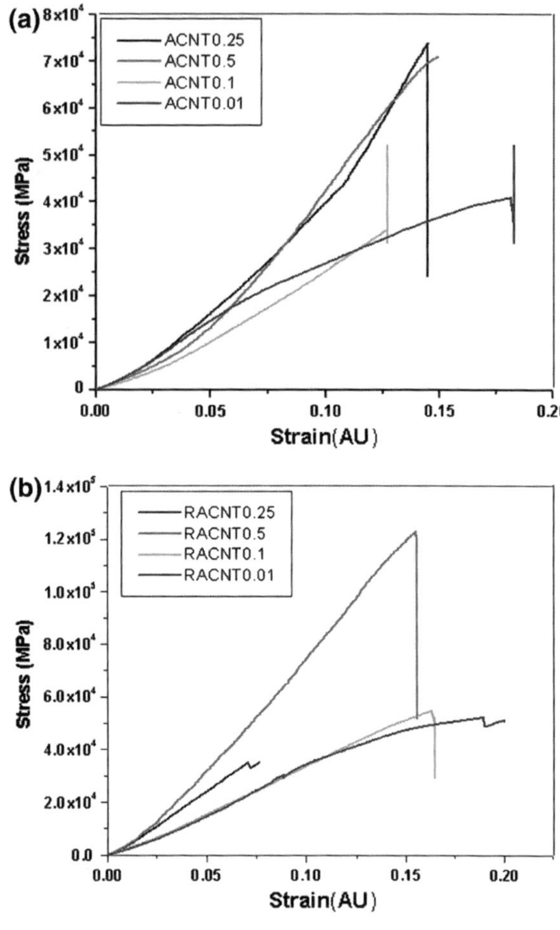

Fig. 4 Stress-strain curves of **a** ACNT composites, **b** RACNT composites

Table 2 Tensile strength and young modulus of ACNT samples

Sample	Strength (MPa)	Modulus (MPa)
ACNT0.01	40.81	2867.5
ACNT0.1	33.87	2879.5
ACNT0.25	73.55	4524.8
ACNT0.5	70.71	4371.7

Table 3 Tensile strength and young modulus of RACNT samples

Sample	Strength (MPa)	Modulus (MPa)
RACNT0.01	52.17	3107.9
RACNT0.1	54.40	3201.7
RACNT0.25	122.73	5836.5
RACNT0.5	35.07	7861.6

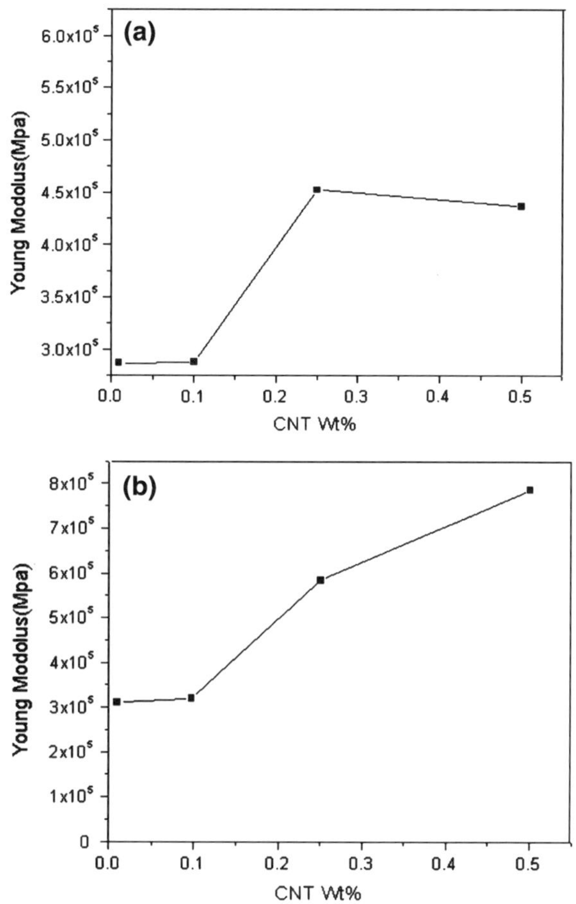

Fig. 5 Young modulus of composites of different wt %. **a** ACNT. **b** RACNT

3.2 Hardness of Nanotube Composites

The hardness of the refrigerated nanocomposites increased continuously with increasing ACNT content (Fig. 7b). The hardness of values were improved by 11 % with 0.1 wt %, 18 % with 0.25 wt % and 46 % with 0.5 wt % infusion of CNTs in comparison to 0.01 wt. %. The high aspect ratio, high modulus, strength of aligned CNTs contributed to the reinforcement thus by improving hardness value. However, the hardness value of room temperature samples are little lower than their refrigerated counterparts (Fig. 7a). Further, it was observed that RACNT0.1 and ACNT0.25 have the same hardness value of 32.3 MPa. These results indicated relatively better curing of epoxy resin at low temperature that allows more interaction with hardener. However, ACNT0.5 has lower value than ACNT0.25. In room temperature samples, voids may be created at the time of

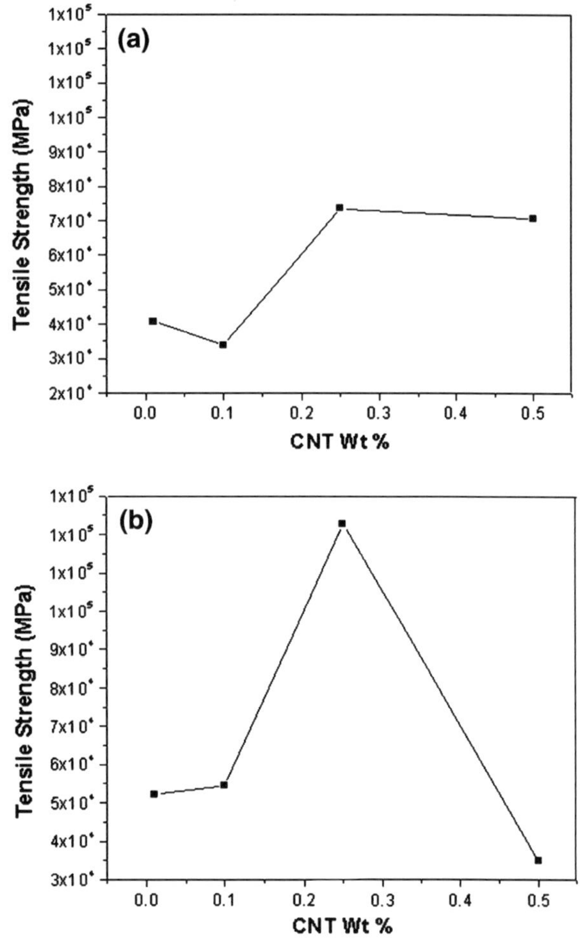

Fig. 6 Tensile strength of aligned CNT composites at different wt %. **a** ACNT. **b** RACNT

fabrication of composites and some amount of unreacted resin may be left due to faster curing that could be the reason for the reduced value of hardness. It was reported that few voids were produced during the fabrication process and that voids increased with the higher nanoparticle contents [21, 22]. Hardness results agreed well with tensile test results with the finding that refrigerated samples have improved mechanical properties than room temperature settled samples. This indicates variation in polymerization process under the action of refrigeration temperature. At low temperatures, polymeric chains get frozen and thermal vibrations and mechanical deformations (moduli) are controlled mainly by the weak interatomic bonding (Van derWaals) forces [23]. Therefore after infusing the dispersion of CNTs that restricts the mobility of polymer chains under loading improved the elastic modulus as well as hardness in refrigerated samples.

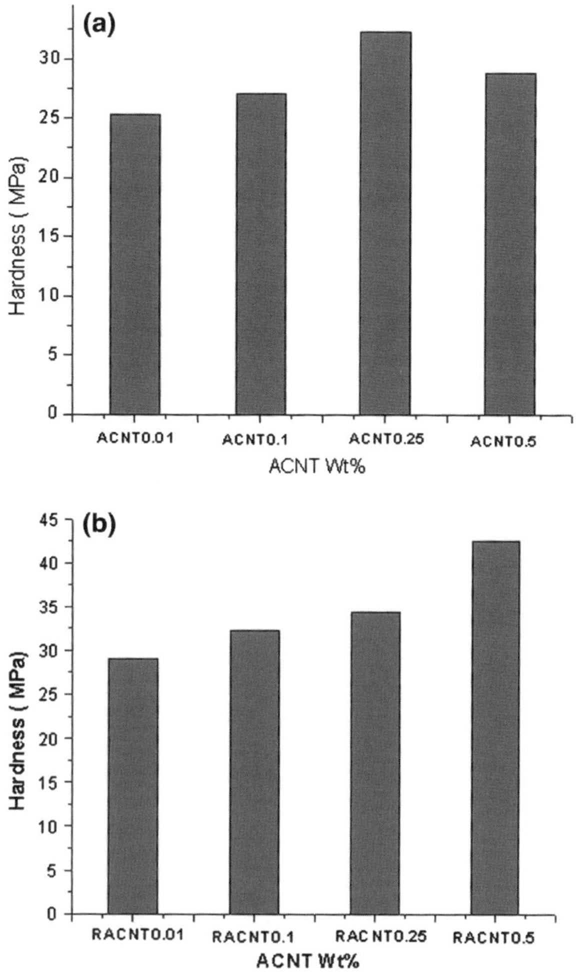

Fig. 7 Hardness values of nanocomposites of different wt %. **a** ACNT. **b** RACNT

3.3 Nanoindentation Result

The nanoindentation was chosen as method of investigation, because this method uses small pieces of sample material, and it does not require specially machined samples. The evaluation of each load curve delivered values of hardness, Young's modulus,and penetration– or contact–depth values. The load curve for ACNT0.25 and RACNT0.25 are given in Figs. 8 and 9 respectively. The corresponding values obtained are (ACNT: Young's modulus = 4.58 GPa, Stiffness = 31.76 mN/μm, Hardness = 0.29 GPa, depth = 1.168) and (RACNT: Young's modulus = 4.66 GPa, Stiffness = 34.53 mN/μm, Hardness = 0.26 GPa, depth = 1.249). The increase in values with penetration depths at low depths is caused by a surface effect in the sample [24]. High value of Young modulus and stiffness of RACNT indicate

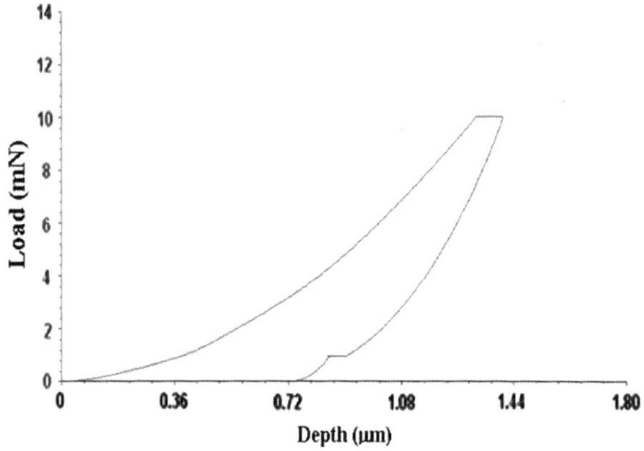

Fig. 8 Nanoindentation load curve for ACNT0.25

toughness of the material that supplements the tensile measurement. But hardness value of ACNT was found to be more than refrigerated sample that is not agreed with the microhardness test.

3.4 Raman Spectroscopy

Figure 10 a, b shows Raman scattering spectra of both types of nanocomposites in high frequency regions. A strong peak around 1,620 cm^{-1} can be assigned to the G-band due to tangential C–C bond stretching motions generically stem from the E_{2g2} mode at 1,580 cm^{-1} in graphite. The weak peak around 1,480 cm^{-1} was assigned to D-line of the graphite arising from disorder induced A (g1) mode and corresponds to the amorphous carbon [25]. The D band is activated in the first-order scattering process of sp^2 carbons by the presence of in-plane substitutional heteroatoms, vacancies, a grain boundary or other defects and by finite size effects, all of which lower the crystalline symmetry of the quasi-infinite lattice. The sharp peak observed for neat ACNTs (Fig. 2) decreases in intensity in all the composite

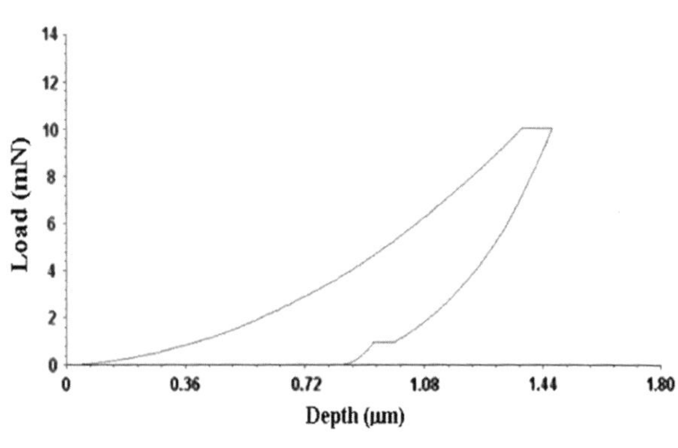

Fig. 9 Nanoindentation load curve for RACNT 0.25

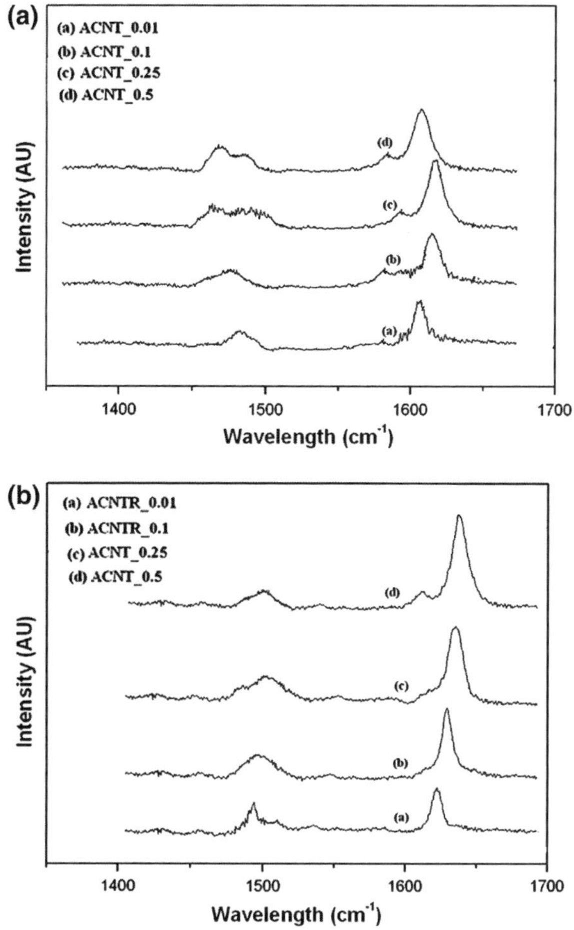

Fig. 10 Raman spectra of nanocomposites with different wt %. **a** ACNT. **b** RACNT

specimens, indicating a dilution effect of the CNTs when blended with epoxy. In case of room temperature samples (Fig. 10a), D-band was seen to increase with order ACNT0.25 < ACNT0.1 < ACNT0.01 i.e. maximum in ACNT0.01. This indicates more no. of carbon nanotubes are converted into amorphous carbons in these samples thus by increasing defect sites. In ACNT0.5, the D-band was observed to be deviate that needs further investigation. Here, the graphite like G modes exhibits a upward shift with increase in nanotube content from 0.01 % (1,613 cm^{-1}) to 0.1 % (1,626 cm^{-1}). However, decrease in intensity and shifting again to 1,612 cm^{-1} was seen in case of 0.5 %. In all refrigerated nanocomposites (Fig. 10b), smaller disorder D-band was observed incomparison to the room temperature ones. Increase in peak intensity and a little shift in G-band were found in refrigerated samples as infusion of CNTs increased. The feature around 1,611 cm^{-1} of RACNT0.01 is shifted to 1,647 cm^{-1} in case of RACNT0.5.

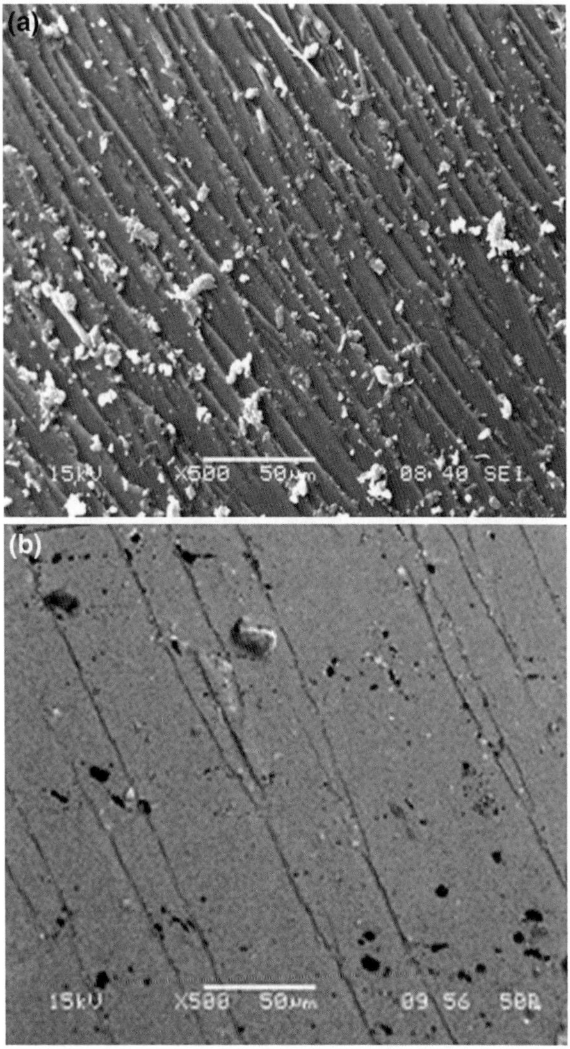

Fig. 11 Fracture surface of **a** ACNT0.5. **b** RACNT0.5

G-band shifting confirms good stress transfer from matrix to nanotube [26]. This result is supporting mechanical test where improvement in modulus was observed in refrigerated case. It implies that low temperature curing facilitates effective reinforcement of matrix and aligned nanotubes that may be possible due to stretching of network. As delay in the settling procedure in these has an impact on the matrix structure and crosslinking ratio and by this way the molecular motions inside the polymer.

3.5 Scanning Electron Microscopic Study

The investigation of the fractured surfaces to analyse the micro deformation and crack propagation mechanism has been carried out using Scanning electron microscope (SEM). The micrographs corresponding to nanocomposites obtained by incorporating ACNT0.5 and RACNT0.5 are presented. It is clearly observed that fracture surface of epoxy/ACNT (Fig. 11a) appears to be rough than epoxy/ RACNT composites (Fig. 11b). The resultant failure mechanism of the epoxy/ ACNT interface was analysed by observing the crack propagation regions within the composite (Fig. 12a). Some nanotubes were observed to be pulled out which

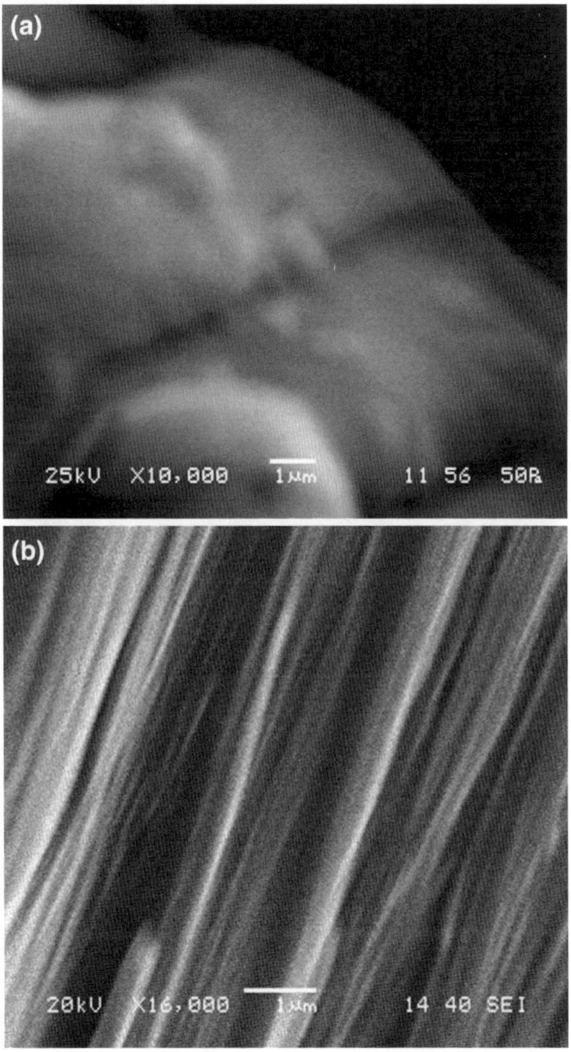

Fig. 12 **a** Nanotubes on ACNT surface. **b** Fracture surface showing crack features in ACNT

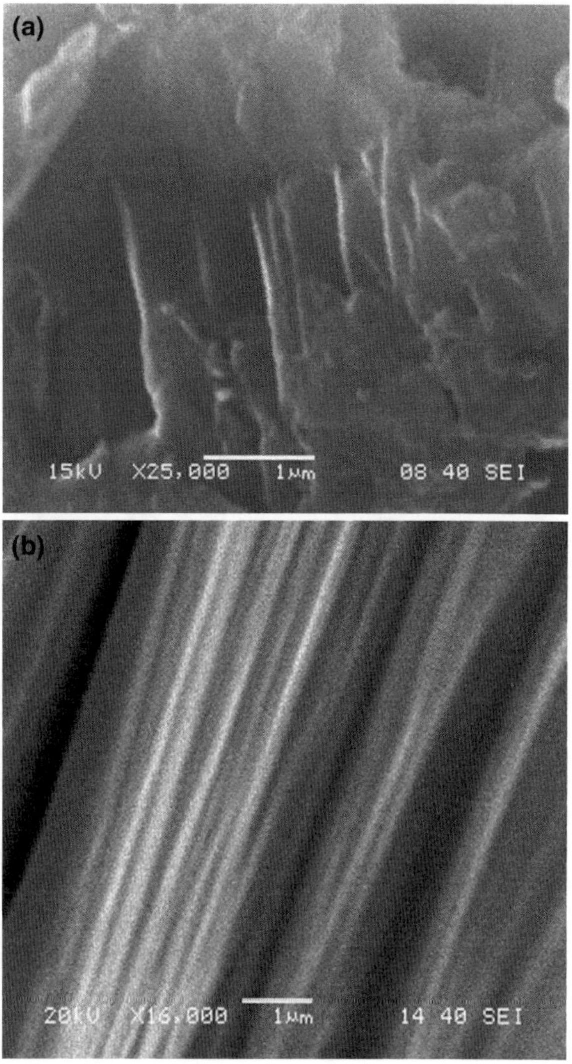

Fig. 13 **a** Bridging mechanism of nanotubes in RACNT composites. **b** crack features in RACNT composites

might be the result of a poor interfacial bonding between the nanotubes and matrix [27]. Therefore, the nanotubes inside the composites could not take up the load, which resulted in the decrease of tensile modulus of the nanotube composite beams [28]. Higher magnification showed crack though the composite surface with smaller crack lengths in vertical direction as well as breaking of nanotubes (Fig. 12b). In comparison to that in case of epoxy/RACNT composites (Fig. 13a), extension and bridging of aligned CNTs across the crack was seen. In these composites, the cracks were spanned by the nanotubes causing enhanced resistance to the crack propagation process (Fig. 13b). The bridging of the nanotubes as a mechanism of inhibiting the crack initiation in polymer and ceramic based

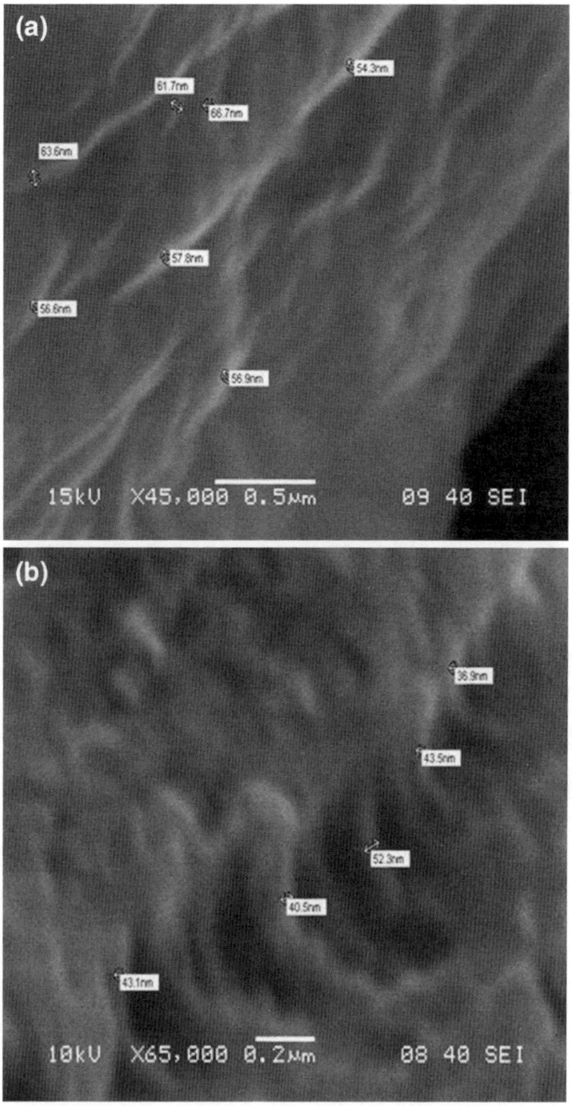

Fig. 14 Dispersion behavior of nanotubes in composites. **a** ACNT. **b** RACNT

nanocomposites has been well illustrated in literature [29, 30]. Figure 14 demonstrates the well dispersion behavior of aligned nanotubes in both composites. Nanotube diameters showed greater value than as received confirms good wetting of polymer around them. However, refrigerated sample (Fig. 14b) revealed better reinforcement between nanotube and matrix than room temperature sample (Fig. 14a). A more thorough study on the evaluation of surface roughness and the toughening mechanisms is necessary and should be further investigated in future works.

4 Conclusion

Addition of very low (up to 0.5 wt %) amount of aligned CNTs brought improvement in mechanical properties of polymer composites. Further, measurements for the Young's modulus of aligned CNT–polymer nanocomposites showed that the low temperature curing process led to a significant improvement. This facilitated better dispersion by optimizing adhesion between epoxy and aligned CNT confirmed from Raman studies as well as SEM. Hardness value and Nano-indentation analysis of the resulting composites showed that loadings as low as 0.25 % of ACNT are effective reinforcing agents, yielding tougher composite.

References

1. E.W. Wong, P.E. Sheehan, C.M. Lieber, Nanobeam mechanics: elasticity, strength, and toughness of nanorods and nanotubes. Science **277**, 1971 (1997)
2. C.G. Zhao, G.J. Hu, R. Justice, D.W. Schaefer, S. Zhang, M.S. Yang, C.C. Han, Synthesis and characterization of multi-walled carbon nanotubes reinforced polyamide 6 via in situ polymerization. Polymer **46**, 5125 (2005)
3. S. Kim, T.W. Pechar, E. Marand, Poly(imide siloxane) and carbon nanotube mixed matrix membranes for gas separation. Desalination **192**, 330 (2006)
4. F.H. Gojny, J. Nastalczyk, Z. Roslaniec, K. Schulte, Surface modified multi-walled carbon nanotubes in CNT/epoxy-composites. Chem. Phys. Lett. **370**, 820 (2003)
5. J.B. Donnet, Nano and microcomposites of polymers elastomers and their reinforcement. Compos. Sci. Technol. **63**, 1085 (2003)
6. H. Mahfuz, A. Adnan, V.K. Rangari, S. Jeelani, B.Z. Jang, Carbonnanoparticles/whiskers reinforced composites and their tensile response. Compos. Part A: Appl. Sci. Manuf. **35**, 519 (2004)
7. Y.H. Liao, M.T. Olivier, Z.Y. Liang, C. Zhang, B. Wang, Investigation of the dispersion process of SWNTs/SC-15 epoxy resin nanocomposites. Mater. Sci. Eng. A **385**, 175 (2004)
8. P. Farhana, Y.X. Zhou, V. Rangari, S. Jeelani, Testing and evaluation on the thermal and mechanical properties of carbon nano fiber reinforced SC-15 epoxy. Mater. Sci. Eng. A **405**(1–2), 246 (2005)
9. S. Bal, Experimental study of mechanical and electrical properties of carbon \nanofiber/ epoxy composites. Mater. Des. **31**, 2406 (2010)
10. M. Moniruzzaman, K.I. Winey, Polymer nanocomposites containing carbon Nanotubes. Macromolecules **39**(16), 5194 (2006)
11. N.R. Raravikar, L.S. Schadler, A. Vijayaraghavan, Y. Zhao, B. Wei, P.M. Ajayan, Synthesis and characterization of thickness-aligned carbon nanotube–polymer composite films. Chem. Mater. **17**(5), 974 (2005)
12. J.N. Coleman, U. Khan, Y.K. Gun'ko, Mechanical reinforcement of polymers using carbon nanotubes. Adv. Mater. **18**(6), 689 (2006)
13. R.H. Baughman, A.A. Zakhidov, W.A. de Heer, Carbon nanotubes–the route toward applications. Science **297**(5582), 787 (2002)
14. A. Allaoui, S. Bai, H.M. Cheng, J.B. Bai, Mechanical and electrical properties of MWNT/ epoxy composite. Compos. Sci. Tech. **62**, 1993 (2002)
15. K. Lau, S. Shi, H. Cheng, Micro-mechanical properties and morphological observation on fracture surfaces of carbon nanotube composites pre-treated at different temperatures. Compos. Sci. Technol. **63**, 1161 (2003)

16. E.T. Thostenson, T.W. Chou, On the elastic properties of carbon-nanotube based composites: modeling and characterization. J. Phys. D: Appl. Phys. **36**, 573 (2003)
17. H. Darmstadt, L. Summchen, J.M. Ting, U. Roland, S. Kaliaguine, C. Roy, Effects of surface treatment on the bulk chemistry and structure of vapor grown carbon fibers. Carbon **35**(1581), 10–11 (1997)
18. M. Endo, Y.A. Kim, T. Takeda, S.H. Hong, T. Matusita, T. Hayashi, Structural characterization of carbon nanofibers obtained by hydrocarbon pyrolysis. Carbon **39**(13), 2003 (2001)
19. R.L. Smith, G.E. Sandland, An accurate method of determining the hardness of metals with particular reference to those of a high degree of hardness. Proc. Inst. Mech. Eng. **1**, 623 (1922)
20. S. Bal, S.S. Samal, U.K. Mohanty, Mechanical and microstructural analysis of carbon nanotube composites pretreated at different temperatures. Am. J. Mater. Sci. **1**(1), 1 (2011)
21. Y.K. Choi, K.I. Sugimoto, S.M. Song, Y. Gotoh, Y. Ohkoshi, M. Endo, Mechanical and physical properties of epoxy composites reinforced by vapor grown carbon nanofibers. Carbon **43**, 2199 (2005)
22. F.H. Gojny, K. Schulte, Functionalisation effect on the thermo-mechanical behaviour of multi-wall carbon nanotube/epoxy-composites. Compos. Sci. Technol. **64**, 2303 (2004)
23. G. Hartwig, *Polymer Properties at Room and Cryogenic Temperatures* (Plenum Publishing Corporation, New York, 1994)
24. P.M. Nagy, D. Aranyi, P. Horváth, P. Pötschke, S. Pegel, E. Kálmán, Nanoindentation investigation of carbon nanotube–polymer composites. Internet Electron. J. Mol. Des. **5**, 135 (2006)
25. R. Saito, G. Dresselhaus, M.S. Dresselhauss, *Physical Properties of Carbon Nanotubes* (Imperial College Press, London, 1998)
26. L. Bokobza, Multiwall carbon nanotube elastomeric composites: a review. Polymer **48**, 4907 (2007)
27. Y.X. Zhou, F. Pervin, S. Jeelani, P.K. Mallik, Improvement of mechanical properties of carbon fabric-epoxy composite using carbon nanofibers. J. Mater. Process. Technol. **198**, 445 (2008)
28. K.T. Lau, D. Hui, Effectiveness of using carbon nanotubes as nano-reinforcements for advanced composite structures. Lett. Carbon **40**, 1605 (2002)
29. P.M. Ajayan, L.S. Schadler, P.V. Braun, *Nanocomposite Science and Technology*, Wiley-VCH Verlag. (2003)
30. C.A. Cooper, S.R. Cohen, A.H. Barber, H.D. Wagner, Detachment of nanotubes from a polymer Matrix. Appl. Phys. Lett. **81**(20), 3873 (2002)

Novel Silicone Rubber/Layered Double Hydroxide Nanocomposite: Preparation and Characterization

Bratati Pradhan and Suneel Kumar Srivastava

Abstract Silicone rubber (SR)/stearate ion modified Mg–Al–LDH (St-LDH) with 1, 3, 5, 8 % loading are obtained by the solution intercalation and characterized. Transmission electron microscope (TEM) of 3 wt % filler loaded SR nanocomposite confirms the exfoliation of St-LDH layers (thickness:1–2 nm, size:75–100 nm). Tensile strength, elongation at break and Young's modulus of the corresponding nanocomposite are found to be maximum improved compared to neat SR by 97, 43 and 55 % respectively. SEM micrographs of the fractured surfaces show the uniform failure of the matrix without involving any crack initiation in neat SR, whereas roughly fractured surface with large number of microvoids are observed in the corresponding of SR nanocomposite. The improvement in the solvent-resistance properties of the SR/St-LDH nanocomposites is attributed to the excellent barrier properties of St-LDH layers and the exfoliated structure of St-LDH. The developed SR/St-LDH nanocomposites are expected to be a novel material with unique combination of mechanical and solvent resistance property.

Keywords Silicone rubber · Transmission electron microscopy · Mechanical property · Solvent resistant property

1 Introduction

The recent development of polymer/inorganic material nanocomposites has been fruitful for industrial applications because of their improved unique properties [1–5] (e.g., mechanical, thermal, and gas-barrier properties and reduced

B. Pradhan · S. K. Srivastava (✉)
Department of Chemistry, Indian Institute of Technology, Kharagpur 721302, India
e-mail: sunit@chem.iitkgp.ernet.in

P. K. Giri et al. (eds.), *Advanced Nanomaterials and Nanotechnology*,
Springer Proceedings in Physics 143, DOI: 10.1007/978-3-642-34216-5_36,

flammability etc.) in comparison to pristine polymer as well as the conventional microcomposites. In this regard, most research work has been focused on the naturally occurring clay minerals such as montmorillonite due to its high aspect ratio, high cation exchange capacity and for their low cost. In recent years, layered double hydroxide (LDH), known as anionic clays, have been replaced in place of cationic clay (layered silicate) due to easy control of the particle size and aspect ratio simply by changing the reaction conditions. The LDH structure consists of stacks of positively charged mixed metal oxide/hydroxide layers that require the presence of interlayer exchangeable anions to maintain overall charge neutrality. The general chemical formula:

$$\left[M^{2+}_{(1-x)}M^{3+}_{x}(OH)_2\right]^{x+}A^{m-}_{x/m}\cdot\, nH_2O$$

where, M^{2+} is divalent cation such as Mg^{2+}. Zn^{2+}etc., M^{3+} is trivalent cataion such as Al^{3+}, Cr^{3+} etc. and A is an anion of valence m (e,g. Cl^-, ${SO_4}^{2-}$ etc.). It is necessary for appropriate surface modification due to short interlayer distance (0.76 nm) and strong electrostatic interaction between LDH layers [6]. LDHs can be organomodified by the replacement of the interlayer anions, e.g. ${CO_3}^{2-}$, ${NO_3}^-$or Cl^- with organic anionic surfactants, e.g. sodium dodecyl sulphate, sodium dodecyl benzene sulphonate, sodium stearate which are packed in mono- and bi-layer form in between LDHs sheets increasing thereby the gallery height.

Silicone rubber (SR) is one of the most widely used synthetic rubber due to unique properties such as excellent weather ability, good chemical stability, oxidation resistance, thermal stability, low-temperature toughness, electrical-insulating properties, low surface energy, low toxicity, and high optical transparency. However, an unfilled silicone elastomer with poor mechanical properties and low thermal/electrical conductivity is demand to improvement these properties for better application. According to the available literature, it can be seen that different type of nanofillers, such as organomodified montmorillonite, hyperbranched organomontmorillonite, nanosilica carbon black, carbon nanotubes, graphite, and nanohydroxyapatite [7, 8] are used to fulfill this purpose. However, the development of SR nanocomposites with other available nanofillers must to be explored. Therefore, present work deals the formation of exfoliated silicone rubber SR/Mg–Al layered double hydroxide (LDH) nanocomposites by solution intercalation method using stearate ion exchanged layered double hydroxide (St-LDH) [9] and observed improved mechanical and solvent barrier properties.

2 Experimental Part

2.1 *Materials*

Commercially available SR (Baysilone U10, vinyl content 0.05 mmol/g), vinyl terminated, linear polydimethylsiloxane base polymer and Baysilone U cross-linking agent 430, polysiloxane containing silicone bonded hydrogen atoms, and Pt catalyst complex were supplied by Momentive Performance Materials, Bangalore, India. Ethynyl cyclohexanol (Inhibitor) was purchased from Sigma to Aldrich. $Mg(NO_3)_2\ 6H_2O$, Al $(NO_3)_3\ 9H_2O$, Na_2CO_3 and carbon tetrachloride (CCl_4) were purchased from E. Merck, India. Sodium hydroxide and stearic acid were obtained from Quest Chemicals, Kolkata, India and SRL, Mumbai, India respectively.

2.1.1 Preparation of Mg–Al LDH and Stearate Ion Modified LDH (St-LDH) and SR/St-LDH Nanocomposite

Mg–Al LDH precursors were prepared according to the co-precipitation method [2, 10]. Stearate ion modified LDH were prepared using earlier reported method [9, 11] where 5 g of calcined Mg–Al LDH (500 °C for 6 h) was dispersed in 120 mL of the standard sodium stearate solution (8×10^{-3} mol/dm^3), stirred for 12h at 80 °C, followed by further refluxing at 100 °C for 6h to yield a white powder of organophilic LDH (St-LDH). Then, the product was filtered and washed several times with distilled water to make it free from the surfactant. The absence of surfactant was ensured spectrophotometrically in the presence of methylene blue [12].

Neat SR and SR/St-LDH nanocomposites were prepared by the typical procedure as reported in our earlier work [9] where desired amount of St-LDH (1, 3, 5 and 8 wt %) in 25 mL of CCl_4 at 80 °C for 6 h was dispersed and added to the earlier prepared solution of SR (10 g) in 25 mL CCl_4. The resultant solution was stirred at 80 °C for 8 h followed by addition of an appropriate amount of catalyst, inhibitor and cross-linker at room temperature. Finally, the resultant mixture is caste to the Teflon Petri dish and subjected to cure at 160 °C for 15 min followed by post curing at 200 °C for 4 h in a hot air oven.

3 Characterisation

The changes in the gallery distance of St-LDH and its SR nanocomposites were measured with Wide angle X-ray diffraction (WAXD) at room temperature in the range of 3–40° on a PANalytical PW3040/60 X'Pert Pro (Holland) with Cu Ka radiation ($\lambda = 0.1542$ nm) at a scanning rate of 3/min. The d value was obtained from the Bragg's equation:

$$n\lambda = 2d \sin\theta \quad (1)$$

Where, n = 1, λ = wavelength of the X-ray used and θ = half the Bragg's angle. The crystallite size was calculated according to the Scherrer equation [13]:

$$L_{hkl} = \frac{K\lambda}{\beta \cos\theta} \quad (2)$$

where, L_{hkl} refers to the size of the crystallites at reflection of hkl, K is the Scherrer factor (0.9) and β is the full width at half maximum.

The nature of dispersion of the St-LDH particles in SR matrix was examined by JEOL 2100 200 kV transmission electron microscope (TEM). The samples for TEM analysis were prepared via the ultramicrotoming of the SR/St-LDH nanocomposites in the form of 80–100 nm-thick slices cut at 150 °C with a diamond knife. The mechanical properties were measured on Tinius Olsen h10KS, ASTMD 412-98 at 25 °C with a crosshead speed of 300 mm/min. Volume fraction and crosslink density of silicone rubber and silicone rubber nanocomposites were measured by following techniques. Previously weighed samples were allowed to swell in toluene at 30 °C for 48 h to obtain equilibrium swelling. The equilibrium swelling was confirmed by this way. At a prescribed time interval, the polymer and polymer nanocomposite films were removed from toluene and weighed after blot-drying the excess toluene at the surface of polymer. After 48 h, the weight of the swollen sample was constant. Then, the samples were taken out from toluene, weighed and dried to a constant weight in a vacuum oven at 70 °C for 6 h.

The volume fraction (V_r) of polymer and polymer nanocomposite in the swollen mass was calculated using the following relation:

$$V_r = \frac{W_r \rho_r^{-1}}{W_r \rho_r^{-1} + W_s \rho_s^{-1}} \quad (3)$$

where W_r, W_s, ρ_r and ρ_s were the weight and densities of polymer/nanocomposite and solvent respectively.

The crosslink density (n_c) and average molecular weight between crosslink (M_c), plays an important role in the improvement of mechanical properties of polymer and polymer nanocomposite under equilibrium swollen conditions, have been calculated according to Flory-Rehner equation [14]:

$$n_c = \frac{-\left[\ln(1 - V_r) + V_r + \chi V_r^2\right]}{V_0\left(V_r^{\frac{1}{3}} - \frac{V_r}{2}\right)} \quad (4)$$

$$M_c = \frac{\rho_r}{2n_c} \quad (5)$$

The morphology of the gold-coated tensile-fractured surface was obtained with a JEOL JSM-5800 scanning electron microscope with an acceleration voltage of 20 kV.

4 Result and Discussion

Figure 1 shows the wide-angle XRD patterns of neat SR and SR/St-LDH nanocomposite and all particulates are listed in Table 1. In Fig. 1, the neat SR exhibits two diffraction peak at $2\theta = 12.20°$ (011, sharp) and $2\theta = 22.86$ (020, broad) [15]. However, the diffraction peaks of SR/St-LDH nanocomposites are shifted to lower diffraction angle, slightly increasing the d spacing between crystal planes. The crystallite sizes of SR, which have been calculated using Eq. (2), in the SR/St-LDH nanocomposites are somewhat lower than the neat SR, suggesting the consequence of imperfect crystalline formation. The formation of imperfect crystallite could be attributed to the incorporation of the filler in polymer nanocomposites, which hinders the crystalline domain of the polymer and could account for the decrease in its crystallite size [15]. The absence of basal peak of St-LDH at $2\theta = 1.7$ in the St-LDH filled nanocomposites suggests the exfoliation of St-LDH in SR matrix [9], which is further confirmed by TEM analysis.

TEM technique confirms the exfoliation of LDH nanolayers in the polymer matrix. Figure 2a–b displays the TEM images of SR/St-LDH (3 wt %) and SR/St-LDH (8 wt %) nanocomposites respectively. It clearly suggests that the LDH layers, (average thickness and lateral sizes around 1–2 and 75–100 nm), are homogeneously dispersed in the SR matrix at 3 wt % filler loading. However, Fig. 2b shows aggregation of St-LDH at 8 wt % filler loading.

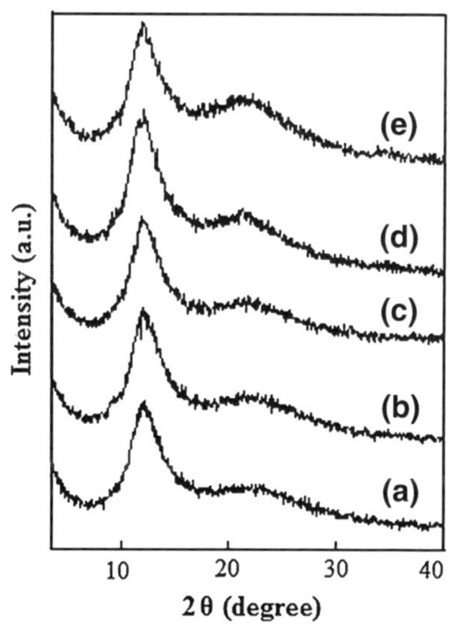

Fig. 1 WAXD spectra of **a** neat SR, **b** SR/St-LDH (1 %), **c** SR/St-LDH (3 %), **d** SR/St-LDH (5 %), **e** SR/St-LDH (8 %)

Table 1 2θ (°), d-spacing (nm), FWHM (°), crystallite size (nm) of (011) plane of neat SR and SR/St-LDH nanocomposites

Sample details	2θ (°)	d spacing (nm)	FWHM (°)	Crystallite size (nm)(011)
SR	12.20	0.72	3.09	2.6
SR/St-LDH (1 wt %)	12.07	0.73	3.31	2.48
SR/St-LDH (3 wt %)	11.93	0.74	3.24	2.47
SR/St-LDH (5 wt %)	12.07	0.73	3.12	2.56
SR/St-LDH (8 wt %)	12.07	0.73	3.12	2.56

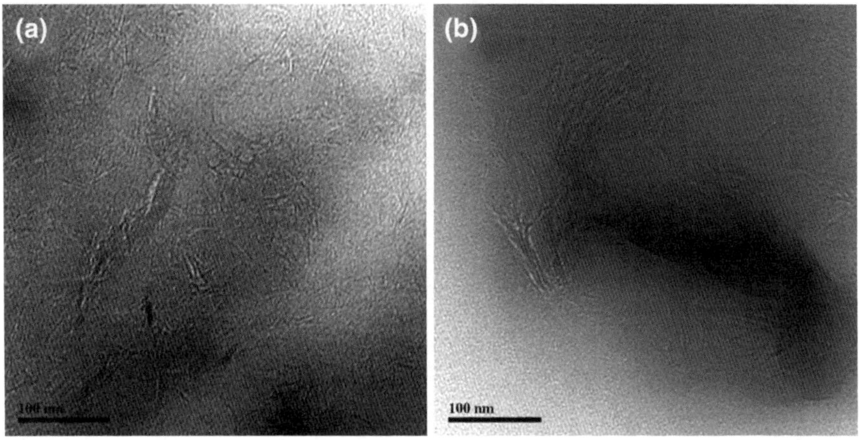

Fig. 2 **a** TEM image of SR/St-LDH (3 wt %) nanocomposite. **b** TEM image of SR/St-LDH (8 wt %) nanocomposite

The measurement of mechanical studies have been shown that the tensile strength and elongation at break of pure SR, SR/St-LDH nanocomposites containing 1, 3, 5 and 8 wt % St-LDH are 0.32, 0.45, 0.63, 0.50, 0.43 MPa and 192, 226, 281, 220, 190 % respectively [9].

These findings clearly demonstrate that SR/St-LDH (3 wt %) nanocomposite shows the maximum improvement of about 97 and 43 % in tensile strength and elongation at break respectively due to the larger aspect ratio and exfoliation of LDH layers as well as polymer-filler interaction [9]. The elongation at break for 8 wt % is 190 % is found to be less than that of pure SR and is in probability is due to the aggregation of St-LDH [2].

Mechanical properties and equilibrium solvent uptake of neat SR and its nanocomposites can be correlated using classic Flory-Rehner approach by calculating polymer volume fraction (V_r), crosslink density (n_c), and average molecular weight between crosslink (M_c). The corresponding data is presented in Table 2 and the variation of crosslink density (n_c) with St-LDH content in SR are displayed in Fig. 3. It is observed that crosslink density and volume fraction of SR/St-LDH nanocomposite is higher than the neat SR. TEM and FT-IR studies indicated that the homogeneous dispersion of St-LDH layers in SR matrix and interaction

Table 2 Density (ρ_r), volume fraction (V_r), crosslink density (n_c) and average molecular weight between crosslink (M_c) of neat SR and its St-LDH nanocomposites

Sample details	ρ_r (g/cm^3)	V_r	$n_c \times 10^{-5}$ (gcc^{-1})	$M_c \times 10^5$ (gmol^{-1})
SR	1.06	0.14	3.67	0.14
SR/St-LDH (1 wt %)	1.12	0.20	8.56	0.065
SR/St-LDH (3 wt %)	1.11	0.22	11.76	0.047
SR/St-LDH (5 wt %)	1.11	0.19	7.59	0.073
SR/St-LDH (8 wt %)	1.13	0.18	7.13	0.079

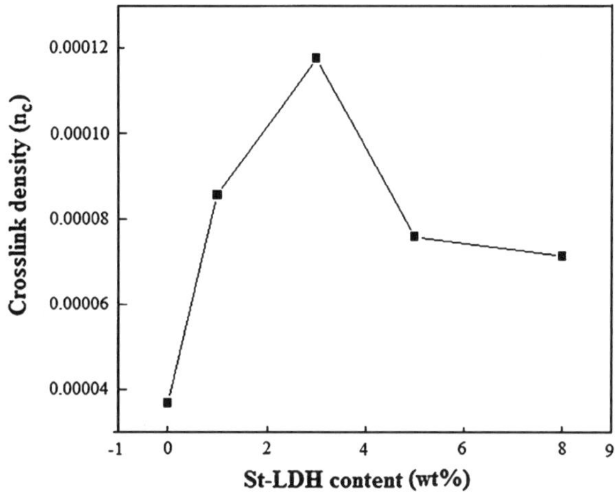

Fig. 3 Variation of crosslink density (n_c) with St-LDH content for neat SR and its SR/St-LDH nanocomposites

between –OH functionality (St–LDH layers) and Si-O groups (SR matrix) could account for such improvement in the mechanical properties of SR/St-LDH (3 wt %) nanocomposite [9].

The optimum values of n_c for SR/St-LDH (3 wt %) nanocomposites could be attributed to the presence of the additional crosslinking sites at the filler-matrix interface [16, 17]. Alternatively, the possibility of the homogeneous dispersion of St-LDH layers in SR matrix and interaction between –OH functionality of St-LDH layers and Si–O polar groups of SR matrix in the improvements of n_c cannot be ruled out. Table 2 also shows the decrease in crosslink density of SR at higher filler loadings (5 and 8 wt %), due to the aggregation of the St-LDH.

SEM micrographs of tensile fractured surfaces of SR and SR/St-LDH (3 wt %) nanocomposites are displayed in Fig. 4a, b respectively. It is clearly observed that the presence of a larger smooth area for neat SR suggesting uniform failure of the SR matrix without any mechanically weaker region for crack initiation. On the contrary, the fracture surface consists of large number of microvoids in the SR/St-LDH nanocomposites. This may also be attributed to the entrapment of the polymer matrix within LDH layers initiating secondary cracks at the organic–inorganic interface and fracture steps before unification with the propagating primary crack [5]. The rough

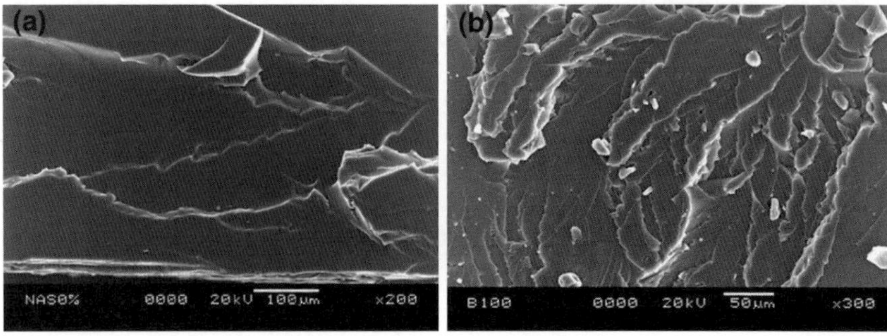

Fig. 4 SEM images of tensile fracture surface of **a** neat SR and **b** SR/St-LDH (3 %) nanocomposite

surface morphology of the SR nanocomposites also accounts for the better mechanical properties similar to those of many other organic polymer nanocomposites.

The toluene uptake (mole %) of neat SR and SR/St-LDH(1, 3, 5, 8 wt %) nanocomposites during swelling are 4.25, 2.79, 2.45, 2.98, 3.0 respectively. It demonstrates that the solvent-uptake capacity is decreased from the surface to the bulk region of the SR nanocomposites and crosslink density (n_c) is increased with up to a 3 wt % St-LDH loading. This may be attributed to the high-aspect-ratio, exfoliated structure of St–LDH layers and the strong interaction between SR and St–LDH (forming a bound polymer). As a result, the polymer chain remains in close proximity to the reinforcing filler (St–LDH) restricting the solvent uptake in SR/St-LDH nanocomposite [5].

5 Conclusion

The nanocomposites of St-LDH with SR have been successfully prepared by the solution intercalation method. XRD and TEM analysis confirm the homogeneous dispersion of St-LDH nanolayers throughout the SR matrix. The nanocomposites are mechanically stable than the neat SR due to the strong interaction between polar group of SR and –OH group of St-LDH. Swelling property analysis confirms the improvement of the solvent resistance of the SR/St-LDH nanocomposites than the neat polymer. The developed SR/St-LDH nanocomposite is observed to be a novel material with unique combination of mechanical and solvent resistance property.

Acknowledgments The authors are grateful to Council of Scientific and Industrial Research (CSIR) and Defence Research and Development Organisation (DRDO), New Delhi, India, for financial support.

References

1. W. Chen, B. Qu, Structural characteristics and thermal properties of PE-g-MA/MgAl-LDH exfoliation nanocomposites synthesized by solution intercalation. Chem. Mater. **15**, 3208 (2003)
2. H. Acharya, S.K. Srivastava, A.K. Bhowmick, Synthesis of partially exfoliated EPDM/LDH nanocomposites by solution intercalation: Structural characterization and properties. Compos. Sci. Technol. **67**, 2807 (2007)
3. M. Pramanik, S.K. Srivastava, B.K. Samantaray, A.K. Bhowmick, Synthesis and characterization of organosoluble, thermoplastic elastomer/clay nanocomposites. J. Polym. Sci. Part B: Polym. Phys. **40**, 2065 (2002)
4. T. Kuila, H. Acharya, S.K. Srivastava, A.K. Bhowmick, Ethylene vinyl acetate/mg-al ldh nanocomposites by solution blending. Polym. Compos. **30**, 497 (2008)
5. B. Pradhan, S.K. Srivastava, R. Ananthakrishnan, A. Saxena, Preparation and characterization of exfoliated layered double hydroxide/silicone rubber nanocomposites. J. Appl. Polym. Sci. **119**, 343 (2011)
6. K. Chibwe, W. Jones, Intercalation of organic and inorganic anions into layered double hydroxides. J. Chem. Soc. Chem. Commun. **14**, 926 (1989)
7. S.D. Burnside, E.P. Giannelis, Synthesis and properties of new poly (Dimethylsi1oxane) nanocomposites. Chem. Mater. **7**, 1597 (1995)
8. L. Dewimille, B. Bresson, L. Bokobza, Synthesis, structure and morphology of poly(dimethylsiloxane) networks filled with in situ generated silica particles. Polymer **46**, 4135 (2005)
9. B. Pradhan, S.K. Srivastava, A.K. Bhowmick, A. Saxena, Effect of bilayered stearate ion-modified Mg–Al layered double hydroxide on the thermal and mechanical properties of silicone rubber nanocomposites. Polym. Int. **61**, 458 (2012)
10. S. Carlino, The intercalation of carboxylic acids into layered double hydroxides: a critical evaluation and review of the different methods. Solid State Ionics **98**, 73 (1997)
11. E. Mahboobeh, W.M.Z.W. Yunus, Z. Hussein, M. Ahmad, N.A. Ibrahim, Flexibility improvement of poly(lactic acid) by stearate-modified layered double hydroxide. J. Appl. Polym. Sci. **118**, 1077 (2010)
12. R. Patel, K.S. Patel, Flow injection determination of anionic surfactants with cationic dyes in water bodies of central India. Analyst **123**, 1691 (1998)
13. Z. S. Mo, and H. F. Zhang, Science Press: PRC; 2003. p. 232
14. P.J. Flory, J.H. Rehner, Statistical mechanics of cross-linked polymer networks I I. swelling. J. Chem. Phys. **11**, 521 (1943)
15. L. Qu, G. Huang, Q. Wang, Z. Xie, Effect of diphenylsiloxane unit content on aggregation structure of poly(dimethylsiloxane-co-diphenylsiloxane). J. Polym. Sci. Part B: Polym. Phys. **46**, 72 (2008)
16. F. Bueche, Molecular basis for the Mullins effect. J. Appl. Polym. Sci. **4**, 107 (1960)
17. F. Bueche, Mullins effect and rubber-filler interaction. J. Appl. Polym. Sci. **5**, 271 (1961)

Preparation and Characterization of Polymer Films of PMMA and its Composites

Basavaraja Sannakki and Anita

Abstract The polymer films of PMMA and its composites with Fe_2O_3 and with Al_2O_3 of different thickness at various weight percentages are prepared by solution casting method. The films of PMMA for 0.5, 1, 2 and 3 N concentrations are used for optical absorption studies using UV-vis Spectrometer over the range 200–900 nm. The samples are characterized by using X-Ray diffractometer and Raman Spectroscopy. The samples are also characterized by using FTIR over the range of wavenumber 3,500–500 cm^{-1}. Further, variation of the Raman intensity of composite films of PMMA with Fe_2O_3 at different weight percentages such as 20, 40 and 60, and with that of Al_2O_3 at 20 wt% for 1 N concentrations are also observed. The particles shape and size of PMMA and its composite are also obtained by UV-vis absorbance spectra and the same was confirmed through the morphological studies of these films by using FESEM. The particle shapes are not in spherical and the particle sizes of PMMA are obtained in the range 39–150 nm and that of PMMA with Fe_2O_3 are from 53 to 111 nm.

Keywords Polymer composite films · Optical absorbance · Raman bands · Particle shape

B. Sannakki (✉) · Anita
Department of Post Graduate Studies and Research in Physics, Gulbarga University, Gulbarga 585106, Karnataka
e-mail: sannakki@rediffmail.com

Anita
e-mail: anita.ghandhe@gmail.com

P. K. Giri et al. (eds.), *Advanced Nanomaterials and Nanotechnology*,
Springer Proceedings in Physics 143, DOI: 10.1007/978-3-642-34216-5_37,

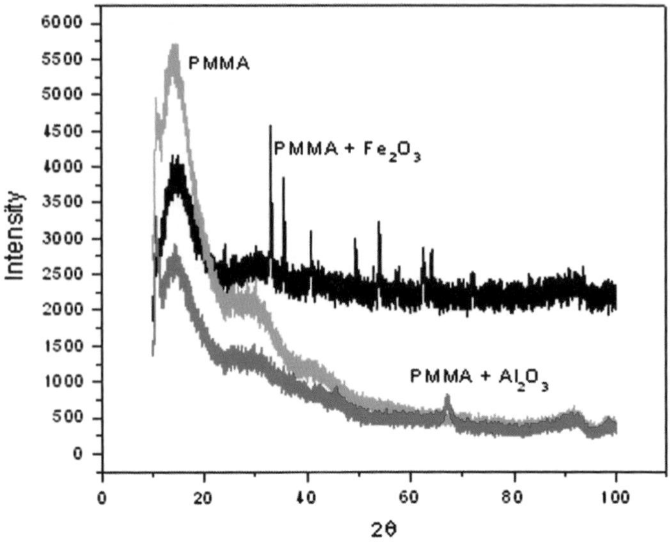

Fig. 1 XRD spectra for PMMA and its composite films

1 Introduction

Developments of Polymer composites and extensive utilization of polymer materials in technology have led to importance in the field of polymer composites and also it is well known that the composites of polymer films exhibit enhanced properties than that of the constituent materials [1–3]. The composite materials of polymers are important to the electronic industry for its optical properties as well as for dielectric properties in the use of capacitors [4]. The development of polymer based composites which exhibit various optical functionalities such as high or low refractive index, tailored absorption and transmission properties. A strong optical non linearity attracts interest because of the potential opto-electronic applications [5, 6]. The polymer composites are typically obtained by the incorporation of functional inorganic particles into a transparent polymer matrix [7]. The application of polymer films include illuminated displays, optical interconnects between electronic components, optical amplifiers and lasers in the visible region, fiber optic components such as filters, optical switches and sensors [8–11]. Poly (methyl methacrylate)-PMMA is an amorphous optically clear thermoplastic polymer material [12]. Hence the present study involves measurement of optical properties of PMMA films and its composites with Fe_2O_3 and with that of Al_2O_3. In the present study particle shape and sizes [13] are also obtained in PMMA films and its composites.

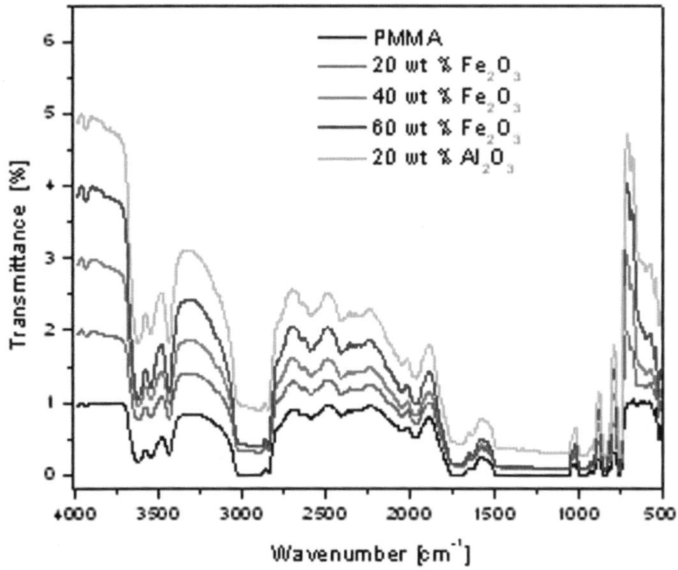

Fig. 2 FTIR spectra for polymer films of PMMA and its composites with Fe_2O_3 and Al_2O_3

2 Experimental Measurements

The substance of Fe_2O_3, Al_2O_3 and Polymethylmethacrylate (PMMA) are obtained from S.d. Fine Chem. Ltd, Mumbai, India. The composite films of PMMA with Al_2O_3 at different weight percentages such as 10 and 20 are obtained by mixing of Al_2O_3 with the PMMA solution, where the PMMA solution is obtained by dissolving substance of PMMA in acetone. Further, the composite films of PMMA with Fe_2O_3 at different weight percentages such as 10, 20, 30, 40, 50 and 60 are also obtained by mixing of Fe_2O_3 with the PMMA solution. The films of PMMA and its composite films with Fe_2O_3 and also with that of Al_2O_3 are prepared by solution casting method. The prepared films are characterized by using XRD, FTIR and Raman spectroscopy. The optical absorbance studies are made using UV-vis spectrometer for determination of particle shape and sizes.

3 Results and Discussion

3.1 Characterization Using XRD

The crystalline or amorphous nature of PMMA and its composite films with Al_2O_3 have been studied using powder X-ray diffractometer (Make: Regaku) at an angle 2θ. The XRD spectra of PMMA and its composite films are given in Fig. 1. It is noticed from Fig. 1 that the diffused and broader peaks are observed for PMMA film indicates that amorphous in nature, whereas in case of composite film of PMMA with Al_2O_3 a peak is observed at angle of 68°. Further, intense sharp peaks are observed for the composite film of PMMA with Fe_2O_3 at 2θ values of 32, 36, 42, 53, 62°, indicating that the presence of crystalline phases in the polymer composite matrix [14]. Unlike composite film of PMMA with Fe_2O_3 sharp peaks are not observed for composite film of PMMA with Al_2O_3 could be the amount of polymer which absorbs on the powder depends on the acid/base characteristics of both polymer and Al_2O_3 samples [15].

The particle size could be easily estimate using the Debye Scherer formula given by

$$D = \frac{0.9\,\lambda}{\beta\,\cos\theta} \tag{1}$$

where D is the particle size, β is FWHM of stronger peak and λ is wavelength of the X-ray.

3.2 FTIR Spectra

The FTIR spectra for optical transmittance in PMMA films and its composites with Fe_2O_3 and with that of Al_2O_3 are observed for various weight percentages and are given in Fig. 2. It is observed that the transmittance of PMMA film is less compared to the composite film with Al_2O_3. It is also noticed that as the weight percentages such as 20, 40 and 60 of Fe_2O_3 with PMMA increases the transmittance also increases. This is due to the interaction between the oxide core and the polymer coating, leading to the intensity with the particle size. Here it is clearly observed that the transmittance intensity of composite films is high compared to the PMMA film. In case of composite films transmittance intensity of the composite film of Al_2O_3 are high compared to that of Fe_2O_3 at 20 wt%. The transmittance intensity depends on surface of the particles and promotes to intensity from interface of ceramic and polymers [16]. The IR studies revealed the nature of the bonding between the oxide cores and the coatings. It is obvious that during process of the PMMA composite film, a substantial number of ester groups is transformed into the corresponding carboxylates and are chemisorbed at the surface of the oxide with an ester like linkage. The transmittance of composite film of PMMA with Fe_2O_3 is lesser to that of the PMMA with Al_2O_3 at 20 wt% due the particles consisting of hydrophobic in PMMA with Fe_2O_3 nano particles.

From Fig. 2 the C=O vibration of the ester carbonyl group appeared at wave number of 1,732 cm^{-1}. The aromatic COO, C–H deformation occurs at 743 cm^{-1}. The C–H, C=O and O–H stretches observed at 2,959, 3,438 and 3,636 cm^{-1} respectively.

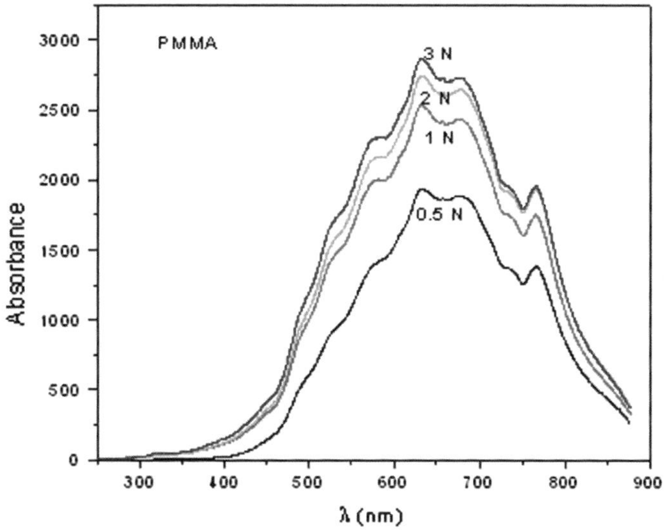

Fig. 3 UV-absorbance spectra for polymer films of PMMA

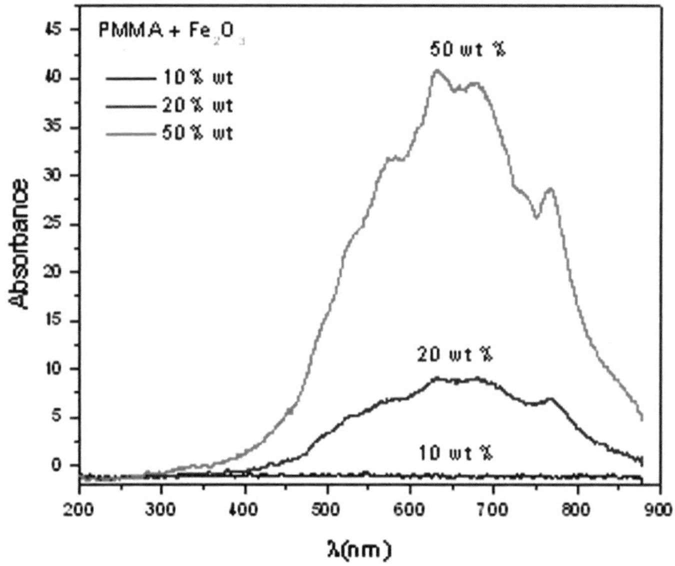

Fig. 4 UV-absorbance spectra for composite films of PMMA with Fe_2O_3

3.3 Optical Spectra

The spectra of optical absorbance for PMMA films and its composite films with Fe_2O_3 are studied for various weight percentages using UV-vis spectrometer

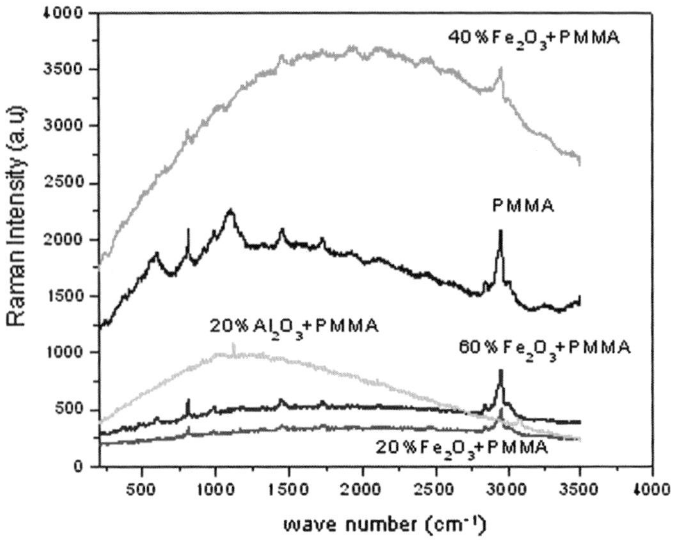

Fig. 5 Raman spectra for PMMA and its composite films with Al_2O_3 and Fe_2O_3

Table 1 Raman bands in PMMA and its composite films

Raman band (cm^{-1})	Assignments
602*	ν(C–COO), ν_s(C–C–O)
812	$\nu(CH_2)$
999*	O-CH_3 rock
1,109	ν(C–C) skeletal mode
1,451*	δ_a(C–H) of α-CH_3, δ_a(C–H) of O–CH_3
1,727*	ν(C=O) of (C–COO)
2,953*	ν_s(C–H) of O–CH_3 with ν_s(C–H) of α-CH_3 and $\nu_a(CH_3)$
3,082*	ν_a(C–H) of O–CH_3 of ν_a(C–H) of α-CH_3
3,455*	($2\nu_2$) Overtone of 1,730 cm^{-1}

* Our observations are almost same with others [22]

(PC2000 ISA). The optical absorbance spectra for films of PMMA are observed at different concentrations such as 0.5, 1, 2 and 3 N and are given in Fig. 3. In Fig. 3 it is observed that the maximum absorbance occurs at 650 and 765 nm for all the PMMA films as well as its composites. The peaks of absorbance spectra for PMMA films occur at 631.94, 632.80, 633.18 and 634.18 nm for 0.5, 1, 2 and 3 N concentrations respectively. Here it is also noticed that slight shift in wavelength at absorbance peak occur with increase in concentration of the film. This indicates there is slight variation in particles size with out much difference. It is clearly observed that as concentration increases the optical absorbance also increases. The size and shape of the particles in the PMMA film can be qualitatively described by

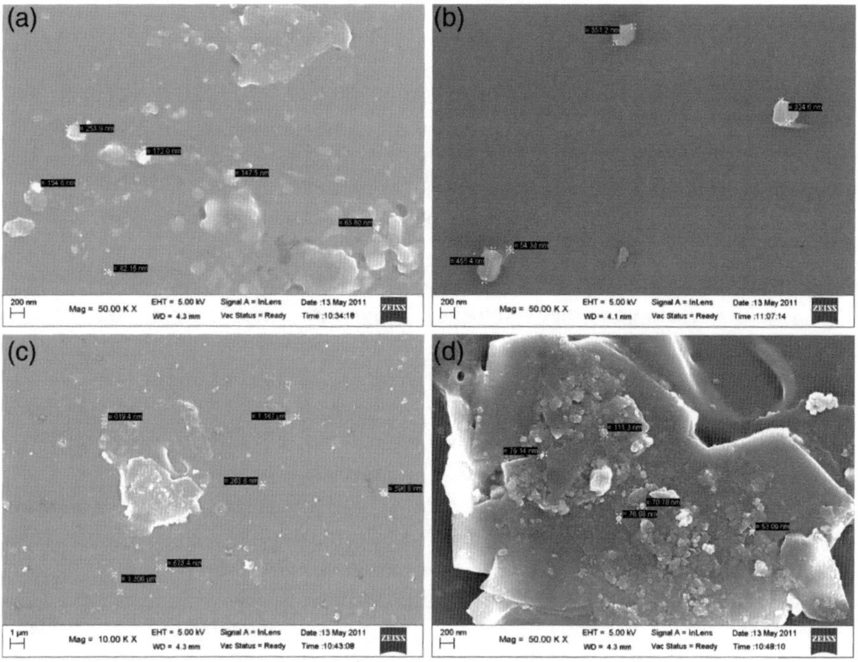

Fig. 6 **a** The micrographs of FESEM for particle sizes in PMMA film at 50 kx magnification. **b** The micrographs of FESEM for particle sizes in PMMA with Al_2O_3 at 50 kx magnification. **c** The micrographs of FESEM for particle sizes in PMMA with Fe_2O_3 at 10 kx magnification. **d** The micrographs of FESEM for particle sizes in PMMA with Fe_2O_3 at 50 kx magnification

the peak position and shape of the observed spectrum [17]. Further the optical absorbance spectra for composite films of PMMA with Fe_2O_3 at 10, 20 and 50 wt% are given in Fig. 4. It is observed from Fig. 4 that as weight percentage increases the absorbance is also increases. It is seen from Figs. 3, 4 for all these samples that the maximum absorbance occurred at wavelength of around 630–700 nm. The optical observation curve for mono-dispersed particles should be narrow, while the optical absorption curve of wide size distribution is broad. The single peak absorbance spectra indicates presence of spherical size of the particles [12], but in Figs. 3, 4 we observed the double-peak absorption spectra that lie between 500 and 800 nm indicate the shape of the particles may be elliptical or cylindrical in PMMA and its composite films. These size and shapes of the particles were confirmed through the surface studies using FESEM images and are given in Fig. 7. From Fig. 3 it is observed that absorbance increases as concentration increases, because at higher concentrations there are more molecules present which in turn more electrons present in the polymer hence more is absorbed at specific wavelength of 631 nm.

3.4 Raman Spectra

The Raman Spectra for PMMA and its composite films with Al_2O_3 and also with that of Fe_2O_3 for various weight percentages over the range 0–4,000 cm^{-1} are studied using Raman Spectrometer (model spex 14018). The plots of Raman intensity as function of wave numbers for 1 N concentration of PMMA and its composite films are given in Fig. 5. Investigations are carried out on the variation of FWHM of the Raman peak at 2,953 cm^{-1} of PMMA polymer films with 20, 40 and 60 wt% of Fe_2O_3 and with 20 wt% of Al_2O_3 composite films. The Raman bands in PMMA and its composite films are given in Table 1. Out of these Raman bands, the one at 2,953 cm^{-1} is the most prominent which is due to the C–H stretching vibration [18, 19]. The other observed bands are at 602, 812, 999, 1,109, 1,264, 1,451, 1,727, 3,082 and 3,455 cm^{-1}. The band at 1,727 cm^{-1} is attributed to the combination band arising out of ν(C=O) and ν(C–COO) modes and the bands at 3,455 cm^{-1} is attributed to the first overtone involving (C=O) of C–COO. The above two Raman bands are observed in PMMA film. The other Raman bands are also seen at same wave numbers for its composite films in Fig. 5 which are shown in Table 1. The observed Raman spectrum agrees fairly well with those available in the literature [20, 21]. It is observed that the spectra of composite film of PMMA at 40 % weight percentage of Fe_2O_3 lies above the spectra of PMMA film and also from other weight percentages.

3.5 Morphology Using FESEM

The morphological studies have been made using the Field Emission Scanning Electron Microscope (FESEM-Hitachi, SU6600) to determine the size of the particles in PMMA films and its composite films with Fe_2O_3 and also with that of Al_2O_3. The micrographs for PMMA and its composites for Fe_2O_3 and for Al_2O_3 are given in Fig. 6. The particle sizes of Fe_2O_3 and Al_2O_3 are in microns [23, 24] prior to using in polymer composites. The particle sizes in PMMA film are 39.68, 44.24, 51.51, 83.5, 86.74 and 149.5 nm and that in composite film of PMMA with Al_2O_3 are 54.38, 324.6, 351.2 and 455.4 nm at 50 kx magnifications. The particle sizes in case of PMMA with Fe_2O_3 are 53.09, 70.78, 76.68, 79.14 and 111 nm are found at 50 k x magnifications. It is clear that the particle size of composite films of PMMA with Al_2O_3 are larger in size with that of PMMA with Fe_2O_3 due to the bonding between oxide ceramic particles with PMMA polymer consisting as one molecule [16]. It also suggests that size of the particles formed in polymer composites are non spherical in shape which could be seen from FESEM image and confirmed by the UV-vis absorbance spectra given in Fig. 4.

4 Conclusion

The XRD spectra of composite films of PMMA with Fe_2O_3 shows sharp peaks for indication of crystalline in nature due to presence of the iron oxide particles. The FTIR spectra for transmittance in PMMA film is less compared to those of its composite films with Fe_2O_3 and with Al_2O_3. The transmittance intensity depends on size and surface area of the particles that promotes to higher intensity, consequently smaller the size of the particles higher the intensity. It is also noticed that as weight percentage of Fe_2O_3 with PMMA increases the transmittance also increases. The UV-vis spectra shows maximum absorbance at specific wavelength of 631 nm for all the concentrations of PMMA and its composites with Fe_2O_3 and it is also noticed that as concentration increases the optical absorbance also increases. In case of composite films of PMMA with Fe_2O_3 as weight percentage increases the absorbance is also increases due to presence of larger number of molecules. The Raman intensity of composite films of PMMA with Fe_2O_3 at 20 and 60 wt% are less than that of PMMA film. The sizes of the particles obtained by FESEM in PMMA film and its composites are shown in nano-particle size and the particles formed in polymer composites are non spherical in shape.

Acknowledgments The authors are thankful to Prof R. D. Mathad, Chairman, Dept of Physics, Gulbarga University, Gulbarga and Prof P. C. Bansal, Dean School of Physics, Central University of Hyderabad for providing facilities and valuable suggestions.

References

1. B. Wenderlinch, *Macromolecular Physics* (Academic, New York, 1973)
2. R.S. Gulalkari, Y.G. Bakale, Electrical conduction mechanism of polyvinyl chloride (PVC) polymethyl methacrylate (PMMA) blend film. Pramana-J. Phy. **69**, 485 (2007)
3. T.K. Vishnuvardhan, V.R. Kulkarni, C. Basavaraj, S.C. Raghavendra, Synthesis, characterization and ac conductivity of polypyrrole/Y_2O_3 composites. Bull. Mater. Sci. **29**, 77 (2006)
4. S.S. Bellad, Composition & frequency dependent dielectric properties of Li-Mg-Ti ferrites. Mater. Res. Bul. **34**, 1099 (1999)
5. W.U. Huynh, J.J. Dittemer, A.P. Alivisatos, Hybrid nanorod-polymer solar cells. Science **295**, 2425 (2002)
6. A.C. Arango, L.R. Johnson, V.N. Bliznyuk, Z. Schlesinger, S.A. Carter, H.H. Horhold, Efficient titanium oxide/conjugated polymer photovoltaics for solar energy conversion. Adv. Mater. **12**, 1689 (2000)
7. L.L. Beecroft, C.K. Ober, Nanocomposite materials for optical applications. Chem. Mater. **9**, 1302 (1997)
8. M.G. Kuzyk, *Polymer Fibre Optics: Materials, Physics and Applications* (Taylor and Francis, London, 2007)
9. H. Yu, Z. Zhang, X. Hao, F. Zhu, A general low-temperature route for large-scale fabrication of highly oriented ZnO nanorod/nanotube arrays. J. Am. Chem. Soc. **127**, 2378 (2005)
10. B. Fei, Z. Xin-Zheng, W. Zhen- Hua, W. Qiang, H. Hao, X. Jing-Jun, Preparation and size characterization of silver nanoparticles. Chin. Phy. Soc. **25**, 4463 (2008). (IOP Publication Ltd)

11. M. Rajesh, M. Sheeba, K. Geetha, C.P.G. Vallabhan, P. Radhakrishnan, V.P.N. Nampoori, Fabrication and characterization of dye-doped polymer optical fiber as a light amplifier. Appl. Opt. **46**, 106 (2007)
12. D. Christopher, K. Shadak Alee, D. Narayana Rao, Synthesis and characterization of silver nanoparticles produced by laser ablation technique in aqueous monomer solution. J. Trends. Chem. **2**, 1 (2011)
13. H.S. Nalwa, *Polymer Optical Fibers* (American Scientific Publishers, Stevenson Ranch, 2004)
14. V. Jagadish Babu, Electrical properties of Electrospun fibers of PANI-PMMA composites. J. Engg. Fibers Fabr. **2**, 25 (2007)
15. E. Papirer, J.-M. Perrin, B. Siffert, G. Philipponneau, Surface characteristics of aluminas in relation with polymer adsorption. J. Coll. Int. Sci. **144**, 263 (1991)
16. D. Vollath, D.V. Szabo, S. Schlabach, Oxide/polymer nanocomposites as new luminescent materials. J. Nanopart. Res. **6**, 181 (2004)
17. H.A. Willis, V.J.I. Zichy, P.J. Hendra, The laser-Raman and infra-red spectra of polymethylmethacrylate. Polymer **10**, 737 (1969)
18. Z. Crnjak, A. Anzlovar, M. Zigon, *Nano Composite as an Effective UV Absorber with Nano Size Zinc Oxide and PMMA* (National Institute of Chemistry, Hajdrihova, 2012)
19. X. Xu, Stimulated Raman spectrum threshold in polymethyl methacrylate optical fibers. Opt. Commun. **279**, 89–93 (1999)
20. M.M.I. Bhuiyan, Y. Haga, M. Esashi, Design and characteristics of large displacement optical fibre switch. IEEE J. Quantum Electron. **41**, 242 (2005)
21. X. Xu, H. Ming, Q. Zhang, Y. Zhang, Properties of Raman spectra and laser induced drift fringes in polymethyl methacrylate optical fibres. J. Opt. A: Pure Appl. Opt. **4**, 237 (2002)
22. K.J. Thomas, M. Sheeba, V.P.N. Nampoori, C.P.G. Vallabhan, P. Radhakrishnan, Raman spectra of polymethyl methacrylate optical fibres excited by a 532 nm diode pumped solid state laser. J. Opt. A: Pure Appl. Opt. **10**, 055303 (2008)
23. A.I.Y. Tok, F.Y.C. Boey, X.L. Zhao, Novel synthesis of Al_2O_3 nano particles by flame spray pyrolysis. J. Mat. Proc. Tech. **178**, 270 (2006)
24. Q. Wrong, S. Yin, C.S. Guo, H.H. Li, N. Kumada, T. Takei, Y. Yonesaki, N. Kinomura, T. Sato, Preparation of α- Fe_2O_3 particles with controlled shape and size via a facile hydrothermal route. J. Phys. Conf. Ser. **339**, 012004 (2012)

A Comparative Performance of Oleic Acid and MWNT-Coated $SmCo_5/Fe$ Nanocomposites Processed by Magnetic Field-Assisted Milling

P. Saravanan

Abstract In the present study, a magnetic field-assisted milling process was adopted for the fabrication of $SmCo_5/Fe$ nanocomposites using oleic acid (OA) and multi-walled carbon nanotubes (MWNT) as surfactants. The bare $SmCo_5/Fe$ powders processed by conventional milling consist of large agglomerates accompanied by amorphization; while both the OA and MWNT-coated $SmCo_5/Fe$ powders possess unusual characteristics such as flake-like structures, reduction in grain size and grain orientation. The OA and MWNT-coated $SmCo_5/Fe$ powders also displayed high degree of texturing in terms of *c*-axis alignment of $SmCo_5$-phase. Enhanced coercivity (H_c) values of 4.37 and 4.67 kOe were obtained for the OA and MWNT-coated powders, respectively and these values significantly higher than that of the bare powders (1.49 kOe). Upon processing the surfactant-coated powders into polymer nanocomposites, relatively higher H_c (12.6 kOe) and M_r/M_{max} (0.82) values were achieved for the MWNT-coated $SmCo_5/Fe$ nanocomposites as compared to that of the OA-coated counterparts ($H_c = 10.2$ kOe; $M_r/M_{max} = 0.70$).

Keywords Nanocomposite magnets • $SmCo_5$ • Magnetic materials • Mechanical milling

1 Introduction

$SmCo_5/Fe$-type nanocomposite exchange coupled magnets which combine both the high anisotropy of hard magnetic phase ($SmCo_5$) with the large magnetization of soft magnetic phase (Fe), have attracted greater attention for their potential to

P. Saravanan (✉)
Defence Metallurgical Research Laboratory, Hyderabad 500058, India
e-mail: psdrdo@gmail.com; ps@dmrl.drdo.in

P. K. Giri et al. (eds.), *Advanced Nanomaterials and Nanotechnology*, Springer Proceedings in Physics 143, DOI: 10.1007/978-3-642-34216-5_38,

offer very high energy products [1, 2]. Most of the preparation techniques reported for the $SmCo_5$/Fe powders mainly uses high energy ball milling; nevertheless, the powder particles produced with this technique possess wide size distribution and, generally, they are magnetically isotropic and hence, not suitable for making anisotropic hard magnets. Consequently, many efforts have been made to transfigure the milling process through introducing either surfactant [3] or magnetic field [4] during conventional ball milling (referred as surfactant- or magnetic field-assisted ball milling) to achieve anisotropic hard magnetic particles with nanosized grains. Wang et al. [5] employed a surfactant-assisted milling technique to synthesize nanoparticles of Sm_2Co_{17} and $SmCo_5$ systems with sizes of about 50 nm, using oleic acid and oleyl amine as surfactants. Similarly, the mechanism of field-induced anisotropy has also been demonstrated in sub-μm size particles of $Nd_2Fe_{14}B$ and Sm_2Co_{17}, prepared by a magnetic field-assisted ball milling [4]. Besides, we have reported the processing of $SmCo_5$/Fe nanocomposite powders by introducing both surfactant and magnetic field simultaneously during the milling process, so as to impart size reduction as well as orientation of the nanocrystallites [6]. Subsequent to the above study, we have also exploited the benefits of multi-walled carbon nanotubes (MWNT) as a surfactant for the processing of $SmCo_5$/Fe nanocomposites by magnetic field-assisted milling [7]. The highly anisotropic powders such as those produced in the above studies would form as an important ingredient for producing high performance resin-bonded magnets. Accordingly, in the present study, efforts have been made to process the oleic acid (OA) and MWNT-coated $SmCo_5$/Fe magnetic powders prepared by field/surfactant-assisted ball milling into resin-bonded magnets. The structural and magnetic properties of such powders as well as the resin-bonded counterparts are compared with that of the conventionally milled one.

2 Materials and Methods

The precursor $SmCo_5$ alloy was prepared by melting of elemental Sm and Co in high purity argon atmosphere. The alloy ingot was crushed into a powder size of $\sim$300 μm and was mixed with 10 wt% of commercially available α-Fe powders having a particle size of less than 10 μm. The starting precursor, i.e. $SmCo_5$ + 10 wt% Fe along with surfactants such as OA (10 wt%) and MWNT (2 wt%) were separately milled in presence of heptane (99.8 % purity) as solvent. The MWNT synthesized by thermal chemical vapour deposition were used in this study. Milling was performed in a planetary ball mill (Fritsch Pulverisette) with the milling vial and balls made of tungsten carbide. For the field-induced milling, blocks of $Nd_2Fe_{14}B$ magnets were fixed around the vial to create a radial magnetic field inside the vial (0.1 T), to synthesize anisotropic powders during milling (referred as magnetic field-assisted milling). The milling time was varied typically from 2 to 10 h at a constant speed of 200 rpm with a ball-to-powder ratio of 10:1. In addition to the preparation of $SmCo_5$/Fe powders by the field/surfactant-assisted

Fig. 1 SEM micrographs of magnetic field-milled $SmCo_5$/Fe nanocomposite powders collected after 10-h of milling time. **a** Without surfactant, **b** with OA and **c** with MWNT

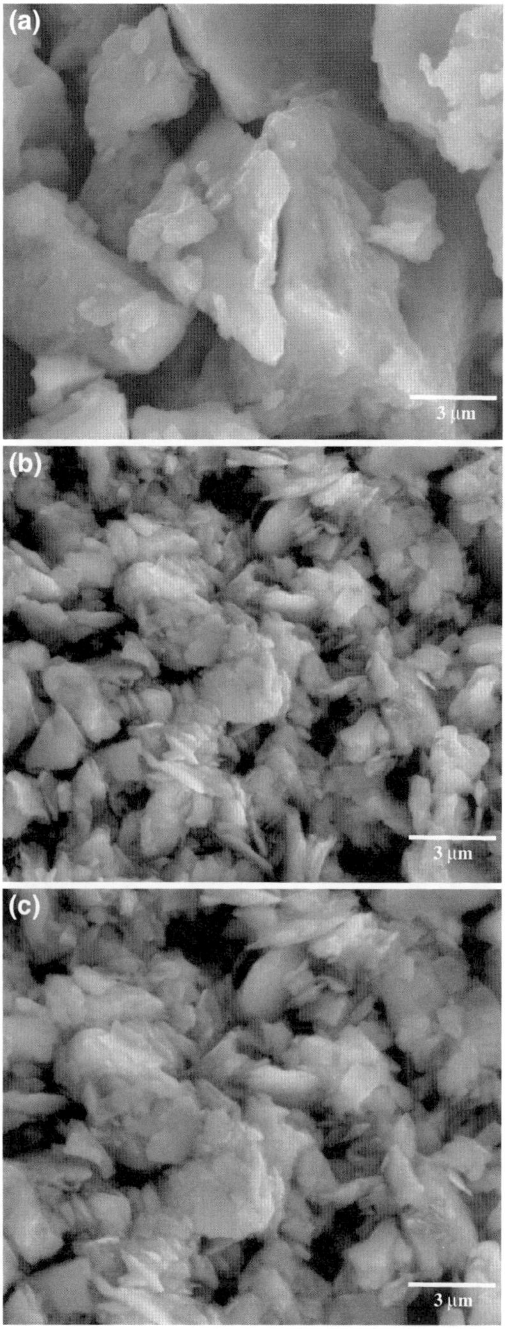

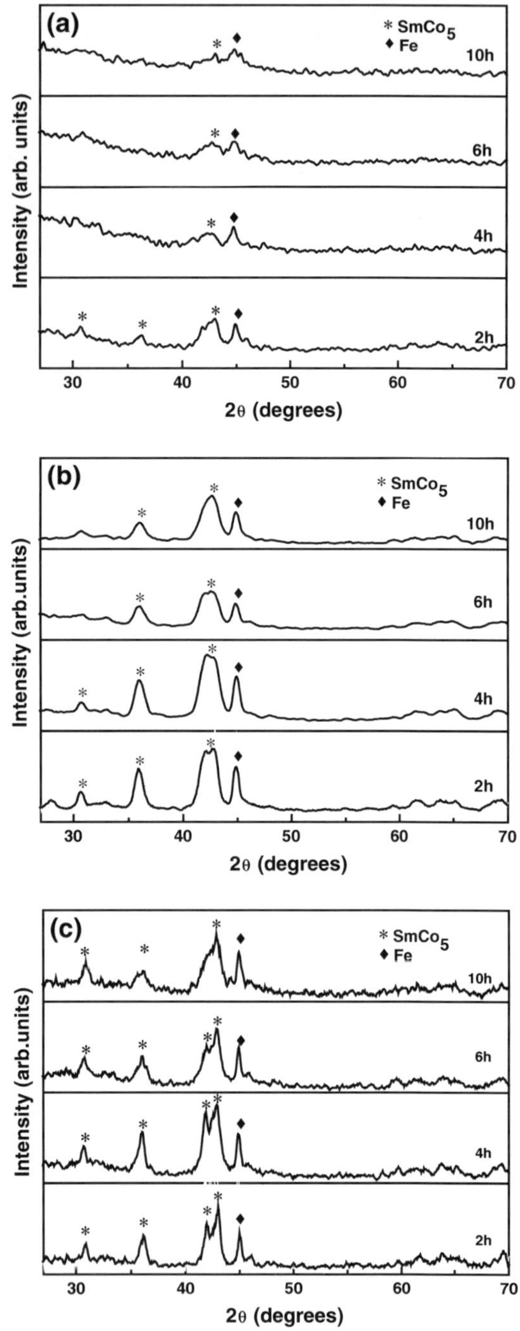

Fig. 2 XRD patterns of $SmCo_5$/Fe nanocomposite powders as a function of milling time. **a** Without surfactant, **b** with OA and **c** with MWNT

Fig. 3 The dependence of **a** coercivity (H_c) and **b** remanence ratio (M_r/M_s) on the milling time for the bare (*dotted line*), OA (*dashed line*) and MWNT-coated (*continuous line*) $SmCo_5/Fe$

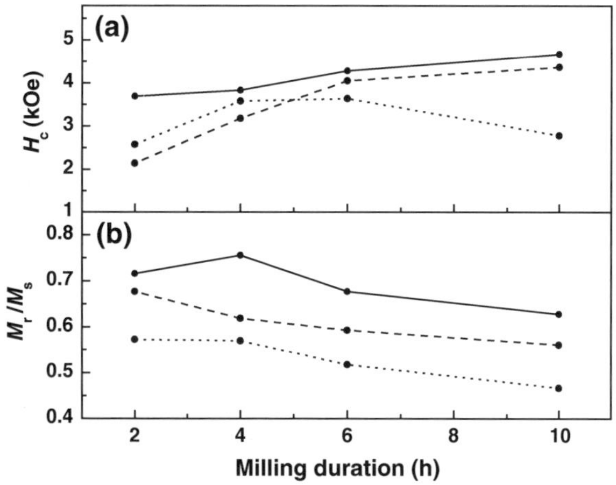

milling, bare powders were processed in a similar manner by conventional milling, using heptane as a solvent (lubricant).

To process bonded magnets, weighed amount of milled $SmCo_5/Fe$ powders were embedded in the epoxy resin (LY556; Hindustan CIBA-GEIGY, Ltd., Mumbai, India) and the mixture was poured into a non-magnetic stainless steel die to form a rectangular block of 10 mm in height and 8 mm in width. The epoxy and magnetic powder mixture were then cured at elevated temperature of 100–120 °C under a pressure of about 10 MPa and aligning field of 2 T. Prior to the magnetic powder loading in the epoxy matrix, they were thoroughly dried in a vacuum oven for several minutes and the powders were mechanically mixed with the resin for 30 min to attain homogeneity. The identification of phases formed in the milled powders was examined by X-ray diffraction (Philips, Cu-Kα radiation). Microstructural studies were conducted using scanning electron microscopy (SEM, Leo 440i) and analytical transmission electron microscopy (TEM, Tecnai 20T G2). The Archimedes method was employed to measure the bulk sample density. Magnetic measurements were performed with a vibrating sample magnetometer (VSM) (ADE make, model EV9) up to a maximum field of 2 T.

3 Results and Discussion

Figure 1 illustrates the typical SEM morphology of 10 h milled nanocomposite powders without and with surfactants. The powders milled without surfactant appear to be large agglomerates with a wide size distribution ranging from 1 to 25 μm; while the surfactant-coated powders show the presence of very fine non-agglomerated particles with sizes in the range 1–5 μm. This indicates that the surfactant-assisted milling is more efficient in reducing the particle size, as well as in dispersing the aggregates into fine particles [6]. In contrast to the irregular shapes

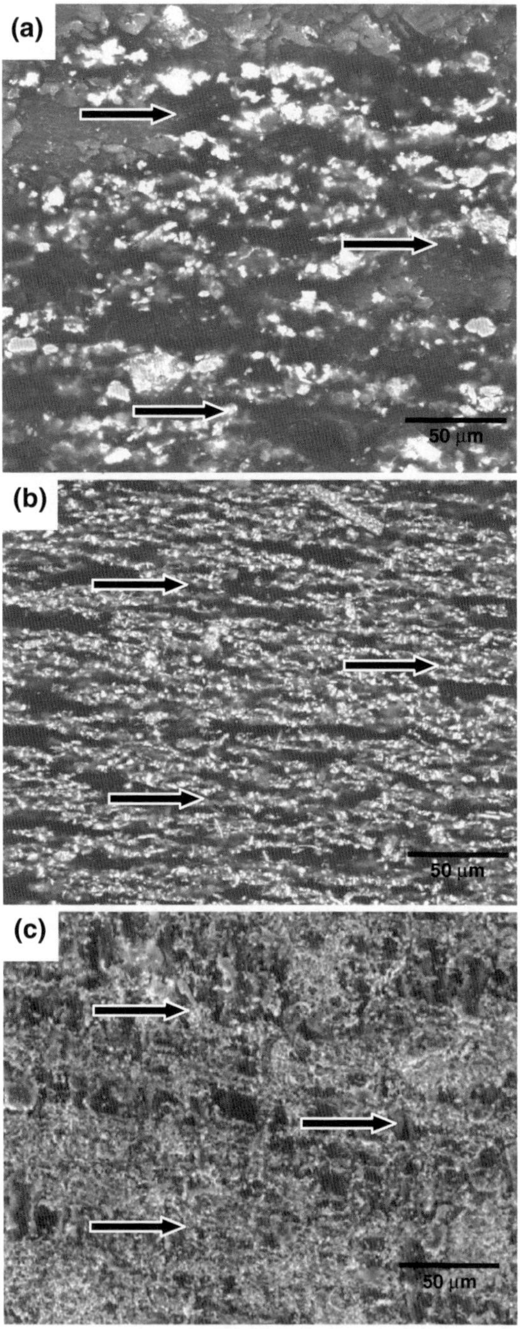

Fig. 4 SEM micrographs of epoxy nanocomposites processed under aligning magnetic field with **a** bare, **b** OA and **c** MWNT-coated $SmCo_5$/Fe powders. Particle alignment along the field direction is indicated by arrows

Fig. 5 XRD patterns of epoxy nanocomposites processed with **a** bare, **b** OA and **c** MWNT-coated $SmCo_5$/Fe powders. The intensity of (002) diffraction peak is the measure of texture or the *c*-axis alignment of $SmCo_5$ phase

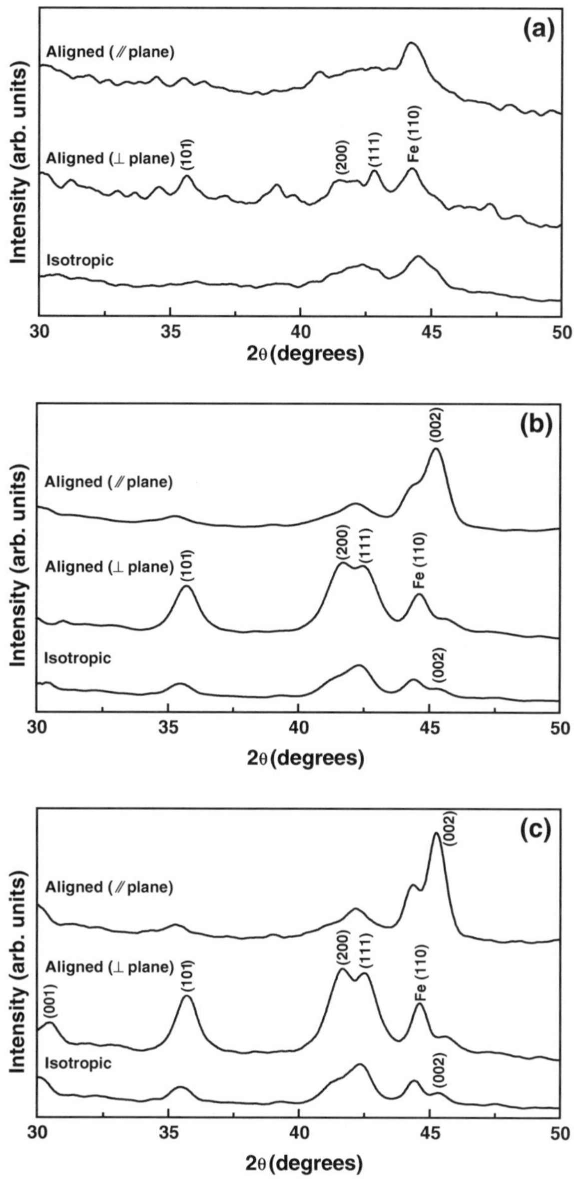

observed in the case of bare powders (Fig. 1a), flake-like structures can be seen both in the case of OA and MWNT-coated powders. Probably, the flake-like structures may be due to the development of anisotropic characteristics during field-ball milling [4]; this presumption can be corroborated in concurrence with the XRD results. The evolution of XRD patterns as a function of milling time for the $SmCo_5$/Fe powders without and with surfactants are shown in Fig. 2. In all the cases, the

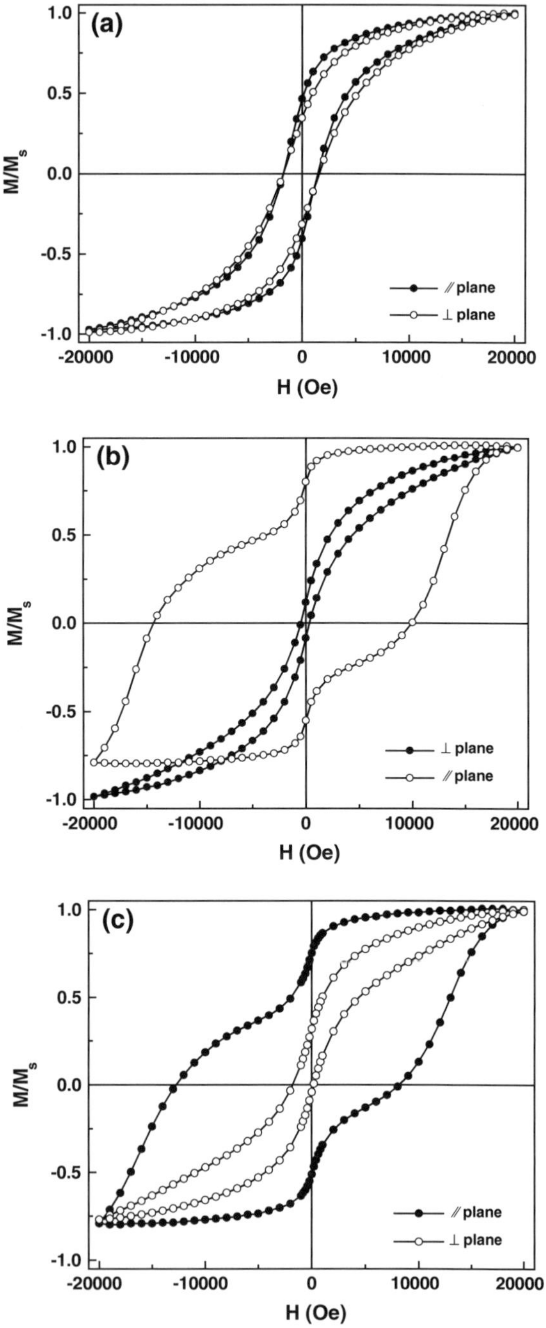

Fig. 6 Magnetic hysteresis loops for the epoxy nanocomposites processed with **a** bare, **b** OA and **c** MWNT-coated $SmCo_5$/Fe powders

XRD peaks are found to be broadened with the progress of milling time. The strains and amorphization induced by the high-energy ball milling may also contribute to the broadening of diffraction peaks. However, for the powders processed with OA and MWNT, the XRD peaks appear to be relatively more intense–showing an indication for possible particle alignment. The bare $SmCo_5$/Fe powder showed a low degree of alignment due to the existence of large size crystallites embedded in the powders, resulting in relatively weak intensity.

Magnetic hysteresis measurement at room temperature showed single-phase behavior for all the $SmCo_5$/Fe powder samples. The coercivity (H_c) values, as well as remanence ratio (M_r/M_{max}) determined from the hysteresis loops were plotted against the milling time and are shown in Fig. 3. The coercivity values for the bare $SmCo_5$/Fe milled powders showed a trend quite similar to the size effect on coercivity in fine particles [8]. It is also possible that the ball milling has led to particle amorphization in the nanoparticles [9] and the smaller particles have more amorphous structure, which leads to the reduced coercivity [10]. On the other hand, the $SmCo_5$/Fe powders processed with OA and MWNT showed continuous increase in coercivity and maximum values of 4.37 and 4.67 kOe, respectively at 10 h of milling time were obtained and these values significantly higher than that of the bare powder obtained at similar milling time. The increase in H_c values may be due to the coating of OA and MWNT over the $SmCo_5$ surface; which acts as a secondary phase in controlling the domain wall motion. Further, on comparing the M_r/M_{max} values of the surfactant-coated powders with those of bare $SmCo_5$/Fe (0.46 %), it is striking to observe that the values M_r/M_{max} ratio for the OA (0.56 %) and MWNT (0.63 %) coated powders are higher than for those milled without surfactants. The increased M_r/M_{max} ratio shows an indication for the existence of anisotropy in the surfactant-coated $SmCo_5$/Fe particles.

In order to have more insight on the particle alignment and anisotropy characteristics, 10 h milled $SmCo_5$/Fe nanocomposite powders were processed in the form of epoxy polymer composites under aligning magnetic field (2 T). Typical SEM micrographs depicting the stacking or alignment of the particles parallel to the field direction are shown in Fig. 4, for the aligned nanocomposites processed with bare, OA and MWNT-coated $SmCo_5$/Fe powders. Though, the particle alignment can be seen in all the cases; it is worth noting that stacking and dispersion of the powder particles within the polymer matrix are more prominent in the case of OA-coated -$SmCo_5$/Fe epoxy composites (Fig. 4b). In the case of MWNT-coated $SmCo_5$/Fe composites, a strong dispersion associated with cementing of fine particles with the polymer matrix can be noticed (Fig. 4c). It is believed that the reorientation of CNTs in a polymeric medium occurs due to the cooperative effect of magnetic torque exerted by the magnetic field directly on the nanotubes and by the hydrodynamic torque and viscous shear (i.e., drag forces) exerted on the nanotubes by the polymer chains, which also respond to the field due to magnetic anisotropy [11].

The above fact on the field-induced particle alignment evidenced in the OA and MWNT-coated $SmCo_5$/Fe epoxy composites can be further corroborated with their corresponding XRD results (Fig. 5). Considering the possible misalignment in the epoxy polymer composites, the isotropic samples had a low degree of alignment; while the aligned samples showed strong peaks–indicating the field-induced texture in the polymer composites. In the case of $SmCo_5$-type magnets, the intensity of (002) plane is considered as measure of texture or the *c*-axis alignment and the degree of texturing for the (002) plane can be calculated in terms of texture coefficient (TC). In the present study, for the aligned samples processed with OA and MWNT-coated $SmCo_5$/Fe powders, a strong (002) diffraction peak can be noticed (Figs. 5b and c) with TC of 2.5 and 2.9, respectively. These values are significantly higher than that of those processed with the bare $SmCo_5$/Fe powder (0.89). The above fact thus apparently confirms that by introducing OA and MWNT during milling improves the *c*-axis alignment of the $SmCo_5$-phase.

The texture effect on the magnetic hysteresis loops of epoxy polymer composites is depicted in Fig. 6. In the case of bare $SmCo_5$/Fe powder, no significant change in their H_c and M_r/M_{max} values was observed in // and $\perp$ planes (Fig. 6a).On the other hand, both OA and MWNT-coated $SmCo_5$/Fe composites showed distinct // and $\perp$ plane anisotropy, in terms of significant change in their H_c and M_r/M_{max} values (Fig. 6b and c). In the case of OA-coated $SmCo_5$/Fe, the measured H_c and M_r/M_{max} values in the // plane are 10.2 kOe and 0.70; while the observed values in the $\perp$ plane are 0.4 kOe and 0.11. As compared to the OA-coated samples, the MWNT-coated $SmCo_5$/Fe showed superior magnetic performance. The measured H_c and M_r/M_{max} values in the // plane are 12.6 kOe and 0.82; while the observed values in the $\perp$ plane are 0.8 kOe and 0.18. The above facts substantiate the existence of better magnetic texture and anisotropy in the case of surfactant-coated $SmCo_5$/Fe epoxy composites. Further, it is interesting to note that a slight step appears on the demagnetization curves of OA and MWNT-coated $SmCo_5$/Fe composites. The two-step demagnetization behavior may result from separate reversal processes of the different parts in the nanocomposites, which is attributed to either the presence of magnetically hard ($SmCo_5$) and soft (α-Fe) phases or the inhomogeneous $_iH_c$ distribution due to the coarse microstructure.

4 Conclusion

In summary, nanostructured anisotropic magnetic particles of OA and MWNT-coated $SmCo_5$/Fe have been produced by magnetic field-assisted milling. Structural and magnetic characterizations reveal that both OA and MWNT-coated $SmCo_5$/Fe particles exhibit field-induced particle alignment and magnetic anisotropy as compared to that of bare $SmCo_5$/Fe nanopowders. Among the two surfactants investigated in this study, MWNT seems to be a promising coating agent for the $SmCo_5$/Fe composites, in terms of providing improved particle alignment and strong dispersability within the polymer matrix. Relatively, higher H_c (12.6 kOe) and M_r/M_{max} (0.82) values were achieved for the MWNT-coated $SmCo_5$/Fe epoxy nanocomposites. The preliminary results of the present study demonstrate that textured nanostructured magnets can be processed by exploiting MWNTs with

$SmCo_5$/Fe nanocomposites. The magnetic properties derived from the nanoscale structure of MWNT-$SmCo_5$/Fe provide opportunities to circumvent traditional performance trade-offs associated with conventional polymer nanocomposites, epitomizing the promise of nano-engineered materials.

Acknowledgments The authors thank the Defence Research and Development Organization (DRDO), Government of India for the financial support to carry out this work. The keen interest shown by the Director, DMRL, in this work is also gratefully acknowledged.

References

1. J. Zhang, S.-Y. Zhang, H.-W. Zhang, B.-G. Shen, Structure, magnetic properties, and coercivity mechanism of nanocomposite SmCo5/α-Fe magnets prepared by mechanical milling. J Appl Phys **89**, 5601 (2001)
2. M. Ito, K. Majima, T. Umemoto, S. Katsuyama, H. Nagai, Magnetic properties and microstructure of $SmCo_5$ + α-Fe nanocomposite magnets prepared by mechanical alloying. J. Alloys Comp. **329**, 272 (2001)
3. M. Chakka, B. Altubcevahir, Z.Q. Jin, Y. Li, J.P. Liu, Magnetic nanoparticles produced surfactant-assisted ball milling. J. Appl. Phys. **99**, 08912 (2006)
4. N. Poudyal, B. Altuncevahir, V. Chakka, K. Chen, T.D. Black, J.P. Liu, Y. Ding, Z.L. Wang, Field-ball milling induced anisotropy in magnetic nanoparticles. J. Phys. D Appl. Phys. **37**, L45 (2004)
5. Y. Wang, Y. Li, C. Rong, J.P. Liu, Sm-Co hard magnetic nanoparticles prepared by surfactant-assisted ball milling. Nanotechnology **18**, 465701 (2007)
6. P. Saravanan, R. Gopalan, N.V. Rama Rao, M. Manivel Raja, V. Chandrasekaran, $SmCo_5$/Fe nanocomposite magnetic powders processed by magnetic field-assisted ball milling with and without surfactant. J. Phys. D Appl. Phys. **40**, 5021 (2007)
7. P. Saravanan, P. Ghosal, V. Chandrasekaran, Multiwalled carbon nanotube-coated $SmCo_5$/Fe magnetic nanocomposites processed by magnetic field-assisted ball milling. Adv. Sci. Lett. **3**, 8792 (2010)
8. D.L.L. Pelecky, R.L. Schalek, Effect of disorder on the magnetic properties of $SmCo_5$. Phys. Rev. B **59**, 457 (1999)
9. C. Suryanarayana, E. Ivanov, V.V. Boldyrev, The science and technology of mechanical alloying. Mater. Sci. Eng., A **304–306**, 151 (2001)
10. R. Gopalan, K. Suresh, D.V. Sridhara Rao, A.K. Singh, N.V. Rama Rao, G. Bhikshamaiah, V. Chandrasekaran, Amorphization, nanocrystallization and magnetic properties of mechanically milled Sm-Co magnetic powders. Int. J. Mater. Res. **99**, 773 (2008)
11. E.S. Choi, J.S. Brookes, D.L. Eaton, M.S. Aal-Haik, M.Y. Hussaini, H. Garmestani, D. Li, K. Dahmen, Enhancement of thermal and electrical properties of carbon nanotube polymer composites by magnetic field processing. J. Appl. Phys. **94**, 6036 (2003)

Novel Porogen Free Porous Hydroxyapatite–Gelatin Nanocomposite: Synthesis and Characterization

K. Sangeetha, S. N. Kalkura, Y. Yokogawa, A. Thamizhavel and E. K. Girija

Abstract A novel method for the preparation of porous Hydroxyapatite–Gelatin (HA-GEL) nanocomposite without the use of any porogen is presented. FT-IR analysis confirmed that the mineral phase of the composite closely resembles that of the natural bone tissue and the chemical bonding between calcium ions of HA and carboxyl ions of GEL molecules induced a gradual blue shift of the 1,634 cm^{-1} band of GEL with decreasing GEL content. XRD pattern showed broad peaks which indicates that the inorganic phase of the composite is poorly crystalline in nature. A higher concentration of GEL induced the formation of very low crystalline HA. SEM observation revealed the formation of homogeneous HA coated porous GEL composite with highly interconnected porous network.

Keywords Hydroxyapatite · Gelatin · Nanocomposite

K. Sangeetha · E. K. Girija (✉)
Department of Physics, Periyar University, Salem 636011, India
e-mail: girijae@yahoo.com

S. N. Kalkura
Crystal Growth Centre, Anna University, Chennai 600025, India

Y. Yokogawa
Graduate School of Engineering, Department of Intelligent Materials Engineering, Osaka City University, Osaka 5588585, Japan

A. Thamizhavel
Department of Condensed Matter Physics, Tata Institute of Fundamental Research, Colaba, Mumbai 400005, India

P. K. Giri et al. (eds.), *Advanced Nanomaterials and Nanotechnology*,
Springer Proceedings in Physics 143, DOI: 10.1007/978-3-642-34216-5_39,

1 Introduction

Bone tissue is a visco-elastic composite made up of long collagen fibrils threading through crystalline hydroxyapatite minerals with highly hierarchical structure [1, 2]. HA is widely used in various biomedical applications such as bone tissue repair and regeneration due to its structural and chemical similarity to bone mineral. HA can be incorporated in the microstructure of gelatin/collagen matrix for mimicking the bone structure. GEL is a product of animal origin biopolymer derived from collagen (the principal intercellular constituent of the white connective tissue of animal skins and bones) by irreversible hydrolytic process. It is a biodegradable natural polymer that exhibits excellent biocompatibility, plasticity, and adhesiveness and also has identical molecular composition and morphological features as that of collagen [3, 4].

In tissue engineering a scaffold serves an important role to deliver biofactors including cells, genes and proteins to regenerate tissues. An ideal scaffold consists of an interconnected macro porous network to provide adequate space for the cell seeding, growth and proliferation. The basic requirements of these scaffolds are biocompatibility, biodegradability, absorbability, expectant mechanical strength, appropriate porous structure and easy processing for the desired shape without unwanted effects [5, 6]. Many synthesis methods have been investigated to prepare porous biodegradable scaffolds for tissue engineering including gas-foaming [7], three-dimensional printing [8], thermal-induced phase separation [9] and electro spinning [10]. Most widely investigated composite for bone tissue engineering scaffolds are composed of polymers reinforced with HA ceramic particles. Traditionally HA-GEL composite have been produced by dispersing the synthesized HA particles into the polymer solution in which it is difficult to obtain a controlled structure with uniform distribution of HA.

In our earlier reports, we had shown that bone like carbonated apatite formed on the reconstituted collagen fibrils [11–13]. In this study we have developed a porogen free novel sol–gel technique for the in situ formation of nano-HA and the formation of homogeneously HA coated porous GEL composite.

2 Experimental Detail

2.1 Materials

Gelatin purified, calcium nitrate tetrahydrate [$Ca(NO_3)_2.4H_2O$, 98 %], di-ammonium hydrogen phosphate [$(NH_4)_2HPO_4$, 99 %], ammonia solution [NH_4OH, 25 %] and 5 % solution of glutaraldehyde ($C_5H_8O_2$) were purchased from Merck. All reagents were used without further purification. Milli-Q water was employed as the solvent.

2.2 Composite Preparation

The GEL solution was prepared by dissolving the required amount of gelatin in the milli-Q water at 40 °C. The Ca and P precursors [0.5 and 0.3 M respectively] were dissolved in water and mixed with the GEL solutions separately. The pH of these solutions was brought to above 10 by the addition of ammonia solution. The solutions of Ca-Gel and P-Gel were stirred separately for 6 h and the reaction temperature was maintained at 50 °C. Then the P-Gel solution was added drop wise into the Ca-Gel solution and stirred for an additional 5 h at the same temperature. The obtained mixture was allowed to mature at room temperature for 12 h. This mixture was cross-linked by 30 μl of Glutaraldehyde (GA) solution and the foams formed were freeze dried for 3 h. The concentration of GEL was changed in the following series: 1.5, 5 and 8 % w/v and the samples were named as HG1.5, HG5 and HG8.

2.3 Characterization

The interaction between HA and GEL were analysed by Fourier transform infrared spectroscopy (FT-IR, Perkin Elmer RXI1). The phase purity and crystallographic structure of the samples were characterized by X-ray diffraction (PANalytical X' Pert PRO diffractometer) by the use of CuKα radiation (1.5406 Å) with voltage and current setting of 40 kV and 30 mA respectively. The lattice parameters were calculated by method of least square and the crystallite size was calculated using the Debye–Scherrer approximation [14], $D_{hkl} = K\lambda/\beta\cos\theta$. where D_{hkl} is the crystallite size as calculated for the (002) reflection, K is the shape factor equal to 0.9, λ is the X-ray wavelength (i.e. 1.5406 Å for CuKα radiation), θ is the Bragg's diffraction angle (in degree) and β is the full width at half maximum (FWHM). The morphology of the composite was observed by Scanning Electron Microscopy (JEOL JSM-6060 model, Japan).

3 Result and Discussions

The FT-IR spectrum (Fig. 1a) of pure GEL exhibited typical bands of C=O stretching at 1,636 and 1,706 cm^{-1} (amide I) and N–H deformation at 1,226 cm^{-1} (amide III). The bands at 2,929 and 1,448 cm^{-1} were assigned to amino acids of GEL structure. The FT-IR spectra of HA-GEL composite show the characteristic PO_4 (v_4) vibrations of HA spectral bands at 566 and 605 cm^{-1} and the broad band around 1,040 cm^{-1} belongs to the bending vibrational mode of PO_4 (v_3). In HG 5 and HG 8, the v_4 peaks of PO_4 around 566 and 605 cm^{-1} cannot be detected individually, instead a broad peak around ~560 cm^{-1} was observed with

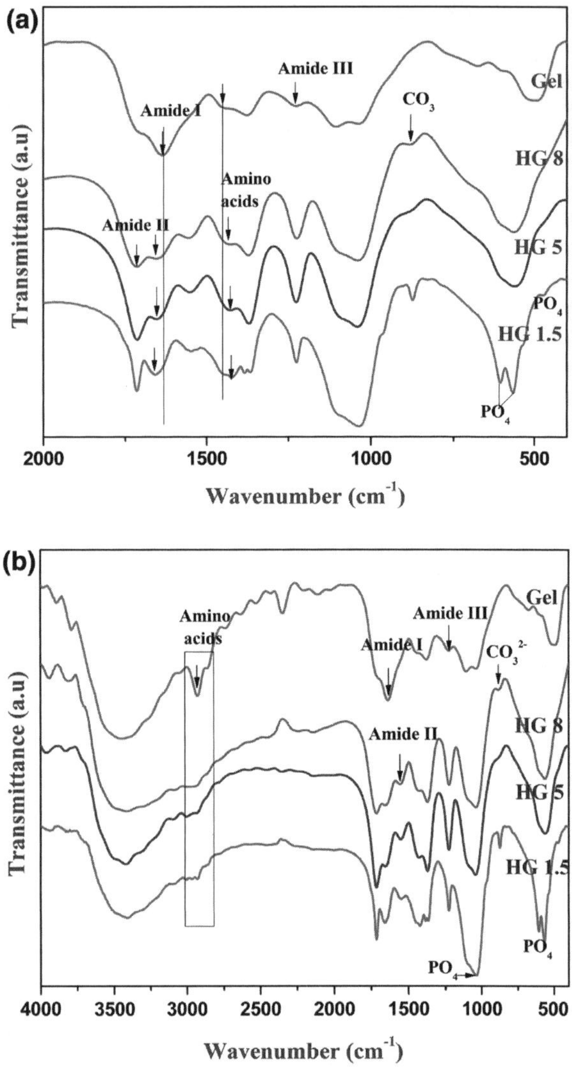

Fig. 1 **a** and **b** FT-IR spectra for pure GEL and HA-GEL composite samples

increasing GEL content. The bands around 875 cm^{-1} appeared for all the samples indicating that during the nucleation of HA, carbonation of inorganic phase also occurs. Moreover, the carbonation can be assigned only to B position; the obtained B-type carbonated apatite closely resembles that of the natural bone tissue [15].

We can obviously see that the typical amide I band of GEL around 1,655 cm^{-1} (peptide bond C=O stretch) indicates that the HA-GEL composite has a predominantly α-helical configuration and this is further confirmed by the presence of amide II band at $\sim$1,551 cm^{-1} (mixed C–N stretch and N–H in-plane bending) which is not observed in the pure GEL [16, 17]. These amide bands of GEL

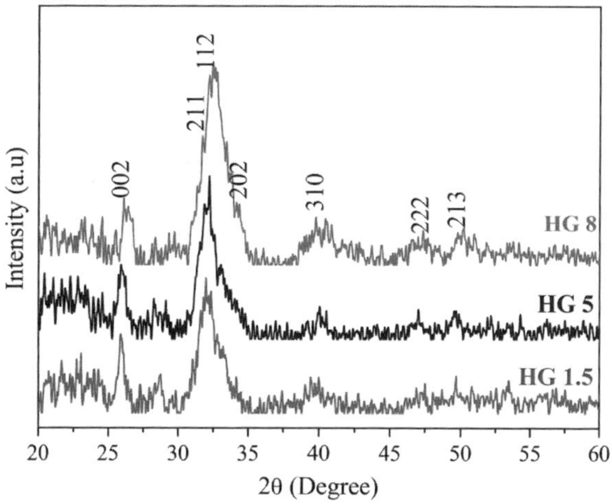

Fig. 2 XRD pattern of the HA-GEL composite samples

resemble that of the collagen bands in the human cortical bone [18]. Also N–H deformation at ~1,224 cm^{-1} for the amide III bands is observed.

The peak at 1,375 cm^{-1} indicates the chelation of Ca^{2+} ions by the carboxylic group of GEL [19, 20]. When GEL content was decreased (Fig. 1b), amide band at 1,636 cm^{-1} revealed a gradual blue shift in wavenumber from 1,634 to 1,659 cm^{-1} which confirmed the chemical interaction between the mineral phase and the organic matrix. In addition the red shift from 1,448 to 1,424 cm^{-1} corresponding to amino acids of GEL indicates the chemical interaction between the GEL amino acids and the calcium and phosphate ionic groups of HA.

XRD pattern of all the samples showed (Fig. 2) the characteristic peaks of HA which are in perfect agreement with the JCPDS data of HA (card no.09-0432). No impurity peaks were observed. The broad peaks centered at ~26 and ~32° corresponds to (002) and (211) plane of HA respectively. These broad peaks indicate that the inorganic phase of the composite is poorly crystalline in nature. When increasing the GEL content the supply of R-COO- coordination sites were abundant for complexation with Ca^{2+} which leads to the formation of very poorly crystalline HA and hence the crystallite size of the sample HG 8 could not be calculated. The lattice parameters and crystallite size calculated for the composite are given in Table 1.

Table 1 Lattice parameters and crystallite size of HA-GEL composite samples

Sample	Lattice parameters (nm)		Crystallite size (nm)
	a	c	
HG 1.5	0.9464	0.6877	48
HG 5	0.9464	0.6874	27
HG 8	0.9385	0.6829	–

Fig. 3 SEM micrographs of HA-GEL composite samples

SEM image of the pure HA (Fig. 3a) shows the spherical nanoporous agglomerates of approximately 1 μm in size, which contains large number of nano-sized particles. Figure 3b–d reports the SEM micrographs of the HA-GEL composite with different GEL contents. When the GEL content was low, non-homogeneously dispersed granular deposits of HA has formed on GEL matrix (Fig. 3b). When increasing the GEL concentration (5 and 8 %), homogeneously HA coated porous GEL composite with highly interconnected porous network was obtained (Fig. 3c and d). It shows apatite has favorably coated on the porous GEL matrix and we cannot distinguish HA and GEL separately. Whereas in the maximum GEL concentration (8 %), highly aligned channel like porous composite was observed. From these results it may be concluded that in terms of porosity the composite can satisfy the required criteria for the body fluids circulations.

The morphological observation of the composite showed the distinct mineral and matrix phases for low concentration of GEL. When GEL content was increased to 5 and 8 %, matrix and mineral phases could not be distinguished. Correspondingly XRD patterns of these samples showed the changes in the characteristic peak intensities. As observed from Fig. 4a–c intensity of (002) plane decreased gradually (211) plane increased and (112) plane increased significantly. This is further evidenced by the blue and red shifts observed for the amide and amino acid groups of the HG 5 and HG 8.

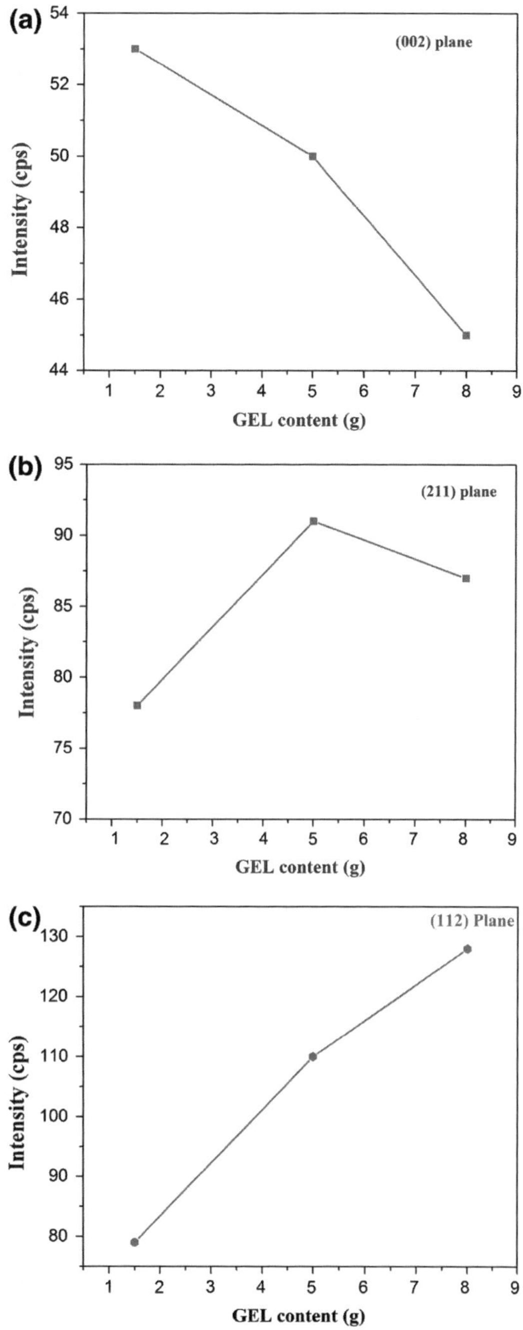

Fig. 4 **a**, **b** and **c** XRD intensity of (002), (211) and (112) planes vs. GEL content respectively

4 Conclusions

HA-GEL nanocomposite was successfully synthesized to mimic both physical architecture and chemical composition of the natural bone. The amount of GEL greatly influenced the nucleation and morphology of the composites. The HA-GEL nanocomposite showed chemical bond formation between HA nanocrystals and GEL macromolecules. Based on these findings, the synthesized nano-HA-GEL composite may be useful for hard tissue regeneration and tissue engineering fields.

Acknowledgments The authors express their sincere thanks to Department of Science and Technology (DST), New Delhi, India for financial support. (Project Ref. No: SR/FTP/PS-24/2009) & (Ref. No. SR/WOS-A/PS-15/2011).

References

1. S. Weiner, H.D. Wagner, The material bone: Structure-Mechanical function relations. Ann. Rev. Mater. Sci. **28**, 271 (1998)
2. M. Jarcho, Calcium phosphate ceramics as hard tissue prosthetics. Clin. Orthop. **157**, 259 (1981)
3. C.L. Tseng, S.Y. Wu, W.H. Wang, C.L. Peng, F.H. Lin, C.C. Lin, T.H. Young, M.J. Shieh, Targeting efficiency and biodistribution of biotinylated-EGF-conjugated gelatin nanoparticles administered via aerosol delivery in nude mice with lung cancer. Biomaterials **29**, 3014 (2008)
4. W.F. Harrington, P.H. Vonhippel, The structure of collagen and Gel. Adv. Protein Chem. **16**, 1 (1961)
5. G.X. Shi, Q. Cai, C.Y. Wang, N. Lu, S.G. Wang, J.Z. Bei, Fabrication of cell scaffold of poly(L-lactic acid) and poly(L-lactic-co-glycolic acid) and biocompatibility. Polym. Adv. Technol. **13**, 227 (2002)
6. Y.S. Nam, T.G. Park, Porous biodegradable polymeric scaffolds prepared by thermally induced phase separation. J. Biomed. Mater. Res. **47**, 8 (1999)
7. D.J. Mooney, D.F. Baldwin, N.P. Suh, J.P. Vacanti, R. Langer, Novel approach to fabricate porous sponges of poly(D, L-lactic-coglycolicacid) without the use of organic solvents. Biomaterials **17**, 1417 (1996)
8. J.K. Sherwood, S.L. Riley, R. Palazzolo, S.C. Brown, D.C. Monkhouse, M. Coates, L.G. Griffith, L.K. Landeen, A. Ratcliffe, A threedimensional osteochondral composite scaffold for articular cartilage repair. Biomaterials **23**, 4739 (2002)
9. P.X. Ma, R.Y. Zhang, Microtubular architecture of biodegradable polymer scaffolds. J. Biomed. Mater. Res. **56**, 469 (2001)
10. J.T. McCann, M. Marquez, Y. Xia, Highly porous fibers by electrospinning into a cryogenic liquid. J. Am. Chem. Soc. **128**, 1436 (2006)
11. E.K. Girija, Y. Yokogawa, F. Nagata, Bone like apatite formation on collagen fibrils by biomimetic method. Chemistry Lett **702**, 7 (2002)
12. E.K. Girija, Y. Yokogawa, F. Nagata, T. Saito, In vitro crystallization of Apatite on type I collagen fibrils: Role of albumin. J. Ceram. Soc. Japan **112**, S853 (2004)
13. E.K. Girija, Y. Yokogawa, F. Nagata, Apatite formation of collagen fibrils in the presence of polyacrylic acid. J. Mater. Sci. Mater. Med. **15**, 593 (2004)
14. R. Jenkins, R.L. Snyder, *Introduction to X-ray powder diffractometry* (Wiley, New York, 1996)

15. C. Du, F.Z. Cui, X.D. Zhu, K. De Groot, Three-dimensional nano-HAp/collagen matrix loading with osteogenic cells in organ culture. J. Biomed. Mater. Res. **44**, 407 (1999)
16. K.J. Payne, A. Veis, Fourier transform IR spectroscopy of collagen and gelatin solutions: deconvolution of the amide I band for conformational studies. Biopolymers **27**, 1749–1760 (1988)
17. B.B. Doyle, Infrared spectroscopy of collagen and collagen-like polypeptides. Biopolymers **14**, 937 (1975)
18. E.P. Paschalis, F. Betts, E. DiCarlo, R. Mendelsohn, A.L. Boskey, FTIR Microspectroscopic Analysis of Human Iliac Crest Biopsies from Untreated Osteoporotic Bone. Calcif. Tissue Int. **61**, 480 (1997)
19. S. Itoh, M. Kikuchi, Y. Koyama, H.N. Matumoto, K. Takakuda, K. Shinomiya, Development of a novel biomaterial, hydroxyapatite/collagen (HAp/Col) composite for medical use. J. Biomed. Mater. Eng **15**, 29 (2005)
20. K.B. Masanori, N. Hiroko, C. Matsumoto, Y. Takeki, K. Yoshihisa, T. Kazuo, T. Junzo, Glutaraldehyde cross-linked hydroxyapatite/collagen self-organized nanocomposites. Biomaterials **25**, 63 (2004)

Effect of γ-Irradiation on Thermal Properties of MWCNTs Reinforced HDPE

Nitturi Naresh Kumar, P. S. Rama Sreekanth and S. Kanagaraj

Abstract Crosslinking by gamma irradiation is a key technique employed in industrial and medical application of polymers to improve their mechanical and thermal properties. The objective of the present work is to investigate the effect of gamma irradiation on thermal properties of high density polyethylene (HDPE) reinforced with multiwalled carbon nanotubes (MWCNTs). The chemically treated MWCNTs were coated on HDPE pellets using nanofluid based deposition technique under continuous stirring, which were processed in an injection moulding machine to obtain nanocomposites having 0.25, 0.50, 0.75 and 1.0 wt% of MWCNTs. The nanocomposites thus obtained were exposed to γ-irradiation in the presence of air at 50 and 100 kGy doses. Thermal characterization of nanocomposites was studied using DSC, TGA and TMA techniques. The results obtained from the DSC studies revealed that the melting point of nanocomposites was not significantly influenced by the presence of MWCNTs, while γ-irradiation resulted a marginal change. The crystallinity of the nanocomposites was increased from 37.4 % for unirradiated HDPE to 55.8 % for 1.0 wt% and 100 kGy irradiated sample, which corresponds to 49 % enhancement. The onset temperature of degradation of nanocomposites was found to be influenced by both MWCNTs and irradiation dosage, which was increased by 16 °C for 100 kGy irradiated 1 wt% nanocomposite compared to unirradiated pure polymer. It is also found that the dimensional stability of the nanocomposites was increased by the reinforcement of MWCNTs. It is concluded from the above studies that thermal properties of nanocomposites are significantly influenced by the irradiation process and the concentration of MWCNTs.

N. N. Kumar · P. S. Rama Sreekanth · S. Kanagaraj (✉)
Department of Mechanical Engineering, Indian Institute of Technology Guwahati,
Guwahati, Assam 781039, India
e-mail: kanagaraj@iitg.ernet.in

P. K. Giri et al. (eds.), *Advanced Nanomaterials and Nanotechnology*,
Springer Proceedings in Physics 143, DOI: 10.1007/978-3-642-34216-5_40,

Keywords Gamma irradiation · MWCNTs · Crystallinity · Thermo-mechanical analysis

1 Introduction

High density polyethylene (HDPE) is a semicrystalline polymer and ideal for industrial and medical applications because of its light weight, ease of processing, resistance to chemical corrosion, toughness and low cost of production. Due to biocompatible and bioinert nature of HDPE, it is explored in bone plates, bone screws, facial implants, orbital implants, artificial limbs and low load-bearing devices. Even though a wide variety of materials is being used as a cervical disc, HDPE has significant relevance due to its characteristics in addition to high cushioning effect [1].

During the last decade, researchers have attempted to improve physical, mechanical and thermal properties of HDPE for various applications. Different types of reinforcements such as carbon nanotubes (CNTs), graphite, hydroxyapatite (HA), calcium carbonate ($CaCO_3$), wollastonite etc., have been used in order to improve the properties of HDPE [2–6]. Among the above mentioned reinforcements, CNTs are an ideal filler material due to their high Young's modulus combined with low density, good electrical and thermal conductivity to prepare polymer nanocomposites [7]. Fouad et al. [8] prepared MWCNTs (up to 8 %) with HDPE via twin-screw extruder and injection moulding. It was observed that MWCNTs had no influence on melting temperature of nanocomposites; however, the crystallinity of the test material was reported to increase. Trujillo et al. [9] studied thermal properties of HDPE/MWCNTs/Al_2O_3 nanocomposites and reported that the onset temperature of degradation and melting temperature of nanocomposites were found to be decreased with an increase of concentration of reinforcement. Chrissafis et al. [10] studied fumed silica nanoparticles (SiO_2)-HDPE composites filled with up to 5 wt% of reinforcement. Though the heat of fusion and the degree of crystallinity of composites were found to be decreased with filler content, no significant improvement on melting temperature was observed. Sahebian et al. [11] reported that the filler content had no influence on melting temperature of HDPE/$CaCO_3$ composites but thermal expansion of them was restricted by the reinforcement of $CaCO_3$. Kodjie et al. [12] prepared HPDE/SWCNTs (0.1, 0.25, 0.50 and 1.0 wt%) nanocomposites by solution blending technique and studied their characteristics. It showed that melting temperature of the nanocomposites was increased by 2 °C but the crystallinity was reduced by 4 % at 1.0 wt% of SWCNTs loading. In addition to that thermal stability of HDPE was also enhanced by 115 °C.

Gamma irradiation is widely used in industries for different purposes including medical sterilization and crosslinking of the polymers, which can be accomplished by chemically or irradiation [13]. As mechanical, thermal and thermo-physical properties of polymers are expected to be influenced by the crosslinking process [14], it is required to study the changes in the properties of polymeric material induced by the irradiation. The effects of gamma irradiation up to 1,000 kGy on HDPE have been studied by Puig et al. [15]. It was reported that the melting temperature of polymer was increased at low irradiation dose up to 150 kGy thereafter it was reduced gradually. Cataño et al. [16] reported that the onset temperature of degradation and melting temperature of HDPE/$CaCO_3$ composites remained constant at different concentration of reinforcement and irradiation dose, however, the activation energy and degree of crystallinity was increased with filler content but not varied significantly with irradiation dose. Carmen et al. [17] studied the effect of gamma irradiation on HDPE/HA composites. It was noted that melting temperature of the composites was remained constant irrespective of concentration of the reinforcement. It was also observed that the crystallinity of composites was increased with filler concentration but it was decreased by 17 % when the test materials were exposed to γ-irradiation.

Though mechanical and thermal properties of MWCNTs reinforced HDPE have been reported, the influence of irradiation effect on thermal properties of the nanocomposites with varying concentration of MWCNTs have been not found in detail in the literature. Hence, the proposed study is focused on thermal characterization of irradiated HDPE/MWCNTs nanocomposites.

2 Materials and Methods

2.1 Materials

The nanocomposites with different concentrations, such as 0.25, 0.50, 0.75, 1.00 wt% of MWCNTs were received in the form of hemispherical cup representing a acetabular cup from Prof. Jose Simoes, University of Aveiro, Portugal. The materials used and processing techniques followed to prepare the nanocomposites were described in detail by Kanagaraj et al. [18].

2.2 Methods

The test materials were gamma irradiated with a ^{60}cobalt source at room temperature in the presence of air at a dosage rate of 2.5 kGy/hr up to 50 and 100 kGy dosage level at M/s Microtrol sterilization Pvt. Ltd, Bangalore. Thermal analysis of the test sample was performed using a NETZSCH simultaneous thermal analyser (DSC & TGA, model STA 449F3). Samples (5.0 $\pm$ 0.1 mg) were heated in Argon atmosphere from room temperature to 550 °C. The thermo-mechanical analysis (TMA) of the nanocomposites was carried out using a Thermo-mechanical analyzer (TMA), Model Exstra 6000 (TMA/SS6000, Seiko Instruments Inc. Japan). The test sample was cylindrical in shape with 8 mm diameter and 2 mm thickness and it was heated from 35 to 85 °C in air under a static load of 10 mN applied on the sample through the expansion probe. All the experiments were performed at a heating rate of 10 °C/min. As recommended by ASTM F2625-10 standard.

3 Results and Discussion

3.1 Differential Scanning Calorimetry Analysis

Melting endotherms of unirradiated and irradiated nanocomposites were analyzed by Differential Scanning Calorimetry (DSC) thermograms and the results are presented in Fig. 1a and b, respectively.

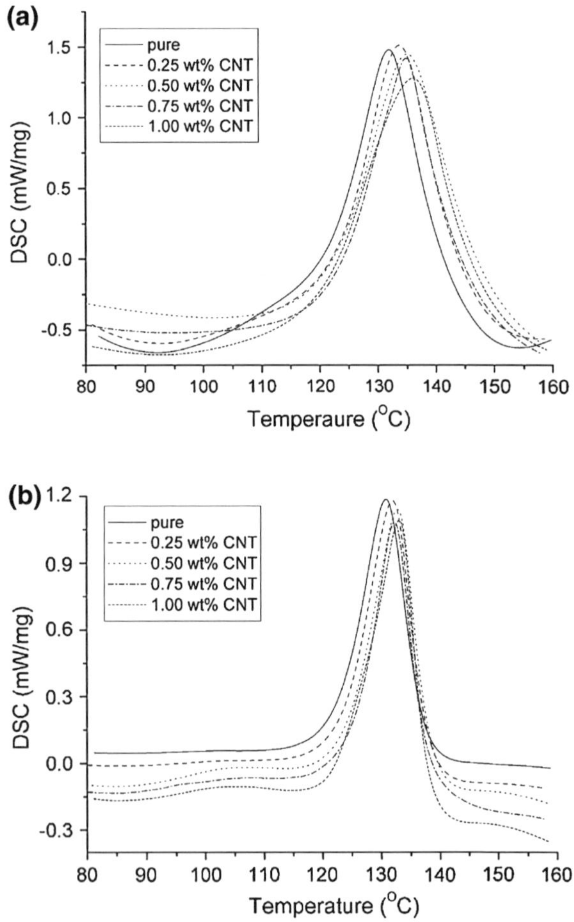

Fig. 1 Endotherms of **a** unirradiated nanocomposites **b** nanocomposites irradiated at 100 kGy

It is observed from Fig. 1a that the melting temperature of unirradiated nanocomposites was found to be slightly varied for different concentrations of MWCNTs. Upon irradiation of the nanocomposites, the melting peak still shifted to higher values, as observed in Fig. 1b. It is observed that the melting temperature of HDPE was raised by increasing the concentration of MWNCTs and γ-irradiation doses. Typical parameters such as melting temperature and degree of crystallinity, were calculated from the DSC plots.

Figure 2a shows the variation of melting point of nanocomposites before and after irradiation. It is observed that the melting point of pure polymer was found to be 131.3 °C, which was increased to 133.8 °C by incorporation of 1.0 wt% MWCNTs corresponding to 1.8 % increase. When the nanocomposites were exposed to gamma irradiation at 100 kGy, the melting temperature of 1 wt% MWCNTs was raised 136.7 °C, which corresponds to a raise of 4 % compared to unirradiated pure polymer. The increase of melting temperature of nanocomposites with irradiation dosage confirmed that the melting process of the test samples was delayed due to the formation of crosslinks produced during the irradiation process, Premnath et al. [19, 20]. Furthermore, an increase of irradiation dose led to higher degree of crosslinking, which resulted in long branched molecules with infinite mass [21] and thus it is expected to demand higher heat input during the melting process resulting an increase of melting temperature of the nanocomposites.

The crystallinity of the nanocomposites was obtained by normalizing the heat of fusion of nanocomposites with that of a 100 % crystalline material, 289.4 J/g in this case [22]. Figure 2b shows the degree of crystallinity of test materials against irradiation dosage and MWCNTs concentration. The crystallinity of nanocomposites was found to be increased with concentration of MWCNTs and gamma irradiation. It was found to be increased from 37.4 to 51.8 % for 1 wt% MWNCTs in HDPE, which was then raised to 55.8 % at 100 kGy irradiation dose. The enhancement of crystallinity was 38.5 % for 1 wt% of HDPE/MWCNTs nanocomposites and 49 % for the same sample irradiated at 100 kGy dose. The increase of crystallinity of the test sample with irradiation dose was due to the fact that chain scission occurred in addition to crosslinking during the irradiation process and these two processes progressed to a higher degree with an increase of dosage level. It was reported by Bhateja et al. [22] that the broken chains formed by the chain scission process were rearranged into the existing crystals, thus an increase of the crystal perfection in an irradiated nanocomposite was occurred. According to Kanagaraj et al. [18], MWCNTs acted as sites for nucleation of crystallization and thus it enhanced the crystallinity of the nanocomposites.

3.2 Thermogravimetric Analysis

Thermal stability of nanocomposites before and after gamma irradiation was studied and it is shown in Fig. 3a and b.

Fig. 2 Irradiation effect on **a** melting temperature and **b** % crystallinity of nanocomposites

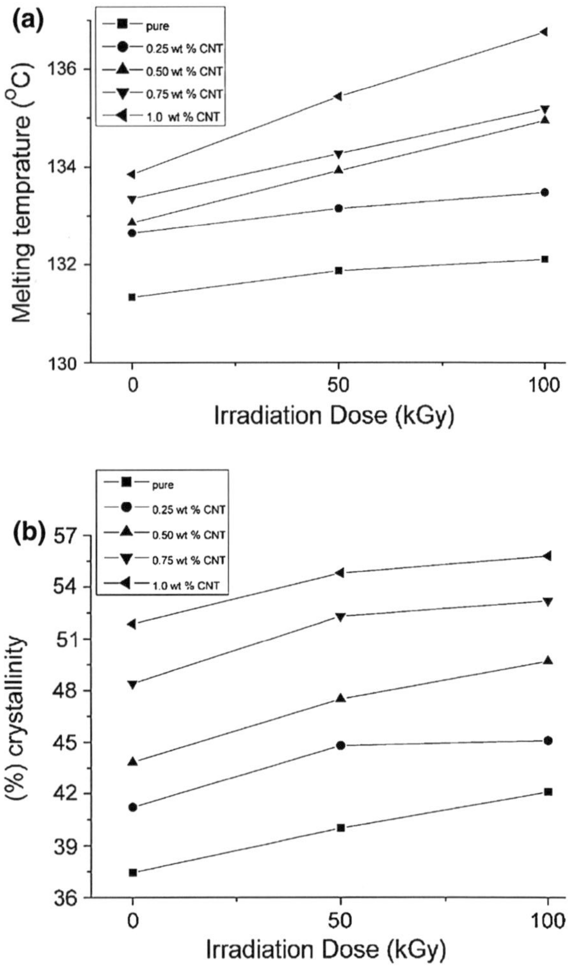

It is observed from Fig. 3 that all unirradiated and irradiated nanocomposites showed higher thermal stability than pure HDPE. It is observed from the results that all nanocomposites are having relatively good thermal stability up to 325 °C at 0 kGy and it was increased to 400 °C at 100 kGy irradiated dose. It is observed from Fig. 4 that the onset temperature of decomposition of unirradiated nanocomposite at 1 wt% of MWNCTs was increased from 457.2 to 467.7 °C, while at 100 kGy the corresponding value was raised from 465 to 473.1 °C. It is due to high thermal stability of MWCNTs that restrict the thermal degradation effect of polymer chains.

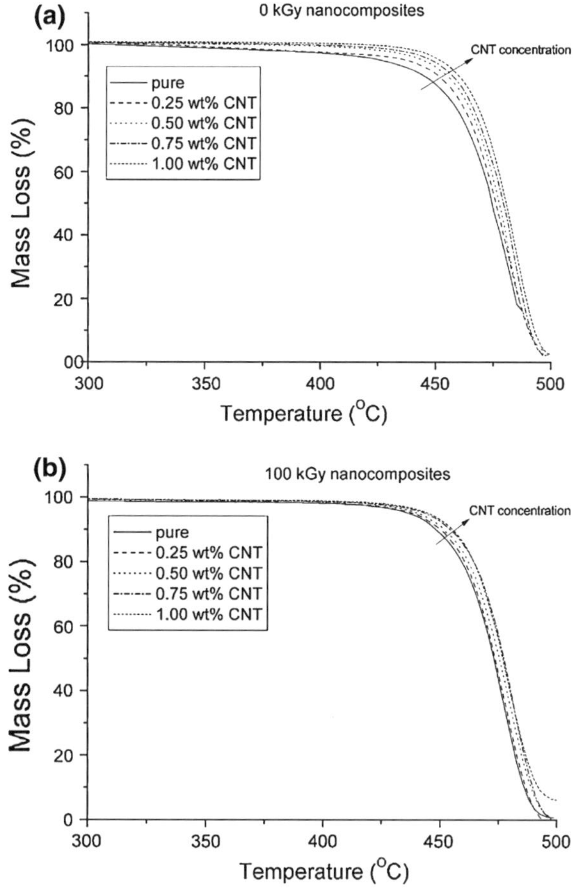

Fig. 3 Thermal stability of **a** unirradiated nanocomposites and **b** irradiated nanocomposites at 100 kGy

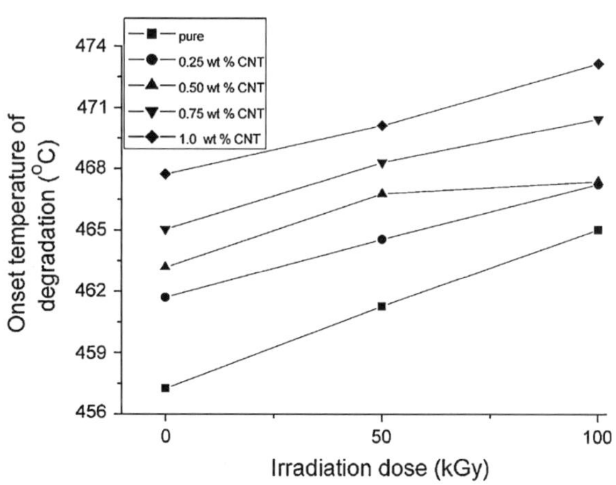

Fig. 4 Effect of irradiation dose on onset temperature of degradation of nanocomposites

3.3 Thermo-mechanical Analysis

In order to investigate the dimensional stability of nanocomposites, thermal expansion of the unirradiated and irradiated samples was studied in the temperature range of 35–85 °C and the results are shown in Fig. 5a and b.

It is observed from Fig. 5 that thermal expansion of both unirradiated and irradiated samples was decreased with an increase of MWCNTs and irradiation dosage but increased with temperature. Thermal expansion of pure HDPE was higher than that of nanocomposites at any concentration and irradiation dose. At 0 kGy, the linear strain of pure polymer was 5.87×10^{-3} m/m and it was reduced to 3.45×10^{-3} m/m at 1 wt% nanocomposite. When the nanocomposites were exposed to gamma irradiation at 100 kGy, the linear strain of 1.0 wt% nanocomposite was further reduced to 2.64×10^{-3} m/m whereas the same for pure HDPE was 3.90×10^{-3} m/m.

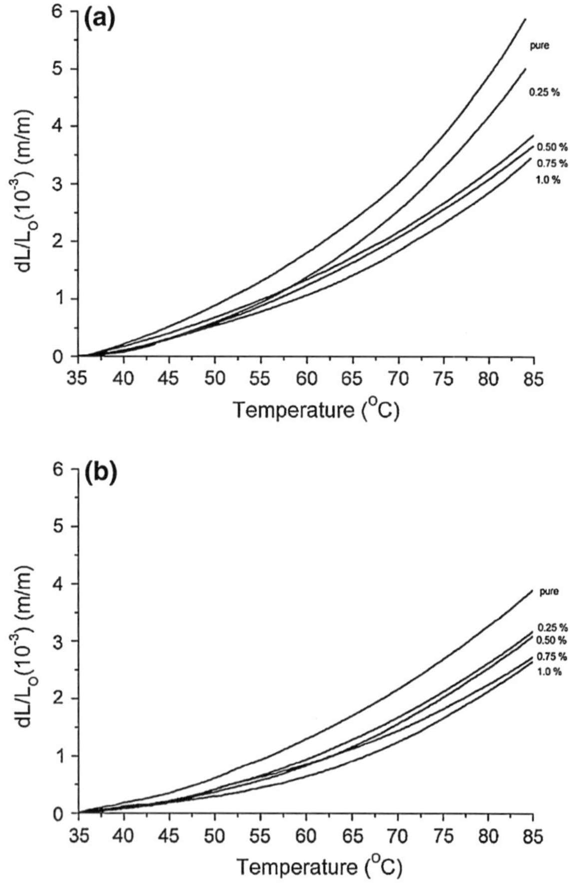

Fig. 5 Linear strain of **a** unirradiated nanocomposites and **b** irradiated nanocomposites at 100 kGy

It is observed from Fig. 6 that thermal expansion of HDPE was decreased by 41.2 and 55 % for 1 wt% of MWNCTs and 100 kGy irradiation, respectively. In general, it is observed that the percentage decrease of linear strain of pure polymer is decreased by increasing the concentration of MWNCTs and the irradiation dosage level. The presence of MWCNTs and crosslinking induced by the irradiation resulted in long molecular chains with infinite mass restricting the mobility of polymer chains, which led to the reduction of linear thermal expansion of polymer and nanocomposites [23].

The coefficient of thermal expansion (CTE) of both unirradiated and irradiated nanocomposites was calculated in the temperature range of 40–80 °C at an interval of 10 °C, where the reference temperature was considered as 35 °C. Figure 7a and b show the CTE of nanocomposites at 0 and 100 kGy irradiation dose, respectively.

It is observed from Fig. 7a, b that (1) CTE of all test samples increased with temperature, (2) CTE of the sample was significantly restricted with an increase of the concentration of MWCNTs, (3) γ-irradiation also influenced to restrict CTE of test samples. It is observed that the CTE of unirradiated pure polymer at 40 and 80 °C was calculated to be 44.9 and 108.6 µm/m-K, respectively, whereas the same for 1 wt% nanocomposite, it was reduced by 62 and 41 %, respectively. Similarly at 100 kGy irradiation dose, the CTE for the pure polymer at same temperatures was noted to be 37.6 and 72.6 µm/m-K, and it was reduced by 57 and 35 %, respectively for 1 wt% nanocomposites. The reduction of CTE of nanocomposites could be attributed to very low intrinsic thermal expansion of nanotubes, which is 26 nm/m up to 800 K [24] and thus it is expected to restrict CTE of pure polymer. Moreover, MWCNTs have very high specific surface area and hence the interfacial area between MWCNTs and polymer matrix is of very high order resulting to the reduction of CTE of the composites.

Fig. 6 Effect of irradiation dose on percentage decrease of linear strain of nanocomposites

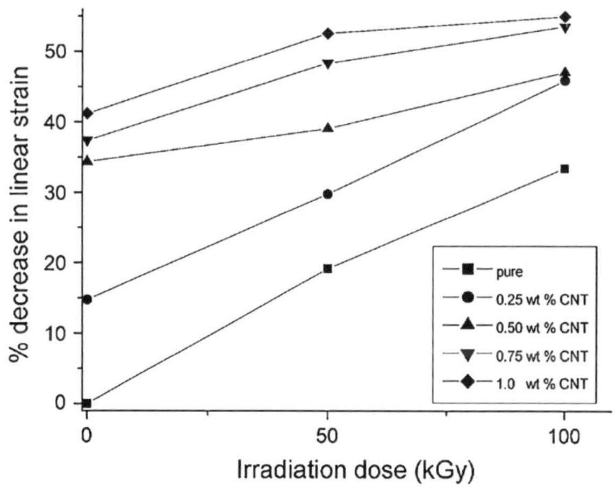

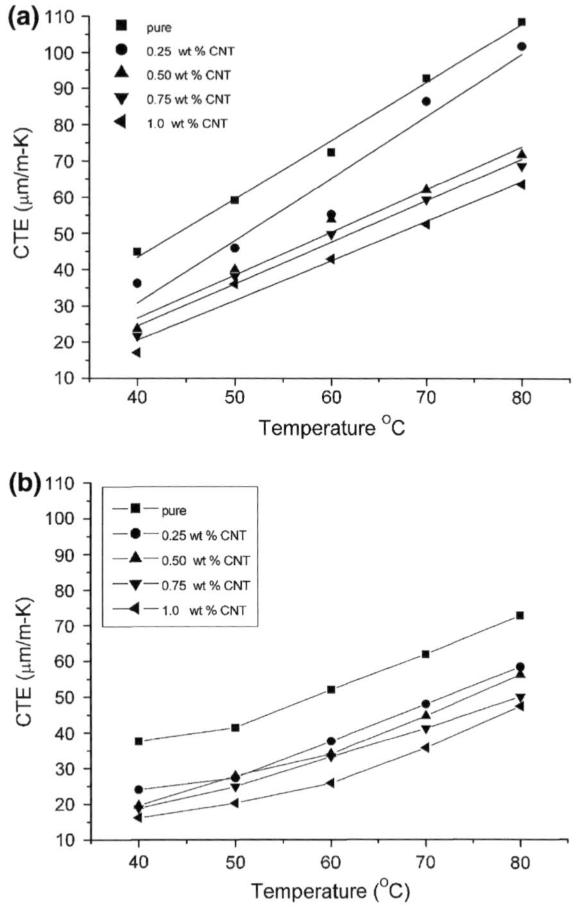

Fig. 7 Coefficient of thermal expansion of nanocomposites at **a** 0 kGy and **b** 100 kGy irradiation dose

4 Conclusions

Thermal characterization of gamma irradiated high density polyethylene reinforced with multiwalled carbon nanotube nanocomposites was studied and the following conclusions are drawn for nanocomposites having 1 wt% MWCNTs and irradiated at 100 kGy.

- The melting temperature and crystallinity of HPDE was enhanced both by presence of MWCNTs and gamma irradiation as well. At 1 wt% MWCNTs in HDPE and 100 kGy irradiation dose, the melting temperature and crystallinity were increased by 5.4 °C and 49 % respectively.
- The onset temperature of degradation of HDPE was delayed by MWCNT reinforcement and irradiation. At 1 wt% MWCNTs and 100 kGy irradiation dose, the degradation was delayed by about 75 °C.

- Dimensional stability of HDPE greatly enhanced by the presence of MWCNTs. It was improved by 56 % upon irradiating 1 wt% nanocomposite at 100 kGy.

Acknowledgments Authors gratefully acknowledge the Department of Science and Technology, India, for funding through a project IDP/MED/2009/04. Technical support given by M/s Microtrol sterilization Pvt. Ltd, Bangalore, Prof. J. A. O. Simoes, University of Aveiro, Portugal, staff from Central Institute Facility and Materials science laboratory at IITG is kindly acknowledged.

References

1. J.B. Park, R.S. Lakes, *Biomaterials: An Introduction*, 2nd edn. (Plenium press, New York, 1992)
2. S.-H. Yu, J.-T. Yeh, B.-C. Huang, K.-S. Huang, Preparation of HDPE/Carbon nanotube composite. Polym. Plast. Technol. Eng. **49**, 15, 1534 (2010)
3. H. Fouad, R. Elleithy, High density polyethylene/graphite nano-composites for total hip joint replacements: Processing and in vitro characterization. J. Mech. Behav. Biomed. Mater. **4**, 1376 (2011)
4. M. Wang, S. Deb, W. Bonfield, Chemically coupled hydroxyapatite-polyethylene composites: processing and characterization. Mater. Lett. **44**, 119 (2000)
5. Z.H. Liu, K.W. Kwok, R.K.Y. Li, C.L. Choy, Effects of coupling agent and morphology on the impact strength of high density polyethylene/$CaCO_3$ composites. Polymer **43**, 2501(2002)
6. R. Yang, J. Yu, Y. Liu, K. Wang, Effects of inorganic fillers on the natural photo-oxidation of high-density polyethylene. Polym. Degrad. Stab. **88**, 333 (2005)
7. P.M. Ajayan, O. Stephan, C. Colliex, D. Trauth, Aligned carbon nanotube arrays formed by cutting a polymer resin–nanotube composite. Science **265**, 212 (1994)
8. H. Fouad, R. Elleithy, S.M. Al-Zahrani, M.A. Ali, Characterization and processing of high density polyethylene/carbon nano-composites. Mater. Des. **32**, 1974 (2011)
9. M. Trujillo, M.L. Arnal, A.J. Müller, E. Laredo, Bredeau, D. Bonduel, Ph. Dubois, Thermal and morphological characterization of nanocomposites prepared by in situ polymerization of high-density polyethylene on carbon nanotubes. Macromolecules **40**, 6268 (2007)
10. K. Chrissafis, K.M. Paraskevopoulos, E. Pavlidou, D. Bikiaris, Thermal degradation mechanism of HDPE nanocomposites containing fumed silica nanoparticles. Thermochim. Acta **485**, 65 (2009)
11. S. Sahebian, S.M. Zebarjad, J.V. Khaki, S.A. Sajjadi, The effect of nano-sized calcium carbonate on thermodynamic parameters of HDPE. J. Mater. Process. Technol. **209**, 1310 (2009)
12. S.L. Kodjie, L. Li, B. Li, W. Cai, C.Y. Li, M. Keating, Morphology and crystallization behavior of HDPE/CNT nanocomposite. J. Macromol. Sci., Part B **45**, 2, 231 (2006)
13. M. Dole, Cross-linking and crystallinity in irradiated polyethylene. Polym. Plast. Technol. Eng. **13**, 41 (1979)
14. S.K. Bhateja, E.H. Andrews, R.J. Young, Radiation-induced crystallinity changes in linear polyethylene. J. Polym. Sci. Part B **21**, 523 (1983)
15. C.C. Puig, C. Albano, E. Laredo, E. Quero, A. Karam, Thermal characterization of the HDPE/LDPE blend (10/90) irradiated using γ-rays. Nucl. Instrum. Methods Phys. Res, Sect. B **268**, 1466 (2010)

16. L. Cataño, C. Albano, A. Karam, N. Domínguez, Y. Sánchez, J. González, Effect of gamma irradiation on mechanical, thermal and rheological behavior of HDPE filled with seaweed residues. Nucl. Instrum. Methods Phys. Res., Sect. B **236**, 348 (2005)
17. A. Carmen, K. Arquímedes, P. Rosestela, G. Gema, D. Nohemy, G. Jeanette, S. Yanixia, HDPE/HA composites obtained in solution: Effect of gamma radiation. Nucl. Instrum. Methods Phys. Res., Sect. B **247**, 331 (2006)
18. S. Kanagaraj, F.R. Varanda, T.V. Zhil'tsova, M.S.A. Oliveira, J.A.O. Simões, Mechanical properties of high density polyethylene/carbon nanotube composites. Compos. Sci. Technol. **67**, 3071 (2007)
19. V. Premnath, W.H. Harris, M. Jasty, E.W. Merrill, Gamma sterilization of UHMWPE articular implants: an analysis of the oxidation problem. Biomaterials **17**, 1741 (1996)
20. V. Premnath, A. Bellare, E.W. Merrill, M. Jasty, W.H. Harris, Molecular rearrangements in ultra high molecular weight polyethylene after irradiation and long-term storage in air. Polymer **40**, 2215 (1999)
21. P.C.P. Watts, P.K. Fearon, W.K. Hsu, N.C. Billingham, H.W. Kroto, D.R.M. Walton, Carbon nanotubes as polymer antioxidants. J. Mater. Chem. **13**, 491 (2003)
22. S.K. Bhateja, R.W. Duerst, J.A. Martens, E.H. Andrews, Radiation-induced enhancement of crystallinity in polymers. J. Macromol. Sci. RMC **4**, 581 (1995)
23. C.F. Deng, Y.X. Ma, P.Zhang, X.X. Zhang, D.Z. Wang (2008), Thermal expansion behaviors of aluminum composite reinforced with carbon nanotubes. Mater. Lett. **62**, 2301 (2008)
24. M.N. Prakash, Master's thesis, Department of mechanical engineering, College of engineering, The Florida State University (2005)

Evidence of Surface Spin Glass Like Ordering and Exchange Bias Effect in Phase Separated $Sm_{0.5}Ca_{0.5}MnO_3$ Nanomanganites

S. K. Giri and T. K. Nath

Abstract In this article, we report that the charge/orbital order state of bulk antiferromagnetic $Sm_{0.5}Ca_{0.5}MnO_3$ is suppressed and confirms the appearance of weak ferromagnetism below 65 K followed by a low temperature spin glass (SG) like transition at 41 K in its nano metric counterpart. Exchange anisotropy effect has been observed in the nanomanganites and can be tuned by the strength of the cooling magnetic field (H_{cool}). The values of exchange bias fields (H_E), coercivity (H_C), remanence asymmetry (M_E) and magnetic coercivity (M_C) are found to strongly depend on cooling magnetic field and temperature. H_E increases with increasing H_{cool} but for larger H_{cool}, H_E tends to decrease due to the growth of finite size ferromagnetic clusters. A proposed phenomenological core–shell type model is attributed to an exchange coupling between the spin-glass like shell (surrounding) and antiferromagnetic core of $Sm_{0.5}Ca_{0.5}MnO_3$ nanomanganites mainly on the basis of uncompensated surface spins. Results suggest that the intrinsic phase inhomogeneity due to the surface effects of the nanostructured manganites may cause exchange anisotropy, which is of special interests for potential application in multifunctional spintronic devices.

Keywords Samarium compound · Nanomanganites · Magnetic anisotropy · Spin glass

---

S. K. Giri (✉) · T. K. Nath
Department of Physics and Meteorology, Indian Institute of Technology, Kharagpur, West Bengal 721302, India
e-mail: samir@phy.iitkgp.ernet.in

T. K. Nath
e-mail: tnath@phy.iitkgp.ernet.in

P. K. Giri et al. (eds.), *Advanced Nanomaterials and Nanotechnology*, Springer Proceedings in Physics 143, DOI: 10.1007/978-3-642-34216-5_41,

1 Introduction

The phase inhomogeneity in nanometer scale is now well established to exists in doped perovskite manganites which exhibits various fundamental features of current interests such as colossal magnetoresistance (CMR), charge ordering (CO), exchange bias effect (EB) and first-order ferromagnetic (FM) transition etc. [1–3]. When a system consisting of ferromagnetic–antiferromagnetic (AFM) [3], FM-spin glass [4], AFM- ferrimagnetic [5] and FM- ferrimagnetic [6] is cooled with the static magnetic field through the Néel temperature (T_N) of the AFM or spin glass temperature (T_{SG}) of the SG, EB is induced showing a shift of the magnetic hysteresis loop [M(H)] along the magnetic field axis. In particular, the magnetic properties of a nanostructure material usually differ from their bulk counterpart due to the surface and size effect. The surface actually leads to a breaking of sub lattice paring in the antiferromagnet and thus to uncompensate surface spin and the appearance of net magnetic moment. Recently, EB effect has been observed in different spontaneously phase separated manganites [7], and cobalites [8], where intrinsic interface exchange coupling between the FM nanodroplets and AFM matrix or SG regions was proposed.

Charge ordered (CO) state is usually observed in half doped manganites with relatively narrow bandwidth system. It is well-known that the antiferromagnetic insulating charge ordered state observed in bulk manganites is suppressed and weak ferromagnetism appears in nano sized manganites. Recently, a few experimental and theoretical studies about the size effect of the CO manganites were reported, in which some interesting effects were revealed when the size reduced to nanoscale [9]. One of the effects is that the antiferromagnetic insulating charge ordered state observed in bulk manganites is suppressed and weak ferromagnetism appears in nano sized manganites, where the key rule of surface effect was argued.

In this present work, we report the suppression of charge ordered state and existence of strong exchange bias effect below T_N in a field cooled magnetic state in a typical nanocrystalline manganites of $Sm_{0.5}Ca_{0.5}MnO_3$, showing spin glass like behavior. This result has important implication for multifunctional nano-device applications of manganites.

2 Experimental Details

Polycrystalline samples of $Sm_{0.5}Ca_{0.5}MnO_3$ have been synthesized through chemical "pyrophoric reaction route" [10]. High purity Sm_2O_3 (99.99 %), $CaCO_3$ (99.99 %) and $Mn(CH_3COO)_2$ (99.99 %) were dissolved in water with appropriate amount of HNO_3. Triethanolamine was then added to this solution and stirred continuously while heating the mixture at 180 °C. Finally, combustion takes place producing a black fluppy powder which, on calcinations, produces polycrystalline samples of $Sm_{0.5}Ca_{0.5}MnO_3$. To synthesize bulk and nano particles of same stoichiometry, we have calcinated one part of the as prepared sample at high temperature (above 1,250 °C) and at 750 °C, respectively. *Thus, we obtained finally two samples of the same stoichiometric formula* $Sm_{0.5}Ca_{0.5}MnO_3$ *having different particle size namely, S17 and SB, where the digits in the code names indicate the corresponding particle sizes. The structural characterization of the samples has been carried out through High Resolution x-ray Diffractometer (HRXRD) from Panalytical (PW3040/60), High Resolution Transmission Electron Microscopy (HRTEM) from JEOL Ltd., Japan, Field Emission Scanning Electron Microscopy (FESEM) and energy dispersive x-ray spectroscopy (EDAX) from Carl Zeiss, Germany*. The ac and static (dc) magnetization, M (T) and M (H), were measured in a superconducting quantum interference device (SQUID) from Quantum Design.

3 Results and Discussion

3.1 Structural Characterization

The high resolution x-ray data (HRXRD) data of the samples have been analyzed through Rietveld refinement using Maud programme. Figure 1a and b shows Rietveld refinement (HRXRD) of single-phase SB and S17 with orthorhombic structure. The average particle size (Φ) estimated using the standard Debye-Scherrer formula ($\Phi = n\lambda/\beta_{\mathrm{eff}} \cos\theta$) for the most intense peak of S17 sample is found to be $\sim$17 nm. Bright field image of S17 [upper inset of Fig. 1b] also shows polydisperse Φ's whose distribution is given by the histogram [inset of Fig. 1b] having average particle size (Φ_{av}) $\sim$ 17 nm. *The histogram is obtained by analyzing several frames of similar bright field images using IMAGE J software and the dispersion in particle size is found to obey log-normal distribution function (black curve in the inset of* Fig. 1b). Inset of Fig. 1a shows the FESEM micrograph of bulk SB sample with a particle size of $\sim$few hundred nm. *The nominal elements of the composition are determined by the EDAX analysis (figure not shown here) of both sample (S17 and SB). We have taken the EDAX spectra at different location in a single sample and they appear to be identical, thus confirming quantitatively good chemical homogeneity of the sample. This confirms that the change in grain size does not affect the stoichiometry of our samples to a considerable extent. The final composition of samples which is determined from EDAX is* $Sm_{0.5}Ca_{0.5}MnO_3$. A details analysis of their crystallographic structure has been carried out on the basis of high resolution x-ray (HRXRD) data. All the samples crystallize in orthorhombic structure with *Pnma* space group. We observe a slight systematic expansion of the unit cell with the decrease of grain size (for S17, V = 219.84 Å^3 and for SB, V = 218.41 Å^3). From the refinement, it is evident that the change in unit cell volume is not significant with the change in grain size. We also notice that the maximum

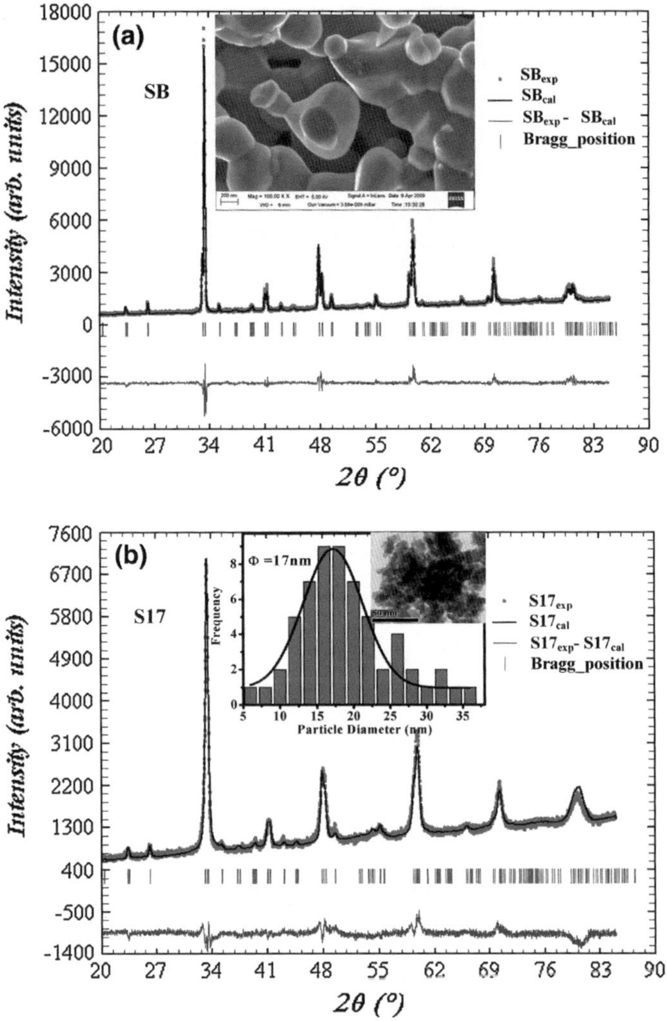

Fig. 1 **a** Rietveld refinement (HRXRD) of XRD patterns of the SB sample. Inset shows the FESEM image of SB sample. **b** Rietveld refinement (HRXRD) of XRD patterns of the S17 sample. Inset shows the histogram showing the size distribution of S17 nanoparticles and TEM micrographs of S17 sample

change in unit cell volume is only $\sim 1\ \%$ for S17 which translates to an insignificant resultant pressure on the grains, calculated from the Birch-Murnaghan (BM) equation of state [11]. This implies that there is no significant role of pressure [12] in the nanomanganites and one has to look for alternatives.

3.2 Dc Magnetization Study

Temperature and magnetic field dependent magnetization [M(T, H)] measurements of all the samples has been done using a SQUID magnetometer (Quantum design, MPMS) from 4 to 300 K at a field of 0.005 T. Both ZFC (zero field cooled) and FC (field cooled) magnetization measurement of all the samples are shown in the Fig. 2a and b. The nature of temperature dependence of M over the whole temperature (T) range is very different from one another for these nano and bulk samples mainly due to the different kind of magnetic ordering. In case of SB [Fig. 2a], a complete charge ordering transition temperature is observed at 270 K as evident from M-T (magnetization vs. temperature) nature. This typical charge order transition temperature ($T_{CO} = 270$ K) consistent with the earlier result [13]. Charge ordering in SCMOBULK system is consistent with spin ordering where an antiferromagnetic transition with CE- type ordering occurs at Neel temperature $T_N = 170$ K. The nearly overlapped ZFC and FC magnetization curves around 34–300 K confirms the antiferromagnetic ordering. Strikingly, the CO transition is not detectable in nanosize S17 sample as shown in the Fig. 2b, clearly showing a destabilization of the antiferromagnetic CO state below 270 K in nanosized grains. *These magnetization behaviors of our nanosized S*17 *sample concur well with recent observation of other CO nanomanganites* [10, 14]. Most likely, this destabilization predominantly occurs at the magnetically disordered surface of the nano grains showing a net moment even below 270 K, originating, presumably, from the unsaturated super exchange [as well as enhanced double exchange (DE)] interactions at the surface region. The strong anisotropic behavior is reflected in the irreversibility in the FC- ZFC magnetization of the nanometric samples. So the large difference between the ZFC and FC curve over a broad temperature range indicates the spin glass like behavior.

3.3 Ac Magnetization Study

To confirm the low temperature glassy behavior of S17 nanomanganite, we have measured the temperature dependent linear ac magnetic susceptibility ($\chi = \chi' + i\chi''$) at a selective frequencies (f) with f = 1, 10, 100, 1,000 and 1,500 Hz in an ac field of 0.0002 T. Temperature dependent real (χ') part of linear ac susceptibility of S17 sample is shown in the Fig. 3. The low temperature peak shifts to a higher temperature with increasing the measuring frequency. Such behavior is a good signature of glassy magnetic state. The frequency dependence peak shift in χ'(T) i.e. $\delta T_f/T_f\ \delta(\log\omega)$ (where T_f is the temperature corresponding to the maxima in χ'in the low temperature regime.) is found to be ~0.01. The value of $\delta T_f/T_f\ \delta(\log\omega)$ for the metallic canonical spin glass system varies from 0.005 to 0.01 and 0.06–0.08 for insulating SG system [15]. So from the result, it is clear that the system S17 behaves more like a metallic spin glass state rather than a

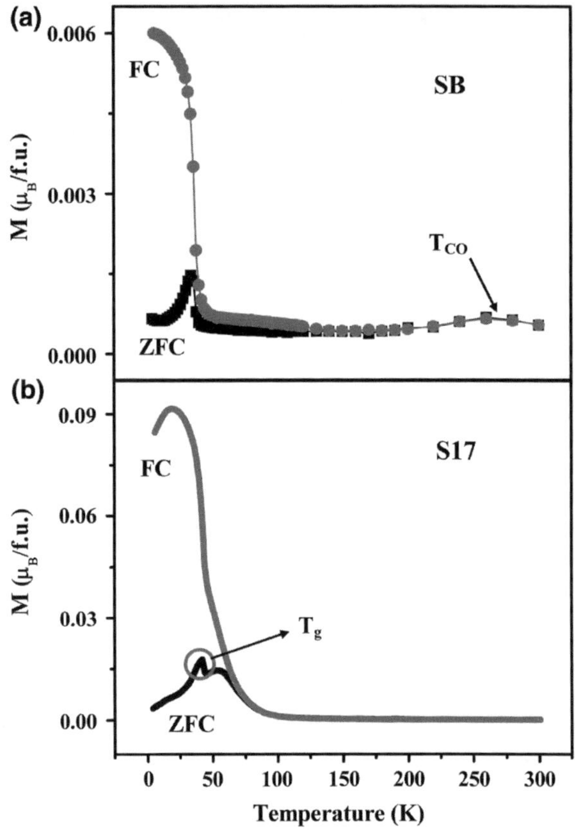

Fig. 2 **a** and **b** Magnetization vs. temperature plots of SB and S17 samples

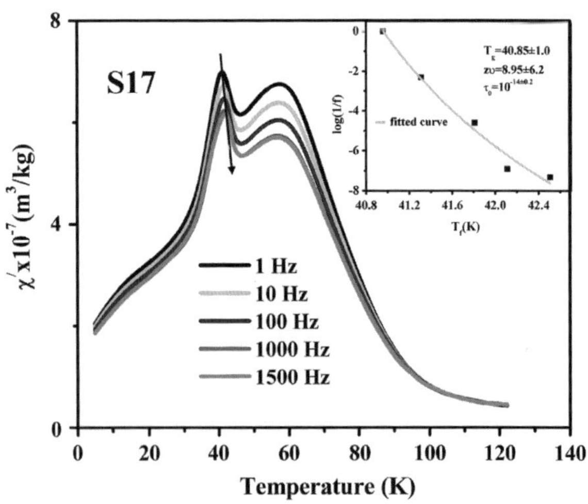

Fig. 3 The real part of linear ac susceptibility vs. temperature plot of S17 sample at different frequencies. Inset shows the plot of log (1/*f*) as a function of freezing temperature T_f

cluster glass state. In case of spin glass system the frequency dependence is very often analyzed by employing the dynamic scaling law or critical slowing down of the form $\tau = \tau_0 (T_f/T_g - 1)^{-z\upsilon}$, where τ_0 is the spin relaxation time, $z\upsilon$ is the critical exponent and T_g is the spin glass transition temperature. T_g is equivalent to the freezing temperature of spin glass in the limit of zero frequency. The best fit from $\chi'(T)$ by dynamical scaling shown in the inset of Fig. 3 produces the value of $T_g = 41$ K which is nearly equal to the dc value found from the temperature dependent dc magnetization [Fig. 2b]. The fitted values of $z\upsilon$ and τ_0 are 9 and 10^{-14} s, respectively. The value of $z\upsilon$ holds satisfactory as generally observed in a canonical SG system in the range 4–12. For a conventional spin glass system typical value of τ_0 is of the order of 10^{-12}–10^{-14} s.

3.4 Exchange Bias Effect

The asymmetric magnetic hysteresis loop (M-H) shifts in both field and magnetization axes, is generally called the exchange bias (EB) phenomena. We observe shift in magnetic hysteresis loop while the S17 sample was cooled in a static dc magnetic field as shown in the Fig. 4. The shift is absent while the cooling magnetic field is almost zero (inset of Fig. 4). The EB field H_E and H_C are generally defined as $-(H_1 + H_2) / 2$ and $-(H_1 - H_2) / 2$, where H_1 and H_2 are left and right coercive field. The remanence asymmetry M_E and the magnetic coercivity M_C are defined as $M_E = -(M_{R1} + M_{R2}) / 2$ and $M_C = -(M_{R1} - M_{R2}) / 2$, M_{R1} and M_{R2} are the positive and negative remanant magnetization, respectively. Figure 5a and b show the cooling field dependence of exchange bias parameter H_E, H_C, M_E and M_C at 5 K for S17 sample. *The value of error in determination of H_E and H_C are* ~1 %. All the above parameters increase sharply up to a field of 3 T as evident from the above

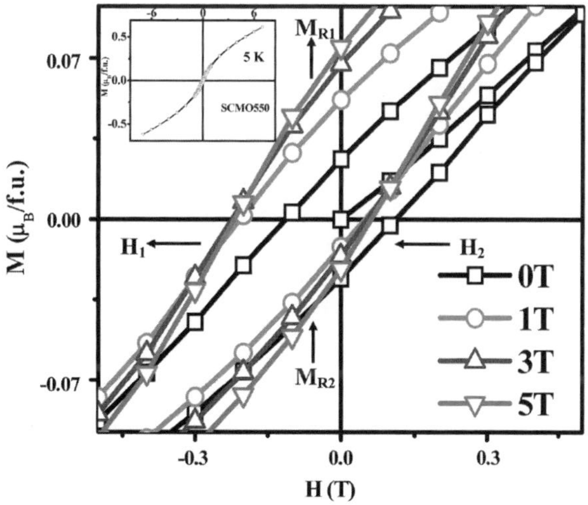

Fig. 4 M-H curves of S17 sample at different cooling fields at 5 K. Upper inset shows the M-H plot of S17 sample at 5 K

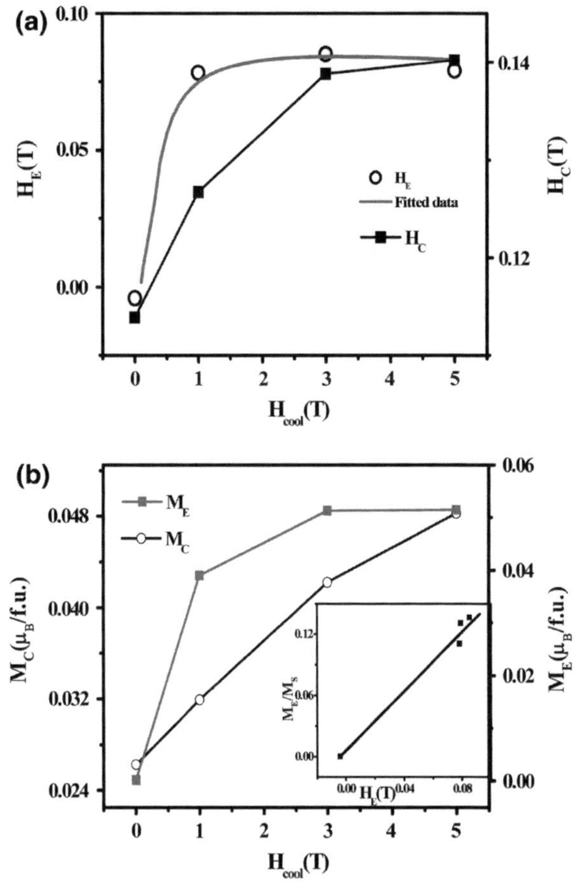

Fig. 5 **a** Variation of H_E and H_C vs. cooling field (H_{cool}) plot of SCMO550 sample. **b** The M_E and M_C vs. cooling field plot of SCMO550 sample. Inset shows the plot of scaled *vertical shift* with *horizontal shift* described in the text

figures. At the higher cooling field M_E tends to saturate, but H_E decreases with increasing the cooling field. M_S is the saturation magnetization, determined using the extrapolation of high field magnetization at $1/H \rightarrow 0$. M_E/M_S is found to increase with H_E. Again the decrease in H_E is accompanied with increase in M_C. So, this suggests that the growth of FM cluster at higher cooling field could contribute partly to the decay of H_E. Consequently, at the high cooling field not only the alignment of degree of FM cluster increases but also the size of FM cluster increases. So EB value reduces due to enhancement of size of FM cluster. According to Bean's relation [3] in case of FM/AFM thin films, the exchange bias field is inversely proportional to the FM layer thickness. The relation is given by

$$H_E = -J_{ex} \frac{S_{AFM} S_{FM}}{\mu_0 t_{FM} M_{FM}} \tag{1}$$

J_{ex} is the exchange constant across the FM/AFM interface per unit area. S_{FM} and S_{AFM} are the interface ferromagnetic and antiferromagnetic magnetization. t_{FM} and M_{FM} are the thickness and magnetization of the FM layer, respectively. So for the strong cooling fields, FM clusters become very large. So, S_{FM} and S_{AFM} decrease and t_{FM} increase, which decrease H_E with increase of cooling magnetic field (H_{cool}). *Due to the existence of glassy behavior of surface spin at low temperature regime, including EB response, Niebieskikwiat and Salamon* [16] *have shown that the H_E depends on H_{cool} by the following relation:*

$$-H_E \propto \frac{M_E}{M_S} \propto J_i \left[\frac{J_i \mu_0}{(g\mu_B)^2} L\left(\frac{\mu H_{cool}}{k_B T_f}\right) + H_{cool} \right] \quad (2)$$

where *L(x)* is the Langevin function, k_B is the Boltzmann constant, and $\mu = N_\nu\ \mu_0$. In Fig. 5a, the red color solid line corresponds to fit of Eq. (2) with an overall scale factor, considering J_i and N_ν are the adjustable parameter. From this equation, the competition between the exchange interaction and the cooling field becomes evident. At low cooling field the first term dominates and depends on J_i^2. However, for large cooling fields the second term becomes important and proportional to J_i, and for $J_i < 0$ the absolute value of H_E would decrease or even more, H_E can change its sign. The exchange constant determined from the best fit of the experimental result, shows negative i.e. $J_i < 0$. This clearly indicates that there is an AFM coupling between the FM domains and the AFM host, explaining the tendency of M_E/M_S towards a reduction at high H_{cool}. So, with increasing H_{cool}, the magnetic coupling (zeeman coupling) between the spin of FM cluster and non-FM matrix at the interface increases as well, tending to orient them along the field direction.

We propose a phenomenological picture (core/shell structure) based on the surface disorder to explain all the above observed results on nanometric samples. *The EB effect observed through the unidirectional shift of the hysteresis loops is induced by the interface exchange coupling between the disordered FM/SG surface and the CO AFM core in this nanomanganite.* A schematic drawing of the nano structure is shown in the Fig. 6. The EB effect can be expected based on the

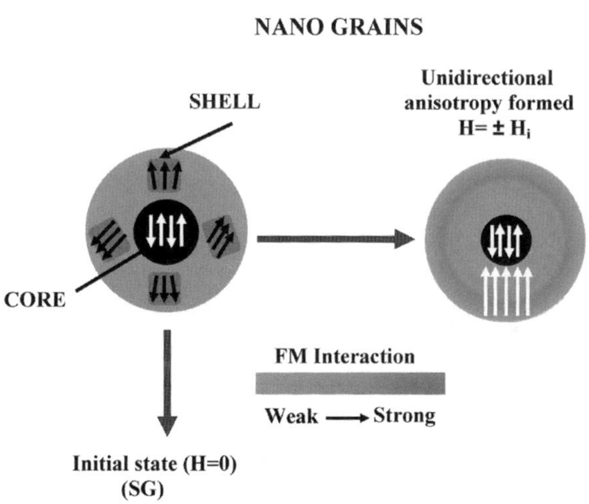

Fig. 6 A schematic phenomenological model of S17 nanograins

fact that there exists a coupling between the SG shell and the AFM core in the nanomanganites, where the SG shell acts as a "FM" component. In case of EB parameter, a low cooling field causes the spins of the SG shell to align partially along the external field, inducing the enhancement of unidirectional exchange interaction between the AFM core and the SG shell. This will contribute to the increase in H_E and M_E. *For lower cooling field, the spins of almost all SG shell align along the cooling field, and the interaction between the AFM core and the SG shell approaches the maximum and, hence,* H_E *and* M_E *tend to saturate up to* 3 *T magnetic field. The decrease in* H_E *at much higher cooling field is due to change of AFM spin structure of the core. There is no change of AFM state at low cooling field as it has CO AFM ordering at ground state. We emphasize the fact that there is a gradual change in the magnetic order from the center to the boundary of these nanograins.* Thus a disordered shell with a gradually ordered core (AFM) region, with the surface effect or the proportion of shell thickness increases as the size of the grain decreases.

4 Conclusions

In summary, the robust charge ordered state in bulk $Sm_{0.5}Ca_{0.5}MnO_3$ is weakened in S17 nanomanganites, accompanied with the emergence of spin glass like phase with ferromagnetic state. Frequency dependent ac magnetic susceptibility clearly suggests the spin glass like transition at 41 K in the nanometric form. Furthermore, after cooling the S17 sample through the Neel temperature with an applied static dc magnetic field, this nanomanganite exhibits that the exchange bias effect can be effectively tuned by the cooling field, which has very useful role for multifunctional application. Our phenomenological hypothesis (core–shell model) is clearly very general and explains qualitatively the exchange bias effects in charge ordering destabilized S17 manganites nanoparticles.

Acknowledgments Authors would like to acknowledge gratefully the financial support of Department Science and Technology (DST), New Delhi, India, through Project No.–IR/S2/PU-04/2006.

References

1. E. Dagotto, Complexity in strongly correlated electronic systems. Science **309**, 262 (2005)
2. Y. Tokura, Critical features of colossal magnetoresistive manganites. Pep. Prog. Phys **69**, 851 (2006)
3. W.H. Meiklejohn, C.P. Bean, New magnetic anisotropy. Phys. Rev. **102**, 1413 (1956)
4. M. Ali, P. Adie, C.H. Marrows, D. Greig, B.J. Hickey, R.L. Stamps, Exchange bias using a spin glass. Nat. Mater. **6**, 75 (2007)

5. G. Salazar-Alvarez, J. Sort, S. Suriñach, M.D. Baro', J. Nogue's, Synthesis and size dependent exchange bias in inverted core-shell MnO ← MnO3 nanoparticles. J. Am. Chem, Soc. **129**, 9108 (2007)
6. W.C. Cain, M.H. Kryder, Investigation of the exchange mechanism in NiFe-TbCo bilayers. J. Appl. Phys. **67**, 5724 (1990)
7. T. Qian, G. Li, T. Zhang, T.F. Zhou, X.Q. Xiang, X.W. Kang, X.G. Li, Exchange bias tuned by cooling field in phase separated $Y_{0.2}Ca_{0.8}MnO_3$. Appl. Phys. Lett. **90**, 012503/3 (2007)
8. Y.-K. Tang, Y. Sun, Z.-H. Cheng, Cooling field dependence of exchange bias in phase-separated $La_{0.88}Sr_{0.12}MnO_3$. J. Appl. Phys. **100**, 023914/3 (2006)
9. Z. Jirák, E. Hadová, O. Kaman, K. Knížek, M. Maryško, E. Pollert, Ferromagnetism versus charge ordering in the $Pr_{0.5}Ca_{0.5}MnO_3$ and $La_{0.5}Ca_{0.5}MnO_3$ nanocrystals. Phys. Rev. B **81**, 024403/8 (2010)
10. S.K. Giri, T.K. Nath, Suppression of charge and antiferromagnetic ordering in $La_{0.5}Ca_{0.5}MnO_3$ nanoparticles. J. Nanosci. Nanotechnol. **11**, 4814 (2011)
11. D.P. Kozlenko, L.S. Dubrovinsky, I.N. Goncharenko, B.N. Savenko, V.I. Voronin, E.A. Kiselev, N.V. Proskurnina, Pressure-induced monoclinic distortion and charge and orbital ordering in $La_{0.5}Ca_{0.5}MnO_3$. Phys. Rev. B **75**, 104408/6 (2007)
12. T. Sarkar, A.K. Raychaudhuri, T. Chatterji, Size induced arrest of the room temperature crystallographic structure in nanoparticles of $La_{0.5}Ca_{0.5}MnO_3$. Appl. Phys. Lett. **92**, 123104/3 (2008)
13. Y. Tokura, Y. Tomioka, Colossal magnetoresistive manganites. J. Magn. Magn. Mater. **200**, 23 (1999)
14. S.S. Rao, S. Tripathi, D. Pandey, S.V. Bhat, Suppression of charge order, disappearance of antiferromagnetism, and emergence of ferromagnetism in $Nd_{0.5}Ca_{0.5}MnO_3$ nanoparticles. Phys. Rev. B **74**, 144416/5 (2006)
15. J.A. Mydosh, *Spin Glasses: An Experimental Introduction* (Taylor and Francis, London, 1993)
16. D. Niebieskikwiat, M.B. Salamon, Intrinsic interface exchange coupling of ferromagnetic nanodomains in a charge ordered manganite. Phys. Rev. B **72**, 174422/6 (2005)

Spin Valve Effect in $Mn_{0.05}Ge_{0.95}$/*p*-Si Structure

S. Bhaumik, S. K. Ray and A. K. Das

Abstract Polycrystalline Mn doped Ge films have been grown on *p*-type Si substrates. The films show hysteresis loop at room temperature (RT) and saturation magnetization increases with decrease in temperature. The presence of multi-magnetic phases having blocking temperature of $\sim$112 K and Curie temperature above RT is observed. $Mn_{0.05}Ge_{0.95}$/*p*-Si heterostructure shows rectification property under forward and reverse bias like a semiconducting diode at RT and works like a spin valve at lower temperature. For a fixed forward bias and at a low temperature, huge change in junction magneto resistance (JMR) is observed.

Keywords Mn doped Ge • Blocking temperature • Semiconducting diode • Spin valve

1 Introduction

The revolutionary effects of spintronics, i.e., spin controlled electronics have started with the invention of giant magnetoresistance (GMR) and have a great impact on modern memory storage devices [1]. GMR based technology depends on the spin scattering between the ferromagnetic and the antiferromagnetic layers. To get the spin amplification along with the spin polarization, one has to choose the other route of spintronics, namely, semiconductor based spintronics, especially diluted magnetic semiconductor (DMS) where magnetic impurity is incorporated

S. Bhaumik · S. K. Ray · A. K. Das (✉)
Department of Physics and Meteorology, IIT Kharagpur, Kharagpur,
West Bengal 721302, India
e-mail: amal@phy.iitkgp.ernet.in

P. K. Giri et al. (eds.), *Advanced Nanomaterials and Nanotechnology*,
Springer Proceedings in Physics 143, DOI: 10.1007/978-3-642-34216-5_42,

into the semiconductor to get the simultaneous effect of magnetic and semiconducting property [2]. Oxide based DMSs [3] and groups III–V based DMSs [4] have been studied extensively, but still are far away from any technological application.

Among various semiconductors, Ge based DMSs have got a great attention in modern research, not only due to its compatibility with the modern Si based technology, but also due to its high mobility. A *p*-type behaviour with carrier concentration of 10^{20} cm^{-3} with Curie temperature (T_C) of 116 K has been reported by Park et al. [5] in Mn doped Ge, whereas a T_C of 285 K has been observed by Cho et al. [6] for 6 % Mn doping. Stroppa et al. [7] found MnGe system to be half metallic with 100 % spin polarization from a theoretical comparison between MnGe and MnSi, although both systems were stabilized in ferromagnetic states, which was also predicted by Zhou et al. [8] from first principle band structure calculations. Self organized GeMn nanocolumns with high Curie temperature has been achieved by Jamet et al. [9]. Ottaviano et al. [10] also found the room temperature ferromagnetism in Mn implanted Ge.

The basic elements of semiconductor spintronics [11] would be spin (magnetic) diodes [12–14], transistors [15–17] and FETs [18]. Successful demonstration of spin light emitting diode [19] and resonant tunneling diode [20] has confirmed the injection of spin polarized carriers into semiconductors. Few experimental demonstrations on magnetic *p-n* junction diodes [21–23] have been made. Here we report on the fabrication of magnetic/nonmagnetic junction diode composed of *p*-type Mn-doped Ge film grown on *p*-type silicon substrate. In the absence of magnetic field, $Mn_{0.05}Ge_{0.95}$/*p*-Si heterostructure behaves like a conventional semiconducting diode under forward and reverse biases and also works like a spin valve in the absence (H = 0) and in presence (H = 150 Oe) of magnetic fields under forward bias at low temperatures.

2 Experimental Details

MnGe film with 5 % Mn concentration was grown on a p-Si substrate having resistivity of 1–10 Ω-cm. The Pulsed Laser Deposition (PLD) technique using KrF excimer laser having wavelength of 248 nm was used under 6×10^{-6} mbar pressure. Two different targets of Ge and Mn with high purity were used for the deposition. The substrate temperature was kept at 60 °C. The grown film was then in situ annealed at 600 °C for 30 min. X-ray diffraction (XRD) measurements were performed using X-pert pro MRD cradle for structural characterization along with atomic force microscopy (AFM) and cross sectional field emission scanning electron microscopy (FESEM). Magnetic measurements were carried out using a superconducting quantum interference device (SQUID) magnetometer. For magnetotransport measurements, cooled head of closed cycle He cryostat in Brucker electromagnet of 2 T was used.

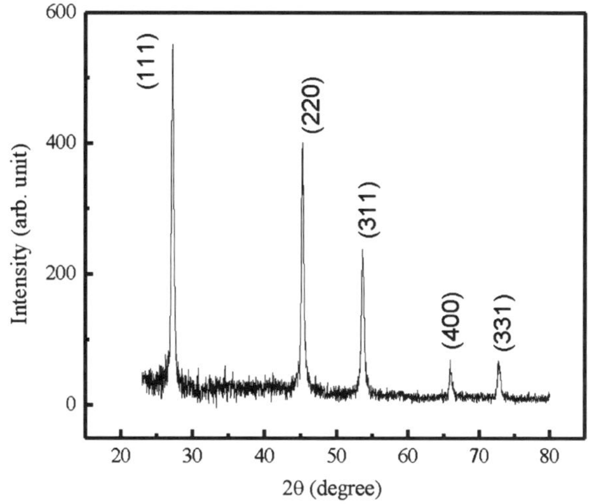

Fig. 1 XRD pattern for $Mn_{0.05}Ge_{0.95}$ film annealed at 600 °C

3 Results and Discussions

XRD plots in Fig. 1 reveal that the film is polycrystalline with Ge peaks corresponding to (111), (220), (311), (400), (331) lattice planes. There is an improvement in crystallinity upon annealing. We have not found any additional peaks due to precipitation of Mn. Cross-sectional FESEM measurements showed that the film thickness is ~50 nm. Energy dispersive spectrometry of the film revealed that the film has a Mn concentration about 5 % (s are not shown here). Typical surface images of the film taken in AFM are shown in Fig. 2. It reveals that the average surface roughness of the film is ~5 nm.

Magnetization (M) as a function of magnetic field (H) at different temperatures is shown in Fig. 3. It is observed that $Mn_{0.05}Ge_{0.95}$ is ferromagnetic with Curie

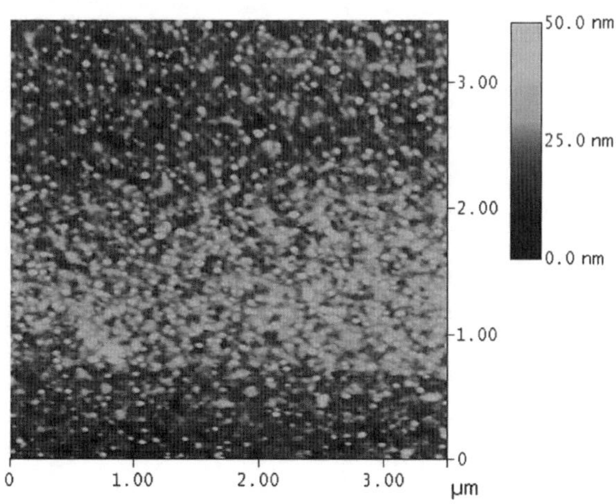

Fig. 2 AFM image of the $Mn_{0.05}Ge_{0.95}$ film

Fig. 3 M-H curves at different temperatures. Inset shows the coercive field and remnant magnetization at different temperatures

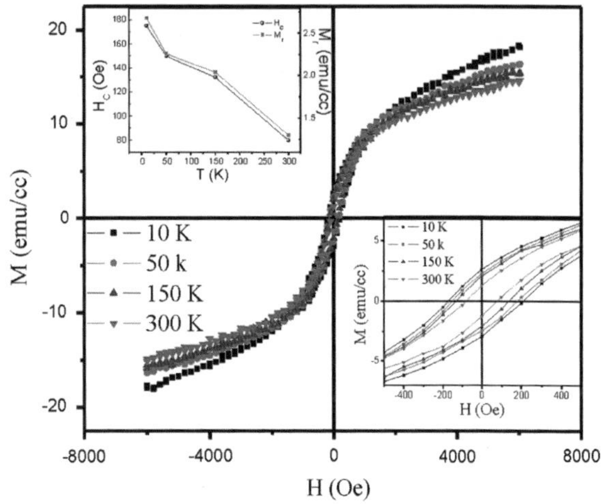

temperature above 300 K. The saturation magnetization (M_S), remnant magnetization (M_r) and coercive magnetic field (H_C) at RT are found to be $M_S = 14.65$ emu/cc, $M_r = 1.30$ emu/cc, and $H_C = 80$ Oe respectively. The M_S increases with decreasing temperature of measurements (T_m). The M-H data near $H = 0$ are magnified and shown in the inset (bottom). The variation of H_C and M_r with T_m is also shown in the inset (top). It clearly shows that both H_C & M_r increase with the decrease of temperature. The temperature (T_m) variation of magnetizations under field cooled (FC) and zero field cooled (ZFC) conditions are shown in Fig. 4. In addition to the ferromagnetic phase, the splitting of FC-ZFC curves indicates the presence of superparamagnetic phase with blocking temperature around 300 K. One can also see the sudden change of slope ($\frac{dM}{dT}$) at T ≈ 120 and T ≈ 12 K,

Fig. 4 FC-ZFC curves for $Mn_{0.05}Ge_{0.95}$

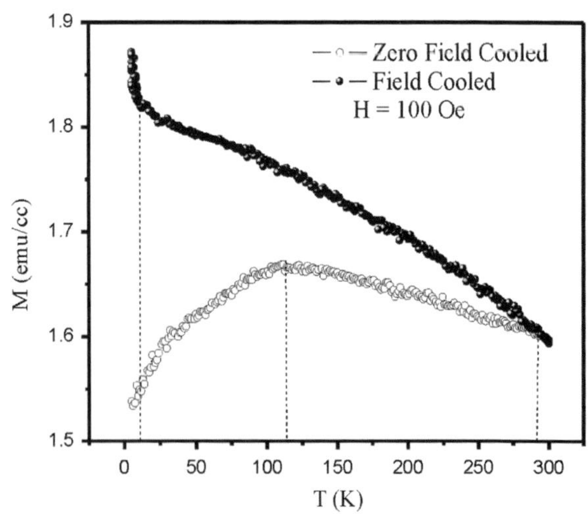

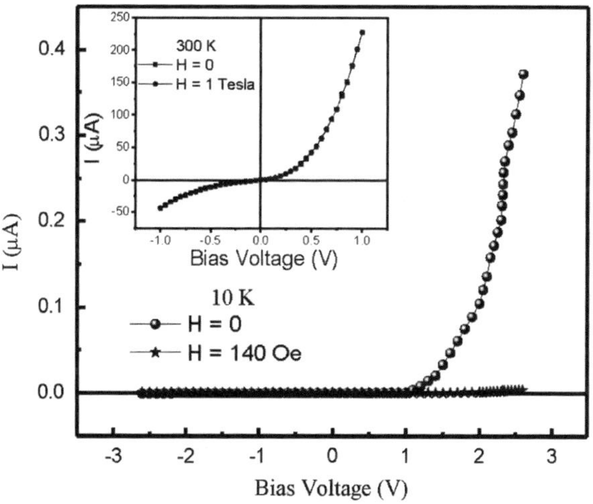

Fig. 5 I–V characteristic for $Mn_{0.05}$ $Ge_{0.95}$/p-Si with and without magnetic field at 10 K, inset shows the I–V characteristic for $Mn_{0.05}Ge_{0.95}$/p-Si at room temperature (300 K)

which we attribute to the other magnetic phases such as Mn-rich clusters and spin glass.

As $Mn_{0.05}Ge_{0.95}$ film was grown on *p*-Si, we have performed magneto-electrical characterization of this heterostructure. In top–bottom configuration we have measured current (I) under forward and reverse bias at RT and 10 K in on–off condition of magnetic field. The results are shown in Fig. 5. The I–V characteristics at H = 0 in Fig. 5 clearly shows the normal semiconducting diode behavior at both temperatures (RT and 10 K). But at H = 140 Oe, a sharp reduction in forward current at 10 K is observed, whereas no effect of magnetic field on forward current is seen at RT. Hence the heterostructure behaves like a spin valve at low temperatures. We have also measured junction magneto resistance (JMR) at 10 K fixed at a forward bias of 3 V. The electrical connection in top–bottom

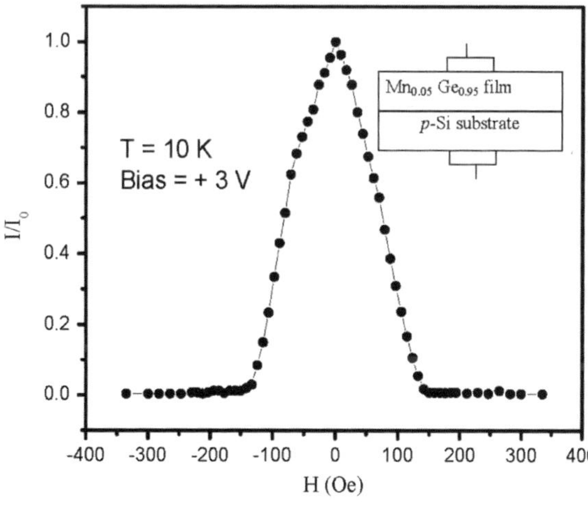

Fig. 6 Change of current as a function of magnetic field normalized to current (I_0) at H = 0 for $Mn_{0.05}$ $Ge_{0.95}$/p-Si in *top–bottom* configuration. Inset shows the *top–bottom* electrical connections

geometry is shown in inset of Fig. 6, where JMR is plotted as a function of magnetic field. The variation of JMR is symmetric with respect to H = 0 and saturates near H = 150 Oe, which is close to the coercive field at T = 10 K. This variation can be correlated to the variation of square of magnetization (M^2).

4 Conclusion

Mn doped Ge films were grown on *p*-Si and structural, magnetic and magneto electric characterizations were performed on the heterostructures. The film of $Mn_{0.05}Ge_{0.95}$ is polycrystalline in nature and contains multimagnetic phases including ferromagnetic phase. The $Mn_{0.05}Ge_{0.95}$/*p*-Si heterostructure behaves like a normal semiconducting diode at all temperatures, but in addition spin valve behaviour is observed at low temperature.

Acknowledgments One of the authors (S. Bhaumik) acknowledges CSIR, Govt. of India for awarding Senior Research Fellowship.

References

1. P. Grunberg, R. Schreiber, Y. Pang, M.B. Brodsky, H. Sowers, Layered magnetic structures: Evidence for antiferromagnetic coupling of Fe layers across Cr interlayers. Phys. Rev. Lett. **57**, 2442 (1986)
2. H. Ohno, Making nonmagnetic semiconductors ferromagnetic. Science **281**, 951–956 (1998)
3. Y. Matsumoto, M. Murakami, T. Shono, T. Hasegawa, T. Fukumura, M. Kawasaki, P. Ahmet, T. Chikyow, S.-Y. Koshihara, H. Koinuma, Room-temperature ferromagnetism in transparent transition metal-doped titanium dioxide. Science **291**, 854–856 (2001)
4. A.C. Mofor, A. El-Shaer, A. Bakin, A. Waag, H. Ahlers, U. Siegner, S. Sievers, M. Albrecht, W. Schoch, N. Izyumskaya, V. Vrutin, Magnetic property investigation on Mn doped ZnO layers on sapphire. Appl. Phys. Lett. **87**, 062501 (2005)
5. Y.D. Park, A.T. Hanbicki, S.C. Erwin, C.S. Hellberg, J.M. Sullivan, J.E. Mattson, T.F. Ambrose, A. Wilson, G. Spanos, B.T. Jonker, A group-IV ferromagnetic semiconductor: Mn_xGe_{1-x}. Science **295**, 651–654 (2002)
6. S. Cho, S. Choi, S.C. Hong, Y. Kim, J.B. Ketterson, B.J. Kim, Y.C. Kim, J.H. Jung, Ferromagnetism in Mn doped Ge. Phys. Rev. B **66**, 033303 (2002)
7. A. Stroppa, S. Picozzi, A. Continenza, A.J. Freeman, Electronic structure and ferromagnetism of Mn-doped group-IV semiconductors. Phys. Rev. B **68**, 155203 (2003)
8. X.H. Zhou, X.S. Chen, X.G. Guo, L.Z. Sun, Y.L. Sun, W. Lu, Ferromagnetism of 3D-impurities substituted in Ge. J. Magn. Magn. Mater. **284**, 253 (2004)
9. M. Jamet, A. Barski, T. Devillers, V. Poydenot, R. Dujardin, P. Bayle- Guillemaud, J. Rothman, E. Bellet-Amalric, A. Marty, J. Cibert, R. Mattana, S. Tatarenko, High-Curie-temperature ferromagnetism in self-organized $Ge_{1-x}Mn_x$ nanocolumns. Nat. Mater. **5**, 653 (2006)
10. L. Ottaviano, A. Continenza, G. Profeta, G. Impellizzeri, A. Irrera, R. Gunnella, O. Kazakova, Room-temperature ferromagnetism in Mn-implanted amorphous Ge. Phys. Rev. B **83**, 134426 (2011)

11. I. Žutić, J. Fabian, S.C. Erwin, Bipolar spintronics: From spin injection to spin-controlled logic. J. Phys.: Condens. Matter **19**, 165219 (2007)
12. M.E. Flatte, G. Vignale, Unipolar spin diodes and transistors. Appl. Phys. Lett. **78**, 1273 (2001)
13. I. Žutić, J. Fabian, S. Das Sarma, Spin-Polarized transport in inhomogeneous magnetic semiconductors: Theory of magnetic/nonmagnetic p-n junctions. Phys. Rev. Lett. **88**, 066603 (2002)
14. J. Fabian, I. Zutic, S. Das Sarma, Theory of spin-polarized bipolar transport in magnetic p-n junctions. Phys. Rev. B **66**, 165301 (2002)
15. M.E. Flatte, Z.G. Yu, E.J. Halperin, D.D. Awschalom, Theory of semiconductor magnetic bipolar transistor. Appl. Phys. Lett. **82**, 4740 (2003)
16. J. Fabian, I. Zutic, S. Das Sarma, Magnetic bipolar transistor. Appl. Phys. Lett. **84**, 85 (2004)
17. J. Fabian, I. Zutic, S. Das Sarma, Spin-polarized current amplification and spin injection in magnetic bipolar transistors. Phys. Rev. B **69**, 115314 (2004)
18. S. Datta, B. Das, Electronic analog of the electro-optic modulator. Appl. Phys. Lett. **56**, 665 (1990)
19. R. Fiederling, M. Kelm, G. Rauscher, W. Ossau, G. Schmidt, A. Waag, L.W. Molenkamp, Injection and detection of a spin-polarized current in a light-emitting diode. Nat. London **402**, 787 (1999)
20. A. Slobodskyy, C. Gould, T. Slobodskyy, C.R. Becker, G. Schmidt, L.W. Molenkamp, Voltage-controlled spin selection in a magnetic resonant tunneling diode. Phys. Rev. Lett. **90**, 246601 (2003)
21. F. Tsui, L. Ma, L. He, Magnetization dependent rectification effect in Ge-based magnetic heterojunction. Appl. Phys. Lett. **83**, 954 (2003)
22. S. Majumdar, A.K. Das, S.K. Ray, Magnetic semiconducting diode of p-$Ge_{1-x}Mn_x$/n-Ge layers on Si substrate. Appl. Phys. Lett. **94**, 122505 (2009)
23. R. Adhikari, A. Sarkar, G.R. Patta, A.K. Das, Magnetic diode exploiting giant positive magnetoresistance in ferrite/semiconductor heterostructures. Appl. Phys. Lett. **98**, 183504 (2011)

Herringbone Nanostructure and Composition Dependent Irreversibility in Martensite Transition Parameters in $Ni_{39+x}Mn_{50}Sn_{11-x}$ ($x \leq 2.0$) Heusler Alloys

A. A. Prasanna and S. Ram

Abstract The ferromagnetic Heusler $Ni_{39+x}Mn_{50}Sn_{11-x}$ ($x \leq 2.0$) alloys have been developed in tailoring the martensite transition with functional magnetocaloric properties. A single martensite phase of a herringbone nanostructure grows in a specific composition $Ni_{41}Mn_{50}Sn_9$ (x = 2.0) which exhibits large irreversibility in the enthalpy change (ΔH), heat-capacity change (ΔC_P), and magnetization (σ) during heating and cooling in the martensite $\leftrightarrow$ austenite transition. As large value as $\Delta H_{M\leftarrow A}$–$\Delta H_{M\rightarrow A} = 0.12$ J/g has been observed with $\Delta C_{P(M\leftarrow A)}$–$\Delta C_{P(M\rightarrow A)}$ $\sim$22.5 mJ/kg-K from a DSC thermogram of a broad peak over from 225 K to 375 K. Thermomagnetic curves measured at a low magnetic field $B = 5$ mT reveal $\sim$58 % lowered σ-value on heating a field cooled sample from a super paramagnetic martensite state at 250 K to a ferromagnetic austenite state with a distinct peak at 320 K. This irreversibility is highly sensitive to the alloy composition. As a result, a small change in the Sn-content $\sim$9.5 at % (x = 1.5) gives reasonably much lowered $\Delta H_{M\leftarrow A}$–$\Delta H_{M\rightarrow A} = 0.1$ J/g and $\Delta C_{P(M\leftarrow A)}$–$\Delta C_{P(M\rightarrow A)}$ = 10 mJ/kg-K in a martensite-austenite composite phase.

Keywords Enthalpy irreversibility · Granular nanostructure · Heusler alloys · Magnetocaloric properties

1 Introduction

The off-stoichiometric Heusler alloys $Ni_{50}Mn_{25+x}Z_{25-x}$ (Z = Ga, Sn, In, or Sb, with $9 \leq x \leq 14$) offer a composition sensitive martensite transition in which a structural transition coincides with a magnetic transition in the form of a first order

A. A. Prasanna · S. Ram (✉)
Materials Science Centre, Indian Institute of Technology, Kharagpur 721302, India
e-mail: sram@matsc.iitkgp.ernet.in

P. K. Giri et al. (eds.), *Advanced Nanomaterials and Nanotechnology*,
Springer Proceedings in Physics 143, DOI: 10.1007/978-3-642-34216-5_43,

magnetostructural transition (FOMT) [1–3]. The FOMT gives rise to unique multifunctional properties, viz., large magnetocalory, magnetic shape memory, and giant magnetoresistance [4, 5], which can be exploited for several applications of magnetic cooling, magnetic actuators and sensors, or spintronic devices. Thermoelastic and/or magnetoelastic character of growth and reversion of the martensite phase is responsible for the useful properties. The thermoelasticity, which arises when a local balance between chemical and nonchemical forces incurs in the interfaces between the martensite strips, results in a reversible thermal process in the martensite (M) $\leftrightarrow$ austenite (A) phase transition [6]. The chemical forces, which arise from a difference in the Gibbs free energy between the martensite and austenite states, drive the forward M $\leftarrow$ A phase transition. The non-chemical forces have two primary origins; the first contribution arises in elastically accommodating the structural change in the lattice so that an elastic energy stored during the M $\leftarrow$ A transition releases reversibly in reverting the transition and the second contribution is the result of an internal work done in the transition in overcoming the frictional barriers during the interfacial motions. The second contribution represents irreversible part of the non-chemical energy that results in a thermal hysteresis in the transition in heating and cooling cycles in a calorimeter [6–8].

A cautious study of the thermal and the magnetic processes in the phase transitions can reveal an irreversible character of the transition parameters that is different from the one that originates from the frictional forces. Coexistence of substantial hysteresis losses and large magnetocalory/magnetoresistance in the M $\leftrightarrow$ A transition regime also demands revival in the understanding of micro-structural, thermal, and magnetic properties. Notwithstanding the crucial knowledge in devising useful shape memory/magnetic refrigerant alloys, no much attention has been given on the irreversible behavior or it was ignored in simplifying the processes. Limited studies are available on the microstructural irreversibility that arises from plastic deformations during the transition [9, 10], but the irreversibility in the other physical properties is rarely studied. A recent study reports an irreversible switch-over of inverse magnetocalory to a conventional magnetocalory in $Ni_{50}Mn_{36}Co_1Sn_{13}$ alloy in the vicinity of reverse M $\rightarrow$ A phase transition under cyclic application of a magnetic field (B) [11]. In this background, we have developed a $Ni_{39+x}Mn_{50}Sn_{11-x}$ ($x \leq 2$) alloy series in tailoring the martensite transition with functional magnetocaloric properties for possible applications. A herringbone nanostructure incurs in a specific $Ni_{39+x}Mn_{50}Sn_{11-x}$ ($x = 2.0$) composition of a single martensite phase and it exhibits reasonably large irreversibility in the transition parameters during the heating followed by cooling cycles in a calorimeter. In a highly composition sensitive thermodynamic property, this kind of the irreversibility depreciates very sensitively in a small compositional change of the martensite phase in the Sn-content from 9.0 at % ($x = 2.0$). As a result, this is an important experimental parameter to attenuate the functional magnetocaloric and other microstructure related physical properties for possible applications.

2 Experimental

The Heusler $Ni_{39+x}Mn_{50}Sn_{11-x}$ ($x \leq 2$) alloys were grown in a herringbone nanostructure by cooling the melts in a controlled shape of a cylindrical disc (15 mm diameter and 8 mm width) in a copper mold in a tungsten inert gas arc melting furnace. A master alloy was prepared by melting and casting a stoichiometric mixture of pure metals in small discs which were flipped and remelted three-to-four times to ensure a homogeneous alloying. The final chemical compositions in these alloys were confirmed in terms of inductively coupled plasma optical emission spectroscopy (performed with a JY- ULTIMA spectrophotometer, France) and energy dispersive X-ray analysis (performed on a scanning electron microscope of JEOL-model JSM-5800). The final compositions show an average 0.1 % depreciation from the initial compositions considered for the alloying. Notwithstanding a cautious alloying, such compositional loss is inevitable due to evaporation of the constituent metals during melting process. For a critical 9.0 at % Sn-content ($x = 2.0$), a single martensite phase turns-up in a $L1_0$ tetragonal crystal structure at room temperature, with an average crystallite size 11 nm and lattice parameters $a = 0.7808$ nm and $c = 0.6954$ nm as characterized by the X-ray diffraction (XRD). A martensite–austenite composite phase transforms at a higher Sn-content ~9.5 at % ($x = 1.5$). An austenite cubic $L2_1$ phase develops only when the x-value is kept below 1.5. High resolution images were recorded using a field emission scanning electron microscope (FESEM) of *ZEISS SUPRA-40* with an accelerated voltage 10 kV. The phase transitions were studied in terms of the heat-outputs measured during heating followed by cooling a specimen over 200–475 K at 10 K/min in a differential scanning calorimeter (DSC Q100, TA Instruments). Temperature dependence of magnetization (σ) over 2–360 K at different B-values such as 5 mT and 5 T was studied using a SQUID magnetometer (Quantum Design).

3 Results and Discussion

In Fig. 1a, b the FESEM images (a) portray a well-resolved pattern of martensite strips with an average 20 nm thickness grown in a specific herringbone nanostructure in a particular composition with $x = 2.0$ in the $Ni_{39+x}Mn_{50}Sn_{11-x}$ alloy series. These FESEM images were recorded from a freshly fractured cross-section surface of an alloy disc. The martensite strips consist of an average 10 nm wide granules arranged in a mirror-image-like stack signifying a twinned arrangement of the atoms as described in a model structure proposed in Fig. 1b. Growth of a granular herringbone nanostructure is possible only when the local strains are effectively relieved in the martensite phase that develops on cooling a sample from the austenite state [12]. Upon cooling the sample through the M $\leftarrow$ A phase transition, a shear stress develops in a cubic crystal structure of the austenite phase,

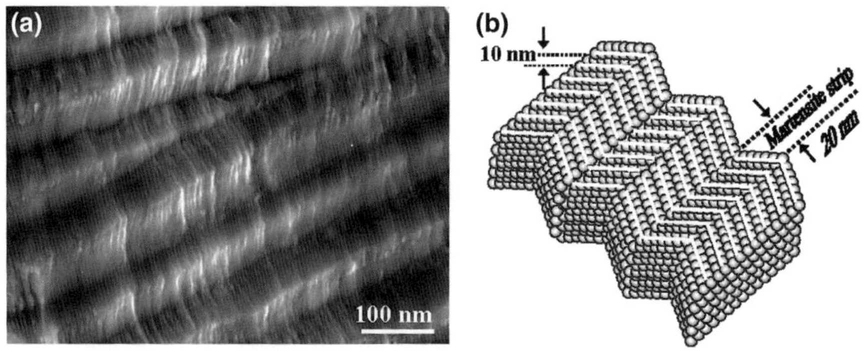

Fig. 1 **a** FESEM images from a $Ni_{39+x}Mn_{50}Sn_{11-x}$ (x = 2.0) alloy consisting of parallel arrays of twinned martensite strips in a herringbone nanostructure, with. **b** a model mirror-image-like stacking of the atoms in a granular herringbone nanostructure

which eventually results in a twinning arrangement of the atoms (Fig. 1b) forming a herringbone nanostructure at the expense of the local strains in the martensite phase of a tetragonal crystal structure. As a result, average thickness in the individual strips developed in a strain-free herringbone nanostructure varies sharply at a narrow scale of 20 ± 2 nm. A low local strain ~1.7 % estimated for the $Ni_{39+x}Mn_{50}Sn_{11-x}$ alloy at x = 2.0 relative to a value 3.5 % found in a composite alloy (x = 1.5) from the XRD patterns using the William-Hall plot supports the observed microstructure [13]. An important characteristic feature of the herringbone structure is that it has a correlation with the martensite magnetic domain structure. A granular boundary which builds-up in this nanostructuring can serve to be a pinning barrier to the single magnetic domains. An average granular width δ in the $Ni_{39+x}Mn_{50}Sn_{11-x}$ (x = 2.0) alloy of nanocrystallites is thus found to be very small of ~10 nm (Fig. 1b) in the form of the thin plates. A pioneering work of Frenkel and Dorfman predicts that a single particle in a ferromagnetic (FM) state can consist of a single magnetic domain only below a critical size δ_c [14]. In the experimental work of Kittel, an approximate δ_c-value varies in a range from 10 nm to 1 μm [15]. Following this conjuncture, a δ–value ~10 nm we observed from the FESEM images in the $Ni_{39+x}Mn_{50}Sn_{11-x}$ (x = 2.0) nanocrystallites is characteristic of a superparamagnetic (SPM) state that has been ascertained with studying the ac susceptibility, thermomagnetic curves, and magnetic hysteresis loops over elevated temperatures [13]. The SPM state is known to exist in the martensite phase in the alloys $Ni_{43.5}Co_{6.5}Mn_{39}Sn_{11}$ and $Ni_{48}Co_2Mn_{39}Sn_{11}$ [16, 17].

In order to establish a correlation of the SPM herringbone nanostructure with the caloric transition processes, we studied the heat-flow over 225–375 K temperature range during heating followed by cooling the $Ni_{39+x}Mn_{50}Sn_{11-x}$ nanocrystallites at a fixed rate 10 K/min, and the results so obtained on the typical alloys (x = 2.0 and 1.5) are compared in Fig. 2. Upon heating the martensite alloy (x = 2.0), an exothermic peak appears with a peak value at point A_p (austenite peak) = 327 K in the M → A transition that begins at the point A_s (austenite start)

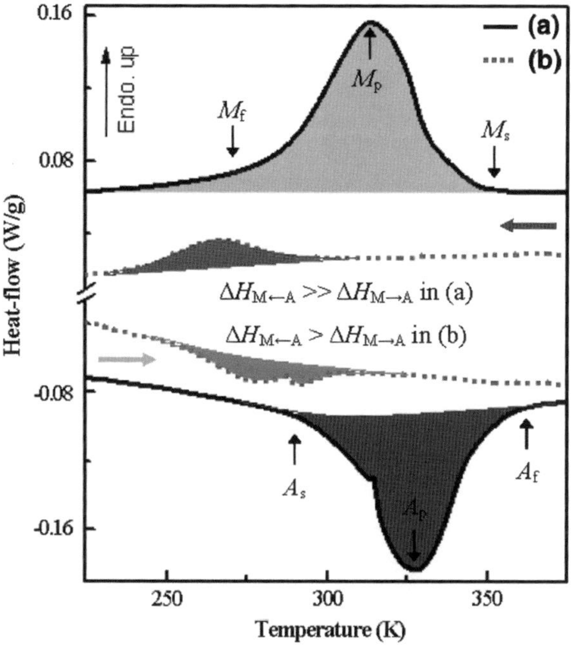

Fig. 2 DSC thermograms during heating followed by cooling (marked with the *arrows*) a $Ni_{39+x}Mn_{50}Sn_{11-x}$ alloy, showing an irreversibility in the ΔH in the martensite transition; **a** x = 2.0 and **b** x = 1.5

~297 K and turns-up at the point A_f (austenite finish) ~348 K. This transition incurs with an enthalpy change $\Delta H_{M \rightarrow A} = 3.205$ J/g, or entropy change $\Delta S_{M \rightarrow A} =$ 9.8 J/kg-K, heat-capacity change $\Delta C_{P(M \rightarrow A)} = C_P(A_f) - C_P(A_s) \approx 67.5$ mJ/kg-K and full-width at half maximum $\beta_{M \rightarrow A} = 28.7$ K. During cooling, the transition starts to revert back at the point M_s (martensite start) ~334 K in a modified peak at the point M_p (martensite peak) ~312 K. The transition finishes at the point M_f (martensite finish) ~287 K, with a large thermal hysteresis $\Delta T = A_p - M_p \approx 15$ K. The M ← A phase transition incurs with the enhanced values $\Delta H_{M \leftarrow A} = 3.324$ J/g, or $\Delta S_{M \leftarrow A} = 10.603$ J/kg-K, $\Delta C_{P(M \leftarrow A)} = C_P(M_f) - C_P(M_s) \approx 90$ mJ/kg-K, and $\beta_{M \leftarrow A} = 29.7$ K. The local strains relieved in regularly displaced martensite strips eventually mediate an enhanced value for $\Delta H_{M \leftarrow A}$ and the related parameters, demonstrating marked irreversibility in this magnetostructural transition process, viz., a large enthalpy irreversibility $\Delta H_{M \leftarrow A} - \Delta H_{M \rightarrow A} = 0.12$ J/g, or heat-capacity irreversibility $\Delta C_{P(M \leftarrow A)} - \Delta C_{P(M \rightarrow A)} = 22.5$ mJ/kg-K. At a nominally higher Sn-content ~9.5 at % in the composite alloy (x = 1.5), the irreversibility in the transition parameters is effectively lowered to the values $\Delta H_{M \leftarrow A} - \Delta H_{M \rightarrow A} = 0.1$ J/g and $\Delta C_{P(M \leftarrow A)} - \Delta C_{P(M \rightarrow A)} = 10$ mJ/kg-K.

Such irreversibility can arise as the martensite strips divide the sample into ideal single magnetic domains in the herringbone nanostructure with the spins pinned down at the boundaries so that they adapt effectively large stresses on a large frictional energy change in the local spin–lattice reconfiguration, absorbing a large $\Delta H_{M \leftarrow A}$ (3.324 J/g) in the transition. Relatively high electronic thermal conductivity and lower stress in the austenite state eases a spin ← lattice heat-transfer in an exchange-coupled spin–lattice relaxation that drives in situ SPM martensite ← FM austenite transition by absorbing the large $\Delta H_{M \leftarrow A}$ over a large span ($\beta_{M \leftarrow A}$ ~29.7 K), with a considerable $\Delta C_{P(M \leftarrow A)}$ (90 mJ/kg-K). On reheating, an apparently lowered $\Delta H_{M \rightarrow A}$ (~3.205 J/g) evolves in this transition that reverts back to the initial FM state and electronically more conducting regime in a rather poor spin–lattice dynamic coupling. This transition coincides with an intermartensite transition (IMT), as evident from a kink in the exothermic peak. The IMT that appears at 315 K in the sample with x = 2.0 shifts to 288 K in the martensite transition peak for the 9.5 at % Sn-content (x = 1.5). Part of the internal stress, which builds-up during the local atomic redistribution, relieves in this IMT signal [18]. Characteristically faster ΔH-change in the martensite ← austenite transition (more conductive) invades the IMT signal in the cooling cycle.

In order to examine how the thermal irreversibility is related with the dynamics of spins in the martensite transition in the $Ni_{39+x}Mn_{50}Sn_{11-x}$ (x = 2.0) nanocrystallites, we studied thermomagnetic curves at an effectively low B-value such as 5 mT over 250–350 K temperature (T) during a zero field-cooled heating (ZFC), a field-cooling (FC), and a field-heating (FH) cycles as shown in Fig. 3. The ZFC curve reveals a spin-disorder (SPM martensite) → order (FM austenite) transition that starts at the point A_s ~297 K and finishes at the point A_f ~322 K with a peak value σ – 0.8 emu/g. This thermal controlled process of the spins follows a FM austenite → paramagnetic austenite transition at thecurie point T_C^A ~334 K. In the FC cycle, the M ← A transition begins to revert with an enhanced peak value σ ~1.9 emu/g at point M_s ~319 K, which finishes at point M_f ~287 K. The FH curve retraces the FC curve up to the M_f-point and then deviates along the ZFC curve. A significant fall in the σ-value by 58 % at the A_f-point in the FH or ZFC curve, relative to the value at the M_s-point in the FC curve, signifies an irreversible spin reordering in the M → A transition. Such

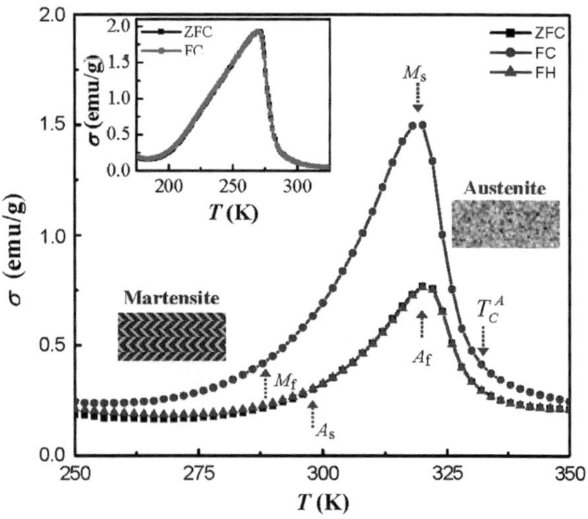

Fig. 3 Thermomagnetic curves in a $Ni_{39+x}Mn_{50}Sn_{11-x}$ (x = 2.0) alloy showing a significant fall in the magnetization ~58 % at A_f–point relative to the magnetization at M_s-point, signifying a large irreversibility in spin reorientation in the martensite → austenite transition at a low bias field B = 5 mT. No such difference in the magnetization in this transition is found in the martensite-austenite composite alloy (x = 1.5) as shown in the inset

irreversibility arises due to hindrance of domain-wall motion during the transition, which slows down the martensite strips transformation into the FM austenite phase. At a sufficiently large B-value, the irreversibility in the σ-value can effectively drop down as the domain-wall moves readily so that the martensite strips coalesce and the spins reorder easily in an applied field direction in the absence of the local pinning barriers. The peak σ-values at the points M_s and A_f remain equal (see the plots in the inset in Fig. 3) in the alloy (x = 1.5) having a residual austenite fraction ~65 % in the martensite phase as found from relative intensities of the XRD peaks.

4 Conclusion

A series of $Ni_{39+x}Mn_{50}Sn_{11-x}$ ($x \leq 2$) alloys of granular nanostructure has been developed in devising the functional magnetocaloric properties for possible applications. At a specific 9.0 at % Sn-content (x = 2.0), this alloy series accomplishes a single martensite phase with a herringbone nanostructure showing large irreversibility in the thermal parameters, such as $\Delta H_{M \leftarrow A} - \Delta H_{M \rightarrow A} = 0.12$ J/g, in the M → A transition with a significant fall in the σ-value by 58 % at 5 mT field. An incomplete reordering of the spins at low fields such as 5 mT during this transition from an electronically low conductive martensite phase to a more conductive austenite phase causes the irreversibility. Such irreversibility is considerably lowered on a nominal increase in the Sn-content to 9.5 at % in the composite alloy (x = 1.5) having a 65 % residual austenite fraction in the martensite phase. Apart from developing the basic physics, a study of the irreversible character of a physical process in the caloric transitions is important in devising newer alloys with functional properties for possible applications.

Acknowledgments Authors are thankful to Dr. D. Das, Scientist F, UGC-DAE Consortium of Scientific Research, Kolkata, India, for providing the facility for the magnetic measurements. A. A. P. is thankful to All India Council for Technical Education (AICTE), New Delhi, and Bahubali College of Engineering, Shravanabelagola, for a financial support.

References

1. T. Krenke, E. Duman, M. Acet, E.F. Wassermann, X. Moya, L. Mañosa, A. Planes, Inverse magnetocaloric effect in ferromagnetic Ni–Mn–Sn alloys. Nat. Mater. **4**, 450 (2005)
2. I. Babita, S. Ram, R. Gopalan, V. Chandrasekaran, Dynamic inverse-magnetocaloric and martensite transition in $Ni_{49}Mn_{38}Sn_{13}$ nanocrystals in low magnetic fields. Philos. Mag. Lett. **89**, 399 (2009)
3. T. Krenke, M. Acet, E.F. Wassermann, Ferromagnetism in the austenitic and martensitic states of Ni–Mn–In alloys. Phys. Rev. B **73**, 174413 (2006)

4. Y. Sutou, Y. Imano, N. Koeda, T. Omori, R. Kainuma, K. Ishida, K. Oikawa, Magnetic and martensitic transformations of NiMnX (X = In, Sn, Sb) ferromagnetic shape memory alloys. Appl. Phys. Lett. **85**, 4358 (2004)
5. T. Krenke, M. Acet, E.F. Wassermann, X. Moya, L. Mañosa, A. Planes, Martensitic transitions and the nature of ferromagnetism in the austenitic and martensitic states of Ni–Mn–Sn alloys. Phys. Rev. B **72**, 014412 (2005)
6. J. Ortin, A. Planes, Thermodynamic analysis of thermal measurements in thermoelastic martensitic transformations. Acta. Metall. **36**, 1873 (1988)
7. C. Segui, E. Cesari, J. Pans, Phenomenological modeling of the hysteresis loop in thermoelastic martensitic transformations. Mater. Trans. JIM **33**, 650 (1992)
8. W.H. Wang, J.L. Chen, Z.H. Liu, G.H. Wu, W.S. Zhan, Thermal hysteresis and friction of phase boundary motion in ferromagnetic $Ni_{52}Mn_{23}Ga_{25}$ single crystals. Phys. Rev. B **65**, 012416 (2001)
9. K. Bhattacharya, S. Conti, G. Zanzotto, J. Zimmer, Symmetry and reversibility of martensite transformations. Nature **428**, 55 (2004)
10. P.J. Shamberger, F.S Ohuchi, Hysteresis of the martensitic transition in magnetocaloric-effect Ni–Mn–Sn alloys. Phys. Rev. B **79**, 144407 (2009)
11. V.V. Khovaylo, K.P. Skokov, O. Gutfleisch, H. Miki, R. Kainuma, T. Kanomata, Reversibility and irreversibility of magnetocaloric effect in a metamagnetic shape memory alloy under cyclic action of a magnetic field. Appl. Phys. Lett. **97**, 052503 (2010)
12. Y.M. Jin, A. Artemev, A.G. Khachaturyan, Three-dimensional phase field model of low symmetry martensitic transformation in polycrystal: simulation of ζ'_2 martensite in AuCd alloys. Acta. Mater. **49**, 2309 (2001)
13. A.A. Prasanna, S. Ram, Local strains, calorimetry, and magnetoresistance in adaptive martensite transition in multiple nanostrips of $Ni_{39+x}Mn_{50}Sn_{11-x}$ ($x \leq 2$) alloys. Sci. Technol. Adv. Mater. (In press)
14. J. Frenkel, J. Dorfman, Spontaneous and induced magnetization in ferromagnetic bodies. Nature **126**, 274 (1930)
15. C. Kittel, Theory of the structure of ferromagnetic Domains in films and small particles. Phys. Rev. **70**, 965 (1946)
16. D.Y. Cong, Q. Luo, S. Roth, J. Liu, O. Gutfleisch, M. Potschke, C. Hurrich, L. Schultz, Sequence of structural and magnetic transitions in $Ni_{48}Co_2Mn_{39}Sn_{11}$ shape memory alloy. J. Magn. Magn. Mater. **323**, 2519 (2011)
17. D.Y. Cong, S. Roth, J. Liu, Q. Luo, M. Pötschke, C. Hürrich, L. Schultz, Superparamagnetic and superspin glass behaviors in the martensite state of $Ni_{43.5}Co_{6.5}Mn_{39}Sn_{11}$ magnetic shape memory alloy. Appl. Phys. Lett. **96**, 112504 (2010)
18. V.V. Kokorin, A.O. Perekos, A.A. Tshcherba, O.M. Babiy, T.V. Efimov, Intermartensitic phase transitions in Ni–Mn–Ga alloy, magnetic field effect. J. Magn. Magn. Mater. **302**, 34 (2006)

Magnetic, Electrical and Optical Properties of $Nd_{0.85}$ $K_{0.15}$ MnO_3 Thin Film

Tribedi Bora, A. Nandy, R. K. Bhuyan, D. Pamu and S. Ravi

Abstract K doped Nd–Mn–O based thin film was deposited on Si substrate by using a target material of composition $Nd_{0.85}$ $K_{0.15}MnO_3$ and a RF magneton sputtering unit. The film was found to be in single phase form with average crystallite size of 54 nm. The band gap and the refractive index values were determined by analyzing the transmission data. They exhibit paramagnetic to ferromagnetic transition along with a signature of antiferromagnetic transition. The magnetic properties such as saturation magnetization, ferromagnetic transition temperature and Curie temperature were determined by analyzing the dc magnetization data. The sizes of grain were determined from FE-SEM images. The film exhibits metal insulator transition in the vicinity of ferromagnetic transition temperature.

Keywords K-doped Nd–Mn–O thin film · Ferromagnetism

1 Introduction

The alkaline earth or alkali ion doped rare earth manganite have been widely studied for about two decades now [1–3]. Depending upon the doping concentration and other structural parameters, they exhibit interesting physical properties such as metal–insulator, ferromagnetic, antiferromagnetic, spin glass or charge order transition and some compounds exhibit large colossal magnetoresistivity (CMR).

T. Bora · A. Nandy · R. K. Bhuyan · D. Pamu · S. Ravi (✉)
Department of Physics, Indian Institute of Technology Guwahati, Guwahati 781039, India
e-mail: sravi@iitg.ernet.in

P. K. Giri et al. (eds.), *Advanced Nanomaterials and Nanotechnology*, 449
Springer Proceedings in Physics 143, DOI: 10.1007/978-3-642-34216-5_44,

The Nd–Mn–O series falls in the category of medium size e_g electron bandwidth of Mn ions, where one can easily tune the system from double exchange ferromagnetic metallic state to robust charge ordered insulating state by varying the average A site ionic size. In addition to that, the presence of two magnetic sublattices, Nd and Mn makes the system as an interesting one. Paramagnetic (PM) insulator to ferromagnetic (FM) metallic transition has been reported in several alkaline earth doped $Nd_{1-x}A_xMnO_3$ (A-alkaline earth) series with a maximum T_C of 250 K in $Nd_{0.7}$ $Sr_{0.3}$ MnO_3 [4–7].

The reports on thin film samples of Nd–Mn–O series are limited and we have not come across any such report on (Nd, K)-Mn–O based thin films. It would be interesting to study the magnetic and electrical transport properties of thin film samples of (Nd, K)-Mn–O series. In the present work, we report the deposition of $Nd_{0.85}$ $K_{0.15}MnO_3$ thin films on Si substrates and study of their magnetic and electrical transport properties.

2 Experimental Details

The K-doped Nd–Mn–O based target material with a composition $Nd_{0.85}$ $K_{0.15}MnO_3$ was prepared by sol gel technique. The stoichiometric ratio of starting compounds such as Neodymium Oxide (Nd_2O_3), Potassium Carbonate (K_2CO_3) and Manganese Acetate ($C_4H_6MnO_4.4H_2O$) with 99.9 % purity were weighed. The weighed compounds were dissolved in nitric acid and then by adding excess citric acid to convert the solution into citrate form. A homogeneous mixture of the solution was prepared by using a magnetic stirrer with hot plate and by adding ethylene glycol such that uniform gel was formed after the slow evaporation. The gel thus obtained was dried and fired at 200° C. The precursor obtained from the sol–gel technique was presintered at 800 °C for 24 h. The presintered powder was pressed into a disc shaped target of 2″ diameter by using a 56 mm die and a hydraulic press with a maximum load of $6Ton/cm^2$. The target was annealed at 900 °C for 24 h with an intermediate grinding and repressing into target disc followed by a final sintering at 1000 °C. Thin film on Si substrate was deposited by using the above target material and RF magneton sputtering system. The deposited film was directly annealed at 800 °C for 2 h. The sputtering was done under Ar/O_2 flow of 15/5 sccm for a duration of 1 h 45 min. The thickness of the film was found to be 618 nm.

X-ray diffraction patterns were recorded at room temperature using a Seifert 3003-TT XRD machine by employing Cu-Kα radiation. The micro structural images were recorded by using Zeiss make field emission scanning electron microscope (FESEM). The temperature variations of zero field cooled (ZFC) and field cooled (FC) magnetization (M) were measured for an applied field of 0.1 T by using a Lakeshore model no. 7410 vibrating sample magnetometer in the temperature range 20 to 300 K. The magnetization loops were measured by varying the field up to ±1.2 T. The transmission spectrum was recorded by using a UV–VIS-NIR spectrometer.

Temperature variation of two probe electrical resistivity was measured by using a commercial Helium closed cycle refrigerator cryostat.

3 Results and Discussion

The XRD patterns of $Nd_{0.85}$ $K_{0.15}MnO_3$ target is shown in Fig. 1. It is observed that the compound is found to be in single phase form. The pattern could be refined to Pbnm space group by Rietveld refinement technique with the help of the FullProf program. The deposited thin films were also found to be in single phase form with typical lattice parameters a = 5.4865 Å, b = 5.4393 Å and c = 7.6945 Å.

The average crystallite size was estimated using Scherrer's formula, $S = k\,\lambda/(\beta cos\Theta)$, where, the constant k depends on the shape of the grain size and is taken as 0.89 by assuming the circular grain, β is full width at half maximum (FWHM) of intensity versus 2θ profile, λ is the wavelength of the Cu–K_α radiation and θ is Braggs diffraction angle. The experimental β value was corrected for instrumental broadening using the relation $\beta^2 = \beta_m^2 - \beta_{ins}^2$. Here β_m is the measured FWHM of the XRD peak and β_{ins} is the instrumental broadening. The average crystallite size of target was found to be 54 nm. Fig. 2 shows the typical FESEM image of thin film sample. One set of grains are found to be in nanocrystalline form with grain size in the range of 50–100 nm and another set in the range of 160–190 nm.

The transmission spectrum of thin film sample recorded at room temperature in the wavelength range of 300 to 2800 nm is shown in Fig. 3. The disappearance of fringes and the fall in transmission values at shorter wavelength is mainly due to the fundamental absorption of the film due to band gap energy. The optical band

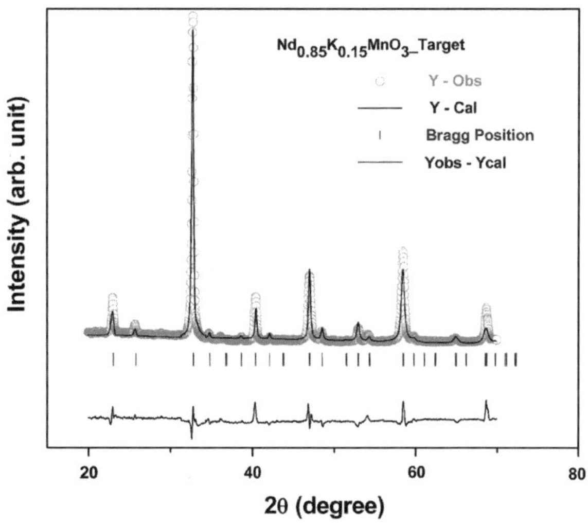

Fig. 1 XRD pattern of $Nd_{0.85}K_{0.15}MnO_3$ target along with Rietveld refinement

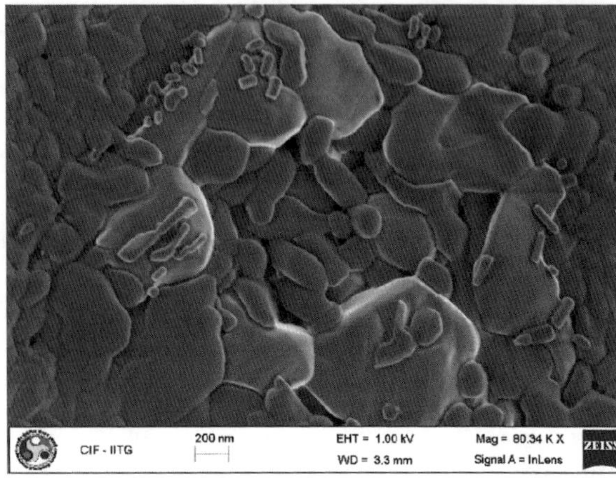

Fig. 2 FESEM image of $Nd_{0.85}K_{0.15}MnO_3$ thin film

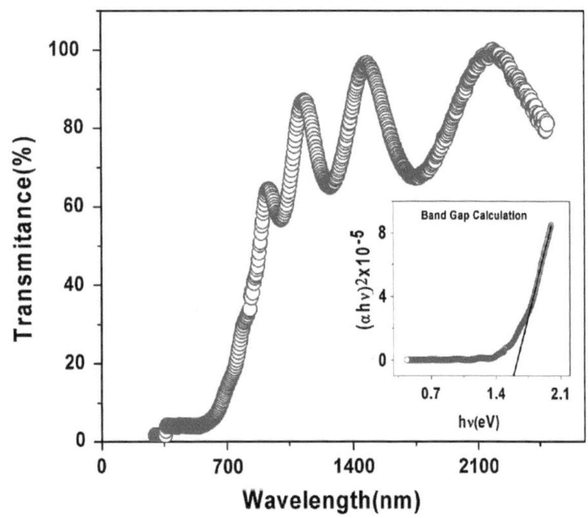

Fig. 3 Transmission spectrum of $Nd_{0.85}K_{0.15}MnO_3$ film. Inset shows $(\alpha h\upsilon)^2$ versus $h\upsilon$

gap energy E_g was calculated by using the equation [8], $\alpha hv = \beta\ (hv\text{–}E_g)^m$, where, hv is the photon energy, $\beta = 1$ (constant) and m $= 0.5$ i.e. for allowed direct band. Here α is absorption coefficient and was determined using the relation, $\alpha =$ -[ln(T)]/d, where d is the thickness of the film and T is the transmittance corresponding to the particular wavelength. The inset of Fig. 3 shows the plot of $(\alpha hv)^2$ versus hv and it could be fitted to the linear relation as shown as solid line. The fitted data closely follow the experimental data and the E_g value determined from the above fit is found to be 1.65 eV.

The refractive index was determined by using the relations [9],

$$n(\lambda) = [N(\lambda) + (N(\lambda)^2 - s(\lambda)^2)^{1/2}]^{1/2} \tag{1}$$

and

$$N(\lambda) = 2s(\lambda)[(T(\lambda)_{max} - T(\lambda)_{min})/((T(\lambda)_{max} T(\lambda)_{min})] + (s^2(\lambda) + 1)/2 \quad (2)$$

Here s is the refractive index of substrate, T_{max} is the maximum transmission and T_{min} is the minimum transmission at particular wavelength (λ). The refractive index value is found to be 2.33.

Magnetization as a function of temperature for the target and film are shown in Fig. 4. The target material shows a long range FM ordering with T_C around 122 K. The paramagnetic susceptibility could be fitted to Curie–Weiss law with a Curie temperature at 134 K. Unlike the target material, the M versus T plots of thin film sample shows PM to FM transition followed by peak effect at around 68 K due to the presence of competing antferromagnetism (AFM). The FC curve shows the suppression of AFM peak. The observed T_C value of thin film is found to be smaller than the target material and it can be understood in terms of lack of oxygenation and hence the reduced concentration of Mn^{4+} ions.

The dc susceptibility data determined from the measured magnetization are found to follow the Curie–Weiss law in the PM region. The plots of inverse susceptibility ($1/\chi$) versus T are shown in the inset of Fig. 4. They exhibit linear behavior and fitted data are shown as solid line. Both target and thin film samples exhibit positive Curie temperature and thus they exhibit basically FM transition.

Typical plot of M versus H recorded at 40 K for the thin film sample is shown in Fig. 5 and it shows the FM behavior and however, the magnetization could not be saturated for the maximum applied field of 1.2 T. The significant linear region observed at higher field can be attributed to the presence of competing AFM or PM matrix. We have estimated the saturated magnetization M_S, after subtracting the contribution from the linear magnetization and it is found to be 13 emu/g.

Temperature variation of electrical resistivity was measured on the target and the thin film samples. The target material shows metal–insulator transition at

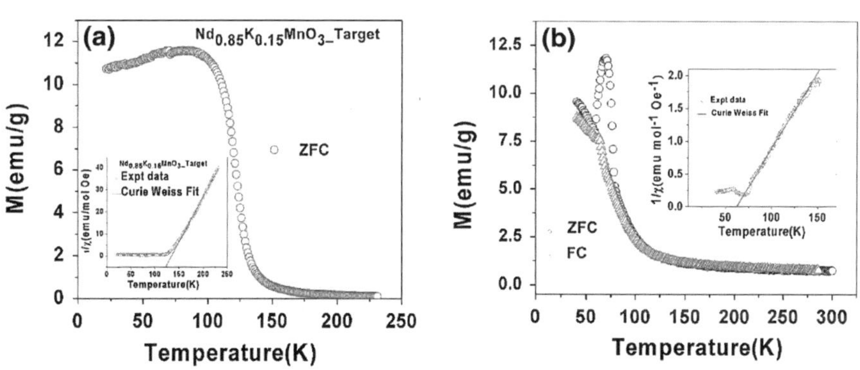

Fig. 4 Temperature variation of magnetization for (a) the target and (b) the thin film sample of $Nd_{0.85}K_{0.15}MnO_3$. $1/\chi$ versus temperature plot along with Curie—Weiss law fit

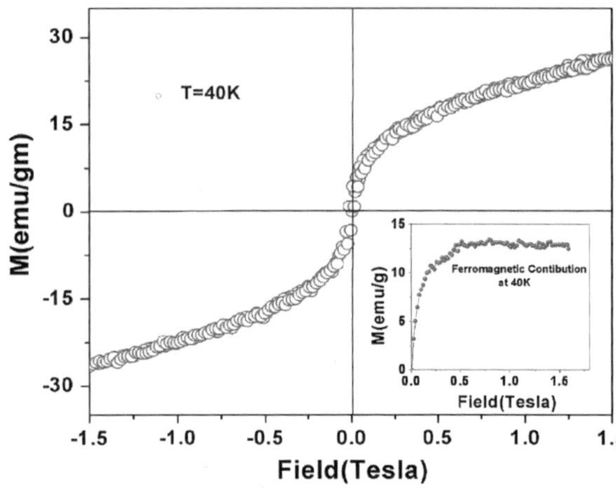

Fig. 5 Magnetization versus magnetic field for $Nd_{0.85}K_{0.15}MnO_3$ thin film at 40 K

around 120 K. Typical plot of ρ versus T for the thin film is shown in Fig. 6. We can see a clear metal–insulator transition at 58 K and it is quite close to the observed Curie temperature of 62 K. For x > 60 K, the resistivity is found to increase by two orders of magnitude with a sudden jump and for clarity the plot is restricted up to 60 K. It clearly demonstrates that observed FM in the present thin film is mainly due to double exchange ferromagnetic interaction. In addition to the above metal insulator transition, we can see an anomaly at around 40 K and it is in the vicinity of the low temperature magnetization peak due to the AFM or other competing interaction.

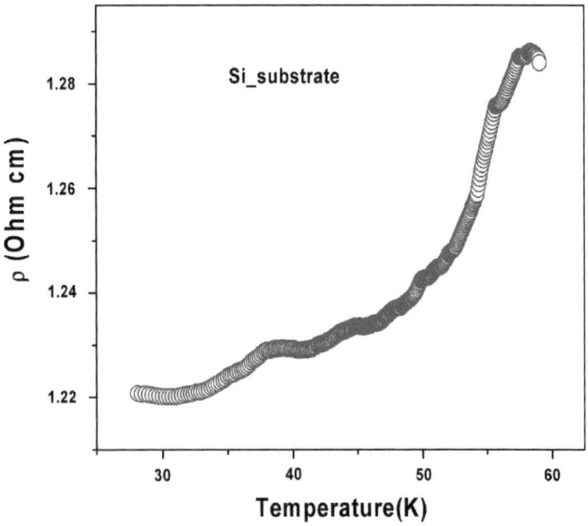

Fig. 6 Temperature variation of electrical resistivity of thin film sample

4 Conclusion

K –doped Nd–Mn–O based CMR thin film was prepared in single phase form. The thickness of the film was found to be 618 nm. They exhibit paramagnetic to ferromagnetic transition temperature $T_C = 68$ K followed by a low temperature AFM behavior. The M–H loop recorded on the film show FM behavior with a considerable linear behavior at higher field due to the presence of AFM. The electrical resistivity data show metal–insulator transition in the vicinity of Curie temperature. The energy band gap value determined from the transmission spectra of thin film samples was found to be 1.65 eV.

Acknowledgments Authors are thankful to BRNS and DST for financial support.

References

1. C.N.R. Rao, B. Raveau, *Colossal Magnetoresistance, Charge Ordering and Related Properties of Manganese Oxides* (World Scientific Publishing Company Incorporated, Singapore, 1998)
2. Y. Tokura, Colossal magneto-resistive Oxides. Gordon and Breach Science Publishers, (2000)
3. A.P. Ramirez, J. Phys.:Condens. Matter, **9,** 8171–8199 (1997)
4. A. Poddar, P. Muruguraj, R. Fischer, E. Gmelin, K. BÃ¤rner, L. Haupt, P. Mandal, G.H. Rao, Physica B, **254,** 21–27 (1998)
5. M. Pattabiraman, P. Murugaraj, G. Rangarajan, Solid State Commun. **132**, 7–11 (2004)
6. I.O. Troyanchuk, D.D. Khalyavin, S.V. Trukhanov, H. Szymczak, J. Phys.: Condens. Matter **11**, 8707–8717 (1999)
7. V.A. Ryzhov, A.V. Lazuta, V.P. Khavronin, I.I. Larionov, I.O. Troaynchuk, D.D. Khalyavin, Solid State Commun. **130**, 803–808 (2004)
8. J.C. Tauc, Optical Properties of Solid, North–Holland, (1972)
9. R. Swanepoel, J. Phps. E: Sci. Instrum **16**, 1214–1222 (1983)

Magnetic Properties of Ultra-Thin FePt Films Grown on Oxidized Si Substrates

Anabil Gayen, P. Saravanan and A. Perumal

Abstract We report the effects of thickness and post annealing temperature on the structural and temperature dependent magnetic properties of FePt (x = 5–50 nm) thin films deposited at ambient temperature on the oxidized Si substrate with an MgO (y = 5, 10 nm) underlayer. All the as-deposited samples show face centered cubic structure, but transformed into $L1_0$ ordered structure upon annealing. The formation of $L1_0$ ordered structure depends strongly on the films thickness and annealing temperature. The coercivity of the FePt films with thickness lower than 20 nm was obtained to be less than 3 kOe, but increases rapidly to above 6 kOe for more thicker films. Room temperature magnetic properties and high temperature magnetic properties of the FePt films were studied by analyzing the magnetization data as a function of temperature to understand the stability of the $L1_0$ ordered structure. High temperature coercivity variation exhibits a linear variation of coercivity up to 540 K and transforms into soft magnetic phase above 640 K. The obtained results are discussed in correlation with the improvement in the $L1_0$ ordering with annealing and the reduction in the temperature dependence of magnetocrystalline anisotropy energy.

Keywords FePt · MgO · $L1_0$ ordered structure · Coercivity · Curie temperature

A. Gayen · A. Perumal (✉)
Department of Physics, Indian Institute of Technology Guwahati, Guwahati 781039, India
e-mail: alagarsamy.perumal@gmail.com

P. Saravanan
Advanced Magnetic Group, Defence Metallurgical Research Laboratory, Kanchanbagh PO, Hyderabad 500058, India

P. K. Giri et al. (eds.), *Advanced Nanomaterials and Nanotechnology*, 457
Springer Proceedings in Physics 143, DOI: 10.1007/978-3-642-34216-5_45,

1 Introduction

Magnetic data storage density has been advancing beyond Tbits/in^2 with remarkable speed supported by the progress in both magnetic recording media and read heads [1]. Perpendicular magnetic recording (PMR) including heat assisted magnetic recording and bit patterned media have been considered as a promising way to achieve the recording areal density beyond 1 Tbits/in^2 [2–4]. Granular type $L1_0$ ordered equi-atomic FePt based nanogranular thin films with (001)-oriented fine FePt particles received much attention for ultra-high density perpendicular magnetic recording medium due to their large magnetocrystalline anisotropy ($K_u = 7 \times 10^7$ ergs/cc) [5–7], which permits to the use of smaller sized particles down to 4 nm before the onset of superparamagnetism. However, from a technological point of view, the fabrication of such smaller particles uniformly with size below 7 nm and the c-axis perpendicular to the film plane, narrow particle size distribution, the reduction of $L1_0$ ordering temperature, and the optimization of coercivity within the writable values are very much important. Hence, extensive studies were reported on FePt based thin films to optimize their microstructure and magnetic properties suitable for recording applications. However, it is important to note that most of the studies were carried out on FePt films fabricated directly on MgO(100) single crystal substrates [8–10] using various fabrication processes. However, the use of the single crystal MgO substrates and the fabrication of the films other than the sputtering process are not suitable for practical applications because of their cost. In recent years, considerable efforts have been devoted to fabricate the $L1_0$ ordered FePt films on economically viable substrates such as glass or oxidized Si [7, 11] with various underlayers [12, 13]. However, controlling c-axis perpendicular to the film plane in nano-sized particles (<7 nm) without expensive single crystal substrates remains a challenge. On the other hand, the development of thick FePt based continuous films with high perpendicular anisotropy on low cost substrates has been explored for the application of bit patterned media. However, no detailed work on the effect of thickness and processing temperature on the magnetic properties of continuous FePt films grown on the low cost substrates is reported. Hence, in this work, we report the fabrication of FePt continuous films at different thicknesses on the oxidized Si substrate with an MgO underlayer and investigate the effects of thickness and post annealing temperature on the structural and room temperature and high temperature magnetic properties of FePt (x = 5–50 nm)/MgO (y = 5, 10 nm) films.

2 Experimental Details

FePt (x nm)/MgO (y nm) (x = 5, 10, 15, 20, 30, and 50; y = 5 and 10) films were prepared in a high vacuum chamber on low cost substrates (thermally oxidized Si or glass substrate). At first, the MgO layer with a particular thickness was deposited at ambient temperature by RF sputtering an MgO target directly on the substrate. Subsequently, the FePt films with various thicknesses were deposited by DC sputtering a FePt alloy target. The nominal thicknesses of the FePt and MgO films were controlled based on the pre-calibrated sputtering rates of FePt and MgO. The base pressure of the chamber was better than 1×10^{-6} Torr. All the films were deposited at 10 mTorr Ar gas pressure. The composition of the FePt films was estimated by energy dispersive x-ray spectroscopy. The crystal structure of the films and the degree of $L1_0$ ordering were examined using X-ray diffraction (XRD) patterns. The plane-view and cross sectional microstructures of the films were characterized by transmission electron microscopes (TEM, JEOL 2100), respectively. The room temperature and high temperature magnetic properties of the films in the temperature range 300–800 K were analyzed by measuring magnetic hysteresis loops at different temperatures by using Vibrating Sample Magnetometer (VSM, Model: LakeShore 7410, USA).

3 Results and Discussion

Figure 1 shows the XRD patterns of the FePt(x)/MgO(y) films with $x = 50$ nm and $y = 10$ (a) and 5 (b) prepared at ambient temperature and post annealed at different temperatures for 30 min. All the peaks in the XRD patterns could be indexed to the $L1_0$ ordered FePt [(001) around $2\theta = 23^{\circ}$ and (002) peak around $2\theta = 49^{\circ}$] and (002) textured MgO underlayer films. It is also to be noted that the XRD peaks observed at $2\theta = 33.03^{\circ}$ represents Si(200) peaks resulting from the thermally oxidized Si substrates [14, 15]. A careful investigation of the XRD patterns reveals that (1) the MgO underlayer showed only the (002) reflection and the intensity of the MgO(002) peak increases with increasing annealing temperature, revealing the development of strong texture in the MgO underlayer with annealing, (2) the ratio of the integrated intensities between the superlattice (001) peak and the fundamental (002) peak increases and (3) the peak position of the (001) superlattice peak [(002) fundamental peak] shifts slightly to higher angles with increasing post annealing temperature suggesting that the degree of $L1_0$ order is enhanced with increasing the post annealing temperature.

In order to understand the effect of increase in degree of $L1_0$ ordering caused by the post annealing on the thickness dependent magnetic properties of FePt films, the room temperature magnetic hysteresis loops were measured for the FePt (x nm) with $y = 5$ (see Fig. 2) and 10 (see Fig. 3). The magnetic hysteresis loops depicted in the Figs. 2 and 3 reveal the following features: (1) the coercivity of the films increases with increasing the film thickness even for the films annealed at 450 °C, (2) the relative variation of the coercivity with the film thickness depends on the annealing temperature, (3) the films thicker than 20 nm exhibits high coercivity above 5 kOe and the loops are not saturated even at 20 kOe applied fields, (4) while the coercivity of the FePt films does not strongly depend on the MgO underlayer thickness (5 or 10 nm) for the FePt film thicknesses less than 50 nm,

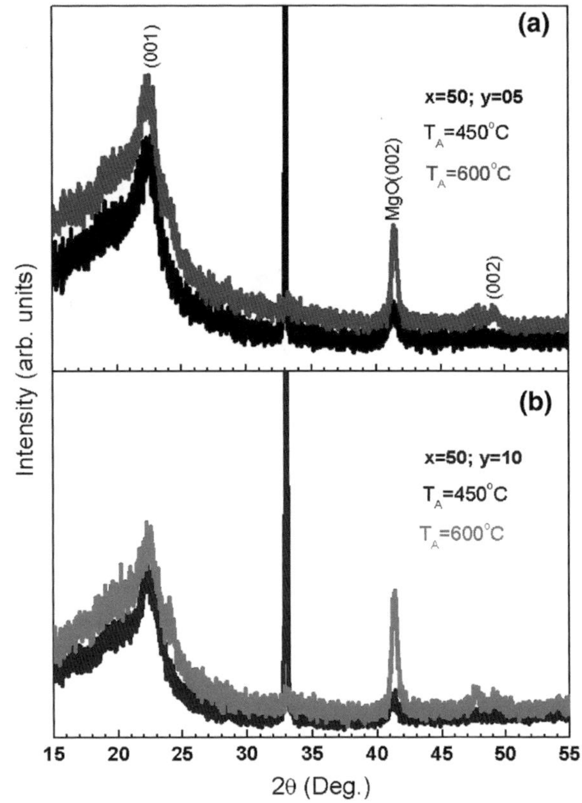

Fig. 1 XRD patterns of the FePt (50 nm) films with MgO underlayer ($y = 5$ nm, **a** and 10 nm, **b**) prepared on thermally oxidized Si substrates at ambient temperature and annealed at different temperatures

the 50 nm thick FePt films shows a large variation in the coercivity depending on the underlayer thickness, (5) In addition, the thicker FePt films deposited on the 10 nm thick MgO underlayer depicts a finite kink around the zero field region while reversing from the saturation, indicating the possibility of presence of both the ordered and disordered magnetic phases. Figure 4 summarizes the coercivity of the FePt films at different thickness with different underlayer thickness annealed at different temperatures. While the nature of coercivity variation exhibits a similar behavior for all the samples with different MgO underlayer thickness, the increase of annealing temperature increases the coercivity in the films. All the samples with 50 nm thick FePt films exhibit large variation in coercivity, which depend both annealing and the underlayer thickness. A maximum of about 12 kOe was observed for the FePt continuous films annealed at 600 °C for the 10 nm MgO underlayer. Note that the curves are not saturated even after applying the field of 20 kOe.

These observations suggest that the coercivity variation in the presently investigated FePt samples is not only affected directly by the ordered volume fraction [16], but also by a microstructural factor that is related to the ordered volume fraction and that scales similarly. Defects in the magnetic materials, such

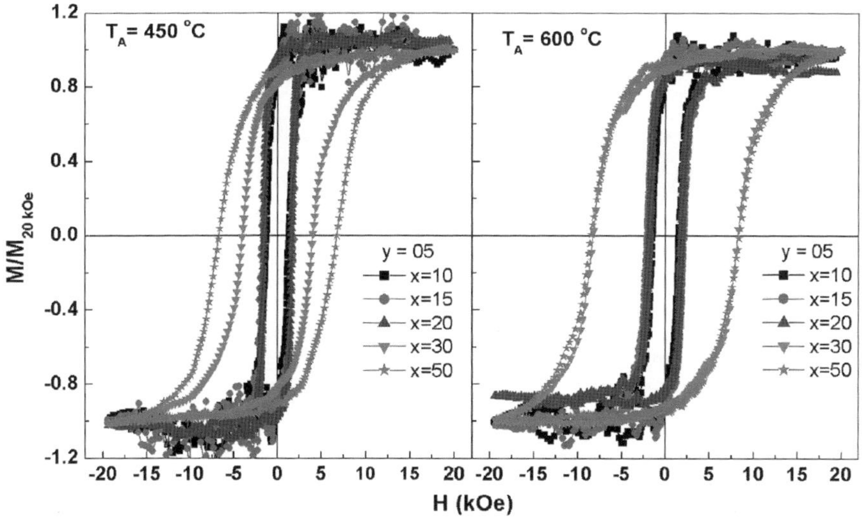

Fig. 2 Room temperature M–H loops of FePt (x nm)/MgO (5 nm) films prepared on thermally oxidized Si substrates at ambient temperatures and annealed at different temperatures

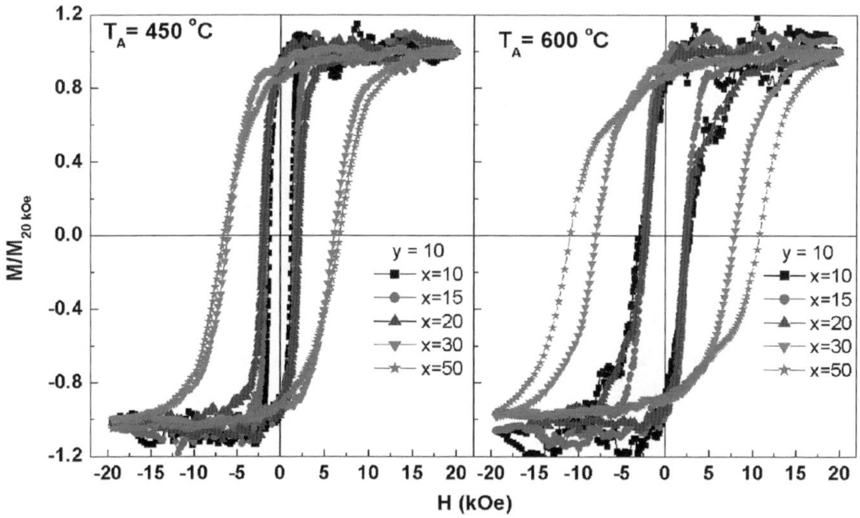

Fig. 3 Room temperature M–H loops of FePt (x nm)/MgO (10 nm) films prepared on thermally oxidized Si substrates at ambient temperatures and annealed at different temperatures

as grain boundaries and phase boundaries, can form pinning sites, which impede the domain-walls and obviously enhances the coercivity. In addition to these, the boundaries between the ordered and disordered regions might also represent

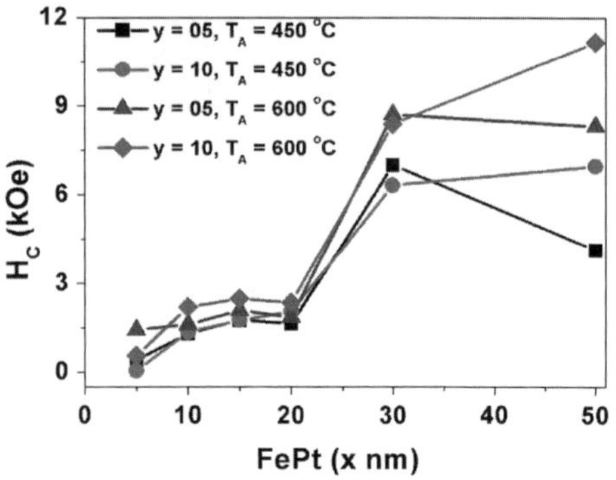

Fig. 4 Plots of variations of coercivity (H_C) as a function of FePt film thickness for the FePt (*x* nm)/MgO (*y* nm) films prepared on thermally oxidized Si substrates at ambient temperatures and annealed at different temperatures

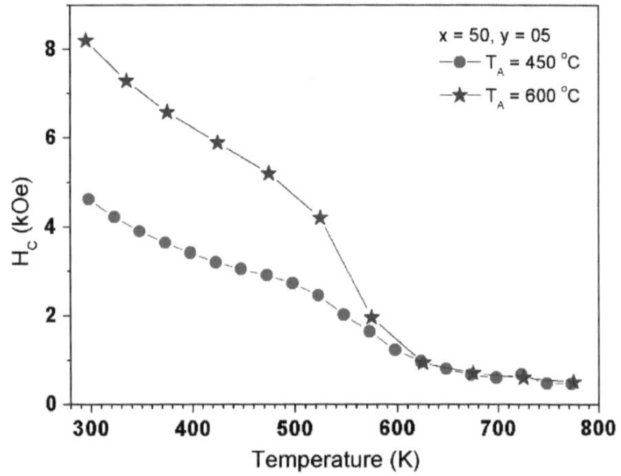

Fig. 5 Plots of variations of coercivity (H_C) as a function of temperature in the temperature range 300 and 800 K for the FePt (50 nm)/ MgO (5 nm) films prepared on thermally oxidized Si substrates at ambient temperatures and annealed at different temperatures

domain-wall pinning sites, i.e., the presence of finely dispersed mixture of the ordered and disordered phases is also responsible for the high coercivity, with pinning of the magnetic domain-walls due to the short range coherency strains between the cubic disordered phase and the tetragonal ordered phase [17].

To study the stability of the $L1_0$ ordered phase in the FePt films at higher temperatures, the magnetic hysteresis loops were measured at different temperatures between 300 and 800 K and the extracted values of coercivity are plotted as a function of temperature in Fig. 5 for FePt(50 nm)/MgO (5 nm) annealed at both 450 and 600 °C. With increasing temperature, the coercivity variation depicts the following features: (1) the temperature dependence of coercivity exhibits clearly a three different zones: (a) 300–540 K and (b) 540–640 K and (c) 640–800 K, (2) In the first zone (300–540 K), the coercivity decreases almost linearly at a rate of 9 (16.5) Oe/K for the samples annealed at 450 °C (600 °C). Highly ordered FePt films exhibits a larger reduction in the coercivity as compared to the films with low $L1_0$ ordering. This could be attributed mainly to the reduction in the magneto-crystalline anisotropy energy with increasing temperature, (3) A rapid variation in coercivity was observed in the second zone mainly due to the instability of the ordered $L1_0$ phase, and (4) finally, the films depicts the soft magnetic properties above 640 K, where the coercivity decreases at a relatively slower rate ($\sim$2.5 Oe/K), as expected for a typical soft magnetic material. These results confirm that the thicker FePt films have higher coercivity which is stable up to about 550 K. However, a careful analysis of the temperature dependent coercivity study at different FePt and MgO underlayer thickness is necessary to reveal the detailed information on the effect of high temperature on the magnetocrystalline anisotropy energy.

4 Conclusion

We studied the effects of thickness and post annealing temperatures on the structural and temperature dependent magnetic properties of FePt/MgO thin films deposited at ambient temperature on the oxidized Si substrate. The development of $L1_0$ ordering depends strongly on the films thickness and annealing temperatures. The MgO underlayer thickness plays a significant role only on the samples annealed at 600 °C. The average coercivity of the FePt films was found to be lower than 3 kOe for the film thickness less than 30 nm, but observed a maximum of about 12 kOe at higher FePt film thickness. The hard magnetic phase, existing due to the presence of the $L1_0$ ordering in the FePt films, transforms into soft magnetic phase above 640 K. The obtained results were correlated with the effects of post annealing on the variation of $L1_0$ ordering and temperature dependent reduction in magnetocrystalline anisotropy energy.

Acknowledgments One of the authors, AG, acknowledges the financial support from the DRDO, New Delhi in the form of research fellowship for the Ph.D. program. This work was financially supported by the DRDO, New Delhi through a research project [ERIP/ER/0900363/M/01/1185]. Infrastructure facilities provided by DST, New Delhi [SR/S2/CMP-19/2006, SR/FST/PII-020/2009] are gratefully acknowledged.

References

1. S.D. Badar, Rev. Mod. Phys. **78**, 1 (2006)
2. A.Yu. Dobin, H.J. Richter, Appl. Phys. Lett. **89**, 062512 (2006)
3. R.C. Victora, X. Shen, IEEE Trans. Magn. **41**, 537 (2005)
4. J.P. Wang, W. Shen, J. Bai, IEEE Trans. Magn. **41**, 3181 (2005)
5. D. Weller, A. Moser, IEEE Trans. Magn. **35**, 4423 (1999)
6. T. Suzuki, Z. Zhang, A.K. Singh, J. Yin, A. Perumal, H. Osawa, IEEE Trans. Magn. **41**, 555 (2005)
7. A. Perumal, L. Zhang, Y.K. Takahashi, K. Hono, J. Appl. Phys. **108**, 083907 (2010)
8. T. Shima, K. Takanashi, Y.K. Takahashi, K. Hono, Appl. Phys. Lett. **85**, 2571 (2004)
9. Z.L. Zhao, J.S. Chen, J. Ding, J.B. Yi, B.H. Liu, J.P. Wang, Appl. Phys. Lett. **88**, 052503 (2006)

10. J. Buschbeck, S. Fahler, M. Weisheit, K. Leistner, J. McCord, B. Rellinghaus, L. Schultz, J. Appl. Phys. **100**, 123901 (2006)
11. Y.C. Wu, L.W. Wang, C.H. Lai, Appl. Phys. Lett. **91**, 072502 (2007)
12. C.C. Chiang, C.-H. Lai, Y.C. Wu, Appl. Phys. Lett. **88**, 152508 (2006)
13. Y.C. Lai, Y.H. Chang, Y.C. Chen, C.H. Liang, W.C. Chang, C.M. Chiou, G.J. Chen, J. Appl. Phys. **101**, 053913 (2007)
14. M.-H. Cho, D.-H. Ko, Y.G. Choi, K. Jeong, I.W. Lyo, D.Y. Noh, H.J. Kim, C.N. Whang, J. Vac. Sci. Technol. A **19**, 192 (2001)
15. Y.-M. Wu, J.-T. Lo, Jpn. J. Appl. Phys. **37**, 4943 (1998)
16. J.D. Livingston, J. Appl. Phys. **52**, 2544 (1981)
17. J.B. Newkirk, A.H. Geisler, D.L. Martin, R. Smoluchowski, Trans. AIME **188**, 1249 (1950)

Substrate Bias Influenced Physical Characteristics of Nanocrystalline Silver Copper Oxide Films

A. Sreedhar, M. Hari Prasad Reddy and S. Uthanna

Abstract The silver-copper-oxide ($Ag_2Cu_2O_3$) films were prepared by reactive magnetron sputtering of equimolar silver copper alloy target ($Ag_{50}Cu_{50}$) onto the glass substrates in an oxygen partial pressure of 2×10^{-2} Pa and at different substrate bias voltages in the range from 0 to -90 V. The X-ray photoelectron spectroscopic studies on the films formed at substrate bias voltage of -60 V showed the characteristic core level binding energies of $Ag_2Cu_2O_3$. The films formed on unbiased substrates were X-ray amorphous. The mixed phase films of $Ag_2Cu_2O_3$ and Ag were deposited at substrate bias voltage of -30 V, where as nanocrystalline and single phase $Ag_2Cu_2O_3$ films were achieved at substrate bias voltage of -60 V. The atomic force micrographs of the unbiased films showed spherical shaped grains while those deposited at higher substrate bias voltages exhibited grains with pyramidal shape. The electrical resistivity of the films increased from 1.2×10^{-5} to 8.6×10^{-3} Ωcm with increase of substrate bias voltage from 0 to -60 V respectively. The optical band gap of the films increased from 1.78 to 1.92 eV with increase of substrate bias voltage from 0 to -60 V.

Keywords Thin films · Reactive sputtering · Silver copper oxide · Structural properties · Optical properties

A. Sreedhar · M. Hari Prasad Reddy · S. Uthanna (✉)
Department of Physics, Sri Venkateswara University, Tirupati 517502, India
e-mail: uthanna@rediffmail.com

P. K. Giri et al. (eds.), *Advanced Nanomaterials and Nanotechnology*,
Springer Proceedings in Physics 143, DOI: 10.1007/978-3-642-34216-5_46,

1 Introduction

The silver-copper-oxygen (Ag–Cu-O) system consists of various ternary compounds of $Ag_2Cu_2O_3$, $Ag_2Cu_2O_4$ and $AgCuO_2$ [1–3]. The ternary silver-copper oxides find novel applications such as positive electrode in button cell batteries, absorber layer in photovoltaic cells, gas sensors, and photocatalytic and antibacterial coatings [4–6]. Gomez-Romero et al. [2] first synthesized silver-copper oxide of $Ag_2Cu_2O_3$. Curda et al. [3] prepared $Ag_2Cu_2O_4$ powers by electrochemical oxidation of suspension of the precursor $Ag_2Cu_2O_3$ [3]. Silver copper oxide in thin film form were prepared through reactive sputtering process using sputter targets such as copper covered by silver chips [7], co-sputtering of silver and copper targets [8] and composite target of $Ag_{50}Cu_{50}$ [9, 10], $Ag_{60}Cu_{40}$ [11] and $Ag_{70}Cu_{30}$ [12, 13] at different sputter process parameters. The physical properties of the sputter deposited films will be highly influenced by the sputter parameters such as oxygen partial pressure, substrate temperature, substrate bias voltage and sputter power. Our earlier report [14] indicated that the $Ag_2Cu_2O_3$ formed at 423 K at oxygen partial pressure of 2×10^{-2} Pa and subsequently annealed in air at 498 K were of single phase $Ag_2Cu_2O_3$ films. Biasing the substrate induced the growth of crystalline films. In the present investigation, the influence of substrate bias voltage on the structural, morphological, electrical and optical properties of $Ag_2Cu_2O_3$ films formed with equimolar target of $Ag_{50}Cu_{50}$ using RF reactive magnetron sputtering was systematically studied.

2 Experimental

$Ag_2Cu_2O_3$ films were deposited on glass substrates using RF magnetron sputtering method. Equimolar ($Ag_{50}Cu_{50}$) target of 50 mm diameter and 3 mm thick was used as sputter material for deposition of thin films. The base pressure of 5×10^{-4} Pa was achieved in the sputter chamber using diffusion pump—rotary pump combination. The required quantities of reactive gas of oxygen and sputter gas of argon were admitted into the sputter chamber through fine controlled needle valves and monitor individually using Aalborg mass flow controllers. The $Ag_2Cu_2O_3$ films were deposited on unheated glass substrates at a constant oxygen partial pressure of 2×10^{-2} Pa, sputtering pressure of 4 Pa and at different substrate bias voltages in the range from 0 to -90 V. The RF power fed to the sputter target using Advanced Energy power source was 65 W. The duration of deposition was 20 min. The core level binding energies of the deposited films was analyzed with X-ray photoelectron spectroscopy (XPS). The crystallographic structure of the films was determined with X-ray diffraction (XRD) taken on a Siefert Diffractometer (model 3003 TT) using monochromatic $CuK_{\alpha 1}$ radiation. The surface morphology of the deposited films was analyzed with atomic force microscope (AFM) taken on Seiko Instruments. The electrical resistivity of the films was measured at room temperature using four point probe method. The optical transmittance of the films was recorded in the wavelength range 300–2500 nm using Perkin-Elmer double beam spectrophotometer.

3 Results and Discussions

The deposition rate of silver copper oxide films was highly influenced by the substrate bias voltage maintained during the growth of the films. The deposition rate of the films was calculated from the film thickness and duration of deposition time. The deposition rate of the films increased from 8 to 9.2 nm/min with the increase of substrate bias voltage from 0 to −60 V and it remains almost constant at higher bias voltages. The increase of deposition rate with the substrate bias voltage was due to the attraction of positively charged molecules and/or clusters of sputtered species in the plasma which enhance the reaction of sputter species on the substrate surface.

Figure 1a, b and c shows the X-ray photoelectron spectroscopic narrow scan core level binding energy spectra of the Ag 3d, Cu 2p and O 1 s at different substrate bias voltages. The core level binding energies of about 368 and 374 eV were related to Ag $3d_{5/2}$ and Ag $3d_{3/2}$ respectively due to the spin–orbit splitting of the energy levels of unbiased films, which shifted to 367.7 and 373.7 eV with increase of substrate bias voltage to −60 V. The peak observed at about 530 eV was related to core level binding energy of oxygen O 1 s at unbiased voltage, shifts towards 529.5 eV with the increase of substrate bias voltage to −60 V. The peaks located at about 573.0 and 603.9 eV were connected to the core level binding energies of Ag $3p_{3/2}$ and Ag $3p_{1/2}$ respectively [14]. The peaks seen around 934.0 and 954 eV were the characteristics core level binding energies of Cu $2p_{3/2}$ and Cu $2p_{1/2}$ respectively at substrate bias voltage of 0 V, which were shifted to 933.4 and 953.7 eV with increase of substrate bias voltage to −60 V, which indicated the growth of single phase $Ag_2Cu_2O_3$ films.

X-ray diffraction profiles of $Ag_2Cu_2O_3$ films formed at different substrate bias voltages were shown in Fig. 2. X-ray diffraction profile of unbiased films not contained any diffraction peaks which indicated that the films were of X-ray amorphous. As the substrate bias voltage increased to −30 V the films showed the (202) and (1–12) reflections of $Ag_2Cu_2O_3$ and Ag respectively [15] indicated the growth of polycrystalline nature of the films. The films formed at substrate bias voltage of −60 V, the intensity of (202) reflection of $Ag_2Cu_2O_3$ increased while (1–12) reflections of Ag disappeared. It revealed that the films formed at substrate bias voltage of −60 V were of single phase $Ag_2Cu_2O_3$ [16]. At higher substrate bias voltage of −90 V, the intensity (202) reflection of $Ag_2Cu_2O_3$ decreased and (1–12) reflection of Ag again noticed due to deformation of $Ag_2Cu_2O_3$ phase. The crystallite size of the films was evaluated form the full width at half maximum intensity of X-ray diffraction peaks using the Debye-Scherrer's relation. The crystallite size of the films increased from 6 to 15 nm with the increase of

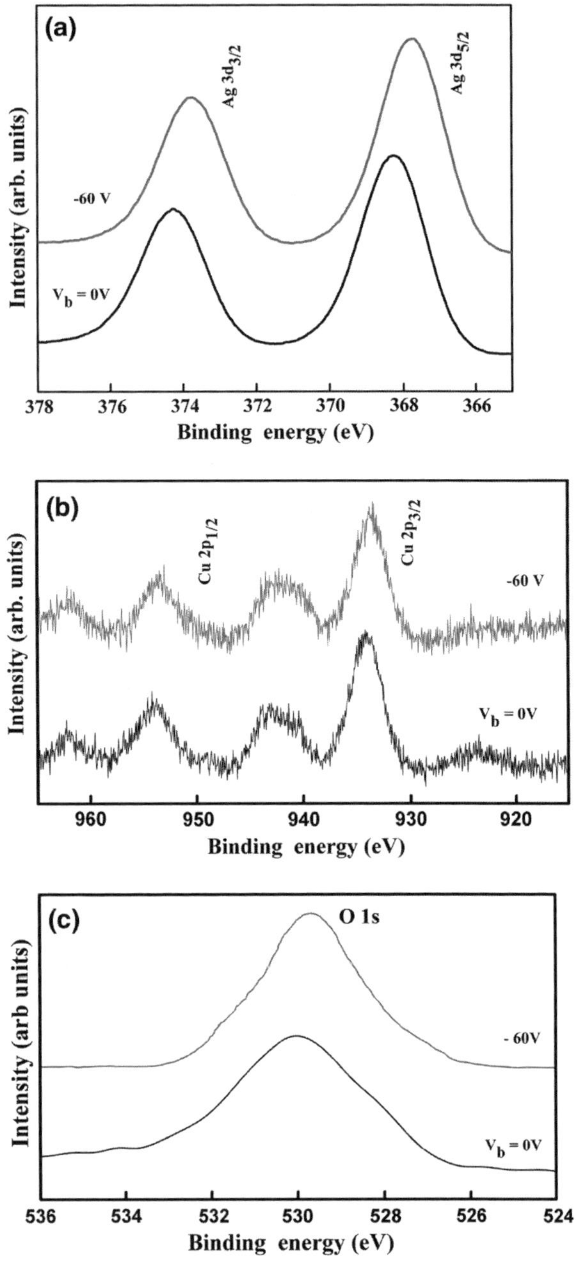

Fig. 1 The narrow scan XPS spectra of **a** Ag 3d, **b** Cu 2p, and **c** O1 s of $Ag_2Cu_2O_3$ films formed at different substrate bias voltages

substrate bias voltage from −30 to −60 V. At higher substrate bias voltage of −90 V, the crystallite size of the films decreased to 12 nm due to the mixed phase of $Ag_2Cu_2O_3$ and Ag. The X-ray diffraction studies revealed that the deposited films were of nanocrystalline with crystallite size in the range 6–15 nm.

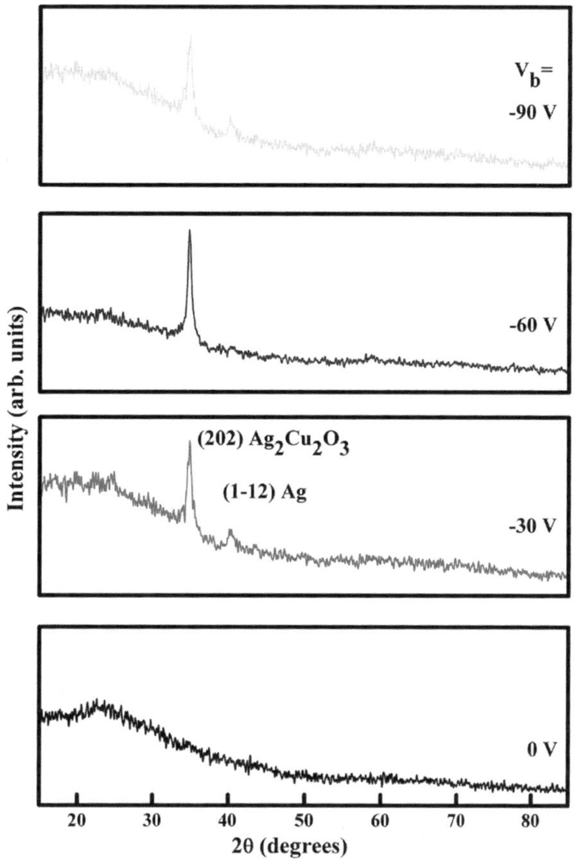

Fig. 2 X-ray diffraction profiles of $Ag_2Cu_2O_3$ films formed at different substrate bias voltages

The AFM three dimensional and two dimensional micrographs of $Ag_2Cu_2O_3$ films formed at different substrate bias voltages were shown in Fig. 3. The films deposited at unbiased substrates showed spherical shaped grains while those deposited at higher substrate bias voltages exhibited pyramidal shape grains. Figure 4 shows the variation of grain size and RMS surface roughness with substrate bias voltage. The grain size of the films increased from 42 to 115 nm with the increase of substrate bias voltage from 0 to −90 V. The root mean square surface roughness of the films increased from 4.60 to 5.80 nm with increase of substrate bias voltage from 0 to −60 V and at higher substrate bias voltage of −90 V it decreased to 5.19 nm.

The electrical resistivity of the films deposited at unbiased substrate was 1.2×10^{-5} Ωcm. The electrical resistivity increased to 8.6×10^{-3} Ωcm with increase of substrate bias voltage to −60 V. The low electrical resistivity of unbiased films was due to presence of mixed phase of $Ag_2Cu_2O_3$ and silver in amorphous phase. The increase of electrical resistivity to 8.6×10^{-3} Ωcm at substrate bias voltage of −60 V was due to the presence of single phase of

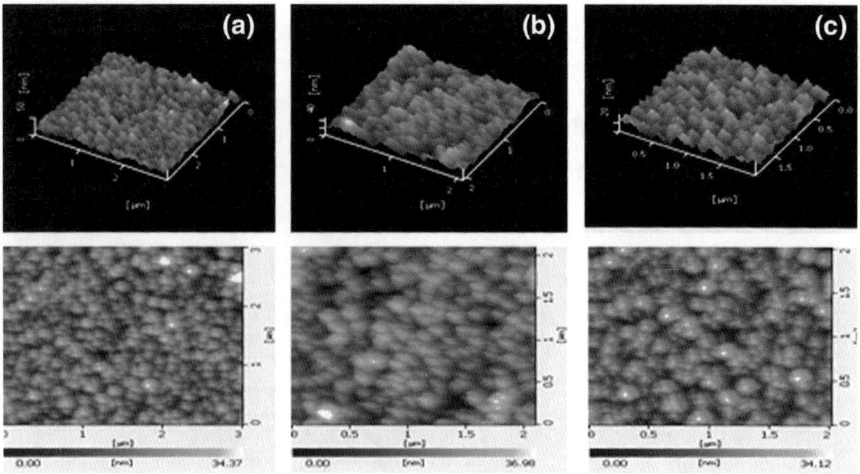

Fig. 3 AFM two and three dimensional micrographs of $Ag_2Cu_2O_3$ films formed at substrate bias voltages: **a** 0 V, **b** −60 V, and **c** −90 V

Fig. 4 Variation in the grain size and RMS roughness of $Ag_2Cu_2O_3$ films with substrate bias voltage

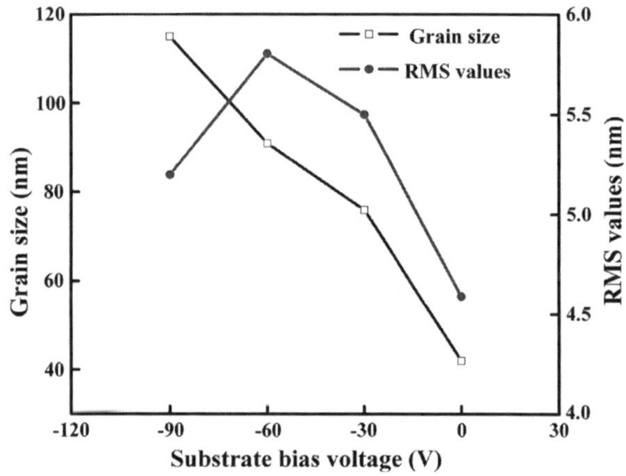

$Ag_2Cu_2O_3$ film. Further increase of substrate bias voltage to −90 V the electrical resistivity decreased to 6.2×10^{-4} Ωcm due to the growth of $Ag_2Cu_2O_3$ along with silver as confirmed by XRD data. Petitjean et al. [1] achieved the electrical resistivity of 1.5×10^{-2} Ωcm in DC reactive sputtered films formed on unheated substrates using equimolar target of $Ag_{50}Cu_{50}$ and at oxygen flow rate of 25 sccm and argon flow rate of 20 sccm.

The optical transmittance spectra of $Ag_2Cu_2O_3$ films formed at different substrate bias voltages are shown in Fig. 5. The optical transmittance of the films (at wavelength $\lambda = 1{,}000$ nm) formed on unbiased substrates was about 31 %. The optical transmittance of the films was increased to 66 % at substrate bias voltage of −60 V while at higher substrate bias voltage of −90 V, it was

Fig. 5 Optical transmittance spectra of $Ag_2Cu_2O_3$ films formed at different substrate bias voltages

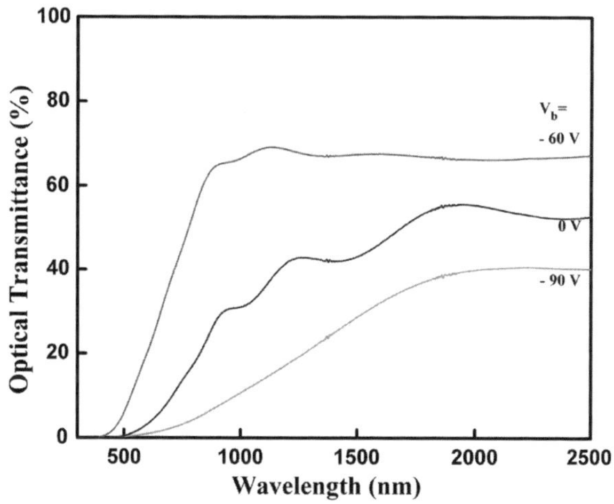

decreased to 11 %. The optical absorption edge of the films shifted towards lower wavelength side with increase of substrate bias voltage up to −60 V, while at higher voltage of −90 V it shifted towards higher wavelength side. The low optical transmittance of substrate bias voltage to < −60 V and at −90 V was due to the presence of metallic silver phase along with $Ag_2Cu_2O_3$. The atoms of metallic silver scatters light hence decrease in the optical transmittance. The optical absorption coefficient (α) of the films was evaluated from the optical transmittance data. The optical band gap (E_g) of the films was estimated from the plots of $(\alpha h\nu)^2$ versus photon energy ($h\nu$) using Tauc's relation assuming that the direct transitions takes place in these material. Extrapolation of the linear portion of these plots to $\alpha = 0$ resulted the optical band gap of the films. The optical band gap of the films increased from 1.78 to 1.92 eV with the increase of substrate bias voltage from 0 to −60 V. The optical band gap of the films decreased to 1.12 eV at higher substrate bias voltage of −90 V due to the presence of metallic silver. Recently, Lund et al. [10] reported that RF magnetron sputtered $Ag_2Cu_2O_3$ films formed with $Ag_{50}Cu_{50}$ target showed a direct band gap of 2.2 eV.

4 Conclusions

This paper is focussed on the feasibility of the growth of nanocrystalline $Ag_2Cu_2O_3$ films using RF bias magnetron sputtering technique. Thin films of $Ag_2Cu_2O_3$ were deposited on to unheated glass substrates by sputtering of equi-molar ($Ag_{50}Cu_{50}$) target in an oxygen partial pressure of 2×10^{-2} Pa, sputter pressure of 4 Pa and at various substrate bias voltages in the range from 0 to −90 V. The influence of substrate bias voltage on the structural, electrical and optical properties of the films was systematically studied. The deposition rate of the films increased from 8 to 9.2 nm/min with the increase of substrate bias voltage from 0 to −60 V and remains almost constant at higher bias voltages. The X-ray photoelectron spectroscopic data showed that the core level binding energies related to the $Ag_2Cu_2O_3$ were obtained in the films formed at substrate bias voltage of −60 V. The X-ray diffraction studies revealed that the films formed on unbiased substrates were of X-ray amorphous, while mixed phase of $Ag_2Cu_2O_3$ and Ag films were achieved at substrate bias voltage of −30 V. Nanocrystalline and single phase $Ag_2Cu_2O_3$ films were grown at substrate bias voltage of −60 V. The crystallite size of the films increased from 6 to 15 nm with the increase of substrate bias voltage from −30 to −60 V. At higher substrate bias voltage of −90 V, it decreased to 12 nm. The atomic force micrographs showed that the films deposited at unbiased substrates showed spherical grains while those formed at higher substrate bias voltages exhibited the grains with pyramidal shape. The grain size of the films increased from 42 to 115 nm with the increase of substrate bias voltage from 0 to −90 V. The electrical resistivity of the films deposited at unbiased substrate was 1.2×10^{-5} Ωcm. The electrical resistivity of single phase $Ag_2Cu_2O_3$ films formed at substrate bias voltage to −60 V was 8.6×10^{-3} Ωcm. The optical band gap of the films increased from 1.78 to 1.92 eV with the increase of substrate bias voltage from 0 to −60 V. In conclusion, nanocrystalline $Ag_2Cu_2O_3$ films with crystallites size of 15 nm, electrical resistivity of 8.6×10^{-3} Ωcm and optical band gap of 1.92 eV were grown at substrate bias voltage of −60 V.

Acknowledgments One of the authors (ASR) thankful to DST—Promotion of University in Research Science and Excellence (PURSE) programme for financial support to carried out this work.

References

1. C. Petitjean, D. Horwat, J.F. Pierson, Effect of annealing temperature on the decomposition of reactively sputtered $Ag_2Cu_2O_3$ films. Appl. Surf. Sci. **255**, 7700 (2009)
2. P. Gomez-Romero, E.M. Tejada-Rosales, M. Rosa Palacin, The $Ag_2Cu_2O_3$: The first silver copper oxide. Chem. Int. Ed. **38**, 524 (1999)
3. J. Curda, W. Klein, M. Jansen, $AgCuO_2$—synthesis, crystal structure, and structural relationships with CuO and $Ag^{I}Ag^{III}O_2$. J. Solid State Chem. **162**, 220 (2001)
4. C.D. May, J.T. Vaughey, New cathode materials for silver-based primary batteries: $AgCuO_2$ and $Ag_2Cu_2O_3$. Electrochem. Commun. **6**, 1075 (2004)
5. F. Sauvage, D. Munoz-Rojas, K.R. Poeppelmeier, N. Casan-Pastor, Transport properties and lithium insertion study in the p-type semiconductors $AgCuO_2$ and $AgCu_{0.5}Mn_{0.5}O_2$. J. Solid State Chem. **182**, 374 (2009)
6. J. Feng, B. Xiao, J.C. Chen, C.T. Zhou, Y.P. Du, R. Zhou, Optical properties of new photovoltaic materials: $AgCuO_2$ and $Ag_2Cu_2O_3$. Solid State Commun. **149**, 1569 (2009)
7. J.F. Pierson, D. Wiederkehr, J.M. Chappe, N. Martin, Reactive sputtering: A method to modify the metallic ratio in the novel silver–copper oxides. Appl. Surf. Sci. **253**, 1484 (2006)
8. J.F. Pierson, D. Horwat, Influence of the current applied to the silver target on the structure and the properties of Ag–Cu–O films deposited by reactive co sputtering. Appl. Surf. Sci. **253**, 7522 (2007)

9. C. Petitjean, D. Horwat, J.F. Pierson, Structure-properties relationship in reactively sputtered Ag–Cu–O films. J. Phys. D Appl. Phys. **42**, 025304 (2009)
10. E. Lund, A. Galeckas, E.V. Monakhov, B.G. Svensson, Structural, optical and electrical properties of reactively sputtered $Ag_2Cu_2O_3$ films. Thin Solid Films **520**, 230 (2011)
11. J.F. Pierson, E. Rolin, C.C. Gendarme, C. Petitjean, D. Horwat, Effect of the oxygen flow rate on the structure and the properties of Ag–Cu–O sputtered films deposited using a Ag/Cu target with eutectic composition. Appl. Surf. Sci. **254**, 6590 (2008)
12. S. Uthanna, M. Hari Prasad Reddy, P. Boulet, C. Petitjean, J.F. Pierson, Effect of deposition temperature on the physical properties of RF magnetron sputtered Ag-Cu-O films with various Cu to Ag ratios. Phys Status Solidi (a) **207**, 1655 (2010)
13. M. Hari Prasad Reddy, P. Narayana Reddy, B. Sreedhar, J.F. Pierson, S. Uthanna, Effect of substrate temperature on the structural, electrical and optical behaviour of reactively sputtered Ag–Cu–O films, Physica Scripta **84**, 045602 (2011)
14. A. Sreedhar, M. Hari Prasad Reddy, S. Uthanna, R. Martins, E. Elangovan, J.F. Pierson, Substrate temperature influenced structural and electrical behaviour of RF magnetron sputtered $Ag_2Cu_2O_3$ films. Acta Physica Polonica A **120**, A-37 (2011)
15. D. Munoz-Rojas, J. Oro, P. Gomez-Romero, J. Fraxedas, N. Casan-Pastor, Electrochemically induced reversible solid state transformations: Electrosynthesis of $Ag_2Cu_2O_4$ by room temperature oxidation of $Ag_2Cu_2O_3$. Electrochem. Commun. **4**, 684 (2002)
16. JCPDS—International centre for diffraction data, ICCD card no. 01-073-6753

Doped ZnO Nanostructured Sensor in Electronic Nose for Detection of Ammonia, Hydrogen and Liquefied Petroleum Gas

Samir Chandra Das, Bipan Tudu, Nabarun Bhattacharyya, R. Bandyopadhyay and Panchanan Pramanik

Abstract In this experimental work, a nanostructured ZnO like nano flowers doped with different elements, X = Pd, Pt, Co, Mn, Ni, Sn are synthesized. X-ray diffraction (XRD), Transmission Electron Microscopy (TEM) and Scanning Electron Microscopy (SEM) are used for characterization of the materials and surface morphology. The materials were synthesized from pyrolysis of $Zn(NO3)_2$ solution with triethanolamine (TEA) with adequate dopant. After complete dehydration of the precursor solution, a black, carbonaceous, mesoporous fluffy mass is obtained, which after calcinations produces the desired nanocrystalline materials. Formation of hexagonal single phase ZnO nanoparticle is confirmed by the careful analysis of XRD data. It follows from non oriented growth for undoped nanostructure to strongly (002) oriented, at intermediate ($\sim$1at. %) doping level. The materials are also characterized through transmission electron microscopy (TEM), and energy dispersive X-ray (EDX) analysis to obtain the shape, size, morphology and elemental compositions. Average crystallite size and particle size

S. C. Das (✉)
Secretary Faculty Councils, University of Kalyani, Nadia 741235 West Bengal, India
e-mail: samirdasju@yahoo.co.in

B. Tudu · R. Bandyopadhyay
Department of Instrumentation and Electronics Engineering, Jadavpur University, Kolkata, India
e-mail: bbt@iee.jusl.ac.in

N. Bhattacharyya
Centre for Development of Advanced Computing (C-DAC), Kolkata, India

P. Pramanik
Department of Chemistry, Indian Institute of Technology, Kharagpur 721302 West Bengal, India

P. K. Giri et al. (eds.), *Advanced Nanomaterials and Nanotechnology*, Springer Proceedings in Physics 143, DOI: 10.1007/978-3-642-34216-5_47,

are observed to be between 24 and 33 nm, which are analyzed through XRD and transmission electron microscope, respectively. Electrical as well as gas sensing properties of the synthesized materials have been studied by dc measurement.

Keywords: Nanostructured sensor · Mesoporous · Fluffy mass · Surface morphology · EDX

1 Introduction

The semiconductor zinc oxide is recognized as one of the most important semiconductor materials which exhibits numerous characteristics that may enable its efficient utilization for various technological applications such as antireflection coatings, transparent electrodes in solar cells, piezoelectric devices, varistors, surface acoustic wave devices, electro-and photo luminescent devices [1], gas sensors, and others. Recently zinc oxide has attracted worldwide research interest because it is considered a promising material for nanostructured gas sensors in electronic noses [2]. Although ZnO is one of the earliest discovered semiconducting oxide gas sensing materials and there are many reports concerning the sensitivity properties of ZnO, most of these works has been done on powder samples usually pressed in pellets and sintered at high temperatures (900–1300 °C). Nanostructure sensor makes use of the chemical sensitivity of semiconductor surfaces to different adsorbed gases. Appropriate donor doping can produce the electronic defect that increase the influence of oxygen partial pressure on the conductivity. Doped zinc oxide nanostructure have a number of attractive applications, such as: gas sensor devices, transparent electrodes, piezoelectric devices. Several techniques have been used to produce many distinct zinc oxide nanostructure as shown in Fig. 1. In this work we use the chemical synthesis to preparation of aqueous solution of zinc oxide and we have investigated the sensing properties of a gas injected ZnO-doped nanostructure sensor, to hydrogen, liquefied petroleum gas and ammonia [3, 4].

2 Experimental

2.1 Raw Materials

The chemicals used are zinc nitrate [$Zn(NO_3)_2$], nickel nitrate [$Ni(NO_3)_2$. $6H_2O$], Manganese acetate [$(CH_3COO)_2Mn$. $4H_2O$], cobalt nitrate [$Co(NO_3)_2$. $6H_2O$], triethanolamine (TEA), distil water and nitric acid (70 %). All the materials were procured from M/S Aldrich (USA).

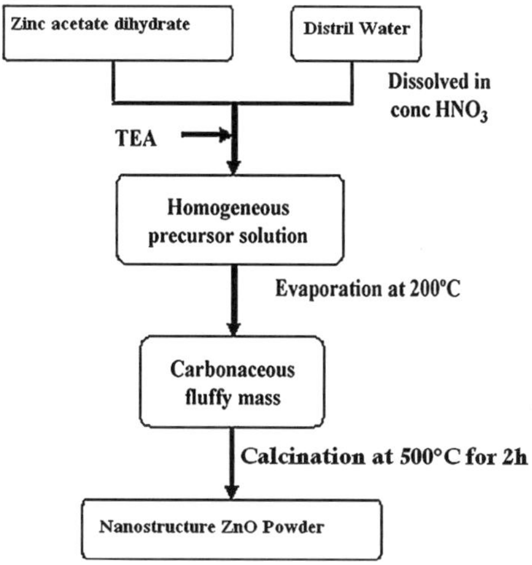

Fig. 1 Schematic representation of the preparation of nanocrystalline powders of ZnO

2.2 Preparation of Aqueous Solution of Doped Zinc Oxide Complex

Prepare 0.25 (M) aqueous solution of zinc nitrate in a 500 ml volumetric flask. Then one by one prepare 0.025 (M) aqueous solution of cobalt nitrate, nickel nitrate, and manganese acetate in a 250 ml volumetric flask.

2.3 Synthesis of Nanosized Co–ZnO Doped, Ni–ZnO Doped, Mn–ZnO Doped Pellet

Use 0.25 (M) aqueous solution of zinc nitrate of 200 ml was mixed with a 0.025 (M) aqueous solution of cobalt nitrate of 20.48 ml. Then the mixture put on a heater for heating. After few time added 14 ml of TEA along with constant stirring of mixture. Then added nitric acid drop wise to avoid precipitation. The pH of the final solution mixture was maintained at almost acidic (pH = 3–4). This mixture solution was completely dehydrated by heating at about 200 °C and then black fluffy mass was formed after decomposition of the metal—organic complex with evolution of fumes. It was a highly exothermic reaction. Then calcinations of the black fluffy at 650 °C for 2 h and then produce carbon free powder. Then this powder was pasted by mortar and collected for prepare pellet. Again, in another beaker 0.25 (M) aqueous solution of zinc nitrate of 200 ml was mixed with 0.025 (M) aqueous solution of nickel nitrate of 20.48 ml and also in another beaker 0.25 (M) zinc nitrate of 200 ml mixed with 0.025 (M) of aqueous solution of manganese acetate of 20.48 ml. Then repeat the above same procedure for these two mixtures.

3 Electrical Measurements and Gas Sensing Studies

Electrical as well as gas sensing properties of the synthesized material were studied by dc measurement. For the experiment, a pellet (10 mm diameter and 2.5 mm thickness) was made from the powder material using 1–2 drops of polyvinyl alcohol (PVA) (5 %). The pellet was then heat-treated at 450 °C for 4 h to remove the residual polymer which helped to produce porous solid. The porous pellet with thin platinum electrodes was housed inside a quartz test-chamber, which was heated by a resistance furnace shown in Fig. 2. The temperature of the test-chamber was controlled using a temperature controller. Resistance of the sensor was measured using auto ranging microvolt DMM (Model 197A, M/s Keithley Instruments, USA) for low temperatures (when the resistance is higher than 100 M) and Agilent Data Acquisition/Switch Unit (model no. 34970A) for higher temperatures. At each sensor operating temperature, the test gas was injected into the chamber through an injection port after a steady base line resistance in air was established. Average crystallite size and particle size are observed to be between 24 and 33 nm, which were analyzed through XRD and transmission electron microscope (TEM) as shown in Figs. 3, 4, respectively. Electrical as well as gas sensing properties of the synthesized materials have been studied by dc measurement. The response of the material to the test gases have been calculated using the equation:

$$response(\%) = \left[\frac{R_a - R_g}{R_a}\right] \times 100$$

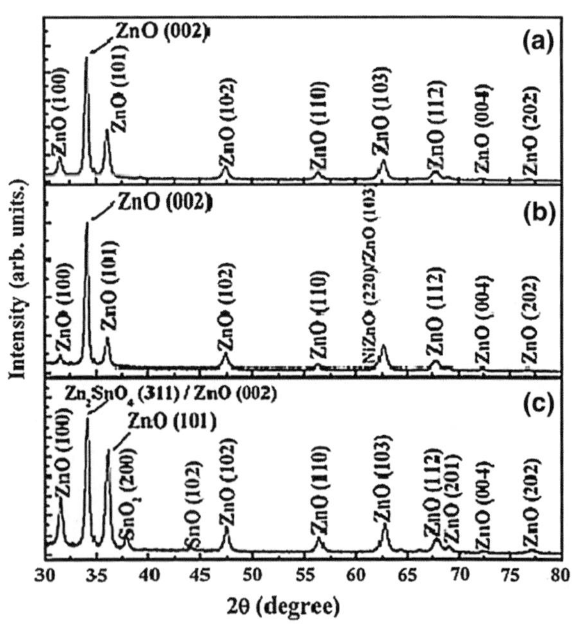

Fig. 2 The XRD pattern of the nanostructured ZnO **a** undoped, **b** Ni-doped, and **c** Sn-doped

Fig. 3 SEM image of pellet surface derived from nanosized ZnO powder

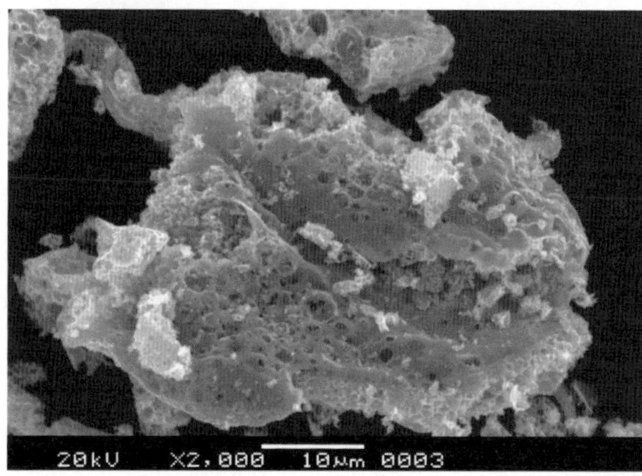

Fig. 4 SEM images showing ZnO(Pt) doped nanostructures obtained on Zn substrate after holding at 400 °C for 2 h

Where, R_a denotes the resistance in air and R_g the resistance in the presence of a test gas. The response of the material is examined with H_2, LPG and NH_3 in the temperature range of 200–400 °C as shown in Figs. 5, 6 respectively and the sensitivity have been observed to be quite appreciable.

4 Results and Discussion

4.1 Structural Analyses

Figure 3 shows the XRD diffraction pattern of the (a) pure, (b) Ni-doped, and (c) Sn-doped ZnO nanostructured. The diffraction peaks in the pattern can be indexed to hexagonal wurtzite structured ZnO (space group: $P6_3mc$ (186); a = 0.3249 nm, c = 0.5206 nm) and diffraction results are in agreement with JCPDS card for ZnO (JCPDS 036–1451) [5]. The intensity of the peaks relative to the background

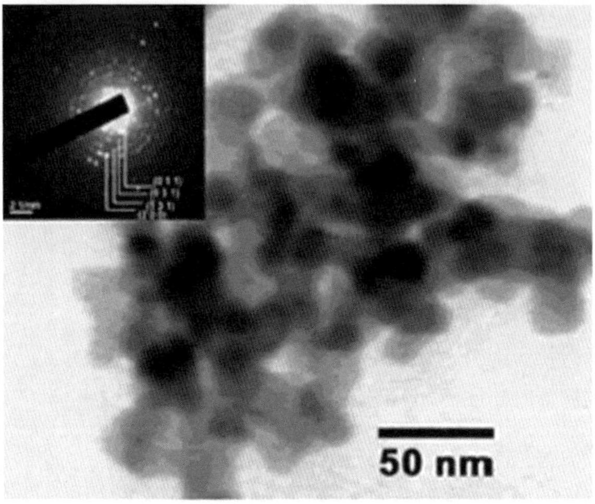

Fig. 5 Bright field TEM micrograph of the calcined (at 500 °C for 2 h) ZnO. (Inset) SAED pattern of the calcined (at 500 °C for 2 h) ZnO

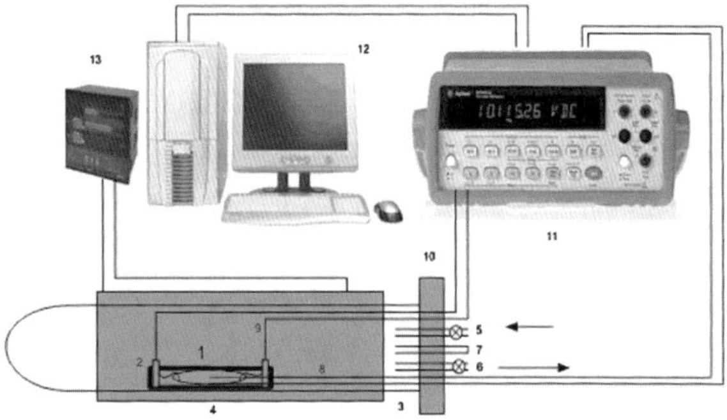

Fig. 6 Setup equipment for in-site resistance measurements. *1* Pellet sample. *2* Two probe sample holder. *3* Quartz test chamber. *4* Furnace. *5* Gas inlet. *6* Gas outlet. *7* Gas injection port. *8* Thermocouple. *9* Platinum wires. *10* Flung. *11* Digital multimeter. *12* Personal computer. *13* Temperature controller (PID controller)

signal demonstrates hexagonal phase of the products and high crystallinity of the ZnO samples grown by chemical synthesized method. The characteristic peaks of $Zn(OH)_2$ was not observed, which indicated a single phase hexagonal ZnO. The nanocrystallites are oriented along the c axis, [002] direction. The doping with Ni (Fig. 3b) seems to take place through substitution of Zn cations by Ni ions, because the modification of ZnO crystal structure was minimum. The doping of Sn (Fig. 3c) on the other hand seems to be problematic due to the appearance of two new peaks of SnO_2 (200) and SnO (102). Also these diffraction results (Fig. 3b and c) are in agreement with JCPDS cards (JCPDS 072–1464, and 024–1470, 041–1445), respectively [6]. The mean grain size (d) of the ZnO nanostructured samples, were calculated using Scherrer's equation:

$$d = \frac{k\lambda}{\left(\beta^2 - \beta_0^2\right)^{1/2}} \cos\theta$$

Where, β is the measured broadening of a diffraction line peak, full width at half its maximum intensity in radians, β_O - the instrumental broadening, $K = 180/\pi$, λ is the X-rays wavelength (0.154056 nm) and θ is the Bragg diffraction angle. The average grain size determined from XRD patterns was 240, 220 and 265 Å for ZnO, Sn–ZnO and Ni–ZnO nanostructured, respectively. For surface morphology and estimation of the nanostructured, SEM technique has been used. Using EDX, the formation of the ZnO and the doping with Sn and Ni in ZnO was confirmed.

4.2 Electrical Properties

The pure nanostructured ZnO have lower stability in corrosive, humid ambient due to large amount of O vacancies. The properties of such zinc oxide films are often altered by adsorption of CO_2, O_2, and water. The electrical resistance decreases with doping with elements Sn, Ni and with rapid photothermal processing temperature up to 650 °C and take place simultaneously with the elimination of voids inside the pure and doped-ZnO nanaostructured. Irreversible changes in the electrical characteristics have been observed when the ZnO nanaostructured were post-growth rapid photothermal processed at temperatures higher than 300 °C for 20 s duration. In the temperature range of 550–650 °C, improvement of the quality and stability of the ZnO sensor samples has been observed [7]. The measurement has been carried out during one year and it was determined that the doped ZnO based sensor elements have higher conductance stability, which ensures a stable zero level for gas sensor applications [1].

4.3 Gas Sensing Properties

The gas sensing phenomenon involves the change of electrical resistance of the sensor material in the presence of gases due to the oxidation of the reducing gases by the surface chemisorbed oxygen species (like O^-, O^{2-},$O_2{}^-$, etc.), which releases the electrons and thus increases the charge in the conduction band of the n-type oxide and hence the conductivity gets increased (i.e. decrease of the potential barrier). The extent of adsorbed oxygen ions and existence of their different chemical forms (O^-, $O^{2\,-}$, $O_2{}^-$, etc.) on the sensor surface are controlled by the sensor operating temperature [4, 8]:

$$O_2(g) + e^- \rightarrow O_2(ads)$$
$$O_2(ads)^- + e^- \rightarrow 2O(ads)$$

The oxidation reactions of H_2 and LPG could be represented as follows [8]:

$$C_pH_{2p+2} + (3p+1)O_{(ads)}- \rightarrow pCO_2 + (p+1)H_2O_{(g)} + (3p+1)e$$

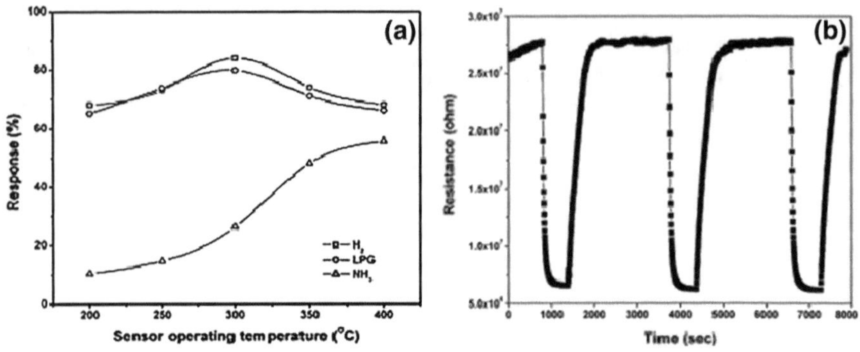

Fig. 7 **a** Sensing characteristics of porous pellet of ZnO(Pd) in alternating environments of air and 500 ppm of H_2, LPG and NH_3. **b** Electrical resistance of porous pellet of ZnO(Pd) in alternating environments of air and 500 ppm LPG at 300 °C

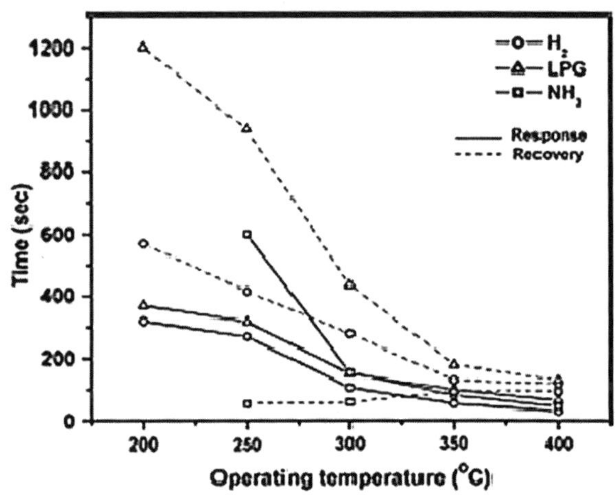

Fig. 8 Temperature variation of response and recovery times for ZnO(Pd) for 500 ppm H_2, LPG and NH_3

LPG is a mixture of hydrocarbons like n-propane and n-butane, so P = 3 or 4. This equation is valid for H_2 when p = 0. The oxidation reactions of NH_3 could be represented as follows [9, 10]:

$$2NH_3 + 3O_{(ads)}- \rightarrow N_2 + 3H_2O + 3_e-$$

Sensing Characteristics of nanosized ZnO(Pd) in the form of porous pellet towards 500 ppm of H_2, LPG and NH_3 in air as function of temperature is shown in Fig. 7a, electrical resistance in alternating environments of air and 500 ppm LPG at 300 °C shown in Fig. 7b. Figure 8 is the temperature variation of response and recovery times for ZnO(Pd) for 500 ppm H_2, LPG and NH_3. Where as the response of porous pellet of nanocrystalline ZnO(Pd) towards 500 ppm of NH_3 in air at different temperatures is shown in Fig. 9.

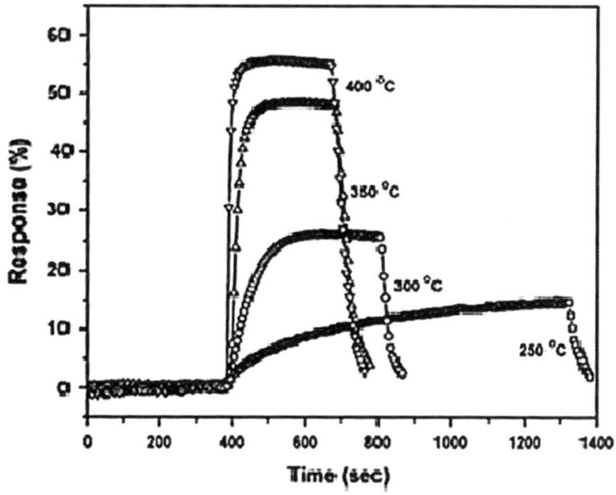

Fig. 9 Response of porous pellet of nanocrystalline ZnO(Pd) towards 500 ppm of NH_3 in air at different temperatures

5 Conclusion

Nanocrystalline ZnO powders have been synthesized by aqueous based metal ion-ligand complex precursor route for the first time. The wurtzite structure of ZnO(Pd) and ZnO(Pt) was realized at 500 °C. The low processing temperature in this method prevented grain growth in the powder. The n-type semiconducting nanocrystalline ZnO(Pd) and ZnO(Pt) showed the sensitivities towards H_2, LPG and NH_3. The maximum sensing responses of Pt and Pd doped ZnO towards H_2, LPG and NH_3 at 350–400 °C in alternating environments of air.

Acknowledgments The authors express thanks to the editor and anonymous referees for their extremely valued comments.

References

1. A. Mitra, R.K. Thareja, V. Ganesan, A. Gupta, P.K. Sahoo, V.N. Kulkarni, Synthesis and characterization of ZnO thin films for UV laser. Appl. Surf. Sci. **174**, 232 (2001)
2. H.K. Hong, C.H. Kwon, S.R. Kim, D.H. Yun, K. Lee, Y.K. Sung, Sens. Actuators, B **66**, 49 (2000)
3. S.C. Das, B. Tudu, N. Bhattacharyya, R. Bandyopadhyay, P. Pramanik, Nanostructured ZnO sensor in electronic nose for environmental monitoring on industrial premises, int proceedings, pp. 165–167. (ISBN: 978-93-80813-14-1) on ICLMSC-2011 (2011)
4. S.K. Biswas, P. Pramanik, Studies on the gas sensing behaviour of nanosized CuNb2O6 towards ammonia, hydrogen and liquid petroleum gas. Sens. Actuators B **133**, 449–455 (2008)
5. R. Cabala, V. Meister, K. Potje-Kamloth, J. Chem. Soc. Faraday Trans. **93**, 131 (1997)
6. P. Mitra, A.P. Chatterjee, H.S. Maiti, Mater. Lett. **35**, 33 (1998)
7. S.C. Ko, Y.C. Kim, S.S. Lee, S.H. Choi, S.R. Kim, Micromachined piezo- electric membrane acoustic device. Sens. Actuators, A **103**, 130 (2003)

8. S.K. Biswas, A. Sarkar, A. Pathak, P. Pramanik, Studies on the sensing behaviour of nanocrystalline CuGa2O4 towards hydrogen, liquid petroleum gas and ammonia. Talanta **81**, 1607–1612 (2010)
9. P. Dhak, U.K. Gayen, S. Mishra, P. Pramanik, A. Roy, J. Appl. Phys. **106**, 063721 (2009)
10. P. Dhak, D. Dhak, K. Pramanik, P. Pramanik, Solid State Sci. **10**, 1936 (2008)
11. S.C. Das, B. Tudu, N. Bhattacharyya, R.Bandyopadhyay, P. Pramanik, Development of nanostructured ZnO based gas sensors to use in electronic nose towards hydrogen, ammonia and LPG proceedings, pp. 204–207. (ISBN: 978-93-80663-36-4) on MDCCT-2012 (2012)
12. S.Y. Sheng, Z.T. Shu, Preparation, structure and gas-sensing properties of ultramicro $ZnSnO_3$ powder. Sens. Actuators, B **12**, 5–9 (1993)

Electron Beam Evaporated Nano-Crystalline V_2O_5 Thin Films for Electrochromic and Electrochemical Applications

P. Rosaiah and O. M. Hussain

Abstract Vanadium pentoxide is one of the most widely investigated multifunctional materials for electro-chromic and electro-chemical device applications. Its layered and open structure makes this material in thin film form well suited for electro-chemical insertion reactions with the Li ions. In the present investigation, V_2O_5 thin films have been prepared by electron beam evaporation technique at a substrate temperature of 523 K in an oxygen partial pressure of 2×10^{-4} mbar. The XRD patterns exhibited predominant (001) diffraction orientation along with (200) and (400) characteristic peaks corresponding to orthorhombic phase of V_2O_5 with P_{mnm} space group. The estimated average grain size is about 90 nm. The optical absorption coefficient 'α' determined from the experimentally measured transmittance and reflectance data for V_2O_5 films was fund to give a better fit for the exponent $n = 3/2$ suggesting the direct forbidden transitions with an estimated optical band gap of 2.36 eV. Two sets of films were prepared with thickness of 300 nm on ITO coated glass for electrochromic analysis and 600 nm on metalized silicon for electrochemical insertion studies. The electron beam evaporated V_2O_5 thin films with thickness 300 nm showed a good reversible cyclic voltammogram on sweeping the potential downward and upward. A two step electrochromism i.e. yellow to green and green to deep blue was observed from reversible current—voltage curves. The electrochemical characteristics of V_2O_5 thin films with thickness 600 nm were examined in non-aqueous region. The film exhibited a discharge capacity of 70 μAh/(cm^2–μm) at a current density of 10 μA/cm^2. Annealing of these films at 673 K exhibited a better discharge capacity of 90 μAh/(cm^2–μm).

P. Rosaiah · O. M. Hussain (✉)
Thin Film Laboratory, Department of Physics, Sri Venkateswara University, Tirupati 517 502, India
e-mail: hussainsvu@gmail.com

P. K. Giri et al. (eds.), *Advanced Nanomaterials and Nanotechnology*, Springer Proceedings in Physics 143, DOI: 10.1007/978-3-642-34216-5_48,

Keywords Nano-crystalline V_2O_5 thin films · Electron beam evaporation · Structural and electrochemical properties

1 Introduction

Transition metal oxides constitute an important group of inorganic solids, exhibiting novel properties for a wide variety of applications in the field of micro-electronics, opto-electronics, display systems, memory switching devices and electrochemical devices [1]. V_2O_5 is one of the potential candidate among the transition metal oxides for application in thin films micro-batteries [2, 3] and electrochromic display devices [4, 5]. It has orthorhombic layered structure, high electrochemical activity, topotactic insertion reaction with lithium, high theoretical energy density for lithium insertion and high stability, the possibility to insert up to 3 lithium ions per V_2O_5 leading to a high specific capacity, which led to its use as a highly promising intercalation material in solid state micro-battery applications. Thin films of V_2O_5 exhibit multi-colored electrochromism and have high potential for use in electrochromic display devices, color filters and other optical devices. V_2O_5 thin films are also used in variable reflectance mirrors smart windows and surfaces with tunable emittance for temperature control of space vehicles. In addition, it is one of the most important optoelectronic materials used in optical switching devices, gas and humidity sensors. Several investigators prepared V_2O_5 thin films by using different techniques and studied physical, electrical, electro-chemical and electrochromic properties [6–8]. Quinzeni et al. [9] studied the electrochemical properties of nano-crystalline V_2O_5 thin films deposited by means of radiofrequency magnetron sputtering. Julien et al. [10] deposited V_2O_5 thin films by thermal evaporation and studied Spectroscopic and electrochemical properties. Similarly, Avellaneda [11] deposited V_2O_5 thin films by sol-gel process and studied the electrochromic performance. Iida et al. [12] deposited V_2O_5 thin films by pulsed laser deposition and studied the electrochromic performance.

However the properties of V_2O_5 depends on structure, composition and morphology of the films, which in turn depend on the type of deposition technique and the process parameters involved during the growth. Electron beam evaporation is one of the potential technique to deposit metal oxide thin films especially V_2O_5 thin films as it offers excellent material utilization, the possibility of precise control of deposition rate and film composition. High deposition rates can be achieved with good structural and morphological control. Very few investigators have employed electron beam evaporation for the growth of the V_2O_5 thin films and studied microstructural, optical and electrical [13–15]. To the best of our knowledge, reports are not available on the electrochemical and electrochromic properties of electron beam evaporated V_2O_5 thin films though this technique has industrial viability. Hence in the present investigation, electron beam evaporation technique has been employed for the deposition of nano-crystalline V_2O_5 thin films on ITO coated glass substrates and metalized silicon substrates and studied electrochemical and electrochromic properties.

2 Experimental

Thin films of V_2O_5 were deposited by electron beam evaporation technique using Balzer BA 510 high vacuum coating unit. Pure V_2O_5 powder (99.9 % Sigma Aldrich) pressed into one inch diameter pellet and subsequently annealed at 873 K was employed as target material. The vacuum chamber was evacuated to a base pressure 1×10^{-6} mbar. Pure oxygen was introduced into the chamber and an oxygen partial pressure of about 2×10^{-4} mbar was maintained during deposition. The power density of electron beam spot was maintained at 1.2 kW/m^2. The source to substrate distance was maintained at 15 cm. Two sets of films were deposited one on gold coated silicon and another on ITO coated glass substrates maintained at 523 K. The structure of the films was studied by a Siefert X-ray Diffractometer with CuK_α radiation ($\lambda = 0.15406$ nm). The surface morphology of the films was studied with Scanning Electron Microscope (Carl Zeiss EVO50) and Atomic Force Microscope (Digital Instrument: Dimension 3,100 series). Compositional studies were performed by EDAX (Oxford Instruments, UK) and The X-ray Photo-electron Spectroscopy (Phoibos 100MCD Energy Analyzer). The optical transmittance and reflectance of the films were recorded in the wavelength range 300–1,500 nm using Perkin Elemer UV–Vis-NIR double beam spectrophotometer. The electrochromic coloration studies were performed by dry lithiation method. The electrochemical properties were carried out on the Li/V_2O_5 cell using EC-Lab Instruments Inc. (v9.5 model) and operated in the cut-off voltage between 3.5 and 1.5 V. Three electrode non- aqueous electrochemical (Li/V_2O_5) cells were assembled in glass beakers with Teflon ceiling in a glove box maintained in a controlled argon environment. The cell was composed of lithium foil as both anode and reference electrode, 1 M $LiAsF_6$ desolved in a mixture contains 1:1 ratio of propylene carbonate (PC) and dimethyl carbonate (DMC) as the liquid electrolyte and V_2O_5 thin film coated on metallized silicon as an active cathode.

3 Results and Discussions

3.1 Structural Studies

Figure 1 shows the X-ray diffraction spectra of V_2O_5 thin films deposited at a substrate temperature of 523 K on ITO coated glass substrates. The well resolved characteristic (200) and (400) peaks along with predominant (001) observed from XRD spectra are corresponding to the orthorhombic structure of V_2O_5 with P_{mnm}

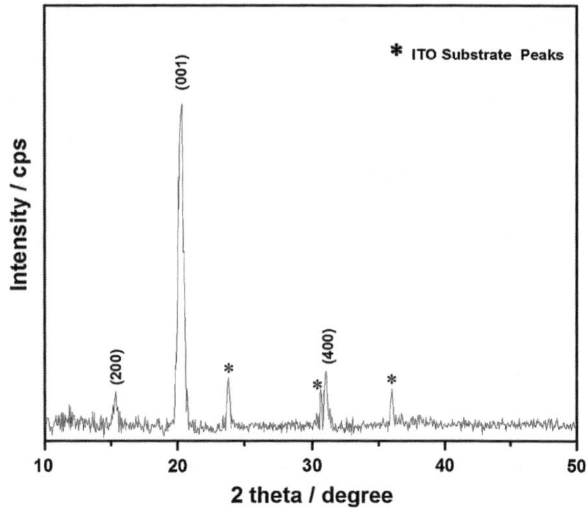

Fig. 1 X-ray diffraction pattern of the V_2O_5 thin films

space group. The XRD results revealed that the structure is such that the crystallographic c-axis is perpendicular to the substrate surface. Debye–Scherrer formula was employed to estimate the crystallite size and found to be 90 nm. The lattice parameters evaluated on the basis of measured 'd' spacing are a = 1.153 nm, b = 0.356 nm, and c = 0.435 nm which are in good agreement with reported valve [16].

3.2 Compositional Studies

The compositional studies were performed by using EDAX and XPS. The compositional analysis of the nano-crystalline V_2O_5 films is carried out using EDAX measurements in order to know the surface elemental atomic percentage and the EDAX spectrum is shown in Fig. 2. Particularly the films deposited at 523 K in an oxygen partial pressure of 2×10^{-4} mbar are observed to be nearly stoichiometric with an estimated O/V ratio of 2.48 from the EDAX spectrum. The XPS spectra of V_2O_5 thin films (Fig. 3) recorded in the binding energy range 510–540 eV exhibited two intense peaks corresponding to vanadium (2P) and oxygen (1s) at binding energies of about 516 eV and 530 eV respectively. The spectra also reveal that characteristic doublets of $V(2P_{3/2})$ and $V(2P_{1/2})$ peaks caused by the Spin-Orbit splitting in vanadium pentoxide. The core level binding energy peak of V $(2P_{3/2})$ and $V(2P_{1/2})$ emission are observed at about 517.10 and 524.50 respectively for the as deposited films. The energy separation $\Delta E = 12.9$ eV of O(1s)–$(2P_{3/2})$ core levels and the energy splitting of the doublet ($\Delta E = 7.40$ eV) shows that the as deposited films are approaches near stoichiometry [17] with vanadium in its highest oxidation state, i.e. V^{+5}.

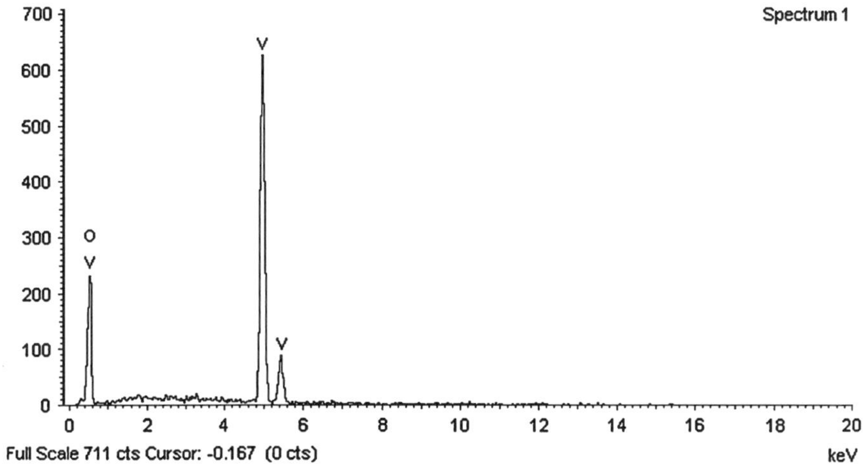

Fig. 2 EDAX spectrum of V_2O_5 thin films

Fig. 3 XPS spectrum of V_2O_5 thin films

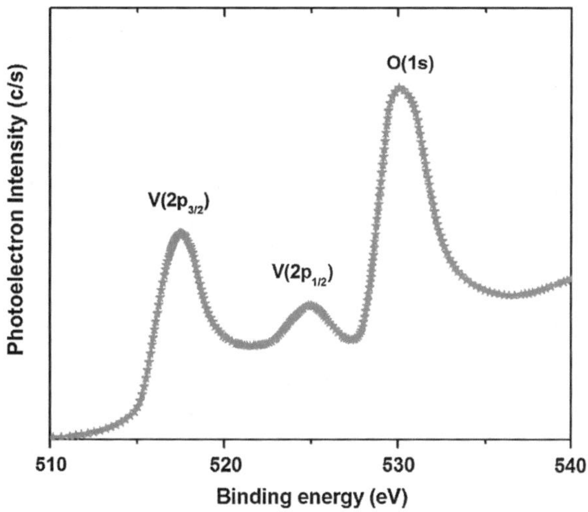

3.3 Optical-Electrochromic Studies

The optical transmittance of nano-crystalline V_2O_5 thin films deposited on ITO glass substrates at deposition temperature 523 K in an oxygen partial pressure 2×10^{-4} mbar is shown in Fig. 4. The as deposited films shows relatively high transmittance around 75 % without near-infrared broad band absorption. This high transmittance reveals that the films were weakly absorbing in the spectral range of investigation. The optical absorption coefficient of the films is evaluated from the transmittance and reflection data using the following equation.

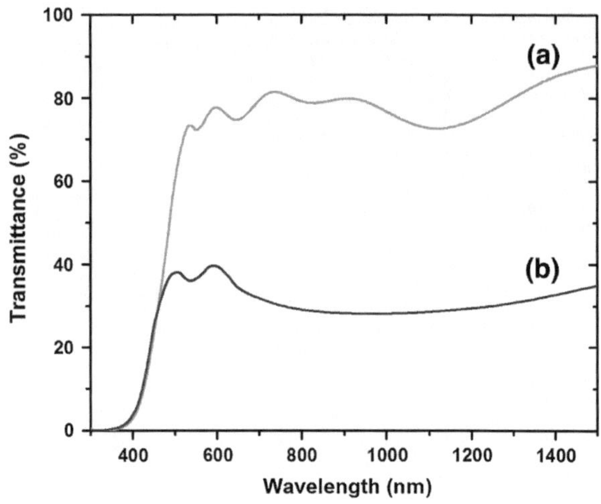

Fig. 4 The optical transmittance of V_2O_5 thin films as a function of wavelength

$$T = (1 - R)^2 \exp(-\alpha d)/\left[1 - R^2 \exp(-2\alpha d)\right], \quad (1)$$

where T is the transmittance, R the Reflectance, α the absorption coefficient, d the film thickness. The optical absorption coefficient (α) is found to be exponentially dependent on the photon energy. Since optical absorption coefficient (α) obeys the Urbach's rule, the exponential dependence of α on the incident photon energy may be expressed as

$$\alpha = \alpha_o \exp(h\nu/E_e) \quad (2)$$

where α_o is experimentally determined constant and E_e often interpreted as the width of the localized states within the band of the material. The width of the localized states within the band gap was found to be 0.28 eV.

The optical band gap of V_2O_5 thin films can be evaluated by the relation as follows.

$$(\alpha hv) = B(hv - E_g)^n \quad (3)$$

where hν the incident photon energy, B the edge width parameter, E_g optical band gap, n_e the exponent determiner the type of electronic transition causing the absorption and can take valves ½, 3/2,2 and 3 for different electronic transition such as direct allowed, direct forbidden, indirect allowed and indirect forbidden transitions respectively. Generally, the conduction band in V_2O_5 is formed by vanadium 3d bands and the valancy band is formed by 2p bands of oxygen. The fundamental absorption is mainly due to the transition from oxygen p-type wave functions to 3d type wave functions following direct forbidden transitions, where $n = 3/2$. Figure 5 shows the plot of $(\alpha h\nu)^{2/3}$ versus hν for as deposited V_2O_5 thin films. The optical band gap of the films is evaluated by extrapolating the linear region of the plot to zero and found to be 2.36 eV and is in good agreement with the reported value for the films [18]. The variation of refractive index and

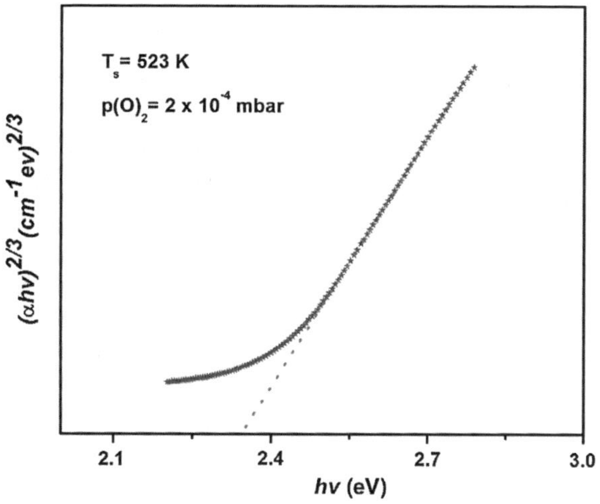

Fig. 5 The optical band gap of the V_2O_5 thin films deposited at 523 K

extinction coefficient as a function of wavelength is shown in Fig. 6. The functional dependence of the refractive index on wavelength was found to obey cauchy's dispersion relation. The refractive index of the films decreased with the increase of wavelength and is almost constant at higher wavelengths. The extinction coefficient (k) was also calculated from the optical absorption coefficient (α) using the relation

$$\alpha = 4\pi k/\lambda, \tag{4}$$

where λ is the wavelength of the incident photon.

The extinction coefficient (k) also decreases with increasing wavelength of incident photon and at higher wavelengths it is nearly constant [19].

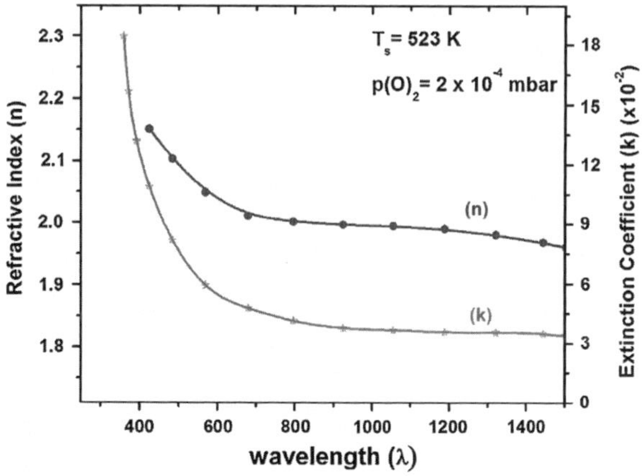

Fig. 6 The variation of refractive index and extinction coefficient with wavelength of V_2O_5 thin films

3.3.1 Cyclic Voltametric Studies

The as deposited V_2O_5 thin films were found to be pale yellow in color. The linear sweep cyclic voltammogram of V_2O_5 thin films at sweep rate 100 mV/sec is shown in Fig. 7. The electron beam evaporated V_2O_5 films gave good reversible cyclic voltammogram curves on sweeping potential downward and upward. A two step electrochromism i.e. yellow to green and then green to deep blue was observed for reversible current–voltage curves. The color changes are accompanied by the appearance of two well defined reduction/oxidation peaks. The similar behavior was also observed for sputtered V_2O_5 films [20]. In the first reduction peak (peak A), only a fraction of the V^{+5} ions are reduced to V^{+4} ions and the remaining V^{+5} are reduced to V^{+4} in the second step (peak B). Similar is the explanation for the remaining two oxidation pairs. The increase in current density and the corresponding color change has been attributed to the progressive reduction of pentavalent vanadium ions by the double injection of electrons and Li^+ ions into the V_2O_5 matrix.

3.3.2 Dry Lithiation Studies

The electrochromic properties of V_2O_5 thin films are also investigated using dry lithiation method [21]. In this method, lithium niobate ($LiNbO_3$) powder was heat treated at 1,113 K under high vacuum to give off lithium atoms for insertion in the exposed V_2O_5 films. The quartz crystal monitor was used for the measurement of film thickness during the lithiation process to control and calibrated the thickness of the deposited film against the electrochemical insertion. In the present investigation, 30 nm effective mass thickness of lithium layer is considered for maximum coloration, which corresponds to approximately 12.7 mC/cm^2 as verified

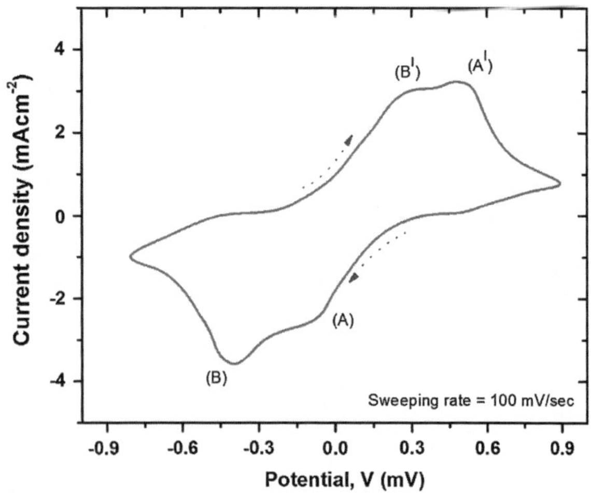

Fig. 7 Cyclic voltammogram (CV) of the V_2O_5 thin films

from the electrochemical method. In the present investigation, the as deposited nano-crystalline V_2O_5 thin films deposited on ITO coated glass substrates demonstrated an optical modulation of 33.4 % at 550 nm with an estimated coloration efficiency of 29.8 cm^2/C. The results are comparable with the reported optical modulation for sputtered V_2O_5 thin films [22].

4 Electrochemical Studies

The second set of the films has been prepared on metalised silicon substrates ($Si/SiO_2/Ti/Au$) with higher thickness of 600 nm under the same conditions. The AFM image of as deposited films is shown in Fig. 8a. The films deposited at (T_s) 523 K are homogeneous and with fine nano particles distributed uniformly over the substrate surface. The estimated average grain size of V_2O_5 thin films is 94 nm. To enhance the crystalline properties, the as grown films are subsequently annealed at 673 K in controlled oxygen atmosphere. The annealed films exhibited predominant (001) orientation along with the other characteristic (200), (400) and (002) orientations, which are corresponding to the orthorhombic structure of V_2O_5 with P_{mnm} space group.

The surface topography recorded by atomic force for the annealed films deposited on metalized silicon substrates is shown in Fig. 8b. The surface of the films is observed to consists of homogeneous distribution of roughly spherical crystallites with an estimated grain size of 220 nm. The effect of heat treatment was to cause the motion of the small islands, which then coalesced with larger grains. The mobility of the islands on the substrates and the number of islands with sufficient energy to move are determined by a thermally activated process. The

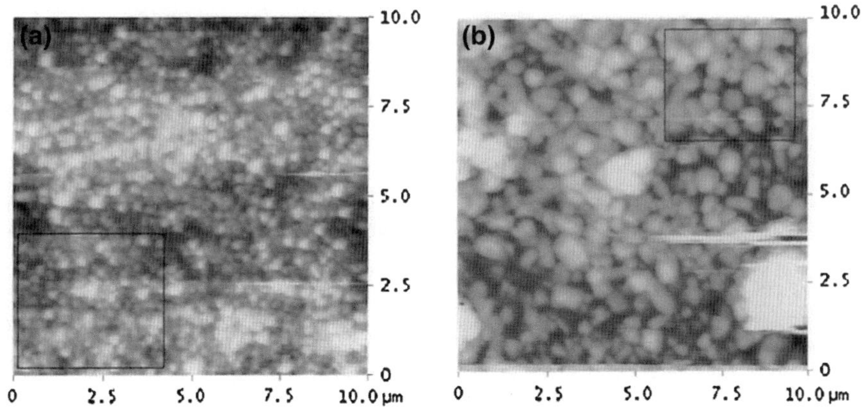

Fig. 8 AFM image of V_2O_5 thin films deposited on metalized silicon substrates. **a** As deposited. **b** Annealed at 673 K

films annealed at higher temperatures has a higher coalescence speed and islands combines, nucleation occurs and the coalescence with each other to form continuous films with larger grain sizes. The enhancement of grain size for the annealed films may leads for better electrochemical properties.

The electrochemical measurements were performed on Li/V_2O_5 cells with Lithium metal foil as an anode and V_2O_5 thin film coated (thickness 600 nm) on metalized silicon substrate as cathode of 1.5 cm^2. The electrochemical measurements are carried at a rate of C/100 in the potential range 1.50–3.7 V at scan rate of 10 mV/s. Typical discharge plots of Li/V_2O_5 cells are shown in Fig. 9. The discharge profile of as deposited films (Fig. 9a) exhibited two plateau regions and showed a discharge capacity of 70 μAh/(cm^2–μm). The cycliabilty of the films tested for 30 cycles and observed a constant discharge capacity up to 10 cycles and degraded gradually as shown in Fig. 10a. The discharge capacity of the annealed films (Fig. 9b) at 673 K for 4 h, is approximately 90 μAh/(cm^2–μm) and also exhibited good cyclic stability even after 25 cycles. Yoon et al. [23] reported discharge capacities in the range 50–120 μAh/(cm^2–μm) for the films deposited by using DC and RF sputtering methods. Park et al. [24] reported discharge capacity of 130 μAh/(cm^2–μm) for RF sputtered V_2O_5 thin films. Navone et al. [25] reported 70 mAh/g discharge capacity for DC sputtered V_2O_5 films of thickness 800 nm. In comparison to the previous reports, the electrochemical performance of electron beam evaporated nano–crystalline V_2O_5 thin films (with thickness of 600 nm) deposited on metalised silicon substrates with subsequent annealing is also good and can be used as binder free cathode material in the fabrication of microbatteries.

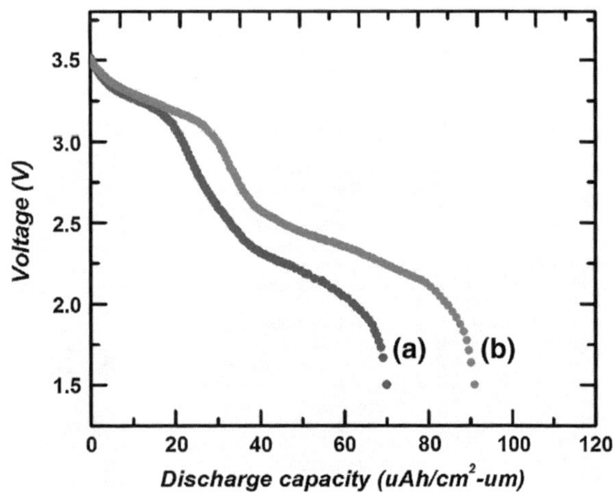

Fig. 9 Discharge curves for V_2O_5 thin films. **a** As deposited. **b** Annealed at 673 K

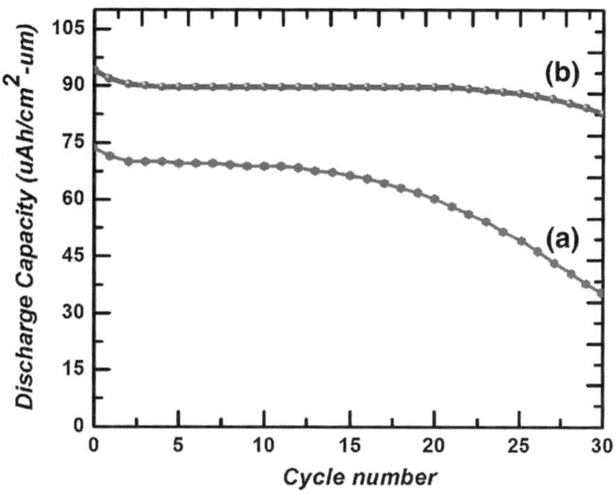

Fig. 10 Discharge capacity as a function of cyclic number for V_2O_5 thin films. **a** As deposited. **b** Annealed at 673 K

5 Conclusions

Nano–crystalline V_2O_5 thin films were deposited on ITO glass substrates and metalised silicon substrates maintained at (T_s) 523 K in an oxygen partial pressure of 2×10^{-4} mbar using electron beam evaporation technique. The films deposited on ITO glass substrates are near stoichiometric with orthorhombic layered structure. The estimated optical band gap of the films is 2.36 eV. The as deposited films deposited ITO coated glass substrates demonstrated better optical modulation of 33.4 % with an estimated coloration efficiency of 29.8 cm^2/C. The as deposited films deposited on metalised silicon substrates showed as lower discharge capacity of 70 $\mu Ah/(cm^2–\mu m)$ with high capacity fading rate due to poor crystallinity. The films annealed at 673 K for 4 h in controlled oxygen atmosphere were found to have orthorhombic structure with an average grain size of about 220 nm. The charge—discharge profiles exhibited two plateau regions and showed a discharge capacity of 90 $\mu Ah/(cm^2–\mu m)$ with constant cyclic rate even after 25 cycles. These results concludes that the V_2O_5 thin films deposited by electron beam evaporation technique at moderate substrate temperature with subsequent annealing at 673 K exhibited better electrochemical properties and may be used as binder free cathode in the fabrication of microbatteries.

Acknowledgments The authors would like to thank the University Grants Commision, New Delhi for providing the financial support.

References

1. A.A. Akl, Thermal annealing effect on the crystallization and optical dispersion of sprayed V_2O_5 thin films. J. Phys. Chem. Sol. **71**, 223 (2010)
2. M.B. Sahana, C. Sudakar, C. Thapa, G. Lawes, V.M. Naik, R.J. Baird, G.W. Auner, R. Naik, K.R. Padmanabhan, Electrochemical properties of V_2O_5 thin films deposited by spin coating. Mater. Sci. Eng. **143**, 42 (2007)

3. S. Oukassi, R. Salot, J.P. Pereira-Ramos, Elaboration and characterization of crystalline rf deposited V_2O_5 positive electrode for thin film battery. Appl. Sur. Sci. **256**, 149 (2009)
4. L. Ottaviano, A. Pennisi, F. Simone, A.M. Salvi, Rf-sputtered electrochromic V_2O_5 films. Opt. Mater. **27**, 307 (2004)
5. G.J. Fang, Z.L. Liu, Y.Q. Wang, H.H. Liu, K.L. Yao, Oriented growth of V_2O_5 electrochromic thin films on transparent conductive glass by pulsed excimer laser ablation technique. J. Phys. D: Appl. Phys. **33**, 3018 (2000)
6. S. Chakraborty, H. Sakata, E. Yokoyama, M. Wakaki, D. Chakravorty, Laser induced forward transfer technique for maskless patterning of amorphous V_2O_5 thin films. Appl. Sur. Sci. **254**, 638 (2007)
7. V.V. Atuchin, B.M. Ayupov, V.A. Kochubey, L.D. Pokrovsky, C.V. Ramana, Y.M. Rumiantsev, Optical properties of textured V_2O_5/Si thin films deposited by reactive magnetron sputtering. Opt. Mater. **30**, 1145 (2008)
8. M. Losurdo, D. Barreka, G. Bruno, E. Tondello, Spectroscopic ellipsometry of V_2O_5 nanocrystalline thin films. Thin Solid Films **384**, 58 (2001)
9. I. Quinzeni, S. Ferrari, E. Quartarone, P. Mustarelli, Structural, morphological and electrochemical properties of nanocrystalline thin films deposited by means of radiofrequency magnetron sputtering. J. Power Sources **196**, 10228 (2011)
10. C. Julien, E. Haro-Ponia towski, L. Eskobar-Alarcon, Growth of V_2O_5 thin films by pulsed laser deposition and their applications in lithium microbatteries. Mater. Sci. Eng. B, **65**, 170 (1999)
11. C.O. Avellaneda, Electrochromic performance of sol-gel deposited V_2O_5Ta films. Mater. Sci. Eng. **138**, 118 (2007)
12. Y. Iida, Y. Kaneko, Y. Kanno, Fabrication of pulsed laser deposited V_2O_5 thin films for electrochromic devices. J. Mater. Proc. Tech. **197**, 261 (2008)
13. C.V. Ramana, O.M. Hussain, S. Uthanna, B.S. Naidu, Influence of oxygen partial pressure on the optical properties of electron beam evaporated vanadium pentoxide thin films. Opt. Mater. **10**, 101 (1998)
14. C.V. Ramana, O.M. Hussain, B.S. Naidu, Growth and structure of electron beam evaporated V_2O_5 thin films. Mater. Chem. Phys. **50**, 195 (1997)
15. C.V. Ramana, O.M. Hussain, B.S. Naidu, P.J. Reddy, Spectroscopic characterization of electron beam evaporated V_2O_5 thin films. Thin Solid Films **305**, 219 (1997)
16. K.V. Madhuri, K.S. Rao, B.S. Naidu, O.M. Hussain, Characterization of laser ablated V_2O_5 thin films. J. Mater. Sci.: Mater. Elec. **13**, 426 (2002)
17. C.V. Ramana, O.M. Hussain, R. Pinto, C. Julien, Microstructural feature of pulsed laser deposited V_2O_5 thin films. Appl. Sur. Sci. **207**, 135 (2003)
18. A.M. Abo El Sound, B. Mansour, L.I. Soliman, Optical and electrical properties of V_2O_5 thin films. Thin Solid Films **247**, 142 (1994)
19. A.A. Akl, Effect of solution molarity on the characteristics of vanadium pentoxide thin films. Appl. Sur. Sci. **252**, 8749 (2006)
20. Y. Fujita, K. Miyazaki, T. Tatsuyama, On the lectrochromism of evaporated V_2O_5 films. Jpn. J. Appl. Phys. **24**, 1082 (1985)
21. K. Harikrishna, O.M. Hussain,C.M. Julien, Electrochromic properties of nanocrystalline WO_3 thin films grown on flexible substrates by plasma assisted evaporation technique. Appl. Phys. A. **99**, 922 (2010)
22. Y.-S. Lin, C.-W. Tsai, Reactive sputtering deposition of V_2O_{5-x} on flexible PET/ITO substrates for electrochromic devices. Electrochem. Acta. **202**, 5645 (2008)
23. Y.S. Yoon, J.S. Kim, S.H. Choi, Structural and electrochemical properties of vanadium oxide thin films grown by dc and rf reactive sputtering at room temperature. Thin Solid Films **460**, 43 (2004)

24. Y.J. Park, K.S. Ryu, K.M. Kim, N-G. Park, M.G. Kang, S.H. Chang, Electrochemical properties of vanadium oxide thin films deposited by rf sputtering. Solid State Ionics **154**, 234 (2002)
25. C. Navone, R. Baddour-Hadjean, J.P. Pereira-Ramos, R. Salot, A kinetic study of electrochemical lithium insertion into oriented V_2O_5 thin films prepared by rf sputtering. Electrochim. Acta **53**, 3331 (2008)

Boron Doped Fullerene C_{60} as Catalyst for $NaAlH_4$

Deepak Agnihotri and Hitesh Sharma

Abstract We report results of our first principles calculation using density functional theory to investigate the effect of boron doping in complexes $C_{60-n}B_nNaAlH_4$; n = 1–4. $NaAlH_4$ molecule interacts weakly with pure C_{60} and the interactions are enhanced for $C_{60-n}B_n$. The hydrogen release energy for $NaAlH_4$ decreases with boron doping in C_{60} cage and is minimum for n = 1. The H release energy is found to decrease significantly from 3.82 eV in $NaAlH_4$ molecule to 3.20, 2.05, 2.23, 2.09 and 2.35 eV in $C_{60}NaAlH_4$, $C_{59}BNaAlH4$, $C_{58}B_2NaAlH_4$, $C_{57}B_3NaAlH_4$ and $C_{56}B_4NaAlH_4$ respectively.

Keywords Ab initio · Catalytic properties · Hydrogen storage · Metal hydrides · DFT

1 Introduction

In hydrogen economy, the storage of hydrogen poses the biggest challenge as the storage medium must meet the requirements of high gravimetric density and volumetric density, fast kinetics and favorable thermodynamics [1]. There are various methods for storing hydrogen such as liquid hydrogen, compressed hydrogen, metal hydrides and hydrogen adsorption on activated carbon which are

D. Agnihotri (✉)
Department of Applied Sciences, RBIEBT, Sahauran, Mohali, Punjab 140104, India
e-mail: deepakphysicist@yahoo.co.in

D. Agnihotri · H. Sharma
Department of Physics, Punjab Technical University, Jalandhar, Punjab 144601, India

P. K. Giri et al. (eds.), *Advanced Nanomaterials and Nanotechnology*,
Springer Proceedings in Physics 143, DOI: 10.1007/978-3-642-34216-5_49,

not optimal. Hydrogen stored in solid materials is a safe alternative to liquid and compressed form and is a convenient fuel with zero emission.

The solid state storage of hydrogen takes place by (1) physisorption as in porous carbon and zeolite materials (2) chemisorption as in metal hydrides. The hydrogen storage capacity of metal hydrides is ~6–18 wt % and these systems have been found to be useful but the operational conditions and the recovery process make them less ideal for most of the applications [2]. The success of metal hydrides has been hindered due to strong covalent bonds between metal and H atoms and high dissociation temperatures. Therefore, the kinetics of the metal hydrides needs to be improved significantly for its commercial success. Among the type (2) materials, sodium alanate ($NaAlH_4$) is the most widely studied complex metal hydride material because of its high hydrogen weight capacity and low cost. The kinetics of hydrogen uptake and release of these materials is such that catalysts are needed to improve the reaction rates. The use of carbon materials as possible catalysts for $NaAlH_4$ has been investigated [3–5]. In the present work, we investigate the possibility of using boron doped fullerene C_{60} as a novel catalyst to release hydrogen from $NaAlH_4$.

2 Computations

Our first-principles calculations were performed using Spanish Initiative for Electronic Simulation with thousands of atoms (SIESTA) computational code based on the generalized gradient approximation to the density functional theory [6]. Core electrons are replaced by non-local, norm-conserving pseudopotentials factorized in the Kleinman–Bylander form [7], whereas valence electrons are described using linear combination of numerical pseudo atomic orbitals of the Sankey–Niklewski type [8] but generalized for double-ζ and polarization functions. For self consistent calculation of Hamiltonian matrix elements, the charge density is represented in a real space grid with a cutoff energy of 200 Ry. The optimized atomic positions were obtained by relaxation until the Hellmann–Feynman force on each atom was less than 0.01 eV/Å. Supercell calculations were employed throughout where the atoms on adjacent nanostructures are separated by 15 Å that is large enough to avoid any significant spurious interactions with periodically repeated images.

We have performed test calculations for C_{60}, B_2 and C–B. The C = C and C–C bond lengths are found to be 1.40 and 1.46 Å, which are in agreement with the experimental values [9, 10]. The B–B and C–B bond lengths come out to be 1.95 and 1.44 Å. Also, the values of ionization potential and electron affinity (6.90 and 2.70 eV) for C_{60} are in excellent agreement with the reported values [11, 12]. The energy needed to remove one H atom from $NaAlH_4$ cluster is 3.82 eV, which is also in agreement with the reported value [13].

3 Results

To study the effect of boron doping in C_{60} cage, we modeled our problem by placing single unit $NaAlH_4$ molecule at different positions on surface of the pure C_{60} fullerene cage and relaxing it till the desired accuracy was achieved. The minimum energy position was taken as starting optimized structure for further investigation. Then we performed the doping of boron in C_{60} cage with boron number varying from 1 to 4 and relaxing the structures again till the desired accuracy was achieved. $NaAlH_4$ molecule interacts with the fullerene structure with Na atom facing towards the carbon substrate. The distances between Na and Al and Al and H are of the order of respectively, 2.62–2.83 and 1.60–1.69 Å in $C_{60-n}B_nNaAlH_4$ complexes for n = 1–4 and are almost alike the corresponding values in the isolated $NaAlH_4$ cluster. The hydrogen release energies for the complexes $C_{60-n}B_nNaAlH_4$ have been calculated by substituting the boron atoms in C_{60} cage varying from n = 1 to 4 using the equation:

$$\Delta E = [E(C_{60-n}B_nNaAlH_3) + E(H)] - E(C_{60-n}B_nNaAlH_4) \quad (1)$$

In the above Eq. 1, $E(C_{60-n}B_nNaAlH_3)$, E(H), and $E(C_{60-n}B_nNaAlH_4)$ are the total energies of $C_{60-n}B_nNaAlH_3$, H, and $C_{60-n}B_nNaAlH_4$ respectively. The H release energies have been mentioned in Table 1.

For pure C_{60} cage, the H release energy for $NaAlH_4$ is 3.20 eV. On single B doping i.e., for n = 1, it is 2.05 eV. Also, for n = 2, 3 and 4, the H release energies for $NaAlH_4$ are 2.23, 2.09 and 2.35 eV respectively. The minimum value of H release energy exists for n = 1. Thus, it is clear from Table 1 that H release energies for the complexes $C_{60-n}B_nNaAlH_4$ show an oscillatory pattern. The decrease in H release energy may be explained on the basis of interaction of $NaAlH_4$ with B doped C_{60} substrate, which effects the ability of Na to donate the electron to form $(AlH_4)^-$ anion and hence covalent bond between Al and H is weakened leading to lowering of desorption energy and improved kinetics.

In summary, we have investigated the effect of boron doping in the complexes $C_{60-n}B_nNaAlH_4$ for n = 1–4. The hydrogen release energy for $NaAlH_4$ decreases with boron doping in C_{60} cage. Therefore, low binding energy of $NaAlH_4$ molecule with B doped C_{60} cage and significant decrease in hydrogen release energy

Table 1 The H release energy (in eV) for the complexes $C_{60-n}B_nNaAlH_4$; n = 1–4

S.No	Number of boron atoms in C_{60} (n)	H release energy for $NaAlH_4$ (eV)
1	0	3.20
2	1	2.05
3	2	2.23
4	3	2.09
5	4	2.35

presents the possibility of boron doped C_{60} to be used as a novel catalyst to improve kinetics of metal hydrides.

Acknowledgments We are thankful to the SIESTA group for providing us the computational code. HS acknowledges the financial support from DST, New Delhi.

References

1. J. Alper, Water splitting goes au naturel. Sci. **299**(5613), 1686–1687 (2003)
2. A.M. Seayad, D.M. Antonelli, Recent advances in hydrogen storage in metal-containing inorganic nanostructures and related materials. Adv. Mater. **16**, 9–10 (2004)
3. C. Cento, P. Gislon, M. Bilgili, A. Masci, Q. Zheng, P.P. Prosini, How carbon affects hydrogen desorption in $NaAlH_4$ and Ti-doped $NaAlH_4$. J. Alloy. Comp. **437**, 360–366 (2007)
4. A. Zaluska, L. Zaluski, J.O.J. Strom Olsen, Sodium alanates for reversible hydrogen storage. J. Alloy. Comp. **298**, 125–134 (2000)
5. J. Wang, A.D. Ebner, J.A. Ritter, Kinetic behavior of Ti-doped $NaAlH_4$ when co-catalyzed with carbon. J. Phys. Chem. B **110**, 17353–17358 (2006)
6. P. Ordejon, E. Artacho, J.M. Soler, Self consistent order-N density-functional calculations for very large systems. Phys. Rev. B **53**, R10441–R10444 (1996)
7. L. Kleinman, D.M. Bylander, Efficacious form for model pseudopotentials. Phys. Rev. Lett. **48**, 1425 (1982)
8. O.F. Sankey, D.J. Niklewski, Ab initio multicenter tight-binding model for molecular-dynamics simulations and other applications in covelent systems. Phys. Rev. B **40**, 3979 (1989)
9. T. Pradeep, V. Vijayakrishnan, A.K. Santra, C.N.R. Rao, Interaction of nitrogen with fullerenes: nitrogen derivatives of C_{60} and C_{70}. J. Phys. Chem. **95**, 10564 (1991)
10. H.B. Burgi, E. Blanc, D. Schwarzenbach, S. Liu, Y. Lu, M.M. Kappes, J.A. Ibers, The structure of C_{60}: Orientational disorder in the low-temperature modification of C_{60}. Angew. Chem. Int. Ed. Engl. **41**, 640 (1992)
11. I.V. Hertel, H. Steger, de Vries, B. Weisser, C. Menzel, B. Kamke, W. Kamke, Giant plasmon excitation in free C_{60} and C_{70} molecules studied by photoionization. Phys. Rev. Lett. **68**, 784–787 (1992)
12. X.C. Wang, C.F. Ding, L.S. Wang, High resolution photoelectron spectroscopy of C_{60}. J. Chem. Phys. **110**, 8217–8220 (1999)
13. P.A. Berseth, A.G. Harter, R. Zidan, A. Blomqvist, C.M. Araujo, R.H. Scheicher, R. Ahuja, P. Jena, Carbon nanomaterials as catalysts for hydrogen uptake and release in $NaAlH_4$. Nano. Lett. **9**, 1501(2009)

Synthesis and Characterization of Photoanode of Nanostructured TiO_2 Sensitized by Natural Dyes for Application in DSSC

N. Goswami, D. Saikia and P. K. Gogoi

Abstract Dye-sensitized solar cells (DSSC), which is a third generation solar cell, have been receiving much attention as a new alternative over to conventional p-n junction devices. The key difference of the DSSC over traditional single junction solar cells is the use of a dye, which broaden the absorption frequency of the wide band gap semiconductor and thereby increase the efficiency of the working electrode. The natural dyes as sensitizers are getting prime importance in recent years due to their availability, low cost and environmental friendliness. Moreover, absorption spectrum of the dye and their anchorage to the surface of the nanostructured TiO_2 are important parameters related to the efficiency of the cell. In the present study we report the synthesis of nanoporous TiO_2 thin film on chemically clean glass substrates by a simple sol–gel technique using titanium isopropoxide as the precursor. The as synthesized films were annealed at 100 °C and were characterised by UV-Vis absorption, X-ray diffraction (XRD) and scanning electron microscope (SEM) for the investigation of their optical, structural and morphological properties. Aqueous and alcoholic extracts of the natural dyes were prepared for the fabrication of TiO_2 working electrode. The sensitization efficiency of the as fabricated photoelectrode or photoanode was investigated using UV-Vis spectroscopy. The effects of changing extraction temperature and extracting solvent are also investigated. The SEM analysis revealed that the as synthesized TiO_2 thin film are mesoporous in nature and favourable for anchorage of the dye on it.

N. Goswami (✉)
Department of Chemistry, Sibsagar College, Joysagar 785665, India
e-mail: nandita_10@rediffmail.com

D. Saikia
Material Science Laboratory, Department of Physics, Sibsagar College, Joysagar, India

P. K. Gogoi
Center for Nanoscience and Composite Material, Department of Chemistry,
Dibrugarh University, Dibrugarh 786004, India

P. K. Giri et al. (eds.), *Advanced Nanomaterials and Nanotechnology*,
Springer Proceedings in Physics 143, DOI: 10.1007/978-3-642-34216-5_50,

The XRD of the films confirmed the presence of anatase structure in the film. In conclusion, we have investigated the electronic coupling and the light harvesting properties of natural-dye sensitized TiO_2 thin film for the fabrication of photoanode for application in DSSC.

Keywords DSSC · Natural dye · Nanoporous TiO_2 · Anchorage

1 Introduction

After the breakthrough discovery by O'Regan and Gratzel, the dye sensitized solar cells (DSSC) have attracted wide attention as a cheaper alternative for converting solar energy to electricity than photovoltaic cells made from silicon and conventional p–n junction [1]. The most important parts in a DSSC are the sensitizer dye and the nanoporous wide band gap semiconductor which works as a photo anode in the device. In recent years, many efforts have been focussed on the synthesis of dyes (both organic and inorganic) as it plays a key role in harvesting sunlight [2]. Although, in the past few years, the most efficient cells have been obtained by using Ru-metal complexes as sensitizer dye, but owing to the high cost and unavailability of the noble metals as well as for their non-benign nature, an alternative would be highly appreciated [3]. As an alternative, pure organic dyes have been used; but, they have also limitations in terms of complicated synthetic routes, low efficiency and use of non-benign chemicals in their synthetic process [4]. Hence, in a quest for green alternatives, natural dyes (containing natural pigments) extracted from fruits, flowers and plants have been used which has great advantages over the organic and inorganic dyes in terms of easy availability, cost efficiency, nontoxicity and most importantly due to their eco friendliness [5]. Numerous efforts have been made by several research groups all over the world to utilize these natural dyes as sensitizers in a DSSC [6–14].

The efficiency of the sensitizing dye is critically dependent on electron injection from a photo excited state of the dye into the conduction band of the semiconductor. As a wide band gap semiconductor, TiO_2 has been used most frequently in a DSSC due to its favourable properties like high photo catalytic activity, anticorrosion, chemical stability, high adsorption capability and eco-friendliness [15–18]. It has attracted tremendous attention among researchers worldwide due to its potential applications in environmental protection, energy generation and many other fields which are closely related to its optical properties [18]. TiO_2 can exist in three main phases: anatase, brookite and rutile. As a bulk material, rutile is the stable phase; however, solution-phase preparation methods for TiO_2 generally favour the anatase structure [17]. The band gap of bulk TiO_2 lies in the range 3.0–3.2 eV (3.0 eV for rutile, 3.2 eV for anatase) and hence are sensitive only to the UV region of the visible spectrum. One of the goals for improvement of the performance of TiO_2 nanomaterials is to increase their optical activity by shifting the response from UV region to the visible region. One of the ways to achieve this goal is the sensitization of TiO_2 by inorganic or organic dyes. Moreover, the properties of TiO_2 nanoparticle are a function of the crystal structure, particle size, and morphology and as such are strongly dependent on the synthesis process [15–24]. Various techniques are available for the synthesis of titania nanoparticle such as sol gel [25–27], chemical vapour deposition (CVD) [28, 29], sputtering [30–32], hydrothermal [33] and peptization [34]. However, sol–gel process is a suitable method for preparing TiO_2 nanoparticle in thin film form since the composition, particle size, film thickness, and porosity can be easily controlled by controlling the parameters like concentration, growth temperature, and sintering condition. [35]. The peptization is also an equally advantageous method for the preparation of TiO_2 nanoparticle in thin film form, in which the reaction can be controlled at a molecular level by using a peptizing agent.

In the present study, thin films of TiO_2 have been prepared by sol–gel and peptization techniques and a comparison between the two were made by observing their response to the natural dyes and incident light. The optical, structural and morphological properties of the as synthesized TiO_2 thin films were investigated by UV-Vis absorption, X-ray diffraction (XRD), and scanning electron microscopy (SEM). Two natural dyes extracted from locally available plant leaves, namely, *lawsonia inermis linn* (Assamese name Jetuka) and *Hibiscus rosa-sinensis* (Asssemese name Jaba) were then adsorbed on the as synthesized TiO_2 thin films for the formation of the working electrode (or photoanode) for application in DSSC. The performance of the as fabricated photoanode as well as the interaction between TiO_2 nanoparticle and the natural dyes were investigated through UV-Vis absorption spectroscopy with an aim to evaluate the optimised film for better anchorage of the dyes.

2 Experimental

All reagents, viz. titanium isopropoxide, isopropyl alcohol, acetic acid, nitric acid, ammonia, ethanol etc. were of analytical grade obtained from Aldrich and Merk (India) Ltd. and used as received without further purification. Deionised water was used throughout the experiments. The crystal structures of the films were analyzed by using a X-ray diffractometer (Make: Seifert, Model 003 T/T) with CuKα radiations operated at 40 kV and 30 mA. For optical studies, optical absorbance spectra in the vicinity of absorption edge were recorded using Perkin–Elmer spectrophotometer. Surface morphology and particle size was examined by LEO 1430VP Scanning Electron Microscope.

Two synthesis approach viz. peptization and sol–gel techniques were adopted for the preparation of TiO_2 thin films.

2.1 Synthesis of TiO_2 Thin Film by Peptization Method

Thin films of TiO_2 were deposited onto chemically clean glass substrates by hydrolysis and peptization of titanium isopropoxide (TIP) solution followed by continuous dip coating technique. The synthetic technique presented here is a modified form from Mahshid et al. [34] in which TiO_2 nanoparticles were prepared in powder form. In a typical reaction 5 ml of TIP and 15 ml of isopropanol were mixed thoroughly with the help of a magnetic stirrer for 30 min and then added drop wise into a round bottom flask containing 250 ml of deionised water. In to this solution, ammonia was added drop wise to adjust the pH at 2 and then kept under vigorous stirring for another 1 h. The resultant turbid solution was subjected to heat treatment at 60–70 °C for almost 20–24 h. The final solution was white–blue and opaque with high viscosity. For the deposition of TiO_2 thin film, chemically clean glass substrates were dip vertically into a bath containing TiO_2 colloidal solution with the help of a substrate holder. The films were withdrawn after 24 h of deposition time and allowed to dry itself at room temperature. A light blue–white transparent film was obtained indicating the formation of TiO_2 nano-particle in thin film form.

2.2 Synthesis of TiO_2 Thin Film by Sol–Gel Method

The synthesis of TiO_2 thin film by sol–gel technique involves the hydrolysis of titanium isopropoxide in presence of ethanol and acetic acid. The synthetic technique presented here is a modification of the method reported by Mohammad et al. [36]. In a typical reaction, 5 ml acetic acid and 6.3 ml of TIP was added drop wise to 50 ml ethanol solution and stirred continuously until a homogeneous gel was obtained. The gel containing the TiO2 particles was coated onto a chemically cleaned glass substrate by dip- coating technique using a single dip coater (Model N0.-SDC 2007C, Apex Instruments Co.). The films were dried in air and then annealed at 450 °C for 30 min. A white transparent film was obtained which indicates the formation of TiO_2 film.

2.3 Extractions of Natural Dyes

Two dyes were extracted from locally available plant leaves namely *lawsonia inermis linn* (Assamese name Jetuka) and *Hibiscus rosa-sinensis* (Asssemese name Jaba) using water and alcohol as solvents.

For aqueous extraction of the dyes, 100 gm of finely crushed fresh leaves of *lawsonia inermis linn* (sample A) and 100 gm of finely crushed fresh flowers of *Hibiscus rosa-sinensis* (sample B) were transferred into two conical flasks containing 100 ml of deionised water and then boiled for 30 min at desired temperatures. For both the sample-A and B, the extraction was done at three different temperatures viz. 50, 75 and 100 °C. The extracts were kept in dark for 24 h and then filtered with a whatman 42 no. filter paper. For alcoholic extraction of the dyes, 100 gm of finely crushed fresh leaves of *lawsonia inermis linn* (sample C) and 100 gm of finely crushed fresh flowers of *Hibiscus rosa-sinensis* (sample D) were mixed with 100 ml ethanol solution and the extraction was done again at three different temperatures using soxhlet apparatus for 2 h. The extracts were kept in dark for 24 h and then filtered with a whatman 42 no. filter paper.

Finally, all the filtrates were collected separately and stored in an air tight bottle at 0 °C.

2.4 Fabrication of Photoanode

For the fabrication of the photoanode, the extracted dyes (sample-A at 50 °C, sample-B at 100 °C, sample-D at 100 °C) were adsorbed onto the surfaces of as synthesized TiO_2 thin films by continuous dip coating technique. The films were dipped vertically with the help of a substrate holder into the baths containing the extracted dyes. The films were withdrawn after 24 h of dip time, washed in distilled water and dried in an oven at 40 °C for 5 h. It was observed that, after immersion in the extracts, the colors of TiO_2 film turned to brown colour for *lawsonia inermis linn* while in case of *Hibiscus rosa-sinensis* it has turned into purple for alcoholic extract and dark pink for aqueous extract indicating the anchorage of the dyes on the surface of TiO_2 thin film.

3 Results and Discussion

3.1 Surface Morphology and Structural Studies

The SEM micrographs of the as synthesized TiO_2 thin films prepared by peptization and sol–gel techniques are shown in Fig. 1a and b respectively. From the micrographs it is observed that surface of both the films are mesoporous in nature and thus suitable for adsorbing the dye on its surface. However, the film obtained by the peptization method is found to be suitable for application in DSSC in comparison to the film prepared from sol gel method in terms of homogeneity and particle size.

Figure 1c shows the XRD patterns of the TiO_2 thin film prepared under peptizing conditions. The XRD patterns of the sample consisted of five distinctive TiO_2 peaks at 25.38, 38.98, 48.28, 54.68 and 55.02° corresponding to (1 0 1), (0 0 4), (2 0 0), (1 0 5) and (2 0 4) crystal planes (JCPDS 21-1272) of anatase phase respectively.

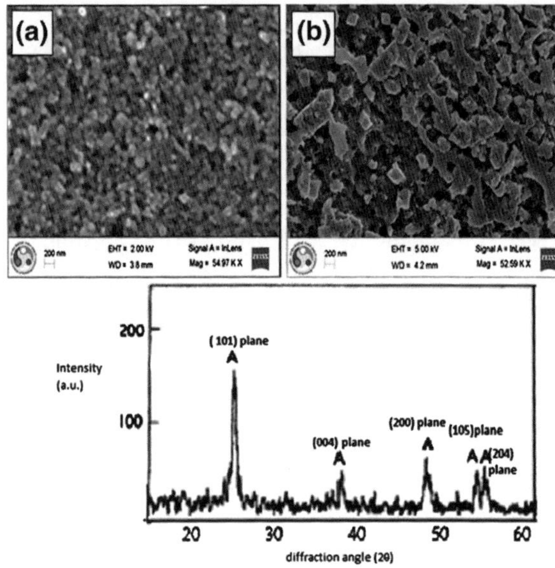

Fig. 1 **a** SEM images of TiO_2 thin film prepared by peptization technique. **b** SEM images of TiO_2 thin film prepared by sol gel technique. **c** X-ray diffraction pattern for TiO_2 thin film prepared by peptization technique

The broadening of the XRD peaks compared to the bulk TiO_2 indicates the formation of small sizes TiO_2 particles in the film.

4 Optical Studies

4.1 Absorption Spectra of Natural dyes

Two different natural dyes extracted from the flowers and leaves of locally available *lawsonia inermis linn* and *Hibiscus rosa sinensis* were used in the present study. The optical absorption spectra of the dyes extracted at three different temperatures (25, 50 and 100 °C), from the fresh leaves of *lawsonia inermis linn* using water and alcohol as solvent are shown in the Fig. 2a and b respectively. From the spectra, it is observed that, absorption pattern of all the samples of *lawsonia inermis linn* are similar, showing good absorption in the visible region with an absorption edge at around 520 nm. However, the dye extracted at 50 °C exhibited a little higher absorption in both the cases (aqueous and alcoholic extracts) in comparison to the others.

The UV-Vis absorption spectra of the aqueous and alcoholic extracts of *Hibiscus rosa-sinensis* flowers, extracted at three different temperatures (25, 50 and 100 °C) are depicted in Fig. 2c and d respectively. The aqueous extract prepared at 100 °C showed better absorption in the visible region in comparison to that extracted at 25 and 50 °C. Form the spectra it is clear that, for both aqueous and alcoholic extracts, the absorption peak was observed at around 544 nm.

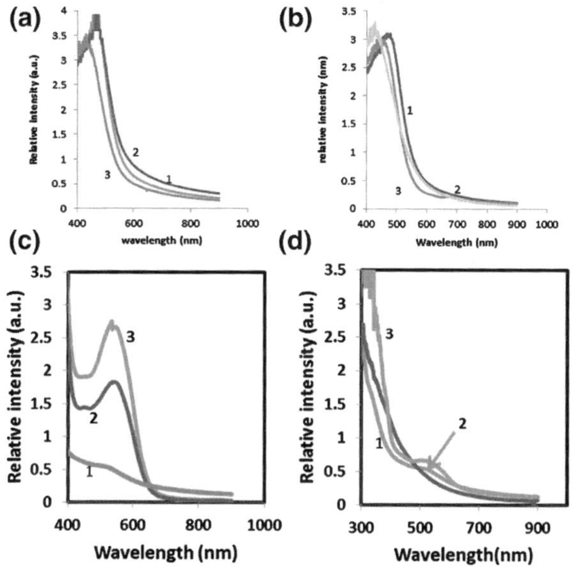

Fig. 2 UV-Vis absorption spectra of **a** aqueous extraction of *lawsonia inermis linn* at three different temperatures *1* 25 °C, *2* 50 °C and *3* 100 °C and **b** alcoholic extraction of *lawsonia inermis linn 1* 25 °C *2* 50 °C and *3* 100 °C **c** UV-Vis spectra of the dye extracted from *Hibiscus rosa-sinensis* water extract) at three different temperatures viz. *1* 25 °C, *2* 50 °C and *3* 100 °C; Fig. 2d: UV-Vis spectra of the dye extracted from *Hibiscus rosa-sinensis* (alcoholic extract) at three different temperatures viz. *1* 25 °C, *2* 50 °C and *3* 100 °C

However, a much higher intensity was observed for the alcoholic extract as shown in Fig. 2d. The difference in the absorption characteristics is attributed due to the complicated mixed composition (viz. chlorophyll, anthocyanins etc.) and colours of the extracts.

4.2 Optical Studies for TiO_2 Films

The optical absorption spectra of the as synthesized TiO_2 thin films prepared by peptization and sol–gel techniques are shown in Fig. 3a and b respectively. Both the film shows absorption in the UV-region indicating the formation TiO_2, as it is sensitive to the UV-region.

The optical band gaps of the films were obtained by using the following equation [37] for a semiconductor

$$A = K\,(h\nu - Eg)^{m/2}/h\nu \tag{1}$$

where 'A' is the absorbance, 'K' is a constant and 'm' is equal to '1' for direct transition and '2' for indirect transition. Linearity of the plots of $(Ah\nu)^2$ versus photon

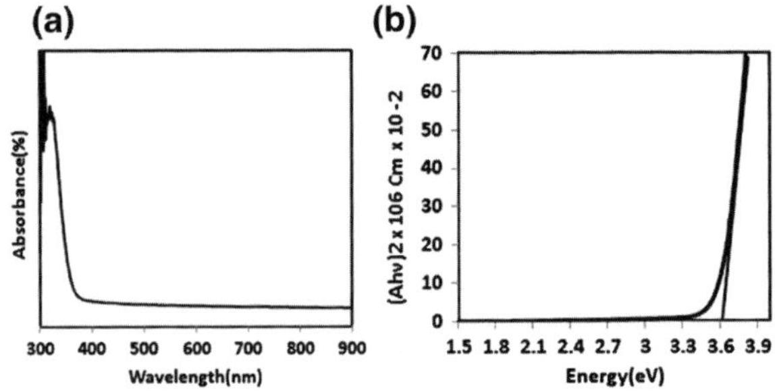

Fig. 3 **a** UV-Vis absorption spectra of TiO_2 thin film prepared by peptizationl technique. **b** Band gap calculation from absorption spectra of TiO_2 thin film prepared by peptization technique. **c** UV-Vis absorption spectra of TiO_2 thin film prepared by sol–gel technique. **d** Band gap calculation from absorption spectra of TiO_2 thin film prepared by sol–gel technique

Table 1 Band gap, shift in band gap and particle sizes calculated from absorption spectra

Sample of TiO_2	Deposition technique	Band gap from UV-Vis (eV)	Shift in band gap (eV)	Grain size from EMA (nm)
Sample A	Peptization	3.6	0.4	4.312
Sample B	Sol–gel	3.4	0.2	6.21

energy $h\nu$ for the TiO_2 thin films indicates that the material is of direct band gap nature. The extrapolation of the straight line to $(Ah\nu)^2 = 0$ axis (Fig. 3c, d) gives the energy band gap of the material. The band gap of the films was found to be 3.6 and 3.4 eV and is depicted in Table 1. From the band gap information, the size of the TiO_2 nanoparticles were calculated using effective mass approximation (EMA) method and following the equation [37] for a semiconductor.

$$E_{gn} - E_{gb} = [(h^2\pi^2/2R^2)1/m*] \quad (2)$$

In the above equation m* is the effective mass of the specimen, R is radius of the particle, Egb is the bulk band gap and Egn is the band gap of the sample. From the calculations it was observed that, the size of TiO_2 particles for the film prepared by sol–gel technique is of 6.21 nm which is slightly greater than the size of 4.312 nm obtained from peptization method and is in agreement with the results observed from SEM studies.

4.3 Efficiency of the Photoanode

Three samples of natural dyes were adsorbed on the surface of the TiO_2 semiconducting nanoparticle in order to fabricate the photoanode. The performance of the natural dyes as a sensitizer for TiO_2 was evaluated by investigating the absorption spectra of as fabricated dye soaked TiO_2 thin films. The chemical

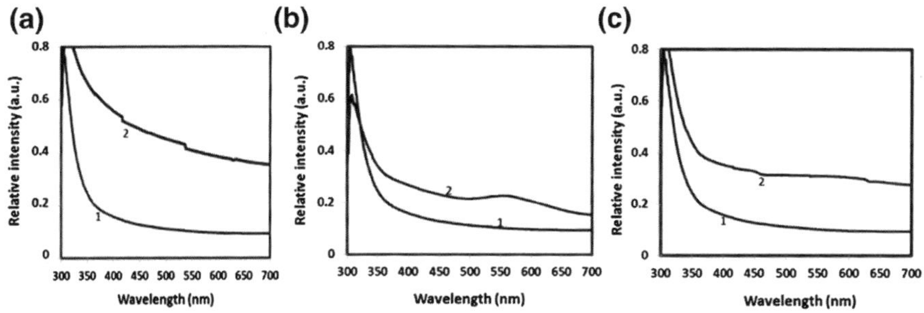

Fig. 4 **a** UV-Vis spectra of TiO_2 thin film unsoaked (1) and soaked (1) with *lawsonia inermis linn* (aqueous extract at 50° C) **b** UV-Vis spectra of TiO_2 thin film unsoaked (1) and soaked (2) with *Hibiscus rosa-sinensis* (Aqueous extract at 100 °C) **c** UV-Vis spectra of TiO_2 thin film unsoaked (1) and soaked (2) with *Hibiscus rosa-sinensis* (Alcoholic extract at 100 °C)

adsorption of these dyes is generally attributed to the carbonyl and hydroxyl groups present in the anthocyanins and other such components present in the extract which bound to the surface of the porous TiO_2 film. The photoanode soaked with the dyes extracted from aqueous solution of *lawsonia inermis linn* at 50 °C exhibited better performance in comparison to the others and as such found to be suitable for application in DSSC.

The UV-Vis absorption spectra of TiO_2 and dye loaded TiO_2 thin films prepared by peptization technique are shown in Fig. 4a–c. The films were soaked with three different dyes: (1) the dye extracted from aqueous solution of lawsonia *inermis linn* at 50° C (Fig. 4a); (2) the dye extracted from aqueous solution of Hibiscus *rosa-sinensis* at 100 °C (Fig. 4b; (3) the dye extracted from Hibiscus *rosa-sinensis* in an alcoholic solution at 100 °C (Fig. 4c). It was observed that all the three set of films exhibited enhanced absorption after being soaked with the dyes indicating the anchorage of the dye on the surface of TiO_2. However, it was found that the film soaked with the dye extracted from aqueous solution of lawsonia *inermis linn* at 50° C showed much higher absorption compared to the other two counter parts and as such may be more suitable for application in DSSC.

5 Conclusion

Nanostructured TiO_2 thin films were prepared by sol–gel and peptization techniques by the hydrolysis and peptization of titanium isopropoxide (TIP). The XRD and SEM studies reveal that the film prepared by peptization technique is suitable for application in DSSC due to its good crystallinity and better surface. Three natural dyes were adsorbed on the surface of TiO_2 for the fabrication of photoanode. However, the dye extracted from aqueous solution of lawsonia *inermis linn* at 50° C exhibited good anchorages to the surface of TiO_2 and hence better absorption in the UV-visible region.

Acknowledgments The authors are thankful to Physics Department and Central Instrumentation facility, IIT Guwhatai for XRD data and SEM images. One of the authors (N.Goswami), acknowledges UGC, New Delhi for financial support under CPE scheme to Sibsagar College, Joysagar.

References

1. B. O'Regan, M. Grätzel, A low-cost, high-efficiency solar cell based on dye sensitized colloidal TiO_2 films. Nature **353**, 737–740 (1991)
2. M. Gratzel, Dye-sensitized soldar cell. J. Photochem. Photobiol. C **4**, 145 (2003)
3. N. Shekhar, G.Y. Vishal, Metal complex dyes for dye sensitized solar cell. General article, Resonance (Sep 2010)
4. K. Pattarith et al., Synthesis of metal free organic dye for dye sensitized solar cell. J. Korean chem Soc. **55**(2) (2011)
5. S. Hao, J. Wu, Y. Huang, J. Lin, Natural dyes as photo sensitizers for dye-sensitized solar cell. Sol. Energy **80**, 209–214 (2006)
6. Y. Amao, T. Komori, Biosensors Bioelectron **19**, 843 (2004)
7. A.S. Polo, N.Y. Murakami Iha, Blue sensitizers for solar cells: natural dyes from Calafate and Jaboticaba. Sol. Energ. Mat. Sol. Cell **90**, 1936–1944 (2006)
8. C.G. Garcia, A.S. Polo, N.Y. Iha, J. Photochem. Photobiol. A **160**, 87 (2003)
9. P.M. Sirimanne, M.K.I. Senevirathna, E.V.A. Premalal, P.K.D.D.P. Pitigala, V. Sivakumar, K. Tennakone, Utilization of natural pigment extracted from pomegranate fruits as sensitizer in solid-state solar cells. J. Photochem. Photobiol. A **177**, 324–327 (2006)
10. K. Tennakone, A.R. Kumarasinghe, G.R.R.A. Kumara, K.G.U. Wijayantha, P.M. Sirimanne, Nanoporous TiO_2 photoanode sensitized with the flower pigment cyaniding. J. Photochem. Photobiol. A **108**, 193–195 (1997)
11. M.S. Roy, P. Balraju, M. Kumar, G.D. Sharma, Dye-sensitized solar cell based on Rose Bengal dye and nanocrystalline TiO_2. Sol. Energ. Mat. Sol. C **92**, 909–913 (2008)
12. J.M.R.C. Fernando, G.K.R. Senadeera, Natural anthocyanins as photosensitizers for dye-sensitized solar devices. Curr. Sci. **95**, 663–666 (2008)
13. N.J. Cherepy, G.P. Smestad, M. Grä tzel, J.Z. Zang, J. Phys. Chem. **B101**, 9342 (1997)
14. Q. Dai, J. Rabani, Photosensitization of nanocrystalline TiO_2 films by anthocyanin Gdyes. J. Photochem. Photobiol. A **148**, 17–24 (2002)
15. T. Sugimoto, Adv. Colloid Interface Sci. **28**, 65 (1987)
16. J. Livage, M. Henry, C. Sanchez, Prog. Solid State Chem. **18**, 259 (1988)
17. M.A. Anderson, M.J. Gieselmann, Q. Xu, J. Membr. Sci. **39**, 243 (1988)
18. E.A. Barringer, H.K. Bowen, Langmuir **1**, 414 (1985)
19. J.H. Jean, T.A. Ring, Langmuir **2**, 251 (1986)
20. J.-L. Look, C.F. Zukoski, J. Am. Ceram. Soc. **75**, 1587 (1992)
21. J.-L. Look, C.F. Zukoski, J. Am. Ceram. Soc. **78**, 21 (1995)
22. X. Bokhimi, A. Morales, O. Novaro, T. Lopez, E. Sanchez, R. Gomez, J. Mater. Res. **10**, 2788 (1995)
23. D. Vorkapic, T. Matsoukas, J. Colloid Interface Sci. **214**, 283 (1999)
24. R.L. Penn, J.F. Banfield, Geochim. Cosmochim. Acta **63**, 1549 (1999)
25. C.J. Barb′e, F. Arendse, P. Comte, M. Jirousek, F. Lenzmann, V. Shklover, M. Gratzel, J. Am. Ceram. Soc. **80** 3157 (1997)
26. T. Yoko, A. Yuasa, K. Kamiya, S. Sakka, Sol-gel-derived TiO_2 film semiconductor electrode for photocleavage of water.Preparation and effects of post heating treatment on the photoelectrochemical behavior. J. Electrochem. Soc. **138**(8), 2279–2285 (1991)
27. K.A. Vorotilov, E.V. Orlova, V.I. Petrovsky, Sol-gel TiO_2 films on silicon substrates. Thin Solid Films **207**(1–2), 180–184 (1992)

28. V.J. Nagpal, R.M. Davis, S.B. Desu, Novel thin films of titanium dioxide particles synthesized by a sol-gel process. J. Mater. Res. **10**(12), 3068–3078 (1995)
29. K.S. Yeung, Y.W. Lam, A simple chemical vapour deposition method for depositing thin TiO films. Thin Solid Films **109**(2), 169–178 (1983)
30. J.P. Lu, J. Wang, R. Raj, Solution precursor chemical vapour deposition of titanium oxide thin films. Thin Solid Films **204**(1), L13–L17 (1991)
31. S. Schiller, G. Beister, W. Sieber, G. Schirmer, E. Hacker, Influence of deposition parameters on the optical and structural properties of TiO films produced by reactive d.c. plasmatron sputtering. Thin Solid Films **83**(2), 239–245 (1981)
32. M.H. Suhail, G.M. Rao, S. Mohan, Dc reactive magnetron sputtering of titanium-structural and optical characterization of TiO_2 films. J. Appl. Phys. **71**(3), 1421–1427 (1992)
33. K. Okimura, N. Maeda, A. Shibata, Characteristics of rutile TiO films prepared by magnetron sputtering at a low temperature. Thin Solid Films **281–282**(1–2), 427–430 (1996)
34. X. Chen, S.S. Mao, Titanium dioxide nanomaterials. Chem Rev **107**(7), 2905 (2007)
35. S. Mahshid, M. Sasani Ghamsari, M. Askari, N. Afshar, S. Lahuti, Synthesis of TiO_2 nanoparticles by hydrolysis and peptizationof titanium isopropoxide solution. Semicond. Phys. Quantum Electron. Optoelectron. **9**(2), 65–68 (2006)
36. Y.-C. Liu, Y.-F. Lu, Y.-Z. Zeng, C.-H. Liao, J.-C. Chung, T.-Y. Wei, Research article nanostructured mesoporous titanium dioxide thin film prepared by sol-gel method for dye-sensitized solar cell. Int. J. Photoenergy
37. H. Mohammed et al., Preparation of titanium dioxide (TiO_2) via the sol-gel Process, on line

First-Principles Study of Adsorption of Alkali Metals (Li, Na, K) on Graphene

B. D. Oli, C. Bhattarai, B. Nepal and N. P. Adhikari

Abstract We have performed the first-principles study of adsorption of alkali metals like lithium, sodium and potassium on graphene. As first-principles methods we have chosen Hartree–Fock method and density functional theory (DFT) method. We have studied the variation of energy of grapheme clusters on sizes of the graphene clusters. Our findings show that the energy of graphene clusters increases with the size of the clusters and for the cluster size more than 30 carbon atoms the energy per carbon atom remains almost constant. This shows that the graphene clusters are stable with respect to energy for the larger size of clusters. The adsorption of alkali metals on the H-grpahene clusters have been studied. The adsorption energy of alkali metal atoms (Li, Na and K) on H-graphene increases on increasing the size of H-graphene cluster. The dependence of the computed adsorption energies of Li, Na and K atoms on the size of H-graphene clusters and the presence of large dipole moments in the adatoms-H-graphene system exhibit the ionic character of the alkali metal adatom-H-graphene interaction. Our findings on the adsorption energy of the alkali metals on H-graphene clusters agree with the previously reported data.

Keywords Graphene · H-graphene · Hartree–Fock · Density functional theory · Alkali metal · Adsorption

B. D. Oli · C. Bhattarai · B. Nepal · N. P. Adhikari (✉)
Central Department of Physics, Tribhuvan University, Kirtipur, Kathmandu, Nepal
e-mail: npadhikari@gmail.com

P. K. Giri et al. (eds.), *Advanced Nanomaterials and Nanotechnology*,
Springer Proceedings in Physics 143, DOI: 10.1007/978-3-642-34216-5_51,

1 Introduction

Graphene is a flat monolayer of carbon atoms tightly packed into a two-dimensional (2-D) honeycomb lattice and is a basic building block for graphitic materials of all other dimensionalities [1]. It can be wrapped up into 0-D fullerenes, rolled into 1-D nanotubes and stacked into 3-D graphite. It's intriguing electric and transport properties have attracted wide attention since its discovery in 2004 by the group led by Geim and Novoselov [2]. The Nobel Prize in Physics 2010 was awarded jointly to Geim and Novoselov "for groundbreaking experiments regarding the two dimensional graphene". After its discovery there was a short period of time when graphene was seen either as mere curiosity or even with distrust, since it was believed that such a material could not exist in crystalline form. The field of graphene research only really took over in 2005 with the measurement of the anomalous Hall Effect [3, 4]. These experiments showed, without any question that the electrons in this material behaved in a 'relativistic' way although they move at velocities 300 times smaller than the speed of light. By showing that the electrons behave as so called Dirac particles [5], it have proved that Pauling's "giant molecules" is actually a crystal since Dirac electrons can only exist on the honeycomb lattice.

Graphene is just one atom thick and the mobility of graphene is very high because the quantum Hall effects can be observed even at room temperature [6]. Graphene has the potential of the material for novel devices because of this high mobility. The energy gap of graphene is zero [2], so graphene cannot be applied to semiconductor devices such as transistors, LEDs, etc. Graphene is the best theoretically studied allotrope of carbon for more than 60 years. It had already been studied theoretically in 1947 by Wallace [7] as a text book example for calculations in Solid State Physics. He predicted the electronic structure and noted the linear dispersion relation.

Graphene is a wonder material with many superlatives to its name. It is the thinnest known material in the universe; it's the strongest material ever measured; it's the stiffest material we know and the most stretchable crystal [8, 9]. The unit hexagonal cell of graphene contains two carbon atoms and has an area of 0.052 nm^2 [10]. So, the density of graphene is 0.77 mg/m^2. It is almost transparent; it absorbs only 2.3 % of the light intensity, independent of the wavelength in the optical domain. It exhibits a breaking strength of $\sim$42 N/m [9] and is about 100 times stronger than the strongest steel. It conducts heat 10 times faster than copper [9] and can carry about 1,000 times the current density copper wire [11]. Graphene is expected to be a revolutionary material that could change the technology of semiconductors and Liquid Crystal Display (LCD) touch screens and monitors, create super small transistors and super dense data storage, increase energy storage and solar cell efficiency, and will transform many other applications.

After the discovery of graphene, it has been extensively investigated, and the idea of studying structural and electronic changes that are induced by the adsorption of foreign atoms [12–16] and molecules [17, 18] was exploited by researchers who were motivated both by technological and academic interest.

Alkali metals (highly reactive chemical species) have a fairly simple electronic configuration and the adsorption on a graphite surface [highly oriented pyrolytic graphite (HOPG)] have been investigated for the past 20 years [19]. The metal-graphite and metal-nanotube structures are of particular interest for catalysis, fuel cell technology and hydrogen storage [20–24]. The doping of single wall carbon nanotubes (SWCNT's) by Li and K is known to enhance conductivity and hydrogen storage capacity [25]. Potassium (K) has also been used to chemically dope graphene, allowing graphene quasiparticle dynamics [26] and minimal conductivity [27] to be studied. Because of the diversity of properties of the metallic elements, the variety of structures formed, and the availability of experimental techniques at the nanolevel, the adsorption of metals on the surface of this material is a promising approach to controllably modify graphene.

In the present work, the first- principles calculations have been carried out in order to study the stability of graphene as a function of the number of carbon atoms; and the structural and electronic properties of alkali metal atoms (Li, Na, and K) adsorbed hydrogen passivated graphene clusters. To study the stability of graphene, the graphene clusters (C_N) having the number of carbon atoms (N) = 6, 10, 13, 16, 19, 22, 24, 27, 30, 32, 35, 37, 40, 42, 45, 47, 50, 52, and 54 are considered and the calculations have been carried out using Hartree–Fock (HF) and DFT (B3LYP) levels of approximation with 3-21G basis set. The adsorption of alkali metal atoms (Li, Na, and K) on only side of the perimeter passivated graphene clusters $C_{16}H_{10}$, $C_{30}H_{14}$, $C_{48}H_{18}$ and $C_{70}H_{22}$ have been considered in the DFT (B3LYP) level of approximation with the choice of basis sets 3-21G and 3-21G*. The adsorption energy, adatom height, the dipole moments, and the population analysis using Natural Bond Orbital (NBO) method have been studied.

In Sect. 2, we briefly describe methods of the calculations. In Sect. 3, we present the results of binding energy and stability of graphene; and the results of adsorption of Li, Na and K atoms on hydrogen passivated graphene (H-graphene). Finally, we conclude in Sect. 4.

2 Computational Methods

We have performed first-principles calculations using the Hartree–Fock (HF) Cluster procedure [28] employing for its implementation the variational Roothaan approach and Density Functional Theory (DFT) [29, 30] with Gaussian orbital functions for the basis sets. In DFT, the gradient-corrected functional B3LYP (i.e. Becke's 3-parameter hybrid exchange functional [31] and Lee, Yang, and Parr correlational functional [32]) has been employed for our calculations. The minimum nature of the structure has been confirmed by frequency calculations at both computational levels. The electronic structure and total energies are calculated using Gaussian 03 set of programs [33] in the present work. Gaussian is widely used quantum chemical calculation software, to calculate the electronic states of isolated molecules. Gaussian was improved in 2003, and Gaussian 03 has been able to calculate the electronic states using periodic boundary conditions. It can calculate the electronic states of not only three-dimensional crystals, but also two-dimensional or one dimensional materials. We can specify the combination of method and basis set in Gaussian 03 set of programs. The basis set 3-21G is, as dependent as the size of graphene clusters, has been chosen and for some cases we used 3-21G* too. All the clusters analyzed in this work have been fully optimized and the convergence criterion was 10^{-8} hartree for the energy, 0.0018 Å for the displacement and 0.00045 hartree for the force.

The present work describes the first-principles study of (1) the equilibrium configurations of the grapheme clusters (C_N) having the number of carbon atoms N = 6, 10, 13, 16, 19, 22, 24, 27, 30, 32, 35, 37, 40, 42, 45, 47, 50, 52, 54 and (2) the adsorption of alkali metal atoms (Li, Na, K) on hydrogen passivated graphene (H-graphene) sheets $C_{16}H_{10}$, $C_{30}H_{14}$, $C_{48}H_{18}$ and $C_{70}H_{22}$. The binding energy of the pure graphene cluster is calculated as,

$$BE = N\, E_{total}(C) - E_{total}(C_N) \tag{1}$$

where N is the number of carbon atoms of a pure graphene sheet considered in the present work, $E_{total}(C_N)$ is the ground state energy of graphene sheet having N number of carbon atoms and $E_{total}(C)$ is the ground state energy of carbon atom. Similarly, the binding energy per carbon atom is estimated by using formula

$$B.\, E._{(percarbon)} = \frac{N\, E_{total}(C) - E_{total}(C_N)}{N} \tag{2}$$

Our findings show that the adsorption of alkali metal atoms (Li, Na and K) on pure graphene sheet of any size is highly unstable for the size of graphene clusters considered in the present work. Therefore, the surface of graphene in real space calculations have been modeled by parallelogram/circular (depending upon their structure) fragments of a graphene sheet passivated by hydrogen along the perimeter. The hydrogen passivated graphene sheet has been termed as H-graphene. The first-principles calculations have been carried out to study the adsorption of alkali metal adatoms (Li, Na, K) on H-graphene clusters $C_{16}H_{10}$, $C_{30}H_{14}$, $C_{48}H_{18}$ and $C_{70}H_{22}$ in the DFT (B3LYP) level of approximation with the choice of basis sets 3-21G and 3-21G*.

We consider the binding of the adatoms (Li, Na, K) on three sites of high symmetry on the H-graphene sheets: the hollow (H) site at the center of a hexagon, the bridge (B) site at the midpoint of a carbon–carbon bond, and the top (T) site directly above a carbon atom. For each adsorption site for the adatom-H-graphene system, the adatoms are relaxed along the z direction considering H-graphene sheet in xy plane. The H-graphene sheets dimension is kept fixed for all these three calculations. Our calculations show that the adsorption of alkali metals on H-graphene is most favourable in the hollow (H) position and hence in the present work we discuss only the adsorption of alkali metals in the hollow site in the present work.

The optimization of the alkali metal-H-graphene system is performed in the following steps. First, we optimize the geometries of H-graphene clusters. After

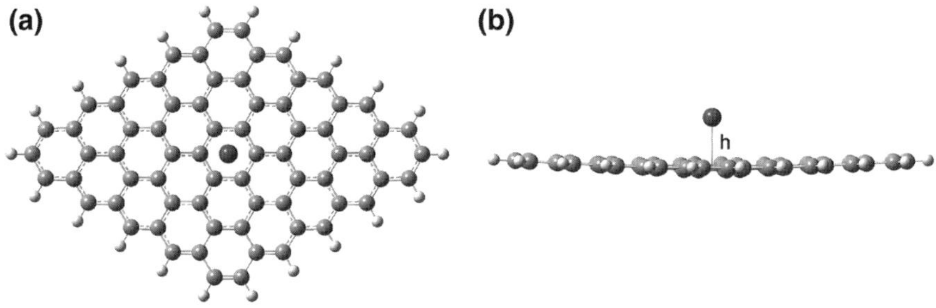

Fig. 1 The optimized geometry of potassium (K) atom adsorbed H-graphene sheet $C_{70}H_{22}$ at the hollow site at the centre of a hexagon in the DFT (B3LYP) level of approximation with the choice of basis set 3-21G*: **a** the *top view*, and **b** the *side view*. The adatom height (h) is equal to 2.62 Å

that, the position of the alkali metals (adatoms) on H-graphene sheet at the hollow (H) site at the center of a hexagon is fixed above certain height and the distances between alkali metal (adatom) and H-graphene sheet have been set to values that were slightly smaller than the sum of the adsorbate and carbon atoms covalent radii. And the final geometries of the adatom-H-graphene systems have been optimized. The optimized H-graphene sheet $C_{70}H_{22}$ with adsorbed potassium atom at the hollow (H) site at the centre of hexagon in the DFT (B3LYP) level of approximation using 3-21G* basis set is shown in Fig. 1. The adatom height (h) (the equilibrium distance between the alkali metals and the surface of H-graphene) is obtained from these optimized structures. The adatom height (h) is defined as the distance between alkali metals (adatoms) and the centre of mass of the carbon ring just below the alkali metals (Li, Na or K) i.e. the plane of the H-graphene sheet.

The binding (adsorption) energy of the alkali metal atoms on H-graphene has been estimated as,

$$\Delta E_{\mathrm{adsorption}} = E(\mathrm{adatom}) + E(\mathrm{H-graphene}) - E(\mathrm{system}) \quad (3)$$

where E(adatom) is the ground state energy of alkali metal (Li, Na or K), E(H-graphene) is the ground state energy of hydrogen passivated graphene sheet and E(system) is the ground state energy of the alkali atom adsorbed H-graphene sheet. The adsorption energy, equilibrium adatom-H-graphene distance (adatom height), and population analysis using Natural Bond Orbital (NBO) method have been studied for all alkali metals-H-graphene systems.

3 Results and Discussion

3.1 The Binding Energy and Stability of Graphene

We have performed the optimization of graphene clusters (C_N) having the number of carbon atoms N = 6, 10, 13, 16, 19, 22, 24, 27, 30, 32, 35, 37, 40, 42, 45, 47, 50, 52, 54 and obtained the ground state energies. The variation of ground state

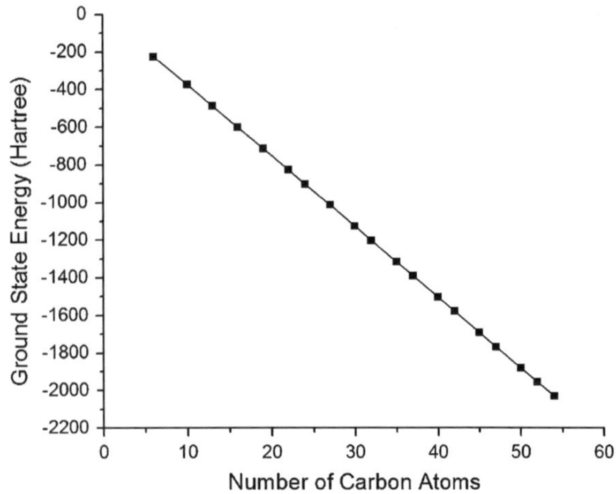

Fig. 2 The variation of the ground state energy of the graphene clusters (C_N) with the number of carbon atoms (N) in the HF/3-21G level of approximation. The best fitted line is: $E(C_N) = 0.43154 - 37.63798 \times N$

energy of graphene clusters (C_N) with the number of carbon atoms (N) in the HF and DFT (B3LYP) levels of approximation using basis set 3-21G are shown in Figs. 2 and 3 respectively. As the number of carbon atoms increases in the graphene sheets the ground state energy decreases linearly in both the levels of approximation with the basis set 3-21G. The linear equations for the ground state energy of graphene clusters (C_N) as a function of number of carbon atoms (N) are $E(C_N) = 0.43154 - 37.63798 \times N$ in the HF level of approximation and $E(C_N) = 0.38972 - 37.87991 \times N$ in the DFT(B3LYP) level of approximation with 3-21G basis set. The energy has been measured in units of Hartee.

Figure 4 shows the variation of the binding energy of graphene clusters (C_N) with the number of carbon atoms (N) using the HF and DFT (B3LYP) levels of approximation with the choice of basis set 3-21G. The binding energy of graphene cluster increases on increasing the size of graphene cluster and the binding energy values obtained in the DFT (B3LYP) level of approximation are larger than the corresponding HF values. This is expected as HF does not take into account of the electron–electron correlation.

We have analyzed all the bond lengths, bond angles and dihedral angles of the symmetric graphene sheet having number of carbon atoms N = 24. The calculation has been performed in the DFT (B3LYP) level of approximation with the choice of basis set 3-21G. The optimized geometry of graphene sheet having 24 number of carbon atoms in the DFT (B3LYP) level of approximation with the choice of basis set 3-21G is shown in Fig. 5. From the optimized geometry of graphene sheet (C_{24}) in the DFT (B3LYP) level of approximation the C–C single bonds (3C–7C, 8C–9C, and 9C–10C) has the bond length of 1.45 Å, whereas the C–C single bonds (2C–3C, 4C–5C, 7C–19C, 8C–14C, 9C–11C, and 10C–22C) have the bond lengths of 1.50 Å. It is seen from the Fig. 5 that the aromatic bonds (shown by a single line and dotted line) are also formed. The aromatic bonds (3C–4C, 4C–8C, and 7C–10C) has the bond lengths of 1.45 Å whereas, the aromatic bonds (1C–2C, 2C–17C, 5C–6C, 5C–16C, 11C–12C, 11C–23C, 13C–14C, 14C–15C, 18C–19C, 19C–20C, 21C–22C,

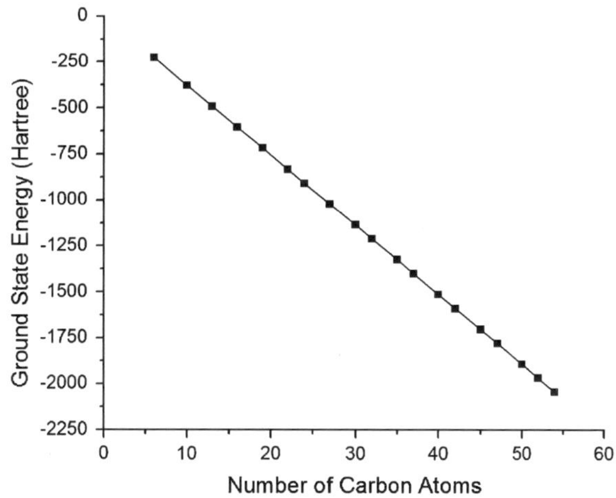

Fig. 3 The variation of the ground state energy of the graphene clusters (C_N) with the number of carbon atoms (N) in the DFT(B3LYP)/3-21G level of approximation. The best fitted line is: $E(C_N) = 0.38972 - 37.87991 \times N$

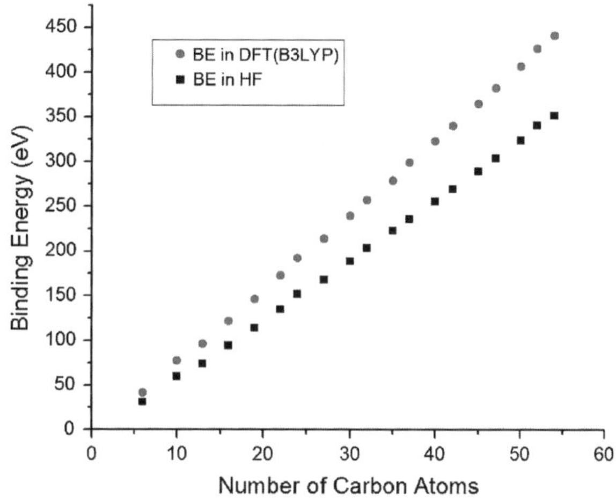

Fig. 4 The variation of the binding energy of the graphene clusters (C_N) with the number of Carbon atoms (N) using the HF and DFT(B3LYP) levels of approximation with 3-21G basis set

and 22C–24C) have the bond lengths of 1.39 Å. This means the aromatic bond gets shortened by 0.06 Å in the edge carbon atoms of the graphene sheet. This may be due to the finite size of the graphene sheet. There are six triple bonds (1C–6C, 12C–13C, 15C–16C, 17C–18C, 20C–21C, and 23C–24C) at the edge of the graphene sheet having bond length of 1.24 Å. The bond angles among the carbon atoms (3C–4C–8C, 4C–8C–9C, 8C–9C–10C, 9C–10C–7C, 10C–7C–3C, and 7C–3C–4C) inside the graphene cluster are 120.0°. The bond angles formed by the single bonds (14C–8C–9C, 8C–9C–11C, 11C–9C–10C, 9C–10C–22C, 19C–7C–3C, 7C–3C–2C) in the graphene cluster has the bond angles 120.0°. The bond angles formed by aromatic and single bonds (2C–3C–4C, 5C–4C–8C, 5C–4C–3C, 14C–8C–4C, 22C–10C–7C, 10C–7C–19C) are 120.0° whereas the bond angles formed by aromatic bonds and single bonds (3C–2C–1C, 6C–5C–4C, 4C–5C–16C, 15C–14C–8C, 8C–14C–13C, 12C–11C–9C, 9C–11C–23C, 24C–22C–10C, 10C–22C–21C, 20C–19C–7C,

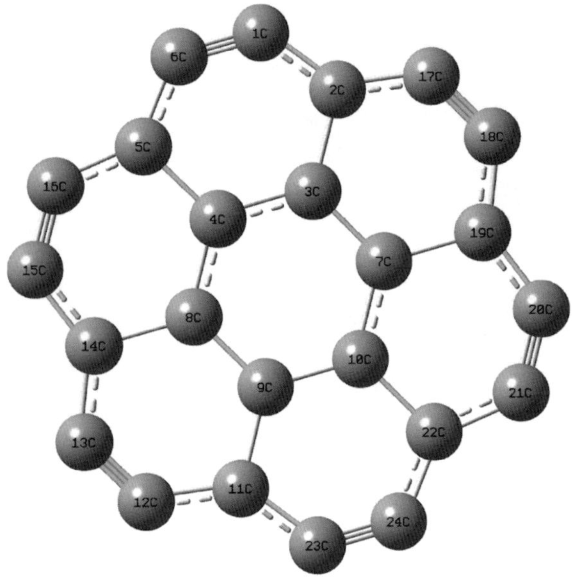

Fig. 5 The optimized geometry of graphene sheet having 24 number of carbon atoms in the DFT (B3LYP) level of approximation using basis set 3-21G

7C–19C–18C, 17C–2C–3C) are 111.9°. The bond angles formed by aromatic bonds and triple bonds (2C–1C–6C, 1C–6C–5C, 5C–16C–15C, 16C–15C–14C, 14C–13C–12C, 13C–12C–11C, 11C–23C–24C, 23C–24C–22C, 22C–21C–20C, 21C–20C–19C, 19C–18C–17C, and 18C–17C–2C) are 128.0°. The dihedral angles are found to be 0.0°.

To study the stability of graphene the binding energy per carbon atom is estimated using Eq. (2). The variation of binding energy per carbon atom of graphene clusters (C_N) with the number of carbon atoms (N) in the HF and DFT (B3LYP) levels of approximation with 3-21G basis set are shown in Figs. 6 and 7 respectively.

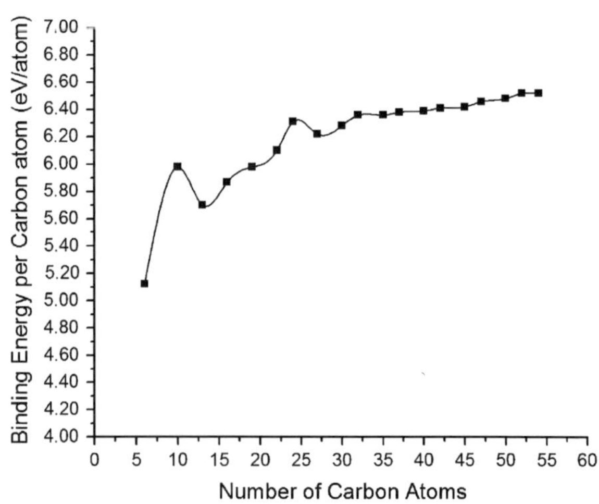

Fig. 6 The variation of binding energy per carbon atom (eV/atom) of graphene clusters (C_N) with the number of carbon atoms (N) in the HF/3-21G level of approximation

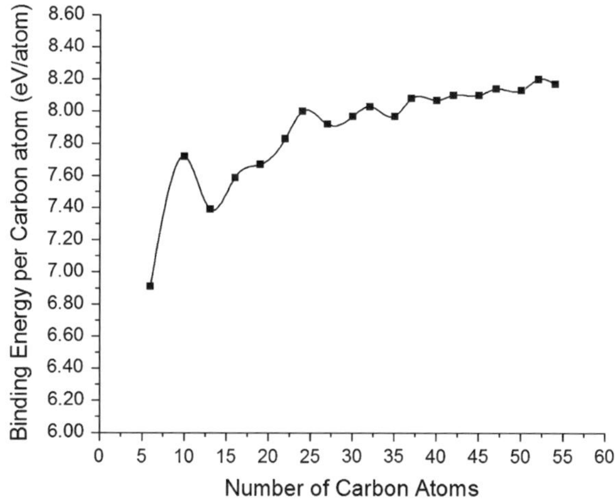

Fig. 7 The variation of binding energy per carbon atom (eV/atom) of graphene clusters (C_N) with the number of carbon atoms (N) in the DFT(B3LYP)/3-21G level of approximation

To our best knowledge in literature there are no systematic studies of binding energy or binding energy per carbon atom as a function of number of carbon atoms in the graphene sheet, which gives stability of graphene as a function of number of carbon atoms. The binding energy per carbon atom of graphene clusters C_{10} and C_{24} (Fig. 4) are more than those of the neighboring graphene clusters. This is because of the fact that for $N = 10$ and 24 graphene sheets are in closed form and more symmetric structures are formed than theirs neighboring graphene clusters having number of carbon atoms $N = 6, 13$ and $N = 22, 27$ respectively. For the graphene sheet having number of carbon atoms 30 and/or more, the change (increase or decrease) in the value of binding energy per carbon atom is within 1 % in the HF and DFT (B3LYP) level of approximation in the same basis set 3-21G. It is concluded that the pure monolayer graphene sheet having 30 and/or more number of carbon atoms is more stable than the lower size graphene clusters. The binding energy of a pure graphene sheet having 32 number of carbon atoms (C_{32}) is estimated using Eq. (2) is 8.03 eV/atom, in the DFT(B3LYP) level of approximation with the choice of basis set 3-21G. This value of the binding energy of the graphene sheet for 32 carbon atoms (C_{32}) agrees within 1.52 % to the previously reported value 7.91 eV/atom [34].

3.2 Alkali Metal (Li, Na, and K) Adsorption on H-Graphene

We investigate the adsorption of the alkali metal atoms (Li, Na, and K) on flat H-graphene clusters. For this study, we select four perimeter passivated graphene clusters: $C_{16}H_{10}$, $C_{30}H_{14}$, $C_{48}H_{18}$ and $C_{70}H_{22}$. The alkali metals (Li, Na, and K) were kept at the distance of 1.75, 2.20 and 2.70 Å respectively above the H-graphene surface at the hollow (H) site at the centre of hexagon. The structures of perimeter-passivated graphene clusters in our study are shown in Fig. 8. The adsorption energy of each adatom H-graphene system is computed as the

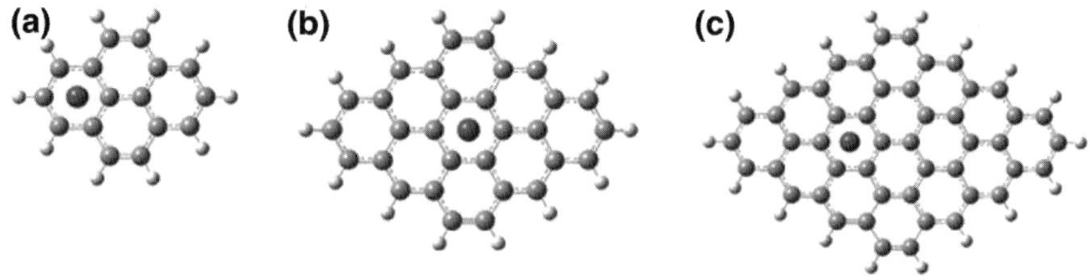

Fig. 8 Structures of hydrogen-passivated graphene (H-graphene) clusters used in the study of alkali metal atoms (Li, Na and K) adsorption on graphene: **a** $C_{16}H_{10}K$, **b** $C_{30}H_{14}K$, and **c** $C_{48}H_{18}K$

difference between the total energy of the isolated alkali metals (Li, Na, and K atom), the optimized H-graphene cluster and the optimized adatom-cluster systems (Eq. 3). In Table 1, we summarize the values of the Li-, Na-, K-cluster adsorption energies ($\Delta E_{adsorption}$), the equilibrium distances between the adatoms and graphene sheet (the adatom height) h, the distance between adatom and nearest carbon atom ($d_{A\text{-}C}$), the dipole moments (μ), and the effective amount of charge transferred (Δq) computed using the DFT(B3LYP) formalisms implemented by the Gaussian 03 set of programs with the choice of basis set 3-21G. In the adsorption of K atom the basis set 3-21G* is used. The adsorption energy of alkali metals (Li, Na, and K) on H-graphene clusters has been corrected by the inherent Basis Set Superposition Error (BSSE) using the 'Counterpoise BSSE Correction' method included in the Gaussian 03 set of programs.

Figures 9 and 10 show the variation of adsorption (binding) energy of Li atom and Na atom on the H-graphene sheets as a function of the number of carbon atoms in the H-graphene sheets in the DFT (B3LYP) level of approximation with the choice of basis set 3-21G with and without BSSE using counterpoise method. Figure 11 shows the variation of adsorption (binding) energy of K atom on the H-graphene sheets as a function of the number of carbon atoms in the DFT(B3LYP) level of approximation with the choice of basis set 3-21G* with and without BSSE using counterpoise method. Our real space DFT calculations reveal a strong dependence of the computed adsorption energies on the size of the H-graphene cluster (Figs. 8, 9, 10). As the size of the H-graphene cluster increases the adsorption energy also increases. From Figs. 8, 9 and 10 it is also seen that after $N = 30$ the adsorption energy of alkali metal atoms on H-graphene clusters becomes almost constant which is expected from our results of the graphene clusters (Sect. 3.1). Our findings for the adsorption of K atom on the H-graphene sheet $C_{30}H_{14}$ is 0.67 eV in the DFT(B3LYP) level of approximation with the choice of basis set 3-21G* taking account of BSSE with the counterpoise method. This value of computed adsorption energy agrees with the adsorption energy for the K-C_{32} sheet within 5.0 % to the previously reported value 0.80 eV [12].

From Table 1, the observed value of charge transferred (Δq) using Natural Bond Orbital (NBO) method from Li, Na and K atom to the H-graphene clusters $C_{16}H_{10}$, $C_{30}H_{14}$, $C_{48}H_{18}$ and $C_{70}H_{22}$ indicate that the effective amount of

Table 1 The DFT (B3LYP) calculations for the 3 alkali metals (Li, Na and K) adsorption on H-graphene clusters considered in the present work

Atom	H-graphene	$\Delta E_{adsorption}$	h	$d_{A\text{-}C}$	μ	Δq
Li[a]	$C_{16}H_{10}$	0.87(0.47)	1.72	2.18	4.8230	0.852
	$C_{30}H_{14}$	1.40(0.97)	1.75	2.26	4.7886	0.912
	$C_{48}H_{18}$	1.76(1.33)	1.76	2.28	4.8473	0.919
	$C_{70}H_{22}$	1.86(1.43)	1.78	2.29	5.9564	0.924
Na[a]	$C_{16}H_{10}$	0.32(0.02)	1.92	2.64	1.9432	0.479
	$C_{30}H_{14}$	0.72(0.36)	1.95	2.57	7.1412	0.958
	$C_{48}H_{18}$	1.03(0.65)	1.97	2.55	7.3255	0.962
	$C_{70}H_{22}$	1.13(0.80)	1.99	2.60	7.2083	0.967
K[b]	$C_{16}H_{10}$	0.17(0.08)	2.60	2.82	5.7864	0.713
	$C_{30}H_{14}$	1.00(0.76)	2.58	2.95	9.0376	0.957
	$C_{48}H_{18}$	1.45(1.20)	2.60	2.92	9.2737	0.960
	$C_{70}H_{22}$	1.64(1.44)	2.62	2.98	9.0286	0.965

The properties listed are the adsorption energy of alkali metals on H-graphene clusters $\Delta E_{adsorption}$, the adatom height h, the distance between adatom and nearest carbon atom $d_{A\text{-}C}$, the dipole moments μ, and the effective amount of the transferred charge Δq are given in eV, Å, Å, D and elementary charge units e respectively. The values in parentheses show the adsorption energy of alkali-metal adatoms on H-graphene clusters with BSSE using 'counterpoise' method

[a] The geometry optimization is done in the DFT (B3LYP)/3-21G

[b] The geometry optimization is done in the DFT (B3LYP)/3-21G*

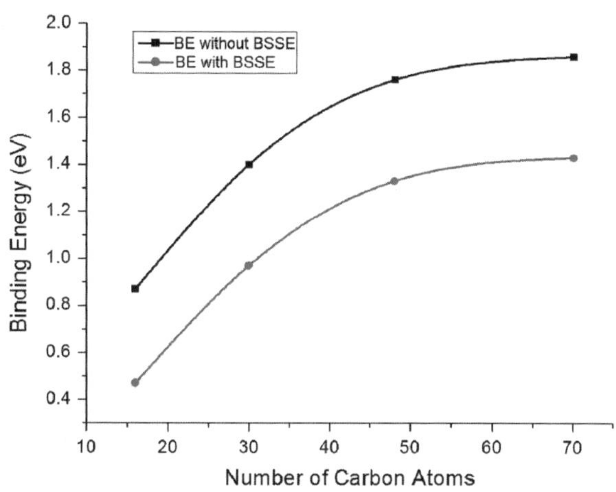

Fig. 9 The variation of binding energy of Li atom on H-graphene sheets as a function of the number of carbon atoms in the H-graphene sheets in the DFT (B3LYP) level of approximation using the basis set 3-21G with and without BSSE using counterpoise method

transferred charge increases with increasing the size of H-graphene. A relatively large fraction of electronic charge transferred from Li, Na and K atom to H-graphene confirms the ionic character of alkali metal-H-graphene interaction. Our study has revealed the significant role of long-range ionic forces in the interaction between Li, Na, K atoms and H-graphene which is demonstrated by the presence of large dipole moments in the alkali metal adatoms-H-graphene clusters and the strong dependence of the computed adsorption energies of Li, Na and K atoms on the size of H-graphene clusters. The strong dependence of the computed

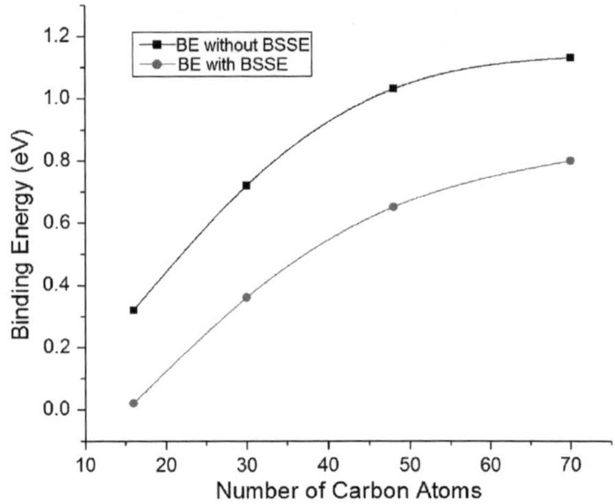

Fig. 10 The variation of binding energy of Na atom on H-graphene sheets as a function of the number of carbon atoms in the H-graphene sheets in the DFT (B3LYP) level of approximation using the basis set 3-21G with and without BSSE using counterpoise method

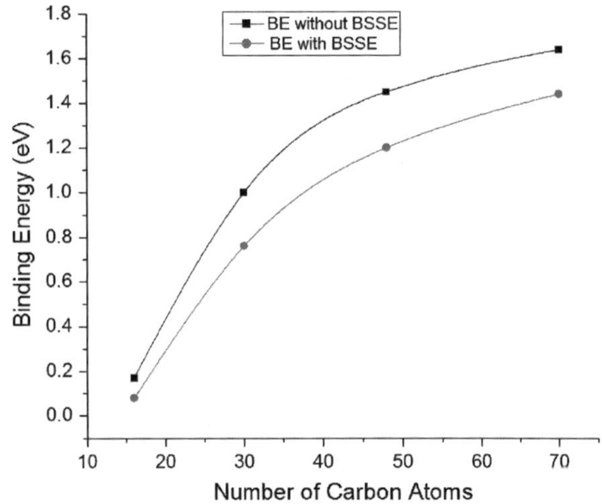

Fig. 11 The variation of binding energy of K atom on H-graphene sheets as a function of the number of carbon atoms in the H-graphene sheets in the DFT(B3LYP) level of approximation using the basis set 3-21G* with and without BSSE using counterpoise method

adsorption energies of Li, Na and K atoms on the size of H-graphene clusters and the large dipole moments of the alkali metal adatoms-H-graphene system suggest that the Li, Na and K atoms are partially ionized and bonded to the surface of H-graphene by electrostatic forces.

4 Conclusions and Outlook

We have studied the stability of graphene considering the graphene clusters (C_N) having the number of carbon atoms N = 6, 10, 13, 16, 19, 22, 24, 27, 30, 32, 35, 37, 40, 42, 45, 47, 50, 52, 54; and adsorption of alkali metal atoms (Li, Na, and K) on H-graphene clusters $C_{16}H_{10}$, $C_{30}H_{14}$, $C_{48}H_{18}$, and $C_{70}H_{22}$ using first-principle calculations. It is found that as the number of carbon atoms increases in the graphene cluster the ground state energy decreases in the HF and DFT (B3LYP) level of approximations with the basis set 3-21G. The linear equations for the ground state energy of graphene clusters (C_N) as a function of number of carbon atoms (N) are $E(C_N) = 0.43154 - 37.63798 \times N$ in the HF level of approximation and $E(C_N) = 0.38972 - 37.87991 \times N$ in the DFT(B3LYP) level of approximation in the same basis set 3-21G. The binding energy of graphene increases on increasing the size of graphene sheet and the binding energy values obtained in the DFT (B3LYP) level of approximation are larger than the corresponding HF values. The pure monolayer graphene sheet having 30 and/or more number of carbon atoms is more stable than the lower size graphene clusters, which is demonstrated by the variation of binding energy per carbon atom of graphene clusters (C_N) with the number of carbon atoms (N) in the graphene cluster in HF/3-21G and DFT (B3LYP)/3-21G (Figs. 6, 7). The adsorption energy of alkali metal atoms (Li, Na and K) on H-graphene increases on increasing the size of H-graphene cluster. The dependence of the computed adsorption energies of Li, Na and K atoms on the size of H-graphene clusters and the presence of large dipole moments in the adatoms-H-graphene system exhibit the ionic character of the alkali metal adatom-H-graphene interaction. The present work can be extended to study the stability of graphene clusters (C_N) and adsorption of metals on graphene having more number of carbons atoms in the graphene with the different and larger size of basis sets.

Acknowledgments The authors acknowledge the partial support for this work by The Abdus Salam International Center for Theoretical Physics (ICTP), Trieste, Italy through Office of External Activities Net-56 and National Academy of Science and Technology (NAST), Khumaltar, Lalitpur, Nepal.

References

1. A.K. Geim, K.S. Novoselov, The rise of graphene. Nat. Mater. **6**, 183–191 (2007)
2. K.S. Novoselov, A.K. Geim, S.V. Morozov, D. Jiang, Y. Zhang, S.V. Dubonos, I.V. Grigorieva, A.A. Firsov, Electric field effect in atomically thin carbon films. Science **306**, 666–669 (2004)
3. K.S. Novoselov, A.K. Geim, S.V. Morozov, D. Jiang, M.I. Katsnelson, I.V. Grigorieva, S.V. Dubonos, A.A. Firsov, Two-dimensional gas of massless Dirac fermions in graphene. Nature **438**, 197–200 (2005)
4. Y. Zhang, Y.W. Tan, H.L. Stormer, P. Kim, Experimental observation of the quantum hall effect and Berry's phase in graphene. Nature **438**, 201–204 (2005)
5. A.H. Castro Neto, F. Guinea, N.M.R. Peres, K.S. Novoselov, A.K. Geim, The electronic properties of graphene. Rev. Mod. Phys. **81**, 109–162 (2009)
6. K.S. Novoselov, Z. Jiang, Y. Zhang, S.V. Morozov, H.L. Stormer, U. Zeitler, J.C. Maan, G.S. Boebinger, P. Kim, A.K. Geim, Room-temperature quantum hall effect in graphene. Science **315**, 1379 (2007)
7. P.R. Wallace, The band theory of graphite. Phys. Rev. **71**, 622–634 (1947)

8. Brumfiel G (2010) Andre Geim: in praise of graphene. Nature. doi:10.1038/news.2010.525
9. A.K. Geim, Graphene: status and prospects. Science **324**, 1530–1534 (2009)
10. M. Gibertini, A. Tomadin, M. Polini, A. Fasolino, M.I. Katsnelson, Electron density distribution and screening in rippled graphene sheets. Phys. Rev. B **81**, 125437–125449 (2010)
11. R. Murali, Y. Yang, K. Brenner, T. Beck, J.D. Meindl, Breakdown current density of graphene nanoribbons. Appl. Phys. Lett. **94**, 243114–243117 (2009)
12. K.T. Chan, J.B. Neaton, M.L. Cohen, First-principles study of metal adatom adsorption on graphene. Phys. Rev. B **77**, 235430 (2008)
13. Y. Mao, J. Yuan, J. Zhong, Density functional calculation of transition metal adatom adsorption on graphene. J. Phys.: Condens. Matter **20**, 115209 (2008)
14. L. Sheng, Y. Ono, T. Taketsugu, Ab initio study of Xe adsorption on Graphene. J. Phys. Chem. C **114**(8), 3544–3548 (2010)
15. M. Vanin, J.J Mortensen, A.K. Kelkkanen, J.M. Garcia-Lastra, K.S. Thygesen, K.W. Jacobsen, Graphene on metals: a van der Waals density functional theory. Phy. Rev. B **81** 081408(R) (2010)
16. R. Varns, P. Strange, Stability of gold atoms and dimmers adsorbed on graphene. J. Phys.:Condens. Matter **20**, 225005 (2008)
17. O. Leenaerts, B. Partoens, F.M. Peeters, Adsorption of H_2O, NH_3, CO, NO_2, and NO on graphene: a first-principles study. Phys. Rev. B **77**, 125416 (2008)
18. M. Zhou, Y.H. Lu, Y.Q. Cai, C. Zhang, Y.P. Feng, Adsorption of gas molecules on transition metal embedded graphene: a search for high –performance graphene-based catalysts and gas sensors. Nanotechnology **22**, 385502 (2011)
19. M. Caragiu, S. Finberg, Alkali metal adsorption on graphite: a review. J. Phys.: Condens. Matter **17**, R995–R1024 (2005)
20. A. Chambers, C. Park, T.K. Baker, N.M. Rodriguez, Hydrogen storage in graphite nanofibers. J. Phys. Chem. B **102**, 4253–4256 (1998)
21. G.E. Froudakis, Hydrogen interaction with carbon nanotubes: a review of ab initio studies. J. Phys.: Condens. Matter **14**, R453 (2002)
22. O. Gulseren, T. Yildirim, S. Ciraci (2002) Effects of hydrogen adsorption on single-wall carbon nanotubes: metallic hydrogen. Phys. Rev. B **66**, 121401(R)
23. Q. Wang, J.K. Johnson, Molecular simulation of hydrogen adsorption in single-walled carbon nanotubes and idealized carbon slit pores. J. Chem. Phys. **110**, 577 (1999)
24. Y. Ye, C.C. Ahn, C. Witham, B. Fultz, J. Liu, A.G. Rinzler, D. Colbert, K.A. Smith, R.E. Smalley, Hydrogen adsorption and cohesive energy of single-walled carbon nanotubes. Appl. Phys. Lett. **74**, 2307 (1999)
25. P. Chen, X. Wu, J. Lin, K.L. Tan, High H_2 uptake by alkali-doped carbon nanotubes under ambient pressure and moderate temperatures. Science **285**, 91–93 (1999)
26. A. Bostwick, T. Ohta, T. Seyller, K. Horn, E. Rotenberg, Quasiparticle dynamics in graphene. Nat. Phys. **3**, 36–40 (2007)
27. J.H. Chen, C. Jang, S. Adam, M.S. Fuhrer, E.D. Williams, M. Ishigami, Charged-impurity scattering in graphene. Nat. Phys. **4**, 377–381 (2008)
28. S.M. Blinder, Basic concepts of self-consistent-field theory. Am. J. Phys. **33**(6), 431–443 (1965)
29. P. Hohenberg, W. Kohn, Inhomogeneous electron gas. Phys. Rev. **136**, B864–B871 (1964)
30. W. Kohn, L.J. Sham, Self-consistent equations including exchange and correlation effects. Phys. Rev. **140**, A1133–A1138 (1965)
31. A.D. Becke, Density functional thermochemistry. III. The role of exact exchange. J. Chem. Phys. **98**, 5648 (1993)
32. C. Lee, W. Yang, R.G. Parr, Development of the Colle-Salvetti correlation-energy formula into a functional of the electron density. Phys. Rev. B **37**, 785–789 (1988)
33. M. J. Frisch, G. W. Trucks, H. B. Schlegel, G. E. Scuseria, M. A. Robb, J. R. Cheeseman, J. A. Montgomery, Jr., T. Vreven, K. N. Kudin, J. C. Burant, J. M. Millam, S. S. Iyengar, J. Tomasi, V. Barone, B. Mennucci, M. Cossi, G. Scalmani, N. Rega, G. A. Petersson,

H. Nakatsuji, M. Hada, M. Ehara, K. Toyota, R. Fukuda, J. Hasegawa, M. Ishida, T. Nakajima, Y. Honda, O. Kitao, H. Nakai, M. Klene, X. Li, J. E. Knox, H. P. Hratchian, J. B. Cross, V. Bakken, C. Adamo, J. Jaramillo, R. Gomperts, R. E. Stratmann, O. Yazyev, A. J. Austin, R. Cammi, C. Pomelli, J. W. Ochterski, P. Y. Ayala, K. Morokuma, G. A. Voth, P. Salvador, J. J. Dannenberg, V. G. Zakrzewski, S. Dapprich, A. D. Daniels, M. C. Strain, O. Farkas, D. K. Malick, A. D. Rabuck, K. Raghavachari, J. B. Foresman, J. V. Ortiz, Q. Cui, A. G. Baboul, S. Clifford, J. Cioslowski, B. B. Stefanov, G. Liu, A. Liashenko, P. Piskorz, I. Komaromi, R. L. Martin, D. J. Fox, T. Keith, M. A. Al-Laham, C. Y. Peng, A. Nanayakkara, M. Challacombe, P. M. W. Gill, B. Johnson, W. Chen, M. W. Wong, C. Gonzalez, and J. A. Pople, Gaussian, Inc., Wallingford CT, 2004

34. A. Bhattacharya, S. Bhattacharya, C. Majumder, G.P. Das, Transition-metal decoration enhanced room-temperature hydrogen storage in a defect-modulated graphene sheet. J. Phys. Chem. C **114**, 10297–10301 (2010)

FEM Simulation Study on Mass Loading Sensitivity of High Aspect Ratio Nano-Pillar Sensing Medium in Saw Sensors

N. Ramakrishnan, Harshal B. Nemade and Roy Paily Palathinkal

Abstract Mass loading effect of a thin film is one of the prime sensing principles in surface acoustic wave (SAW) sensors. Recently we reported mass loading effect of a high aspect ratio resonant structure such as a nano-pillar attached normal to the surface of the SAW resonator. When resonance frequency of the pillar is closer to SAW resonator frequency the resonance frequency shift caused by mass loading of the pillars tends to a negligible value. The high aspect ratio structure and the SAW resonator together in unison form a system of coupled resonator. More recently we reported employing high aspect ratio nano-pillars as sensing medium to design high sensitive SAW sensors such as a SAW hydrogen sensor. In this paper we present finite element method (FEM) simulation study of a SAW resonator with nano-pillars as sensing medium and report the significance of cross-section of resonant pillars on sensitivity. The simulation results shows the mass loading sensitivity of the smaller cross-section pillars are significantly higher. Further the results show that if the pillars cross-section is in nano scale, there is more chance of positive resonance shift at smaller heights and the sensitivity is higher for small change in dimensions of the pillar.

Keywords SAW sensor · Nano-pillars · Sensitivity · FEM

N. Ramakrishnan (✉)
School of Engineering, Monash University Sunway Campus,
46150 Bandar Sunway, Malaysia
e-mail: ramakrishnan@monash.edu

H. B. Nemade
Electrical and Electronics Engineering and Centre for Nanotechnology,
Indian Institute of Technology, Guwahati 781039, India

R. P. Palathinkal
Electrical and Electronics Engineering, Indian Institute of Technology,
Guwahati 781039, India

P. K. Giri et al. (eds.), *Advanced Nanomaterials and Nanotechnology*,
Springer Proceedings in Physics 143, DOI: 10.1007/978-3-642-34216-5_52,

1 Introduction

Surface acoustic wave (SAW) devices are widely employed in sensing applications. Mass loading effect of thin film in SAW sensors is one of the important sensing phenomenon's [1]. In the past few years, SAW sensors employing nanostructured sensing medium have been reported for their improved sensitivity [2–6]. In the all of these reports the improved sensitivity is obtained by the higher surface to volume ratio of the nano sized sensing medium or by the advantage in the material properties. We were first in reporting application of high aspect ratio structures such as a nano or micro pillars as sensing medium in SAW sensors. Initially we worked on a SAW resonators consisting of nano-pillars attached normal to the surface of a SAW resonator [7]. We observed, when the resonance frequency of the nano-pillar (f_r) is equal to the resonance frequency of the SAW resonator ($f_0|_{h=0}$), the nano-pillar offer negligible mass loading to the SAW device and result in positive resonance frequency shift (Δf_0) [7]. When such resonating nano or micro structures are used as sensing medium in SAW sensors, a small change in the properties of the resonating structure will alter the fundamental frequency of the structure and the corresponding Δf_0 will be significantly high. Accordingly, high sensitivity SAW sensors gas sensors can be developed [7]. Further we demonstrated significance of nano-pillars as sensing medium and its potential in developing high sensitive SAW hydrogen sensor [8]. More recently we found that the resonating structure attached to a SAW resonator and SAW resonator together in unison form system of coupled resonator. We reported experimental results on mass loading effect of SU-8 micro pillars attached to SAW resonator [9]. However, piezoelectric transducer based coupled resonating systems are well known [10]. In the present work we have extended our study in investigating the significance of cross-section of nano-pillar sensing medium in SAW sensors. It is more convenient and quicker to study inside physics of micro and nano systems through finite element method (FEM) simulation than experimental analysis. Further one can optimize the experimental procedure from simulation study. Thus we are motivated to perform FEM study of SAW resonator attached with high aspect ratio nano-pillars and study their effect on Δf_0 for different case of pillar cross-section. Eigen frequency analysis of SAW resonator for with and without pillars is performed and Δf_0 is calculated for different cross-sections and heights of the nano-pillar. We observed that the sensitivity of the SAW sensor increases for the pillar with smaller cross-section. The simulation methodology, mass loading results obtained for different cross-section of nano-pillars, discussion on the sensitivity of the nano-pillars for different cross-section of nano-pillars are presented in the following sections.

2 Simulation Setup

A SAW resonator made of single interdigital transducer (IDT) with large number of electrodes is considered for the FEM simulation. As the electrodes in IDT are periodically arranged, the entire SAW resonator is modeled as a segment of one period length (2 μm) of the SAW resonator. As shown in Fig. 1, the segment consists of single electrode of length 1 μm, and appropriate periodic boundary conditions are provided to the walls of the segment to simulate a complete SAW resonator. The dimensions of the segment are chosen to simulate an 871 MHz SAW resonator. Lithium Niobate crystal constants are assumed to the substrate. As the surface waves amplitude are negligible as it travels into the depth of the substrate, the height of the substrate is assumed to be 40 μm (10 times of the SAW wavelength) and the bottom surface of the substrate is fixed in its position. Periodic boundary conditions are provided as follows, the displacement and voltage of the right boundary of the segment is set negative to the displacement and voltage at the left side boundary [9]. Eigen frequency analysis of the model is performed and resonance frequency ($f_0|_{h=0}$) of the SAW resonator without pillar is recorded. Further to study the mass loading effect of nano-pillars of different cross-section, nano-pillars made of arbitrary material is placed in the active area of the segment substrate (see Fig. 1) and resonance frequency (f_0)of the SAW resonator with pillars is recorded.

The mass loading effect caused by the nano-pillars is studied in six different cases of pillar size (cross-section), active area dimension and number of pillars

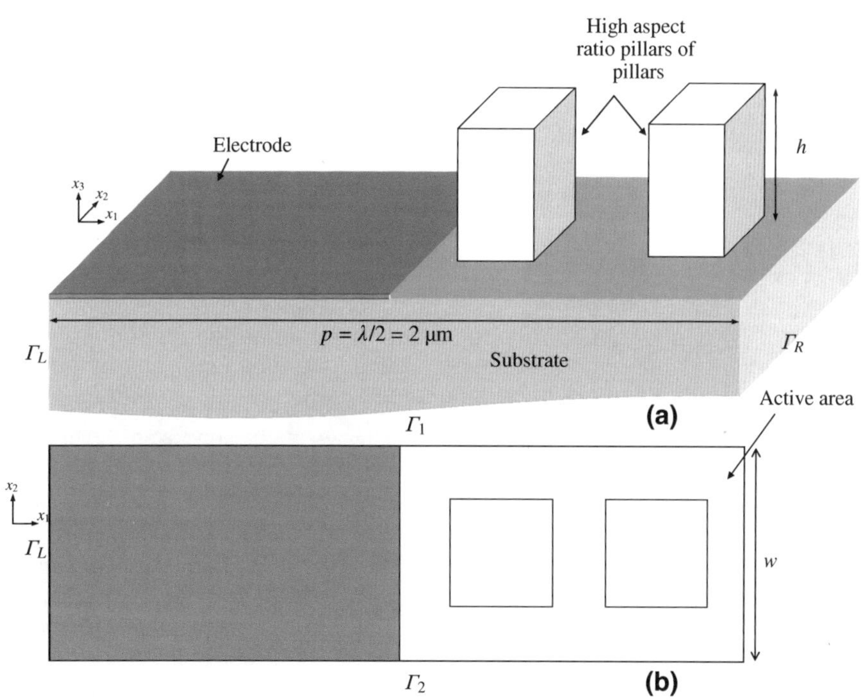

Fig. 1 **a** SAW resonator geometry considered for the simulation of hydrogen sensor using palladium pillars with cross-section of 250 × 250 nm, **b** *Top view* showing the boundaries

Table 1 Variation in resonance frequency for change in *w*

w (nm)	Resonance frequency (MHz)
500	850.3962
250	850.4236
125	850.4342
62	850.4347
31	850.4351
8	850.4348

attached to the substrate as tabulated in Table 1. In order to study the significance of cross-section alone, the mass of the pillars per unit area is kept equal for every case of study. For example in case 1, two pillars of cross-section 250 × 250 is placed in middle of the active area, hence to simulate the same mass per area of the pillars with cross-section 8 × 8 nm the active area is reduced to 16 × 1,000 nm and 64 pillars of cross-section 8 × 8 nm is placed over the active area. It should be noted that if the active area of the substrate has to be maintained constant, then for the case of pillars with cross-section 8 × 8 nm, 4,096 such pillars are to be placed within the active area of 500 × 1,000 nm. This situation will increases the number density of the mesh elements and subsequently the number of degrees of freedom to solve, which will be beyond the capacity of the computer to solve. Hence it was decided to reduce the size of *w* proportionally with respect to the cross-section of the pillar in each case of study and maintain the mass per unit area of the pillars same for all cases.

3 Results and Discussions

3.1 Observation on the Resonance Frequency

Initially the eigen frequency analysis of the SAW resonator without pillars is performed for different aperture widths (*w*) and the resonance frequency ($f_0|_{h=0}$) of the SAW resonator is recorded and it is shown in Table 1. The $f_0|_{h=0}$ for different *w* is found to be within the range of 850 MHz In the second stage, the eigen frequency analysis is performed for the SAW resonator with high aspect ratio pillars attached to the middle of the active area as shown in Fig. 1. Each pillar is assumed to have Young's modulus of 128 GPa and density of 12,020 kg/m^3. The height of the nano-pillars is varied and the corresponding resonance frequency f_0 of the SAW resonator with the pillar structure attached is recorded for each case. Figure 2a–f shows the plot of Δf_o versus height, *h* of pillars for the case 1, 2, 3, 4, 5, and 6 respectively. As expected for every case, increase in mass loading (increase in *h*) the resonance frequency shift decreases. However, at heights $h = 320, 264, 180, 140, 99$, and 72 nm of the pillar for the case 1, 2, 3, 4, 5, and 6 respectively, Δf_0 tends to zero. The resonance frequencies of the pillar alone are

Table 2 Different case of pillar dimensions considered for sensitivity study

Case	Cross section of the pillar (nm^2)	Active area (nm^2)	Number of pillar ($x_2 \times x_3$)	Top view of active area
				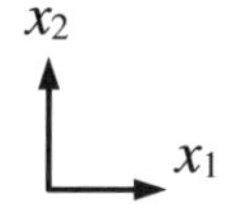 Pillars
1	250 × 250	500 × 1,000	2	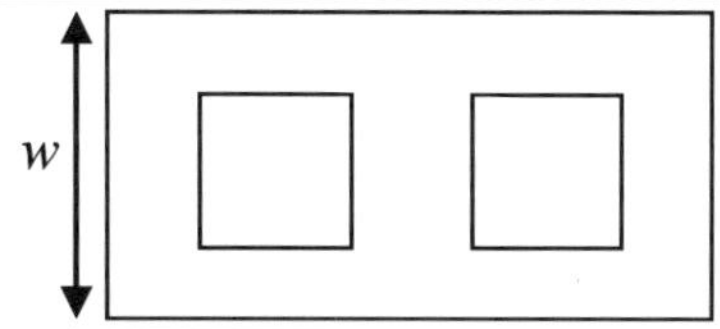
2	125 × 125	250 × 1,000	4	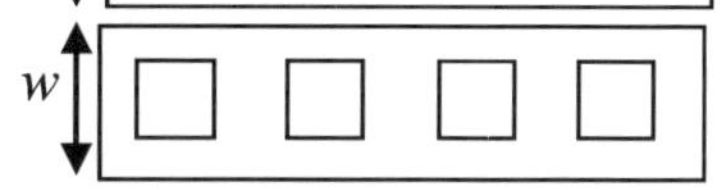
3	62 × 62	125 × 1,000	8	
4	31 × 31	62 × 1,000	16	
5	16 × 16	31 × 1,000	32	
6	8 × 8	16 × 1,000	64	

calculated. Figure 3 shows the plot of resonance frequency versus height of the pillar. It can be seen from Figs. 2 and 3, when the heights of pillar are such that they resonate with SAW resonator, the Δf_0 tends to zero.

3.2 High Aspect Ratio Nano-Pillars as Sensing Medium

SAW sensors widely employ mass loading effect of sensing film coated over its substrate. One of the early reports on such sensors is a SAW hydrogen sensor reported by Amico et al. [11] in 1982. The SAW hydrogen sensor consists of a sensing medium made of thin palladium film coated over its acoustic path. It's reported that in the presence of hydrogen the palladium film material density decreases and Young's modulus increases. Further the volume of the film increases due to the addition of hydrogen molecules to the film [11]. These changes in the sensing film properties introduce mass loading to the SAW device and alters the

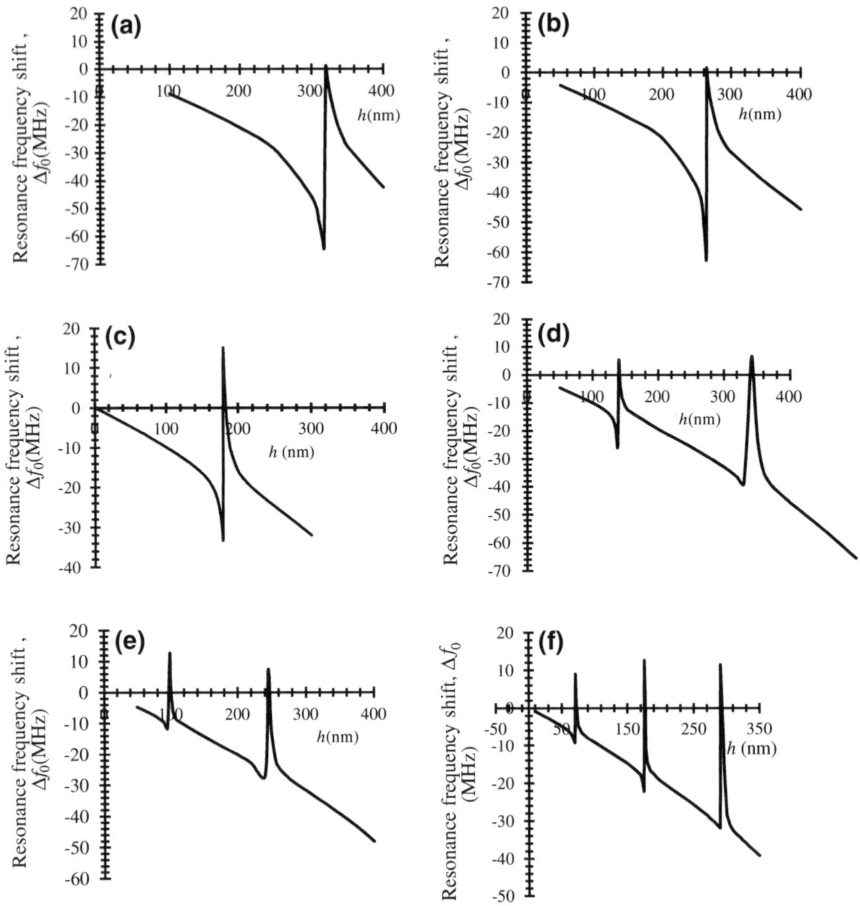

Fig. 2 Plot of resonance frequency shift versus height of the pillars for **a** case 1, **b** case 2, **c** case 3, **d** case 4, **e** case 5, and **d** case 6 of Table 2

SAW properties and corresponding hydrogen concentration is measured. Thus if the sensing medium is made of nano-pillars such that they resonate at $(f_0|_{h=0})$ will offer negligible mass loading and further small change in properties of material similar to the SAW hydrogen sensor [11] will offer significant change in mass loading to the SAW device. Accordingly, high sensitive SAW sensor can be developed. Thus it is decided to calculate the mass loading effect of the pillars for small change in their height and estimate the mass loading sensitivity of the pillars from the slope of Δf_0 curve (Fig. 2) obtained for every cross-section of pillars. The sensitivity in this case can be defined as change in Δf_0 for 1 nm increases in height of the pillars at the maximum Δf_o of the curve in Fig. 3. In other words, Δf_0 obtained for 1 nm height change from their respective resonant height of the pillar. Figure 4 shows the maximum sensitivity that can be obtained by using the

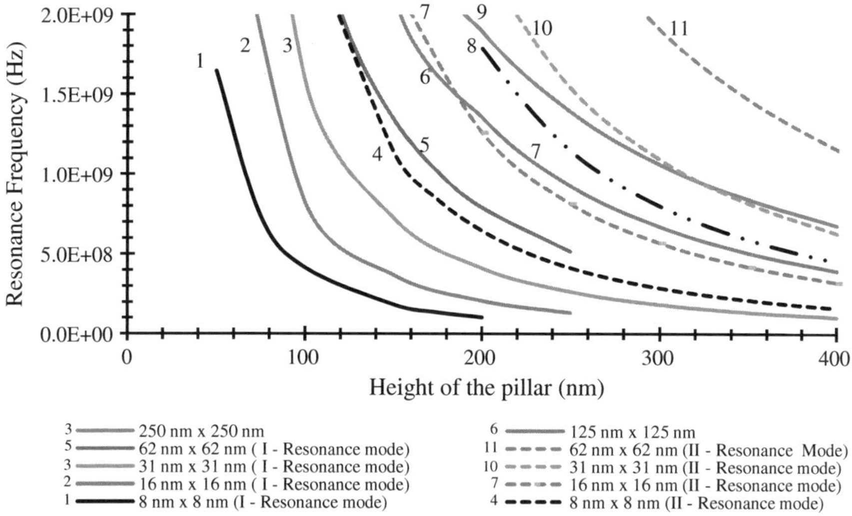

Fig. 3 Resonance frequencies of the pillar of cross-section consider for the study

Fig. 4 Bar graph showing sensitivity obtained for different case mentioned in Table 2

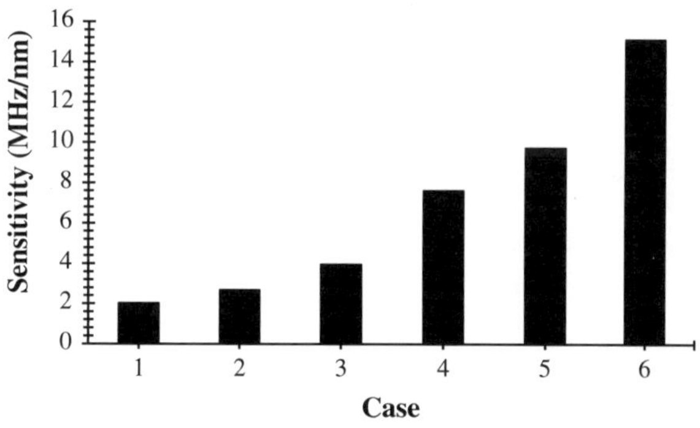

resonant nano-pillars used in the present study as sensing medium. It can be seen from Fig. 4 that for smaller the cross-section of the pillar the mass loading sensitivity is significantly high. It is also evident that if the pillars are in nano scale cross-section, the resonant heights can be obtained at smaller heights of the pillars.

4 Conclusion

FEM simulation of SAW resonator with high aspect ratio nano-pillars attached to normal to their substrate is performed. The significance of pillar size in deciding the mass loading sensitivity is studied in six different cases of nano-pillar cross-sections. Mass loading sensitivity is estimated from resonance frequency shift curves obtained for every case of study. It was observed that smaller cross-section pillar show higher sensitivity to small changes in the pillar dimension. Thus choosing resonant structures with nano scale dimensions will certainly aid in designing high sensitive SAW sensors.

Acknowledgments The authors thank Monash University Sunway Campus internal seed grant E-15-12 and PSCT research strength support.

References

1. J.W. Gardner, V.K. Varadan, O.O. Awadelkarim, *Microsensors MEMS and Smart Devices* (Wiley, Chichester, 2002)
2. M. Penz, A.Z. Sadek, H.D. Zheng, P. Aversa, D.G. McCulloch, K. Kalantar-Zadeh, W. Wlodarski, SAW gas sensors with titania nanotubes layers. Sens. Lett. **9**, 925 (2011)
3. D.T. Phan, G.S. Chung, Surface acoustic wave hydrogen sensors based on ZnO nanoparticles incorporated with a Pt catalyst. Sens. Actuators B **161**, 341 (2012)
4. S. Liu, H. Sun, R. Nagarajan, J. Kumar, Z. Gu, J. Cho, P. Kurup, Dynamic chemical vapor sensing with nanofibrous film based surface acoustic wave sensors. Sens. Actuators A **167**, 8 (2011)
5. A.I. Buvailo, Y. Xing, J. Hines, N. Dollahon, E. Borguet, TiO2/LiCl-based nanostructured thin film for humidity sensor applications. ACS Appl. Mater. Interfaces **3**, 528 (2011)
6. N. Ramakrishnan, T. Vamsi, A. Khan, B. Harshal, B. Nemade, R.P. Palathinkal, Humidity sensor using NIPAAm nanogel as sensing medium in SAW devices. Int. J. Nanosci. **10**, 259 (2011)
7. N. Ramakrishnan, R.P. Palathinkal, H.B. Nemade, Mass loading effects of high aspect ratio structures grown over SAW resonators. Sens. Lett. **8**, 253 (2010)
8. N. Ramakrishnan, H.B. Nemade, R.P. Palathinkal, Finite element simulation of a surface acoustic wave hydrogen sensor with palladium nano-pillars as sensing medium. Sen. Lett. **8**, 824 (2010)
9. N. Ramakrishnan, H.B. Nemade, R.P. Palathinkal, Mass Loading in coupled resonators consisting of SU-8 micropillars fabricated over SAW devices. IEEE Sens. J. **11**, 430 (2011)
10. G. L. Dybwad, A sensitive new method for determination of adhesive bonding between a particle and a substrate. J. Appl. Phys. **58**, 2789 (1985)
11. D. Amico, A. Palma, E. Verona, Surface acoustic wave hydrogen sensor. Sens. Actuator **3**, 31 (1982)

Quantum Breathers in Nonlinear Metamaterials

Bijoy Mandal, Reshmi Basu, Kamal Chowdhury, Arindam Biswas, A. K. Bandyopadhyay, A. K. Bhattacharjee and D. Mandal

Abstract The goal of this paper is to explore whether SRR system, through engineering of their geometry (i.e. change of coupling) and permittivity, shows any sensitivity on the femtosecond response of quantum breathers by quantum calculations hitherto not done in metamaterials. This study is quite realistic to understand quantum localization by nonlinearity that is essential for many small-structured nano-devices. Therefore, this study should be viewed as working towards understanding of many nano structured devices.

Keywords Metamaterials · SRR · Quantum breathers · K-G lattice

B. Mandal · A. K. Bandyopadhyay
Faculty of Engineering and Technology, NSHM Knowledge Campus,
Durgapur 713212, India

R. Basu · K. Chowdhury
Ideal Institute of Engineering, West Bengal University of Technology, Kalyani,
Nadia 741235, India

K. Chowdhury
Department of Materials Science and Engineering, University of Florida, FL,
Gainesville 32611, USA

A. Biswas · A. K. Bhattacharjee (✉) · D. Mandal
Department of Electronics and Communication Engineering, N. I. T,
Durgapur 713209, India
e-mail: asisbanerjee1000@gmail.com

P. K. Giri et al. (eds.), *Advanced Nanomaterials and Nanotechnology*,
Springer Proceedings in Physics 143, DOI: 10.1007/978-3-642-34216-5_53,

1 Introduction

With respect to a voluminous work done on classical breathers [1] on metamaterials which are theoretically and experimentally very important with a lot of interesting applications [2–4], not much work has been done on their quantum perspective. This brings us to the new and evolving field of quantum breathers that have already been studied in nonlinear optical materials with charge defect [5]. Before we go into some details in this field, let us briefly describe metamaterials: Substances with both negative dielectric constants (ε) and magnetic permeability (μ) possess a negative index of refraction and consequently they exhibit a variety of interesting optical properties. These negative indexed materials (NIM) called "metamaterials" (MM) do not occur in nature, and only recently it has been possible to artificially fabricate them as split-ring-resonators (SRRs) for antenna arrays, which are embedded in a nonlinear Kerr-medium so that the entire system can be treated as nonlinear. Based on a theory of Pendry et al. [2], Smith et al. [3] engineered a practical metamaterial with negative refraction property that can even cross the diffraction limit for various applications [4]. Next let us look at discrete breathers (DBs).

Discrete breathers (DBs) also known as intrinsic localized modes (ILM) are nonlinear excitations that are produced by the nonlinearity and discreteness of the periodic lattice [1, 6]. They are formed as a self-consistent coupling between the mode and the system nonlinearity. Thus, DB modifies the local property that provides the environment for the DB to exist. The appropriate length scale drives us to nano-range in MM containing SRR elements. Thus, localization assumes more significance that arises due to interplay of nonlinearity and discreteness.

To characterize classical DBs, the bulk system was the right tool [7], but for very small (nano) systems, the laws of classical mechanics are not valid that brings us to the quantum breathers (QBs) [8, 9]. These are observed in many interesting fields, viz. BEC in photonic lattices, antiferromagnetics, DNA, metamaterials (SRR), Josephson junctions, ferroelectrics, etc. (see Ref. [5] for all the relevant references).

The signature of QBs is manifested in two-phonon bound state. The interplay of nonlinearity and discreteness with respect to charge density in terms of both space and time domains gives rise to a K-G model through variational principle based on our discrete Hamiltonian that embodies both coupling and linear permittivity [1]. On the one hand, the 'geometry' of such engineered metamaterials that are modeled as R-L-C circuit can be varied in terms of slit size and other parameters to change the coupling of the system. This coupling may be a sensitive variable to the formation of quantum breathers and hence on their lifetime in femtoseconds that has important implications for THz applications. On the other hand, the linear permittivity that is present in our Hamiltonian of the system may also be considered to play a role in the femtosecond response of SRR based metamaterials for antenna applications.

2 Theoretical Aspect

Quantum localization in K-G system has been studied by many researchers with four atom lattice, particularly by Proville [10]. Based on the discrete Hamiltonian [11], by using the 'variational principle', the governing equation in relation to the charge density in the SRR assembly can be expressed as nonlinear Klein-Gordon dynamical equation that is valid for metamaterials. This gives rise to dark and bright solitons, dark and bright discrete breathers as well as breather pulses [1]. The latter cannot be obtained by discrete nonlinear Schrodinger equation (NLSE). So, our approach on a Klein-Gordon lattice model is quite richer in the theoretical context as well as in the practical applications. Such characteristics of classical discrete breathers are mentioned here, as they also involve 'localization' due to nonlinearity that is our focus in the present work.

In this context, it is to be noted that our discrete Hamiltonian that was developed for another important nonlinear optical material, such as ferroelectrics, can also be used [6] for the purpose of second quantization by Bosonic operators to reveal the presence of quantum breathers. However, in the present work, for the temporal evolution spectra of quanta, we have adopted a generalized Hamiltonian that is quantized for our purpose. It is worth mentioning that Proville [10] took a 'four-atom' lattice for studying quantum breathers, but our method is capable of taking any number of quanta on an arbitrary number of sites in a non-periodic boundary condition.

Let us consider a simple case, as shown in Fig. 1. This is a typical SRR assembly due to Kerr dielectric filling between the SRR slits with linear permittivity (ε_l). A one-dimensional discrete, periodic array of identical nonlinear.

SRRs are considered here for a device-oriented model for an antenna-array application having a varying inductance (L) and mutual inductance (M) values [1, 11].

The generalized Hamiltonian for the Klein-Gordon system for order parameter (y_n) at nth site can be expressed as:

$$H = \sum_n \frac{p_n^2}{2m} + \frac{A}{2}y_n^2 + \frac{B}{4}y_n^4 + k(y_n - y_{n-1})^2 \tag{1}$$

The first term is momentum at nth site (p_n), the second and third terms are nonlinear potential and the last one includes a coupling term (λ); the third term includes both linear permittivity (ε_l) and focusing nonlinearity ($\alpha = +1$). Here A and B are constants.

The first term is momentum at nth site (p_n), the second and third terms are nonlinear potential and the last one includes a coupling term (λ); the third term

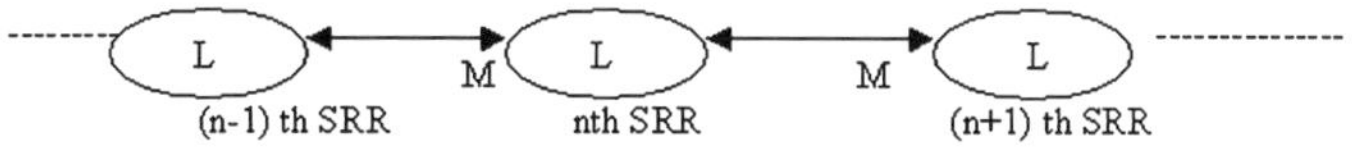

Fig. 1 Three adjacent split-ring resonators

includes both linear permittivity (ε_l) and focusing nonlinearity ($\alpha = +1$). Here A and B are two constants. From Eq. (1), we deduced the equation of motion after rescaling of time with the timescale as 3.048 fs for a coupling value of $\lambda = 0.01$ for SRR based metamaterials.

Now, for second quantization, we need to make the creation and annihilation Bosonic operators that act on the above Hamiltonian. In order to characterize quantum breathers, let us resort to $\langle n_i \rangle(t) = \langle \psi_t | \hat{n}_i | \psi_t \rangle$ for the temporal evolution of quanta at each site of the system. We take i-th eigenstate of the Hamiltonian, and then make it time dependent:

$$|\Psi_i(t)\rangle = \sum_i b_i \exp(-iE_i t/\hbar)|\psi_i\rangle \tag{2}$$

where ψ_i and E_i are the i-th eigenvector and eigenvalue respectively, t is time, the Planck's constant (h) taken as unity and $b_i = \langle \psi_i \mid \Psi(0) \rangle$ for each site i and for a given range of t, where $\Psi(0)$ stands for initial state.

3 Results and Discussion

For the nonlinear Klein-Gordon lattice, in contrast with discrete nonlinear Schrodinger equation, the energy is not completely transferred between the anharmonic oscillators and there is a critical redistribution time of quanta [9] that is proportional to the quantum breathers' lifetime in femtoseconds (fs) which has an important implication in THz devices. Unlike natural materials, metamaterials show a large magnetic response at THz frequency and hence their THz applications make significant advances [12].

In a periodic boundary condition with a Bloch function, the two-phonon bound state has already been observed in ferroelectrics [5] that is a signature of quantum breathers. However, as said earlier, here we are dealing with non-periodic boundary condition for the temporal evolution spectra of quanta by Eq. (2). A typical spectrum is shown in Fig. 2. This is for 12 quanta for a focusing nonlinearity $\alpha = +1$ and linear permittivity $\varepsilon_l = 0.2$ that is less than that taken by Lazarides et al. [11] for a coupling value of 0.01. Here, the initial localization is mainly at the first site and then there is a fast redistribution of quanta between the other two sites until they become equal or almost equal. Here, the critical time for redistribution (t_{re}) is around 1.2 that is proportional to quantum breathers' lifetime. This is about 3.6 fs (278 THz) for this value of coupling. This value seems to be on the very much lower side of that compared to ferroelectric materials, such as lithium niobate, with 12 quanta, i.e. 562 fs (about 2 THz) for a coupling of 0.1, but with a much higher nonlinearity.

The temporal evolution spectra are quite sensitive to both coupling and permittivity values that are used in our computation as input data. For a change of three orders of magnitude of permittivity, the lifetime changes from 16.4 to

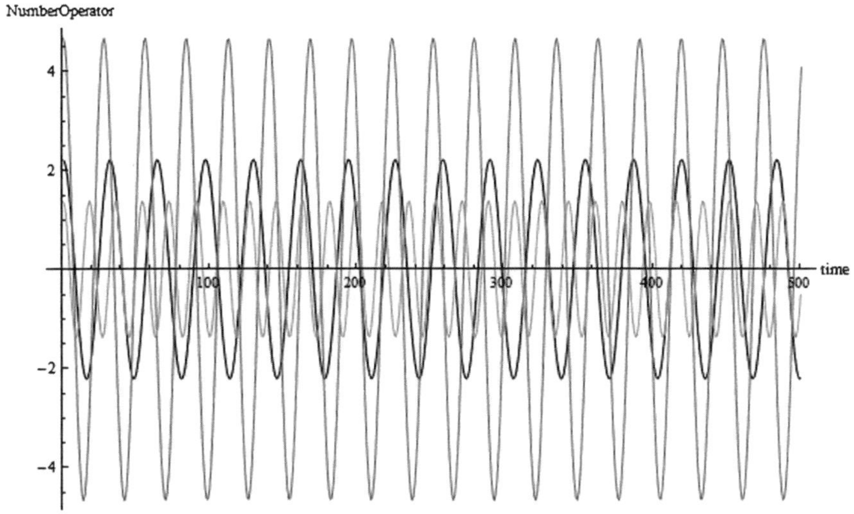

Fig. 2 Spectra for SRR based metamaterials

17.2 THz that is not significant, but it is quite higher for lower number of quanta. However, for a change of coupling within the SRR assembly of more than two orders of magnitude, the lifetime changes from 62.5 to 157.5 THz (higher THz response with lower permittivity).

4 Conclusion

In an extension to our work on the quantum breather (QB), whose signature was evident in two-phonon boundstate (TPBS) in ferroelectrics containing charge defects that was done in a periodic boundary condition, here, we have employed a non-number conserving approach with a non-periodic boundary condition, although there are various other methods to characterize QBs. In a non-periodic boundary condition, the temporal evolution spectra show important variation with permittivity that could have important consequences for future THz and other applications of quantum breathers. The time of redistribution of quanta in femtoseconds shows interesting behaviour with permittivity from design viewpoint in THz applications.

Acknowledgments The authors would like to give special thanks to Professor A. J. Sievers of Department of Physics at Cornell University (USA) for clarifying the role of localization in the discrete breathers.

References

1. P. Giri, K. Choudhary, A. Sengupta, A.K. Bandyopadhyay, A.R. McGurn, Klein-gordon equation approach to nonlinear split-ring resonator based metamaterials: One-dimensional systems. Phys. Rev. B. **84**, 155429 (2011)
2. J.B. Pendry, A.J. Holden, D.J. Roberts, W.J. Stewart, Low frequency plasmons in thin-wire structures. J. Phys. Condens. Matter. **10**, 4785 (1998)
3. D.R. Smith, W.J. Padilla, D.C. Vier, S.C. Nemat-Nasser, S. Schultz, Composite medium with simultaneously negative permeability and permittivity. Phys. Rev. Lett. **84**, 4184 (2000)
4. A.D. Boardman, K. Marinov, Nonradiating and radiating configurations driven by left-handed metamaterials. J. Opt. Soc. Am. B. **23**, 543 (2006)
5. A. Biswas, K. Choudhary, A.K. Bandyopadhyay, A.K. Bhattacharjee, D. Mandal, Quantum pinning transition due to charge defect in ferroelectrics. J. Appl. Phys. **110**, 024104 (2011)
6. A.K. Bandyopadhyay, P.C. Ray, L. Vu-Quoc, A.R. McGurn, Multiple-time-scale analysis of nonlinear modes in ferroelectric LiNbO3. Phys. Rev. B. **81**, 064104 (2010)
7. P. Giri, K. Choudhary, A. Sengupta, A.K. Bandyopadhya, P.C. Ray, Discrete breathers in nonlinear LiNbO3-type ferroelectrics. J. Appl. Phys. **109**, 054105 (2011)
8. S. Flach, A.V. Gorbach, Discrete breathers—advances in theory and applications. Phys. Rep. **467**, 1 (2008)
9. M. Sato, B.E. Hubbard, A.J. Sievers, Colloquium: Nonlinear energy localization and its manipulation in micromechanical oscillator arrays. Rev. Mod. Phys. **78**, 137 (2006)
10. L. Proville, Dynamical structure factor of a nonlinear Klein-Gordon lattice. Phys. Rev. B. **71**, 104306 (2005)
11. M. Eleftheriou, N. Lazarides, G.P. Tsironis, Magnetoinductive breathers in metamaterials. Phys. Rev. E. **77**, 036608 (2008)
12. R. Singh, E. Plum, W. Zhang, N.I. Zheludev, Highly tunable optical activity in planar achiral terahertz metamaterials. Opt. Exp. **18**, 13425 (2010)

Quantum Breathers in Lithium Tantalate in Klein-Gordon Lattice

Arindam Biswas, K. Choudhury, A. K. Bandyopadhyay, A. K. Bhattacharjee and D. Mandal

Abstract We present the mathematical model for TPBS parameters via Fourier Grid Hamiltonian method after second quantization on a K-G lattice is done with Bosonic operators. This study is realistic to understand quantum localization by nonlinearity that is essential for many small-structured nano-devices. An eigen-spectrum is shown that indicates two-phonon QB state separated at upper segment of energy from a single-phonon continuum. Hence, the objective of this paper is to explore whether the switching is easier at a particular point of nonlinearity by quantum calculations hitherto not done on lithium tantalite.

Keywords Ferroelectrics · Quantum breathers · Two phonon bounce state (TPBS) · K-G lattice

1 Introduction

One of the well-studied materials is ferroelectric in the field of solid state physics, which has several important applications as memory switching, in nonlinear optical communications, non-volatile memory devices, and many others [1].

A. Biswas (✉) · A. K. Bhattacharjee · D. Mandal
Department of Electronics and Communication Engineering, N. I. T, Durgapur 713209, India
e-mail: asisbanerjee1000@gmail.com

K. Choudhury
Department of Materials Science and Engineering, University of Florida, Gainesville, FL 32611, USA

K. Choudhury · A. K. Bandyopadhyay
Ideal Institute of Engineering, West Bengal University of Technology, Kalyani 741235, India

P. K. Giri et al. (eds.), *Advanced Nanomaterials and Nanotechnology*,
Springer Proceedings in Physics 143, DOI: 10.1007/978-3-642-34216-5_54,

Lithium tantalate with a low poling field appears to be the most interesting candidate as a key material for a variety of devices: such as optical parametric oscillators, nonlinear frequency converters, electro-optics, second-order nonlinear optical material, holography, etc. [2]. Many of such devices include important nano devices.

Below Curie point, certain perovskites show spontaneous polarizations that are called ferroelectrics which show nonlinear hysteresis in terms of polarization (*P*) and electric field (*E*) vectors. Ferroelectrics occur as collections of domains, where the ferroelectric domains, as with ferromagnetic domains, are created and oriented by a need to minimize the fields as well as the free energy of the crystal. The dynamic properties and domain structure of these materials have been extensively studied [2, 3 and 4] (see references therein). However, recently they have gained renewed interest for potential applications in nano science and the design of nano devices, where the focus is on properties exhibited at small length scales. These have interesting application in various nano devices, such as surface mounted capacitors, electromechanical sensors, actuators, and transducers, infrared sensors, tunable thermostats, nonlinear dielectric materials, etc. [1]. Such a small length scale (nano region) thus brings us to discrete breathers (DBs).

Discrete breathers (DBs) also known as intrinsic localized modes (ILMs) are nonlinear excitations that are produced by the nonlinearity and discreteness of the periodic lattice [1, 5]. There is an importance of domains and domain walls in ferroelectrics that can be described by a soliton solution, i.e. nonlinear localized traveling waves that are robust and propagate without change in shape giving the polarization profile [4, 6]. On the other hand, discrete breathers (DBs) are discrete solutions, periodic in time and localized in space and whose frequencies extend outside the phonon spectrum [4]. This is described by our discrete Hamiltonian [1]. The question is about the appropriate length scale, which drives us to nano-ferroelectrics in the context of discreteness. Thus, localization assumes more significance.

To characterize classical DBs, the bulk system was the right tool [6, 7], but for very small systems, such as nano systems, the laws of classical mechanics are not valid that brings us to the quantum breathers (QBs), which have been described in lithium niobate ferroelectrics in terms of Klein-Gordon lattice [4] that are observed in many interesting fields. For QBs, it is important to consider detailed information on two-phonon bound state (TPBS) that is sensitive to the degree of nonlinearity. The branching out of the QB state from the single-phonon continuum is known in nonlinear systems with charge defects [4]. Hence, the objective of this paper is to explore whether the switching is easier at a particular point of nonlinearity by quantum calculations hitherto not done on lithium tantalate. Here, we present the mathematical model for TPBS parameters via Fourier Grid Hamiltonian method after second quantization on a Klein-Gordon lattice is done with Bosonic operators. This study is realistic to understand quantum localization by nonlinearity that is essential for many small-structured nano-devices. A brief description is given here of our Hamiltonian that gives a good description of the mode dynamics by considering an idealized model of one-dimensional array of N identical rectangular slab (ferroelectric) domains arrayed in the x-direction.

2 Theoretical Development

The nearest neighbor domains, i.e. the polarization in the ith domain (P_i) with that in the (i-1)th domain (P_{i-1}) were taken to interact by a harmonic potential with a spring constant k so that the resulting Hamiltonian for the polarization is given by:

$$H = \sum_{i=1}^{N} \left(\frac{p_i^2}{2m_d} + \frac{k}{4}(P_i - P_{i-1})^2 - \frac{\alpha_1}{2}P_i^2 + \frac{\alpha_2}{4}P_i^4 - EP_i \right) \tag{1}$$

Equation (1) gives a general treatment of the mode dynamics in the array, particularly for modes, which are strongly localized over a small number of domains in the array. Here, α is the nonlinearity parameter. For extended modes and modes which are localized, and slowly range over a large number of consecutive domains, our discrete Hamiltonian [Eq. (1)] can be split as:

$$\tilde{H} = H_0 + H_1 \tag{2}$$

Then, a general 'basis' can be written for n particles. The numerical analysis was carried out with Fourier grid Hamiltonian method [8] with 1,000 grids and 0.006 spacing to calculate various eigenvalues and eigenvectors. We restrict ourselves to two phonon states, since at the working temperature the number of phonon is smaller. In order to reduce the computer memory requirement, we take the advantage of translational invariance by periodic Bloch wave formulation, as detailed in Ref. [4, 8]. For a two-phonons case, the non-zero hopping coefficients (D_{mn}) are: $D_{01} = D_{10}, D_{12} = D_{21}$. The energy gap between the single phonon continuum and a bound state is given by:

$$E_g = E_2 - E_0 - 2(E_1 - E_0) \tag{3}$$

where E_0, E_1 and E_2 are three eigenvalues at different points of wave vector (k) that are calculated from our computation to generate $E(k)$ vs. k curve, which gives the signature of quantum breathers (QB) in terms of two-phonon bound state. The width of the single-phonon in the eigenspectrum is given by the magnitude of 4σ, where σ is expressed as:

$$\sigma = -\frac{\lambda}{2}D_{01}^2 \tag{4}$$

where λ is an interaction term (i.e. equivalent to k). D_{01} represents the coefficient for zero to single phonon generation. The variation of the single phonon spectrum width (W_{ph}) represents (through $D_{01} = D_{10}$) the creation of a new phonon or annihilation of an existing phonon. Again, the hopping coefficient for a single phonon to become a two-phonon bound state is given by:

$$\mu = -\frac{\lambda}{2} D_{01} D_{12} = -\frac{\lambda}{2} D_{10} D_{21} \tag{5}$$

All the above calculations were done for 51 sites or domains and $\lambda = 12$. More data points could be used in our present simulation, but here we are primarily focused to study nonlinearity/impurity induced critical behavior of discrete breather motion and its quantum origin. To treat the problem analytically we take the help of second-quantization method. By quantizing the Hamiltonian in Eq. (1) in a number conserving quantized form with $N \to \infty$ [9] with $\lambda_1 = \frac{\lambda}{(\lambda-\alpha)}$, $\eta = \frac{2\alpha}{(\lambda-\alpha)}$ and $E = 0$ in Eq. (1) leads to the equation for two-phonon amplitudes as:

$$\hat{H} = \sum_n a_n^+ a_n + \frac{3}{8} \eta a_n^{+2} a_n^2 + \frac{\lambda_1}{2} \{a_n^+ (a_{n+1} + a_{n-1}) + h.c.\} \tag{6}$$

where a_n^+ and a_n are the creation and annihilation operators with h as planck's constant and c is the velocity of light; the terms η and λ_I are already defined above.

3 Results and Discussion

An eigenspectrum is shown Fig. 1 that indicates two-phonon quantum breather state separated at upper segment of energy from a single-phonon continuum. As an input data, the coupling is taken as 12 and a moderate poling field as 13.9 kV/cm for a nonlinearity value of 514. The Landau or nonlinearity parameter (α) is inversely proportional to the impurity content or poling field, but the impurity content of this sample is not known. In the absence of the impurity data, we could operate through nonlinearity parameter, which should also give a correct picture of quantum breathers in terms of various TPBS parameters, as nonlinearity and discreteness give rise to quantum localization or rather to the formation of quantum breathers.

As the nonlinearity decreases in the system, i.e. the poling field increases the width of the single-phonon continuum increases [Eq. (4)]. This has also an effect on the energy gap between quantum breather state (upper branch in Fig. 1) and the single phonon continuum [Eq. (3)] as well as on hopping coefficient [Eq. (5)]. It may be considered quite significant for lithium tantalate, which is gaining popularity as an important candidate as device materials that are also supported by its lower value of poling field for switching. This might indicate that whatever be the impurity content in lithium tantalate, there does not seem to be an appreciable effect of pinning in the system. It should be pointed out here that 'pinning' is a

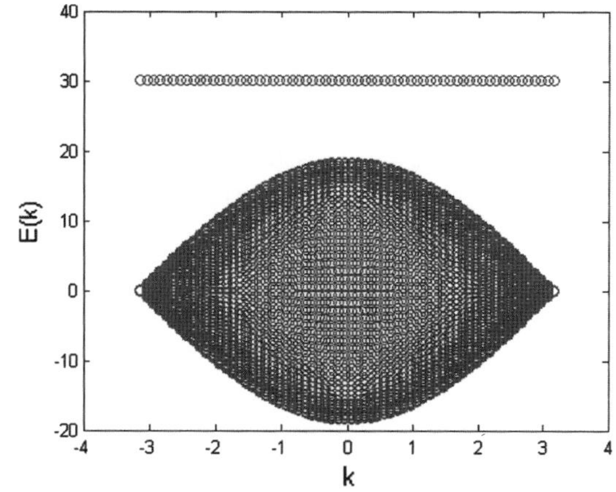

Fig. 1 Eigenspectrum for lithium tantalate

macroscopic phenomena involving the charge defects in the system [4], whereas TPBS parameters are microscopic in nature that are obtained after detailed quantum calculations by Fourier Grid Hamiltonian method.

4 Conclusion

Various TPBS parameters, such as energy gap, single-phonon width and hopping coefficient show some type of functional dependence on nonlinearity that is congenial for the formation of quantum breathers. Although the Eigen spectra of lithium tantalate show interesting behavior against nonlinearity in the lithium tantalate nano-ferroelectric, it is necessary to use another powerful technique, such as density function theory (DFT), as done by Phillpot and coworkers [10]. The quantum calculations shown in this paper can be considered important for various applications of lithium tantalate, such as photonics and switches.

Acknowledgement The authors would like to thank Dr. Sergej Flach of the Max Planck Institute (MPI) at Dresden, Germany, for his valuable guidance during the computational work. K.C. would also like to thank MPI for warm hospitality.

References

1. A.K. Bandyopadhyay, P.C. Ray, L. Vu-Quoc, A.R. McGurn, Multiple-time-scale analysis of nonlinear modes in ferroelectric LiNbO3. Phys. Rev. B **81**, 064104 (2010)
2. V. Gopalan, Defect–domain wall interactions in trigonal ferroelectrics. Ann Rev. Mater. Res. **37**, 449 (2007)
3. A.K. Bandyopadhyay, P.C. Ray, V. Gopalan, Dynamical systems analysis for polarization in ferroelectrics. J. Appl. Phys. **100**, 114106 (2006)

4. A. Biswas, K. Choudhary, A.K. Bandyopadhyay, A.K. Bhattacharjee, D. Mandal, Quantum pinning transition due to charge defect in ferroelectrics. J. Appl. Phys. **110**, 024104 (2011)
5. M. Sato, B.E. Hubbard, A.J. Sievers, *Colloquium*: Nonlinear energy localization and its manipulation in micromechanical oscillator arrays. Rev. Mod. Phys. **78**, 137 (2006)
6. A. Biswas, K. Chowdhury, A.K. Bandyopadhyay, A.K. Bhattacharjee, D. Mandal, in *Dissipative Intrinsic Localized Modes in Nano-Ferroelectrics, Techconnect World 2011,* at the Hynes Convention Center in Boston, vol. 2. Massachusetts, pp. 635–638, 13–16 June 2011
7. P. Giri, K. Choudhary, A. Sengupta, A.K. Bandyopadhyay, P.C. Ray, Discrete breathers in nonlinear LiNbO3-type ferroelectrics. J. Appl. Phys. **109**, 054105 (2011)
8. J.P. Nguenang, R.A. Pinto, S. Flach, Quantum *q*-breathers in a finite Bose-Hubbard chain: The case of two interacting bosons. Phys. Rev. B **75**, 214303 (2007)
9. S. Flach, A.V. Gorbach, Discrete breathers—advances in theory and applications. Phys. Rep. **467**, 1 (2008)
10. H. Xu, D. Lee, S.B. Sinnott, V. Gopalan, V. Dierolf, S.R. Phillpot, Structure and energetics of Er defects in LiNbO3 from first-principles and thermodynamic calculations. Phys. Rev. B **80**, 144104 (2009)

Effect on Shell Thickness on Intersubband Transition Energies in GaAs/$Al_xGa_{1-x}As$ Inverted Core–Shell Nanodot

Arpan Deyasi, S. Bhattacharyya and N. R. Das

Abstract Energy eigenvalues of an inverted core–shell spherical quantum dot is numerically computed using finite difference technique, and corresponding intersubband transition energies are evaluated for the lowest three eigenstates. GaAs/$Al_xGa_{1-x}As$ material composition is considered for simulation purpose. Al mole fraction of the heterostrucutre is varied within type-I limit to see the effect on the eigenstates and transition energies. Dimensional effect is also studied by varying thickness of constant dot size as well as by varying inner radius of constant dot thickness. Fine optical tuning may be obtained by suitably applying this effect on quantum dot laser. The analysis is carried out by taking into consideration conduction band discontinuity and effective mass mismatch at core–shell junction following BenDaniel Duke boundary condition.

Keywords Finite difference technique · Energy eigenvalue · Inverted core–shell nanodot · Intersubband transition energies

A. Deyasi (✉)
RCC Institute of Information Technology, Canal South Road, Beliaghata, Kolkata 700015, India
e-mail: arpan.deyasi@gmail.com

S. Bhattacharyya
Asansol Engineering. College, Asansol, Burdwan, West Bengal 713305, India
e-mail: swapanbhattacharyya@ieee.org

N. R. Das
Institute of Radio Physics and Electronics, University of Calcutta, Kolkata 700009, India
e-mail: nrd@ieee.org

P. K. Giri et al. (eds.), *Advanced Nanomaterials and Nanotechnology*,
Springer Proceedings in Physics 143, DOI: 10.1007/978-3-642-34216-5_55,

1 Introduction

With emergence of nanostructure devices, it has been already established by intense research activities that the existing problems of present VLSI-based electronics can be solved by quantum-confined semiconductor structures where miniaturization is possible beyond the existing saturation point. Advancement in the technology of fabrication by MOCVD or MBE techniques for growth of complex low-dimensional heterostructures [1, 2] has opened new dimension in experimental nanoscience; simultaneously providing theoretical researchers a great opportunity to propose new nanostructures for novel electronic and photonic applications. Confinement of electrons along reduced dimensions that are comparable to the electron wavelength leads to the formation of quantum wells, wires and dots; which are already been used in designing transistors [3, 4], optical transmitters [5], receivers [6] and modulators [7]. Several geometrical variations of quantum dots have already been proposed, and it is recently established that spherical quantum dots with multiple layers can be used as efficient optical emitters [8], tailoring of material parameters as well as core dimension and shell thickness lead towards better and accurate optical tuning, desired for modern technological applications, precisely for spintronic nanodevices [9].

Core–shell quantum dot is one with 'core material' at the centre of the dot, where dimension of dot is comparable to or less than the de-Bröglie wavelength. In inverted spherical core–shell nanodot, core layer is made by higher bandgap material, whereas shell material is made by lower bandgap material. This particular type of structure has attracted physicists in recent past due to its high photoluminescence quantum efficiency and ease of optical tuning in less than meV range.

Understanding of physical properties of these devices is based on knowledge of quantum-transport processes. The transport property for semiconductor nanostructure can be predicted by appropriate mathematical model using reliable numerical methods. Different numerical approaches are already suggested by researchers [10, 11]; as analytical solutions are difficult, and if at all possible. Accurate solution thus requires the incorporation of material parameters and their relative interdependency in adopted numerical technique. Finite Difference method (FD-Q) is a technique, which can provide results with better accuracy compared to other well-known techniques [12–14], and thus is considered for computation of eigenstates of finite barrier core–shell quantum dot structure, as shown in Fig. 1.

Photoluminescence characterization in visible spectrum region was first carried out on spherical core–shell structure [15] and first excited state was modeled for different shell thickness [16]. Hole wave functions and its impact on band alignment and quantum confinement for both type-I and type-II heterostructures were also studied, and optical transition matrix were calculated for possible optoelectronic applications. Energy levels and wave functions of PbSe/PbS quantum dots were calculated using envelope function approximations [17] and location property of LUMO and HOMO was analyzed. The impact of shell structure on exciton

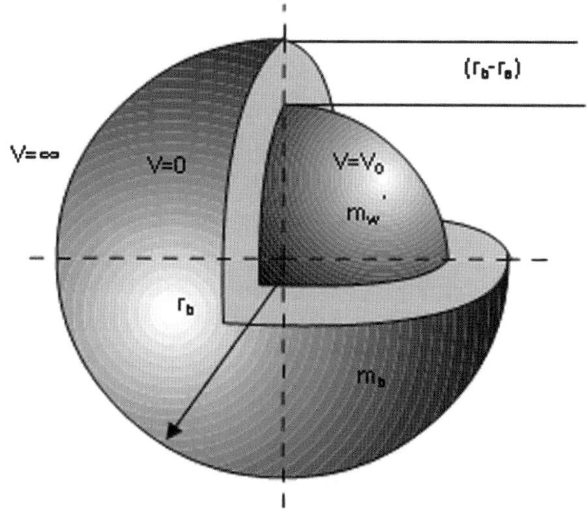

Fig. 1 Inverted Core–Shell Nanodot

and biexciton binding energies considering ZnSe/ZnS material composition was studied [18] using WKB approximation.

Recently, chemical bonding properties and optical gap shifts of core–shell quantum dots for different material compositions [19] were computed, whereas finite difference approach was incorporated for computation of electron states [20] of a CdTe/CdS core–shell quantum dot. Experimental work on inverted core–shell structure [21] and radiative recombination lifetime was estimated by varying shell thickness. It is now established that inverted core–shell exhibits a better lasing performance than other type of quantum dots [22].

In this paper, energy eigenvalues of inverted core–shell spherical nanodot are calculated using finite difference technique, and corresponding intersubband transition energies are computed for the first three lowest states considering GaAs/ $Al_xGa_{1-x}As$ material composition, where core layer is made by higher bandgap $Al_xGa_{1-x}As$ material, and shell is made by lower bandgap GaAs material. Analysis is carried out considering conduction band discontinuity and effective mass mismatch at junctions where both are functions of material parameters where later is considered through BenDaniel Duke boundary conditions [23]. Dimensional variation and material composition disparity are studied together to see the effect on eigenstates and transition energies. This emphasizes the importance of this device as optical tuner.

2 Modeling Procedure

Schrödinger's equation for the spherical quantum dot having circular cross-section can be written in Cartesian co-ordinate system as

$$-\frac{\hbar^2}{2m^*}\left[\frac{\partial^2}{\partial x^2}+\frac{\partial^2}{\partial y^2}+\frac{\partial^2}{\partial z^2}\right]\psi(x,y,z)+V(x,y,z)\psi(x,y,z)=E_{x,y,z}\psi(x,y,z) \quad (1)$$

Transforming into cylindrical coordinate system and due to independence of wave function on angular ordinate owing to circular symmetry, Eq. (1) may be re-written as

$$-\frac{\hbar^2}{2m^*}\left[\frac{2}{r}\frac{\partial}{\partial r}+\frac{\partial^2}{\partial r^2}\right]\psi(r)+V(r)\psi(r)=E_r\psi(r) \quad (2)$$

where E_r is the eigenvalue associated with confined cross-sectional motion, m^* is effective mass of electron which is different for different materials. It is assumed that BenDaniel Duke boundary condition is satisfied at all junctions.

Introducing finite difference technique, Eq. (1) may be modified as

$$\begin{aligned}\psi_r(j+1)\left[-\frac{\hbar^2}{2m^*r(\delta r)}-\frac{\hbar^2}{2m^*(\delta r)^2}\right]+\psi_r(j)\left[\frac{\hbar^2}{m^*(\delta r)^2}+V(r)\right]\\+\psi_r(j-1)\left[\frac{\hbar^2}{2m^*r(\delta r)}-\frac{\hbar^2}{2m^*(\delta r)^2}\right]=E_r\psi_r(j)\end{aligned} \quad (3)$$

Solution of Eq. (3) is used to determine the eigenenergy of core–shell structure, where core and shell regions are defined as

$$\begin{aligned}0<r_a\leq a\\ a<(r_b-r_a)\leq b\end{aligned} \quad (4)$$

a and b are radii of core and shell structures respectively. For analysis purpose, effective mass dependence on material parameter and conduction band discontinuity are also taken into account to obtain near accurate result.

3 Results and Discussion

First three eigenstates are numerical calculated to determine GaAs/$Al_xGa_{1-x}As$ spherical quantum dot using finite difference technique. Here BenDaniel Duke boundary condition has been incorporated for considering effective mass mismatch at the junctions along with conduction band discontinuity.

The above analysis is carried out by differing dimension and thickness of inverted nanodot. Figure 2 represents the energy profile of the three lowest eigenstates considering GaAs/$Al_{0.3}Ga_{0.7}As$ dot for 60 nm outer radius.

It is observed that with increase of thickness, energy eigenvalue decreases, and the effect is more pronounced for lower dimensions. This is quite expected, as energy becomes higher with lowering of the dimension of the structure. For higher eigenstates, the shift of energy values becomes more prominent with change of outer radius.

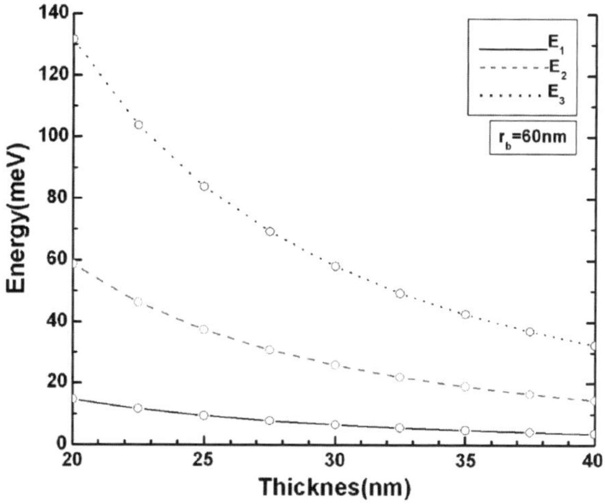

Fig. 2 Profile of thickness vs. energy for *three lowest* eigenstates for *outer radius* ($r_b = 60$ nm) considering $GaAs/Al_{0.3}Ga_{0.7}As$ composition

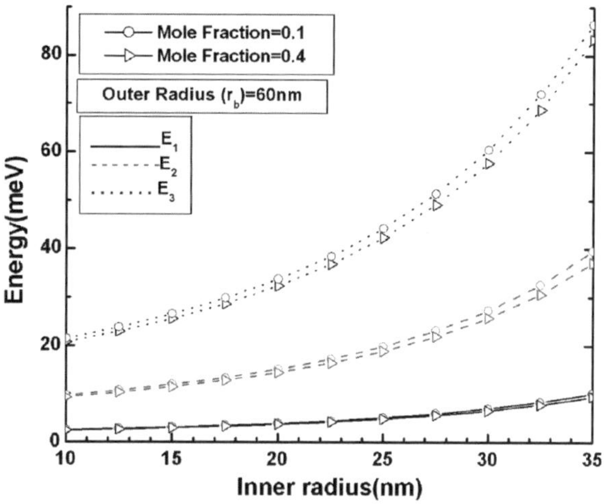

Fig. 3 Profile of *inner radius* vs. energy for *three lowest* eigenstates with constant *outer radius* ($r_b = 60$ nm) for two different material compositions

For constant outer radius, as inner radius is increased, i.e., thickness is decreased, eigenenergy increases. This is plotted in Fig. 3. With change of material composition, i.e., by making higher barrier potential of the core layer, it is observed that eigenvalue decreases. Here barrier height increases by increasing percentage of Al composition, as it increases conduction band discontinuity. The shift is vivid for higher states, and it is negligibly small when ground state is considered having smaller inner radius.

For different material composition of heterostrucutre quantum dot, comparative study reveals the fact that with increase in barrier height, i.e., by increasing mole fraction of Al in core region, eigenenergy decreases slowly for specified dimension. This is due to the fact that higher Al percentage leads to increase in barrier height of the core layer, as well as disparity of effective mass. With increase in

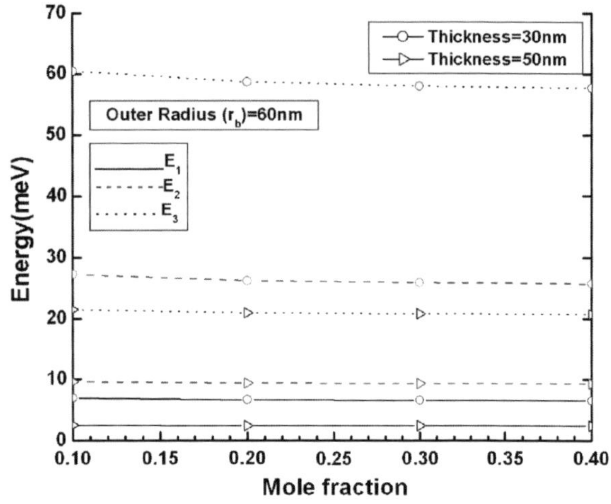

Fig. 4 Energy profile with Al mole fraction for *three lowest* eigenstates with constant *outer radius* ($r_b = 60$ nm) and two different shell thicknesses

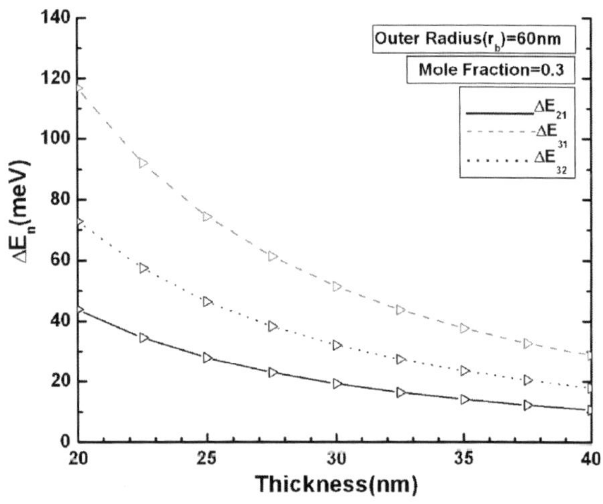

Fig. 5 Profile of *lowest three* intersubbnad transition energies with thickness *outer radius* ($r_b = 60$ nm) considering GaAs/ $Al_{0.3}Ga_{0.7}As$ composition

thickness of the structure, the effect of barrier height becomes less pronounced. This is shown in Fig. 4.

Intersubband energies are calculated for the first three eigenstates, which provides a theoretical understanding about optical tuning so that it can be used for quantum dot laser. Figure 5 shows the variation of intersubband energies with thickness, keeping outer radius and specified material composition constant. It is evident from Fig. 2 that for two different outer radii, difference of energy decreases with increase in thickness of the shell's outer dimension.

By varying thickness of the shell, it can be seen that transition energies increase rapidly with ratio of core radius to shell thickness for constant outer core dimension considering specified material composition. This is shown in Fig. 6.

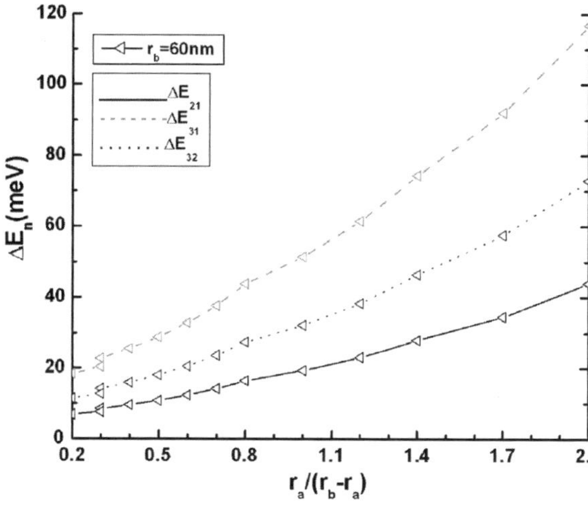

Fig. 6 Profile of *lowest three* intersubbnad transition energies with ratio of *inner radius* and thickness for *outer radius* ($r_b = 60$ nm) considering GaAs/ $Al_{0.3}Ga_{0.7}As$ composition

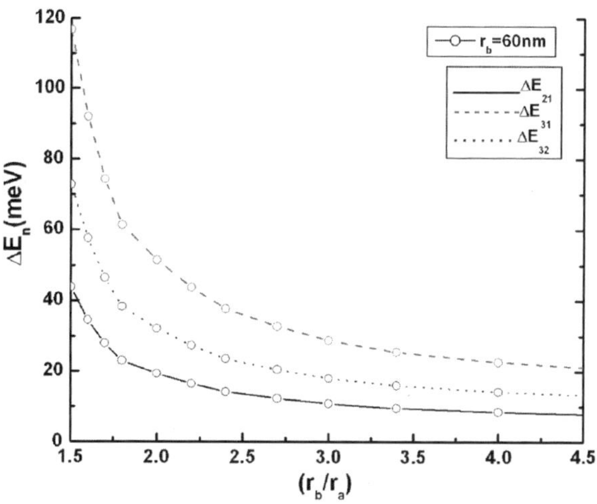

Fig. 7 Profile of *lowest three* intersubbnad transition energies with ratio of radius considering GaAs/ $Al_{0.3}Ga_{0.7}As$ composition

Keeping outer radius constant and by varying the inverted core shell dot's outer to inner radius ratio, it can be seen that transition energies fall rapidly when inner dimension is comparable to the dimension of the entire structure, but it decreases slowly when inner dimension is small. This is plotted in Fig. 7 considering GaAs/ $Al_{0.3}Ga_{0.7}As$ composition.

Increment of Al mole fraction of core layer material increases the barrier height of the core layer, which in turn, decreases the energy value. This also effectively reduces the transition energies, which is plotted in Fig. 8 for two different shell thicknesses and constant outer radius, i.e., dot dimension. Thus, optical tuning may be achieved by varying material composition also together with dimensional variation.

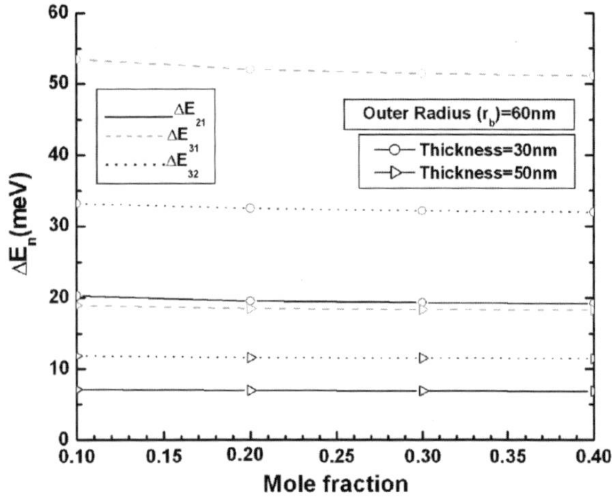

Fig. 8 Profile of *lowest three* intersubbnad transition energies with Al mole fraction with constant *outer radius* ($r_b = 60$ nm) and two different shell thicknesses

4 Conclusion

Eigenenergies along with intersubband transition energies of inverted core–shell quantum dot having spherical geometry are computed for different core dimension and also for different shell thickness, along with suitable variation of material composition. Increasing the shell thickness of constant dot dimension lowers the electron eigenenergies, which signifies that dot's transition energies are reduced as the structure itself is enlarged. This results a blueshift in absorption and emission spectra, which can accurately be controlled at the time of manufacturing from the knowledge of material parameters. Reverse trend is observed when dimension of core layer is increased for constant dimensional configuration, which causes a redshift in absorption and emission spectra. Moreover, role of material composition is also quite important. Changing material composition of core layer modifies the energies and hence intersubband transition energies. Increasing the Al composition enhances the barrier height, which effectively reduces the eigenstates of the nanostructure while dimensional specifications are kept constant. Thus, optical tuning, which is the intersubband transition energies, can be controlled by choosing suitable shell thickness for constant core layer diameter and for particular core layer composition of inverted core–shell quantum dot. This speaks in favor of the candidature of the proposed structure as quantum dot laser.

Acknowledgments The authors wish to acknowledge the help of the colleagues in their research group in the department of Radio Physics and Electronics, University of Calcutta. The authors also acknowledge the constant encouragement of Prof. A. N. Chakravarti during the period of this research work.

References

1. M. Kitamura, M. Nishioka, J. Oshinowo, Y. Arakawa, In situ fabrication of self- aligned InGaAs quantum dots on GaAs multiatomic steps by metalorganic chemical vapor deposition. Appl. Phys. Lett. **66**, 3663 (1995)
2. J. Tersoff, C. Teichert, M.G. Lagally, Self-organization in growth of quantum dot superlattices. Phys. Rev. Lett. **76**, 1675 (1996)
3. V. Vijayakumar, R. Seshasayanan, Optical response of quantum dot transistor with front side illumination. Recent Adv. Space Technol. Serv. Clim. Change **290**, (2010). doi:10.1109/RSCSCC.2010.5712862
4. D. Urban, M. Braun, J. König, Theory of a magnetically controlled quantum-dot spin transistor. Phys. Rev. B **76**(12), 125306 (2007)
5. R. Debusmann, T.W. Schlereth, S. Gerhard, W. Kaiser, S. Hofling, A. Forchel, Gain studies on quantum-dot lasers with temperature-stable emission wavelength. IEEE J. Quantum Electron. **44**(2), 175 (2008)
6. D. Cui, X. Jian, X. Sheng-Yong, G. Paradee, B.A. Lewis, M.D. Gerhold, Infrared photodiode based on colloidal PbSe nanocrystal quantum dots. IEEE Trans. Nanotechnol. **5**(4), 362 (2006)
7. A. Majumdar, N. Manquest, A. Faraon, J. Vuckovic, Theory of electro-optic modulation via a quantum dot coupled to a nano-resonator. Opt. Express **18**(5), 3974 (2010)
8. K.K. Song, S. Lee, Highly luminescent (ZnSe) ZnS core-shell quantum dots for blue to UV emission: synthesis and characterization. Curr. Appl. Phys. **1**(2–3), 169 (2001)
9. Y. Yang, O. Chen, A. Angerhofer, Y.C. Cao, Radial-position-controlled doping in CdS/ZnS core/shell nanocrystals. J. Am. Chem. Soc. **128**(38), 12428 (2006)
10. Z. Heng, S. Jun-Jie, Ground state of excitons in quantum-dot quantum-well nanoparticles: stochastic variational method. Chin. Phys. **13**, 2136 (2004)
11. C. Weng, I. Chen, Y. Tsai, Electron-acoustic-phonon interaction in core/shell nanocrystals and in quantum-dot quantum wells. Phys. Rev. B **76**, 195313 (2007)
12. M. Tsetseri, G.P. Triberis, A study of the ground state of quantum wires using the finite difference method. Superlattices Microstruct. **32**, 79 (2002)
13. D.M. Sullivan, D.S. Critin, Determining quantum eigenfunctions in three-dimensional nanoscale structures. J. Appl. Phys. **97**, 104305 (2005)
14. D.M. Sullivan, Determining a complete three-dimensional set of eigenfunctions for nanoscale structure analysis. J. Appl. Phys. **98**, 084311 (2005)
15. B.O. Dabbousi, J. Rodriguez-Viejo, F.V. Mikulec, J.R. Heine, H. Mattoussi, R. Ober, K.F. Jensen, M.G. Bawendi, (CdSe)ZnS core-shell quantum dots: synthesis and characterization of a size series of highly luminescent nanocrystallites. J. Phys. Chem. B **101**, 9463 (1997)
16. J. Li, L.W. Wang, First principle study of core/shell structure quantum dots. Appl. Phys. Lett. **84**(18), 3648 (2004)
17. A.C. Bartnik, F.W. Wise, A. Kigel, E. Lifshitz, Electronic structure of PbSe/PbS core-shell quantum dots. Phys. Rev. B **75**, 245424 (2007)
18. P. Sen, S. Chattopadhyay, J.T. Andrews, P.K. Sen, Impact of shell thickness on exciton and biexciton binding energies of a ZnSe/ZnS core–shell quantum dot. J. Phys. Chem. Solids **71**, 1201 (2010)
19. F. Gao, Effect of quantum confinement and shape on band gap of core/shell quantum dots and nanowires. Appl. Phys. Lett. **98**, 193105 (2011)
20. C.Y. Woon, G. Gopir, A.P. Othman, Finite difference calculation of electron states in CdTe-CdS core-shell quantum dots. Sains Malaysiana **40**(1), 55 (2011)
21. L.P. Balet, S.A. Ivanov, A. Priyatinski, M. Achermann, V.I. Kilmov, Inverted core/shell nanocrystals continuously tunable between type-I and type-II localization regimes. Nano Lett. **4**(8), 1485 (2004)

22. J. Nanda, S.A. Ivanov, H. Htoon, I. Bezel, A. Priyatinski, S. Tretiak, V.I. Kilmov, Absorption cross-sections and Auger recombination lifetimes in inverted core-shell nanocrystals: implications for lasing performance. J. Appl. Phys. **99**, 034309 (2006)
23. D.J. BenDaniel, C.B. Duke, Space-charge effects on electron tunneling. Phys. Rev. **152**, 683 (1966)

Spin Transfer Torque Driven Magnetic QCA Cells

Nilanjana Pradhan and Debashis De

Abstract In this paper we have proposed spin transfer torque current (STT) driven Magnetic QCA cells. The switching energies dissipated gets significantly reduced when the single domain ferromagnet changes their states ("spin up-logic 0" and "spin down-logic 1") in Magnetic QCA. Spin transfer torque generates spin induced current which manipulates the magnetic moment of the electron and it reaches a value such that the overall energy is larger than the energy loss due to damping. This principal is applied in Magnetic QCA for reducing switching energy dissipation.

Keywords Spintronics · Spin transfer torque (STT) · Magnetic QCA

1 Introduction

Spintronics means "spin electronics". The electrons are spin polarized in spintronic device. As a result of this spin polarization ("spin up" or "spin down") spin current is generated. The spin is manipulated to obtain different states. Some of the things which should be explored are spin polarization, spin dynamics and spin polarized transport [1]. There is a difference between spin transport and charge transport. The term spin stands for either the spin of a single electron switch can be

N. Pradhan (✉) · D. De
Department of Computer Science and Engineering, West Bengal University of Technology, BF-142, Sector-1, Saltlake, Kolkata 700064, India
e-mail: nilanjana.pradhan@gmail.com

D. De
e-mail: dr.debashis.de@gmail.com

P. K. Giri et al. (eds.), *Advanced Nanomaterials and Nanotechnology*,
Springer Proceedings in Physics 143, DOI: 10.1007/978-3-642-34216-5_56,

obtained from its magnetic moment $-g\mu_B s$ (μ_B is the Bohr magnetron and g is the electron g factor, in a solid) or an average spin obtained from an assembly of electrons). The spin polarization can be obtained in several ways. Bohr magnetron μ_B is a physical constant and the natural unit for expressing an electron dipole moment.

In addition to their mass and electric charge, electrons have an intrinsic quantity of angular momentum called spin. A magnetic field is associated with the spin which lines up with the spin axis. Spin is a vector quantity. In a magnetic field electronics with "spin up" and "spin down" have different energies. Spintronic devices create spin polarized currents and use the spin to manipulate the magnetization of ferro-magnets. STT rotates the magnetization axis of a nanomagnet by exerting a torque on it [1]. STT driven MQCA cells suffers from low switching energy dissipation by reducing the saturation magnetization of a magnet (M_s).STT (Is) $\propto M_s$.

$$\mu_B = e\hbar/2m_e c$$

e	elementary charge
ħ	reduced planck constant
m_e	electron rest mass
c	speed of light

Elementary charge means electric charge carried by a single proton. Plank constant is the physical constant reflecting the sizes of quanta in quantum mechanics.

Spintronics actively manipulates the spin degrees of freedom. Spin means the spin of a single electron s, which is detected by its magnetic moment. The goal of spintronics is to understand the interaction between the particle spin. Spintronics include investigation of spin transport in electronics materials, as well as understanding spin dynamics and spin relaxation. Spin polarization means creating a non equilibrium spin population. Spin relaxation mechanisms involves in the process of bringing accumulated spin population back in equilibrium. Spin detection is also a part of spintronics which senses the changes in the signals caused by the presence of non equilibrium spin in the system [1]. The common goal in many spintronic devices is to maximize the spin detection sensitivity to the point it detects not the spin itself but changes in the spin states. Electrons with spin which has magnetic moment opposite to the magnetization of a ferromagnetic layer are scattered more than electrons with parallel magnetic moments, resulting in a spin filtering effect. Spin transport devices are based on the effect that the magnetic spin of electrons can be manipulated in optoelectronic devices by magnetic fields or applied voltage [2]. The domain of spin-transport devices is commonly referred to as magneto electronics or Spintronics. The spin transport effects are most widely used today in magnetic metallic devices such as read heads for Hard Disks Drives and in Magnetic Random Access Memories (MRAM) [3].These device structures comprise magnetic thin films, the magnetic orientation of which can be changed with applied magnetic fields. In case of non-volatile MRAM the orientation will be preserved until the bit is switched again [4]. Hence a locally generated field performs the writing of a magnetic bit, by currents in the vicinity of the magnetic structure. The reading principle of magnetic bits is based on the discrimination between two distinct magnetic devices.

The magnetic field or the presence of magnetic materials not necessarily essential for manipulating spins. GMR-based memory devices and spin valves are elementary spintronic applications where the role of spin, however is passive in dictating the size of the resistance depending on the spin direction controlled by local magnetic fields. In spintronics active control of spin dynamics is envisioned to lead to novel quantum-mechanical enabling technologies such as spin transistors, spin filters and modulators, new memory devices [5]. The inherent quantum mechanical nature of spin as dynamical variable and the inherently long relaxation or coherence time associated with spin states are two important characteristics of spintronics.

2 Generation of Spin Transfer Torque

Landau-Lifshitz-Gilbert equation defines the movement of the magnetic moment. There are three terms of the equation. The first term describes the fact that the magnetic moment will rotate around the applied field and is a procession term. The second term denotes the damping torque. The damping torque aligns the magnetic moment to the applied field. The last term is called the spin transfer torque [6]. This torque acts along with the damping torque or against it depending on the direction of the current. The damping and the spin transfer torque if works together the magnetic moment will align with the applied field. The spin transfer torque and damping can work against each other such that the magnetic moment reaches a steady state.

$$\frac{d\vec{M}_{free}}{dt} = -\gamma\vec{M}_{free} \times \vec{H} + \alpha\vec{M}_{free} \times \frac{dM}{dt} - I\gamma ST\vec{M}_{free} \times \left(\vec{M}_{free} \times \vec{M}_{spin}\right) \quad (1)$$

The Hamiltonian of conduction band electron with momentum $\vec{p}$ and effective mass m is

$$H = \frac{p^2}{2m} + s_\alpha Q_{\alpha\beta} p_\beta \quad (2)$$

$Q_{\alpha\beta}$ characterizes the spin-orbit interaction, s_α is the αth component of electron spin.

The effective magnetic field is given by $B_p = Qp/\mu_B g$ (μ_B is the Bohr magneton, g is the electron g-factor)

The value and direction of the magnetic field is determined by electron momentum $\vec{p}$.

The frequency of the electron spin is given by $\Omega = Qp/\hbar$. The spin orbit interaction is qs[p × n]. The spin density is determined by the following continuity equation: $J_{\alpha\beta}$ is the α component of the flow of electrons along β axis

$$\frac{\partial S_\beta}{\partial x} + \frac{\partial J_{\alpha\beta}}{\partial x_\alpha} = \frac{m}{\hbar}\varepsilon_{\beta j\gamma} Q_{ji} J_{i\gamma} + (\frac{\partial S_\beta}{\partial t}) others \tag{3}$$

The spin current operator is given by:

$$\widehat{J}_\alpha = \left(\widehat{\mathrm{V}_{\alpha\beta}\widehat{\mathrm{S}}_\beta} + \widehat{\mathrm{S}}_\beta \widehat{\mathrm{V}}_\alpha\right)/2 \tag{4}$$

$$\widehat{\mathrm{V}}_\alpha = \frac{\partial y}{\partial x} = \frac{\mathrm{p}_\alpha}{\mathrm{m}} + \mathrm{Q}_{\beta\alpha} + \widehat{\mathrm{S}}_\beta \tag{5}$$

The above equation gives the spin velocity operator.

In magnetic multilayer when current flows the electron becomes spin polarized and the angular momentum which causes reversal of magnetization in the thin magnetic layers. An electron spin carried by the current interacts with magnetic layer the interaction causes torques between the spin and the magnetization. The current when flows through a ferromagnet it becomes spin polarized and hence carried angular momentum. The current remains polarized in neighboring non-magnetic layers so that the angular momentum carried by the current can interact with the magnetization. The spin current exerts a spin transfer torque on the magnetizations in the device. The spin transfer induced magnetization and the generation of spin transfer torque is demonstrated by certain devices which have multiple layers like copper layer(non magnetic), an insulating layer and a cobalt layer(magnetic layer). As current passes have uniform density, uniform magnetization the spin transfer torque is also uniform.

3 Spin Current Derivation

Pure spin current is deprived of any kind of association with electric current. In electric current the spins of charge carriers are random. The electrons have two polarizations spin up and spin down.

$$\mathrm{J}\uparrow(\mathrm{spinup}) = -\mathrm{j}\downarrow(\mathrm{spindown})$$

$$\mathrm{J_c} = -\mathrm{e}(\mathrm{j}\uparrow + \mathrm{j}\downarrow) = 0$$

The spin current $\mathrm{J_S} = \; /2\,(\,\mathrm{j} * - * \varphi) \gtrless$

Vector potential for spin current

$$\mathrm{H} = \sum\nolimits_{\mathrm{t},\sigma} = \frac{1}{2\mathrm{m}}\left(\mathrm{p_t} - \frac{\mathrm{e}}{\mathrm{c}}\mathrm{A}\right)^2 + \mathrm{V} \tag{6}$$

The Dirac equation of an electron in electromagnetic field is

$$\left(\beta mc^2\varphi(x,t) + \sum_{k=1}^{3} \alpha pc\right)\varphi(x,t) = i\hbar\frac{\partial\varphi(x,t)}{\partial t} \tag{7}$$

m	rest mass of an electron
c	speed of light
p	momentum
x, t	space and time constant
ħ	reduced Planck constant
Ψ	wave function

4 × 4 matrices α_k and β. $\alpha_{i=}\ \beta^2 = I_4$.

Schrondinger equation:

$$\frac{-\hbar^2}{2m}\nabla^2\phi = i\hbar\frac{\partial}{\partial t}\phi \tag{8}$$

The vector potential for the electron spin is

$$H = \sum_{i,\sigma}\frac{1}{2m}\left(P_i - \frac{e}{c}A_\sigma\right)^2 + V \tag{9}$$

$$v_{i,\sigma} = \frac{p_i - \frac{e}{c}A_\sigma}{m} = -\frac{c}{e}\frac{\partial H}{\partial A_\sigma}$$

Spin Velocity =

$$j_\uparrow = -\frac{c}{e}\frac{\partial E(A_\uparrow, A_\downarrow)}{\partial A_\uparrow};$$
$$j_\downarrow = -\frac{c}{e}\frac{\partial E(A_\uparrow, A_\downarrow)}{\partial A_\downarrow}.$$

Spin dependent current:

$$J_s = \frac{\hbar}{2}(j_\uparrow - j_\downarrow) \neq 0 \tag{10}$$

The read–write operation in Magnetic QCA cells using Spin Transfer Torque (STT) reduces the switching energies dissipated during the change of state of electron spins [1]. Hence our proposed scheme is a new approach which aims at reducing the switching energy dissipation in magnetic QCA cells.

$$A_\uparrow = -A_\downarrow \Rightarrow j_\downarrow = -j_\uparrow$$

Pauli found that the data can be recovered from the Zeeman's experiment if an extra spin term $\frac{e\hbar\sigma}{2m}$B is appended.

$$i\hbar\partial_t\psi = \left[-\frac{\hbar^2\nabla^2}{2m} - \frac{e\hbar}{2m}\sigma B\right]\varphi \tag{11}$$

Dirac equation was able to derive spin from information contained within fundamental wave equation.

Dirac equation for particles with spin is:

$$\left[(\alpha, p)c + \left(\beta mc^2\right)\right]\varphi = E\varphi \tag{12}$$

$$\left[-\frac{\hbar^2\nabla^2}{2m} - \frac{e\hbar}{2m}\sigma B\right]\varphi = i\hbar\partial_t\varphi$$

Multiplying both sides by $\varphi^\dagger$

$$\left[-\frac{\hbar^2\nabla^2}{2m} - \frac{e\hbar}{2m}\sigma B\right]\varphi\varphi^\dagger = i\hbar\partial_t\varphi^\dagger \tag{13}$$

Since

$$\varphi^\dagger = \begin{pmatrix}\varphi_1 \\ \varphi_2\end{pmatrix}^\dagger = \left(\varphi_1^*\varphi_2^*\right)$$

Spin operator are Hamiltonian

$$\sigma^\dagger = \sigma$$

Solving the above equations and comparing it with the continuity equation we get

$$J_{pauli} = \frac{i\hbar}{2m}\left[\left(\varphi^\dagger\right) - \varphi^\dagger\,\varphi\right]$$

Wave function for spin up is

$$\uparrow\varphi_n = (array{*}20c\varphi_n 0)$$

Wave function for spin down is

$$\downarrow\varphi_n = \begin{pmatrix}0 \\ \varphi_n\end{pmatrix}$$

4 Magnetic QCA Cell

In early 1993, the Quantum dot cellular Automata was introduced by Lent [7]. The prime contribution on QCA based logic circuits are developed and different types of defect analysis is reported in [8–13]. Magnetic QCA is nanometer scale magnetic particles (nano-magnets) which exhibits bi-stable states ("0" or "1") on application of magnetic field presenting binary information [14]. In magnetic QCA cell there are mainly five nanomagnets. There is a central nanomagnet surrounded by four others as shown in Fig. 1.

Three of the neighbors can be used as inputs driven by additional driver nanomagnets oriented along the direction so the clock-field. The nanomagnet to the right of the central magnet is the output. The majority gate is simulated by applications of horizontal clock-field. The three-input majority gate acts as a programmable two input NAND or NOR gate depending upon the state of any one of the three input magnets. Boolean logic functions can be built by a network of majority gates. The input is set by external clock field. The driver magnet helps in realizing the different combinations of input. The intersection of horizontal and vertical wires are common to all structures. Figure 2 shows the switching of MQCA wire by an external input and clocking field [15]. The clocking field H_{ext} is applied in the direction of the hard axis. Before the application of magnetic field the nanomagnets are in metastable states as shown in Fig. 2.

As soon as a horizontal magnetic field is applied the nanomagnets are aligns themselves along their hard axes as shown in Fig. 2. The nanomagnets returns to their ground state and arrange themselves in alternating "up" and "down" direction as the magnetic field is reduced slowly as shown in Fig. 2. The "up" and "down" state is equivalent to logic "0" and "1" respectively (Fig. 3).

If we fabricate the Majority gate by Magnetic Tunnel Junction structure which consists of Cu (Antiferromagnet or nonmaget), Cobalt (Ferromagnet) and Insulator and allow current pass through it, it will create spin polarized current which gives rise to Spin torque [5]. The polarity of the voltage applied to each of the input of the majority gate may be a positive voltage "plus" or logic 1 and a negative voltage or "minus" which indicates logic 0 [16]. The "minus" input voltage drives the current from bottom to top and aligns the magnetization of the free layer to align opposite to the fixed layer. The magnetic tunnel junction logic which we are using over here to construct the majority gate consists of free layer and a fixed layer. The free layer which responds to the current induced torques and the fixed layers which is susceptible to Spin Transfer Torque (STT) [16]. The resistance of

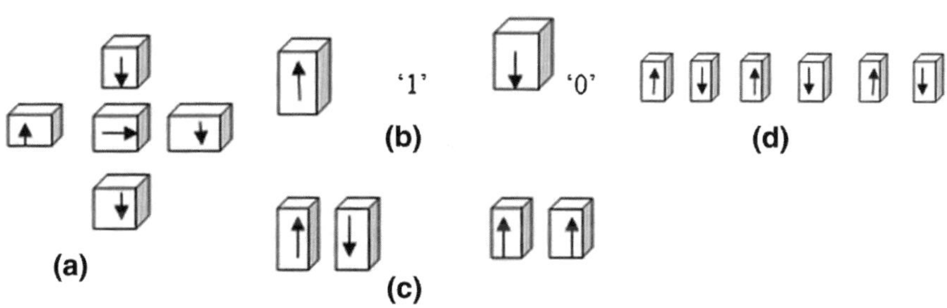

Fig. 1 MQCA Circuits. **a** Majority gate. **b** Logic '1' and '0' for nanomagnets. **c** Ground state, metastable state. **d** MQCA wire

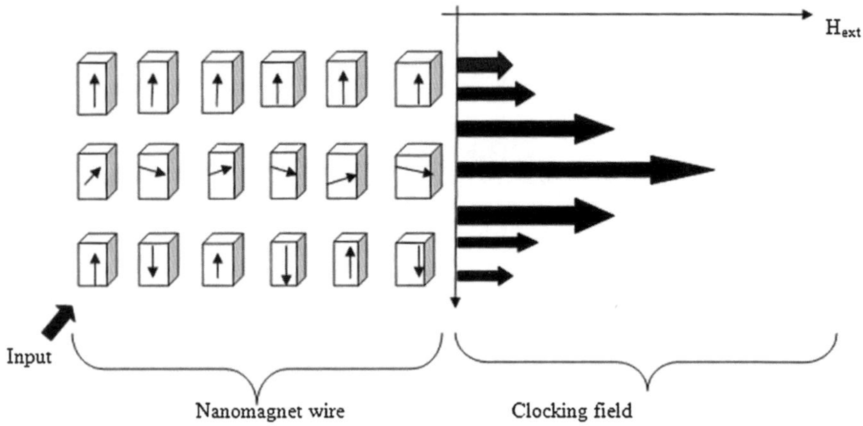

Fig. 2 Clocking of magnetic QCA devices

Fig. 3 Lithographically defined point contact which acts as the majority voter input in magnetic QCA

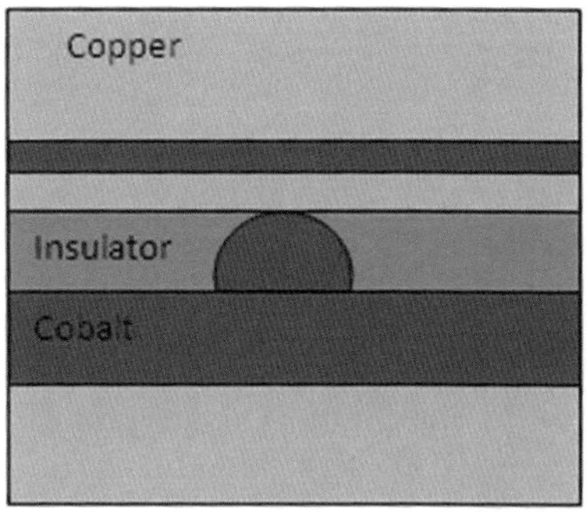

the structure depends on its magnetic state especially their GMR effect. The spin transfer effect changes the resistance of the device when the current is non-uniform. Spin transfer torque is also non-uniform (asymmetric). The device must have a small cross-sectional area because the heat generated by the current may destroy the device if it is not concentrated into a small area. Spin transfer torque effect decreased as the cross sectional area decreases spin transfer torque is uniform when current density and magnetization is uniform.

5 Conclusion

In this paper we have discussed some of the interesting property of nanomagnets. Spin transfer torque (STT) is one of the interesting property. STT is utilized to drive MQCA majority gates. Special layered structure using the concept of Magnetic Tunnel Junction (MTJ) is used over here. This structure actually generates the spin transfer torque which operates the MQCA cells. Nanomagnet QCA is easier to design and operate with proper usage of its magnetic properties. Hence in future we can utilize this property to bring about a versatile change in the field of Quantum mechanics and Nanotechnology.

Acknowledgments Authors are grateful to UGC major project entitled "Study of quantum dot cellular automata for designing circuits and implementing them for high speed and low power fault tolerant computing" under which this paper has been completed.

References

1. I. Zuti, J. Fabian, *Spintronics: Fundamentals and Applications*, vol 76. (Maryland, USA, 2004), pp 323-410
2. N. Gupta, K. Chaudhary et al., *Spintronics-A New Light in Nanoscale Devices*. Proceedings of the 5th National Conference, (INDIACom-2011, India, 2011), pp 1–9
3. P.M. Braganca, J. A. Katine et al., A three-terminal approach to developing spin-torque written magnetic random access memory cells. IEEE **8**, 190–195 (2009)
4. O. Heinonen, *Spintronics in Applications: Hard Drives, MRAM, and Spin Torque Oscillators*. (October 2011), pp 3–36
5. Z. Diao, Z. Li et al., Spin-transfer torque switching in magnetic tunnel junctions and spin-transfer torque random access memory. J. Phys.: Condens. Matter **19**, 165209 (2007)
6. V.L. Korenev, A.F. Ioffe, Bulk electron spin polarization generated by the spin hall current. Phys. Rev **74**, (2006)
7. C.S. Lent, P.D. Taugaw, W. Porod, G.H. Bernstein, Quantum cellular automata. Nanotechnology **4**, 49–57 (1993)
8. K. Das, D. De Characterization, test and logic synthesis of novel conservative and reversible logic gates for QCA. Int. J. Nanosci. **9**(2), 201–214 (2010)
9. K. Das, D. De, *Novel Approach to Design A Testable Conservative Logic Gate for QCA Implementation*. In Proceedings of the IEEE International Conference of IACC'2010, (19–20 Feb 2010), pp 82–87
10. K. Das, D. De, *A Novel Approach of And-Or-Inverter (AOI) Gate Design for QCA*. In Proceedings of the IEEE conference on CODEC-09, (Dec 2009), pp 1–4
11. K. Das, D. De, QCA defect and fault analysis of diverse nanostructure for implementing logic gate. Int. J. Recent Trends Eng. **3**(1), 1–5 (2010)
12. K. Das, D. De Characterization, applicability and defect analysis for tiles nanostructure of quantum dot cellular automata. J. Mol. Simul. **37**(3), 210–225 (2011)
13. K. Das, D. De A study on diverse nanostructure for implementing logic gate design for QCA. Int. J. Nanosci. **10**(1–2), 263–269 (2011)
14. N. Pradhan, K. Das, D. De, *Low Power Logic Design Using MQCA*. IWPSD Conference (2011)
15. N. Pradhan, K. Das, D. De, *Diverse Clocking Strategy in MQCA*. In Proceedings of the IEEE International conference on RAIT (2012)
16. D.E. Nikonov, G.I. Bourianoff, *Proposal of a Spin Torque Majority Gate Logic*, vol 32 (June 2011)

Mechanical Properties of Short Monatomic Gold Chain

Sumali Bansal, Rajiv Bhandari and Keya Dharamvir

Abstract A number of experiments have been conducted over the past decade to study the mechanical properties of a single string of gold atoms suspended between two gold tips, using a scanning tunneling microscope (STM) or an atomic force microscope (AFM). We consider four, five, six, seven, eight and nine atom monatomic chains of gold stretched between two end atoms whose positions remain fixed while those of the rest of the atoms are relaxed. Though there is not any difference in the behavior of these chains but in order to avoid overlapping in graphs we have differentiated chains on the basis of even (4, 6, 8) atom and odd (5, 7, 9) atom chains. We use the Gupta Potential which is a many body Glue Potential describing atom- atom interactions. The distance between the end atoms is the total length of the chain and the total energy of the chain is the sum of interaction energies between all pairs of atoms. The intervening atoms are allowed to adjust their positions till minimum energy is reached. The results from this procedure show that the minimum energy configurations are zig-zag chains which straighten when stretched. The Young's modulus is in the range of a Gigapascal. On further stretching, the chain breaks. The maximum force for these short chains to break is of the order of 1.2, 1.4, 1.3 nN for even (4, 6, 8) chains and 1.7, 1.5 and 0.99 nN for odd (5, 7, 9) chains respectively. The region of length beyond the point of inflexion represents plasticity, and therefore cannot extend to the large strains. Hence this must be unphysical and we take the maximum force to be the breaking force. These results compare well with existing literature qualitatively and quantitatively. However, for long chains, containing 15 or more atoms,

S. Bansal (✉) · K. Dharamvir
Department of Physics, Panjab University, Chandigarh, India
e-mail: sumali.bansal@gmail.com

R. Bhandari
Post Graduate Government College for Men, Sector 11, Chandigarh, India
e-mail: keya@pu.ac.in

P. K. Giri et al. (eds.), *Advanced Nanomaterials and Nanotechnology*, Springer Proceedings in Physics 143, DOI: 10.1007/978-3-642-34216-5_57,

the zigzag configuration is not favoured. Experimental realizations of such long monatomic chains are awaited but experimental results are available for stretching up to eight atom chain.

Keywords Gupta potential • Monatomic chain • Young's modulus • Quasi static structures

1 Introduction

Among the nanostructured materials, nanowires have emerged important structural components in future nanomaterials, mainly due to their high strength and ductility as compared to bulk materials. A number of theoretical and experimental [1] studies have been conducted over the past decade to study the mechanical properties of single atom chains of gold suspended between two tips of a scanning tunneling microscope (STM) [2] or an atomic force microscope (AFM) [3]. Numerical simulations of gold nanowires have been done mainly using Density Functional Theory (DFT) which involves very elaborate calculations [4]. Park et al. [5] used MD simulations to study gold nanowires under tensile loading at several strain rates and wire sizes at room temperature using various forms of embedded atom potential. In the present work we investigate the tensile strength of short monatomic gold chains containing four to nine atoms, in the quasi static framework. We use an effective atom–atom potential, the Gupta potential [6] which is a many body glue potential.

2 Methodology

We consider monatomic chains of gold containing 4, 5, 6, 7, 8 and 9 atoms stretched between two fixed end atoms. The two fixed end atoms were kept at a certain distance (total length of chain) and the intervening atoms were allowed to relax till the minimum energy is reached. Thus we are using a relaxation technique at zero temperature. Computations were done using the GP potential given in Eq. 1.

$$V = \sum_{i-1}^{N}\left[\sum_{\substack{j=1\\ j\neq i}}^{N} A exp\left(-p\left[\frac{r_{ij}}{r_o}-1\right]\right) - \sqrt{\sum_{\substack{j=1\\ j\neq i}}^{N} \xi^2 exp\left(-2q\left[\frac{r_{ij}}{r_o}-1\right]\right)}\right] \tag{1}$$

where r_{ij} is the distance between any two atoms indexed i and j. A, r_o, ξ, p and q are the parameters of the potential and are determined by fitting to experimental values of the cohesive energy, lattice parameters and independent elastic constants for the reference bulk crystal at 0 K [7].

Fig. 1 Zig-zag seven-atom gold chain

3 Results and Discussions

Geometric optimization is done during the relaxation of the chain under the given potential. For the short chains considered by us, using the GP the minimum energy configuration is found to be a zig-zag chain. The optimized zig-zag chains were then stretched by slightly changing the coordinates so that the total length increases. Relaxation was again carried out keeping the two end atoms fixed. As the longitudinal strain on the chains is increased, the zig-zag chain straightens at a particular length. A typical zig-zag chain is shown in Fig. 1.

Table 1 Elongation stages of gold nanowire (quasi static view)

S. no.	Total length of chain (Å)	Energy (eV)	Structure	Remarks
1	10.00	−16.2534	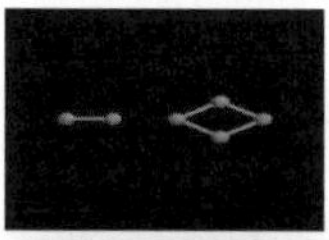	Closed structure bond formation with dimer and tetrahedral structure
2	10.25–10.75	−15.9329 to −15.7561	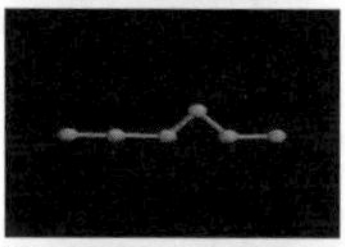	Side kink towards right
3	11.00–11.25	−15.7344 to −15.6983	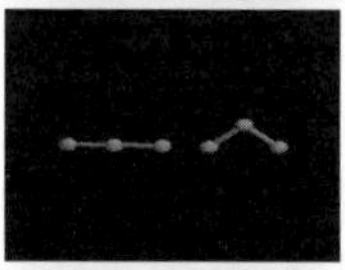	Side kink towards the end of chain
4	11.50–12.00	−15.6825 to −15.6716	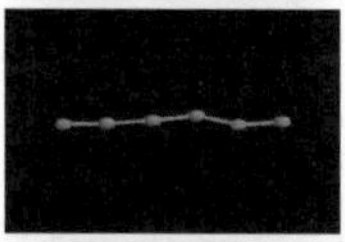	Reaching the zig-zag geometry
5	12.25–13.00	−15.6655 to −15.1898	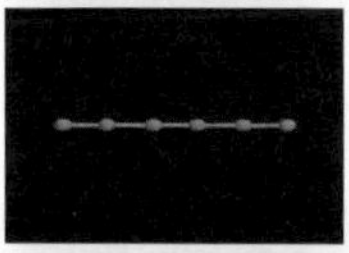	Straight chain
6	13.05–14.30	−15.1717 to −15.0552	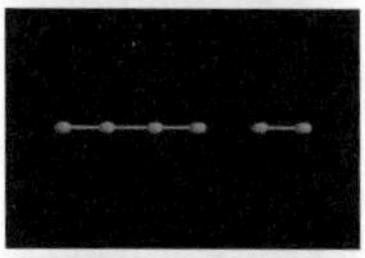	Bond breakage
7	14.35–25.00	−15.0547 to −14.5097	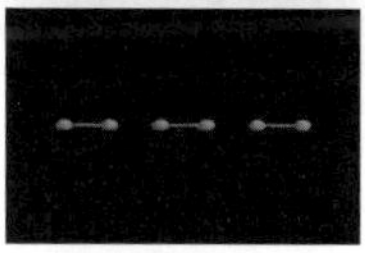	Formation of dimers after bond breakage

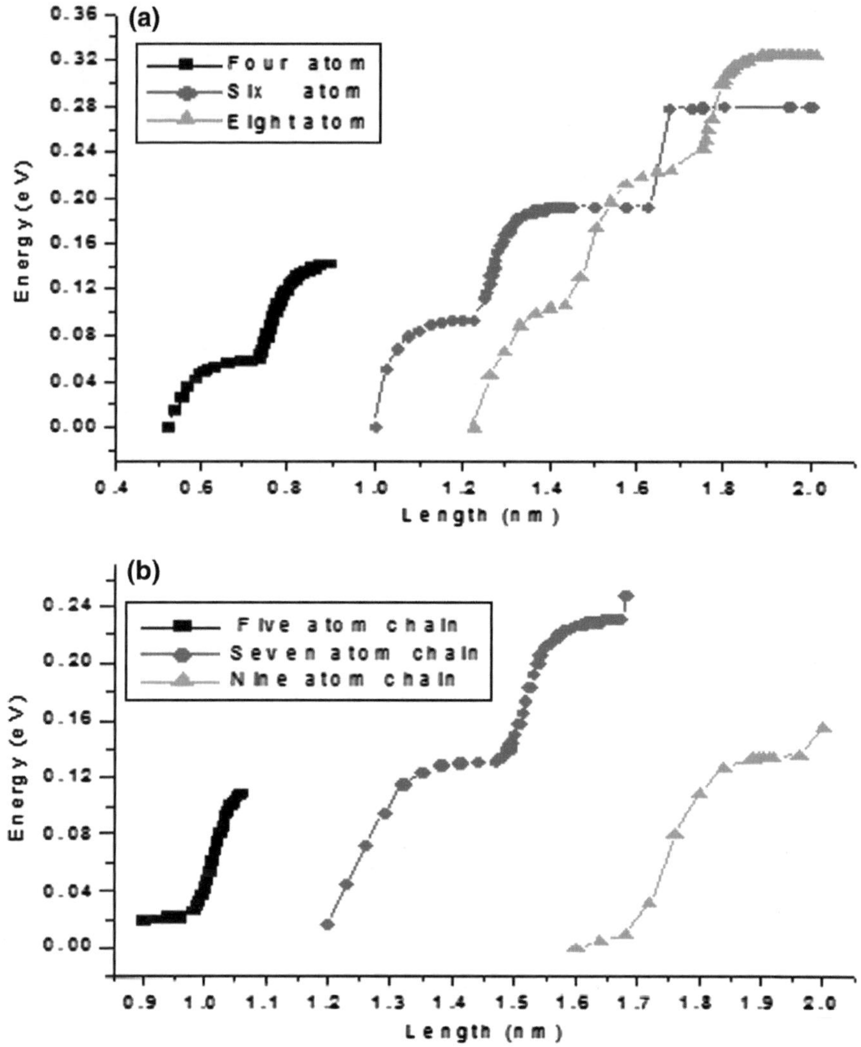

Fig. 2 Variation of total energy (U) with length (L) of short monatomic gold chains **a** 4, 6, 8 **b** 5, 7, 9 atoms

The total energy is a monotonic function of chain length with a plateau whenever there crossover from zig-zag to island-containing to straight character takes place. These qualitative changes in structure can be seen in Table 1.

We find that under the Gupta potential the chains straighten at a lower value of strain as compared with ref. 5 which uses DFT. Figures 2a, b show the total energy (U) vs. total length (L) graphs for chains containing even (4, 6 and 8) and odd (5, 7 and 9) numbers of atoms respectively. The separation into a and b is only to avoid overlapping of curves and there is no qualitative difference between odd- and even-atom chains.

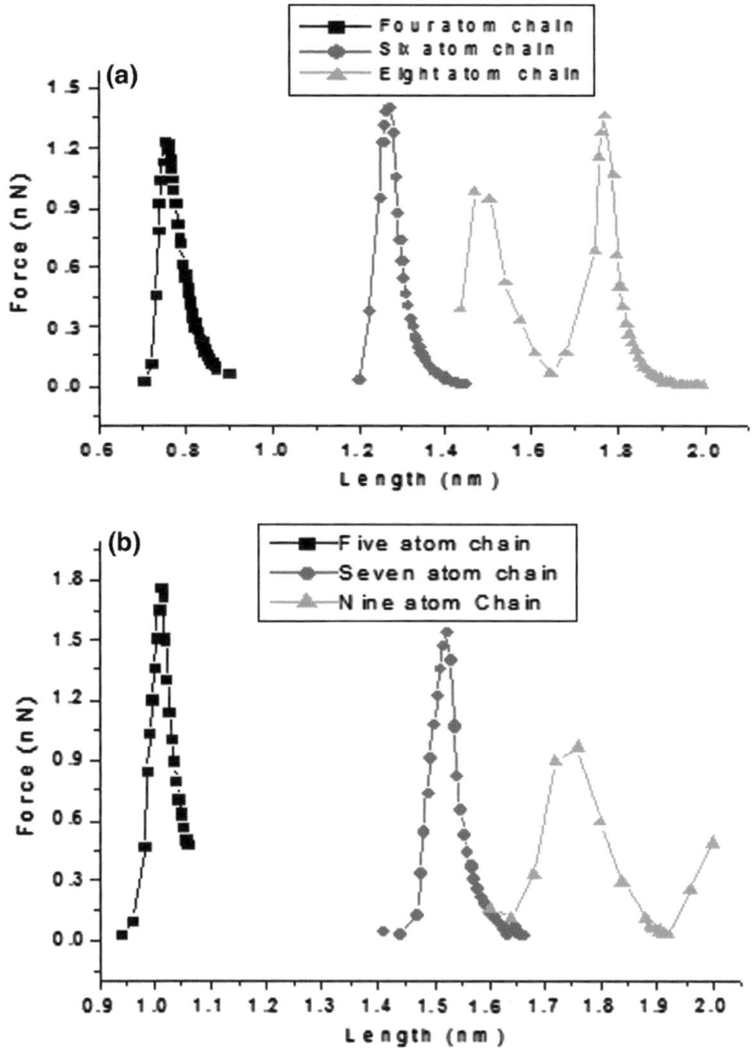

Fig. 3 Variation of force with the total length of a gold chain **a** 4, 6, 8 atoms **b** 5, 7, 9 atoms

We use the U-L curve to find the relation between force and total length of the wire using Eq. 2.

$$F = -\frac{\partial U}{\partial L}. \tag{2}$$

Force as a function of length is shown in Fig. 3. This is nothing but a stress–strain curve. The usual elastic region is where the curve rises with increase in length. However it reaches a maximum and then turns down depicting the onset of

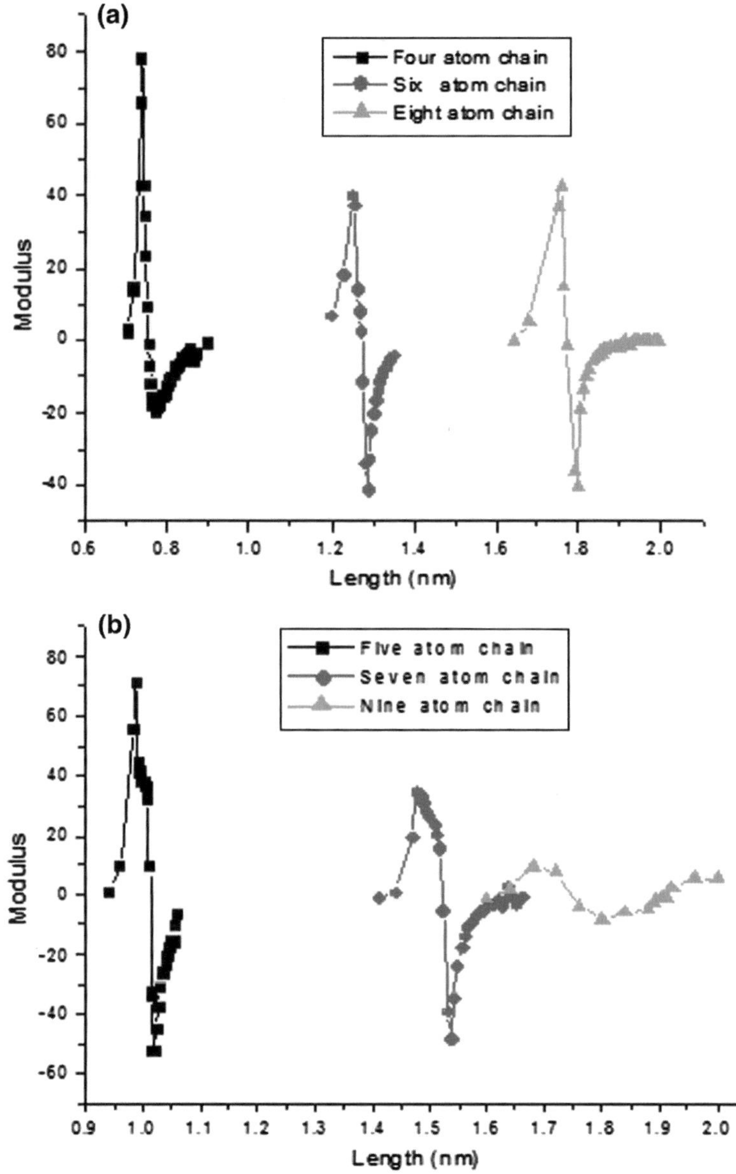

Fig. 4 Modulus vs. total length **a** 4, 6, 8 atoms **b** 5, 7, 9 atoms

plasticity. Thus the peak represents the maximum permissible force, or the breaking force. Figure 3 shows that the breaking forces of the chains lie between 0.99 and 1.7 nN, which is in agreement with existing experimental and theoretical calculations [4]. However there seems to be no systematic way in which the breaking force changes with the number of atoms in the chain. In Fig. 4 we plot 'modulus' vs. total length of the wires, where 'modulus' is defined as follows.

$$Modulus = \frac{\partial F}{\partial l}. \tag{3}$$

The modulus thus defined is proportional to Young's modulus. For a nano cross-sectioned chain as ours, it is difficult to define the area of cross section, more so in the case of zig-zag ones. Hence we use Eq. 3.

The 'unphysical regions in Figs. 3 and 4, where the curves have slopes and values negative respectively, cannot exist as stable states of the wire. However in a dynamic situation (e.g., when the wire is being pulled continuously) they may well exist.

4 Conclusion

We find that qualitatively different structures are possible for any given monatomic wire for different strains. This is a new result. Interestingly, such regions of strain are separated by plastic regions which can be accessed by applying the strain dynamically. The breaking force more or less agrees with experimentally observed values [4]. However, the strain at breaking point is too small.

References

1. P. Zijlstra, A.L. Tchebotareva, J.W.M. Chon, M. Gu, M. Orrit, Acoustic oscillations and elastic moduli of single gold nanorod. Nano Lett. **8**, 3493 (2008)
2. A.I. Yanson, G.R. Bollinger, H.E. van den Brom, N. Agraït, J.M. van Ruitenbeek, Formation and manipulation of a metallic wire of single gold atoms. Nature **395**, 783 (1998)
3. Stalder, U. Durig, Study of yielding mechanics in nanometer-sized Au contacts. Appl. Phys. Lett. **68**, 63 (1996)
4. G.R. Bollinger, S.R. Bahn, N. Agrait, K.W. Jacobsen, S. Vieira, Mechanical properties and formation mechanisms of a wire of single gold atoms. Phys. Rev. Lett. **87**, 2 (2001)
5. H.S. Park, J.A. Zimmerman, Modeling inelasticity and failure in gold nanowires. Phys. Rev. B **72**, 054106 (2005)
6. R.P. Gupta, Lattice relaxation at a metal surface. Phys. Rev. B **23**, 6265 (1981)
7. F. Cleri, V. Rosato, Tight-binding potentials for transition metals and alloys. Phys. Rev. B **48**, 22 (1993)

A DFT Study for the Structural and Electronic Properties of Zn_mSe_n Nanoclusters

Phool Singh Yadav and Dheeraj Kumar Pandey

Abstract An ab-initio study has been performed for the stability, structural and electronic properties of 19 small zinc selenide Zn_mSe_n (m + n = 2–4) nanoclusters. Out of these nanoclusters, one nanocluster is found to be unstable due to its imaginary vibrational frequency. A B3LYP-DFT/6-311G(3df) method is used in the optimization of the geometries of the nanoclusters. We have calculated the zero point energy (ZPE), which is ignored by the other workers. The binding energies (BE), HOMO-LUMO gaps and bond lengths have been obtained for all the optimized nanoclusters. For the same value of 'm' and 'n', we designate the most stable structure the one, which has maximum final binding energy (FBE) per atom. The adiabatic and vertical ionization potentials (IP) and electron affinities (EA), dipole moments and charge on atoms have been investigated for the most stable nanoclusters. For the same value of 'm' and 'n', the nanocluster containing maximum number of Se atoms is found to be most stable.

Keywords DFT study · Semiconductor · Nanoclusters · HOMO-LUMO gap

1 Introduction

The period for more than three decades has drawn significant attention to explore the physics and chemistry of II–VI semiconductor nanoclusters [1–6]. There exists a continuous transition from molecular electronic structure to that of the bulk. A nanocluster is an intermediate phase between the molecule and bulk, whose

P. S. Yadav (✉) · D. K. Pandey
Department of Physics, University of Allahabad, Allahabad 211002, India
e-mail: psycmprl@rediffmail.com

P. K. Giri et al. (eds.), *Advanced Nanomaterials and Nanotechnology*,
Springer Proceedings in Physics 143, DOI: 10.1007/978-3-642-34216-5_58,

electronic and other properties may be exotic. However, two fundamental facts are attributable to the small size of these particles instead of that their physical properties are very different from those of the bulk material. The first fact is that the excitons are confined in a dimension to the order of the particle size, resulting in a compression of the bulk exciton and a corresponding blue shift in the first excited state with decreasing size [7]. This property is known as the quantum confinement effect. According to the technological point of view, the quantum confinement effect has created a great interest from due to the unique ability to tune the optical properties by changing particle size. The second important fact of these quantum-confined nanoparticles (quantum dots or QDs) is that the surface to volume ratio is much greater than that of the bulk semiconductor and thus the nature of the surface influences more on the physical properties of these particles.

Among II–VI semiconducting materials, ZnSe is a direct band gap semiconductor with room temperature band gap energy and an emission at 2.80 eV, which suggests that ZnSe is a potentially good material for short—wavelength lasers and other photoelectronic devices. In addition, ZnSe is of special interest as it exhibits tunable-ultraviolet (UV) luminescence via quantum confinement effects. ZnSe is one of the promising material for fabrication of light emitting devices, such as blue–green laser diodes and tunable mid-IR laser sources for remote sensing applications [8–15].

A number of experimental [16–20] and theoretical studies [21–25] for the physical properties of both the surfaces and the bulk phases have been performed. However, a study of the small size nanoclusters of these selenides is still lacking for their better understanding in terms of the size dependence of the properties. Matxain et al. [21] has calculated ground state geometries of small size Zn_nSe_n (n = 1–9) nanoclusters using B3LYP gradient-corrected density functional method. Deglmann et al. [22] have obtained the atomic structure of Zn_nSe_n nanoclusters up to heptamers (n = 7). Using time-dependent (TD) density functional response theory (DFRT) within the tight-binding approach, the structural and electronic properties of unpassivated Zn_mSe_n ($m + n = 200$) clusters have been calculated by Goswami et al. [23, 24]. Most recently, Sanville et al. [25] produced the ZnSe nanoclusters by direct laser ablation method and analyzed in a time of flight mass spectrometer. They predicted only few physical properties like atomization energy, HOMO-LUMO gap and vertical ionization potential of small stoichiometric Zn_nSe_n (n = 1–16) nanocluster at B3LYP level of theory using SKBJ (d, 2df) basis set.

Earlier, we have performed an ab initio study for Ga_xN_y and Zn_xS_y nanoclusters [26, 27] and the study has been extended to Zn_mSe_n small nanoclusters. In this paper, we report the results of a theoretical study of the structural, stabilities, HOMO-LUMO gap, adiabatic and vertical ionization potential (IP) and electron affinity (EA), charge on atoms and dipole moment of small size Zn_mSe_n (m + n = 2–4) nanoclusters by using the B3LYP-DFT/6-311G(3df) method. We present the method used in the investigations in second section. Third section contains the calculation and results and the last section contains the conclusions.

2 Method

All the geometrical structures are fully optimized using the hybrid gradient-corrected functional (B3LYP) [28–30] within density functional theory frame [31, 32] in the Gaussian-03 code [33]. The harmonic vibrational frequencies of each optimized structures are determined by analytical differentiation of gradients. The polarizable triple split valance basis set, 6-311G(3df) is used as the basic basis set in this study for the most precise calculation after selecting a large number basis set for each atom. The main advantage of this split valance basis set in comparison to others is that the orbital is allowed to change its size without making any change in its shape. The three *d* electrons and one f electrons of Zn and Se atoms are included with the valence electrons as a polarized function due to their importance for the description of the ground state of each atom in three dimensions.

3 Calculation and Results

3.1 Stability of Structures

All the possible two and three-dimensional structures have been considered for optimization to obtain a stable ground state geometry for ZnSe nanoclusters. Minimum energy for each structure is achieved by relaxing the atomic positions. In all fully optimized structures, the convergence of the system energy is obtained up to 10^{-7} meV and the forces of 10^{-3} eV/Å on each atom are achieved. We have obtained the most stable structure for each Zn_mSe_n nanocluster. The stability of nanocluster is considered on the basis of final binding energy (FBE). For a more precise calculation, the harmonic vibrational frequencies and the corresponding zero point energy (ZPE) for all the fully optimized structures have also been calculated. The value of FBE is obtained as,

$$FBE = [(E_i - E_t)/p] - ZPE$$

where E_i, E_t and p stand for the sum of the energies of all the isolated atoms present in the nanocluster, total energy of the nanocluster and the total number of atoms present in the corresponding nanocluster, respectively. All the most stable structures are presented in Fig. 1, while the symmetry, multiplicity of the ground state and the FBEs for all the optimized structures are presented in Table 1. The most stable structures have been depicted by bold letters. For the most stable structures, the calculated Se–Se, Zn–Se and Zn–Zn bond lengths are presented in Table 2. Now, we discuss each nanocluster individually as below:

ZnSe The ground state of ZnSe nanocluster is the singlet state. The present FBE (0.52 eV) is lower than the experimental value (0.89 eV) [34]. Our computed value for Zn–Se bond length of 2.22 Å is very close to the values reported by others [21, 25] as shown in Table 2.

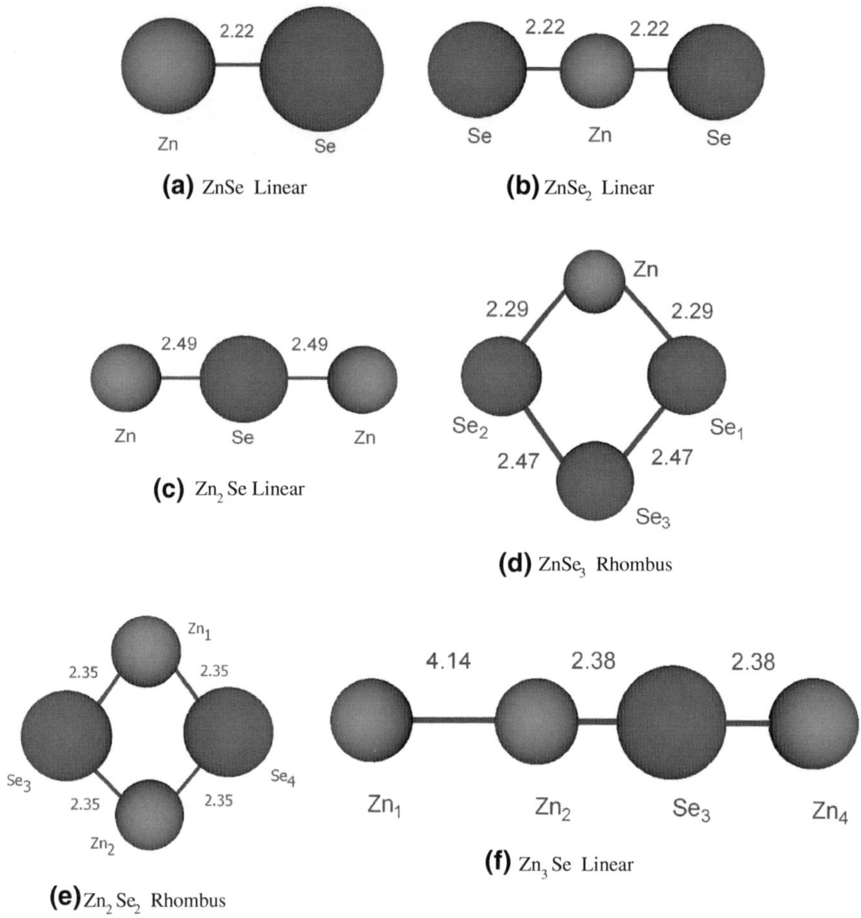

Fig. 1 Most stable structures of Zn_mSe_n (m + n = 2–4) nanocluster (all the bond lengths are in Å)

Zn_mSe_n ($m + n = 3$) The ground state of three atom nanoclusters is either singlet or triplet. For $ZnSe_2$ configuration, the ground states of the linear and triangular structures are triplet and singlet, respectively where as the reverse is true for Zn_2Se configuration.

$ZnSe_2$ We have considered the two linear structures and a triangular structure. Among them, the linear ZnSeSe has maximum FBE (1.24 eV), but it is unstable due to its imaginary frequency. The other linear SeZnSe structure having $D_{\infty h}$ symmetry has FBE of 1.22 eV and is found to be most stable. The bond length of Zn–Se of the linear SeZnSe structure is same as in linear ZnSe structure.

Zn_2Se The investigated structures are similar to $ZnSe_2$. Among them the linear ZnSeZn structure possessing $D_{\infty h}$ symmetry has maximum FBE of 0.60 eV and is

Table 1 Symmetry, multiplicity of the ground state (GS), bindingenergy peratom (BE), zero point energy (ZPE), final binding energy (FBE) and HOMO-LUMO gap for all the configurations of Zn_mSe_n (m + n = 2–4) nanoclusters

Nanocluster	Configuration	Symmetry	GS multiplicity	BE (eV)	ZPE (eV)	FBE (eV)		HOMO-LUMO gap (eV)	
						(Present)	(Others)	(Present)	(Others)
ZnSe	**Linear ZnSe**	$\mathbf{C_{\infty v}}$	**1**	**0.54**	**0.02**	**0.52**	**0.89**[a]	**1.60**	**2.58**[b], **2.41**[c], **2.69**[c]
$ZnSe_2$	Linear ZnSeSe[a]	$\mathbf{C_{\infty v}}$	3	1.27	0.03	1.24		2.37	
	Linear SeZnSe	$\mathbf{D_{\infty h}}$	**3**	**1.27**	**0.05**	**1.22**		**1.14**	
	Triangular SeZnSe	C_s	1	1.18	0.04	1.14		1.57	
Zn_2Se	**Linear ZnSeZn**	$\mathbf{D_{\infty h}}$	**1**	**0.63**	**0.03**	**0.60**		**1.04**	
	Triangular ZnSeZn	C_{2v}	3	0.54	0.03	0.51		1.67	
	Linear ZnZnSe	$\mathbf{C_{\infty v}}$	1	0.45	0.03	0.42		1.98	
ZnSe$_3$	**Rhombus**	$\mathbf{C_{2v}}$	**1**	**1.77**	**0.05**	**1.72**		**2.37**	
	Triangular-planer	C_{2v}	1	1.50	0.06	1.44		2.98	
	Linear SeZnSeSe	$\mathbf{C_{\infty v}}$	3	1.36	0.06	1.30		1.00	
	Linear ZnSeSeSe	$\mathbf{C_{\infty v}}$	3	1.22	0.04	1.18		1.30	
	Pyramidal	C_s	5	1.22	0.05	1.17		1.64	
Zn_2 Se_2	**Rhombus**	$\mathbf{D_{2h}}$	**1**	**1.56**	**0.08**	**1.48**	**0.68**[c]	**2.89**	**2.69**[d]
	Linear ZnSeZnSe	$C_{\infty v}$	3	1.09	0.05	1.04		0.00	
	Linear SeZnZnSe	$D_{\infty h}$	3	0.95	0.07	0.88		1.83	
	Linear ZnZnSeSe	$D_{\infty h}$	3	0.75	0.03	0.72		0.54	
Zn_3Se	**Linear ZnZnSeZn**	$\mathbf{C_{\infty v}}$	**1**	**0.48**	**0.03**	**0.45**		**2.10**	
	Trigonal	C_{2v}	3	0.48	0.03	0.45		1.60	
	Rhombus	C_{2v}	1	0.27	0.04	0.23		1.05	

The most stable configurations are bold-faced ones. The final binding energy (FBE) = BE—zero point energy (ZPE)

[a] Unstable structure due to having at least an imaginary frequency

[b] Ref. [34]

[c] Ref. [35]

[d] Ref. [25]

Table 2 Bond lengths (Å) for all the most stable configurations of Zn_mSe_n (m + n = 2–4) nanoclusters

Nanocluster	Configuration	Bonds	Bond length (Å)	
			(Present)	(Others)
ZnSe	Linear ZnSe (1a)	Zn–Se	2.22	2.20[a], 2.30[b]
$ZnSe_2$	Linear SeZnSe (1b)	Zn–Se	2.22	
Zn_2Se	Linear ZnSeZn (1c)	Zn–Se	2.49	
$ZnSe_3$	Rhombus (1d)	Zn–Se	2.29	
		Se–Se	2.47	
Zn_2Se_2	Rhombus (1e)	Zn–Se	2.35	2.38[a,b]
Zn_3Se	Linear ZnZnSeZn (1f)	Zn–Zn	4.14	
		Zn–Se	2.38	

[a] Ref. [21]
[b] Ref. [25]

found to be most stable. For the linear ZnSeZn structure, the Zn–Se bond length is 2.49 Å as shown in Table 2. The other two structures triangular Zn_2Se and linear ZnZnSe have lower FBEs in comparison to linear ZnSeZn structure.

Zn_mSe_n $(m + n = 4)$ All the four atom structures have the singlet and triplet ground states except the pyramidal $ZnSe_3$ structure which has pentalet ground state.

ZnSe$_3$ We have investigated five different structures like two linear chains SeZnSeSe and ZnSeSeSe, rhombus, triangular-planer and pyramidal ones. The rhombus geometry having C_{2V} symmetry has maximum final binding energy of 1.72 eV and is most stable. The calculated values of Zn–Se and Se–Se bond lengths for most stable structure are 2.29 and 2.47 Å, respectively.

Zn$_2$*Se*$_2$ We have studied three linear chains (ZnSeZnSe, SeZnZnSe and ZnZnSeSe) and a rhombus structure. Among them rhombus geometry having D_{2h} symmetry has maximum FBE of 1.48 eV and is found to be most stable. Sanville et al. [25] have also reported rhombus geometry as a most stable one with binding energy of 0.68 eV. The computed Zn–Se bond length of rhombus structure is very much close to that of others [21, 25] as shown in Table 2.

Zn$_3$*Se* Here, we have considered the three different geometries as linear chain (ZnZnSeZn), trigonal and rhombus. All the geometries have very low FBE in comparison to other structures. Among them linear ZnZnSeZn structure is most stable one with FBE of 0.45 eV and has $C_{\infty V}$ symmetry. The calculated and predicted values of Zn–Zn and Zn–Se bond lengths of linear ZnZnSeZn nanocluster are presented in Table 2.

We observe that the nanocluster containing maximum number of Se atoms for same value of 'm' and 'n' is most stable in comparison with others. The study also reveals that the FBE of most stable nanoclusters increases with the size of the nanocluster.

3.2 *Electronic Structure*

The calculated energy difference between highest occupied molecular orbital and lowest unoccupied molecular orbital (HOMO-LUMO gap) for all the studied structures are presented in Table 1. The predicted HOMO-LUMO gap of ZnSe nanocluster is lower than the others [34, 35] but for the rhombus Zn_2Se_2 structure, it has higher value than others [25] as shown in Table 1. The value of HOMO-LUMO gap increases with number of Se atoms, except for two atom ZnSe nanocluster.

3.2.1 Ionization Potential and Electron Affinity

The ionization potential (IP) is defined as the amount of energy required to remove an electron from a nanocluster. We determine the adiabatic IP by evaluating the energy difference between the neutral and the ionized nanoclusters after finding the most stable state for the ionized nanoclusters by using the optimization procedure. The electron affinity (EA) is defined as the energy released when an electron is added to a neutral nanocluster. We have determined the adiabatic EA by finding the energy difference between the neutral and the anionic nanocluster. The anionic nanocluster has been relaxed to its most stable state. The adiabatic and vertical IPs and EAs for the most stable ones are included in Table 3.

The linear SeZnSe (Fig. 1b) geometry has the highest value of adiabatic IP and vertical IP as well as adiabatic and vertical EA. Sanville et al. [25] have reported the vertical IP for rhombus Zn_2Se_2 structure (Fig. 1e) as 8.11 eV, which is very close to our calculated value of 8.03 eV. The linear Zn_3Se configuration (Fig. 1f) has the lowest value of adiabatic and vertical IP, while Linear ZnSeZn structure (Fig. 1c) has the lowest value of adiabatic EA and the lowest value of vertical EA with Linear Zn_3Se (Fig. 1f) configuration. A zigzag behavior in the variation of both IP and EA with nanocluster size is observed. Neither experimental data nor earlier calculation is available for comparison of IP and EA.

Table 3 Adiabatic and vertical ionization potential (IP) and electron affinity (EA) in eV for all the most stable configurations of Zn_mSe_n (m + n = 2–4) nanoclusters

Nanocluster	Configuration	IP (eV)	EA (eV)
ZnSe	Linear ZnSe (1a)	8.05 (8.09)	1.70 (1.66)
$ZnSe_2$	Linear SeZnSe (1b)	9.52 (9.63)	3.54 (3.43)
Zn_2Se	Linear ZnSeZn (1c)	7.52 (7.53)	1.40 (1.39)
$ZnSe_3$	Rhombus (1d)	7.36 (7.55)	1.75 (1.42)
Zn_2Se_2	Rhombus (1e)	7.97 (8.03)	1.58 (1.46)
Zn_3Se	Linear ZnZnSeZn (1f)	6.94 (7.42)	1.46 (1.39)

The quantities given in brackets are the vertical IP and EA

Table 4 Charge on atoms of the most stable configurations of Zn_mSe_n (m + n = 2–4) nanoclusters and their dipole moment (in Debye units)

Nanocluster	Configuration	Charges on atoms					Dipole moment (debye)
		q_1	q_2	q3	q4	q_5	
ZnSe	Linear ZnSe (1a)	0.789	−0.789				5.95
$ZnSe_2$	Linear SeZnSe (1b)	−0.587	1.174	−0.587			0.00
Zn_2Se	Linear ZnSeZn (1c)	0.442	−0.884	0.442			0.00
$ZnSe_3$	Rhombus ZnSeSeSe (1d)	1.146	−0.623	0.100	−0.623		3.31
Zn_2Se_2	Rhombus ZnSeSeZn (1e)	1.170	−1.170	−1.170	1.170		0.00

The charges are distributed on the atoms in the same order of atoms as given in table

3.2.2 Charge on Atoms and Dipole moment of Nanoclusters

The charge on atoms of the most stable geometries of Zn_mSe_n nanoclusters and their dipole moments are depicted in Table 4. The two atom linear ZnSe nanocluster (Fig. 1a) has highest dipole moment of 5.95 Debye due to large charge transfer between the atoms. The linear SeZnSe (Fig. 1b), linear ZnSeZn (Fig. 1c), and Zn_2Se_2 rhombus (Fig. 1e) structures have zero dipole moment due to their symmetry. The Linear ZnZnSeZn (Fig. 1f) structure has quite small dipole moment. No experimental data for charge on atoms and dipole moment are available for comparison.

4 Conclusions

The occurrence of the most stable configurations of the various ZnSe nanoclusters has been established in the present investigations. For Zn_mSe_n nanoclusters, we have predicted the bond lengths, binding energies, HOMO-LUMO gaps, adiabatic and vertical ionization potentials and electron affinities, charge on atoms and dipole moments.

We observe that the nanocluster containing maximum number of Se atom for the same value of ‘m’ and ‘n’ is most stable in comparison with others. The FBEs of most stable nanoclusters also increases with the nanocluster size. The nanoclusters with high (low) FBEs have large (small) number of Se atoms. The value of HOMO-LUMO gap increases with number of Se atoms, except for two atom ZnSe nanocluster. We observe that the IP and EA both show zigzag behavior. The imaginary frequency causes a linear ZnSeSe structure as unstable structure, although it has the highest FBE among the $ZnSe_2$ nanoclusters. The studied most stable nanoclusters need to be grown experimentally.

Acknowledgments The authors are thankful to University Grants Commissions, New Delhi for its financial assistance.

References

1. A. Alivisatos, Perspectives on the physical chemistry of semiconductor nanocrystals. J. Phys. Chem. **100**, 13226–13239 (1996)
2. C.B. Murray, C.R. Kagan, M.G. Bawendi, Synthesis and characterization of monodisperse nanocrystals and close-packed nanocrystal assemblies. Ann. Rev. Mater. Sci. **30**, 545–610 (2000)
3. T. Trindade, P. O'Brien, N.L. Pickett, Nanocystalline semiconductor: Synthesis, properties and perspectives. Chem. Mater. **13**, 3843–3858 (2001)
4. A.L. Efros, M. Rosen, The electronic structure of semiconductor nanocrystals. Ann. Rev. Mater. Sci. **30**, 475–521 (2000)
5. A. Hagfeldt, M. Gratzel, Light-induced redox reactions in nanocrystalline systems. Chem. Rev. **95**, 49–68 (1995)
6. L.E. Brus, Electron–electron and electron hole interactions in small semiconductor crystallites: The size dependence of the lowest excited electronic state. J. Chem. Phys. **80**, 4403 (1984)
7. R. Rosetti, J.L. Ellison, J.M. Gibson, L.E. Brus, Size effects in the excited electronic states of small colloidal CdS crystallites. J. Chem. Phys. **80**, 4464 (1984)
8. M.A. Hasse, J. Qiu, J.M. Depuydt, H. Cheng, Blue-green laser diodes. Appl. Phys. Lett. **59**, 1272–1574 (1991)
9. H. Jeon, J. Ding, W. Patterson, A.V. Nurmikko, W. Xie, D.C. Grillo, M. Kobayashi, R.L. Gunshor, Blue-green injection laser diodes in (Zn, Cd)Se/ZnSe quantum wells. Appl. Phys. Lett. **59**, 3619–3621 (1991)
10. T. Tawara, S. Tanaka, H. Kumano, I. Suemune, Growth and luminescence properties of self-organized ZnSe quantum dots. Appl. Phys. Lett. **75**, 235 (1999)
11. D. Sarigiannis, J.D. Peck, G. Kioseoglou, A. Petrou, T.J. Mountziaris, Characterization of vapour-phase-grown ZnSe nanoparticles. Appl. Phys. Lett. **80**, 4024–4026 (2002)
12. H. Rho, H.E. Jackson, S. Lee, M. Dobrowolska, J.K. Furdyna, Raman scattering from CdSe/ZnSe self-assembled quantum dot structures. Phys. Rev. B **61**, 15641 (2000)
13. B. Xiang, H.Z. Zhang, G.H. Li, F.H. Yang, F.H. Su, R.M. Wang, J. Xu, G.W. Lu, X.C. Sun, Q. Zho, D.P. Yu, Green-light- emitting ZnSe nanowires fabricated via vapour phase growth. Appl. Phys. Lett. **82**, 3330 (2003)
14. G.N. Karanikolos, P. Alexandridis, R. Mallory, A. Petrou, T.J. Mountziaris, Synthesis of ZnSe nanostructures using lytropic liquid crystals. Nanotechnology **16**, 2372 (2005)
15. S.L. Cumberland, K.M. Hanif, A. Javier, G.A. Khitrov, G.F. Strouse, S.M. Woessner, C.S. Yun, Inorganic clusters as single- source precursors for preparation of CdSe, ZnSe and CdSe/ZnSe nanomaterials. Chem. Mater. **14**, 1576–1584 (2002)
16. L.M. Kukreja, A. Rohlfing, P. Misra, F. Hillenkamp, K. Dreisewerd, Cluster formation in UV laser ablation plumes of ZnSe and ZnO studied by time-of-flight mass spectrometry. Appl. Phys. A Mater. Sci. Process. **78**, 641–644 (2004)
17. J. Zhu, Y. Koltypin, A. Gedanken, General sonochemical method for the preparation of nanophasic selenides: Synthesis of ZnSe nanoparticles. Chem. Mater. **12**, 73 (2000)
18. N. Kumbhojkar, S. Mahamumi, V. Leppert, S.H. Risbud, Quantum confinement effects in chemically grown, stable ZnSe nanoclusters. Nanostruct. Mater. **10**(2), 117–129 (1998)
19. P. Reiss, ZnSe based colloidal nanocrystals: Synthesis, shape control, core/shell, alloy and doped systems. New J. Chem. **31**, 1843–1852 (2007). doi:10.1039/b712086a
20. V.V. Nikesh, A.D. Lad, S. Kimura, S. Nozaki, S. Mahamuni, Electron energy levels in ZnSe quantum dots. J. Appl. Phys. **100**, 113520 (2006)

21. J.M. Matxain, J.M. Mercero, J.E. Fowler, M. Ugalde, Small clusters of group-II–VI. materials: ZniXi, X = Se, Te, i = 1–9. Phys. Rev. A **64**, 053201 (2001)
22. P. Deglmann, R. Ahlrichs, K. Tsereteli, Theoretical studies of ligand-free cadmium selenide and related semiconductor clusters. J. Chem. Phys. **116**, 1585–1597 (2002)
23. B. Goswami, S. Pal, P. Sarkar, G. Seifert, M. Springborg, Theoretical study of structural, electronic, and optical properties of Zn_mSe_n clusters. Phys. Rev. B **73**, 205312 (2006)
24. B. Goswami, S. Pal, P. Sarkar, Theoretical studies of the effect of surface passivation on structural, electronic, and optical properties of zinc selenide clusters. Phys. Rev. B **76**, 045323 (2007)
25. E. Sanville, A. Burnin, J. Belbruno, Experimental and computational study of small (n = 1–16) stoichiometric zinc and cadmium chalcogenide clusters. J. Phys. Chem. A **110**(7), 2378–2386 (2006). doi:10.1021/jp056218v
26. P.S. Yadav, R.K. Yadav, B.K. Agrawal, Structural, electronic and vibrational properties of small GaxNy (x + y = 2–5) nanoclusters: A B3LYP-DFT study. J. Phys. Condens. Matter. **19**, 076209 (2007)
27. P.S. Yadav, D.K. Pandey, S. Agrawal, B.K. Agrawal, Ab initio study of structural, electronic, optical and vibrational properties of Zn_xS_y (x + y = 2 to 5) nanoclusters. J. Nanopart. Res. **12**, 737–757 (2010)
28. A.D. Becke, Density–functional thermochemistry. III. The role of exact exchange. J. Chem. Phys. **98**, 5648 (1993). doi:10.1063/1.464913
29. A.D. Becke, Density-functional exchange-energy approximation with correct asymptotic behaviour. Phys. Rev. A **38**, 3098 (1988)
30. C. Lee, W. Yang, R.G. Parr, Development of the Colle-Salvetti correlation-energy formula into a functional of the electron density. Phys. Rev. B **371**, 785 (1988)
31. P. Hohemberg, W. Kohn, Phys. Rev. **136**, 3864 (1964)
32. B. Miehlich, A. Savin, H. Stoll, H. Preuss, Results obtained with the correlation energy density functionals of Becke and Lee, Yang and Parr. Chem. Phys. Lett. **157**, 200 (1989)
33. Gaussian Inc. Gaussian 03, *Revision C.03* (Gaussian, Pittsburgh, 2003)
34. Weast RC, Lide DR, Astle MJ, Beyer WH, in *CRC Handbook of Chemistry and Physics*, 70th edn. (Chemical Rubber, Boca Raton, 1990)
35. W. Luo, S. Ismail-Beigi, M.L. Cohen, S.G. Louie, Quasiparticle band structure of ZnS and ZnSe. Phys. Rev. B **66**, 195215 (2002)

Novel Waste Water Treatment Strategy Using Titanate Nanofibers in Fixed Bed Reactor

A. Arumugam, V. Ponnusami, C. Narendhar and T. S. Bhuvaneshvari

Abstract In this paper, the preparation of TiO_2 nanorods by sol–gel-template process has been considered. The prepared sols were characterized by using SEM images show that TiO2 nanorods with uniform diameter of about Uniformly sized TiO2 nanorods were obtained of approximately 10–50 nm in diameter. Effects of pH, concentration are systematically studied. Results show that the adsorption is highly pH-dependent and the removal is almost complete (99 %) for pH 3. Equilibrium batch experiments were conducted and the results were found to be fit with the Langmuir isotherm equation. Langmuir constants were obtained $b = 0.0273$ L/mg and $q_m = 147$ mg/g. This study is to evaluate the suitability of the titanate nano fibres as adsorbent media in a continuous flow packed bed reactor for removing arsenate from water. The desorption studies shows that the percentage desorption has reached 92 % at 75 min of desorption. Since the equipment required for TNs synthesis is simple and cheap, and alkali solutions are reusable, TNs can be regarded as an efficient, low-cost adsorbent for the removal of arsenic.

Keywords Arsenic · Adsorption · Packed bed reactor · Titanate nanotubes

1 Introduction

Arsenic contamination in drinking water is a global problem. National Research Council (NRC) classified arsenic as a human carcinogen (prone to cancer of the skin, lungs, kidney, bladder, liver, and prostate). The Maximum Contamination

A. Arumugam (✉) · V. Ponnusami · C. Narendhar · T. S. Bhuvaneshvari
Department of Chemical Engineering, School of Chemical and Biotechnology, SASTRA University, Thanjavur, Tamilnadu 613401, India
e-mail: aruchemxl@scbt.sastra.edu

P. K. Giri et al. (eds.), *Advanced Nanomaterials and Nanotechnology*, Springer Proceedings in Physics 143, DOI: 10.1007/978-3-642-34216-5_59,

Level (MCL) for arsenic in drinking water was lowered from 50 to 10 μg/L in 2002 by the USEPA followed by WHO, the European Union and several other countries. This lowering of MCL has led to the increase in need of potable water supply systems and industrial discharges for arsenic treatment processes suitable for water sources.

As_2O_3 was used widely as an herbicide, in the US notably at electric power substations. Many soils underlying cattle dipping vats that used arsenic-containing solutions to kill cattle ticks was found to contain arsenic. Arsenic contamination of drinking water is mainly due to naturally occurring arsenic. Industrial arsenic contamination is not a cause but agricultural or industrial activity can mobilise arsenic. Most soils contain at least one hundred times the MCL since arsenic is present in typical soils at roughly 2–20 ppm (10 ppb) [1].

There are several stable oxidation states of arsenic (−III, 0, +III, +V) but in natural systems, +III and +V are the most common. Arsine (−III) is a very toxic compound formed under reducing conditions. But its occurance in nature is very rare. Arsenic is present in both organic and inorganic forms in the environment. Inorganic forms are more abundant in freshwater. Mobility and toxicity of arsenic strongly depends on the structure and oxidation state. The inorganic forms are found to be more acutely toxic than the organic compounds. Also As (III) is found to have higher human mammalian toxicity than As (V). The effects of oxidation state on chronic toxicity are confounded by the redox conversion of As (III) and As (V) within human cells and tissues [2].

A variety of methods including coagulation and filtration with or without sedimentation, coagulation assisted micro-filtration membrane separation, lime softening, adsorption on granular activated alumina, adsorption on ion exchange resins, and reverse osmosis, and electro-dialysis, are available for arsenic removal from contaminated water. However, each of these traditional methods of treatment involves substantial costs and is often difficult to properly control, especially for small water treatment systems [3].

The process of accumulation or concentration of a substance at a surface or interface is known as adsorption. The adsorbing phase is known as the adsorbent and the substance is called adsorbate. Adsorptive processes are easy to operate and they provide high concentration efficiency. This is because these processes have low capital and operating costs, and they do not require any chemical feed. Arsenic adsorbents are prepared using iron and aluminium. Activated alumina is more versatile than the ion exchange process and is more cost effective when compared to other membrane separation techniques. In order to ensure effective treatment, it requires pH adjustment. Activated alumina requires an optimum pH of around 5.5–6 for adsorption process. Since pH of most ground waters is 6–9, it is required to reduce the pH for effective arsenic removal [3].

Titanate nanotubes is used in fuel cells, bone implant materials, pigment for paints, for photo catalysis in solar cells, as it is an oxide material. Titanate nanotubes are used as a catalyst in most of these applications as it provides a physically and chemically stable platform for catalytic processes [4].

Titanate nanotubes are synthesised using different procedures. Two main methods are mainly used for titanate synthesis; a sequential process of synthesis of titanate nanotubes nanoparticles from titanium alkoxide compounds which are then converted to nanotubes by alkaline treatment, and the method is a direct hydrothermal synthesis process which involves anatase nanoparticles to form nanotubes [5].

Nanotubes are fabricated by the hydrothermal treatment of TiO_2 (anatase, rutile brookite or amorphous forms) in a highly concentrated NaOH aqueous solution at moderate temperatures of around 90–180 °C. The main disadvantages of hydrothermal process are preparation conditions such as reaction time, alkali concentrations and temperatures. In the present work we have investigated the formation of TiO_2 nanorods using template assisted sol–gel method. CTAB is used as a structure directing agent. Thus toxicity of the process is overcome by using the sol–gel method.

2 Materials and Method

2.1 Chemicals

All chemicals were of analytical grade and were purchased from sigma Aldrich. Titanium isopropoxide is used as precursor and CTAB as a template. Ammonia Solution was purchased from Merck Chemicals Pvt Ltd.

2.2 Preparation of Titanate Nanotubes

Titanium isopropoxide 1 mM (100 mL) and 50 mL CTAB (1 g in 50 mL) were mixed for 20 min in magnetic stirrer. 0.1 M Ammonia (10 mL) was added drop wise, and was heated at 110 °C for 6 h in a round bottom flask with continuous water circulation. It was then cooled to room temperature and the solution was centrifuged. The pellet collected from the bottom was mixed with HCL (fill the tube fully) and centrifuged. The pellet was taken and dried at room temperature.

2.3 Packed Bed Systems

The glass column (7 mm ID, 20 cm length) was to be packed with titanate nanotubes. Mean particle size of the nanotubes are 0.127 mm, the packing depth is 103 mm. Bed volume was 3.4 mL. Hydraulic loading rate of 5.2 mm/s was maintained. The Flowrate was maintained at 12 mL/min using peristaltic pump.

3 Experimental Procedure

3.1 Characterization of Titanate Nanotubes

The titanate nanotubes texture was characterized by Scanning Electron Microscopy (SEM).

3.2 Batch Adsorption Studies

Stock solution of As (V) was prepared by dissolving required amount in double distilled water. Effect of initial metal concentration was studied by varying initial nickel concentration between 5 and 40 mg/dm^3. For each run, 25 mL of As (V) solution of known concentration was taken in a 250 mL conical flask. 25 mg of adsorbent was added to the mixture and the flask was agitated in an orbital rotary shaker at constant temperature (303 K), pH (5) and rpm (175). Samples were taken after 4 h and filtered to remove the adsorbent particles. The amount of nickel adsorbed per unit mass of the adsorbent at equilibrium (w_e, mg/g) was calculated using the following expression:

$$q_e = \frac{(C_o - C_e)}{m} \tag{1}$$

where C_o and C_{oe} are initial and equilibrium As (V) concentrations in mg/dm^3 and m is the amount of adsorbent in g/dm^3. The percentage As (V) adsorption (%R) was calculated using the following expression [6]:

$$\%R = \frac{(C_o - C_e)}{C_o} \times 100 \tag{2}$$

3.3 Equilibrium Analysis

Equilibrium data were analyzed with most popularly used Langmuir and Freundlich isotherms models. These models are given by following equations:

Langmuir isotherm:

$$q_e = q_m \frac{K_L C_e}{1 + K_L C_e} \tag{3}$$

Freundlich isotherm:

$$q_e = K_F C_e^{1/n} \tag{4}$$

where w_e, q_m, and K_L stand for equilibrium solid phase concentration (mg/g), Langmuir monolayer adsorption capacity (mg/g), and Langmuir equilibrium constant (dm^3/mg) respectively. K_F and n, Freundlich isotherm parameters, are indicators of adsorption capacity and adsorption intensity respectively. While Langmuir isotherm is applicable for homogenous surface adsorption, Freundlich model is used for heterogeneous surface [7].

3.4 Regeneration

One of the important aspects of designing sorbents is the ease of regeneration. The ease with which the arsenic leaches out of the adsorbents was assessed using the procedure given below. The adsorbent particles containing adsorbed As (V) were re-suspended by stirring in 1 M NaOH solutions of 50 mL for 2 h.

4 Results and Discussion

4.1 Characterisation

Scanning Electron Micrographs (Fig. 1) show that synthesized nanotubes are elongated structures with high aspect ratios was obtained. Individual nanotubes can range from a few hundreds of nanometers to a micron in length.

4.2 Effect of pH on Adsorption of Nickel

The pH value affects the surface charge of absorbents, the metal ionization degree and the metal speciation, all of which can affect the adsorption mechanism and the uptake Capacity Effect of pH is shown in Fig. 2. From the figure it could be observed that adsorption capacity decreased with increase in solution pH. The adsorption of arsenate was more favoured in acid solution. The effect of pH is strongly dependent on the surface charge of the adsorbent material. At low pH, adsorbent surface has more positive charge density. As a result, electrostatic repulsion will be high and hence metal uptake will be lower. The increase of adsorption at pH between 2 and 3 can be attributed to the more negatively charged adsorbent surface and to the decrease competitive adsorption by protons.

Fig. 1 SEM image of TiO_2 nanoarchitectured mesh synthesized by sol–gel process

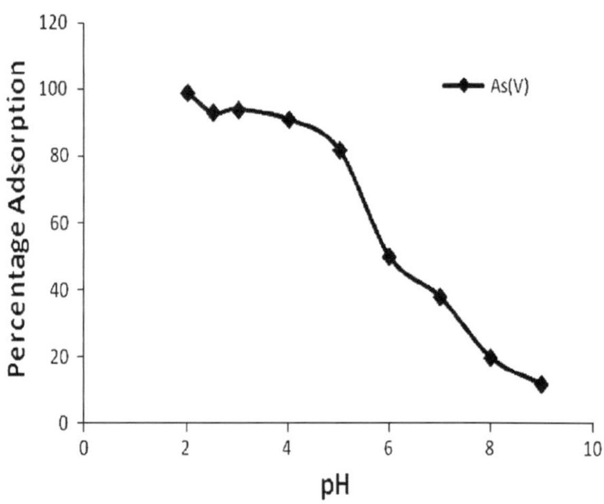

Fig. 2 Effect of solution pH on the adsorption of As (V) on to titanate nano tubes

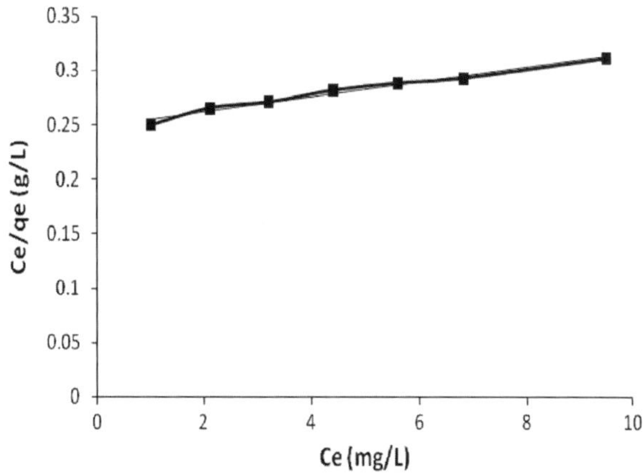

Fig. 3 Langmuir isotherms for adsorption of As (V) on to titanate nano tubes

4.3 Equilibrium Studies

The relation between the initial concentration of As (V) and its percentage removal from solution was studied for all adsorbents included in the study. The initial concentrations of As (V) studied were 5, 10, 15, 20, 25, 30 and 40 mg/L at an adsorbent concentration of 25 mg/L. The adsorption equilibrium data are conveniently represented by adsorption isotherms, which correspond to the relationship between the mass of the solute adsorbed per unit mass of adsorbent w_e and the solute concentration for the solution at equilibrium C_e. From the Fig. 3 Langmuir constants were obtained $b = 0.0273$ L/mg and $q_m = 147$ mg/g with $R^2 = 0.9862$. The equilibrium data for Freundlich equation is fairly satisfactory empirical isotherm can be used for non-ideal adsorption. The Freundlich isotherm constants K_f and n are constants incorporating all factors affecting the adsorption process such as of adsorption capacity and intensity of adsorption. The Fig. 4 shows that Freundlich Constants $K_f = 0.24$ L/g and $n = 10$. The Langmuir Isotherm is to best fit for the adsorption of As (V).

4.4 Packed Bed Experiment

Figure 5 presents column test results. Column tests were conducted at the same loading rate (flow rate = 5.2 mm/s). Arsenic breakthrough began (i.e., $C\mathrm{As} > 5$ g/L) at 3,000 BVs. Titanate nanotubes media can increase the porosity of and expand the packed bed, resulting in significant reduction in head loss and operational problems. Channeling, another efficiency-reducing problem can be ignored.

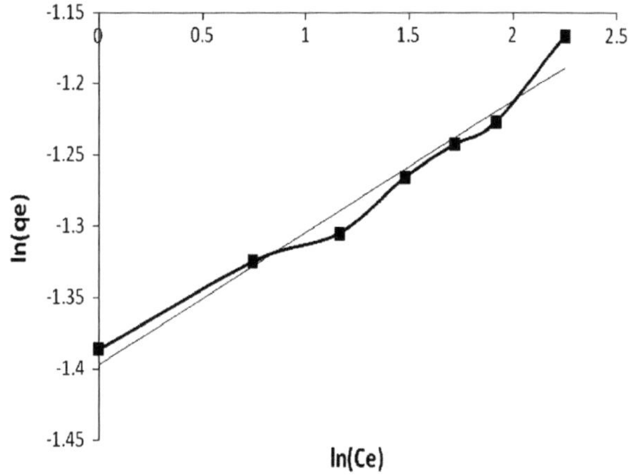

Fig. 4 Freundlich isotherms for adsorption of As (V) on to titanate nano tubes

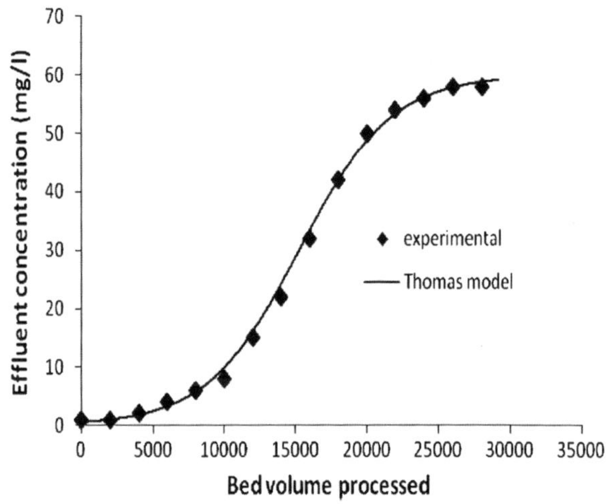

Fig. 5 Column tests (effluent concentration versus bed volumes processed)

4.5 Desorption of Arsenic

Titanate nanotubes adsorbents to be regenerated after adsorption of arsenic, desorption study was carried out. Increased concentration of OH- should compete greatly with the already adsorbed arsenate on Titanate nanotubes, hence sodium hydroxide solution was used to desorbs arsenic from adsorbent. Desorption of adsorbed arsenic on adsorbent was performed using 1.0 M of NaOH solution. From the Fig. 6 the percentage desorption has reached 92 % at 75 min of desorption.

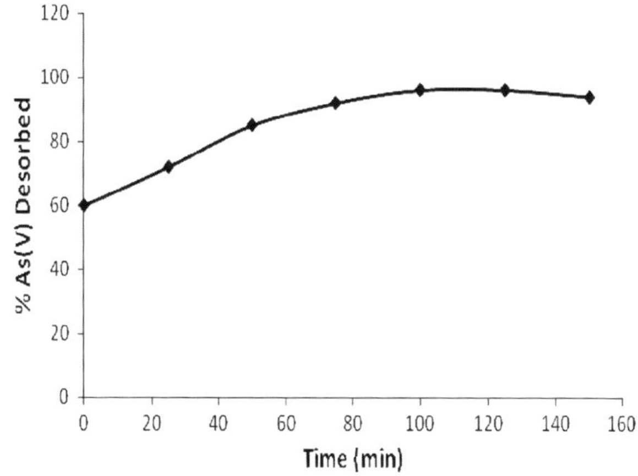

Fig. 6 Desorption of As (V) with 1.0 M NaOH solution

5 Conclusion

TiO_2 nanorods have successfully been prepared by sol–gel template process. Uniformly sized TiO2 nanorods were obtained of approximately 10–50 nm in diameter. The study revealed that titanate nanotubes obtained was efficient adsorbent for As (V) removal. The adsorption strongly depends on pH. Best results were obtained at lower pH. The arsenic was rapidly adsorbed when lower concentrations were used. The adsorption isotherm is consistent with both Freundlich and Langmuir isotherm. This study demonstrated that titanate nanofibers can be synthesized into highly porous adsorbent media that removes arsenic from water in a continuous flow application without release of titanate nanotubes from the packed bed.

Acknowledgments The authors are thankful to the management of SASTRA University for providing necessary facility to carry out this research.

References

1. M. Pena, X. Meng, G.P. Korfiatis, C. Gjing, Adsorption mechanism of arsenic on nanocrystalline titanium dioxide. Environ. Sci. Technol. **40**, 1257–1262 (2006)
2. H. Akbari, A.R. RMehrabadi, A. Torabian, Determination of nanofiltration efficency in arsenic removal from drinking water. Iran. J. Environ. Health. Sci. Eng. **7**(3), 273–278 (2010)
3. D. Mohan, C.U. Pittman Jr., Arsenic removal from water/wastewater using adsorbents—a critical review. (2007). 10.1016/j.jhazmat.2007.01.006
4. M.N. Chongab, H.Y. Zhuc, B. Jin, Response surface optimization of photocatalytic process for degradation of Congo red using H-titanate nanofiber catalyst. Chem. Eng. J. **156**, 278–285 (2010)
5. H.Y. Niu, J.M. Wang, Y.L. Shi, Y.Q. Cai, F.S. Wei, Adsorption behavior of arsenic onto protonated titanate nanotubes prepared via hydrothermal method. Microporous Mesoporous Mater. **122**, 28–35 (2009)

6. K.G. Bhattacharyya, A. Sharma, Azadirachta indica leaf powder as an effective biosorbent for dyes: A case study with aqueous Congo red solutions. J. Environ. Manag. **71**, 217–229 (2004)
7. N. Meunier, J. Laroulandie, J.F. Blais, R.D. Tyagi, Cocoa shells for heavy metal removal from acidic solutions. Bioresour. Technol. **90**, 255–263 (2003)

Structural, Magnetic and Antibacterial Properties of Some Biocompatible Nano-Crystalline Ferrimagnetic Glass/Glass-Ceramics

G. P. Kothiyal, K. Sharma and A. Srinivasan

Abstract Ferrimagnetic glass-ceramics have been prepared from three different glass compositions; (a) $34SiO_2$–(45−x) CaO–$16P_2O_5$–4.5MgO–$0.5CaF_2$–xFe_2O_3 (x = 10–20 wt%), (b) $25SiO_2$–50CaO–$15P_2O_5$–(10 − x)Fe_2O_3–xZnO (x = 0–5 mol %) and (c) $25SiO_2$–(50−x)CaO–$15P_2O_5$–$8Fe_2O_3$–2ZnO−xAg (where x = 2 and 4 mol %). Structural and micro structural properties have been investigated using X-ray diffraction (XRD) and scanning electron microscopy (SEM). Magnetic properties were studied using vibrating sample magnetometry and Mössbauer spectroscopy. X-ray diffraction studies revealed the presence of hydroxyapatite [$Ca_{10}(PO_4)_6(OH)_2$], wollastonite [$CaSiO_3$] and magnetite [Fe_3O_4] crystalline phases in the glass-ceramic samples. The microstructure revealed the formation of nano size crystallites. Magnetic studies showed the evolution of ferrimagnetic property. Increase in magnetization has been observed with an increase of ZnO up to 2 mol % and Ag content up to 4 mol %. The bioactivity of the glass-ceramics samples was investigated by examining apatite formation on the surface of the samples treated in simulated body fluid (SBF). Mössbauer spectroscopy has shown the relaxation of magnetic particles. Bioactive response in SBF was found dependent on iron oxide content. Antibacterial response of Ag containing glasses against with *Escherichia coli (E. coli)* was studied and found to depend on Ag ions concentration in the glass matrix. The glass samples with 4 mol % Ag have shown effective antibacterial activity against *E. coli.*

Keywords Glass-ceramic · Magnetite · XRD · Simulated body fluid (SBF)

G. P. Kothiyal (✉) · K. Sharma
Glass and Advanced Ceramics Division, Bhabha Atomic Research Centre,
Mumbai 400085, India
e-mail: gpkoth@barc.gov.in

A. Srinivasan
Department of Physics, Indian Institute of Technology, Guwahati 781039, India

P. K. Giri et al. (eds.), *Advanced Nanomaterials and Nanotechnology*,
Springer Proceedings in Physics 143, DOI: 10.1007/978-3-642-34216-5_60,

1 Introduction

The use of glass/glass-ceramics has received lot of importance in the recent years not only for biomedical research but also for bio-medical applications [1]. A bioactive glass-ceramic forms a biologically active hydroxyapatite layer on its surface that permits bonding with bone and soft tissues [2]. Nanostructure in glass/glass-ceramic materials can be prepared by controlling the crystal growth process. Recently, glass and glass-ceramics containing Fe ions have been found useful for many important applications like cancer treatment by elimination of cancerous cells in bones [3, 4]. However, magnetic and bioactive properties of these materials depend on the composition. Magnetic properties in the ferrimagnetic glass-ceramics arise from the conversion of iron oxide into magnetite, maghemite and hematite [5]. Magnetic properties also depend on the exchange coupling and arrangement of atoms [6]. When metallic Ag and Fe are added to glass system, local arrangement of atoms is affected. This leads to change in the magnetic properties of materials. Further, higher amounts of Fe_2O_3 may result an improved magnetic properties but the content of iron oxide in glass matrix is limited by the glass forming capability of the system. Thus, desired response can be obtained by controlling the composition or the content of iron oxide in the glass structure. Addition of ZnO improves the magnetic properties and has a stimulatory effect on bone formations [7, 8]. Silver is known to have antibacterial properties [9] thereby affecting the materials behaviour.

In this paper, we present results of our studies on some Fe_2O_3 containing silico-phosphate glasses/glass-ceramics. The effect of ZnO and Ag addition on the structural and magnetic properties of the glass-ceramics is particularly investigated. Bioactive response has also been studied by in vitro tests (reaction with SBF) while the glass containing Ag is also studied for antibacterial activity against *Escherichia coli* (*E. coli*).

2 Experimental

Base glasses with compositions (a) $34SiO_2$–(45–x) CaO–$16P_2O_5$–4.5MgO–$0.5CaF_2$–xFe_2O_3 (x = 10–20 wt%), (b) $25SiO_2$–50CaO–$15P_2O_5$–(10–x)Fe_2O_3–xZnO (x = 0–5 mol %), and (c) $25SiO_2$–(50–x) CaO–$15P_2O_5$–$8Fe_2O_3$–2ZnO–xAg (where x = 2 and 4 mol %) were prepared by melt quenching followed by annealing. The glasses were transformed into glass-ceramics by controlled crystallization performed using the thermal data obtained from differential thermal analysis (DTA). DTA experiments were carried out in Ar environment.

Phase formation and evolution of microstructure in the glass-ceramics were analyzed using X-ray diffraction (XRD, Philips PW 1710) and scanning electron microscopy (SEM, Tescan Vega MV 2300T/40), respectively. Magnetic properties were studied with a superconducting quantum interference device (SQUID,

Quantum Design MPMS) magnetometer. Mössbauer spectra have been obtained using a Mössbauer spectrometer operated in constant acceleration mode. The source employed is ^{57}Co in Rh matrix of strength 50 mCi. The calibration of the velocity scale is done using iron metal foil. Mössbauer spectra are fitted with appropriate doublets and sextets using least square fit program. Bioactivity of the glass/glass-ceramics was studied in vitro by exposure to simulated body fluid (SBF) and the antibacterial response for series C was found out in *E. coli*.

3 Results and Discussion

The bubble free transparent glass samples with brownish or blackish tinge were obtained. Glass transition (Tg) and crystallization temperatures (T_x) of glasses belonging to series A were obtained from DTA plots. Glass transition temperature (T_g) decreased from 698 to 663 °C as Fe_2O_3 content is increased from 10 to 20 wt%. Increase in Fe_2O_3 content results in a decrease in fraction of bridging oxygens. Consequently the network becomes less rigid and requires less energy to break, thereby decreasing the T_g. With the addition of Fe_2O_3, the peak temperature of crystallization (T_p) for glass samples decreased and was found to be 728 °C for the sample having 20 wt% Fe_2O_3. This is due to enhanced nucleation and crystallization in glasses as a result of increase in Fe_2O_3 content from 10 to 20 wt%. Glass (series B) containing 2 mol % ZnO shows one crystallization peak around 715 °C. T_p increases marginally to 725 °C when 5 mol % ZnO was added (Fig. 1). Crystallization of these glasses was carried out by subjecting them to controlled heat treatment at 900 and 1,000 °C for 6 h. Crystallization yielded various bio-compatible and bioactive phases.

XRD patterns of samples of series A after heat treatment at 1,000 °C exhibited the formation of apatite [$Ca_5(PO_4)_3OH$], wollastonite ($CaSiO_3$) and magnetite (Fe_3O_4) as major crystalline phases. The relative magnitude of magnetite phase increased when iron oxide content was increased to 20 wt%. The ZnO containing

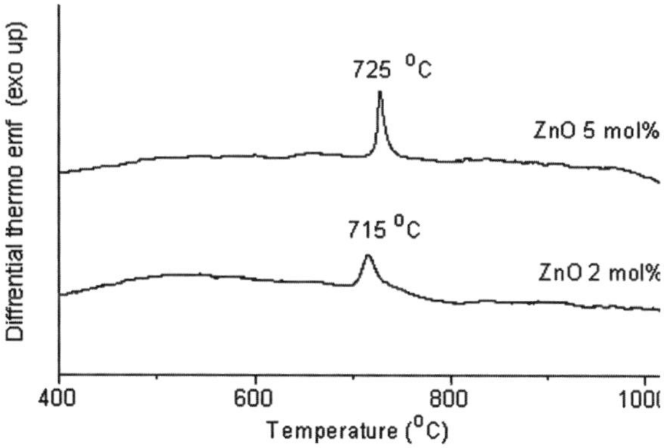

Fig. 1 DTA plots glass samples of series B

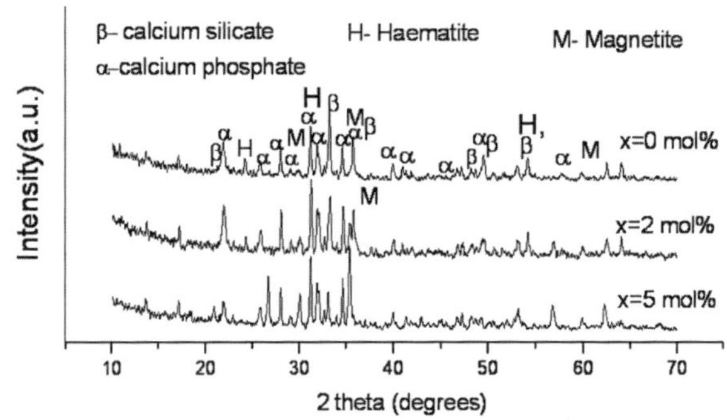

Fig. 2 XRD of glass samples after heat treatment for series B

glass-ceramics show calcium phosphate [$Ca_3(PO_4)_2$], rankinite ($Ca_3Si_2O_7$), magnetite and hematite phases (Fig. 2). The average crystallite size estimated using the Scherrer's formula [10] is found to be around 50 nm.

The microstructures of the glass-ceramics are found to be granular with crystallite size distribution and morphology depending on the amount of Fe_2O_3. In the case of series A, the crystallite size decreased from 70 to 50 nm for change in Fe_2O_3 concentration from 0 to 15 wt% [11]. The microstructure of ZnO based glass-ceramics (series B) also exhibits the granular morphology (Fig. 3). The microstructure without ZnO exhibited the formation of crystallites of average size of 55 nm. The crystallite size marginally decreased to around 50 nm when ZnO content was increased to 5 mol %. The microstructure becomes more homogeneous and granular with the addition of ZnO.

Magnetisation (M) versus applied field (H) curves for series B glass-ceramics samples are given in Fig. 4. Magnetization in glass-ceramics samples is attributed to development of magnetite phase.

We have observed that the magnetization improved with addition of ZnO. Glass-ceramic without ZnO content shows saturation at 5,000 Oe with saturation magnetisation of 1.7 emu/g. The sample shows the typical hysteresis loop with coercive field of 28 Oe. The value of coercive field decreases to 8 Oe with the

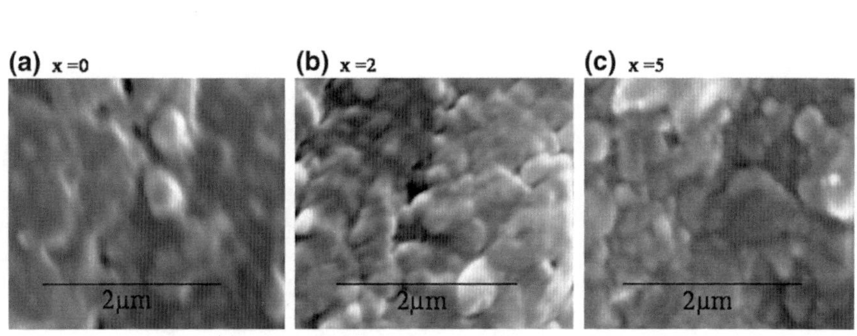

Fig. 3 SEM micrographs of series B glass-ceramics containing ZnO (x) mol %

Fig. 4 M-H plots of glass-ceramics containing different amounts of ZnO (x)

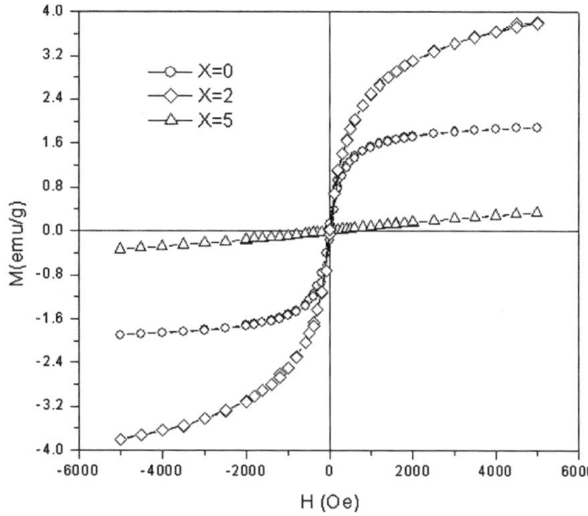

addition of 2 mol % ZnO while the saturation magnetisation increased to 3.2 emu/g. In ferrite like phases, Zn prefers the A-site, which can lead to an overall increase in net magnetic moment. With further increase of ZnO to 5 mol %, concentration of magnetic particles reduced as the content of Fe_2O_3 has decreased.

Figure 5a shows the zero-field cooled (ZFC) and field cooled (FC) curves of glass-ceramics of series B. FC curve follows the ZFC curve and a sharp decrease in the magnetization is observed at around 50 K due to the Morin transition in hematite [12]. Hematite is formed as one of the crystalline phases upon heat treatment. Hematite behaves as a weak ferromagnet above the Morin transition temperature and a perfect antiferromagnet below this temperature. Further, decreasing the temperature to 5 K, we observed an increase in magnetization of the sample under ZFC and FC curves. This behaviour is attributed to the presence of some magnetic particles which behave superparamagnetically even at 5 K and field and zero-field cooling does not have any effect on the magnetic ordering of these particles.

Figure 5a (inset) shows the ZFC and FC curve for sample containing 2 mol % ZnO. FC curve follow the ZFC till a broad peak observed in the ZFC curve around 80 K (which is the blocking temperature, T_B). The divergence between the FC and ZFC curve results from the anisotropy energy barrier of the nanoparticles. In the FC curve, magnetization increases as the sample is cooled as moments of magnetic particles are aligned with the field during the cooling process. The anisotropy energy barrier does not have an effect on the FC curve as field is applied at $T >> T_B$. While in ZFC process, the magnetic particles need to overcome the anisotropy energy barrier, as the moments are oriented along their easy axes. We have observed that the FC curve tends to flatten below T_B as the magnetization is blocked below T_B. It suggests the existence of strong interactions among the particles. The interaction may be of dipole-dipole interaction among the particles

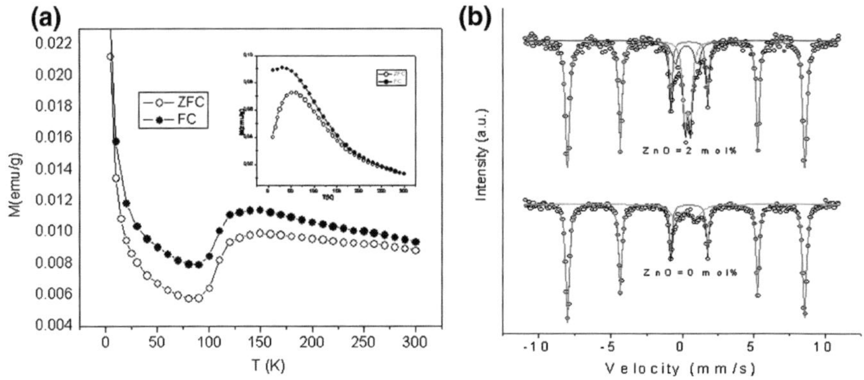

Fig. 5 **a** FC/ZFC curve. **b** Mössbauer spectra; of glass-ceramics containing different amounts of ZnO (series B)

[13]. ZFC/FC curves of samples depend on the volume and spatial arrangement of the magnetic particles and T_B increases with increasing strength of dipolar interaction [14]. We observed that T_B has decreased from 130 to 50 K with the addition of 2 mol % ZnO, indicating the decreasing strength of dipolar interactions. This is explained in terms of replacement of magnetic cations by Zn ions which will reduce the magnetic linkage with neighboring cations. Morin transition was not observed in sample containing 2 mol % ZnO. Morin transition varies considerably with impurities and can be suppressed by doping [15]. Addition of ZnO resulted in suppression of ferromagnetic to antiferromagnetic transition.

The Mössbauer spectra for different series B glass-ceramics samples are shown in Fig. 5b. Hyperfine interaction parameters are derived from the Mössbauer spectra and given in Table 1. The magnetic sextets and doublets have been observed. One sextet has been observed for the sample with no ZnO. The sextet with an isomer shift of 0.46 mm/s and quadrupole splitting of 0.09 mm/s has been

Table 1 Various parameters obtained from room temperature Mössbauer spectra of glass/glass-ceramic samples

Sample (x) mol %	Com	Magnetic (sextet)				Paramagnetic (doublet)			
		H_{int} (kg)	QS	IS	RI (%)	QS	IS	RI (%)	Γ
ZnO(0)	1	515.5	0.096	0.464	85.8	1.45	0.22	14.16	0.892
ZnO(2)	1	516.1	0.096	0.462	69.2	0.37	0.32	20.54	0.341
	2					1.55	0.26	10.22	0.577
Ag(2)	1					1.54	0.85	4.76	0.274
	2					1.78	1.55	28.36	0.695
	3					1.06	0.34	66.88	0.662
Ag(4)	1	472	0.004	0.30	3.2	0.94	0.39	46.26	0.695
	2	418	0.006	0.55	27.1	1.63	1.15	29.30	0.948

IS isomer shift (mm/s), *QS* quadrupolesplitting (mm/s), *RI* relative intensity, *Γ* line width (mm/s), H_{int} internal field (kg), *Com* component

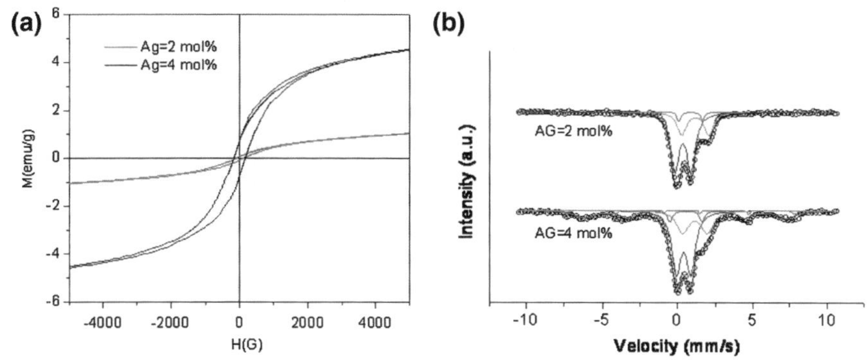

Fig. 6 **a** M-H plots and **b** Mössbauer spectra; of glass-ceramics containing different amounts of Ag (x)

identified to the presence of Fe_2O_3 phase [16]. The doublet is due to the presence of superparamagnetic relaxations of magnetic particles. For the sample having 2 mol % ZnO, one sextet and two doublets are observed. The sextet can be attributed to hematite phase in magnetic ordered state. The doublet with IS = 0.32 and QS = 0.37 is due to Fe^{3+} in octahedral coordination while the doublet with IS = 0.26 and QS = 1.55 is assigned to Fe^{3+} ions in tetrahedral coordination [17]. These two doublets may be due to small clusters whose magnetization appears completely relaxed.

Magnetic response of glass-ceramics containing silver (series C) has been studied. The sample containing 2 mol % Ag shows hysteresis loop with a coercive field of 100 Oe (Fig. 6a). However, the value of coercive field increases to 166 Oe with an increase of Ag content to 4 mol %. In addition, the saturation magnetization also increases from 0.78 to 4.34 emu/g. The increase in saturation magnetization is due increased nucleation and completion of the magnetic structure of magnetic particles when Ag was increased to 4 mol %.

Mössbauer spectrum of glass-ceramic containing 2 mol % Ag could not be fitted with sextet (Fig. 6b). However, the spectra of these samples could be fitted with doublets (Table 1). Superparamagnetic relaxation gives rise to line broadening and a pronounced central peak in Mossbauer spectra [18]. Sample containing 4 mol % Ag shows two magnetic components (Fe^{3+} in octahedral coordination) with the magnetic hyperfine fields H_{hf} = 472 and 418 kg and two doublet components (Fe^{3+} in tetrahedral coordination). The low value of quadrupole splitting of 0.004 mm/s is attributed to cubic symmetry around iron ions. The doublet with IS = 0.39 and QS = 0.94 is assigned to Fe^{3+} in octahedral coordination. This doublet may be due to the presence of isolated ion-clusters showing complete magnetic relaxation. The other doublet with IS = 1.15 and QS = 1.63 is assigned to Fe^{3+} in octahedral coordination. The QS of the doublet is larger than that of the sextet. It was because, the particle sizes represented by the doublet are small and their magnetic structure is not complete. These signify more

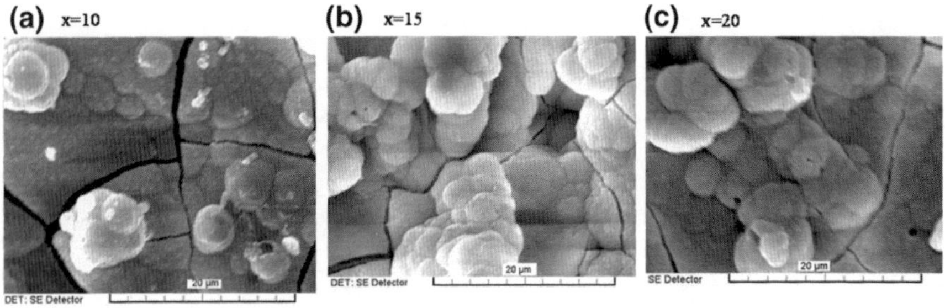

Fig. 7 SEM images of Fe_2O_3 (x wt%) glass-ceramics after 4 weeks of immersion in SBF

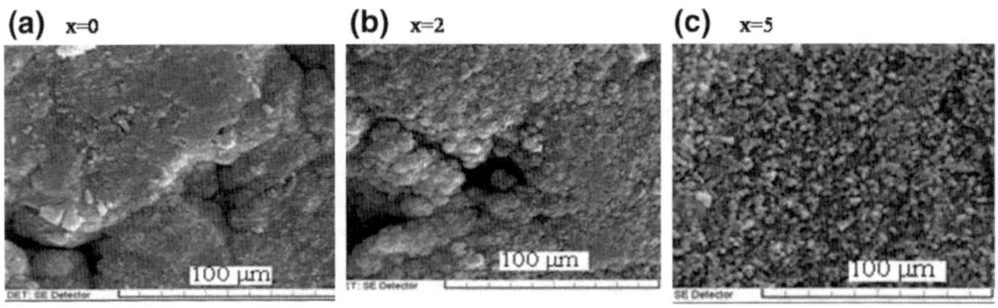

Fig. 8 SEM images of ZnO (x mol %) containing glass-ceramics after 4 weeks of immersion in SBF

crystal defects and lattice distortions, which result in reducing the symmetry around the Fe ions.

We have studied the surface response of different glass-ceramics when immersed in SBF. Formation of bioactive (apatite) layer on the sample surface was observed for all the samples. For iron containing glasses, the best results was obtained with 15 wt% iron oxide (Fig. 7), while for ZnO based glasses, the best morphology and surface coverage were obtained for the glass-ceramics with 5 mol % ZnO. Figure 8 shows the SEM micrographs of ZnO containing glass-ceramics samples treated with SBF. Series A and B glass-ceramics show both magnetic and bioactive properties and thus may be useful as implant where magnetic and bioactive properties are required.

The antibacterial activity of Ag containing glass-ceramics (series C) was evaluated using *E. coli* MG1655, Gram negative coliform bacteria. The glass-ceramic containing 4 % Ag showed complete inhibition to bacterial growth at a concentration of 10 % (w/v). However, the glass-ceramic containing 2 % Ag showed the similar effect at 20 % (w/v).

4 Conclusions

Ferri-magnetic bio-ceramics with different compositions have been prepared. Crystallization of glasses produced nanometer size crystallites of calcium phosphate, hematite and magnetite, which were evenly dispersed in the glassy silica matrix. Blocking temperature decreased with addition of ZnO implying the decrease in strength of dipolar interaction. Morin transition was suppressed while the saturation magnetization increased with ZnO addition. Magnetization also increases with increase in amount of Ag. This trend is attributed to the completion of magnetic ordering. Mössbauer spectroscopy has shown the relaxation of magnetic particles, indicating the size of magnetic clusters is small. The investigated glass-ceramics exhibited formation of apatite like layer when immersed in SBF, optimum response was found with 15 wt% iron oxide. Silver containing glass-ceramics have shown significant antibacterial activity and thus may find use as bactericidal material.

Acknowledgments The authors wish to thank Dr. A. K. Suri for interest and support to this work. They also like to thank Dr. S.K. Gupta for SEM, Mr. M. R. Singh and Mr. Parjapat for magnetic, Dr. A. K. Tyagi for XRD, Dr. S. M Yusuf and Mr. Sher Singh Meena for Mossbauer facilities and measurements. They would also like to thank Shri Sudhansu, Dr. S. Gautam and Dr. A. Sharma for providing antibacterial facilities. Technical assistance of Mr. Anupam Dixit is gratefully acknowledged.

References

1. L.L. Hench, Bioceramics. J. Am. Ceram. Soc. **81**, 1705 (1998)
2. K. Sharma, A. Dixit, S. Singh, C.L. Prajapat, S. Bhattacharya, Jagannath, M.R. Singh, S.M. Yusuf, P.K. Sharma, A.K. Tyagi, G.P. Kothiyal, Preparation and studies on surface modifications of calcium-silico-phosphate ferrimagnetic glass-ceramics in simulated body fluid. Mater. Sci. Eng. C **29**, 2226 (2009)
3. R.K. Singh, A. Srinivasan, Bioactivity of SiO_2–CaO–P_2O_5–Na_2O glasses containing zinc–iron oxide. Appl. Surf. Sci. **256**, 1725 (2010)
4. T. Kokubo, Y. Ebisawa, Y. Sugimoto, M. Kiyama, K. Ohura, T. Yamamuro, M. Hiraoka, M. Abe, Preparation of bioactive and ferromagnetic glass-ceramic for hyperthermia. Bioceramic **52**, 13 (1992)
5. K. Sharma, S. Singh, C.L. Prajapat, S. Bhattacharya, Jagannath, M.R. Singh, S.M. Yusuf, G.P. Kothiyal, Preparation and study of magnetic properties of silico phosphate glass and glass-ceramics having iron and zinc oxide. J. Mag. Magn. Mater **321**, 3821 (2009)
6. D. Cullity, Introduction to magnetic materials. (Addison-Wesley, Reading, 1972)
7. A. Ito, H. Kawamura, M. Otsuka, M. Ikeuchi, H. Ohgushi, K. Ishikawa, Zinc-releasing calcium phosphate for stimulating bone formation. Mater. Sci. Eng. C **22**, 21 (2002)
8. M. Yamaguchi, A. Igarashi, S. Ychiyama, Bioavailability of zinc yeast in rats: Stimulatory effect on bone calcification in vivo. J. Health Sci. **50**, 75 (2004)
9. H.L. Su, C.C. Chou, D.J. Hung, S.H. Lin, I.C. Pao, J.H. Lin, F.L. Huang, R.X. Dong, J.J. Lin, The disruption of bacterial membrane integrity ROS generation induced by nanohybrids of silver and clay. Biomat. **30**, 5979 (2009)
10. B.D. Cullity, Elements of X-ray diffraction. (Addison-Wesley, Reading, 1978)

11. G.P. Kothiyal, K. Sharma, A. Dixit, A. Srinivasan, Structural and magnetic studies on nano-crystalline biocompatible glass/glass-ceramic. AIP Conf. Proc. Inter. Conf. Phys. Emerg. Funct. Mater (PEFM-2010) **1313**, 55 (2010)
12. M.F. Hansen, C. Bender Koch, S. Mørup, Magnetic dynamics of weakly and strongly interacting hematite nanoparticles. Phys. Rev. B **62**, 1124 (2000)
13. M.E. Hilo, K.O. Grady, R.W. Chantrell, Susceptibility phenomenon in fine particles system: 1 concentration dependence of the peak. J. Magn. Magn. Mater **114**, 295 (1992)
14. Z. Mao, D. Chen, Z. He, Equilibrium magnetic properties of dipolar interacting ferromagnetic nanoparticles. J. Magn. Magn. Mater **320**, 2335 (2008)
15. I.V. Golosovsky, I. Mirebeau, F. Fauth, D.A. Kurdyukov, Y.A. Kumzerov, Magnetic structure of hematite nanostructured in porous glass. Solid State Commun. **141**, 178 (2007)
16. J. De Sitter, C. Dauwe, E. De Grave, A. Govaert, On the Mossbauer parameter in $BiFeO_3$. Solid State Commun. **18**, 645 (1976)
17. J.A. Johnson, C.E. Johnson, Mossbauer spectroscopy as a probe of silicate glasses. J. Phys. Condens Matter **17**, 381 (2005)
18. H.N. Ok, M.S. Han, Mossbauer studies on superparamagnetic behaviour of 69–30 at % FeNi fine particles. J. Appl. Phys. **44**, 1931 (1973)

Printed by Printforce, the Netherlands